Genus Cyclamen

in Science, Cultivation, Art and Culture

Edited by BRIAN MATHEW

DEDICATION

In memory of Fred Buglass
(1905–1998)

who suggested this book and instigated the Society's programme of field work.
Founder member of The Cyclamen Society and President 1981–1994

and

Kees Sahin
(1944–2006)

Charismatic plantsman and breeder
With Elisabeth his wife, founded K. Sahin Zaden BV, an innovative seed company in Holland.
Cyclamen enthusiast, breeder and benefactor of this book

Genus Cyclamen

in Science, Cultivation, Art and Culture

Edited by BRIAN MATHEW

Watercolours by Christabel King and Pandora Sellars

Kew Publishing

Royal Botanic Gardens, Kew

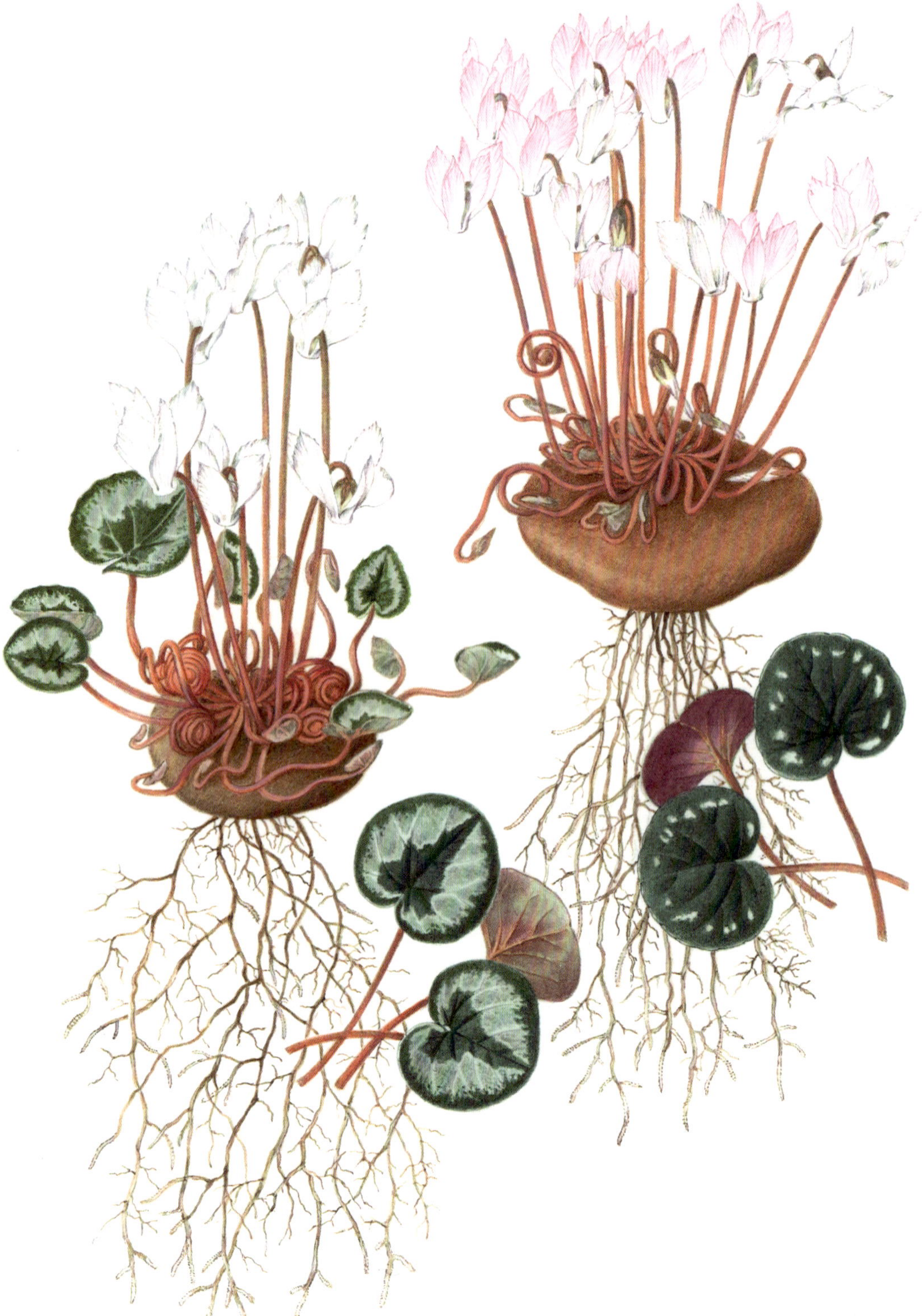

Contents

Detailed contents

Kew
ROYAL BOTANIC GARDENS

The Cyclamen Society

First Edition published in 2013
by Royal Botanic Gardens, Kew
Richmond, Surrey, TW9 3AB, UK
www.kew.org

in association with the Cyclamen Society

Distributed on behalf of the Royal Botanic Gardens, Kew in North America by the University of Chicago Press, 1427 East 60th Street, Chicago, IL 60637, USA

ISBN 978-1-84246-472-4

British Library Cataloguing in Publication Data

A catalogue record for this book is available from the British Library

Project editor: Sharon Whitehead
Page design and layout: Margaret Newman

Cover illustrations: Front – 'Cyclamen Bipenninum Flore Carneo' from Nauwkeurige Beschryving der Aard-gewassen by Abraham Munting, Utrecht and Leiden, 1696
Back – *Cyclamen graecum* subsp. *graecum*, painted by Christabel King
Dedication images: Photograph of Fred Buglass courtesy of the Leverstock Green Village Association
Photograph of Kees Sahin courtesy of Elisabeth Sahin

Printed and bound by Firmengruppe APPL, aprinta druck, Wemding, Germany

FSC www.fsc.org MIX Paper from responsible sources FSC™ C004592

For information or to purchase all Kew titles please visit
www.kewbooks.com or email publishing@kew.org

Kew's mission is to inspire and deliver science-based plant conservation worldwide, enhancing the quality of life.

Kew receives half of its running costs from Government through the Department of Environment, Food and Rural Affairs (Defra). All other funding needed to support Kew's vital work comes from members, foundations, donors and commercial activities including book sales.

Foreword

When the Cyclamen Society was first established at a meeting held at the Royal Horticultural Society Halls in Westminster on 27 January 1977, the founding members agreed that, in addition to bringing together gardeners who were interested in growing and showing *Cyclamen* species and cultivars, consideration should be given to discussing wider issues, including the conservation of the genus in the wild and in cultivation. The need to become involved in the conservation of cyclamen in the wild was one of the subjects highlighted in the first issue of the *Journal of the Cyclamen Society*, published in June 1977, in which attention was drawn to the problems and members were encouraged "to propagate and spread rare species and forms already in cultivation". In later years, the Society developed a "Code of Conduct for cyclamen Growers" as well as a manifesto "promoting the conservation of cyclamen in nature and in cultivation by all acceptable means".

I draw attention to this aspect of the Cyclamen Society's remit early in this foreword to illustrate the enthusiasm and commitment of the Members not only to providing as much help and guidance as possible for gardeners growing their favourite genus but also to recognising the importance of conservation of both wild and cultivated plants. Browse through the Society's journals, which have been published twice a year since the Society was formed, and you will find a wealth of information on all aspects of cyclamen cultivation and conservation, as well as views (sometimes differing) and comments on the taxonomy and nomenclature of cyclamen species and their variants, and accounts of the experiences and observations of the many Members who have travelled to see populations of various cyclamen species in the wild.

Not long after the formation of the Cyclamen Society, one of the founding members, Fred Buglass, who I knew as a great plant enthusiast with particular interests in cyclamen and snowdrops, suggested that one of the aims of the Society should be to prepare a book covering all aspects of the genus *Cyclamen*. During his Presidency, which ran from 1981 to 1994, a publications subcommittee and a research and exploration sub-committee were set up to develop this book and other projects. Subsequently, work initiated by these subcommittees has proved immensely valuable in helping to increase knowledge of the genus both in the wild and in cultivation. Plans were made to develop a series of scientific studies and field trips, to be carried out mainly by members of the Cyclamen Society but also involving other collaborators and scientists working on the genus. The first of these expeditions took place in 1987 in Turkey, where the distribution of *C. coum* was mapped and the variation found both within populations and between populations studied in relation to their habitats. Since that initial foray, there have been 28 further expeditions involving studies of 14 further taxa. In addition, the Society has been fortunate to have the cooperation of research teams from Montpellier and the University of Reading contribute information on their scientific work specifically, in the former case, on *Cyclamen* subgenus *Psilanthum*.

Another of the very important initiatives taken by the Society at this time was to sponsor the translation of Professor Friedrich Hildebrand's seminal monograph, *Die Gattung* Cyclamen *L.* published in 1898, into English. This daunting task was undertaken by Erna Frank, a very knowledgeable and enthusiastic member of the Cyclamen Society. Erna was fluent in German and English and had a wide and detailed knowledge of the genus in the wild and in cultivation. Published in 1999, this translation has proved to be extremely useful in furthering studies of the genus being undertaken by the Cyclamen Society and other researchers.

The aim of the Cyclamen Society has always been to produce a book that, while having the botany of the genus (its morphology, classification and distribution) at its core, is more wide ranging in scope and includes much of the broad array of information and knowledge that is to be found in the Society's journal, as well as recent research and the invaluable fieldwork carried out by Members during the past 25 years.

In this, the Society has succeeded admirably, bringing together information from a range of authors, largely members of the Cyclamen Society but also with input from other collaborators, not only on the botany of the genus but also on aspects of ecology and evolution. This information supplements the accounts of the species, providing details of cultivation and propagation not only in Europe but in other areas of the world where cyclamen are grown. The Society then added accounts of the development of *C. persicum* cultivars as florist's flowers and pot plants, and a summary of investigations into the varying scents of cyclamen flowers; then a further fascinating chapter on the numerous ways cyclamen images have been used in popular culture — on stamps, jewellery, ceramics and glassware — and an invaluable bibliography on the genus; all supported by a wide range of very fine paintings and more than 700 high-quality photographs.

The *Genus Cyclamen in Science, Cultivation, Art and Culture* amply fulfils the original vision of the Members who instigated the project. It will be greatly valued and used not only by current and future cyclamen enthusiasts but also by a much wider audience of gardeners in many parts of the world who grow these infinitely variable and enchanting plants.

Chris Brickell

Introduction

It may seem surprising that a small group of plants, numbering just 23 species, should attract enough attention to support a specialist society of over 1,000 members that is now in its 36th year. In addition to the 70-plus issues of the Society's journal, which contain much varied information, there are several authoritative books on the subject. Some of these are definitive botanical monographs, others literature from one of the specialist nurseries that strive to supply the seemingly endless demand for these highly attractive plants. One might imagine that with such a limited range of species, the subject of cyclamen would become exhausted after a few years and attentions might then be redirected to a more extensive group of plants, but this is not the case and enthusiasm for cyclamen continues unabated. Why is this so?

The aesthetic appeal of cyclamen is without question: their simple, graceful form has attracted the attention not only of devoted gardeners but also of artists, potters, jewellery makers and textile and stamp designers. Yet cyclamen have so much more to offer than their immediately compelling beauty. One of the primary reasons they have fascinated plant enthusiasts over the years is perhaps their extraordinary capacity for variation, providing the aficionado with the on-going excitement of seeking and selecting new forms. This variation is not confined to the flowers, although there is plenty of scope for selection of floral traits and most species are now available in a wide range of shades in the pink-purple region of the spectrum and as white variants. It is in the array of foliage markings that cyclamen have so much appeal: at one extreme the leaves can be plain green, at the other, wholly silver- or pewter-coloured, and in between there is almost every possible variation of light and dark patterns, clearly defined zones and random silver splashes. Cyclamen offer all this diversity in flower and leaf colouring, in addition to variation in other characteristics, such as the size and shape of leaf and flower, fragrance and time of flowering. In this last character, cyclamen have an attribute that is particularly valuable to the enthusiastic gardener: the various species flower over a long period from late summer through autumn and winter into late spring. It has been said that one can have a cyclamen in flower in every month of the year, although this is perhaps more the optimism of an addict than reality! Flowering in ten months of the year is certainly possible. For those who delve deeper, the underground parts, the tubers, have very distinctive characteristics and some species can be easily identified during dormancy. Even the fruiting stages of cyclamen are fascinating, so taking all these facets together, there is no doubt as to why these plants have such an enduring appeal. Furthermore, this morphological variation reflects the wide distribution of cyclamen in the wild and their choice of habitat, another whole field of study.

Some years ago the committee members of the Cyclamen Society decided that a book devoted to their chosen genus should be written. This would not be a solely botanical monograph but would aim to encompass all aspects of cyclamen and draw on the great wealth of knowledge that has appeared in the pages of the Society's journal. Hence the eclectic mix of information contained in this work. Of course, to give the book a logical structure, botany is necessarily at its the core; but the book also contains accounts of some of the most up-to-date research with major contributions from two research teams, one in Montpellier and another at The University of Reading. Other scientific aspects of the genus include a survey of the various floral fragrances emitted by cyclamen.

There is a detailed chapter on cultivation, an area in which the Society's members excel. To avoid too much European parochialism in this chapter, contributions have been sought from other countries where cyclamen species are cultivated extensively, notably Australia, Japan, North America and New Zealand. The 'Florists' cyclamen' could not be ignored and the late Kees Sahin, whose innovative seed business in Holland developed new cyclamen breeding lines, took on the task of preparing a detailed account of the history of the *Cyclamen persicum* cultivars. Sadly he died before this chapter could be completed, but Kees' wife and business partner Elisabeth agreed that Sahin's senior plant breeder, Diony de Bont, should complete the work with the assistance of John Grimshaw.

Cyclamen have long attracted the attentions of botanical illustrators, so we commissioned new paintings of each of the species. We were indeed fortunate to interest two of the country's leading illustrators, Pandora Sellars and Christabel King, to take on the preparation of a set of watercolours. In view of the fact that cyclamen have appealed to artists since the Middle Ages, the Society decided to make a special feature of this by tracing the history of their illustration from early woodcuts through to the present. Cyclamen have also been used in many ways to adorn cards of varying types, stamps, jewellery, ceramics and glassware; and the name cyclamen has even found its way on to ships. It soon became clear that the use of cyclamen in popular culture was so extensive and interesting that a whole section of the book has been devoted to the extraordinary range of art works and designs that cyclamen have inspired.

One of the valuable aspects of the book is the wealth of personal observation volunteered by members of the Cyclamen Society, largely through its journal. Extracts from this information are brought together here in the 'Further observations' section for each of the species. They encompass many topics but focus particularly on plants seen in the wild and so contain much information about habitat and distribution. The book can be seen as a tribute to the work of this remarkable Society and underlines the extent to which a group of specialist enthusiasts can enhance the knowledge of a genus.

Brian Mathew
November 2012

Acknowledgements

For a book such as this, it is almost impossible to list all the contributors; in addition to the authors of individual chapters and sections, there are many whose notes have been extracted from the pages of the Cyclamen Society's journal. Suffice to say that we are indebted to them for their valuable observations, often the result of enjoyable holidays spent in the pursuit of their hobby. In most cases, their names appear along with the quotations or extracts from their articles. In particular, we thank those who have contributed substantially by providing new material constituting whole chapters or sections: Alisdair Aird, Vic Aspland, Richard Bailey, Pedro Bota, Chris Clennett, Robin Clery, Alastair Culham, Heather Dale, Diony de Bont (of K. Sahin Zaden BV), Max Debussche, Martyn Denney, John Fielding, John Grimshaw, Terry Hatch, Ray Johnstone, Eugene N. Kozloff, John T. Lonsdale, Brian Mathew, Peter Moore, Motomu Nagaoka, Neriman Özhatay, Oron Peri, Myrna Semaan, John D. Thompson, Clare Trivedi, Jill Weatherhead, John Whitehead, Trevor Wiltshire and the late Jeremy Wood. We are grateful to Takii Europe BV, into which K. Sahin Zaden BV is now incorporated, for approving the publication of Diony de Bont's extensive work on florists' cyclamen.

We are indebted to many other people and horticultural organisations who have supported the preparation of this book in various ways, providing information or allowing use of their cyclamen artefacts and images, these include: James Armitage, Sema Atay, Jan and Mieke Bravenboer, Chris Brickell, Paul Cumbleton (of the RHS Wisley Alpine Department), Morel Diffusion, Keith Fry, Goldsmith Seeds, Christopher Grey-Wilson, Peter Jones, Jill Marsden, Geoffrey Munn, Yoko Otsuki (translation skills), Neriman Özhatay, Edwin Rose, Elsie Sadler, the late Graham Simpson, Syngenta, Brian and Jo Walker, and Jill White. The late Kees Sahin and his wife Dr Elisabeth Sahin were enthusiastic supporters of the book in its early stages and it was at Kees' suggestion that the chapter 'A celebration of cyclamen in art' came into being.

The Cyclamen Society's Committee and Publications Sub-Committee deserve special thanks for their support and enthusiasm during the long period of preparation; especially deserving of thanks are Alisdair Aird, former Editor of the Society's journal, and Helena Wiesner, Assistant Editor, whose skill in editing the notes extracted from the journals was much appreciated. Helena and Hilary Temple have also assisted greatly in taking on the task of obtaining formal clearance from the many contributors to include their material in this book. Chris Clennett kindly checked through the botanical descriptions of all the species.

As can be seen, many images appear in the book and a list of all the photographers to whom we offer our grateful thanks appears below. Martyn Denney has given invaluable support in the selection of images and is to be congratulated for his diligent research in tracking down many early illustrations from the Old Herbals. Librarians across Europe have helped enrich this chapter on the Old Herbals by responding enthusiastically to requests, usually in a language foreign to them, for images of precious manuscripts of which they are custodians.

We would especially like to thank Peter Prokop of the Österreichische Nationalbibliothek, Vienna, whose willingness to search through the Fuchs Codex for cyclamen paintings saved a trip to Vienna. Librarians in the United Kingdom have without exception welcomed personal visits to consult the precious volumes of manuscript and printed herbals in their care. Amongst them, we would particularly like to thank Anne Marie Townsend of the Sherardian Library of Plant Taxonomy, Oxford; the staff of the Royal Horticultural Society's Lindley Library, London and Garden & Science Libraries, Wisley; and especially Andrea Hart of the Botany Library at the Natural History Museum, London for her cheerful forbearance and enthusiasm in the face of hundreds of requests to produce trolley loads of herbals and florilegia over several days.

For original artwork, we have been extremely fortunate to attract the interest of two leading botanical illustrators, Pandora Sellars and Christabel King. Their superb watercolours have captured the subtleties of each of the species of cyclamen and greatly enhanced this publication.

Thanks go to the staff of Kew Publishing for their constructive help and expertise during the design and production stages.

The Editor wishes to record his thanks to his wife Margaret for her support and remarkable degree of patience, particularly during the later stages of preparation.

Key to photographers

ACC Antique Collectors' Club, **BC** Brian Cresswell, **BM** Brian Mathew, **BW** Brian Walker, **DDB** Diony de Bont, **DK** David Kikodze, **DS** David and Helen Sissons, **DT** Doug Thompson, **EG** Eugene Kozloff, **ET** Evrim Tabur, **EW** Edwin J Rose, **GD** Geneviève Debussche, **GF** Gerald Firak, **GM** Gilbert Mangin, **GN** Graham Nichols, **GSM** Goldsmith, **IR** Ian Robertson, **IWM** Imperial War Museum, **JA** Jim Almond, **JB** Jan Bravenboer, **JF** John Fielding, **JG** John Grimshaw, **JL** John Lonsdale, **JMVW** Jean Marie van der Weghe, **JBC** Jože Bavcon, **JSI** John Ingham, **JW** Jeremy Wood, **JWea** Jill Weatherhead, **JWhe** John Whitehead, **JWhi** Jill White, **KF** Keith Fry, **KMV** Kunsthistorisches Museum, Vienna, **LA** Lynn Almond, **LB** Lesley Butler, **MaxD** Max Debussche, **MD** Martyn Denney, **MELJ** Melvyn Jope, **MH** Marco ten Hope, **MKNGP** Musashi Kyuryo National Government Park, Japan, **ML** Magnus Lidén, **MM** Mat Murray, **MR** Mike Read, **MS** Myrna Semaan, **MTh** Mats Thulin, **NG** Nancy Goodwin, **NO** Neriman Özhatay, **NY** Naoki Yokoyama, **OP** Oron Peri, **PF** Paul Furse, **PH** Paul Hendrikx, **PJ** Peter Jones, **RD** Ray Drew, **RT** Roman Tolordava, **RW** Robert Wallis, **SAH** Sahin collection, **SB** Sheila Brown, **SGN** Syngenta, **TB** Todd Boland, **TBB** Tinatin Barblishvili, **TH** Tony Heaton, **TM** Tim Murphy, **TW** Trevor Wiltshire, **VAA** Vic Aspland, **VC** Viv Condon and **WM** Wayne Mountford, Portmeirion.

About the major contributors

Alisdair Aird is a former Chairman of the Cyclamen Society, and edited its journal for many years. He has a wide interest in other plants, is Vice Chairman of the Royal Horticultural Society Lily Group, and in 2007 was awarded the Lyttel Cup by the R.H.S. for his work on lily species. Alisdair is active in the Mediterranean Garden Society, moderating its internet forum, and organising its trips abroad, sometimes in conjunction with the Cyclamen Society. He gardens in Sussex (UK), in the Peloponnese, and in south-west France growing cyclamen in all three gardens.

Vic Aspland. Although a Chartered Chemist by profession, Vic has been a lifelong gardener, with a particular interest in smaller hardy plants. He has been a very active member of various plant societies for over 40 years. Vic is currently President of the Cyclamen Society, and Practical Gardening Correspondent for the Alpine Garden Society. He has also been Chairman of the Horticultural Advisory Committee of Birmingham Botanical Gardens for more than 12 years, and has recently been elected Chairman of Trustees. He is a popular lecturer and author, with over 100 articles published in various journals. In addition to cultivating his garden, Vic enjoys travelling to mountains with his wife Janet to study and photograph plants. His journeys have incorporated five field studies organised by the Cyclamen Society.

Sema Atay is based in Istanbul and until 2010 worked for DHKD (The Turkish Society for the Conservation of Nature) on conservation biology and rural development projects. She is currently a freelance consultant and is involved with two projects. One of these, called the 'Taste of Fethiye', is developing a system for linking tourism and local agriculture; the aim is to increase the benefit to farmers from tourism and to encourage hotels to use local produce from the area. The other project involves landscape architecture in Bodrum. She is a co-author of the publication *122 Important Plant Areas of Turkey* (2005).

Richard Bailey. Primarily a fruit and vegetable man, Richard is a member of the R.H.S. Fruit, Vegetable and Herb committee, retired Editor of the National Vegetable Society Bulletin and holder of the R.H.S.'s Gold Medal. Richard joined the Cyclamen Society in 1977 and its Committee in 1978, serving as Publications Secretary for 30 years; he was awarded Honorary Life Membership in 1994. He has been on seven of the Society's Expeditions, mostly to Turkey, and remains active in the Society. More generally, Richard is a horticultural lecturer and judge of exhibitions ranging from Chelsea to tiny local shows. His early working life included six challenging years teaching pure science to undergraduate fine art and fashion students, after which Richard went on to teach wildlife conservation in university adult education.

Diony de Bont. "Since my early youth I have been interested in flowers and plants. After a horticultural education, I started as a breeder at K. Sahin Zaden BV, where the first crop I worked on was florists' cyclamen. I was surprised by the variation in colours, flower shapes and leaf patterns of the crop. Our aim was to make F_1 hybrids. The late Kees Sahin and also the late Victor de Troyer were my teachers in plant breeding. Kees had intended to write the chapter about florists' cyclamen for this book but due to illness he asked me to do it. It has been a great honour to do so. Besides the chapter text, we also decided to make a list of florists' cyclamen cultivars that is as complete as possible, sorted by year of introduction. I would like to thank John Grimshaw, who corrected my English, and the many cyclamen growers and flower seed companies who provided me with information."

Pedro Bota is Senior Support Technician (Biochemistry) in the Department of Plant Sciences at the University of Oxford. His main role is plant metabolite profiling using gas chromatography-mass spectrometry, helping and supporting students and researchers who need to use GC-MS and to analyse samples.

Chris Clennett is a professional botanist and horticulturist working for the Royal Botanic Gardens, Kew as garden manager at Wakehurst Place. He has specialised in the Mediterranean flora and bulbous plants, particularly *Erythronium*, and has a long-term interest in cyclamen, having studied the genus for an MSc. Chris has been a member of the Cyclamen Society for 30 years and has travelled with the Society on field work in Greece and Turkey. He is honorary curator of the Society's herbarium and was Editor of the Society's journal from 2000 to 2010.

Robin Clery is a botanist who works as a research scientist in Natural Products Research group at Givaudan, the leading global flavour and fragrance company. After obtaining a degree in Botany and a PhD in essential oil chemistry, Robin spent 20 years researching the chemical composition of flower scents and essential oils and recreating these for use in perfumes; he has contributed to several books and papers on the subject and is a member of the editorial board of the *Journal of Essential Oil Research*. From 1992 to 2007 Robin worked for Quest International, developing new techniques for the extraction and analysis of perfumes from a wide variety of natural products. During this time, he undertook a survey of the floral scent of cyclamen using specimens from the collections provided by Wisley and Trevor Wiltshire.

Alastair Culham is a professional botanist and plantsman with a keen interest in communicating the excitement of botanical and horticultural science to others. He is based at the Centre for Plant Diversity and Systematics, School of Biological Studies, University of Reading, where he organises the MSc in Plant Diversity, curates the Reading University Herbarium and runs the World's first Web 2.0-enabled greenhouse. He has collaborated with the Cyclamen Society for many years, publishing in their journal and in the scientific press.

Max Debussche is a retired Research Director at the Centre d'Ecologie Fonctionnelle et Evolutive (CNRS, Montpellier, France). His research focuses on plant population and community dynamics in the context of land-use and climate change, with emphasis on rare and endemic plant species and their conservation. He has published 22 articles concerning cyclamen, in international academic as well as in popular journals, many of them with John Thompson. He is a member of several scientific councils of national and regional agencies devoted to nature conservancy and environmental resource management. As a hobby, Max cultivates Mediterranean plants, including cyclamen, in his garden near Montpellier, carries out wildlife surveys and prospects Cévennes torrents for brown trout.

Martyn Denney has been growing cyclamen in north-east Hampshire since 1983 and exhibits regularly at Cyclamen Society shows. He has studied cyclamen in the wild for over 20 years and has participated in several of the Society's field studies. Martyn has an interest in the history of cyclamen, specialising in historical botanical illustrations. He has been the Secretary of the Cyclamen Society since 2002, is chairman of the Society's Research Committee and writes regularly for the Society's journal *Cyclamen*. As an antidote to cyclamen, Martyn has travelled widely in the eastern Himalaya studying the plants (*Arisaema* in particular) and the diverse cultures of the region.

John Grimshaw is Director of the Castle Howard Arboretum, North Yorkshire. Until 2012, he was Gardens Manager at Colesbourne Park, Gloucestershire, responsible for developing the snowdrop collection and display there — enhancing it with many cyclamen in the process. John has been a keen gardener all his life and joined the Cyclamen Society in 1982, aged 14. He began to develop a collection of *Cyclamen* soon afterwards, which was used in part for cytological investigations into the *C. hederifolium*–*C. africanum* complex during 1989. In addition to snowdrops and trees, John is particularly interested in African plants, having studied the Afromontane forests of Mt Kilimanjaro for his doctorate.

Ricardus M. Haber and **Myrna Semaan Haber**. Ricardus is father of the environmental movement in Lebanon, spreading awareness and founding the first environmental NGO in the Middle East in 1972. He created the first two Lebanese nature reserves and lobbied for their legal recognition, thus establishing a legislative precedence for nature conservation that preceded the Biodiversity Convention. Ricardus inspired the creation of the Lebanese Ministry of Environment by lobbying the political establishment.

Walking the natural landscape with his camera, Ricardus studied the flora of Lebanon, continuing the work of botanist P. Mouterde. Myrna, his student at the American University of Beirut and PhD holder in molecular systematics from the University of Reading, later joined him, and together they amassed a collection of photographs of over 3,000 species. They described three novelties: *Ballota byblensis, Salvia fairuziana* and *Phlomis tathamiorum* as well as several new records. Their books include *Orchids of Lebanon* (one of the series *Natural Heritage from the Mediterranean — Spaces and Species*), which covers 86 taxa that are illustrated by outstanding photography; *Lebanon, Land of the Four Seasons*; and *Floral Enchantment to Lebanon*, which presents 110 taxa bearing the name of Lebanon and narrates the history of botanical explorations from the 1500s through to the later visits of prominent botanists.

Terry Hatch has been a plantsman for over 60 years. Born in England, he completed his horticultural apprenticeship in a parks department before emigrating to seek adventure and new challenges in New Zealand in the early 1960s. Once there, he continued his passion for growing and breeding plants and, with his wife Pam, established Joy Plants Nursery, which has now been running for 47 years. Terry has exhibited at many New Zealand horticultural shows including 10 years in Auckland's Ellerslie flower show; over the years, his exhibits have received gold, silver and bronze medals, including the Horticultural Excellence award in 2004. More recently, Terry has been a judge at the shows. He has a keen practical interest in the conservation of New Zealand's native flora, and one of the vegetation restoration projects in which he is involved has been on-going for 30 years. In 1992, Terry was awarded the prestigious Plant Raisers' Award from the R.N.Z.I.H.. He is a member of the International Plant Propagators Society (IPPS) of which he was World President in 2008 and has been awarded life membership.

Ray Johnstone. "I live in Ryton, Tyne & Wear, and have been an Alpine Garden Society member since 1966. I am also a member of the Scottish Rock Garden Club, the R.H.S., the Cyclamen Society and the Crocus Group. My main interests are alpines and bulbs, vegetables and fruit. I am very keen on raising plants from seed and enjoy taking part, donating and receiving seed from the various societies. My interest in cyclamen developed in 1974 after finding *C. creticum* in western Crete. Holidays to the former Yugoslavia, Italy, Greece, Cyprus and Turkey cemented this interest, leading me to join the Cyclamen Society in 1978. I took part in the Society's 1992 Peloponnese field study and have visited this area many times since. My cyclamen are kept in a cool greenhouse; the more-tender species are kept frost-free in winter."

Christabel King has an Honours Degree in Botany from University College London and studied scientific illustration at Middlesex Polytechnic. Since taking up freelance botanical illustration, she has worked mainly at Kew and has painted many plant portraits for *Curtis's Botanical Magazine* and its associated monographs *Echinocereus* (1985), *Pleione* (1989), *Lewisia* (1989) and *Galanthus* (1999) and some artwork for *Lavandula* (2004). She also supplied illustrations for *Flowering Plants of the World* by V. H. Heywood (1978) and *All Good Things Around Us* by Pamela Michael (1980). Christabel has travelled widely; during an expedition to Uganda's Ruwenzori Mountains, she prepared plant portraits and landscapes for Guy Yeoman's *Mountains of the Moon* (1989). She has also tutored young artists from Brazil, as part of the Margaret Mee Amazon Trust scholarship scheme, as well as many students at Capel Manor and in Turkey. Her awards include the Jill Smythies Award from the Linnean Society of London.

Eugene N. Kozloff is an American marine biologist and botanist, who worked for many years at Friday Harbor Laboratories, University of Washington, and is now at Shannon Point Marine Center, Anacortes, Washington. He has written several books, some in collaboration with others, on the marine, freshwater, and terrestrial plants and animals of the Pacific Northwest. He has also had a long-standing interest in growing cyclamen. His collection of postage stamps that feature cyclamen can be seen in a section of our book devoted to this topic.

Dominique Lelièvre. "I was born in the South of France, but as my parents were born in Bourbon island (Ile de la Réunion), I had the good fortune to spend holidays there from time to time. There, I could savour tropical fruits and experience local plants and flowers such as vanilla, geranium and a rich and varied palette I could not name. During my first professional experience in Mexico, where I embarked on the long path of studying perfumery, I could also enjoy and extend these delicious sensations. Then, continuing my career in perfumery in England, I joined Robin Clery in a visit to Kew Gardens, where cyclamen were awaiting. Through direct olfaction, and with the help of headspace technique, we were able to get a little closer to understanding and reproducing these subtle and delicate odours."

John Lonsdale has grown cyclamen for more than 25 years. Born in Sheffield and living in the south-east of England, John's focus for more than 10 years was the propagation and cultivation of Primulaceae, particularly *Androsace*, *Cyclamen*, *Dionysia* and *Primula*. He enjoyed growing and showing plants in pots and was awarded the Alpine Garden Society's Gold Merit Medal. Following a move to the USA in 1995, John gardens at Edgewood Gardens (www.edgewoodgardens.net), a property of a little over 1.5 acres in Exton, south-east Pennsylvania. Working in a wonderful climate for growing in-character hardy plants, particularly bulbs, John has developed a garden focused mainly on woodland plants, especially *Cyclamen* and *Trillium*. Two greenhouses contain many *C. graecum* selections and facilitate the propagation of several thousand cyclamen from seed each year. More than a dozen cyclamen species are grown without protection in the garden, including tens of thousands of *C. hederifolium*.

Brian Mathew's career with plants began as a student at the Royal Horticultural Society's Garden at Wisley in 1960. He spent 25 years as a botanist in the Herbarium at the Royal Botanic Gardens, Kew, where he specialised in the study of the petaloid monocots (loosely known as 'bulbs'). He has a keen interest in growing bulbs, has travelled extensively to study them in the wild and has lectured widely on both the botanical and gardening aspects of the subject. His 20 books include monographs of *Crocus*, *Iris*, *Helleborus*, *Lewisia* and *Daphne* (with Chris Brickell). For 10 years, Brian was Editor of *Curtis's Botanical Magazine*, the longest-running periodical containing watercolours of plants. In 1992, the R.H.S. awarded him their highest award, the Victoria Medal of Honour.

Peter Moore. "My introduction to cyclamen was while walking in Austria when I was a boy. The plant seemed to have the combined merits of lovely scented flowers and beautiful patterned leaves. Over the following years, I built up a comprehensive collection of the different species. In 1986, I was able to take early retirement from my job as a printer in London and started Tile Barn Nursery. The nursery specialised in cyclamen and for the next 25 years provided me, and I hope my customers, with much pleasure. I was for many years an Officer of the Cyclamen Society, serving a term as its President, and was privileged to go on some of the Society's field studies."

Neriman Özhatay is a Director of the Department of Pharmaceutical Botany and ISTE Herbarium, Istanbul University. She is a taxonomic botanist and lectures as professor on medicinal plant systematics in the Faculty of Pharmacy of Istanbul University. Neriman is particularly interested the Turkish flora and in geophytes including *Cyclamen*; she has been a member of the Cyclamen Society since 1990. Neriman is a keen conservationist and was President of the Turkish Nature Protection Society from 2003 to 2006. She has written books and booklets to raise public awareness about the richness of flowering plants in Turkey and how this diversity needs protection. She has travelled extensively around Turkey recording and collecting specimens, in particular of the genera *Allium*, *Fritillaria*, *Muscari* and *Bellevalia*, describing new species such as *Allium ilgazense*, *A. sivasicum*, *Muscari anatolicum* and *Bellevalia anatolica*.

Pandora Sellars came to botanical illustration following her early careers in textile design and teaching. She is now acknowledged as one of the world's leading botanical illustrators and has contributed paintings to several books including the *Flora of Jersey* (1984), *The Genus Paphiopedilum* (1987), *The Flowering of Kew* by Richard Mabey (1988) and *The Genus Arum* (1993); some of her illustrations have appeared in the prestigious *Curtis's Botanical Magazine*. Pandora painted the portraits for a set of five British postal stamps depicting orchids and designed the commemorative plate and presentation picture for Diana, Princess of Wales to mark the occasion of the opening of the new Conservatory at Kew in 1987.

Graham Simpson was a great supporter of the Cyclamen Society's activities and was one of its first members, thereafter writing many articles for the Society's journal and serving as Treasurer for 10 years from 1987. He then continued as a committee member and took on the role of the Secretary of the Panel of Experts, assisting many Members and cyclamen enthusiasts with their queries and problems. Graham's profession as a solicitor provided great benefits when it came to writing the Society's rules. If any species could be said to be his favourite, it was probably *C. purpurascens*, which he travelled to see in its wild state on many occasions. As can be seen by the section in our book, cyclamen stamps were another interest that he addressed in a methodical manner. Graham died in 2003.

John Thompson is a research director with the CNRS at the "Centre d'Ecologie Fonctionnelle et Evolutive" in Montpellier. He has worked for more than 20 years on the ecology, adaptation and conservation ecology of Mediterranean endemic plants, both in relation to environmental variation and habitat fragmentation in southern France and in studying the processes of differentiation and evolution across the Mediterranean region. In 2005, John published the book *Plant Evolution in the Mediterranean* (Oxford University Press). His present research concerns the conservation status of rare species populations and their habitats, methods to establish conservation priorities for rare species, reintroduction projects for rare species and the evaluation of the attenuation and compensation measures associated with development projects that impact rare species populations.

Mats Thulin is a botanist and professor in the Department of Systematic Biology, Uppsala University. He specialises in the study and classification of tropical plants, in particular those of north-east Africa, and has for the last 18 years been preparing a *Flora of Somalia*. This work, often involving arduous and hazardous journeys, has contributed to Somalia now being acknowledged as one of the world's biological 'hotspots'. Mats has described and collected many new species, including the surprising *C. somalense*, and more recently a pink-flowered species of *Acacia* which proved to be widespread although previously unrecorded and very distinctive.

Jill Weatherhead is a garden designer in Victoria, Australia who has grown cyclamen for over 20 years, both in Melbourne and in the cooler Dandenong Ranges to the east. After obtaining a Bachelor's degree in Science at Melbourne's Monash University, she studied horticulture at Burnley College of Agriculture and Horticulture (now part of the University of Melbourne). Jill established the mail-order nursery Possum Creek Perennials which, amongst other bulbs, sold 15 Cyclamen species and varieties within Australia. Further studies included Landscape Architecture at the University of Melbourne. Jill also writes articles for magazines, newspapers and internet sites.

John Whitehead is a Kew-trained arboriculturist with a wide interest in all types of plants from wherever they occur in the world. He has travelled extensively to see and photograph rare plants and, fortunately for us, visited Cyrenaica in Libya where he observed *C. rohlfsianum*. One of his more recent projects has been to establish a tree trail and guide to the Grade 1 listed Glynllifon Gardens in Gwynedd, Wales, where unusual trees were first planted over 200 years ago. The range now includes Himalayan rhododendrons, bamboo-leaved oaks, Japanese cedars and Californian redwood; native trees have also been reintroduced to help preserve the habitat of the lesser horseshoe bat.

Trevor Wiltshire has been a Cyclamen Society member since 1979. He was selected to take part in the first Cyclamen Society field trip to Northern Turkey in 1987 and participated in Society field trips to Crete and Turkey. Trevor changed his career from finance to horticulture, becoming Rock Garden Superintendent at the Royal Horticultural Society gardens at Wisley in Surrey. He instigated a huge increase in the use of cyclamen across all the R.H.S. gardens. Every year, tens of thousands of cyclamen were planted. Trevor has travelled extensively, even as far as Japan, to lecture on cyclamen. He has been involved with indigenous plant propagation projects at Başkonuş and Dümlügöze in Turkey. Now retired in Cornwall, Trevor is still growing, studying and lecturing on cyclamen.

Jeremy Wood was a keen plantsman with several specialist interests including cyclamen and hellebores; he held one of the UK's National Collections of the latter genus. For many years, Jeremy chaired the Cyclamen Society's Research and Exploration Sub-committee and was instrumental in organising and leading the first of a long series of field studies. In 1990, he headed the team that studied *C. coum* in Turkey; we are fortunate that Jeremy completed the interesting account of this fieldwork, which is published here, well before his death in 2008.

CHAPTER ONE

The botany of cyclamen

1.1 STRUCTURE OF THE CYCLAMEN PLANT

Brian Mathew & Chris Clennett

Tubers and roots

Our initial impressions of a cyclamen are understandably focused on the above-ground parts, for they are what attract people to these lovely plants, but there are also plenty of interesting characteristics below ground in the form of a substantial storage organ: the tuber. This organ is the key to the success of cyclamen, and a great many other 'bulbous' plants, in coping with the alternating wet–dry climate of the Mediterranean. In this climate, many plants survive long dry inclement periods by existing underground as bulbs, corms or tubers.

Tubers may be modified stems, as in the potato, or swollen roots, as in *Dahlia*. The tuber of cyclamen is thought to be formed in the seedling stage by the swelling of the hypocotyl, the region between the roots and the stem of a young plant. In fact, some of the most diagnostic characters that allow us to distinguish cyclamen species are to be found in the shape of the tuber, the type of outer layer it possesses, the position of the roots on its surface and the thickness of the roots themselves. In some species, notably *C. graecum*, the roots are thick and fleshy, whereas in others, they are very fine and much-branched. Although it is sometimes difficult to identify a cyclamen species by its tubers alone, some species have unique features and can be identified at a glance. Enthusiastic growers who frequently handle cyclamen tubers when re-potting will be aware of just how diverse and interesting cyclamen tubers can be. The descriptions in this book

Tuber of *C. hederifolium* with underside devoid of roots (BM)

Tuber of *C. graecum*, with a surface like cracked bark and with thickened roots (BM)

Tuber of *C. coum*, smooth with roots from the centre underside (BM)

Tuber of *C. mirabile* showing thick roots produced in a ring above the base (BM)

apply to tubers that have reached their mature shape and are of a size that can support a flowering plant.

Some time after a cyclamen tuber becomes buried in the soil, it is likely to form a stem-like extension on its upper surface, thereby bringing the growing point to the surface while the tuber itself remains deep underground. This extension is often referred to as a 'floral trunk', although it also bears the leaves; it is usually fairly short but in extreme cases may be 20 cm or more long. All species are probably capable of forming a floral trunk.

Leaves

The foliage of cyclamen species appears at varying times of the year, depending upon the overall growth cycle. The autumnal species produce leaves during or soon after flowering; in the wild, the actual time of leaf production depends largely upon whether rain has arrived; in cultivation, it is dependent on watering. The spring-flowering cyclamen mostly have leaves already present when they bloom. In the majority of species, the leaves die away in summer, but three species, *C. parviflorum*, *C. purpurascens* and *C. colchicum*, tend to retain their leaves throughout the year, producing fresh ones in the growing season before the old ones have died away.

The leaves arise directly from the growing point of the tuber, or from the 'floral trunk' if one has developed, and usually several leaves form from each point, arising in a dense cluster. If a floral trunk has formed, it is possible to see that the leaves are actually produced alternately, not in an opposite arrangement or all in a cluster from one point (i.e. in a rosette). Their long petioles are either more or less erect, as in *C. persicum*, or grow horizontally beneath the soil surface before turning upwards to bring the leaf blade into the light. The latter is the most common state in the genus and is no better exemplified than in *C. hederifolium* and *C. repandum*. The petioles of the separate leaves tend to radiate from the tuber in different directions, thus spacing out the foliage over a wider area. Those who cultivate cyclamen will be familiar with the situation in which a plant in a pot produces its foliage all around the rim leaving a bare centre above the tuber.

The leaf blade (lamina) is extraordinarily variable in shape and markings and this is, of course, one of the main reasons for cyclamen's great appeal. The lamina have an instantly recognisable basic shape, which is generally heart-shaped to rounded with a markedly bilobed base. The margins are smooth, dentate, scalloped or distinctly and coarsely lobed. In general, it is possible to recognise each cyclamen species from its leaves, even though there is a great range of variability within most species. The photographs within the species descriptions in this book (Chapter 2) show just how remarkable and interesting the foliage can be. The markings on the upper surface of the leaves attract particular attention from enthusiasts. These most often form zonal patterns of light and dark green, pewter or silver, but may also take the

Leaves of *C. hederifolium*, a familiar form of the ivy-leaved cyclamen (BM)

The rounded leaves of *C. coum* (BM)

Cyclamen mirabile has coarsely toothed leaves (BM)

The lobed, jagged leaves of *C. rohlfsianum* (BM)

Cyclamen graecum has heart-shaped, very variable leaves with a satin-like overlay (BM)

Cyclamen graecum with long 'floral trunks' showing how leaves are attached alternately to the stem. A specimen from Attica, Greece (MD)

form of netted patterns of pale or silvery veins. Sometimes there are no markings at all, the leaves being wholly green (as in *C. parviflorum*) or wholly silvery-grey (as in *C. graecum* 'Glyfada'). The lower leaf surface is not patterned and is green or, more commonly, a shade of beetroot purple.

Inflorescence

The flowers of cyclamen are solitary, occurring on long pedicels that are each produced in the axil of a leaf. This is, however, almost impossible to observe because of the normally very compact nature of the growing point and the lack of a stem.

Pedicels

The pedicels, like the leaves, either are erect or suberect from the tuber (as in *C. persicum*) or grow horizontally before turning upwards and emerging from the soil (as, for example, in *C. hederifolium*, *C. coum* and *C. repandum*). At the apex, the pedicel turns over to form a crook shape, so that the flower is held in a pendent position. The pedicels are usually covered with minute glandular hairs. After successful pollination and fertilisation, the whole corolla falls off and the pedicel either starts to coil like a watch spring (in most species) or bends over towards the ground to bring the developing ovary close to the soil (in only *C. persicum* and *C. somalense*). In the former strategy, coiling starts at the top of the pedicel, just beneath the calyx, except in *C. graecum* and *C. rohlfsianum*, whose pedicels begin to coil in the middle or lower part. Either way, the effect is the same: to bring the young capsule down to ground level where it will mature over the course of several months, protected by the leaves from inclement weather and perhaps also from physical damage by grazing animals. It might seem curious that the seeds reach maturity on top of the parent tuber, but they are attractive to ants and are normally soon distributed away from the parent. Growers of cyclamen are well aware of the need for vigilance when the capsules are nearly ripe, as their crop of seeds can soon vanish. If the flowers are not fertilised, they shrivel while still attached to the pedicel, which becomes flaccid and withers away without coiling.

Flowers

The structure of cyclamen flowers is remarkably similar to that of a *Primula*, a fact that is more obvious if a cyclamen flower is 'caught' when the five lobes are spread out horizontally, halfway between the bud stage and the fully open stage during which the corolla lobes ('petals') are reflexed. Interestingly, although they are dicotyledonous plants, cyclamen have a flower shape that occurs not infrequently among petaloid monocotyledons ('bulbs'): *Erythronium* being the classical example but also *Narcissus cyclamineus*, many *Lilium* species and *Conanthera bifolia*. This strategy has almost certainly come about in plants that have evolved to flower during a rainy season after a long dry period; these condition are typical of the winter-rainfall Mediterranean region in which cyclamen grow naturally. Two factors are involved: the need to protect the pollen from rain and the need to attract pollinators when it is not raining. There are various ways of achieving this. 'Rainy-season' flowers such as *Crocus*, although upward-facing, have evolved the ability to open and close in response to light intensity and/or temperature; some plants that grow in this climate, including many *Lilium* species, have downward-slanting funnel-shaped flowers; others, such as *Fritillaria*, have pendent bell-shaped flowers. Cyclamen flowers are a sophisticated variation of this third type, with a subtle difference. They are pendent with the lower part forming a small bell that protects the stamens from the elements, but the larger free part of the corolla (the five obvious lobes) reflexes sharply upwards and acts as a showy advertisement to passing insects.

Looking in more detail at the cyclamen flower, we find a five-lobed calyx that is usually minutely glandular-pubescent; the lobes are united at the base to form a short tube or cup and most commonly they are ovate and pointed, sometimes also toothed at the margins. The base of the five-lobed corolla is also in the form of a short tube or cup, which can be seen most clearly when a flower falls off. This rather globular corolla tube nestles neatly within the cup-shaped calyx, with the five corolla lobes alternating with the lobes of the calyx. At the point where the corolla tube ends, the five lobes separate and bend sharply outwards and

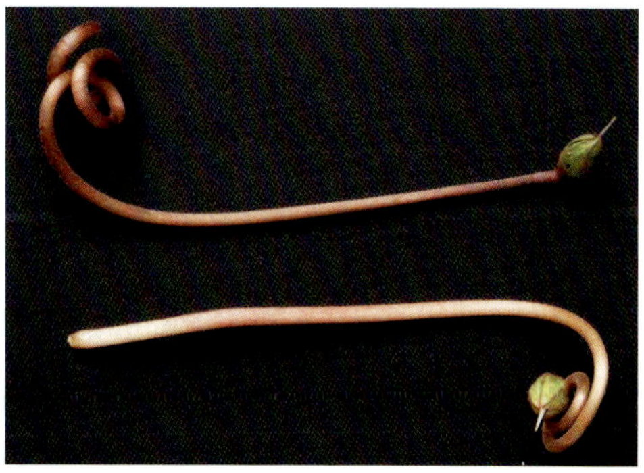

Pedicels of cyclamen after fertilisation of flowers: lower: *C. hederifolium* coiling from top; upper: *C. graecum* from the bottom (BM)

Flower of *C. hederifolium* from the back to show the globular corolla tube (BM)

backwards (except in *C. alpinum* in which they are often held more or less horizontally), thereby producing the characteristic cyclamen shape.

Often, there are conspicuous marks at the base of each lobe that act as a homing device for pollinators; the flowers are sometimes scented (see also 'Pollination', pages 6–7). Most species have their five corolla lobes twisted longitudinally. This might make the flowers more conspicuous when approached from different directions; alternatively, it might be quite coincidental and arise purely as a result of the very tight convolute vernation of the flower buds (i.e. like a rolled umbrella). Members of the Apocynaceae have similar vernation and, in some genera (e.g. *Vinca*), these too have a tendency to have corolla lobes that are slightly twisted as they unfurl.

Bud of *C. persicum* showing the 'rolled umbrella' effect (BM)

The corolla lobes, especially those of *C. mirabile*, might also be toothed at their tips. A useful diagnostic feature occurs at the mouth of the corolla; here, there might be an expanded portion that takes the form of two 'ears' (or auricles as they are usually referred to), one on either side of each corolla lobe base resulting in a ring of ten such swellings around the mouth of the corolla. These are present in the flowers of *C. hederifolium*, *C. africanum*, *C. graecum*, *C. rohlfsianum* and *C. cyprium*. The flowers of other species have a smooth transition from the corolla mouth into the lobes with no markedly expanded portion.

Face-on view of flower of *C. hederifolium* showing auricles and cone of anthers (BM)

Flower of *C. hederifolium* cut open to show 1-veined calyx and cone of anthers (BM)

Cyclamen hederifolium with swollen appendages (auricles) at the mouth (BM)

Cyclamen coum has no appendages (auricles) but very strong markings (BM)

The small small flowers of *C. intaminatum* are unmarked at the mouth (BM)

The protruding anthers of *C. rohlfsianum* are a unique characteristic in the genus (BM)

In the centre of the flower and converging to form a cone are the five stamens (the male part of the flower). Each of these consists of a very short filament, which is attached to the base of the corolla, and a narrowly triangular anther, which splits longitudinally to release the pollen. The anther slits face inwards to the centre of the flower (providing 'introrse dehiscence'); the outer faces of the stamens are furnished with minute wart-like papillae. A detailed study of the stamens by Clennett (2002) has shown that anther morphology is particularly interesting in delimiting cyclamen species.

The female part of the flower, the gynoecium, consists of a superior ovary (i.e. it is encompassed within the corolla rather than being outside it as in, for example, a *Narcissus*), which is usually ovoid and tapers upwards into the style, which continues to taper to an awl-like apex with a minute stigma. The ovary consists of five completely fused carpels and has free-central placentation with numerous ovules. The style and its stigma overtop the stamens but usually do not extend beyond the mouth of the flower; in some species (e.g. *C. repandum*), however, the style is exserted from the corolla tube, while in *C. rohlfsianum*, both the style and the cone of stamens protrude beyond the mouth (see photograph page 6). Cyclamen flowers are normally protandrous, that is the stamens reach maturity and release their pollen before the stigma is receptive, this being a method of encouraging cross-fertilisation between plants.

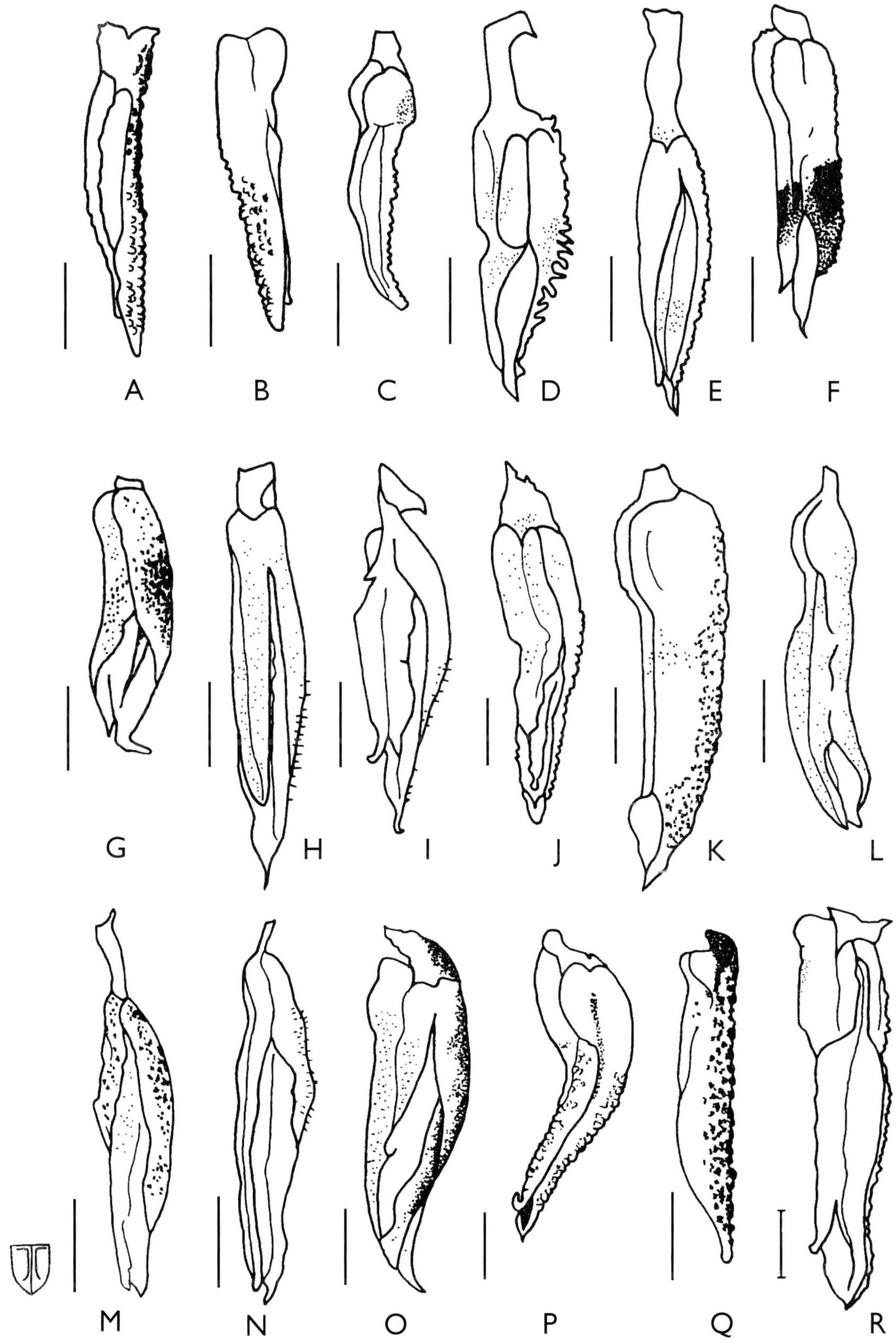

Stamen morphology in cyclamen (all species are shown except *C. intaminatum*, which is visually identical to *C. cilicium*). **A**, *Cyclamen repandum*; **B**, *C. creticum*; **C**, *C. balearicum*; **D**, *C. coum*; **E**, *C. elegans*; **F**, *C. alpinum*; **G**, *C. parviflorum*; **H**, *C. cilicium*; **I**, *C. mirabile*; **J**, *C. pseudibericum*; **K**, *C. libanoticum*; **L**, *C. cyprium*; **M**, *C. hederifolium*; **N**, *C. africanum*; **O**, *C. graecum*; **P**, *C. purpurascens*; **Q**, *C. persicum*; **R**, *C. rohlfsianum*. Scale bars = 1 mm. From Clennett, *Analysis and Revision of Cyclamen*. © *Botanical Journal of the Linnean Society* 138: 477 (2002) (CC)

Seed capsules of *C. hederifolium* showing coiling of pedicels (BM)

Ripe capsule of *C. hederifolium* showing method of splitting and large seeds (BM)

Flower of *C. rohlfsianum* showing relative position of stamens and style (BM)

Fruit and seeds

The cyclamen fruit is a subglobose capsule that is carried on a coiled pedicel. Only in *C. persicum* and *C. somalense* is the pedicel straight and arches downwards to reach the ground where the fruit matures. On ripening, a process that can take up to 9 months, the capsule splits unevenly at the apex, usually into five teeth that tend to curl outwards as they dry out, making the seeds visible and accessible. In each capsule, there are many large, irregularly shaped sticky seeds that are sugary and attractive to foraging ants, which facilitate their distribution away from the parent plant. The slowest cyclamen fruit to reach maturity after fertilisation are autumn-flowering species such as *C. hederifolium*, *C. graecum* and particularly *C. purpurascens*, which usually is the slowest of all cyclamen in forming mature capsules.

Pollination

The cyclamen flower has evolved to attract insect pollinators, of that we can be sure. First, it is relatively large and showy, so it is visible from a distance. Upon closer inspection, it can be seen that most species have, to the human eye, darker markings at the mouth of the corolla; these are often an indication that there is a roughly equivalent 'signal patch' in the UV spectrum, which humans cannot see but which is visible to a range of insects. Additionally, several cyclamen species have a noticeable fragrance that emanates from the interior of the corolla tube; this varies considerably (see 'The scent of cyclamen', Chapter 3.2) from species to species and might well act as advertisement to a variety of potential pollinators.

Other factors that are relevant to pollination are: (1) the pendent position of the flower and the position of the stamens and stigma within it (which allow shed pollen to fall due to gravity); (2) the small, lightweight nature of the pollen; and (3) the fact that the style is longer than the stamens, so that its apical stigma is held clear of the anthers (i.e. it lies beneath the stamens as the flower is pendent). This arrangement is most easily observed in *C. rohlfsianum* in which the stamens and style are clearly visible. These factors might suggest various things: pollen could fall under its own weight directly onto the stigma, thus effecting self-pollination; similarly, any insect visitor could readily be dusted with pollen on its head and other upper parts; and, being very light in weight, cyclamen pollen might also be distributed by wind.

So, there are three possibilities: insect pollination and wind pollination (both of which could result in cross-fertilisation) or self-pollination. The desirability of cross-pollination between different plants of the same species is clear: to encourage genetic diversity as opposed to in-breeding. Many plant species have evolved ways to encourage or even ensure this, for example by being genetically self-incompatible or by being dioecious (i.e. with male and female flowers on separate plants). The genus *Primula* has flowers with differing lengths of style (heterostyly) but this feature is not present in *Cyclamen*. Although insect and wind pollination might lead to cross-pollination, it is not necessarily certain; cyclamen plants often have several flowers open at the same time and insects often visit several of them before moving on to those of another plant, so selfing is a distinct possibility. Self-pollination is also a strong possibility given the flower shape and the relative positions of the stamens and stigma; but are cyclamen self-compatible? Although it makes evolutionary sense to encourage cross-pollination, it makes survival sense to have a back-up system if cross-pollination fails. Personal experience, and a fact known to most cyclamen growers, is that cyclamen are readily self-compatible. It is not unusual to have a cyclamen plant with a solitary flower that is fertilised and produces viable seeds. It is a long-standing joke that if you wish to get good seed production, you should take the plant to a show by car as shaking during the journey ensures pollination! Failing that, just gently tapping the flowers daily throughout their flowering period has the same effect.

Self-compatibility presumably accounts for the ability of certain selections to reproduce reasonably uniformly by seed, a factor that enables growers to select and name cultivars that are relatively stable in their characters, a very useful attribute. In many genera, it is not possible to obtain uniform cultivars by seed as they vary so much. As yet, no satisfactory method of increasing cyclamen by vegetative means has been found so it has been impossible to select individual clones as is possible for many cultivars. Good examples of acceptably uniform (seed-raised) cultivars in cyclamen are *C. graecum* 'Glyfada' and several of Peter Moore's 'Tilebarn' selections of various species.

It may be that not all cyclamen are self-compatible. Mats Thulin's account of *C. somalense* in cultivation certainly seems to suggest that this species is not self-compatible (Chapter 2, page 127). Nevertheless, one must be careful in reaching conclusions; sometimes a lack of seed production in early- or late-flowering geophytes is brought about by inclement weather (e.g. temperatures that are too low for pollen germination) or unsuitable growing conditions that result in weakened stressed plants.

Returning finally to insect pollination, bees and hoverflies seem to visit cyclamen regularly and the following have also been observed: butterflies, beetles and, in the case of *C. persicum* in Israel, a small moth (*Micropterix elegans*) and several species of *Thysanoptera* thrips (Schwartz-Tzachor *et al.* in *Journal of the Cyclamen Society* 29: 64–68, 2005). Pollination can be effected in the obvious way, by physical transfer of pollen from one flower to another on an insect's body (leading to cross-pollination). It can also result from the high-frequency movement of the insect's wings, causing disturbance of the pollen (i.e. 'buzz pollination') some of which might well fall onto the stigma resulting in self-pollination.

What is clear is that there is still plenty of scope for detailed observations of and studies into the whole matter of pollination and compatibility in each cyclamen species. The range of flower colours, markings and scents suggests that the different species could have subtly differing pollination mechanisms. Anyone who has sniffed the flowers of a wide range of cyclamen will be aware of the marked differences between them, from the sweet, heavy scent of *C. purpurascens* to the distinctive coconut smell of *C. mirabile* and the more honey-like perfume of *C. cilicium*. The scientific investigation of fragrance in cyclamen is dealt with in Chapter 3.2 in considerable depth.

Cyclamen hederifolium with honey bee visitor in the wild (BM)

Cyclamen purpurascens visited by hover fly (BM)

Cyclamen purpurascens with bumble bee visitor (BM)

Cyclamen coum shedding copious amount of pollen, activated by tuning fork to simulate 'buzz' pollination (BM)

Chromosome numbers

Repandum Group (= Subgenus *Psilanthum*)
Cyclamen repandum 2n = 20
Cyclamen creticum 2n = 20, 22
Cyclamen rhodium 2n = 20
Cyclamen balearicum 2n = 20
Coum Group (= Subgenus *Gyrophoebe*, Series 1)
Cyclamen coum 2n = 30
Cyclamen elegans 2n = 30?
Cyclamen parviflorum 2n = 30
Cyclamen alpinum 2n = 30
Pseudibericum Group (= Subgenus *Gyrophoebe*, Series 2)
Cyclamen pseudibericum 2n = 30
Cilicium Group (= Subgenus *Gyrophoebe*, Series 3)
Cyclamen cilicium 2n = 30
Cyclamen intaminatum 2n = 30
Cyclamen mirabile 2n = 30
Cyprium Group (= Subgenus *Corticata*)
Cyclamen cyprium 2n = 30
Cyclamen libanoticum 2n = 30
Persicum Group (= Subgenus *Persicum*)
Cyclamen persicum 2n = 48 (72, 96, 136 in some cultivars)
Cyclamen somalense 2n = c. 96 (Thulin 2006, a new count, reported here on page 128)
Hederifolium Group (= Subgenus *Cyclamen*, Series 1)
Cyclamen hederifolium 2n = 34, 68
Cyclamen confusum 2n = 102
Cyclamen africanum 2n = 34, 68
Purpurascens Group (= Subgenus *Cyclamen*, Series 2)
Cyclamen purpurascens 2n = 34
Cyclamen colchicum 2n = 34 (as reported by Grey-Wilson in *Curtis's Botanical Magazine* 15(3): 181 (1998))
Graecum Group (= Subgenus *Cyclamen*, Series 3)
Cyclamen graecum 2n = 84 (85, 86, 87)
Rohlfsianum Group (= Subgenus *Cyclamen*, Series 4)
Cyclamen rohlfsianum 2n = 96

1.2 CYCLAMEN CURIOSITIES AND ABERRATIONS

Vic Aspland

One of the enduring questions to the Cyclamen Society Panel of Experts begins in one of two different ways: "This year, one of my cyclamen has..." or "I recently bought a cyclamen that had...". A description of the plant follows, and the question is rounded off with: "Is this unusual?"

One of the greatest fascinations of growing cyclamen is the apparently infinite variation in leaf shape and pattern, within the confines of a more-or-less 'standard' design. A lifetime can be spent collecting different leaf forms of a single species. But this is just normal 'variation', and cyclamen are capable of much more extreme behaviour, which we label aberrations. Aberrant forms can be found of all parts of the plant: flower, sepal, leaf, tuber and floral trunk.

A plant of *C. purpurascens* with long 'floral trunks' and new tubers forming above the old (JB)

Tuber and roots

The tubers of an individual species have a fairly standard form: in *C. graecum*, for example, younger tubers of up to 10–20 years old are more-or-less spherical, have a thick brownish skin that breaks into strips of flakes, and thick roots growing from the base of the tuber. Nevertheless, there is considerable variation between species. With some experience, many growers can distinguish between the dormant tubers of most species, but even here, aberrations can be found. Staying with *C. graecum*, I have two spindle-shaped tubers with smooth skins (origin unknown). Could these be detached trunks that are gradually turning into tubers?

Cyclamen repandum and *C. creticum* are well-known for their ability to form floral trunks. In nature this allows them to locate the growth-point from which their flowers and leaves develop at the optimum depth. Less commonly, *C. graecum*, *C. rohlfsianum*, *C. hederifolium* and *C. cilicium* also form floral trunks, and in the Lake Garda area of Italy, populations *of C. purpurascens* exist that appear to consist of floral trunks only: no tuber! In *C. cilicium* and *C. purpurascens*, the trunks might also show brownish nodules, which may be incipient tubers.

In general, roots are the least-variable part of the cyclamen plant. For each species, they generally emerge from a particular area of the tuber, but here also, variation can occur. The roots of younger *C. graecum* and *C. mirabile* plants emerge from a circular area at the base of the tuber. If the roots of older tubers are lost from the base, replacements usually re-grow at the outside edge of this area. Repeated errors in cultivation, and root loss, will lead to the production of a 'bald' central area with an annulus of replacement roots around it.

The search for aberrant forms can add a new angle to your cyclamen collection. You may encounter the pretty, the strange, the weird and the ugly. Perhaps you may care to reserve a corner of your greenhouse for such things? If you do, you will be able to follow an idea of the great gardener, E. A. Bowles, who had a corner of his garden reserved for odd plants, which he christened 'The Lunatic Asylum'.

Leaves

Perhaps the most familiar aberration in leaves is the development of colour. *Cyclamen mirabile* was so named (*mirabile* = wonderful) by Hildebrand for the red-purple colour present on the young

A remarkable situation recorded in *C. repandum* in which roots are being produced from leaf stalks, observed by Jean Marie van der Weghe (JMVW)

Pinkish-leaved forms of *C. mirabile* are particularly desirable (MD)

leaves of the single specimen that he possessed. It is unfortunate that the colour fades as the leaves develop. Other species can also show colour development, though perhaps not to the same extent. In *C. hederifolium*, a proportion of the plants bearing the name Bowles Apollo Group can show a red or pink flush over part or the whole of the leaf surface. *Cyclamen rohlfsianum* sometimes shows the same effect, although in this case, the colour tends to be concentrated in and near the leaf patterns. More recently, I have seen specimens of *C. cilicium* showing the same phenomenon. Some forms of *C. graecum* hailing from mainland Greece and some of the islands have an underlying pinkish-brown cast, often completely masking the normal green colour. In all of the plants I have seen, the red colours fade as the leaves develop and age. Microscopic examination of *C. mirabile* leaves has shown that the colour is produced by glandular cells on the upper leaf surface. I am aware of no published examinations of the leaf colour of other species, but have been shown a fascinating trick developed by a Society member's *C. purpurascens*. The young leaves have a red-purple flush over the upper surface, and the underside is pale green. As the leaves age, the colour is lost from the upper surface and the underside becomes beetroot red. When I did a microscopic examination of leaf sections, I observed that the colour resulted from the presence of a beetroot red dyestuff held within individual leaf cells, but only those in the outermost layer in both leaf surfaces. Throughout the colour-change process, all cells within the main body of the leaf remained green. What a fascinating metabolic trick!

Rarely, leaves are partly or wholly marked with yellow. A number of *C. hederifolium* plants showing this effect have been

A variegated form of *C. coum* raised by Jan Bravenboer (JB)

found by Society expeditions, and a seed-raised plant was exhibited by a member at the Wisley show in 2005. A variegated form of *C. coum* has been selected by Jan Bravenboer and a plant of *C. rhodium* seen by Marco ten Hoope has almost wholly cream-coloured leaves. The effect is usually curious rather than attractive, although who knows what might be achieved by careful breeding?

The colour changes described above are usually shown by all, or most, of the leaves on an individual plant, whereas many aberrations of leaf shape are one-off affairs. A not-uncommon variation is the production of a leaf with two, or even three, points, instead of the normal one at the tip. In the most interesting cases, the leaf can even become V-shaped. The cause is a 'genetic hiccup' of some kind, in which the normal well-ordered replication of the rapidly increasing cells in the growing leaf is disrupted in some way. In the opposite direction, plants might produce leaves that are truncate: with a straight rather than a pointed tip, as if cut off with a pair of scissors. The Society expeditions that studied *C. alpinum* in Turkey found several plants that had this leaf form. In this case, their shape was genetically determined, as all the leaves on the individual plants were truncate.

The leaves of this plant of *C. rhodium* seen by Marco ten Hoope contain very little chlorophyll at all (MH)

A fine silver leaf form of *C. coum*, selected by Jan Bravenboer (JB)

A form of *C. alpinum* seen in populations near Marmaris, Turkey, with permanently pendent petals; this feature is retained in cultivation (MD)

To encompass the many variations in 'normal' leaf shapes would be impossible, as no two plants are *exactly* the same, but some fashion trends are developing at present. In terms of leaf shapes, very long narrow leaf forms seem to be gaining a keen following, although really weird mutations such as that recorded by A. P. de Candolle in 1808 are very unusual (see Chapter 7, page 495). In the opposite direction, an enthusiasm for holly-like leaves seems to be developing. Not so many years ago, leaves with a silvery or pewter-coloured upper surface were regarded as 'unusual', or aberrant if you wish. Now, forms of *C. coum* and *C. hederifolium* with this type of leaf are regarded as relatively commonplace (although 'better' forms are always eagerly snapped up) and the hunt is on for 'silvers' in other species. I recently saw offered seed under the name *C. cilicium* Silver Leaf. Perhaps careful breeding using selected *C. creticum*, *C. balearicum* and *C. rhodium* subsp. *peloponnesiacum* could lead to the next developments in this area, as there are forms that are already heavily patterned with silver.

Flowers

Flower shape

The flower contains some of the most rapidly developing cells in the whole cyclamen plant, so it is not surprising that the occasional flower develops in an abnormal way. The opening of a cyclamen bud is a fascinating process in itself. The petals, which are tightly wrapped around each other rather like a rolled umbrella, gradually elongate and unwrap; once separated, they move through almost 180 degrees to adopt the normal position.

The exception is, of course, *C. alpinum*, in which the petals only move to the horizontal. Experience shows that the final move to the vertical can be arrested at any stage, or may not take place at all. I have grown a specimen of *C. purpurascens* that produced flowers of a propeller-like shape similar to those of *C. alpinum*. During the Cyclamen Society expeditions to south-western Turkey to study *C. alpinum*, a number of plants were seen in which the whole of each petal drooped down below the level of the calyx. One expedition member (T. Wiltshire) became so excited when each one was spotted that these were rapidly christened 'Trevor's drooping'. Variations such as these are usually a genetically determined characteristic of the individual plant, and will occur consistently each growing season.

The reflexing of the petals can, however, be arrested for very obvious physical reasons. Although not an everyday occurrence, it is not uncommon for the petals to separate only partly, or for only one pair of petals to separate. In such a case, the flower adopts a fan shape, and remains in a downward-facing position. A drawing of a flower of this type produced by a *C. persicum* plant was once published in the Cyclamen Society's journal and, more recently, I have been sent a photograph of a *C. coum* on which all of the flowers remained as flared funnels, creating the overall shape of a *Soldanella*. On one of the Society's field trips, another curious form was observed: a *C. rhodium* that looked more like a snowdrop. Whether this characteristic is genetically determined remains to be seen in future seasons.

An aberrant flower of *C. rhodium* with petals failing to reflex. Seen in the wild in Greece (MD)

Flower colour

Most cyclamen enthusiasts will, at some time, have suffered the tired old joke made by the uncommitted, that the colour range of cyclamen flowers runs all the way from white to pink. This is very easy to refute of course, by pointing out the magenta strains of *C. coum* or *C. pseudibericum*, *C. hederifolium* Ruby Group and so on. But more subtle colour variations also occur. *Cyclamen hederifolium* is well-known for its ability to produce white flowers in one season and pink in the next. Despite numerous studies, the crucial factor that causes this variation it is still unknown because it happens so unpredictably; summer temperature, drought or moisture, feeding or starving remain possible causes. More perplexing still, a plant will very occasionally produce both white and pink flowers at the same time. At the time of writing, I have a *C. coum* performing this trick; last year, all of its flowers were white! I have no reports of other species changing colour in this way; although the reason for this might be simply that there is greater opportunity for this phenomenon to be observed in *C. hederifolium*, as it is grown in enormous numbers, than in other species. There is more fun in store with all of this spontaneous variation. The occasional flower has individual petals that are either divided into pink and white zones or have stripes. This feature is highly unstable and might be prominent for a single year and never appear again. It is most often seen in *C. persicum* (especially in the highly bred florists' forms) and in *C. hederifolium*.

A variation that has only been reported for *C. persicum* is the formation of crested petals. In this form, the central area of the petal running upward from the mouth of the flower can be thickened, ribbed or fluted, producing the crest. A number of older strains of florists' hybrids still exist, but are rarely offered for sale. Occasionally forms arise in which the petals are distinctly fimbriate or fringed at the margins. Such a plant was raised by Jan Bravenboer.

Cyclamen hederifolium with fimbriate-margined petals (JB)

To round off the discussion on colour variation, I will stay with florists' *C. persicum*. Plants are occasionally seen in which the flowers have an all-over flecking, the spots being of a deeper shade of pink than the main body of the petal. This feature is usually consistent and has been known to growers for a very long time. A chromolithograph published in *The Floral Magazine* in 1869 shows two white-flowered plants with pale pink flecks. Another has been illustrated in Vilmorin-Andrieux's *Album de Clichés*. You should be warned, however, that such flecks could also be an indication of an infestation by cyclamen bud mite. Fortunately, this pest seems to be much less common than it was 30 years ago, and it is quite a number of years since I have seen a plant that might have suffered an attack.

Twin flowers, double flowers

Individual species have particular foibles when it comes to aberrations, but the speciality of *C. coum* seems to be the production of two flowers on the end of a stem. These flowers are almost invariably very close together, and each has its own calyx. The effect can be quite attractive, as usually each flower develops normally, but this oddity is usually confined to the occasional flower stem. The stems bearing such flowers may be fasciated: wider than normal and flattened, or having a cross-section like a figure eight.

A florists' *C. persicum* with speckled petals, from Vilmorin-Andrieux *Album de Clichés* (DDB)

The doubling of flowers is an interesting phenomenon, and can occur in a number of ways. Most frequently, it is produced by the conversion of anthers into extra petals. In the case of fully double florists' auriculas, this conversion can be so complete that a plant cannot be used for breeding. In almost any cyclamen species, the odd anther or two can undergo this transformation, producing a petal that sticks down from the mouth of the flower. The effect is decidedly odd, but not attractive and rarely appears in subsequent years. The cause is unknown; presumably a cell in the microscopic developing bud malfunctions or receives the wrong chemical message, a gene is switched off (or on?), and the cell divides to form the wrong organ.

Perfect doubling can be found in florists' *C. persicum*, but is very unusual among the 'undeveloped' species. Nevertheless, in John Parkinson's *Paradisi in Sole*, published in 1629, there is a clear woodcut illustration and a careful description of what he called 'Double flowred Sowebread of Antioch'. His description reads as follows:

> "This Sowebread of Antioch with double flowers hath his leaves fomewhat round, like vnto the leaues of the Summer Sowebread, but with leffe notches or corners, & full of white fpots on them: it beareth flowers on ftalks, like vnto others, & likewife fome ftalks that haue two or three flowers on them, which are very large, with ten or twelue leaues a peece, of a faire Peach colour, like vnto the flowers of purple Sowebread of the Spring, and deeper at the bottome.
>
> There are of this kinde fome, whofe flowers appeare in the Spring, and are as large and double as the former, but of a pure white colour."

Cyclamen creticum near Spili with seven petals (JF)

An unusual multipetalled form of *C. coum* selected by Jan Bravenboer (JB)

A double form of *C. persicum* selected in Japan (NY)

This aberrant *C. hederifolium* produced three seed heads joined together (JB)

A white-flowered 'Stargazer' (JB)

Flowers of *C. hederifolium* of the 'Stargazer' type (JB)

Other oddities of flowers and stems

A further oddity is shown in Parkinson's double-flowered *C. persicum*: the pedicel (the stem bearing the flower) is branched, and also bears two leaves. An identically branched stem of florists' *C. persicum* was sent to me by a Society member a few years ago. A more recent break occurred in *C. hederifolium*: the pedicel fails to bend through 180° in normal shepherds-crook fashion, the flowers face upwards rather than downwards. These aberrants have been named 'Stargazer'.

The calyx is a fruitful source of aberrations. It consists essentially of five modified leaves that form a cup or sheath around the base of each flower. Occasionally, specimens of *C. coum* will show reversion of the calyx segments to the ordinary circular or kidney-shape of normal leaves. Unfortunately, this is an impermanent feature; a plant might show this effect for one or several years, and then it disappears, never to happen

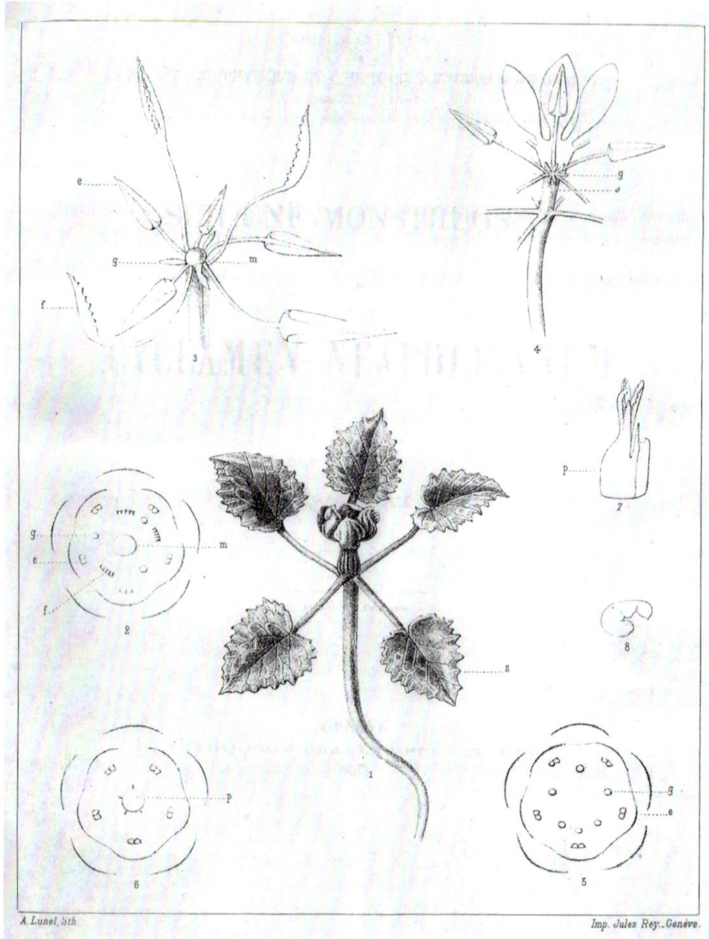

Flower wholly mutated into small leaves. This was noted in 1887 by C. de Candolle and published as: '*sur une monstruosité de Cyclamen neapolitanum*' (BM)

A plant of *C. hederifolium* with the same mutation as that recorded by de Candolle; raised by J. Lever (BM)

again. There are suspicions that exposure to critical levels of herbicide or (more likely) pesticide might cause this to happen. In the case of *C. hederifolium*, the flower itself may disappear completely, leaving only the calyx. In 1887, C. de Candolle published a description and illustration of a 'monstruosité' in which the flower — or what remained of it — was surrounded by small, stalked leaves. Jim Lever of Bolton once raised three seedlings having this characteristic, which is interesting but not immediately attractive! This can also happen in other species. During some field work in 1996, the Cyclamen Society found a strange plant of *C. creticum* in which the petals were absent and the enlarged leaf-like calyx formed the dominant feature of the flower. Recently, Martyn Denney saw another similar malformation in a wild population of *C. cyprium*, which was not unattractive when seen from the reverse side.

A form of *C. creticum* seen in the wild with a leaf-like calyx (JF)

Cyclamen cyprium with leaf-like calyx mutation. Seen by M. Denney in Cyprus November 2011 in the Pentadaktylos/Beşparmak mountains c. 10 km west of Kantara Village, altitude approximately 600 m (MD)

1.3 USES, LEGENDS, CULINARY VALUE AND TOXICITY OF CYCLAMEN

Brian Mathew

Much has been recorded over a period of two millennia about the herbal properties and toxicity of cyclamen, and extracts from some of the Old Herbals can be found in our chapter on early botanical illustration (Chapter 6). Some traditions are clearly based on superstition, such as the recommendation that a pregnant woman should not step over a cyclamen plant for fear of terminating her pregnancy. There is, however, no doubt that cyclamen, in particular the tubers, contain toxic substances and should certainly not be eaten; conversely, there are recent accounts of the leaves being used in cooking. The safe option is probably to note the various records with interest but to continue growing cyclamen only for their aesthetic properties! This succinct verse recorded by Professor D. Soban, an anaesthetist from Slovenia, is a salutary reminder of the toxic nature of cyclamen, which are perhaps even as poisonous as the notorious toadstools:

Amanita or Cyclamen

Dangerous Amanita (BM)

Toxic cyclamen (BM)

or vice versa
Cyclamen and Amanita —
Amen

The following notes give an outline of the wide range of uses that have been attributed to cyclamen including, somewhat surprisingly, their value in the kitchen. Dr Joseph Rupreht of the Faculty of the Department of Anaesthesiology, Erasmus University Rotterdam, provides a scientific account of the chemistry behind these alleged properties. I recommend that the two articles are read in conjunction as the latter might act as a deterrent for any inexperienced chefs!

Perhaps one of the more practical uses of cyclamen was recorded by Richard Bailey in northern Turkey: the juice from cyclamen — presumably *C. coum* — is rubbed on the hands as a grease remover. This ties in well with Dr Rupreht's chemical explanation below concerning the saponins contained in cyclamen plants. Similarly, Melvyn Jope and Peter Moore learned that, in the Peloponnese, cyclamen tubers are put into streams to stun and poison fish: this can be explained by the presence of the toxin cyclamine. This seems to be an easy way to catch fish, although one hopes the corpses are safely edible afterwards. This method of fishing was noted by Pliny (d. 79 AD) who recorded it in his monumental *Historia Naturalis* (c. 77–79 AD) and is still in practise in some places.

As to legends, there is the advice for pregnant women mentioned above, which is probably based on the known effects of ingesting just a small amount of cyclamen tuber: severe internal illness that could lead to an abortion. Antonia Fraser in her book, *The Weaker Vessel*, notes the deliberate use of 'sowbread' tubers as an abortifacient.

A quite unrelated tale, recounted to Alisdair Aird in Corfu, is that cyclamen are dug out of vineyards as their presence is thought to result in poor fruit. Some plants are suspected of having a deleterious effect on the surrounding soil in order to deter competition, but could that explain this practice? Or could it possibly be that the situation and soil conditions in which the cyclamen thrive are not ideal for growing grape vines. This legend dates back at least to the time of Pliny, who advised that a vine would wither away if planted near a cyclamen. The great herbalist Dioscorides, in the first century AD, recommended cyclamen as a cure for very diverse problems: snake bite, smallpox, jaundice, loss of hair and constipation. The list was added to in the many herbals of the Middle Ages and more detail can be found in our account of early botanical illustration in Chapter 6. Even in present times, one can find cyclamen used as an additive in the homeopathic treatment of migraine and tinnitus and in a fungicidal skin cream.

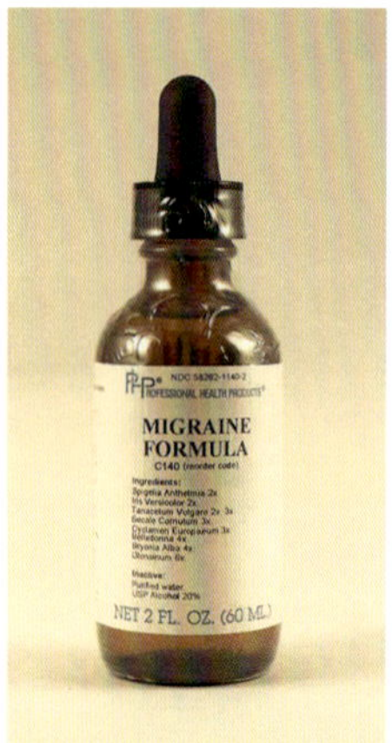

Homeopathic migraine treatment

In view of these various medicinal attributes, it is somewhat surprising that cyclamen have been used in the kitchen — but only in a well-cooked state. Melvyn Jope wrote in the Cyclamen Society's journal for December 2008 that he had come across references to cyclamen leaves being used like vine leaves as the wrapping for meat or vegetable rolls. He also found a recipe for 'cyclamen leaves stuffed with split peas', which originated from the Greek island of Simi and involving peas, rice, onions, tomatoes and olive oil. Prompted by this article, Anne Killingback and Michael Pritchard looked further into the culinary value of cyclamen leaves and were able to add some information that is worth repeating as it indicates a current usage in Lebanon and is in considerable detail.

Lebanese cooking with cyclamen

Anne Killingback and Michael Pritchard

Our daughter Leda is married to Awdat, a Lebanese Druze. He is a chef and they run a Lebanese restaurant in Norwich. Over a long dinner with them, Awdat told us about cooking with cyclamen during his childhood and as a chef. This was a small part of a much wider discussion of stuffed leaves and vegetables in Lebanese cuisine; a vast topic in its own right. Considerable discussion ensued to arrive at an English translation from Arabic, as ingredients have to be identified and an English spelling assigned. Leda then wrote up several pages of notes on which the following is based.

Pickled cyclamen tubers

These are not very palatable and would be made only in times of great hardship, such as the civil war which was only 15 years ago. During the war, there was some food aid but growing crops in the fields was dangerous as shelling from across the valley was very likely. So crops such as turnips became impractical.

To pickle the cyclamen tubers, it is necessary to top and tail them and remove any damaged areas. They are then immersed in brine with the addition of wine vinegar. The concentration of the brine should be such that an egg will float in it. Leave for at least four months. They will keep for years and it is probably best to do so as they can be very hard and horrible. Not recommended as it is very difficult to get them right. Pickled turnips are a hundred times better!

Stuffed cyclamen leaves

Stuffed cyclamen leaves are known in Lebanon as *Dwayk ak Jabal* (which translates as 'Cock of the Mountain') and are still to be found on the menus of some restaurants in Lebanon.

Just before our discussion in mid-January 2009, Awdat had spent some time in Lebanon. The cyclamen leaves were abundant and much larger than usual, being up to 20 cm (8 in) in length in some cases. People were coming out from Beirut to the mountains to harvest them for cooking. The larger leaves are found in shady places, often under pine nut trees. The season for the leaves is November to February when other local vegetables are scarce.

Stuffed vegetables and leaves are the centre-piece of traditional Lebanese cooking. They are time-consuming to make but are a way of making expensive scarce ingredients go a long way. Stuffed vegetables, known as *me'hershi* or *mashi*, have two types of stuffing. The first contains meat, usually lamb, and is served hot as a main course. The second stuffing is made entirely from vegetables and is served cold as a starter. The most prized leaves for stuffing are cyclamen, Swiss chard and vine leaves, in that order. The large cyclamen leaves are easier to stuff than vine leaves, which have serrated edges to cope with. The vegetable stuffing is preferred with cyclamen leaves as it brings out the subtle flavour of the leaves.

Before starting, it is necessary to have a large, heavy-bottomed pan. A pan of about 3.5 litres (6 pints) will hold 40–50 leaves.

The stuffing can now be made. The recipe given by Melvyn Jope in the Cyclamen Society's journal is satisfactory, but Awdat considers it to be too bland. His recipe is basically the same, but the split peas are replaced by chick peas that have been soaked overnight and then rolled in a tea towel to remove their skins. In addition, there would be a finely chopped red or red-green pepper, a third of a cup of chopped parsley, two tablespoons of finely chopped mint, a tablespoon of fresh lemon juice, a quarter teaspoon of ground cinnamon, a similar quantity of seven peppers spice mix (black pepper, allspice etc.), salt, pepper and chilli to taste. Mix all the ingredients and leave to stand. Note that there are no strict proportions for this type of stuffing; it all depends on the chef's preferences. Now the pot is lined with a layer of sliced potatoes (with their skins on) and tomatoes. This is covered with vine leaves if available. When finished, this layer is in itself a tasty dish.

The leaves are blanched and rolled to produce the typical dolmades shape that Melvyn describes. These are packed into the lined pan with the seam downwards. Two tablespoons of olive oil, lemon juice and water are mixed and poured over the stuffed leaves to just cover them. Inverted saucers or small plates are put on top to hold everything down.

The pan is brought to simmer and allowed to cook for an hour or so until the stuffed leaves are tender and the filling fully cooked. The water must be kept topped up throughout to avoid burning. When cool, the *me'hershi* can be unpacked from the pan. They can be frozen for later use and will keep for a year in the freezer. They can be served cold with lemon wedges.

Toxicity of *Cyclamen purpurascens*

Dr Joseph Rupreht

Tubers of *C. purpurascens*, and probably of other species as well, contain considerable quantities of a very toxic glycoside, cyclamine, which is responsible for the bitter and burning taste. According to several sources, as little as 0.3 g of ingested tuber can cause vomiting, diarrhoea and abdominal cramps. Abortions may occur in women as a result of the severity of the illness. Larger quantities of the fresh tubers cause severe intoxication that can end lethally. The cause of death is thought to be paralysis of the breathing centre. In practice though, the supra-regional Toxicological Centre of Zurich has no record on file of any known serious intoxication with *C. purpurascens*.

However, two children and two cows are known to them to have vomited after ingestion of the green parts of cyclamen. Whether this concerned florists' cyclamen (*C. persicum*) or *C. purpurascens* is not known. Cyclamine is extremely toxic to fish and even minute quantities cause excessive permeability of their breathing surfaces; essential electrolytes are then lost into the surrounding water and the fish die. By contrast, it is interesting that pigs can tolerate considerable quantities of cyclamen tubers without apparent ill-effects.

Properties of cyclamine

Cyclamine is a saponin, a class of chemical compounds that frequently occur in plant tissues. True to their name, which is derived from the Latin word for soap, saponins form foam in water solutions. Poisoned animals may produce urine that forms a remarkable amount of foam, so much so that horses poisoned by a saponin can produce a carpet of foam up to their hocks. Chemically, saponins are glycosides and consist of one or several sugars bound to an organic base. One group of saponins comprises cardiac glycosides, which are usually described separately because of their specific actions on the heart. The presence of saponins in herbal material is determined by foam tests, a haemolysis test, thin layer chromatography or photometrical means. Cyclamine is a potent haemolytic (blood-destroying) compound, the organic base part of which is called aglycon. Aglycons determine the capacity for haemolysis, whereas the attached sugar molecules determine the magnitude of this effect. Cyclamine owes its potency as a haemolytic poison to the fact that it has four sugar molecules, attached at one position only (see figure), and the whole substance has a neutral pH in a water solution (Steinegger & Hansel, *Lehrbuch der Pharmakognosie und Phytopharmazie*, Springer-Verlag, 1988).

There seems to be ample chemical evidence that cyclamine is very toxic. Books on toxic plants support this knowledge (Habermehl, *Mitteleuropäische Giftpflanzen und ihre Wirkstoffe*, Springer-Verlag, 1985; MAFF, *British Poisonous Plants*, Bulletin 161, 1979). It is therefore surprising that no documented

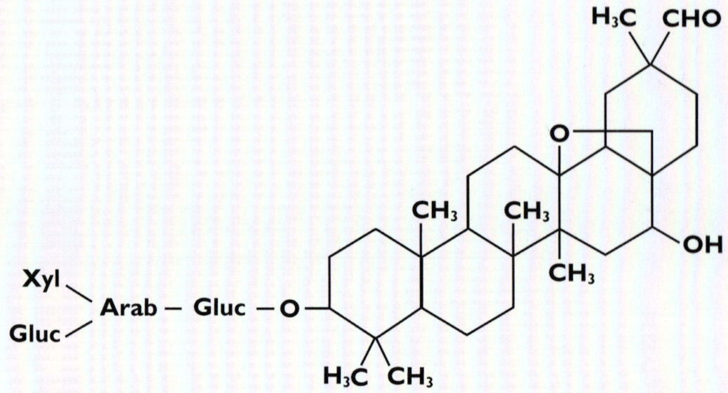

Cyclamine, a 3-terpen-glycoside with four sugars and an esteric conformation

recent cases of serious intoxications resulting from cyclamen ingestion have been reported. From the available scientific data about cyclamine, it is difficult to deduce why the tubers of C. *purpurascens* and *C. persicum* would initiate abortion and why the final stage of poisoning should end in central respiratory arrest. The effects of cyclamine on the cell membrane, with the exception of the haemolytic action, are not well understood. Substances from the tuber might act on smooth muscles, causing intestinal cramps, but this symptom could also be brought about by damage to the mucosa caused by saponin. Final respiratory arrest would most probably be the end stage of a hypovolaemic shock. Treatment of cyclamen poisoning would probably include an attempt to empty the stomach, treatment of haemolysis, and intensive cardiovascular and ventilatory support when necessary.

Cyclamen belongs to the family Primulaceae, other members of which also store saponins in their underground parts. Some species of *Primula* (*P. obonica* and *P. malacoides*) regularly cause a nasty rash or a serious dermatitis. This contact dermatitis develops if the substance primin is liberated from special hairy glands. This raises the question: are any similar glands known in cyclamen and are there any records of allergic phenomena caused by cyclamen?

CHAPTER TWO

Cyclamen in nature

2.1 INTRODUCTION, DISTRIBUTION AND CLASSIFICATION OF CYCLAMEN SPECIES

Brian Mathew & Chris Clennett

Introduction

The following section describes the 23 species of *Cyclamen*, arranged in their perceived natural groupings (i.e. their classification). We follow a standard format for each species: after the introductory remarks, we provide botanical information — the place of first publication of the species, the type locality and specimen, and a botanical description. This is followed by a broad outline of the species' natural distribution and habitat. The 'Further observations' sections cover articles and extracts written by cyclamen enthusiasts that provide valuable first-hand information, mostly about plants observed during their travels. For some species, there are longer accounts of fieldwork carried out by members of the Cyclamen Society, whose purpose is not only to enjoy growing and promoting these wonderful plants but also to extend scientific knowledge of them.

In instances where information was lacking on matters such as distribution, habitat and conservation, or where there were questions of taxonomic status, the Society has allocated funds and expertise for field studies, which were mostly carried out in collaboration with local scientists and conservationists. Such studies have added much valuable information to the pool of knowledge about cyclamen. Examples of this work include extension of the known distributions of several Turkish species that were previously thought to be rather restricted, and clarification of the taxonomy of the *C. hederifolium* group in Crete and the subsequent recognition of *C. confusum* as a species. Accounts of the more extensive of these field surveys, and reports on the ongoing molecular studies in which the Society is involved, are given in Chapter 3 'Scientific studies and fieldwork'. Also included there is a description of the collaborative research work carried out by Robin Clery on the wide range of fragrances encountered in cyclamen.

Distribution

Cyclamen species are distributed widely through southern Europe, western Asia and North Africa. Looking more closely at the distribution, one realises that on the whole they are never far from the influence of the seas: the Mediterranean, Black Sea and Caspian. They do not penetrate far inland into the dryer interiors of countries such as Turkey, where they would be away from moderate coastal climates. The one exception is *C. purpurascens*, which ranges from sea level on the Adriatic coast of the Balkans up into the mountains of Italy and Switzerland and then eastwards as far as Poland and Romania. Even *C. mirabile*, which does extend far inland in western Turkey, is usually found in areas that have Mediterranean enclaves: for example near the large lake of Eğirdir. Nowhere, do cyclamen extend into what would be classed as an Irano–Turanian climatic zone, with harsh winters and long dry summers.

The total distribution of the genus is from the Balearic Islands and southern France, east through the Mediterranean to Turkey

(which is especially rich in species), the Caucasus and the Talysh and Elburz ranges, west and south of the Caspian Sea as far as north-eastern Iran, where the more arid conditions again hold sway and prevent any further eastern expansion. Cyclamen occupy most of the Mediterranean and Aegean islands, certainly all the larger ones and probably many of the smaller ones too. Northwards, they extend to Austria, southern Germany, southern Poland and the Crimea; southwards, they occur through Syria, Lebanon and Jordan as far as Israel. In North Africa, cyclamen are present in Tunisia, Algeria and Libya; there is one outlier in north-east Africa, *C. somalense* from the north-eastern tip of Somalia, which also has a Mediterranean-type climate.

Some species, like *C. hederifolium* and *C. purpurascens*, are very widespread; others, such as *C. intaminatum*, *C. parviflorum*, *C. pseudibericum*, *C. cyprium*, *C. rohlfsianum* and *C. somalense*, are local endemics.

Classification of *Cyclamen* species

There have been many suggested systems of classification for the genus *Cyclamen* at levels above that of species: subgenera, sections and series. It is difficult to choose which system to follow for a book such as this, which is intended primarily for cyclamen enthusiasts. The situation will no doubt continue to change as techniques of analysis become ever more sophisticated and hopefully nearer to revealing 'the truth'. What doesn't change very much is the classification at the species level, which is comforting for the cyclamen aficionado. Yes, there has been some shuffling between members of the *C. repandum* alliance, which continues, and some reassessment of *C. hederifolium* and its variants, but on the whole, the species are distinct and easily recognised. With regard to the *C. repandum* and *C. hederifolium* complexes, much more will be found later in the book under the species accounts below and in Chapter 3, 'Scientific studies and fieldwork', which includes two recent in-depth examinations of subgenus *Psilanthum* (the *C. repandum* group) by Debussche & Thompson and by Culham & Bota.

For those who wish to learn more about the history of cyclamen subgeneric classification, there is much available: Hildebrand's monograph of 1898; Glasau, 1939; Legro, 1959; Meikle, 1978, 1985, 1989; Schwarz, 1955, 1964; Anderberg, 1994; Grey-Wilson, 1997, 2002, Clennett, 2002 (the full references can be found in Chapter 9 'Cyclamen literature'). The systems proposed by Grey-Wilson (1997, 2002) and by Clennett (2002) have much in common and take into account the results of previously published studies. These systems are also based on a thorough knowledge of the species, in cultivation, in the field and in the laboratory. For those who know and grow cyclamen, their classifications of the species have a practical approach that we follow in the layout of this book.

Groups of *Cyclamen*
following Grey-Wilson (2002), with modifications

Repandum group (= Subgenus *Psilanthum*)
- *Cyclamen repandum*
  - var. *baborense*
- *Cyclamen balearicum*
- *Cyclamen creticum*
- *Cyclamen rhodium*
  - subsp. *peloponnesiacum*
  - subsp. *vividum*

Coum group (= Subgenus *Gyrophoebe*, Series 1)
- *Cyclamen coum*
  - subsp. *caucasicum*
- *Cyclamen elegans*
- *Cyclamen parviflorum*
  - subsp. *subalpinum*
- *Cyclamen alpinum*

Pseudibericum group (= Subgenus *Gyrophoebe*, Series 2)
- *Cyclamen pseudibericum*

Cilicium group (= Subgenus *Gyrophoebe*, Series 3)
- *Cyclamen cilicium*
- *Cyclamen intaminatum*
- *Cyclamen mirabile*

Cyprium group (= Subgenus *Corticata*)
- *Cyclamen cyprium*
- *Cyclamen libanoticum*

Persicum group (= Subgenus *Persicum*)
- *Cyclamen persicum*
  - var. *autumnale*
- *Cyclamen somalense*

Hederifolium group (= Subgenus *Cyclamen*, Series 1)
- *Cyclamen hederifolium*
- *Cyclamen confusum*
- *Cyclamen africanum*

Purpurascens group (= Subgenus *Cyclamen*, Series 2)
- *Cyclamen purpurascens*
- *Cyclamen colchicum*

Graecum group (= Subgenus *Cyclamen*, Series 3)
- *Cyclamen graecum*
  - subsp. *candicum*
  - subsp. *anatolicum*

Rohlfsianum group (= Subgenus *Cyclamen*, Series 4)
- *Cyclamen rohlfsianum*

2.2 THE SPECIES OF *CYCLAMEN*

Brian Mathew

with contributions from Members of The Cyclamen Society

2.2a The Repandum group (= Subgenus *Psilanthum*)

Comprising *C. repandum*, *C. repandum* var. *baborense*, *C. balearicum*, *C. creticum*, *C. rhodium*, *C. rhodium* subsp. *peloponnesiacum* and *C. rhodium* subsp. *vividum*

A group of four spring-flowering species that have ovate or triangular leaves and elegant flowers with long narrow petals. The corolla is white or pink to reddish-purple, sometimes with a darker stain around the mouth but with no obvious dark blotch or jagged markings. There are no swellings ('auricles') around the mouth. The stamens do not protrude from the corolla but the style is often clearly exserted. In fruit, the pedicels start to coil downwards from the apex. Chromosome number 2n = 20, also with a record of 2n = 22 in *C. creticum*.

The 'Repandum group', or Subgenus *Psilanthum*, probably presents the most taxonomically difficult challenge within the genus *Cyclamen*, probably one to which there is no absolutely definitive answer. The group includes a collection of similar but slightly different entities that are distributed in the Mediterranean region, from the Balearic Islands and southern France, through Italy and the Adriatic part of the Balkans, to Greece and the islands of Crete and Rhodes. Surprisingly, they are not recorded on the nearby mainland of Turkey, but there is also a North African representation in Algeria. The present distribution is most likely the outcome of a process whereby populations of a formerly widespread cyclamen became physically separated and isolated during periods of changing climate and sea levels. Over time, this has led to a degree of morphological and genetic differentiation but, as evolution is a continuous process, the story is in a state of flux, particularly so in the case of *C. repandum* and its allies. Taxonomists try their utmost to put everything into discrete 'boxes', but there is often no neat 'answer' to questions regarding classification for the simple reason that it is subject to continual change. In view of this state of affairs, this book includes both our standard layout of the *C. repandum* group (in this chapter) and learned accounts by two research teams (in Chapter 3). These two teams are not in opposition, in fact their findings are largely complementary, but there is a difference in the outcome of their deliberations, although surely both would agree that this might well change as their studies continue.

First, it should be noted that there is no disagreement between researchers as to which cyclamen belong within the 'Repandum group' (subgenus *Psilanthum*), but the status of each of these is another matter. This problem has been addressed several times in the light of increased knowledge and increasingly sophisticated techniques for assessing data, so we are now confronted by a range of opinions as to what constitutes a species, subspecies, variety or form. Extensive studies of the group have been carried out by Max Debussche and John Thompson working at the Centre d'Ecologie Fonctionnelle et Evolutive in Montpellier. Starting with detailed local studies of the small, fragmented southern French populations of *C. balearicum*, they extended their research to include other members of the alliance, combining various techniques to put together a comprehensive account of the whole of this group.

Another study by Alastair Culham and Pedro Bota at Reading University, employing both morphological and molecular techniques, suggests that this assemblage of entities appears to separate into two groups, although even then the situation is still unclear. For this reason, they regard their findings as interim results and suggest that further detailed work is required using both disciplines; an alternative suggestion by Culham is that it might be better to wait for evolution to sort the matter out by causing the entities to diversify further! Although this remark is made in jest, it does underline the fact that much of the plant world is subject to continuous change and cannot be strictly regimented, however sophisticated the techniques. Nevertheless, we do need names in order to communicate, whether it is to convey erudite information, to discuss plants with friends, to offer plants for sale in a nursery or for conservation purposes. In difficult cases such as this, the 'Repandum group' of *Cyclamen*, it may be that a compromise is the best that can be achieved: a system that is practical and based on appearance (morphology) combined with an awareness of perceived lineage (genetics).

We are indebted to the teams of researchers from Montpellier and Reading for making their valuable work available to us and both research papers are included in full in the 'Scientific studies and fieldwork' chapter of this book (sections 3.3 and 3.4). Nevertheless, it will be helpful to provide here an outline of the classifications proposed by the two teams, demonstrating how there is full agreement on the taxa recognised but not on their status.

Both groups of researchers use, for the sake of convenience, an informal means of communication rather than a cumbersome hierarchy. Culham & Bota (2009) give the six taxa 'brief names' and, using morphological data from leaves, showed that the taxa fell into two distinct groups: (a) *balearicum*, *creticum* and *rhodium* and (b) *repandum*, *peloponnesiacum* and *vividum*.

Analysis of DNA sequences also shows them to fall into two distinct groups, although slightly different from those above: (a) *creticum*, *rhodium*, *peloponnesiacum* and *vividum* and (b) *balearicum* and *repandum*. Applying the code of nomenclature to these molecular groups would result in the following combinations: *creticum* subsp.

creticum, *creticum* subsp. *rhodium*, *creticum* subsp. *peloponnesiacum* and *creticum* subsp. *vividum* (*C. creticum* being the earliest of all these epithets at specific level); *repandum* subsp. *repandum*; and *repandum* subsp. *balearicum*. However, Culham & Bota do not yet intend to make these new combinations formally, as further sampling and more sophisticated analysis is needed before nomenclatural changes should be made on the basis of DNA sequence data.

In their work at Montpellier, Debussche & Thompson used common names specially coined for the purpose and usefully based on the plants' origin (see page 213 for the full list). Their research suggests that two species should be recognised in the group, although these are different from those suggested by Culham & Bota: *C. balearicum* and *C. repandum*, the latter with the following subspecific variants: *C. repandum* subsp. *repandum* (with two varieties, var. *repandum* and var. *baborense*), subsp. *peloponnesiacum* (with two varieties, var. *peloponnesiacum* and var. *vividum*), subsp. *creticum* and subsp. *rhodense*.

There is still more research work to be done; here, we take a perspective that will hopefully be simple to follow and will not confuse cyclamen enthusiasts any further. Our treatment recognises *C. repandum* (including var. *repandum* and var. *baborense*), *C. balearicum*, *C. creticum* and, for the Peloponnesian and Rhodian representatives of the group, *C. rhodium* (comprising subsp. *rhodium*, subsp. *peloponnesiacum* and subsp. *vividum*). It should be explained that *C. rhodium*, published in 1975 by Otto Schwarz & Lothar Lepper, is the earliest and therefore valid name at the species level.

Cyclamen repandum var. *repandum* and var. *baborense*

Flowering in spring

This familiar spring cyclamen is certainly one of the best for naturalising in gardens that are situated in relatively mild, moist climates. Some gardens in the south-west of England (e.g. Knightshayes, Killerton) have dramatic drifts of *C. repandum*, which flowers at about the same time as many of the spring bulbs, such as *Fritillaria meleagris*, *Erythronium* and dwarf *Narcissus* species. The vibrant reddish-purple shade of its flowers provides a bright splash of colour with the added advantage of a delicate fragrance. This garden-worthiness has not gone unnoticed: the species been awarded not only the ultimate accolade for a plant of horticultural value, the RHS Award of Garden Merit (AGM), but also an Award of Merit (AM) in 1959 and a First Class Certificate (FCC) in 1973. The white-flowered variant, forma *album*, had been awarded an AM even earlier, in 1952. Interestingly, the FCC and AM plants all originated in Corsica and the comment has been made that Corsican forms are larger-flowered than those in other parts of the species' distribution. It would take extensive field studies to ascertain the validity of such a claim, but it certainly appears to be the case that white forms occur more frequently in Corsica than in other parts of this plant's range. The species has a fairly extensive distribution in Corsica, Sardinia, Sicily, mainland Italy and the Adriatic region of the Balkans. It is to be found in one region of southern France, near St Tropez, and in a rather restricted outlying area in northern Algeria, where it is referred to as var. *baborense*. There have also been reports that *C. repandum* occurs in Corfu, Albania and southern Switzerland; although these seem plausible places for it to exist, there are, to our knowledge,

A fine drift of *C. repandum* at the National Trust's garden Killerton, Devon (BM)

Cyclamen repandum var. *repandum*. Painted by Christabel King (reproduced at 115% of original size)

no positive sightings. The supposition that *C. repandum* is present in Corfu possibly comes from the comment in one of Gerald Durrell's books, that he saw cyclamen growing there in spring. A very full account of the species' distribution and ecology can be found in Max Debussche and John Thompson's very thorough paper, which comprises Chapter 3.4.

The first valid name for this well-known species (i.e. designated after Linnaeus' *Species Plantarum*, 1753) was published by James Edward Smith in the first volume of Sibthorp & Smith's sumptuous *Flora Graeca Prodromus* (1806) and illustrated by no less a botanical artist than Ferdinand Bauer. Not surprisingly, this plant, which occurs in parts of southern Europe and notably in Italy, was known long before this and we can find excellent woodcuts depicting it from the sixteenth century onwards. More can be found about these and other early illustrations in Chapters 6.1 and 7.

The epithet *repandum* refers to the leaf margin meaning wavy or uneven.

C. repandum Smith in Sibth. & Smith, *Fl. Graeca Prodr.* 1: 128 (1806) var. ***repandum***. Lectotype: [designated by Compton & Culham, *Journ. Cyclamen Soc.* 27(2): 70 (2003)] Bauer's original watercolour, tab. 186, page 177 (*MS Sherard* 244, OXF).

Synonyms Probably all of the following are referable to *C. repandum* as minor variants:

C. eucardium Jord. in Jord. & Fourreau, *Icon. Fl. Europ.* 3: 16–17 (1903).

C. ilicetorum Jord. in Jord. & Fourreau, *Icon. Fl. Europ.* 3: 17 (1903).

C. lobospilum Jord. in Jord. & Fourreau, *Icon. Fl. Europ.* 3: 17 (1903).

C. spectabile Jord. in Jord. & Fourreau, *Icon. Fl. Europ.* 3: 16–17 (1903).

C. stenopetalum Jord. in Jord. & Fourreau, *Icon. Fl. Europ.* 3: 16 (1903).

C. rarinaevum Jord. in Jord. & Fourreau, *Icon. Fl. Europ.* 3: 16–17 (1903).

C. repandum Smith var. ***baborense*** Debussche & Quézel, *Acta Bot. Gallica* 144, 1: 30 (1997). Type: Algeria, Petite Kabylie, Djebel Babor, El Ma Berd, 1890, *Battandier* s.n. (holotype MPU).

Description (both varieties) **Tuber** depressed-globose, not more than 7 cm diameter at maturity, brown, usually smooth and velvety but sometimes becoming slightly corky in old age, rooting only from the centre of the underside. **Petioles** prostrate in the lower part below ground, then ascending to erect above ground. **Leaves** well developed before flowering time. Leaf blade broadly ovate-triangular, 2–15 cm long, 1.5–15 cm wide, usually with an irregularly triangular dark green zone in the centre and a dark green marginal zone, separated by a very variable pale green or grey green jagged zone, which is sometimes broken into separate patches on a green background, occasionally almost wholly dark green with few or no markings; margins somewhat lobed and scallop-toothed, the teeth widely spaced and varying from shallow to very prominent; the underside pale green or purplish suffused. **Pedicels** horizontal in the lower part, turning upwards as they emerge above ground, after anthesis coiling from the top downwards. **Flowers** produced in spring, fragrant; petals rather

Cyclamen repandum in Gargano, Italy (TH)

Tuber of *C. repandum* (BM)

Flower of *C. repandum* (BM)

A population of *Cyclamen repandum* in the Gargano region, southern Italy (TH)

In the Foresta Umbra, Gargano National Park, Italy, *C. repandum* growing in oak woods, 19 April 2011 (TH)

long and narrow, swept upwards into an erect position, rich reddish-pink, stained darker reddish-purple towards the mouth of the corolla, narrowly elliptic-oblanceolate, c. 1.5–2.5 cm long (var. *repandum*) up to 3 cm long (var. *baborense*), 0.5–1 cm wide (var. *repandum*), 0.7–1.1 cm wide (var. *baborense*), acute or subacute at the apex, without swollen appendages ('auricles') at the base, strongly twisted lengthways; style apex clearly exserted from the mouth of the corolla. 2n = 20.

Flowering period March–May (in the wild).

Distribution var. *repandum*: Italy, Sicily, southern France (Maures Massif), Adriatic coastal region of the Balkans, Corsica, Sardinia; var. *baborense*: northern Algeria.

Habitat var. *repandum* — shady places in deciduous and evergreen woods and scrub, shady rocky terrain, 0–1,300 m; var. *baborense* — deciduous oak forests, 400–1,800 m.

Variations The naturalised populations of *C. repandum* seen in gardens in Britain are generally remarkably uniform in appearance, with the 'standard' hastate leaf markings in pale or greyish green. In the wild, however, there are variants with foliage ranging from completely green to almost wholly pewter or silver. Apart from in the white-flowered forma *album*, the flower colour is generally a rich reddish-purple, the tone a little variable, but not to the same extent as that of many other cyclamen species.

Cyclamen repandum country, the Solenzara Valley, Corsica (MD)

Cyclamen repandum f. *album* at St. Florent, Corsica (MD)

Named cultivars

'Pelops' was used originally to denote the populations growing on the Peloponnese, which are treated here as *C. rhodium* subsp. *peloponnesiacum* so the cultivar name does not now apply.

Further observations on *Cyclamen repandum*

An early case of mistaken identity Only a year after the publication of James Edward Smith's name *C. repandum* and the accompanying watercolour by Ferdinand Bauer, the species was illustrated again, this time by Sydenham Edwards in *Curtis's Botanical Magazine*. Unfortunately, the application of Latin names to cyclamen at that time was often accompanied by errors and this was the case here, the plate being published in 1807 under the name "*Cyclamen hederaefolium*. Ivy-leafed Cyclamen, or Sow-Bread". The description includes the observations that is a "native of Italy and a very valuable plant on account of its early flowering, sweet scent and beautiful foliage" and that it "could be propagated by cuttings of the root". The inaccurate application of the name seems to have gone largely unnoticed for a while, and it was not until later in the nineteenth century that James Atkins, the snowdrop devotee of Painswick, Gloucestershire, wrote to Joseph Hooker the Director of Kew Gardens and Editor of Curtis's journal that he would have gladly supplied specimens of *C. repandum* to be featured in the magazine "had it not been already figured in the Bot. Mag. t.1001 under the erroneous name of *hederaefolium* – the syns. there given are some of them quite wrong – and I have found in looking into the matter pretty fully that several authorities repeat the same syns. with all their errors one after the other evidently without examination."

We can probably identify all our cyclamen quite accurately nowadays, if only botanists could agree on the classification!

***Cyclamen repandum* and *C. purpurascens* together?** On the face of it, this combination of species growing together in the wild seems a little unlikely, the former being in general a fairly lowland plant whereas the latter is often seen in mountain woodlands. Nevertheless, they can both occur at or near sea level, and the combination of *C. purpurascens* and *C. repandum* together on the Adriatic islands of the balkans was confirmed in an article in the Cyclamen Society's Journal of June 2001 by **József Szendefi** from Hungary. He writes:

> "In mid January 2001, I visited the region of Split in southern Croatia, and then in early February, Krk island in the north of the country. From the overall distribution of cyclamen species, it should at least be possible to find *C. repandum* in both places. On 4 February, I found *C. repandum* all around Krk Island, with the exception of the area around the bridge from Rijeka at the northern tip of the island...... I found a place with both *C. repandum* and *C. purpurascens* growing together under one tree, which was the first time I have seen both species together."

Cyclamen repandum also occurs on the Adriatic coast of the Balkans wherever there is a suitable shaded habitat (BM)

***Cyclamen repandum* on Cyprus? Desmond Meikle** reports in his *Flora of Cyprus* (1985) that in 1900 Post recorded *C. repandum* occurring at Evrykhou ("Ericon"), but adds that Post was probably mistaken and that the plants were likely to have been *C. cyprium*. Martyn Denney visited the area, in the Solea Valley, but notes in the Cyclamen Society's journal of December 2001 that he found neither species in that location.

Cyclamen balearicum

FLOWERING IN SPRING

Although this white-flowered species is one of the smallest cyclamen, it has considerable appeal as the leaves are often beautifully silver-washed and the flowers delicately scented. Gardeners might expect it to be a rather tender plant, as it originates from southern France and the Balearic Islands, but it is not particularly so and is certainly a good subject for an unheated glasshouse or even a protected position outdoors. The detailed ecological account presented by Max Debussche and John Thompson in Chapter 3.4 gives a clear indication of why this is so: the species can be subjected to fairly harsh conditions in at least some of its locations.

Cyclamen balearicum in the Formentor area of Mallorca (DT)

C. balearicum Willk., *Oesterr. Bot. Zeit.* 25: 111 (1875). Type: Lectotype [selected by J. A. Rosselló & L. Sáez in *Collect. Bot. (Barcelona)* 25 (1): 3–203 (2000)]: Mallorca, prope Miramar, 16 April 1873, *Willkomm* Herb. Balear. 20 (COI-Willkomm).

SYNONYM *C. hederifolium* subsp. *balearicum* (Willk.) O. Schwarz, *Gartenflora n.s.* 1: 22 (1938).

DESCRIPTION **Tuber** depressed-globose, usually flattened both above and below, 2–4 cm diameter at maturity, brown, smooth, rooting from one point on the underside. **Petioles** prostrate in the lower part below ground, then ascending to erect above ground. **Leaves** well developed before flowering time. Leaf blade often slightly revolute (turned downwards) at the edges, broadly ovate, 2–6(–9) cm long, 2–6(–9) cm wide, green or grey-green with conspicuous silver marbling, often almost entirely silver-grey; margin entire or very weakly toothed; the underside normally purplish suffused. **Pedicels** horizontal in the lower part, turning upwards as they emerge above ground, after anthesis coiling from the top downwards. **Flowers** produced in spring, scented; petals swept upwards to an erect position, wholly white with semi-transparent veins, rarely with faint pink suffusion at the mouth and margins, narrowly elliptic-oblanceolate with subacute or obtuse tips, c. 8–20 mm long, 0.3–0.5 cm wide, twisted lengthways, without swollen appendages ('auricles') at the mouth; style held within the mouth of the corolla or slightly exserted. 2n = 20.

Tuber of *C. balearicum* (BM)

Flower of *C. balearicum* (BM)

FLOWERING PERIOD February to May

DISTRIBUTION Spain: Balearic Islands of Majorca, Minorca, Ibiza, Cabrera and Dragonera; southern France in the Languedoc-Roussillon region (for more detail see page 213); Corsica (see page 215).

HABITAT Usually in shady places, on north-facing slopes in evergreen woods and scrub (e.g. *Pinus*, *Quercus* or *Buxus*), in gorges and rock crevices, usually on limestone formations, occasionally sandstone or schist; sea level to 1,445 m.

Cyclamen balearicum. Painted by Christabel King (reproduced at 130% of original size)

A habitat for *C. balearicum*, Formentor, Mallorca (DT)

Healthy regeneration by seed in a population of *C. balearicum* at Formentor in Mallorca, growing in pine mould (DT)

C. balearicum in Mallorca growing in a shady humus-rich scree, with fairly dull leaves (DT)

Cyclamen balearicum in France, at Mont Luziers, showing attractively silvered leaves (MD)

A fine plant of *C. creticum*, Aradhena gorge, Crete (JF)

Variations In cultivation, *C. balearicum* is frequently seen with almost wholly silver-suffused leaves, presumably because these forms are considered the most desirable, but in the wild the leaves can vary in the degree of markings to almost completely green. There is only a little variation in leaf shape, but the margin can be nearly entire or with a few shallow teeth. Flower size, as for all cyclamen species, varies considerably between approximately 1 and 2 cm long. Although the flowers are generally white with almost transparent veins, it is possible to find plants with the faintest pink tinge at the mouth of the corolla (see also comments on page 213).

Cyclamen creticum

Flowering in spring

Like *C. balearicum*, which it closely resembles, at least in flower characteristics, this small cyclamen is not one of the most striking. Nevertheless, it has undoubted charm, especially when seen in its native haunts on Crete. There, it is a quite common plant, widespread on the island wherever there are suitably shady conditions. It can also be found in more open situations, usually as a result of woodland or scrub clearance. Although not a showy species, *C. creticum* is, like its Balearic relative, well worth cultivating. It is probably best kept in a glasshouse or frame where the soft leaves get some protection from late frosts but, in mild districts, there is no reason why it should not be as successful as *C. repandum*.

C. creticum (Döerfl.) Hildebr., *Beih. Bot. Centr.* 19 (2): 367 (1906). Type: Crete, Aghios Vasilis, Kedros, *Doerfler Iter Cretica* 902, 1904 (?BP, W)

Synonyms

C. repandum var. *creticum* Döerfl., *Ver. Zool.-Bot. Ges. Wien* 55: 20 (1905).

C. hederifolium subsp. *creticum* (Döerfl.) O. Schwarz, *Gartenflora n.s.* 1: 22 (1938).

C. repandum subsp. *creticum* (Döerfl.) Debussche & Thompson, *Bot. J. Linn. Soc.* 139: 141 (2002).

Description **Tuber** depressed-globose, up to c. 4 cm diameter at maturity, brown, smooth, with slender roots arising in a cluster from the underside. **Petioles** prostrate in the lower part below ground, then ascending to erect above ground. **Leaves** well developed before flowering time. Leaf blade broadly ovate, sometimes angular, 4–12 cm long, 3.5–12 cm wide, often greyish-green with a more conspicuous grey or silver zone and frequently with irregular splashes of silver; margin weakly to strongly toothed; the underside normally purplish-red suffused. **Pedicels** horizontal in the lower part, turning upwards as they emerge above ground, after anthesis coiling from the top downwards. **Flowers** produced

Tuber of *C. creticum* (BM)

Flower of *C. creticum* (BM)

Cyclamen creticum. Painted by Christabel King (reproduced at 120% of original size)

A range of leaves to show variation in *C. creticum* (JF)

in spring, scented; petals swept upwards to an erect position, uniform white with semi-transparent veins or less commonly pale pink, narrowly elliptic-oblanceolate with subacute or acute tips, c. 1.5–2.5 cm long, 0.4–0.7 cm wide, slightly twisted lengthways, without swollen appendages ('auricles') at the mouth; style held within the mouth of the corolla or slightly exserted. 2n = 20, 22 (for additional information see 'Further observations', page 35).

Flowering period Usually March to May.

Distribution *C. creticum* is endemic to the islands of Crete and Kárpathos.

Habitat In partial to deep shade in gullies, streamsides and north-facing banks or in scrub and light woodland, near sea level to c. 1,300 m.

Variations Although the foliage of *C. creticum* is only a little variable in shape, usually rather triangular, the markings make up for this by varying greatly. In general cultivation, the forms tend to have grey-green cordate leaves with only faint markings and little variation. Plants collected by the Cyclamen Society on a field study of the species prove, however, that more well-defined silvery markings on a dark green background are not uncommon. The size and frequency of the silvery splashes that are often present on the upper surface vary greatly. There is usually also a paler grey- or silver-green region in a zone around a green central mark, and the margin may be finely to coarsely toothed. In flower colour, the most noticeable variation from the normal white is a pale pink form that can be found scattered through the populations; this has been formally described by C. Grey-Wilson as forma *pallide-roseum* to distinguish it from white-flowered forma *creticum*. In the wild, however, there appears to be no clear distinction and mixed populations can occur, ranging from plants that have white flowers, some with a slight tinge of pink, through to those with markedly pink petals.

Named cultivars

'Silvery Hope', with bright silvery leaves and white flowers, originated from Marco ten Hoope from the Netherlands.

'Tilebarn Spoa' was found near the village of Spoa on the island of Kárpathos, so it is interesting geographically but not appreciably different from the Cretan forms of the species.

Cyclamen creticum west of Argiroúpoli, Crete (JF)

Habitat of *C. creticum* near Argiroúpoli, Crete (JF)

Cyclamen creticum shows great leaf variation in the wild in Crete (JF)

A beautifully speckled leaf form of *C. creticum* at Omalos, Crete (JF)

Cyclamen creticum showing its long, elegant petals. The white-flowered form near Spili, Crete (JF)

The pale pink *C. creticum* f. *pallide-roseum* in the Aradhena Gorge, Sfakia, Crete (JF)

Cyclamen creticum 'Silvery Hope' a very good silver leaf selection by Marco ten Hoope (MH)

Further observations on *Cyclamen creticum*

Cyclamen in *Flowers of Crete* One of the most recent accounts of *C. creticum* in its native haunts is that of **John Fielding** and **Nicholas Turland** in their impressive and very valuable tome on the flora of Crete, *Flowers of Crete* (2005). Cyclamen is of particular interest to John Fielding, who also took part in the Cyclamen Society's field work to study this species, and the genus is given considerable coverage in the book. The text concerning *C. creticum*, which is enhanced by a very good photographic record, is worth quoting because it contains personal observations of a kind not normally found in a Flora:

> "*Cyclamen creticum* occurs throughout most of Crete with the possible exception of the south-central area south of the Mesará plain. It is most prolific as a woodland floor plant in the moister valley bottoms, but can be found in the majority of gullies. It also grows in less likely habitats such as open, exposed phrygana on the coasts of the far east of Crete and at over 1,100 m in the Lefká Óri in an equally exposed and wind-blasted habitat. *Cyclamen creticum* is endemic to the Cretan area, which also includes Kárpathos, where it appears to be comparatively rare. It is fairly uniform in appearance often with dull leaves, sometimes wholly green but usually with a silver hastate pattern and often with subtle degrees of speckling. Very occasionally one can find a wholly silver leaf, or one with bright silvery speckling. The leaf blade size varies according to habitat, usually about 5 cm long but occasionally as much as 15 cm in lush lowland conditions. Flowering occurs from early February in sheltered, low-altitude habitats such as in the Thériso gorge to as late as early June on the 'E4' footpath route above Askifou. Normally the flowers are pure white and fairly small with a pleasant sweet scent, but sporadically throughout Crete there occurs a soft pink form. Sometimes this may be just an occasional plant, but in the eastern populations these pink forms may represent as much as ten percent of a population. Those observed so far have not been an intense colour, always a 'washed-out' pale pink. Reports of *Cyclamen repandum* from around the town of Kándanos, west of the Lefká Óri, are probably founded on sightings of this pink form of *C. creticum*. The latter species certainly occurs around the town, including occasional pink ones. There is a strange anomaly in that plants from a cluster of populations south of Kándanos show a chromosome count of 2n = 22, whereas the usual count for this species is 2n = 20. To date no explanation has been found for this and it may have no great significance."

Flowers of Crete by John Fielding and Nicholas Turland (2005) in which cyclamen are given considerable prominence!

Cyclamen rhodium subsp. *rhodium*. Painted by Christabel King (reproduced at 95% of original size)

Cyclamen rhodium subsp. *rhodium*, subsp. *peloponnesiacum* and subsp. *vividum*

Flowering in spring

The spring-flowering cyclamen that occurs on Rhodes has been known for at least half a century as a distinct entity and, when brought into cultivation in the early 1960s, was loosely referred to as "*C. repandum* Rhodes form". It was Richard Gorer who drew attention to the fact that this plant might merit botanical recognition. In the *Gardeners' Chronicle* 154: 388 (1963), he mentioned it under the specific name *C. rhodium*, unfortunately without the formal necessities that constitute valid publication; consequently, the name was overlooked or disregarded. Two years later, Desmond Meikle published an article with Lewis Palmer in the *Journal of the Royal Horticultural Society* (now *The Garden*) in which he recognised this eastern Aegean island endemic as a variety (var. *rhodense*) of the much more widespread *C. repandum*. It was not until 1975 that Otto Schwarz and Lothar Lepper took up Gorer's name and validly published *rhodium* at the specific level. Following the International Code of Botanical Nomenclature, this means that if treated as a species, the Rhodian cyclamen takes the epithet *rhodium*, but if given varietal or subspecies status, *rhodense* takes precedence.

In April 1991, a team representing the Cyclamen Society visited Rhodes to carry out a survey of the cyclamen, in particular of *C. rhodium*, which was found to be widely distributed on moist, wooded slopes all over the island. It was confirmed at all the previously reported sites, and new records extended its known distribution to other locations in the extreme south, west and east. The lowest altitude at which it was observed was about 25 m. Kárpathos, a large island to the south-west of Rhodes, was also visited with the expectation that *C. rhodium* might also occur there but without success. The species has, however, been found on another eastern Aegean island, Kos, where it was noted in 1996 by Dr David Thomas of Cardiff University and was further studied during Cyclamen Society fieldwork in 1998. Kos is very close to the Turkish mainland, but in spite of searches, no member of subgenus *Psilanthum* has yet been recorded in those nearby areas of Turkey that face west towards the Aegean islands.

The two fine spring-flowering cyclamen of the Peloponnese have also had a chequered history at the hands of botanists and horticulturists. In the 1960s, when the first tubers of the speckled-leaved, pink-flowered plant came into cultivation, these plants were regarded as a pale variant of *C. repandum*. They were known merely as the Peloponnese form to distinguish them from the 'standard' *C. repandum* and from the even paler form from Rhodes that has almost white, pink-nosed flowers. A pale pink version with a darker carmine 'nose' then acquired the cultivar name *C. repandum* 'Pelops' and received an RHS Award of Merit under that name in 1968 when exhibited by the cyclamen guru Doris Saunders. Twenty years later, botanical recognition was bestowed upon it by Christopher Grey-Wilson who described it

A drift of *Cyclamen rhodium* subsp. *rhodium* on the Island of Rhodes, near Embonas (JF)

Cyclamen rhodium subsp. *peloponnesiacum*, showing considerable leaf variation in one population (MELJ)

Cyclamen rhodium subsp. *peloponnesiacum*. Painted by Pandora Sellars (reproduced at 120% of original size)

The richly coloured *C. rhodium* subsp. *vividum* between Ag. Andreas and Ag. Pardeleimon, Peloponnese (JF)

as a subspecies of *C. repandum* (subsp. *peloponnesiacum*) in his 1988 publication *The Genus Cyclamen*. It ultimately was given species status, as *C. peloponnesiacum*, in Kit Tan's *Endemic Plants of Greece — The Peloponnese* (2001), although this is actually an invalid name according to the requirements of the International Code of Botanical Nomenclature.

Vividum was the epithet given by Christopher Grey-Wilson to a darker, reddish-purple variant that he first regarded as a form of *C. repandum* subsp. *peloponnesiacum* and then upgraded to varietal level; the most recent status of *vividum* is that of a subspecies. It appears that Doris Saunders knew of this darker variant as her account of 'Pelops' at the time it received an AM includes the statement: "The Greek forms of *C. repandum* from the Peloponnese … are extremely variable, ranging from bright red (i.e. *vividum*) through deep rose-pink … to palest lilac… . They would thus seem to be intermediate between the dark red Italian forms and the variety *rhodense* from the Island of Rhodes, which is white or very pale pink...."

In this book, we take an intermediate position between those of the two teams of researchers whose deliberations appear in our introduction to the *Psilanthum* group, largely for convenience's sake until a definitive classification emerges (if it ever does!). We treat these three entities as subspecies, not of the more widespread *C. repandum* but of a species comprising all of the Peloponnesian and Rhodian spring cyclamen. As explained above, *C. rhodium* R. Gorer ex O. Schwarz & Lepper is the earliest name for this species as a whole, and so this name takes precedence, and we have as a result: *C. rhodium* subsp. *rhodium*, *C. rhodium* subsp. *peloponnesiacum* (Grey-Wilson) J. Compton & Culham and *C. rhodium* subsp. *vividum* (Grey-Wilson) J. Compton & Culham.

In the Peloponnese, subsp. *peloponnesiacum* is particularly plentiful in the Sparta region and the northern part of the Mani Peninsula, with a few other populations known in the northern Peloponnese. It has been reported on the islands of Andros, Tinos, Ikaria and Samos, but investigations by Max Debussche and by Peter and Nick Moore (published in the Cyclamen Society's journal for June 2006) show that the records for Andros are a straightforward case of mistaken identity and so, probably, are the others. In 2009, however, *C. rhodium* was located by Vic and Janet Aspland in a gorge on the western side of Kythira island, where it is said to be more akin to the darker subsp. *vividum* than to subsp. *peloponnesiacum*; part of their interesting report is repeated below in 'Further observations'. On the Peloponnese, subsp. *vividum* is, in general, a more easterly occurring plant, reported to inhabit light scrub or dryish woodland and rather more open situations in rocky places. There are, however, places in which the two variants meet, for example in the mountains of the south-eastern part of the Peloponnese.

Cyclamen rhodium subsp. *vividum* in an open rocky situation with *Fritillaria messanensis* near Vaskina, Peloponnese (JF)

Although seldom seen outside specialist collections, the various forms of *C. rhodium* are all attractive plants with bicoloured, fragrant flowers. There is no reason why they should not thrive outdoors in mild climates if grown in a sheltered, shady position. Peter Moore suggests that subsp. *vividum* might need full sun, perhaps with a cool root run to protect the fragile roots.

C. rhodium R. Gorer ex O. Schwarz & Lepper, *Feddes Repert. Sp. Nov. Regni Veg.* 86: 491 (1975) subsp. **rhodium** Type: Greece, Rhodes, Mt Prophitis Elias, 150 m, April 1964, *Palmer* 4 (holotype K).

Synonyms

- *C. repandum* var. *rhodense* Meikle, *J. Roy. Hort. Soc.* 90: 291, Pl. 121 (1965).
- *C. repandum* subsp. *rhodense* (Meikle) Grey-Wilson, *The Genus Cyclamen*: 60 (1988).
- *C. peloponnesiacum* subsp. *rhodense* (Meikle) Kit Tan, *Endemic Plants of Greece — The Peloponnese*: 239 (2001).

C. rhodium subsp. **peloponnesiacum** (Grey-Wilson) J. Compton & Culham, *Journ. Cyclamen Soc.* 27: 72 (2003). Type: Greece, Peloponnese, Taygetos Mts, Langada Gorge, W of Sparta, April 1987, *Grey-Wilson s.n.* (holotype K).

Synonyms

- *C. repandum* subsp. *peloponnesiacum* Grey-Wilson, *The Genus Cyclamen*: 60 (1988) var. *peloponnesiacum*.
- *C. peloponnesiacum* (Grey-Wilson) Kit Tan, *Endemic Plants of Greece — The Peloponnese*: 238 (2001) *nom. illegit.*

C. rhodium subsp. **vividum** (Grey-Wilson) J. Compton & Culham, *Journ. Cyclamen Soc.* 27: 72 (2003). Type: Greece, Peloponnese, Mt Parnon, SW of Kosmas, c. 700 m, April 1987, *Grey-Wilson s.n.* (holotype K).

Synonyms

- *C. repandum* subsp. *peloponnesiacum* forma *vividum* Grey-Wilson, *The Genus Cyclamen*: 60 (1988).
- *C. repandum* subsp. *peloponnesiacum* var. *vividum* (Grey-Wilson) Grey-Wilson, *Cyclamen*: 174 (1997).
- *C. peloponnesiacum* (Grey-Wilson) Kit Tan subsp. *vividum* (Grey-Wilson) Kit Tan, *Endemic Plants of Greece — The Peloponnese*: 239 (2001).

Description (all subspecies) **Tuber** depressed-globose, not more than 7 cm diameter at maturity, brown, usually smooth and velvety but sometimes becoming slightly corky in old age, rooting only from the centre of the underside. **Petioles** prostrate in the lower part below ground, then ascending to erect above ground. **Leaves** well developed before flowering time. Leaf blade

Tuber of *C. rhodium* subsp. *rhodium* (BM)

Flower of *C. rhodium* subsp. *rhodium* (BM)

Flower of *C. rhodium* subsp. *peloponnesiacum* (BM)

Flowers of *C. rhodium* subsp. *vividum* (BM)

The triangular leaves of *C. rhodium* subsp. *rhodium* often have unremarkable markings: here growing among mossy rocks on Rhodes (MH)

Cyclamen rhodium subsp. *vividum*. Painted by Christabel King (reproduced at original size)

Cyclamen rhodium subsp. *rhodium* here in its natural shady habitat in Rhodes in April (MH)

The wooded habitat of *C. rhodium* subsp. *peloponnesiacum* in the Peloponnese (SB)

Cyclamen rhodium subsp. *vividum* often grows in more exposed and rocky places than subsp. *peloponnesiacum* (JF)

broadly ovate-triangular, 2–15 cm long, 1.5–15 cm wide, green or greyish-green with a paler, jagged hastate pattern in the centre and variably splashed and speckled with silvery patches, occasionally with a more distinct silvery zone or with a wholly dull silvery appearance and sometimes wholly plain green; margins somewhat lobed and scallop-toothed, the teeth widely spaced and varying from shallow to very prominent; the underside pale green or purplish suffused. **Pedicels** horizontal in the lower part, turning upwards as they emerge above ground, after anthesis coiling from the top downwards. **Flowers** produced in spring, fragrant; petals rather long and narrow, swept upwards in an erect position, white or very pale pink with a pinkish-purple stain towards the mouth of the corolla (subsp. *rhodium*), pale to deep pink with a darker reddish-pink stain at the mouth (subsp. *peloponnesiacum*) or rich glowing ruby-red (subsp. *vividum*), narrowly elliptic-oblanceolate, c. 1.5–3 cm long, 0.5–1 cm wide, acute or subacute at the apex, without swollen appendages ('auricles') at the base, strongly twisted lengthways; style apex clearly exserted from the mouth of the corolla. 2n = 20.

Flowering period March to May

Distribution Subsp. *rhodium* — Eastern Aegean Islands of Rhodes and Kos; subsp. *peloponnesiacum* — central-southern, eastern and northern Peloponnese; subsp. *vividum* — south-eastern Peloponnese, Kythira Island (see page 44).

Habitat Subsp. *rhodium* — in shade under scrub and in rocky places, often in *Pinus brutia* forest, 50–800 m; subsp. *peloponnesiacum* — humid, shady places in woodland, scrub and base of cliffs, 350–1,500 m; subsp. *vividum* — light woodland, open rocky places and in dryish scrub, 400–1,680 m.

Variations In all three subspecies of *C. rhodium*, variation in leaf markings is considerable, perhaps the most common form having a dark green triangular centre surrounded by a variable jagged-edged silvery-green zone and the whole leaf covered in few–many uneven splashes and blotches of silver-grey. At the extremes, the whole leaf blade can be plain dull, dark green or almost wholly greyish-green with silver flecking.

In flower colour, subsp. *rhodium* is normally near-white with a pinkish 'nose' but flowers with petals slightly tinged pink can be found in wild populations. In 1998 the Cyclamen Society carried out fieldwork in the Dodecanese islands, during which they recorded pure white-flowered forms of subsp. *rhodium*, one of which had upward-facing ('stargazer'-type) flowers. Subsp. *peloponnesiacum* has flower colours that range through shades of pink, normally with a deeper carmine stain around the mouth of the corolla. A pure white form that does not have the darker stain has been found, and this was described in 1999 as *C. rhodium* subsp. *peloponnesiacum* forma *albiflorum*. The flower colour of

Selection of leaves of *C. rhodium* subsp. *peloponnesiacum* (MH)

A form of *C. rhodium* subsp. *rhodium* on Rhodes with markedly silver-splashed foliage (BC)

subsp. *vividum*, which is not as variable as those of the other forms, is an intense deep glowing ruby, which tends to obscure the darker stain at the mouth, although this is present on close inspection. A white form of subsp. *vividum* has also been found.

Named cultivars

Few variants of *C. rhodium* subsp. *peloponnesiacum* have been named. The cultivar name 'Pelops' was at one time applied to Peloponnesian plants that were considered as variants of *C. repandum*. The basis for a number of silvery-leaved variants known as the Silver-Leaf Group was a plant introduced by the Cyclamen Society in 1992; some plants selected by Mr Forster were probably also involved in the parentage of this group and such plants are now not uncommon.

Pure white form of *C. rhodium* subsp. *peloponnesiacum* (BM)

In the Peloponnese the intensely coloured *C. rhodium* subsp. *vividum* is instantly distinguishable from the paler subsp. *peloponnesiacum* (SB)

Variation in flower colour of *C. rhodium* subsp. *rhodium*: pink and white forms intermixed on Rhodes (MH)

A silver-leaved selection of *C. rhodium* subsp. *peloponnesiacum*, this one from Marco ten Hoope (MH)

Further observations on *Cyclamen rhodium*

***Cyclamen rhodium* subsp. *rhodium* on Rhodes. Martyn Denney** gives an interesting account of his observations of *C. rhodium* (as *C. repandum rhodense*) in the Cyclamen Society's Journal of December 1994, part of which is repeated here:

"On Mount Filerimos, to the south-west of Rhodes Town, *C. persicum* was still in flower on the slopes immediately below the monastery… . Further down Filerimos, *C. persicum* gave way to *C. repandum rhodense* (hereafter referred to as *rhodense*). Generally, *rhodense* appeared to have finished flowering, but there was very little evidence of seed having been set. These plants had generally unremarkable leaves, ranging from plain green to lightly marbled hastate pattern. Mount Profitis Ilias has been reported as being the most widely known site. Certainly, *rhodense* is very abundant throughout the area, frequently occurring in huge drifts, often with *Paeonia rhodia*. *Rhodense* is quite clearly a woodland plant, enjoying shaded damp conditions, often in *Pinus brutia* forest, and often growing in deep leaf or pine needle litter, or in silt around the edge of streams. Where a significant depth of needle litter exists, floral trunks can be a considerable length. Plants on Profitis Ilias were in full flower, with some seed having been set, and flower buds still waiting to open. Leaf patterns were generally far more interesting than on Filerimos, and at a couple of sites, leaves with a high proportion of silvering were seen. A minority of flowers had a significant pink flush; the majority being the standard off-white with a purple basal blotch. Although Bailey [in *Cyclamen Society's Journal* 15, 1991] reports that *rhodense* is more widespread throughout Rhodes than other publications might suggest, nothing can prepare one for the sheer volume of plants in any suitable habitat. My time was very limited, and I travelled no further south than Mount Profitis Ilias, but throughout the central spine of the island in the part that I did visit, *rhodense* was 'everywhere' — not always in flower, but at least present in leaf."

High-altitude *C. rhodium* subsp. *peloponnesiacum* The Cyclamen Society's Journal is the source of many interesting and informative articles, often reporting the field observations of its members during their travels. In the June 1998 issue, there was an interesting account of a search by **Alisdair Aird** for the Peloponnesian cyclamen at higher levels in the Taygetos mountains; the following is an extract from his article:

"An April excitement … was the discovery of a large population of *peloponnesiacum* at 1,560 m (5,120 ft), as high as it's ever been recorded previously. This was in the southern part of the Taygetos mountains, on the western slopes but in a semi-alpine bowl where most of the plants actually faced north. They were growing on a 30% gradient among thin woodland and scrub of mixed pine, fir and juniper, and although this particular spot faces quite severe winter conditions, the tubers did not appear to be any deeper-seated than normal for the species. A much higher proportion than normal of plants up here had plain unmarked leaves. Some *C. hederifolium* were scattered among them. Hardly any of these highest-altitude plants were yet starting to show flower buds, though the leaves of some plants here were already showing signs of desiccation (after some days of a drying wind and clear skies). Highest of all, though, was yet another much smaller population of *peloponnesiacum*, found at a record 1,680 m (5,510 ft). This was just across the divide, i.e. on the east-facing side of the range. The site was a steep rocky slope facing south, in thin mixed pine and fir woodland, and the tubers appeared to be very deeply seated."

A mystery solved on Kythira In April 1997, the Cyclamen Society sent a team to investigate the island of Kythira working on the hypothesis that, as there were members of the *Psilanthum* group on both the mainland and in Crete, then logically there might well be a representative on this island, which lies between the two. Nothing was found on that occasion, but the tantalising thought remained and **Vic and Janet Aspland** returned in 2009, this time with success. This extract from their report is taken from the Society's journal of December 2009.

"A steep descent along a south-west-facing slope through a very dry landscape with magnificent sea views gave us our first views of a gorge terminating at the coast, but no cyclamen. The turning point — literally and figuratively — occurred when we continued our descent along a north-west-facing slope, and saw the first *C. graecum*. The whole of this aspect, virtually from the coast up to the hill tops was covered with scrub, mainly of *Quercus coccifera* to a maximum height of about two metres (about six feet). Then, only a few tens of metres further down, in the fallen debris at the foot of a bank, appeared the first leaves of what was undoubtedly a cyclamen of the *rhodium* persuasion. As we descended further, the number of plants increased.

This hillside above the gorge was both very steep and densely covered with *Quercus*, so virtually impossible to climb, but examination with a monocular revealed more cyclamen as far up the hillside as the ground could be seen. Immediately adjacent to the road, the plants grew in soil somewhat resembling terra-rossa, but of a paler pinkish shade. Higher up, the substrate was almost totally mineral, consisting of decomposing limestone detritus with a minimal mulch of very slowly decomposing *Quercus* leaves. The pH of the soil, such as it was, lay in the range 8.0–8.2.

From this point, the cyclamen was abundant, all of the way down the slope, and even into the bottom of the gorge, where it grew in banks of fine white sand. Finding leaves was easy, but to classify our plant, we needed to see flowers too. We soon realised that we were right at the end of the flowering season (the last week in April); the peak would have been in the second half of March, but eventually we found a few plants in flower … [we would] probably place these plants within the envelope of *C. rhodium* subsp. *vividum*, but some caution is needed. It could be that the later-flowering plants seen by us all happen to have darker flowers … . Further inland the gorge opens out but time prevented us from exploring this area. We did, however, examine the opposite, south-east-facing slope, which was completely different: a very much hotter and drier environment, with a sparse covering of *Pistachio*, carob, *Cistus* and the suite of associated shrubs that will be familiar to Mediterranean travellers. In this cyclamen-unfriendly area, we saw only a very few *C. graecum*.

This new location is … scientifically interesting, as it bridges the gap between two well-explored cyclamen areas. The habitat available is extensive, and is very unlikely to be exploited in any way by man. Coupled with this, the very high rate of regeneration from seed of the cyclamen discovered makes it unlikely to be under threat."

Observations on subsp. *vividum* Any observations made by **Peter Moore**, former President of the Cyclamen Society, inveterate traveller in search of cyclamen, a specialist commercial grower of them and one of the Britain's most successful exhibitors, are likely to be valuable. Writing in the Society's journal of June 1996, Peter recounts his encounters and experiences with *rhodium* subsp. *vividum* and subsp. *peloponnesiacum* in the wild:

"Although *peloponnesiacum* is to be found at many sites in the Peloponnese, *vividum* is confined to an area east, south and south-west of Mount Parnon (which lies on the eastern side of the Peloponnese, above Sparta). The most striking difference between the two forms and indeed between *vividum* and all other members of the *C. repandum* group was highlighted by the Society's 1993 Expedition (for an account of this, see Chapter 3, pages 255–257). One of the expedition's aims was to extend the known sites of *vividum*, and also to try to find whether *vividum* and *peloponnesiacum* overlapped in their distribution in the wild.

At the base of the valley by the village of Ag. Petros north-east of Mount Parnon, *vividum* and *peloponnesiacum* were found at a site growing together in an approximate 50/50 ratio. Leading away towards the north-west of Mount Parnon, *peloponnesiacum* gradually took over and *vividum* disappeared completely. On the other sunny side of the valley to the east of Mount Parnon, *vividum* was in the ascendant and soon *peloponnesiacum* was no longer found. Careful observation at further sites emphasised the entirely different growing conditions needed by *vividum*. Here was a form [i.e. a relative — ed.] of *C. repandum* growing in hot sunny situations, with minimum shelter from low scrub and rocks. *Vividum* was usually found in association with *C. graecum*, and this highlights the required growing conditions. Where *vividum* is found growing in the open the cool root is often in leaf mould from the nearby *Quercus coccifera* scrub: as with *peloponnesiacum*, the tubers are generally found at a greater depth than with other species."

Habitat of *C. rhodium* subsp. *vividum* on Mt Parnon, eastern Peloponnese (JF)

2.2b The Coum group

(= Subgenus *Gyrophoebe*, Series 1)

Comprising: *C. coum*, *C. coum* subsp. *caucasicum*, *C. elegans*, *C. parviflorum*, *C. parviflorum* subsp. *subalpinum*, *C. alpinum*

This group of four winter- or early-spring-flowering cyclamen species is characterised by their rounded to heart-shaped leaves and smallish bright carmine to pink flowers, which have a darker blotch at the base of each petal. There are no swellings ('auricles') around the mouth of the corolla. The stamens are not exserted from the corolla but the style is sometimes shortly exserted. When in fruit, the pedicels start to coil downwards from the apex. Chromosome number 2n = 30.

Cyclamen coum growing above Ikizdere, northern Turkey, 21 March 2011 (MD)

Cyclamen coum subsp. *coum* and subsp. *caucasicum*

FLOWERING WINTER AND EARLY SPRING

For a display in mid-winter, *C. coum* is probably one of the mainstays of gardens in cool temperate climates, thriving in areas that do not become too hot or dry in summer. Exactly why this should be so can probably be explained by reference to its natural habitat. Jeremy Wood's extensive account of the Cyclamen Society's field trips to map and record the species in the wild provides a valuable picture of its habitat and variation (see the 'Scientific studies and fieldwork' section on pages 226–237). It is a widespread species growing from Romania and Bulgaria in the west through northern Turkey to the Crimea and Caucasus in the east, and southwards as far as Syria, Lebanon and Israel. The conditions in the greater, more northerly part of its distribution are influenced by the Black Sea. Here, *C. coum* experiences a moderate climate with year-round rainfall and no prolonged hot dry period, unlike that of the Mediterranean region where most cyclamen species occur. *C. coum* is therefore ideally suited to temperate countries that have cool, but not severe, winters and some summer rainfall.

Typical wooded habitat for *C. coum* above Hemsin, northern Turkey (MD)

Cyclamen coum. Painted by Pandora Sellars (reproduced at 135% of original size)

A thriving colony of *C. coum* among the tea bushes near Hemsin, northern Turkey (MD)

Thriving garden populations can be seen in many places in Britain and *C. coum* does especially well in the north-western states of North America; for example, in the Seattle garden of the late Brian and Margaret Mulligan, *C. coum* naturalised itself in drifts that would be dramatic even by Black Sea coastal woodland standards. Although in nature *C. coum* is usually found growing in acidic conditions, often in association with rhododendrons and in the tea (Camellia) plantations along the Turkish coast, acidic soils do not seem to be a prerequisite for successful cultivation. Wessel Marais' garden in the Dordogne is on a limestone hillside and has thousands of *C. coum* plants growing happily and seeding around in the semi-shade of deciduous trees. Elsewhere in this book, there are references to the success of *C. coum* in other parts of the world.

C. coum was the focus of an extensive field survey carried out by the Cyclamen Society in 1987, 1988 and 1990, the purpose of which was to acquire further knowledge about its distribution and variability. In particular, it was hoped that the study would provide definitive information about the correlation between the species' distribution and its leaf shape. Since 1849, when K. Koch first described the eastern-most representatives (as *C. purpurascens* var. *caucasicum* rather than as a variant of *C. coum*), there has been a proposition that there is a change in the outline shape of the leaf from west to east, with more or less round leaves with a smooth margin at the western end of the species' range (Bulgaria and north-western Turkey) and heart-shaped leaves with toothed margins at the eastern end (north-eastern Turkey and the Caucasus). Jeremy Wood's account (pages 226–237) presents the findings of that study although, perhaps not unexpectedly, there are no definitive answers. Many plant species that have extensive distributions show clines of variation. When only the extremes are taken into account, it can seem unhelpful, even nonsensical in some cases, to merge them under one species, but when material from the whole range is considered, the position often becomes blurred and the distinctions imprecise. This is where the rank of subspecies can be useful, allowing us to acknowledge, for the purposes of communication, that a certain pattern of variation exists while at the same time accepting that there is overlap in the characteristics. In fact Otto Schwarz elevated the status of the more easterly occurring populations of *C. coum* from that of a variety to that of subspecies (as *C. coum* subsp. *caucasicum*) in 1955, and this has been followed by subsequent authors such as Desmond Meikle in the *Journal of the Royal Horticultural Society* 90: 297–8 (1965) and Christopher Grey-Wilson in *Cyclamen*: 74–5 (2002).

The populations that have more-rounded leaves are automatically referred to as subsp. *coum* because it was this variant that was known to Philip Miller when he first described the species in 1768. The populations from the Caspian Sea region are distinct from both subsp. *coum* and subsp. *caucasicum* and were described as *C. elegans* by Boissier & Buhse in 1860. Subsequently, this entity was merged with *C. coum* subsp. *caucasicum* by several authors, but is currently recognised once more at specific level as *C. elegans* (see page 57).

This is not the end of the story because *C. coum* also has a southwards extension into the Amanus Mountains of southern Turkey and on into Syria, Lebanon and Israel, although there is a break between this 'southern arm' and the populations in the Black Sea region of Turkey. These southern populations are attributed to subsp. *coum*, although there are some distinctions of note, even if at present these are considered to be of insufficient importance to define a further subspecies.

Much has been written about the origin of the epithet *coum* and no doubt the speculation will continue. A few of the views are repeated in the 'Further observations on *C. coum*' section.

Cyclamen coum in one of its southern locations, photographed in the mountains of Lebanon by Richard and Myrna Semaan (MS)

C. coum Miller subsp. ***coum*** *Gard. Dict.* ed. 8, no. 6 (1768). Type: described from a specimen cultivated at Chelsea Physic Garden, deposited at the Natural History Museum, London (BM).

Synonyms

C. orbiculatum Miller, *Gard. Dict.* ed. 8, no. 5 (1768).
C. hyemale Salisb., *Prodr.* 118 (1796).
C. hiemale Hildebr., *Gartenflora* 53: 70 (1904).
C. durostoricum Pantu & Solacolu, *Bull. Sect. Sc. Ac. Roum.* 9: 23 (1924).
C. orbiculatum var. *coum* (Miller) Doorenbos, *Meded. Landbouwhogeschool* 50: 25 (1950).
C. coum Miller subsp. *hiemale* (K. Koch) Schwarz, *Feddes Repert. Sp. Nov.* 58: 249 (1955).
C. kusnetzovii Kotov & Czernova, *Fl. RSS Ucr.* 8: 521 (1958).

C. coum subsp. ***caucasicum*** (K. Koch) O. Schwarz, *Feddes Repert. Sp. Nov.* 58: 250 (1955). Type: unknown.

Synonyms

C. vernum Sweet, *British Flower Gard.* 1: t. 9 (1823).
C. europaeum var. *caucasicum* K. Koch, *Linnaea* 23: 619 (1849).
C. ibericum T. Moore in Henfrey *et al.*, *Garden Companion* 1: 90, f. 2 (1852).
C. caucasicum Willd. ex Boiss., *Fl. Orient.* 4: 11 (1875).
C. coum var. *abchasicum* Medw. ex Kusn., *Fl. Cauc. Crit.* 4, 1: 167 (1902).
C. vernum var. *caucasicum* Schwarz, *Gartenflora N.S.*, 1: 20 (1938).
C. abchasicum (Medw. ex Kusn.) Kolak., *Fl. Abkhazya* 3: 274 (1948).
C. circassicum Pobed., *Bot. Zhurn.* 33: 226 (1948).

Description (both subspecies) **Tuber** depressed-globose, usually very flattened above, up to c. 4.5 cm diameter at maturity, brown, smooth and velvety-pubescent, rooting from the centre of the underside. **Petioles** prostrate in the lower part below ground, then ascending to erect above ground. **Leaves** well developed before flowering time. Leaf blade orbicular to broadly ovate, 2–8 cm long, 1.5–8 cm wide, most frequently deep dull green with a central triangular green pattern surrounded by a paler green or grey zone that can be continuous or broken into separate patches on a green ground; around the paler zone there is usually a dull green border out to the margins; occasionally the leaves are wholly dark green or wholly silvery or pewter-coloured; the margins are almost entire to sparsely and shallowly toothed; the underside is normally stained purplish-red. **Pedicels** horizontal in the lower part, turning upwards as they emerge above ground, after anthesis coiling from the top downwards. **Flowers** produced in winter or early spring (also a Turkish record of an autumn-flowering population), usually without or sometimes with a slight scent; petals swept back into a suberect or erect position, usually pale pink to deep rich magenta, each with a darker magenta-purple basal stain at the base surrounding a white or pale pink zone thus forming a white or pink rim around the mouth of the corolla, although this may be absent, broadly ovate to suborbicular with obtuse tips, c. 0.5–1.5 cm long, rather weakly twisted lengthways, without swollen appendages ('auricles') at the base; style not exserted from the corolla. 2n = 30.

Flowering period (November) December to April.

Tuber of *C. coum* (BM)

Flower of *C. coum* (BM)

A rather typical leaf form of *C. coum* in the hazel (*Corylus*) plantations of the Black Sea region of Turkey (MD)

This form of *C. coum* from Russia lacks the usual white 'nose' of the species (JB)

This tumbled rocky river bed near Hemsin, Turkey, provides a good habitat for *C. coum* (MD)

Cyclamen coum is often one of the first plants to flower as the snow melts above Ikizdere, Turkey (MD)

A frequent companion plant for *C. coum* is *Helleborus orientalis*. Forest road from Artvin towards Fıstıklı (MD)

Cyclamen coum and *Primula vulgaris* often occur together, here seen near the Sumela Monastery (MD)

Distribution Both subspecies — Bulgaria, north-west, north and central-south Turkey, Crimea, Caucasus, western Syria, northern Lebanon, northern Israel. Records from Cyprus have not been confirmed; subsp. *coum* — Bulgaria, Romania, northern and central-southern Turkey, Crimea, the Caucasus, western Syria, northern Lebanon, northern Israel, Cyprus? (see 'Further observations on *C. coum*, page 56); subsp. *caucasicum* — North-eastern Turkey, southern Caucasus.

Habitat Usually in woodland (coniferous, deciduous or mixed) in humus-rich soil, sometimes in scrub or in rocky places; at higher elevations, may be found in the open in alpine turf; sea level to 2,150 m.

A grassy bank above Ikizdere, Turkey, with a mass of *C. coum* and leaves of emerging *Colchicum* in late March. At the higher altitudes, *C. coum* tends to occur in more open situations (MD)

Variations With a distribution spanning nearly 2,000 km from west to east *C. coum* is, unsurprisingly, very variable in its natural state and even more so in the selections made in cultivation. Apart from the general leaf shape, which varies from approximately orbicular (subsp. *coum*) to ovate (subsp. *caucasicum*), there is an enormous array of variation in the markings of the foliage. At one end of the scale, the foliage can be totally plain green, a variant that at one time was regarded by nurseries and gardeners as the true *C. coum*, through to attractively silver-leaved, these forms having been referred to under names such as *C. orbiculatum*, *C. atkinsii*, *C. hiemale* and *C. ibericum*. The plain green forms are not always the product of horticultural selection and can be found in the wild; the Bowles Scholarship Expedition (1963), for example, recorded entirely dark-leaved plants at Lake Abant in north-western Turkey. Between these two extremes of dark green or wholly silver (or pewter) is a great array of leaf patterns, mostly consisting of a dark green central triangular pattern, aptly described as a 'Christmas-tree' shape, surrounded by a pale contrasting zone in varying degrees of silvery green; the outer margin of the leaf may also have a thin edging of dark green.

C. coum plants that have a range of pewter-coloured leaves with only a very narrow green area of central veining are referred to as the Pewter Group and those with a more silvery appearance, the Silver Group. In some cases, the silvery zone is broken into a ring of irregular blotches. Variants have now been raised in which the green central zone is completely lacking, the only break in colour being a narrow dark green margin. There is also a rather unpleasant variegated-leaved form.

Coupled with these foliage variations, there is also a great range in the flower colour of *C. coum*, so combining these two characteristics provides the potential for a large number of selections — and cyclamen aficionados have taken full advantage of this. If it were possible to say that there is a typical flower colour then, in the wild, it would be a bright magenta with a dark purple stain towards the bottom of the petals and with white markings, which have been referred to as 'eyes', right at the mouth of the corolla. At the extremes, *C. coum* flowers can be an intense reddish-magenta right through to pale pinkish-purple shades or white. The white variants are, most frequently, white with the usual dark purple markings near the mouth, but rare white forms without such markings have been found, for example during the Cyclamen Society's fieldwork. Another plant with pure white flowers is reported to have occurred in the garden of Stephen Blore among a patch of normal self-sown pink-flowering plants and a most attractive form has been raised in New York by Ellen Hornig. Several pure white-flowered cultivars have now been named. Some of the coloured forms have 'picotee' flowers in which the petals have a darker rim of colour. In addition to these colour forms, there are oddities such as plants that have markedly serrated-edged petals. In cases where particular variants have been selected and are being propagated true to form, they have been given cultivar names or sometimes 'group' names if there is a little variation.

Although *C. coum* normally flowers in early spring in its native state, an autumn-flowering population has been recorded in northern Turkey by Neriman Özhatay.

Selection of leaves of *C. coum* grown by Marco ten Hoope, showing great variation (MH)

Scent is not a feature usually associated with *C. coum*, but it has been observed that when many plants in flower are growing together in an enclosed space, there is a distinct fragrance reminiscent of that of *C. cilicium* but with a citrus-like note. Others have indicated a likeness to the scent of primroses. Scent was one of the points of interest noted for the plant described as *C. kusnetzovii* from near Russkoje in the Belogorsk district of the Crimea by M. I. Kotov & N. M. Czernova in 1958. The other distinguishing feature of this plant was said to be its flower size, with petals 10–16 mm long, but this is such a variable character in cyclamen that it is of little significance as a distinguishing feature. The leaves of plants from this area are plain green or only slightly patterned and those in cultivation are rather compact. This is thought to be a very hardy cyclamen, and although the name *C. kusnetzovii* is generally regarded as a straight synonym of *C. coum*, the hardiness of these plants could be a major attraction for gardeners in cold-winter regions.

Two variants of *C. coum* are currently recognised at subspecies level, but it must pointed out that there is much variation within each and their characteristics overlap, but this is an acceptable situation in the concept of subspecies. As Christopher Grey-Wilson sensibly says, the most practical approach for the intermediates is to use the name *C. coum* without qualification. The premise was that there exists an east-west gradation of variation — from rounded or kidney-shaped leaves (as wide as long or wider than long) in the west (subsp. *coum*) to ovate leaves (usually longer than wide) in the east (subsp. *caucasicum*) — and this is to some extent valid (Grey-Wilson, 1988, 1997, 2002). It was for this reason that the Cyclamen Society, in collaboration with Turkish botanists, undertook the fieldwork described on pages 226–237. In addition to the leaf characteristics, it has been pointed out that the flowers of subsp. *coum* usually have a white patch at the corolla mouth, whereas in subsp. *caucasicum* this patch is generally pinkish or may be lacking altogether, the dark stain extending right to the mouth.

The pure white *C. coum* forma *albissimum* near Sumela above Trabzon (MD)

Above Fıstıklı, Artvin Province. Heart-shaped leaves tend to be more prevalent at the eastern end of the range in Turkey (MD)

Variants of subsp. *coum* with white flowers having a carmine stain around the mouth of the corolla are named **forma *pallidum*** C. Grey-Wilson. Pure white-flowered individuals, first recorded during Cyclamen Society fieldwork in north-eastern Turkey, take the name **forma *albissimum*** Bailey *et al.* The 'normal', but very variable, magenta-flowered plants automatically become referable to **forma *coum***.

Although subsp. *caucasicum* probably varies in the wild almost as much as subsp. *coum* there are fewer variants in cultivation, perhaps because it is less widely grown by enthusiasts.

Named cultivars

coum subsp. *coum* forma *coum*

With such a variable species that is easy to grow and raise from seed, it is not surprising there are many named selections.

'Blush': green-edged pewter-coloured leaves; flowers pale rose-pink with rose-magenta blotch at mouth.

Broadleigh Silver Group: silver or pewter leaves; flowers variable shades of pink; possibly *C. coum*–*C. alpinum* hybrids.

Near the Sumela Monastery, above Trabzon, northern Turkey. The mixed forest here is the habitat for both *C. coum* and *C. parviflorum* (BM)

'BSBE': derived from plants that had striking silver and green leaf patterns collected by the Bowles Scholarship Botanical Expedition in 1963 in the Black Sea coastal region; now many generations down the line, these plants may be found listed under the BSBE tag, sometimes with the number 518 attached.

'Crimson King': attractively marked leaves; rich crimson-coloured flowers.

'Dusky Maid': greyish-green leaves; crimson-magenta flowers.

'Heavy Metal': rounded, completely silver-washed leaves, sometimes with thin green margin and midrib; pale pink flowers.

'Linnett Jewel': silver and green patterned leaves; flowers large, deep magenta, the petals paler towards the centre.

'Linnett Rose': leaves deep green with grey pattern; flowers pinkish-rose flowers, the petals deeper at their margins.

'Maçka': originating from the town of that name in Turkey, collected in 1966 by S. Albury, M. Cheese and J. Watson; leaves rounded, silver-washed with a small, central deep green area; flowers deep pinkish-magenta.

'Magenta Surprise': flowers bright magenta.

'Meadens Crimson': plain deep green leaves; deep crimson-magenta blooms.

'Merymana': originating from Merymana (Sumela) near Trabzon.

'Nymans': typical rounded *coum* leaves of a rather yellowish-green, flowers small, deep claret-coloured with pale markings in the form of a V at the base of each petal; this cultivar was selected from a collection by E. K. Balls in 1930s sent to Nymans garden in Sussex.

Pewter Group: leaves pewter-coloured with a green margin and central vein; flowers variable pink to magenta; raised by Kath Dryden.

'Roseum': leaves attractively patterned silver; flowers large, rose-pink with magenta markings at the mouth.

'Rubrum': general name given to deeply coloured forms.

'Silver Star': striking silver-coloured leaves with obvious veining; flowers bicoloured pink and white, roughly shuttlecock-shaped; developed from second-generation seedlings from a plant collected by The Cyclamen Society under the number S21 N397 in northern Turkey.

'Sterling Silver': leaves with very striking luminous quality, silvery-white, unmarked; flowers in various shades of pale magenta.

'Tilebarn Elizabeth': leaves entirely pewter-coloured without green veining; flowers bicoloured, very pale pink or whitish, flushed deeper towards the petal margins, blotched magenta-crimson at mouth.

'Tilebarn Graham': leaves rounded, silver-coloured; flowers rather small with elegant narrow, twisted petals, pink with deep magenta blotch at mouth.

'Turkish Princess': leaves green with overlay of silver; flowers bicoloured pink and magenta.

The commonly cultivated white variant with a purple stain at the mouth is *C. coum* forma *pallidum* (MD)

'Urfa': leaves rounded with a slightly pointed apex, marked with a striking 'Christmas tree' pattern; flowers rather squat, deep magenta-pink with a deeper blotch at mouth.

coum subsp. *coum* forma *pallidum*

'Album': leaves rounded, deep green, shiny and without markings; flowers white with magenta markings at mouth.

'Marbled Moon': handsomely marked leaves; small ('dumpy') white flowers with magenta markings at the mouth.

Cyclamen coum forma *albissimum* 'George Bisson' combines pure white flowers with marbled leaves (MD)

The plain-leaved *C. coum* forma *albissimum* 'Golan Heights' in cultivation. Originally found by members of the Cyclamen Society during field work in Israel in 1990 (MD)

'Maurice Dryden': variable silver or pewter-coloured leaves with green margin; flowers white with a dark purple stain at the mouth.

'White Pewter': the white-flowered version of Pewter Group, the flowers with a purple stain at the mouth.

coum subsp. *coum* forma *albissimum*

'George Bisson': leaves with a pale green zone and central darker green 'Christmas tree' pattern; flowers pure white.

'Golan Heights': plain glossy green leaves; pure white flowers. The original plant of this was found on the Golan Heights near the village of Mas'ada by the Cyclamen Society during fieldwork in Israel in 1990.

'Lake Effect': leaves deep matt green slightly marked silver; flowers pure white with pleated-edged petals giving a rippled appearance; raised by Ellen Hornig of Oswego, New York.

'Tilebarn Fenella': attractively patterned leaves; pure white flowers.

coum subsp. *caucasicum*

'Album': heart-shaped leaves; flowers white with magenta marks at mouth.

'Linnett Charm': leaves heart-shaped, deep green with a hastate pattern in various shades of pewter; flowers mid-pink with a basal darker blotch and pink or lavender rim to the mouth (unlike subsp. *coum* in which this zone is white).

Further observations on *Cyclamen coum*

The curious epithet *coum* The name *C. coum* has puzzled researchers for a long time. If the species were known to grow on the island of Cos there would probably be no further discussion; unfortunately, it has not been found there and it is unlikely that it will be, as this is a comparatively well-known island and anything as obvious as a cyclamen would have been detected long ago. Cos is also a long way from the nearest locations of *C. coum* in Turkey. The species had been referred to in the eighteenth century as the 'Cyclamen de Chio' but the same caveat applies: it has never been recorded on Chios island. There have been various conjectures as to the origin of the name, none of which has really gained particular favour over the others. While browsing through an article on cyclamen by Otto Schwarz in *Gartenflora* 13 (1938), **Erna Frank** came across a location for '*C. europaeum* subsp. *ponticum*' as 'the gorge of Okoum' and wonders whether it could point to the origin of the name *coum*. This gorge is in Abkhasia and is in fact the type locality of the plant now known as *C. colchicum*; *C. coum* probably also grows in the same area.

This suggestion prompted **Charles Hollidge** to consider the matter in detail and he came up with one of the more interesting proposals, which appeared in the Cyclamen Society's journal in 1989:

> "The earliest reference to *coum* appears to have originated in a catalogue published by the Morin family, gardeners and florists, in Paris in the 17th century … Joncquet, a herbalist, knew that '*coum*' meant from Cos. He also knew that *C. coum* was not found in Cos. He offered a tentative suggestion that that particular cyclamen came from Chios. He did not know that '*coum*' could also mean 'from Coa'. If his knowledge of Cilicia came mainly from the Bible and commentaries on the Bible, he may have been acquainted with Kuhe, Kue, Qui, or a mass of other variations; under Syrian, Greek, Roman, Armenian, Seljuk and Ottoman occupation at different times, this polyglot country had no standardised spelling until very recently. Had Joncquet waited another 300 years before rushing into print, he would have discovered that the acceptable transliteration is Coa. This name was used principally for the eastern region of Cilicia. It was the land of fertile plains stretching north to the wooded mountains, where *C. coum* abounds. In short, '*coum*' can mean 'from Cos' or 'from Coa'. No-one suggests that as applied to a cyclamen it means 'from Cos'. Therefore it probably means 'from Coa', which is consistent with all the geographical, botanical and etymological knowledge we have."

Perhaps the only point to be made about this particular explanation is that eastern Cilicia is a rather outlying location for *C. coum*, the main part of its distribution being in the Black Sea region. Any early introductions of the species are much more likely to have been from the latter region.

The mystery is far from settled!

***Cyclamen coum* in Cyprus?** In his *Flora of Cyprus* Vol. 2, **Desmond Meikle** reports two sightings of ?*C. coum* in Cyprus, both by a discerning plantsman, Eliot Hodgkin. In a letter to Meikle dated 16 April 1971, Hodgkin's notes read:

> "*Cyclamen* sp. ?*coum* 1. Kykko Monastery. In dense shade in punky [sic], deep leafmould under bushes of *Quercus alnifolia* and other shrubs. Growing with *C. cyprium*. 2. Below Nikos. Near the road. Corm small and smooth, and roundish leaves unspotted. Roots springing from the base only. Growing with *C. cyprium*."

***Cyclamen coum* on Cyprus: a follow-up** The question of *C. coum*'s occurrence on Cyprus was taken up by **Martyn Denney**. His quest suggests that this is either a case of mistaken identity or that the species no longer occurs there. Writing in the Cyclamen Society's journal of December 2001, Denney says that:

> "With Cyprus showing a certain botanical allegiance to the countries of the Levant to the east, the existence of *C. coum* on the island is at least believable. Unfortunately, neither Jill White (1987) nor myself succeeded in finding evidence of *C. coum* near Kykko, though whereas she found *C. cyprium* there, I did not. 'Below Nikos' could indicate immediately underneath the village (which is perched on the side of the Marathasa Valley), or further down the valley itself. Here also I failed to find *C. coum* or *C. cyprium*, either underneath the village or further down the valley. However, ... since 1971 a new road has been constructed along the Marathasa Valle and, immediately below the village of Nikos, the Kaleopanayiotis Dam has been built thus flooding part of the area. It is therefore highly likely that if *C. coum* ever did exist, this is no longer the case. In addition ... ... a 'smooth tuber with roots springing from the base' is a description which could equally apply to *cyprium* and *coum* and cannot therefore be considered diagnostic."

So whether or not *C. coum* is present on Cyprus remains another mystery to which there is no answer, unless it is relocated on the island.

Cyclamen hunting trip in Romania and Bulgaria. József Szendefi, describes, in the Cyclamen Society's journal for June 2002, a journey made in early March 2001 from Hungary through Romania and Bulgaria, searching for cyclamen. He travelled first to the Danube delta then followed the Black Sea coast southwards to the Turkish border, turned inland towards the Rhodope mountains, then northwards crossing Stara Zagora, Veliko Tarnovo and Ruse, before returning to Hungary via Romania. In the Dobruja area of Romania, the country is hilly, with woods full of snowdrops and this type of country continues across into Bulgaria. Although Szendefi found no cyclamen there, he did encounter them south of Burgas where there was a large population of *C. coum* subsp. *coum* in the woods, all with pink flowers except for one white-flowered plant. So on the basis of the observations made on this journey, *C. coum* certainly occurs in the extreme south-east corner of the Bulgaria, adjacent to Turkey.

***Cyclamen coum* in Romania** Although József Szendefi did not find *C. coum* in the Dobruja region of southern Romania or in northern Bulgaria, it does occur there. In 1924, a 'new' species was described from this area as *C. durostoricum*, named after the town of Silistra, the Roman Durostorum. The description and drawing of the plant show that it is undoubtedly the same as *C. coum*. The authors of the new species were the Romanian botanists Pantu and Solacolu and their description was published in the *Bulletin of the Roumanian Science Academy*. Silistra was granted to Romania by the Treaty of Bucharest of 1913 but then became Bulgarian in 1918. In 1919, it was returned to Romania and remained part of this country until 1940, when it rejoined Bulgaria. So, although the plant stayed in the same place, the border has changed frequently — but when named in 1924, *C. durostoricum* (= *C. coum*) was a Romanian species!

The southern-most populations of *C. coum*: in Israel The southern distribution of *C. coum*, from southern Turkey southwards to Israel, presents something of a dilemma. In his detailed monograph of 2002, Christopher Grey-Wilson includes populations in this region in subsp. *coum* but with a qualifying 'probably' and notes that they "are misfits for they do not conform neatly with the other subspecies". Israel represents the southernmost part of the distribution of *C. coum*, which is known from the Golan Heights, Mt Hermon and Mt Meron.

An isolated population is also known even further south, and in the journal *Cyclamen* of December 2001, **Oron Peri** described this interesting location. He notes that in other places in Israel, *C. coum* is found growing under trees at altitudes above 1,000 m, protected from direct sun and winds, and exposed to a few days of snow almost every year. Then, some 20 years ago, a botanist reported seeing a new colony on Mt Harashim (850 m), a location about 20 km further south than the nearest known colony where climate, altitude and conditions are very different. This colony consisted of about 50 plants growing on

Cyclamen coum photographed growing in the Golan Heights, Israel by Oron Peri

the southern edge of the mountain in a very exposed rocky and windy area. Surprisingly, the colony grew together with *C. persicum*, which was in flower in mid-March (when *C. coum* had already formed seed pods). The plants were growing in rock crevices and under a few low-growing *Pistacia lentiscus* and *Quercus calliprinos*, together with plants usually seen at much lower altitudes: *Ranunculus asiaticus*, *Narcissus tazetta*, *Orchis anatolica* and many others. Although the habitat of this colony is different from that of other Israeli populations, mainly in that it is much more exposed to sun and wind, the plants look perfectly healthy, although the leaves, which were green in colour with no silver patterns, were starting to turn yellow earlier than those of other populations.

Grey-Wilson states that the southern *C. coum* populations (i.e. those from southern Turkey southwards) clearly require further study. In particular, he notes their ovate leaves, which have shallowly scalloped margins, and their small, dark-coloured flowers with rounded petals.

***Cyclamen coum* in southern Turkey** In the central-south of Turkey, there is an interesting area where *C. coum* and *C. pseudibericum* overlap in their distribution. **Harry Gill** and **Pat Nicholls**, who were members of the Cyclamen Society's field survey of the latter species (the full report of which by Vic Aspland is described on page 275), made the following observation in the Cyclamen Society's journal for June 2002.

"We spent the last day searching north of Tarsus which was also the furthest westerly point of our survey. Yet another surprise was in store ... We were returning from the extensive and mostly uninhabited 'holiday home' large village of Çamiyayla on a road going east to rejoin the major road between Tarsus and Pozanti, when Pat yelled "stop" within a few yards of a bridge crossing a lake. A huge colony of *Cyclamen coum* covered the hillside including many with very choice leaf patterns and large triangular leaves, reminiscent of *Cyclamen coum* subsp. *caucasicum*. There were some particularly striking leaves on plants underneath a huge walnut tree. We speculated on the nature of the flowers and how the area would look in January/February, at the same time believing it would be a difficult site to access because of the narrowness of the road and the winter snow. It is possible that it is kept open because of the bridge crossing and the dam nearby which was securing the lake."

Cyclamen elegans

Winter or spring flowering

It is evident from the chosen specific epithet, meaning elegant, that Edmund Boissier and Friedrich Buhse must have known this species as a living plant when they described it in 1860, as it is unlikely that a dried specimen of any cyclamen would prompt the use of the name *elegans*! In fact, it was seen growing in north-eastern Iran by Buhse, who prepared the dried specimens that are deposited in several herbaria, including that in St Petersburg and the Boissier herbarium in Geneva. It is indeed an elegant species, particularly when seen in its natural state in the Caspian forests.

Cyclamen elegans is undoubtedly related to *C. coum* and opinions have varied as to its taxonomic status: some authors have retained its original position as a species, some have reduced it to a mere synonym of *C. coum* subsp. *caucasicum*, whereas others have placed it in an intermediate position as a separate subspecies of *C. coum*. In the latest assessments, *C. elegans* is back to being recognised as a distinct species. Suffice to say that *C. elegans* is a fairly distinctive plant having flowers that are generally larger than those of *C. coum*; its petals are relatively long and often pointed at the apex, contrasting with the rather 'dumpy', more

Cyclamen elegans. Painted by Pandora Sellars (reproduced at 120% of original size)

The flowers of *C. elegans* are longer, more elegant than those of *C. coum* and generally a softer pinkish colour (BM)

rounded petals of the more westerly occurring relative. The overall aspect of the flowers is thus really rather different from that of *C. coum*.

Field notes made in 1963 by members of the Bowles Scholarship expedition to Iran go as far as to suggest that the flowers of *C. elegans* are similar to those of *C. libanoticum* when seen at a distance. This illusion is helped no doubt by the flower colour of *C. elegans*, which in general is paler and of a more pinkish tone than the frequently deeper, purplish-red flower colour of *C. coum*. *Cyclamen coum* and *C. elegans* have probably evolved out of the same ancestral stock, the more easterly ones (now *C. elegans*) in the Caspian Sea region having been cut off geographically from the Black Sea or Caucasian ones (*C. coum*) by climate changes that have created a harsh no-go region for cyclamen in the eastern uplands of Anatolia and Transcaucasia.

The leaves of *C. elegans* are always quite obviously heart shaped, whereas those of *C. coum* have, at least in the west of this species' range, almost round leaves, although these tend to be more heart-shaped in the east. This apparent geographical cline in the leaf shape of *C. coum* formed the subject of a series of field investigations by representatives of the Cyclamen Society in the 1980s; an interesting account and discussion of their work can be found on pages 226–237.

Cyclamen elegans occurs in the damp woodlands on the western, southern and south-eastern margins of the Caspian Sea in Iran and possibly also in Azerbaijan, although the latter needs confirmation as the literature records might refer to *C. coum* subsp. *caucasicum* with which *C. elegans* was at one time merged. In Iran, *C. elegans* has been recorded in the provinces of Gilan, Mazanderan and Gorgan, indicating a distribution that follows much of the Iranian Caspian coastline. *Cyclamen elegans* can be seen in association with violets, primroses, *Parrotia persica* and *Carpinus*, and occasionally in evergreen woods.

The rainfall in this region varies considerably, but can be as much as 2,000 mm (80 inches). Although *C. elegans* can be found at low altitudes, only just above Caspian Sea level (which is c. 28 m below mean sea level), it does occur in higher regions of the Caspian forest and was recently (April 2009) noted by Bob and Rannveig Wallis, John Ingham and others at 1,450 m in the Elburz between Ruyan and Baladeh, north-east of Tehran, on the Caspian side of the watershed. At this altitude, *C. elegans* was growing in open woodland, which was dry under foot with lots of dead leaves, on a fairly steep slope. In the lower part of its range, *C. elegans* can become a pseudo-epiphyte: the members of the 1963 Bowles Scholarship expedition (B. Mathew, S. Baker, D. Pycraft and D. Barter) reported this species growing in humus-rich crevices and on the mossy, fern-bedecked branches of trees and rotting logs, a habitat made possible only by the high level of precipitation. Further east, near Gorgan, another group, visiting the area in 2004, reported lots of cyclamen with many seedlings on a winding mountain track. It is good to know that this attractive Iranian species is alive and well in its type locality.

C. elegans Boiss. & Buhse, *Nouv. Mem. Soc. Imp. Naturalistes Moscou* 12: 145 (1860). Type: N. E. Iran, Gorgan: 'in Persia boreali prope Asterabad', *Buhse* (LE).

Description **Tuber** depressed-globose, up to c. 4.5 cm diameter at maturity, brown, smooth and velvety-pubescent, rooting from the centre of the underside. **Petioles** prostrate in the lower part below ground, then ascending to erect above ground. **Leaves** well developed before flowering time. Leaf blade broadly ovate, 2–7 cm long, 1.5–5.5 cm wide; margin scallop-toothed, dull green with a central triangular green pattern surrounded by a pale green or greyish zone that may be continuous or broken into separate patches on a green ground, then a dull green border out to the margins; occasionally wholly grey-pewter coloured; the underside normally purplish suffused. **Pedicels** horizontal in the lower part, turning upwards as they emerge above ground, after anthesis coiling from the top downwards. **Flowers** produced in winter or

Tuber of *C. elegans* (BM)

Flowers of *C. elegans* (BM)

Note the markedly heart-shaped leaves, typical of this species; this is a good silvery-leaved form in cultivation (JB)

Mixed deciduous woodland including *Parrotia persica* near the Caspian coast, a habitat for *C. elegans* (BM)

early spring, without or with a slight scent; petals swept upwards to an erect position, usually pale to medium pink with a dark purple basal stain and two pale pink marks at the mouth of the corolla, ovate-elliptic with acute tips, c. 2–2.5 cm long, rather weakly twisted lengthways, without swollen appendages ('auricles') at the base; style held within the mouth of the corolla. 2n = 30.

Flowering period December to March (in the wild).

Distribution *Cyclamen elegans* is endemic to the Caspian region, on the Caspian Sea side of the watershed in the foothills and lower mountain regions of the Talysh and Elburz ranges. It is known from north-western, northern and north-eastern Iran and adjacent Azerbaijan.

Habitat Deciduous and evergreen woods, mossy banks and shady cliffs, and in remnants of low altitude woodlands, Caspian Sea level (–28 m) to 1,450 m.

At low levels in the Caspian woods, *C. elegans* occasionally takes to the trees, growing in mossy humus (BM)

Variations *Cyclamen elegans* is not a highly variable species when compared with some of its relatives such as *C. coum* and *C. alpinum*. The flower colour is usually a pale but vibrant shade of pink, and this, combined with the sizeable flowers, means that *C. elegans* is far from being a dull plant. The leaves are, however, not very exciting, with a comparatively sober variable zoning of silvery- or grey-green on a dull green background. A variant collected by the Bowles Scholarship Expedition (BSBE812) had wholly pewter-coloured leaves. There are now some interesting forms in cultivation with pewter or silver foliage.

A fine plant of *C. elegans* with attractive silvery leaves (MD)

Further observations on *Cyclamen elegans*

The Furse expeditions of the 1960s The botanical explorations of **Paul and Polly Furse** in their Land Rovers 'Rose of Persia' and 'Afghan Rose' captured the imagination of many bulb enthusiasts of the day. Although *C. elegans* had been described 100 years earlier, it was more or less unknown, certainly in horticulture, so it was particularly exciting to see material of it introduced into cultivation. In writing up an account of the 1964 expedition, Paul Furse notes that:

> "By mid-March ... On our descent to the Caspian it was still early spring with the snow just gone and the villages in a wild spring fever ... the most fascinating flower was the pink cyclamen (PF5023) which grew under the big trees or on mossy rocks or often in hollow tree-stumps; this is close to the *coum* group of cyclamen and may be a sub-species of it, with pointed leaves and large pale pink flowers with crimson base and no white nose, and it is probably the plant that has in the past been called both *C. elegans* and *C. caucasicum*. It is common in the western part of the forest and gets scarcer as you go east, till it seems to disappear near the south-east corner of the Caspian."

***Cyclamen elegans* in twenty-first century Iran** There are not many articles or notes about *C. elegans* in the Cyclamen Society's journal, but one eye-witness report, published in the June 2006 issue, does contain some useful observations.

The group was travelling in April 2004 through lower altitudes of the Caspian woods to the south of the town of Damavand, heading north toward Qa'emshahr. After the town of Pol-e Sefid, turning north-east through the foothills toward Sari, they reached a section of wooded hillside where, in the still leafless woods, they saw the leaves of cyclamen plants, some of which were still in flower. Nearby on a fairly steep wooded track on clay soil, cyclamen were dotted all over the place.

Cyclamen elegans near Ramsar, as recorded by Paul Furse in 1964 (PF)

Travelling on north and east, climbing into the mountains, occasional forestry scars could be seen near the sides of the road, but mostly there were thick stands of deciduous trees with natural clearings occurring from time to time. Eventually, upon reaching the top of a pass between Pol-e Sefid and Sari at about 1,200 m:

> "there was a cutting through sandy clay with a steep erosion slope and a tangle of roots and branches falling from the edge high above. Something looking like a flattened pebble caught my eye, lying at the edge of the road drainage ditch under the steep bank. I bent down to examine it. A cyclamen tuber! Of course, excited now, I looked around — and there were lots of them. Some hanging from the top of the bank, some growing in little landslips but otherwise undisturbed, some in a jumble at the bottom of a small cutting caused by a seasonal stream … some were in flower and … some even had seed pods on them despite spring being later at this higher altitude."

From there, the group travelled to Gorgan (Asterabad of old) where they found more cyclamen in the mountains a few miles south of Gorgan, in a densely wooded area at the bottom of a cliff of sandy clay soil. Here, there were many seedlings, the result of seeds dropped from the plants on the cliff. They found that this was the extent of the distribution as "to the east of Gorgan the forests peter out and are replaced by scrub and then semi-desert, the habitat of tulips, iris and fritillaries." This last observation underlines the fact that cyclamen are essentially mesophytes, not plants of arid or intensely cold places; they do not grow, for example, in the whole of the upland interior of Anatolia, Iran and Transcaucasia. Even those from Mediterranean climes find shelter in scrub or among rocks and in crevices. *Cyclamen elegans* is perhaps one of the more extreme in its requirement for a humid woodland habitat.

Further observations of *C. elegans* in the Caspian forests

Over the past decade, **Bob and Rannveig Wallis** have visited many sites in both the Talysh and Elburz mountains of Iran, studying the flora and adding further information to our knowledge of this distinctive cyclamen of the Caspian region. The following is an extract from their report for the Cyclamen Society is published in full in the Society's journal *Cyclamen* for June 2011.

> **"The Talysh Mountains**
>
> The northern-most site we have found *C. elegans* in Iran is in the Talysh Mountains just inland from Astara within 1 km (0.6 mile) of the border between Iran and Azerbaijan. Here, the road winds up through the deciduous forest right next to the high wire fence which forms the border. The cyclamen were scattered here, growing under small bushes of *Zelkova* which formed an under-storey under larger trees. It looked as if these bushes had been coppiced over a number of years. The tubers were in deep leaf mould and moss. The altitude here is 190 m (620 ft). We are informed that it also grows further along the Talysh, the other side of the border because, some years ago, we were sent two seedlings of *C. elegans* from a friend in Latvia which had been grown from stock originating on the Azerbaijan side.
>
> We have searched in vain in several places, several times on the road between Asalem and Khalkhal but failed to find any suitable habitat and no leaves.
>
> Another road heads up from the wide wet plains of the Sefid river delta at Fuman to, and beyond, the old mountain village of Masuleh. The delta is a hive of agriculture with numerous large fields of tea and rice often edged with large stands of *Parrotia persica* which still looks good in its spring foliage. Just think what a sight it would be in full autumn glory. *Cyclamen elegans* can be found all along this road as far as Masuleh, growing with *Primula vulgaris* (presumably var. *heterochroma* although we did not see any flowers), *Scilla hohenakeri* and one of the prettiest of stream-side plants, *Rhynchocorus elephas*.The last mentioned forms large colonies beside any water seeps, the large yellow flowers looking, as the name suggests, just like an African elephant complete with large ears, eyes and a trunk.
>
> We did not venture out of the bus once we were beyond the town because we were concerned about the throughway. The road is not surfaced after Masuleh and it is badly eroded by the streams which cross it. The woods and scrub look extremely interesting since there seems to have been little human interference and a second visit is definitely warranted. Above the tree line there were large stands of *Orchis pinetorum*, *Anemone caucasica*, *Crocus speciosus*, *Colchicum speciosum* and *Fritillaria kotschyana*.
>
> **The Elburz Mountains**
>
> Once one ventures east of the Sefid River, one is in the Elburz Mountains. We drove once on the road from Siyahkal to Deylaman. *Cyclamen elegans* was frequent everywhere we stopped up to an altitude of about 240 m (790 ft) but we did not find it despite considerable searching on a stretch of scrubby grassland which was obviously a popular picnic spot at about 1,740 m (5,700 ft) in spite of the habitat looking right.
>
> Perhaps one of the more interesting places is the road from Abbas Abad to Kalardasht because the woods are full of *C. elegans* here. We were very surprised to find them above the line of the large trees, growing under 3 m high beech and hawthorn bushes at an altitude of 1,230 m (4,035 ft), higher, by a long way, than reported in the literature. The flowers had long since gone over by this time, 7 May 2006.
>
> *Cyclamen elegans* is well known on the next road eastwards as this is the one which joins Chalus over the Kandovan Pass to Tehran and has been well reported in the past. It is still found in the low altitude woods in company with *Scilla hohenakeri*, and occasionally, the leaves of *Crocus caspius*.
>
> There is a new asphalt road which joins Alạmdeh to Baladeh which winds its way through the most unspoilt Caspian forests we have seen, until it tops out at a heady and very snowy 3,200 m (10,500 ft). The bus had wheezed its way over three of these monster passes before it was able to descend into the valley where Baladeh is situated, many hours later. The northern approach goes through some very beautiful historic villages and access off the roadsides is not

One associate of *C. elegans* in the Elburz range is *Epimedium pinnatum*, seen growing here at 1,420 m, south of Alamdeh (RW)

too bad, so we spent rather too long searching in the forests. We found *C. elegans* in several places in the woods right up to 1,420 m (4,660 ft). The woods here are somewhat less dense and spring had been late since we found just one *C. elegans* flower on 24 April 2009. It was growing with some nice stands of *Epimedium pinnatum* and *Primula vulgaris* both of which were in flower, both with yellow flowers. We feel that this must be an altitude record for *C. elegans*. Higher up, the habitat enters a scrub phase and then montane moorland, both of which are absolutely full of flowers at this time. We recorded: *Scilla siberica*, *Ornithogalum sigmoideum*, *Anemone caucasica*, gageas, *Iris reticulata*, *Corydalis seisumsiana* and a large-flowered yellow *Adonis* (possibly *A. wolgensis*) just emerging from the ground. The journey over the top is rewarded by the fact that the cliffs in the river gorge at Baladeh are studded with cushions of *Dionysia aretioides*. We have not observed *C. elegans* on the Amol to Tehran road in spite of several searches.

In 2007, we ventured east of the Caspian Sea and entered the forest from Aliabad, some 75 km (45 miles) beyond the eastern shore. This is again another very popular picnic spot owing to its springs and wonderful views of the mountain forests ascending steeply to the south. We found a few *C. elegans* in what we took for a managed forest of *Parrotia persica* at 246 m (807 ft). This is as far east as we were able to search but is already many kilometres further than it has been reported in the past."

Cyclamen parviflorum subsp. *parviflorum* and subsp. *subalpinum*

SPRING-FLOWERING

The northern Turkish *C. parviflorum* is a challenge to cultivate. This fine specimen shows the slightly lavender shade of the flowers (MD)

This diminutive cyclamen is aptly named as it does have the smallest flowers of all cyclamen, save perhaps for some of the smaller forms of *C. intaminatum*. This does not, however, mean that this plant is without its attractions as the whole plant is in proportion, with round, plain-green leaves that stay almost pressed to the ground even in cultivation, certainly in the case of the higher altitude populations. This is not surprising when one considers its habitat in the mountains of Turkey's northern range, the Pontus. Here, it is subjected to cool moist winds from the Black Sea in the summer and, at the higher levels, it lies buried in snow from October to May, so there is every encouragement to stay close to the ground. The Bowles Scholarship Expedition (B. Mathew, S. Baker, D. Pycraft and D. Barter) travelled in early March 1963 over the Zigana Pass, where their Landrover was dwarfed by the snowdrifts, the road having been cleared using bulldozers such was the quantity of snow. Not surprisingly therefore, most of the plants associated with *C. parviflorum* are also relatively dwarf; one of the most exciting of these is the small shrub *Daphne glomerata* but there are many other desirable alpines including *Trollius ranunculinus*, *Viola altaica*, *Crocus vallicola*, *C. scharojanii* subsp. *lazicus* and *Gentiana verna* var. *pontica*. Lower down the mountains, for example in the area of the Sumela Monastery, *C. parviflorum* occurs in wooded situations where the plants are somewhat larger and have taller flower stems and leaf stalks, characters that prompted Christopher Grey-Wilson to distinguish them first as var. *subalpinum* (1997) then as a subspecies (2002). Jill White, who has made a particular

Zigana Pass in summer where *C. parviflorum* inhabits the short turf, often bathed in mist on the seaward side (BM)

Zigana Pass in late March (MD)

Cyclamen parviflorum emerging from its blanket of snow, 24 March 2011 (MD)

Cyclamen parviflorum. Painted by Christabel King (reproduced at 115% of original size)

Daphne glomerata, growing in association with *C. parviflorum* on the Zigana Pass (BM)

Sumela Monastery, the forested habitat of *C. parviflorum* in its lower altitude range (BM)

Another associate of *C. parviflorum* in the eastern Pontus is *Gentiana verna* subsp. *pontica* (JSI)

study of the species, records that there are intermediate habitats for *C. parviflorum*, on open slopes under low scrub, and that in such sites, a mixture of both types of plants can be observed.

In some ways, this tiny species is similar to *C. coum*, but the two are readily distinguishable and appear not to hybridise although both can be found in the same general area. A comment from Richard Bailey about the lower-altitude colonies of *C. parviflorum*, based on fieldwork in Turkey, suggests, however, that there could have been some introgression of genes from *C. parviflorum* into *C. coum* populations: "The flowers range in colour from a deep rose pink to a shade so pale as to be almost white — although with a deep purple blotch at the nose … The leaves vary in shape from round to heart-shaped and while most are plain green some have substantial marbling." In general, though, the plain, dull green, almost circular leaves of *C. parviflorum* provide immediate recognition, supported by the flower colour which is subtly different from that of *C. coum*. In *C. parviflorum* there is a discernible blue element in the colouration of the flowers, giving an almost lavender shade and contrasting with the vibrant carmine pinks and red-purples of *C. coum*. At the base of the corolla of *C. parviflorum*, there is a deep, dull purple blotch that continues right to the mouth, whereas the dark blotch in *C. coum* is situated above the base and is replaced by a paler pink or white zone at the mouth.

In growth pattern, *C. parviflorum* differs from nearly all other cyclamen in that it has no proper dormant season. It has evolved in a region that has no markedly dry period and as soon as flowering and fruiting have finished, the new leaves and flower buds for the next spring are already forming. The plants then go into 'cold storage' until the following spring, when they complete their development as the snow melts.

Tuber of *C. parviflorum* (BM)

Flower of *C. parviflorum*.The dark violet stain continues right into the mouth of the corolla (BM)

Two subspecies are recognised: subsp. *parviflorum* and subsp. *subalpinum*.

Key to the subspecies (after C. Grey-Wilson 2002)

Leaves usually less than 2.5 cm wide; petals only 4–8 mm long, held erect . . . . subsp. *parviflorum*
Leaves up to c. 7 cm wide; petals 8–14 mm long, spreading . . . . subsp. *subalpinum*

C. parviflorum subsp. **parviflorum** Pobed., *Bot. Mater. Gerb. Bot. Inst. Komarova Akad. Nauk SSSR* 9: 250 (1946) and *Syst. Herb. Inst. Bot. Acad. Sci. URSS* 9: 250 (1946)]. Type: Turkey, Artvin (L).

C. parviflorum subsp. **subalpinum** (Grey-Wilson) Grey-Wilson, *Cyclamen* (ed. 2): 207 (2002). Type: Cultivated specimen, 1995, *J. White* (holotype K).

A fine patch of *C. parviflorum* showing the round, plain green leaves characteristic of the species. Near Sumela Monastery, above Trabzon (MD)

Synonym

C. parviflorum var. *subalpinum* Grey-Wilson, *Cyclamen* (ed. 1): 174 (1997).

Description **Tuber** small, depressed-globose, up to c. 2 cm diameter at maturity, brown, smooth and velvety-pubescent, rooting from the centre of the underside. **Petioles** prostrate in the lower part below ground, then ascending to erect above ground. **Leaves** well developed before flowering time, old or new leaves nearly always present and more or less held flat on the ground (subsp. *parviflorum*) or raised on short petioles (subsp. *subalpinum*). Leaf blade orbicular or suborbicular, 1.5–3.5 cm long and wide (to 7 cm in subsp. *subalpinum*), deep dull green with no markings; the margin entire or subentire; underside purplish suffused. **Pedicels** horizontal in the lower part, turning upwards as they emerge above ground, after anthesis coiling from the top downwards. **Flowers** produced in spring, without a scent or slightly fragrant (see note on fragrance below); petals swept back to an erect position or sometimes held almost horizontally, pale to medium pinkish-violet with a dark purple basal stain at the mouth of the corolla, broadly ovate with subacute tips, c. 5–14 mm long, twisted lengthways, without swollen appendages ('auricles') at the base; nectaries are visible inside the corolla near base of petals (according to Grey-Wilson 2002); style held within the mouth of the corolla or just exserted. 2n – 30.

Flowering period April to June in the wild, earlier in cultivation.

Lower altitude *C. parviflorum* near the Sumela Monastery with *Primula vulgaris* and *Galanthus rizehensis* (MD)

At higher altitudes, *C. parviflorum* grows in alpine turf in association with *Trollius ranunculinus* and primroses; leaves of the autumn-flowering *Crocus vallicola* can also be seen (JSI)

Cyclamen parviflorum, accompanied by the superb *Daphne glomerata* (foliage top and bottom) which is found only in alpine regions (JSI)

The aptly named *C. parviflorum* is a true alpine, dwarfed by the harsh conditions in the high Pontus range, northern Turkey (JSI)

Distribution Both subspecies — endemic to north-eastern Turkey, recorded in the vilayets (provinces) of Artvin, Giresun, Gümüshane, Rize and Trabzon; the precise distribution of each of the two subspecies within this range is not known and further detailed field research would be required to ascertain the exact distribution of each, in some places populations are certainly mixed; subsp. *subalpinum* —Turkey, probably represented in the same provinces as subsp. *parviflorum* but at lower altitudes.

Habitat At its lowest altitudes, *C. parviflorum* (subsp. *subalpinum*) occurs in damp situations in mixed coniferous woods (*Picea*, *Abies*, *Pinus* and *Fagus*) that are rich in leafmould with, for example, *Primula vulgaris* and *Galanthus rizehensis*. Higher up, subsp. *parviflorum* occurs in open damp, peaty alpine turf near melting snow, with primulas, gentians (*Gentiana verna* and *G. pyrenaica*), *Daphne glomerata*, crocus (*Crocus scharojanii*, *C. vallicola* and *C. aerius*), *Trollius ranunculinus* and *Caltha polypetala*; altitude range (both subspecies) 730–2,400 m.

Variations Apart from the distinction between the small alpine and larger lower-altitude variants, *C. parviflorum* is remarkably, for a cyclamen, fairly uniform in its characteristics. The absence of any markings on the leaves immediately rules out the great range of possibilities that are apparent in almost every other species. In flower colour, there is variation in the degree of lavender-pink from very pale to a deeper more violet shade, occasionally nearly white. Jill White has made the following observations on the range of flower colour and markings: "Colours vary from nearly white to a deep pink ... Those without a solid blotch at the base of the petal can have dark purple lines running into each other which can look like a solid blotch. In others the lines are much more noticeable, being further apart; or the blotch can have lines the colour of the petal. These lines all curl up into the mouth of the flower (tram-lines for any would-be pollinator)."

Further observations on *Cyclamen parviflorum*

Mixed populations with *C. coum* Although clearly a member of the same group as *C. coum*, it is also apparent that *C. parviflorum* is distinct morphologically. In some places in north-eastern Turkey, *C. coum* and *C. parviflorum* can be found growing in close proximity and indeed have been recorded in mixed populations. **Jeremy Wood** reported in the *Cyclamen Society's Journal* 11 (1987) that:

> "Excitement was provided by finding *C. coum* and *C. parviflorum*, both in flower, growing as mixed populations in two separate wooded valleys at altitudes of around 3,000 to 4,000 ft, accompanied in places by primroses, scillas and snowdrops. These valleys are very far removed from the classical habitat of *C. parviflorum* which is open alpine turf at 6,000 ft and over, such as occurs on the Zigana pass."

These lower-altitude plants are probably referable to subsp. *subalpinum*. It was, however, noted on the Mathew–Tomlinson field trip of 1965 that *C. coum* occurred near *C. parviflorum* among bushes just below the Zigana pass summit (at c. 2,000 m) on the south side of the Pontus, so in this case it is subsp. *parviflorum* that meets up with *C. coum* and might also form mixed populations.

Cyclamen parviflorum and *C. coum* grow intermingled near Sumela, the latter in flower (MD)

Fragrance in *C. parviflorum* Scent is not a notable feature of the related *C. coum* but has been recorded for *C. parviflorum*. **Richard Bailey**, a member of the Cyclamen Society's field survey in the late 1980s, noted that over 80% of the plants of *C. parviflorum* examined bore scented flowers. To Bailey, the scent was reminiscent of honey, whereas **Keith Fry** and **Jill White**, both good growers of this none too easily cultivated species, note the fragrance as being "lilac-like" or "a cross between a violet and a primrose". **Ian White**, writing in the Society's journal in 1989, commented upon the scent of plants in wild populations on a visit to north-eastern Turkey:

> "The cyclamen were at their flowering peak! Tens of thousands of the little beauties carpeting the ground with different shades of pink (and occasional white). Where they grew in the open in close-cropped alpine turf there were very few that had a noticeable scent but where they grew in more sheltered woodland there were far more scented forms."

The habitat of *C. parviflorum*. **Jill White** is particularly successful at cultivating *C. parviflorum* in Essex, and her observations about the species in its Turkish habitat have no doubt provided her with a useful insight into its requirements. Writing in *Cyclamen*, June 1991, of subsp. *parviflorum*, she remarks that:

> "In winter, which can last for eight months of the year, they are covered by a thick layer of snow many metres deep in places. As some plants are still flowering as late as July and even into August, their seeds must ripen while under snow, where they lie alongside the parent plants until the snow melts in the spring… .
>
> The soil is very fibrous, and in May, June and July always on the dry side. They are protected from any drying winds by the turf around them, and by midday what sun there might have been has gone, to be replaced by mists which roll up from the Black Sea coast, through the valleys, and on to the tops of the mountains, to give *C. parviflorum* its daily bath of mist or rain… .
>
> The tubers can be green or brown. The tiny just-germinated tubers, if on the soil surface and exposed to any light, are often green, just like a potato grown too near the surface."

Of subsp. *subalpinum*:

> "At the lower altitude of 1,200–1,700 m on very steep slopes, in woodland under *Picea orientalis* and the sweetly scented *Rhododendron luteum*, *C. parviflorum* grows in deep leafmould, rotting tree branches and moss, with rock debris beneath to give superb drainage. The leafmould is again on the dry side in May, June and July. The plants under the conifers are in heavy shade; only a glint of sunlight

A thriving population of *C. parviflorum* in a woodland scree, with abundant seedlings (JSI)

gets to them for a short time. Although it rains a lot in the Pontic Alps, usually in the afternoon, it isn't a cold rain as in England, and the conifers must keep a lot of it off of the cyclamen. Here most are much bigger in every way than their high alpine brothers, with leaves up to 70 mm across……the flowers are quite different from the alpine type, and very variable. Some have flowers held like *C. trochopteranthum* [now *C. alpinum* — ed.] propeller-like… . It is interesting to note that any plants that strayed out into the open, away from the conifers, were the small alpine type. [Although this suggests that the differences are not stable, merely environmental, it does seem that the two 'types' retain their characteristics in cultivation, at least to some extent — ed.]; *C. coum* was found not many metres away, with *C. parviflorum* under *Rhododendron luteum* and *R. ponticum* in a carpet of thousands of *Primula vulgaris* subsp. *vulgaris*, a very lovely sight — but not a marked leaf of *C. parviflorum* was to be found [in contrast to the statement above by Richard Bailey – ed.]."

Cyclamen alpinum

Flowering in spring

Cyclamen alpinum cultivated to perfection at a Cyclamen Society show (MD)

This pleasing early-spring-flowering *Cyclamen* species, although similar in many respects to *C. coum*, is usually immediately recognisable on account of its musty-fragrant flowers in which the petals are often spread out sideways rather than swept right back in typical cyclamen fashion. Because of this, this species has been known more commonly in recent times by the cumbersome but very descriptive name of *C. trochopteranthum*, meaning propeller-shaped flower (rather more akin to a ship's propeller than an aircraft's).

For a species as distinct as this, *C. alpinum* has a rather confused history. Living specimens of a cyclamen introduced in 1956 by Dr Peter Davis and Oleg Polunin (nos 25368 and 25579) from south-western Turkey were identified as the little-known *C. alpinum*. This had been first described, although very scantily, by Carl Sprenger in 1892 on the basis of plants stated to grow "by the eternal snow in the highest mountains of Asia Minor". Further information obtained soon afterwards by Friedrich Hildebrand was a little more precise: "in the Cilician Taurus near the snow line". Sprenger was closely involved with the Naples nursery of Dammann & Co, which appears to have imported the plant and made it available in commerce. The living specimens grown and studied by Hildebrand in the late nineteenth century were obtained from that company.

Some 80 years later, the cyclamen specialist Otto Schwarz decided that the Davis–Polunin plants were not the same as Sprenger's *C. alpinum*; so in 1975, he described a new species, *C. trochopteranthum*, to accommodate them and cited *Davis* 25368 as the type specimen. This of course created a mystery in that if Davis's cyclamen did not represent *C. alpinum*, where did Sprenger's plant actually come from and what was it like? The matter had in fact been clarified by

Cyclamen alpinum. Painted by Pandora Sellars (reproduced at 150% of original size)

Habitat for *C. alpinum* among the cedars south of Elmalı in south-western Turkey (BM)

Hildebrand in 1898, but his very impressively detailed monograph *Die Gattung Cyclamen* was apparently overlooked by subsequent researchers. In all probability, it was the translation of his work into English by Erna Frank in 1999 that drew attention to the true facts. Hildebrand's very precise study of the plants received from Dammann indicates that he was describing the same species that Davis and Polunin collected all those years later; and that it was they who had at last pinpointed its wild origin — not primarily in the Cilician Taurus but predominantly further west, especially in south-western Turkey.

In 2002, Christopher Grey-Wilson reviewed the facts of the debate as to whether *C. alpinum* and *C. trochopteranthum* were one and the same and decided that "the arguments in favour of this view are very persuasive." The rule of priority dictates that *C. alpinum* is the correct name for this delightful cyclamen. It was perhaps an unfortunate choice of epithet by Sprenger: the Cyclamen Society's fieldwork in the 1990s showed that it is misleading to regard it as an 'alpine' species as it was recorded across a wide altitude range from sea level up to about 1,620 m.

All of this history could have been very different had the dried specimens collected by Edward Forbes in 1842 been identified at the time as representing an unknown species; instead, they were deposited in the Kew herbarium under the name *C. hederifolium* and that is where they remained until the late twentieth century when someone realised the error (see page 77).

C. alpinum seldom 'does well' as an outdoor garden plant, unlike its relative *C. coum* which generally thrives and self-seeds freely. However, when grown in pots for display in an alpine house, it can be superb.

C. alpinum Damman ex Sprenger, *Gartenflora* 41: 526 (1892). Type: no type specimen exists; Grey-Wilson (2002) cites as a lectotype the illustration in Hildebrand, *Die Gattung Cyclamen* t. 111, figs 45–63 (1898).

Synonym

C. trochopteranthum O. Schwarz, *Feddes Repert. Sp. Nov.* 86: 493 (1975).

Description **Tuber** depressed-globose, flattened on top and below, up to 5 cm diameter at maturity, brown, smooth and velvety-pubescent, with slender roots produced only from the centre of the underside. **Petioles** prostrate in the lower part below ground, then ascending to erect above ground. **Leaf** blade broadly

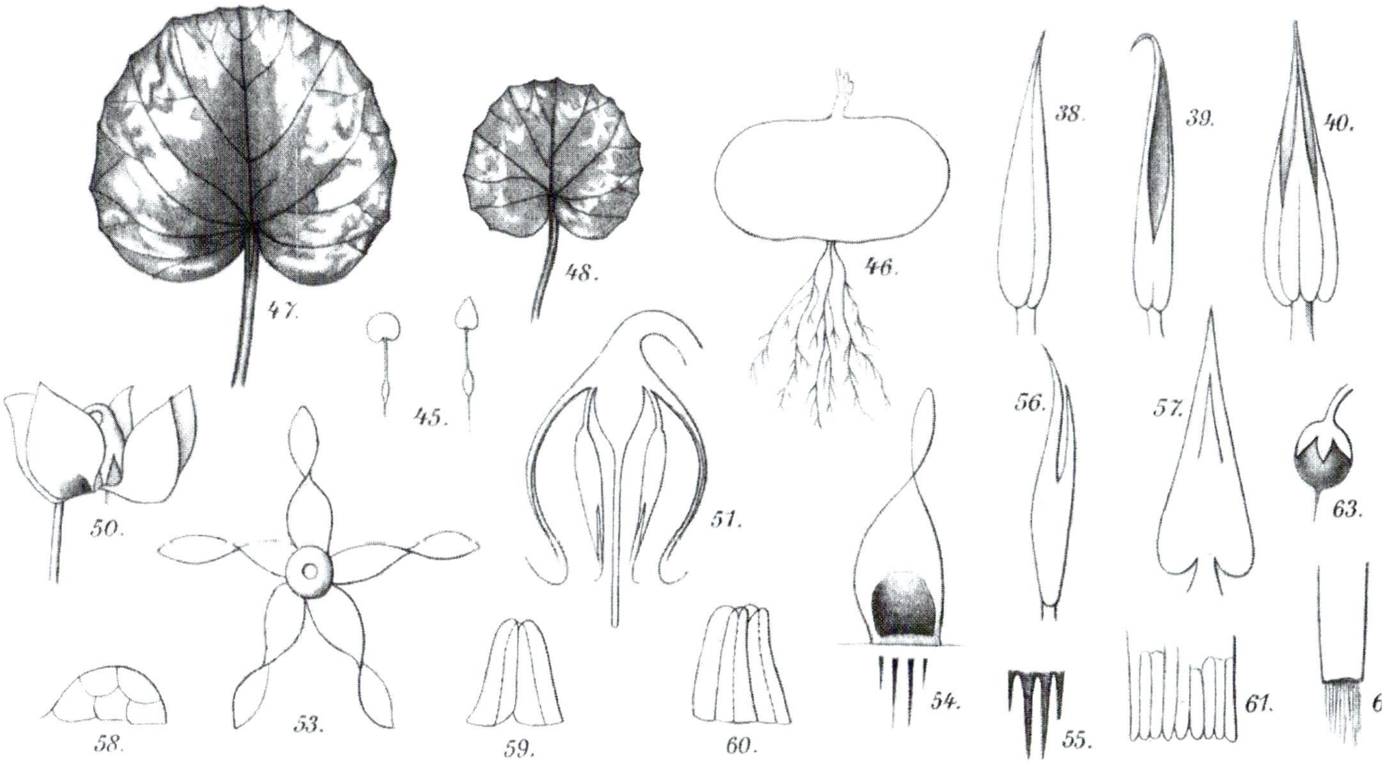

Illustrations of *C. alpinum* from Hildebrand's monograph of 1898

Tuber of *C. alpinum* (BM)

Flower of *C. alpinum* (BM)

ovate to almost orbicular, 1.2–9 cm long, 1–9.2 cm wide; margin slightly toothed or scalloped, dull green, most frequently with a central deep green 'Christmas tree' pattern surrounded by a grey-green or silver-green zone, which breaks up into pale spots and blotches on a green ground towards the margins; the underside normally purple. **Pedicels** horizontal in the lower part, turning upwards as they emerge above ground, after anthesis coiling from the top downwards. **Flowers** produced in early spring, normally with the developing leaves, fragrant (a somewhat musty-sweet perfume); petals usually pale to deep pinkish-purple with a darker purple basal blotch, strongly twisted lengthways and often held horizontally creating a ship's propeller shape, without swollen

A fine, pale form of *C. alpinum* with attractive foliage. Near Çakırlar, south-west of Antalya in limestone rock formations (TW)

A range of leaves of *C. alpinum* showing great variation, observed by Marco ten Hoope (MH)

appendages ('auricles') at the base, 9–16 mm long, 7–15 mm wide; style apex held just within the mouth of the corolla to slightly exserted from it. 2n = 30.

Flowering period February to April.

Distribution *Cyclamen alpinum* is endemic to south-western Turkey, recorded in the vilayets (provinces) of Antalya, Burdur, Denizli, Isparta and Muğla.

Habitat Semi-shade of deciduous trees and shrubs or at edges of evergreen (e.g. *Quercus coccifera*) scrub and in open pine woods; sometimes in shaded or north-facing screes. Usually on limestone formations at 0–1,620 m.

Variations In wild populations, rare leaf variations include plain green leaves and ones in which the silver or grey zones predominate; there is great variation in size from the large-leafed, low-altitude forms to the tiny-leafed high-altitude ones.

The flower colour varies greatly in shades of pinkish-purple and white forms can occasionally be found, although always with the deep crimson or purple blotch at the mouth; a completely white-flowered *C. alpinum* is not currently known to exist. White-flowered variants (with a purple blotch) were named and described collectively as forma *leucanthum* by Christopher Grey-Wilson in 1997.

Perhaps one of the most variable characteristics of *C. alpinum* is in the aspect of the petals. Although they are frequently held out horizontally, resulting in the propeller shape that suggested to Schwarz the name of *C. trochopteranthum* (now reduced to a

In suitable habitats, *C. alpinum* is often abundant. Here at only 100 m in scrub beneath pines between Dalaman and Fethiye, south-west Turkey (TW)

Cyclamen alpinum, showing characteristic propeller-shaped flowers. South-western Turkey, n of Fethiye (TW)

A typical habitat for *C. alpinum* at its higher altitudes among *Cedrus libani* and junipers, here 1,650 m at Çamkyüsü in south-western Turkey (TW)

Although appearing harsh, damp scree conditions can provide a suitable habitat for *C. alpinum* (TW)

synonym), it is not unusual to find plants in which the petals are more erect, in which case the flowers resemble those of *C. coum*. In other variants, the petals droop to the near vertical, so that the flowers look as if they are wilted even when freshly opened; such variants were seen from time to time during the Cyclamen Society's fieldwork and became known to the group as 'Trevor's drooping' as Trevor Wiltshire seemed to be adept at spotting them! Although an interesting variant, this is hardly an attractive attribute.

Compact dark forms of *C. alpinum* with good foliage are arguably among the most attractive of all cyclamen (TW)

Named cultivars

'Album': white-flowered variants have been referred to as 'Album'; these are not completely white, having the purplish stain around the mouth, and come under the umbrella of forma *leucanthum*, named by Christopher Grey-Wilson in 1997.

'Nettleton White': leaves marbled; flowers near-white with purple stain at the corolla mouth.

'Pink Swirl': Large pink flowers; a selection with particularly large and elegant flowers in 'clear pink'.

'Red Devil': Deep red flowers; the two-toned leaves set off well the numerous rather tall blooms of clear red.

'Speckles': flowers quite distinctive, pale pink with carmine speckling.

Further observations on *Cyclamen alpinum*

***Cyclamen alpinum*: extensions to the known range** In autumn 2003, during a study of *C. mirabile*, members of the Cyclamen Society and their Turkish colleagues located a small population of *C. alpinum* already in leaf near the village of Gölçük on the Muğla to Kale road, which is a not surprising addition to the list of known sites. This population was in a fairly conventional sort of habitat for the species, a damp *Pinus–Styrax–Quercus* wood, although the cyclamen tended to grow in crevices on the limestone cliffs. Further to the north-east, beyond Denizli, the country changes considerably and a major feature is the Acigöl salt lake with its harsh, dry steppe country and scattered junipers, not looking at all like 'Cyclamen territory'. North of the Denizli–Dinar road, the group continued along tortuous routes through the hills and mountains, but no *Cyclamen* species was found in spite of some habitats looking not too uninviting. Interestingly, when enquiries were made of local people in two separate villages, they were adamant that there were cyclamen in the hills around but that these were spring-flowering species. There is in fact a published record for *C. coum* in the mountains to the north-west of Dinar, but this is more likely to represent a range extension for *C. alpinum*. This is definitely a location that needs further exploration by someone who knows their *alpinum* from their *coum*!

Edward Forbes and *Cyclamen alpinum*. In 1842, Edward Forbes, in the company of Lt. A. B. Spratt and the Rev. E. T. Daniell, carried out extensive fieldwork in south-western Turkey, which Forbes referred to under the ancient geographical term of Lycia. Forbes' interests were wide-ranging but for this expedition appeared to be centred upon archaeology: his group visited several of the ancient sites that are scattered liberally through the region and made detailed surveys of some of them. Although this focus provided a framework for his travels, Forbes's party also made geological, zoological and botanical observations. With regard to the last of these disciplines, and most importantly, he prepared a collection of dried specimens that are housed in the herbarium at the Royal Botanic Gardens, Kew.

Forbes was born on 2 February 1815 on the Isle of Man and died on 18 November 1854. Although commencing a career in medicine, he had such broad interests that he soon became a very competent naturalist and joined the survey ship *Beacon* in this capacity. Later, he became Professor of Botany at King's College, London, and then held the Chair of Natural History at the University of Edinburgh. In botanical circles, it is perhaps the 1842 visit to Anatolia (while attached to the *Beacon* survey) for which he is best remembered as it resulted

A good form of the white-flowered *C. alpinum* forma *leucanthum*, grown by Tim Murphy (MD)

Prof. Edward Forbes (1815–1854)

in the discovery of several plants unknown to science at the time, among them *Fritillaria forbesii* and *Chionodoxa forbesii*. One of Forbes's dried specimens in the Kew herbarium is a *Cyclamen*. The original label bears the inscription "Lycia 42" and "Prof. Forbes 429" in two different hands. The former label, "Lycia 42", is probably in Forbes's own handwriting. A tentative identification in another hand has been added in pencil: "Cyclamen hederifolium?". From the details of the specimen, this is clearly incorrect, so the question arises as to what it really is. Of the species that are known to occur in south-west Anatolia, there is only one logical possibility, and that is *C. alpinum* (previously *C. trochopteranthum*). In fact, this is exactly what the specimen appears to represent, even without the geographical link. Indeed, someone has added a Kew label with the name *C. trochopteranthum*, but this identification must have been relatively recently, i.e. post-1975 when this name was published by Schwarz.

The account of this journey of 1842 is written up in a fascinating two-volume work: *Travels in Lycia, Milyas and the Cybiratis*, published in London in 1847. Although many of the individual specimens collected on the expedition have no precise data on them — often just "Lycia" — we can trace the course of separate journeys by reading the descriptions provided in the books. One of these journeys was to the mountain range Baba Dag (known to Forbes as Cragus) to the east of Fethiye. This large and imposing mountain clearly fascinated Forbes who wrote about it in detail, describing the climb from sea level through the lush forests to the upper levels. Baba Dag is a botanically very rich mountain with several locally distributed species, probably endemics, notably *Chionodoxa forbesii* (i.e. the true species, not the commonly cultivated "C. forbesii hort.", which is *C. siehei*) and the white *Sternbergia candida*.

Forbes wrote: "... a few days later [i.e. after 21 May] we ascended Cragus, and carefully collected every plant then flowering on the mountains, including *Scilla bifolia* whose exquisitely blue flowers contrasted with the snow masses in the clefts." One would not describe the rather violet-blue *Scilla bifolia* as exquisitely blue, so I am certain that this was the then unknown *Chionodoxa forbesii*. He may have collected his specimens of *Cyclamen alpinum* there, although *Cyclamen* are not in his list of plants seen. The Cyclamen Society's survey of this species in 1997–1999 certainly found it to be plentiful on Baba Dag, and it is very unlikely that Forbes would have missed seeing, or neglected to collect, such a striking plant. It is possible, however, that it was already past flowering. The genus *Cyclamen* does appear in a list of plants from another mountain that I have not yet traced. Forbes wanted to compare the plants he found on the limestone mountain Cragus with those from a similar altitude on a serpentine formation, and this he did. The site of this mountain was reported as Balbura, near the ancient site of Katara. The Society's survey did record *C. alpinum* on serpentine as well as limestone.

This is all very interesting, as it means that Forbes was the first person to record and collect this cyclamen, 50 years before it was first named and described as *C. alpinum* by Sprenger, 133 years before it was named *C. trochopteranthum*, and 114 years before the re-discovery of its wild habitat and introduction to cultivation by Peter Davis and Oleg Polunin. Had Forbes's specimen been identified in 1842 as representing a hitherto unknown species, we might well have a '*Cyclamen forbesii*'!

2.2c The Pseudibericum group

(= Subgenus *Gyrophoebe*, Series 2)

Comprising: *Cyclamen pseudibericum*

The solitary species in this Series has ovate, toothed leaves and large flowers in spring. The corolla is bright purple with a darker blotch around the mouth, which has a white rim. There are no swellings ('auricles') at the mouth. The stamens are not exserted from the corolla; the style may be shortly exserted. In fruit, the pedicels start to coil downwards from the apex. Chromosome number 2n = 30.

Cyclamen pseudibericum

FLOWERING IN SPRING

The very showy spring-flowering *C. pseudibericum* can be grown outside in mild areas but makes a very good alpine house subject for cooler regions (MD)

Cyclamen pseudibericum. Painted by Pandora Sellars (reproduced at 125% of original size)

Among the spring-blooming species, *C. pseudibericum* is one of the showiest with large, richly coloured flowers. It is also one of the more restricted in distribution, endemic to southern Turkey in the Anti-Taurus and Amanus mountains. For this reason, *C. pseudibericum* is subject to a Turkish government collecting ban, unlike its geographical neighbours *C. cilicium* and *C. coum* of which commercial gathering is allowed, although with strict quotas.

The species was described in 1901 by Friedrich Hildebrand and has a rather interesting history; but first, his apparently curious choice of name, which means 'false ibericum', should be explained. This epithet refers back to a time when '*C. ibericum*' (now regarded as a synonym of *C. coum*) was recognised as a distinct species. The name, incidentally, indicates an origin in the Caucasus (Iberia), not the Iberian Peninsula. It was considered that the newly recognised species from southern Turkey was similar in appearance to, although not quite the same as, this Caucasian plant, so the epithet *pseudibericum* was chosen to imply a 'nearly but not quite' relationship.

For 50 years, *C. pseudibericum* was an enigmatic plant with no recorded origin, regarded by some as a variant of *C. libanoticum* and by others as a hybrid that had arisen in cultivation. Then, in 1952, the Turkish botanist Dr Hüsnü Demiriz rediscovered this plant in the Amanus range in southern Turkey. This record was followed by that of Dr Peter Davis in 1957 (No. 26117), made during his extensive botanical expeditions, and as a consequence the species became better known in cultivation. Further locations were added during the travels of Mathew and Tomlinson (1965: MT4141 and MT4146), Albury, Cheese and Watson (1966: ACW664), Archibald (1985) and, Erna and Ronald Frank (1986, 1992 and 1993). Bob and Rannveig Wallis, who travelled with the Franks, made their observations available to the Cyclamen Society. These and the older distribution records provided the background information for some dedicated fieldwork by the Society in 2002, an account of which is presented in the 'Scientific studies and fieldwork' chapter on page 275.

It is not unusual to find more than one cyclamen species in a given area, and this is so in the case of *C. pseudibericum*. In some of its locations, for example near Karsantı, *C. pseudibericum* meets with *C. cilicium*, whereas near Dörtyol and Düziçi village, which is to the north of Osmaniye, it occurs with *C. coum*. Nevertheless, no hybrids have been recorded: these three species are clearly genetically distinct and, in the case of *C. cilicium*, the very different flowering time precludes the possibility of hybridisation.

C. pseudibericum Hildebr., *Beih. Centralbl, Bot.* 10: 522 (1901). Type: Turkey, a specimen cultivated in Germany at Freiburg, *Hildebrand*.

SYNONYM

C. libanoticum Hildebr. subsp. *pseudibericum* (Hildebr.) Glasau, *Planta* 30: 523 (1939).

DESCRIPTION **Tuber** often almost subglobose or somewhat flattened above, up to 4 cm diameter at maturity, brown, smooth but becoming corky with age, with slender roots produced only from the underside. **Petioles** prostrate in the lower part below ground, then ascending to erect above ground. **Leaves** well developed before flowering time. Leaf blade broadly ovate, 2–8 cm long, 2–8.5 cm wide, subacute or obtuse, dark green, variable in markings but often with a central narrowly triangular, jagged-margined green pattern surrounded by an arrow-shaped paler green or grey zone usually forming a continuous band, then the outer edge towards the margins is often deep green variably marbled or splashed silvery-green; margin regularly scallop-toothed; the underside usually suffused purplish-red. **Pedicels** horizontal in the lower part, turning upwards as they emerge above ground, after anthesis coiling from the top downwards. **Flowers** produced in spring, with strong scent; petals swept

Tuber of *C. pseudibericum* (BM)

Flower of *C. pseudibericum* (BM)

Cyclamen pseudibericum in the wild. Richly coloured flowers and striking foliage are the characteristics of this Turkish endemic species (SB)

Variation in leaves of *C. pseudibericum* (BM)

A natural garden in southern Turkey formed by *C. pseudibericum*, primroses and blue scillas (?*Scilla melaina*) (SB)

Cyclamen pseudibericum **in scrub near Baskonus, Kahramanmaraş province, with evergreen oaks (ET)**

upwards to an erect position, usually rich pinkish-purple but sometimes paler pink, always with a dark purple basal stain near the base and a white basal zone forming a white rim at the mouth of the corolla, ovate-elliptic with acute tips, c. 2–2.5 cm long, twisted lengthways but not strongly so, without obvious swollen appendages ('auricles') at the base; style usually held within the mouth of the corolla. 2n = 30.

Flowering period March to May.

Distribution *C. pseudibericum* is endemic to southern Turkey, in the Amanus and Anti-Taurus mountain ranges, recorded in the vilayets (provinces) of Adana, Gaziantep, Hatay, Kahramanmaraş and Osmaniye.

Habitat Often in the shade of deciduous trees such as oak (*Quercus* spp.), beech (*Fagus*) and hop-hornbeam (*Ostrya*), sometimes also in rocky valleys near streams and in light *Pinus brutia* forests, 500–1,650 m.

Variations Although a geographically more restricted species than most, *C. pseudibericum* does display a considerable amount of variation in its leaf markings. Usually there is a narrowly triangular, ragged-edged dark green zone in the centre, surrounded by a grey- or silvery-green zone that varies in intensity, in 'good' forms providing a strikingly contrasting leaf marking; sometimes the leaf is also splashed with silvery-green spots. Living material collected during some of the early field trips displayed a good range of leaf variation and flower colour; for example, some of the Mathew–Tomlinson collections have very richly coloured flowers with well-marked leaves, whereas the Albury–Cheese–Watson collection ACW664 had attractive softer pale to mid rose-pink flowers. The usual or 'typical' form, f. *pseudibericum*, has flower colours in rich magenta or purple shades; paler forms have been described as f. *roseum* by Christopher Grey-Wilson, published in *Cyclamen*: 174 (1997). Some variants of this species are so pale as to be nearly white but pure white forms are a very rare occurrence.

The pink-flowered *C. pseudibericum* forma *roseum* in cultivation at a Cyclamen Society show, 25 March 2006 (MD)

Further observations on *Cyclamen pseudibericum*

An unusual site for *C. pseudibericum* During his very extensive plant-hunting travels, the late **Jim Archibald** made many astute observations. In 1985, he located *C. pseudibericum* in a somewhat isolated situation, away from the main area of distribution and bordering onto a rather different climatic zone:

> "growing among deciduous oaks, along with populations of such other woodland plants as *Anemone blanda*, *Corydalis* and *Doronicum*, on a north-facing slope of Armut Daği, above Tekir, at about 1,500 m. This locality is south of Göksun and about 70 km north of the well-known habitats near Haruniye, but what interested me was the fact that the species had penetrated to the colder, drier inland slopes of the Anti-Taurus. The plants were flowering but first impressions were that the colony was not thriving."

Hunting for C. *pseudibericum* The late **Erna and Ronald Frank** were well known for their remarkable journeys, many of them undertaken in Turkey where they often recorded new locations or made valuable observations. Their hunt for *C. pseudibericum* was described in the December 1993 issue of *Cyclamen*, an extract of which is repeated here:

> "About 10 km north of Düziçi is that distinctive peak, the DülDül Dağ (2,246 m), where we have recorded *C. pseudibericum* up to a height of 1,650 m. In the hamlets skirting the DülDül Dağ, children pick bunches of its flowers in the spring but this need not cause any anxiety to conservationists as it grows abundantly in scrub and hedges, even on rubbish heaps. But the beech forest above provides the happiest biotope for the cyclamen, with its dappled shade and leafmould. *Cyclamen coum* occurs in the same area but we have not seen mixed populations of the two species as there are in the Amanus near Dörtyol. Dr and Mrs Wallis, however, have reported *C. pseudibericum* and *C. cilicium* growing side by side near Karsantı, to my knowledge the first time that those two species have been found together.
>
> The road from Kozan to Feke runs through beautiful mountain scenery and over a couple of passes. *Cyclamen pseudibericum* accompanies the road on the slopes at about 700 m and this is one of the well known areas for the species. We have found the extent of the distribution somewhat larger than expected from reports, starting from about 26 km north of Kozan and continuing for about 18 km towards Feke. Although it would be an arduous task, I think it would be rewarding to clamber

about in those not very accessible mountains, where one would surely discover a much wider spread of the species. Jim Archibald found a location about 50 km east of Feke, as the storks fly, south of the Püren Pass on the Kahramanmaraş to Göksun road. When we visited the site last spring, the mountainside on which the cyclamen grow was very boggy, with many rivulets of snow-melt water. The flowers were quite small and not very plentiful, but there was the added interest of *Fritillaria alfredae glaucoviridis* as one of the companion plants.

Some years ago we happened upon a *C. pseudibericum* location that we had not heard of, though I am sure somebody must have been there before us. This was a stretch of the Yenicekale–Andırın road at 1,180 m. It was a thriving habitat not only for the cyclamen but also for fritillarias, scillas, primulas, corydalis and others.

The colour of the flowers we saw in all these locations seemed to diverge very little from the typical carmine-magenta, but some years ago Alfred Schattanek of Vienna sent seeds of a light pink form which he had discovered at Karatepe, of Hittite fame. As long ago as 1966, Albury, Cheese and Watson found a new location of *C. pseudibericum* in the Amanus Mountains near Dörtyol, where the flowers displayed a remarkable range of colours from white through shades of pink to the standard magenta. Here the pale pink shades of *C. pseudibericum* dominate and even *C. coum*, which intermingles with *C. pseudibericum*, seems to favour the light pink hues."

***C. pseudibericum* growing alongside *C. coum*.** **Harry Gill** and **Pat Nicholls**, who accompanied Vic Aspland on the Cyclamen Society's field survey described on page 275, wrote separately of their own experiences, some of which are repeated here in a selection from the Cyclamen Society's journal for June 2002.

"The outer boundary of the area we had surveyed can be roughly determined by joining points starting 35 km NW of Tarsus at Çamıyayla, to Aladag (Karsantı) 90 km N of Adana, to Andırın 60 km N of Osmaniye, to Kurucaova 15 km SE of Göksun, to Fevzipaşa 40 km E of Osmaniye, and to Belen 5 km S of Iskenderun. To cover such a large area we used several bases: mainly Osmaniye, but also Adana, Karamanmaraş and Iskenderun. Our daily outings would tend to resemble bicycle spokes in and out of the individual bases. All in all, we drove 3,100 km in our search for the plants.

Regrettably, we were unable to survey the southern-most site we found, which was just to the south-east of Iskenderun, not altogether surprising since the road on which we had stopped ultimately led to a complex of telecommunications masts. Although *C. pseudibericum* was found at several locations with the northern-most site being some considerable distance from the southern-most, it was not found growing extensively over large areas but appeared to grow in quite discrete colonies in limestone rocky areas shaded by aspect and overtopped by deciduous trees and shrubs.

In the northern areas, particularly the Aladağ region, the weather and a dangerous muddy track prevented us extending our search, but we were fairly sure that, having found the cyclamen in one area of limestone-type pavement, it would occur elsewhere since the pavement covered many visible hillsides and plateaus. The region is home to many tiny hamlets so the whole area is heavily grazed, leaving the surviving cyclamen to grow in clints and underneath various species of grazed scrub.

There were many memorable occasions probably none more so than finding on our very first day a new site for *C. pseudibericum*, which we surveyed and which ultimately proved to be the best site for plants in leaf and flower… . Other memories include seeing strikingly deep-coloured flowers of *Cercis siliquastrum*, *Daphne sericea* en masse in flower and scenting the surrounding area, *Primula vulgaris* in abundance including almost white flowers and a couple with pink flowers, *Anemone blanda* here, there and everywhere, *Muscari* and *Scilla* species and *Iris persica*. The lovely flowers of the iris were found in two locations, the second one being in an area where we were guided to *C. pseudibericum* by a farmer. There were probably ten or so flowers but imagine our consternation when our guide casually picked the flowers and ate them!

We saw masses of *Colchicum szovitsii* looking like water lilies as many of the plants were in snow-melt water. This was in the Gülpinar Pass south of Göksun, where we also had the pleasure of seeing a beautiful golden crocus species flowering in abundance on the lower slopes where the snow had melted away. On one occasion, we found *C. cilicium* on one side of a road and *C. pseudibericum* on the other! We found a few *Fritillaria alfredae* and one *Fritillaria acmopetala*, and we will not forget the distinctive *Helleborus vesicarius*. *Cyclamen coum* was found growing with *C. pseudibericum* but also in many areas where no *C. pseudibericum* could be found."

***C. pseudibericum* in the Amanus Mountains** The accounts of the 2002 Cyclamen Society's fieldwork interested **Tony Willis**, who wrote in the Society's journal for June 2003:

"Reading the article on the Turkey Expedition 2002 ... and the apparent gap in its distribution I thought these observations might be of interest. I have found this cyclamen species in a very pale flowered form approximately 10 km east from Osmaniye in the direction of Gaziantep in the mountains 15 km to the south of the road. The plants were growing in an area that I think would have been covered in snow at the time the expedition visited, at an altitude I estimate to have been about 600 m. They were growing under a dwarf oak species at its highest altitude where it merged into dense pine forest on a steep northerly-facing slope. Other plants in the area were *Primula vulgaris*, *Iris unguicularis* and several orchids in abundance (two species of *Cephalanthera*, *Limodorum abortivum* and *Orchis purpurea*). This general area of the Amanus Mountains has no roads, over a very extensive area, only logging and donkey tracks and therefore further observation is very difficult. Looking at my records on the colour of the plants found, one plant was a very pale pink, one was white with a faint pink flush to the tips of the petals ... and one was white with a pink picotee edge to the petals, with the rest as indicated above being pale pink."

2.2d The Cilicium group

(= Subgenus *Gyrophoebe*, Series 3)

Comprising: *Cyclamen cilicium*, *C. intaminatum* and *C. mirabile*

A group of three closely related autumn-flowering species with white or pink flowers having no swellings ('auricles') around the mouth of the corolla; there are either no markings at all at the mouth or clearly defined dark blotches or jagged markings. The stamens and style are not exserted from the corolla. In fruit, the pedicels start to coil downwards from the apex. Chromosome number 2n = 30.

Cyclamen cilicium

FLOWERING IN AUTUMN

Along with *C. hederifolium*, this autumn-flowering species is one of the better known in cultivation, although perhaps not for the right reasons. Wild-dug tubers have been traded by the million for decades, many of them ending up in colourful packets, not always correctly identified, on garden centre shelves, over-dried and with only a slim chance of becoming established when planted in the garden. Nevertheless, this is a common plant in its native haunts. It is a small graceful cyclamen with honey-scented flowers in variable shades of pink and, like many other cyclamen, with darker purple markings around the mouth of the corolla. Often, the flowers

A well-grown plant of *C. cilicium* at a Cyclamen Society show, here flowering in September before the leaves have developed (MD)

appear before the leaves emerge but this plant blooms over a long period between September and November, so the foliage is usually present by the end of the flowering season.

This is yet another endemic of Turkey, a country that is so richly endowed with cyclamen species. The species name informs us that it is a native of the ancient Roman region of Cilicia along the Mediterranean coast in the south of the country. In fact, it is confined to the Taurus (Toros) mountains, although it has an extensive distribution within that great chain. The eastern end of *C. cilicium* territory is in Karaman province, and it then occurs in the west as far as the Dedegöl mountain range, which lies between lakes Beyşehir and Eğırdır in the south-western corner of Konya province; there are also populations in the mountains on the north-west side of Lake Eğırdır, where this species occurs not far from the most easterly record for *C. mirabile*, although the two have not yet been recorded in close proximity. Southwards, *C. cilicium* reaches into Antalya province. Although the whole area covered by *C. cilicium* is known to be extensive, its actual distribution has not been well recorded in the past. For this reason, the Cyclamen Society carried out detailed field studies in the autumn of 2008 and 2009 to ascertain exactly where *C. cilicium* grows and to fill the gaps in information about its habitat, natural variation, conservation status and so on. A full report on this work appears in the 'Scientific studies and fieldwork' chapter on pages 296–306.

Those who are familiar with cyclamen species are unlikely to have any difficulty in differentiating *C. cilicium* from other species. Although the flowers are very similar to those of *C. mirabile*, the latter usually has conspicuous teeth at the apex of the petals and the

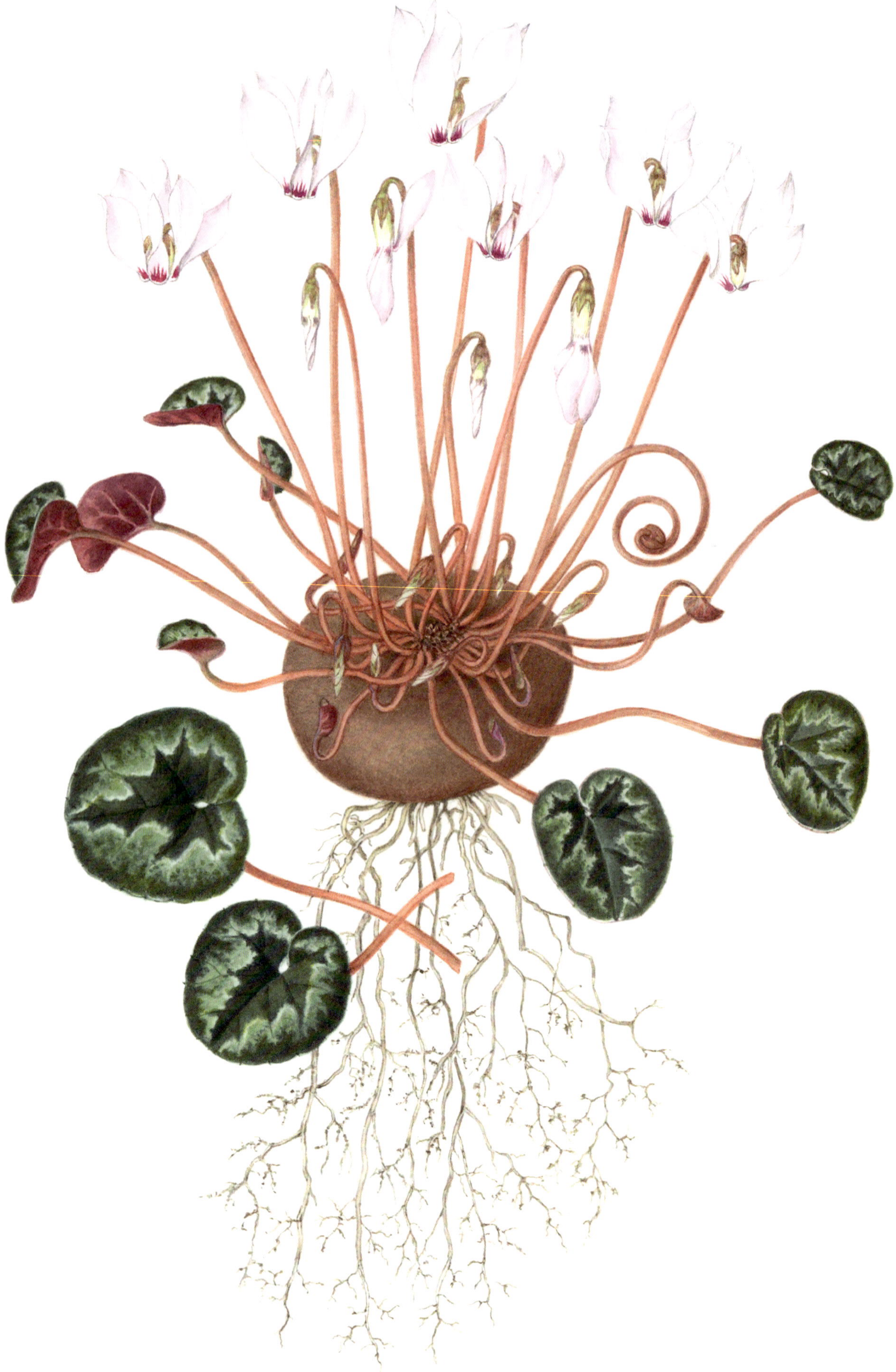

Cyclamen cilicium. Painted by Christabel King (reproduced at 125% of original size)

Cyclamen cilicium country: the Toros mountain range in southern Turkey (MD)

flowers have a distinctive coconut scent, unlike the more honey-like fragrance of *C. cilicium*. The leaves too are different, those of *C. mirabile* having obvious teeth on the margins, but it is the position of the roots on the tuber that is perhaps the most definitive characteristic. In *C. cilicium*, the smooth tuber has fine roots emerging together from a point near the centre of the underside, whereas the much thicker roots of *C. mirabile* are produced in a ring around the lower sides of the rather cork-like tuber.

Cyclamen cilicium was named by Edmund Boissier (1810–1885) and Theodor von Heldreich (1822–1902) based on specimens collected by the latter in the region of the Cilician Gates pass to the north-west of Adana. These two remarkable men travelled and collected widely and were leading authorities on the flora of southern Europe and the Near and Middle East. The Swiss botanist Boissier compiled the monumental five-volume *Flora Orientalis* (1867–1884), while his German colleague Heldreich completed his 13-volume *Herbarium Graecum Normale* between 1856 and 1896. The type specimen of *C. cilicium* and duplicate sheets of it ('isotypes') can be found in several herbaria, including that of Kew.

C. cilicium Boiss. & Heldr. in Boiss., *Diagn. Ser. 1*, 11: 78 (1849). Type: Turkey, Cilicia, 'Gülek Boghas' (Cilician Gates), *Heldreich* (isotype, Kew).

A fine wide-petalled individual of *C. cilicium* (MR)

Tuber of *C. cilicium* (BM)

Flower of *C. cilicium* (BM)

A selection of leaves of *C. cilicium* showing great variation (MD)

Description **Tuber** flattened above, up to 6 cm diameter at maturity, often light brown, with a smooth, puberulous surface; roots produced in the centre of the underside of the tuber. **Petioles** suberect to erect, sometimes tending to be horizontal at the base before turning upwards. **Leaf** blade usually ovate or suborbicular, 1.5–6 cm long, 1.5–5 cm wide, the margin entire or shallowly crenate, apex obtuse; the upper surface usually with a green central jagged or 'fir tree' pattern surrounded by a broken grey-green or paler green zone; the underside is normally a reddish-purple. **Pedicels** suberect or erect directly from the tuber, after fertilisation coiling from the top downwards. **Flowers** produced in autumn, normally just before or with the developing leaves, usually with a honey-like scent; petals without swollen appendages ('auricles') at the base, pale to deep pink with a darker crimson-purple basal blotch, twisted lengthways and held in a suberect position, 15–20 mm long, 4–6 mm wide, acute, margin entire or occasionally slightly toothed; stamens included within the corolla; style apex just exserted from the mouth of the corolla. 2n = 30.

Flowering period September to December.

Distribution Endemic to southern Turkey, the species is found extensively in the Taurus (Toros) mountains, the vast chain that runs along the Mediterranean coast, from Barla Dağ by Eğirdir Gölu to Ala Dağ at the eastern end of its range; it has been recorded in the vilayets (provinces) of Adana, Antalya, Içel, Isparta, Karaman and Konya.

Habitat *Cyclamen cilicium* is essentially a mountain species and has been found growing over a wide altitude range from 174 to over 2,000 m, almost exclusively where the underlying rock is limestone. The habitat is usually in woodland or scrubby hillsides and it has been noted in sparse coniferous woods of *Pinus*, *Abies*, *Juniperus* and *Cedrus* and in scrub formed by species such as *Cotinus*, *Styrax*, *Daphne* and *Quercus* spp. Occasionally, it can be found in the open, growing in rock crevices, but this is usually where the trees and scrub have been felled.

Cyclamen cilicium in the wild in southern Turkey: near Aladag, north of Adana. A plant with well-developed leaves, found in a damp position near a stream under plane trees, alongside *Daphne sericea*, 927 m (MD)

In dry conditions, *C. cilicium* produces flowers before the leaves. Southern Turkey, in sparse forest of cedar, fir and juniper north of Akseki, 1,527 m (MD)

A cemetery about 12 km south-south-west of Ermenek on a limestone slope under sparse coniferous and deciduous trees provides suitable conditions for a thriving population of *C. cilicium*, which is no doubt protected from agricultural development (MD)

Variations Like most *Cyclamen* species, particularly those with a sizeable distribution as this species has, *C. cilicium* has variable flower colour, ranging from pale to deep pinkish-purple. There is also an excellent true-breeding pure white form now widespread in cultivation (forma *album*). The normally ovate leaves have only shallowly toothed margins and they are very variably marked on the upper surface, but frequently have a green triangular 'Christmas tree' pattern in the centre, surrounded by a jagged or broken creamy-green or silvery zone that reaches almost to the margin. The cultivated stocks in Britain are not particularly exciting when compared to those of several other *Cyclamen* species. However, nursery beds seen in Turkey (M. Denney, personal communication) suggest that there is a wide range of attractive forms, including some with particularly striking foliage. The absence of 'good' forms in cultivation might be explained by the fact that many of the tubers imported in the past appear to

Holes in tree trunks and rock crevices give perfect protection for cyclamen tubers. Here *C. cilicium* is in the bole of an oak, north of Aydıncık on the road to Gülnar, 660 m (MD)

A well-flowered deep pink form of *C. cilicium* in the wild growing at the edge of pine woodland in the Toros mountains, about 25 km north-west of Mersin, 1,262 m (MD)

have originated in a limited area in the more westerly part of the species' range. The fieldwork by the Cyclamen Society, reported on pages 296–306, will no doubt result in some interesting new variants being introduced into cultivation.

Named cultivars

'Album': This pure white-flowered variant, the discovery of which is described opposite, can be found under this name but does have a valid botanical status as forma *album*, published by Erna Frank and Manfred Koenen in 1983.

'Bowles variety': Grey-Wilson (2002) notes this as being a rather ordinary *C. cilicium* of no particular merit.

The pure white-flowered *C. cilicium* forma *album* between Alanya and Konya in the Toros range growing among junipers and *Quercus coccifera* (MD)

Further observations on *Cyclamen cilicium*

The white-flowered *C. cilicium* White-flowered variants of most cyclamen species that normally have pink or carmine flowers have now been recorded, either in the wild or occurring in cultivated stocks. Some forms are white with a pink 'nose' to the flower whereas others are wholly white. The all-white version of *C. cilicium* is a particularly fine plant, vigorous, free-flowering and breeding true to type from seed. It was found nearly 30 years ago by **Erna and Ronald Frank** on one of their many excursions in Turkey, on this occasion travelling with Manfred Koenen. It was described by the Franks as forma *album* in the *Quarterly Bulletin of the Alpine Garden Society* of June 1983. They write:

> "During our recent holiday in Turkey, in late October 1982, we were fortunate to find a number of pure white-flowered plants amongst a small population of the usual pink-flowered *C. cilicium* var. *cilicium* (see 'Note' below), growing under *Quercus coccifera* and *Pinus brutia* together with *Biarum bovei*. The leaves were typical ... with a small hastate pattern surrounded by a grey-green area, edged by a darker green band with silvery spots around the slightly wavy margin. As no previous record of a white-flowered *C. cilicium* var. *cilicium* has been traced the botanical rank of forma is here proposed for this newly discovered albino variant."

The Franks visited the site on several subsequent occasions to check on progress, commenting in the Cyclamen Society's journal for December 1995:

> "It seems strange that in these very popular mountains no one seems to have come across another white *C. cilicium* since Ronald found that first *C. cilicium* forma *album* in 1982. We visit what we call the "locus classicus" every time we pass by, and find that there are still always four to five white individuals in that very small population of pinks."

Note: To avoid confusion it should be pointed out that, at the time the Franks were writing in 1983, *C. intaminatum* was regarded as a variety of *C. cilicium*; hence, they use *C. cilcium* var. *cilicium* to indicate the typical variety. As *C. intaminatum* is now treated as a separate species, the 'var. *cilicium*' is meaningless and can be dropped.

And an update ... ... A team from the Cyclamen Society visited the site in November 2009 and reported seeing 15 white-flowered individuals. They considered the population to be stable but not expanding. From the one tuber introduced to cultivation in 1982, the white form is now well established in cultivation and appears regularly at the autumn cyclamen and alpine shows. As a measure of its appeal, a plant exhibited by the Franks was given the RHS Award of Merit in October 1988.

The pure white *C. cilicium* forma *album* is now a familiar plant in cultivation (MD)

Cyclamen intaminatum

AUTUMN-FLOWERING

The specific epithet for this small autumnal species appears at first to be a rather curious one, meaning 'untainted' but was chosen to indicate that the flowers are completely plain, without the conspicuous coloured blotches or marks that are a feature of so many cyclamen species. The name was coined by the Kew botanist Desmond Meikle in 1978 to distinguish this entity from *C. cilicium*, of which it was formerly regarded as a variety. It seems clear, however, that it is best treated as a distinct species. *Cyclamen intaminatum* is known to have a very different distribution and ecology from the Cilician species, occurring as it does in north-western Turkey. It is among the earliest of the autumnal species in gardens, although in the wild, it usually does not bloom until October or November, often before the leaves emerge. In distribution, *C. intaminatum* is one of the more restricted species, found in a limited area to the west of Eskişehir where the vilayets (provinces) of Bilecik, Bursa, Eskişehir and Kütahya meet. It grows in deciduous (mainly *Quercus*) woods and thickets in dappled shade. In recent years, the species has been studied in the wild by a team sent by the Cyclamen Society. This team worked in collaboration with Turkish botanists primarily to ascertain the precise distribution of *C. intaminatum* and the extent of its variability.

The small, delicate-looking *C. intaminatum* is a hardy species but it can be appreciated more closely in a pot. Pinkish forms are not uncommon (MD)

There has been much discussion about the original collection of this species, made by Edward K. Balls under the number 669a, which is said to have been collected in southern Turkey in the Cilician Gates area. This is contrary to the known distribution of the species today and although it is impossible to state categorically that *C. intaminatum* does not occur in the eastern Cilician Taurus, it seems extremely unlikely. This is a much-visited area (where *C. cilicium* is common) so, were it present in the area, one would expect *C. intaminatum* to have been seen again during the 70 years since the E. K. Balls & W. Balfour Gourlay expedition. Furthermore, Balls and Balfour are known to have travelled right through the area where *C. intaminatum* does grow on their way south to the Cilician Taurus. The records show that in early April 1934 they travelled from Istanbul to Bursa and on through Inegöl,

Edward K. Balls 'gone native' in Turkey, May 1934 (RBG Kew)

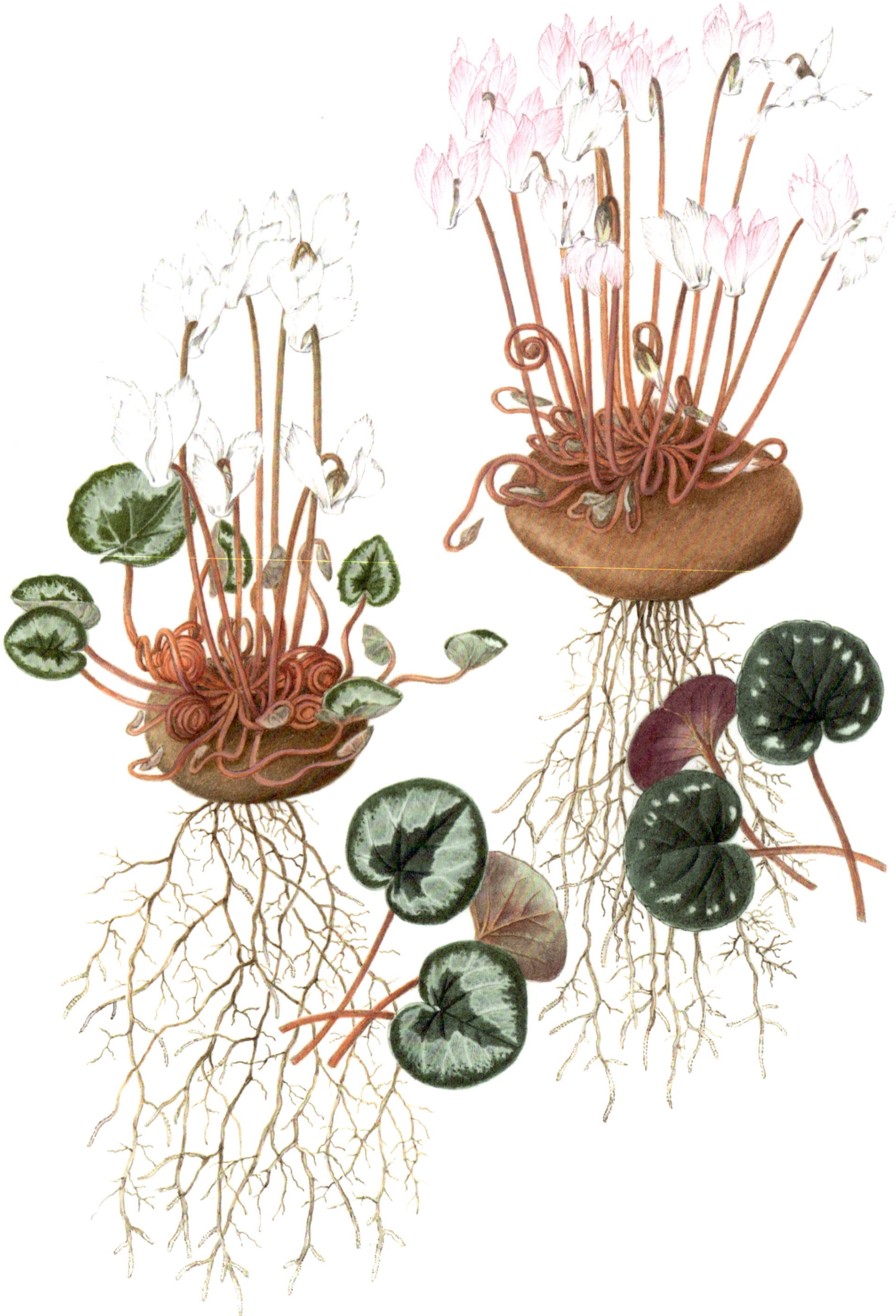

Cyclamen intaminatum. Painted by Christabel King (reproduced at 120% of original size)

Bozüyük, Inönü and Kütahya to Afyon, Konya and the Taurus. The collection numbers in the Inegöl–Bozüyük–Inönü–Kütahya area (where *C. intaminatum* occurs) are from EKB 621 to 636; collection 628a was a cyclamen, and Balls's notes for this are as follows: "Between Inönü and Kütahya. April 3 1934, 3,200 ft. Out of flower. Leaves varying from nearly plain green to strongly marbled. Purplish wash on reverse. Growing under scrub oak in small open copse." Three days after gathering collection 628a, Balls made collection 669a at Burucek near Pozantı in the Cilician Gates area; this collection most probably comprised tubers of *C. cilicium* as no other cyclamen has since been recorded there and this species is common thereabouts. Unfortunately, there is an additional puzzle. Writing in the *Gardeners' Chronicle* ser. 3, 98: 318–319 and 370–372 (1935) Balls refers to two collections: B669a (from Burucek) and B185 (from Namrun, which is further west, north of Mersin). He describes them as having flowered in cultivation from November to March and as having rounded pink flowers, thus these plants sound more like *C. coum* — but this species has never been recorded in the Cilician Taurus range. On the other hand, *C. coum* does grow mixed with *C. intaminatum* in a few of its localities. The most likely explanation for all this confusing information is that the tubers of these cyclamen were mixed up some time after their collection: during their transportation back to England, unpacking, distribution, potting or subsequent cultivation. Anyone who has been involved in cultivating a collection of plants or in field studies will appreciate just how easily this can be done!

On the basis of the presently known facts and the phytogeography of these species, we can safely assume that *C. intaminatum* is, in reality, a rather local species that is confined to north-western Turkey.

Although one of the smallest-flowered cyclamen species, *C. intaminatum* has a certain delicacy and is particularly suitable for growing in pots, which allow it to be more closely appreciated than is the case when growing in the open garden.

For many years, the plants of *C. intaminatum* that were in cultivation gave the impression that the species was not greatly variable and somewhat dull in its leaf markings; one of the most common forms has dull green foliage with just a ring of silvery spots. Recent fieldwork indicates, however, that there is in fact considerable variation within *C. intaminatum* and that there are forms that are more desirable than those in cultivation at present, not all of them as diminutive as Balls's original introduction.

C. intaminatum (Meikle) Grey-Wilson, *The Genus Cyclamen*: 71 (1988). Type: Turkey, Eskişehir, 53 km from Kütahya to Çukurhisar, 1,000 m, October 1975, *T. Baytop* ISTE 33888 (holotype K, isotype ISTE).

Synonyms

'*C. cilicium* var.' Turrill, *Curtis's Bot. Mag. N.S.* t. 307 (1957).

C. cilicium var. *intaminatum* Meikle, *Notes Roy. Bot. Gard. Edinb.* 36: 2 (1978).

Description **Tuber** small, depressed-globose, up to c. 3.5 cm diameter at maturity, brown, smooth and velvety-pubescent, roots slender, produced from the centre of the underside. **Petioles** at base horizontal, then turning upwards from the tuber. **Leaves** not or partially developed at flowering time. Leaf blade reniform or suborbicular, sometimes emarginate at apex, 1–5.2 cm long, 0.5–5.1 cm wide, plain dull green or with a green border and central

Cyclamen intaminatum between Inegöl and Pazargeri at 950 m, 23 October. Although frequently white-flowered, some forms have a pink flush (MD)

Tuber of *C. intaminatum* (BM)

Flowers of *C. intaminatum* showing the lack of markings at the mouth and the conspicuous veins of this small-flowered species (BM)

A range of leaves of *C. intaminatum* from plants growing at Wisley (BM)

A woodland habitat (mainly *Fagus*) for *C. intaminatum* (NO)

irregularly triangular green pattern separated by a very variable pale green or greyish zone, which may be continuous or broken into separate patches on a green ground; margin almost entire to shallowly crenate-toothed; the underside green or slightly purplish suffused. **Pedicels** more or less erect from the tuber, after anthesis coiling from the top downwards. **Flowers** produced in autumn, without a noticeable scent; petals swept upwards to an erect position, white or less commonly pink, with no darker stain at the mouth of the corolla, oblanceolate-elliptic with acute or subacute tips, c. 1.2–2.9 cm long, c. 0.4–0.9 cm wide, twisted lengthways, without swollen appendages ('auricles') at the base; style held within the mouth of the corolla or slightly exserted. 2n = 30.

Flowering period September to November.

Distribution *Cyclamen intaminatum* is a native of north-western Turkey, recorded in the vilayets (provinces) of Bilecik, Bursa, Eskişehir and Kütahya.

Habitat Mainly deciduous (*Fagus* and *Quercus*) woods and thickets in dappled shade, 300–1,796 m.

Variations The small, more or less orbicular leaves show a considerable amount of variation: at one end of the range, a plain, rather dull green, through to forms with just a zone of separate greyish or silvery blotches, and on to more attractive forms with a green triangular central mark surrounded by a continuous silvery zone. Occasionally, there is a pink flush on the upper surface of the leaf, although this is not as pronounced as in some of the forms of *C. mirabile*. The flowers, although small by the standards of most cyclamen species, do vary considerably in size, the Cyclamen Society's field explorations of this species (see pages 270–274) showing that they can be as large as those of the related *C. cilicium*. White is the usual colour for *C. intaminatum* flowers but pink forms also occur in wild populations. The species always appears to have plain-coloured flowers without a dark 'nose' at the mouth of the corolla.

Cyclamen intaminatum in beech woodland, Bursa Province south of Inayet, 1,133 m (MD)

At its higher altitudes *C. intaminatum* grows in less wooded environments, here among sparse *Pinus* and *Abies*. Turkey, Kütahya Province, east of Domaniç at 1,796 m (MD)

Occasionally plants of *C. intaminatum* may be found in which the leaves are overlaid with a pinkish tone. This was seen in a population between Inegöl and Pazargeri (MD)

Cyclamen intaminatum in sparse vegetation under oaks (*Quercus*) in north-western Turkey in Bilecik Province south of Bozüyük

A beautiful soft pink form of *C. intaminatum* in Bursa Province between Inegöl and Pazargeri (MD)

Named cultivars

'E. K. Balls': a name used to denote plants derived from the original collection by E. K. Balls in Turkey, now many generations down the line; the leaves are generally deep green with a zone of silvery-green blotches and the white flowers are at the small end of the range for this species.

Further observations on *Cyclamen intaminatum*

Cyclamen intaminatum **in the wild and in cultivation. Ian Robertson**, a member of a team from the Cyclamen Society, observed *C. intaminatum* and *C. coum* growing together in the wild near the village of Oylat in north-western Turkey. He writes:

> "Just outside the village we looked at an interesting site we had been told about. On a north-facing bank by the path were some *C. intaminatum* in flower with some very nice leaves, so we decided to collect one [they did have the necessary Turkish approval and CITES certificates — ed.!]. Imagine my surprise on digging down, when I found that the flowers and leaves were not attached to the same tuber! The flowers were *C. intaminatum* all right, but the leaves belonged to *C. coum*. This led to the thought that it might be worth trying to grow *C. intaminatum* and *C. coum* together in the garden in a well-drained site."

Ian went on to experiment with the cultivation of this attractive little species, at first in pots but then in the open garden. Of the ones planted in the garden, he writes:

> "I have planted *C. intaminatum* under nut bushes and other deciduous shrubs, in many cases among *C. coum*. There is shade in the form of a high hedge and fence to the south, which helps to keep the site cool. The drainage is always sharp and, as the soil contains a lot of old builder's rubble, it is alkaline. It is early days yet, but so far the plants have come through two winters down here on the Wiltshire–Dorset borders and seem to be growing all right."

And in Surrey... On a visit to Wisley in autumn 2011, it was interesting to come across a small patch of *C. intaminatum* — almost a relief from the thousands of *C. hederifolium*. These had obviously been planted some time ago as they looked to be well established. No such intentional experiments were involved in nearby **Brian Mathew**'s garden. A few years ago, a *C. intaminatum* appeared in a scree bed that had been made up in part of recycled potting compost. The plant must have been derived from seed as the cyclamen are grown singly in pots, making it impossible to overlook and throw out a tuber at re-potting time. The plant looked healthy enough, so it was left to its own devices and has flourished and flowered each year since. Although Surrey is regarded by those living further north as part of the 'banana belt' of Britain, it can have inclement winter weather, as the south-east is often influenced by nearby continental Europe — quite unlike the climate of the damper, more south-western and western parts of the country. This particular garden is also on heavy clay and slopes

The natural conditions can be replicated in cultivation: *C. intaminatum* in an oak woodland situation at Wisley (BM)

to the north-east, so has a cold microclimate. The lowest temperatures of the past few years have been in the –9°C to –12°C range, usually without the snow cover that the plants would certainly have in the wild. Although the winters are not dramatic in their severity, this is often enough to damage the leaves of *C. coum* growing nearby, especially if wet weather has preceded the frost. The good news is that *C. intaminatum* is not only surviving but that seedlings are appearing in several other places. The origin of this particular individual would have been seed from BM 11139 as this was the only plant in the collection at the time; this was collected in 1988 in Bilecik province at an altitude of about 1,000 m.

Cyclamen mirabile

Autumn-flowering

This attractive autumn-flowering species is very similar in the form and colour of its flowers to *C. cilicium*, but there is no doubt that they are distinct, both morphologically and geographically. Their tubers alone are quite different, those of *C. mirabile* having a rough, almost cork-like surface and fairly thick, fleshy roots that appear in a ring around the perimeter of the base; *C. cilicium*, on the other hand, has an apparently smooth tuber that has a very fine, velvet-like pubescence and slender roots that arise at or near the centre of the base. The leaves too are distinctive: in the case of *C. mirabile*, they are rounded and scallop-toothed at the margin, whereas in *C. cilicium* they are more obviously ovate with shallowly toothed margins. When compared with those of *C. cilicium*, the leaves of *C. mirabile* often have a darker green

Cyclamen mirabile. Painted by Pandora Sellars (reproduced at 130% of original size)

A superb specimen of *C. mirabile* at a Cyclamen Society show. (MD)

ground colour with an almost metallic sheen; and not infrequently, the young foliage has a pink or purple suffusion on the upper side, a feature much enhanced by Peter Moore (of Tile Barn Nursery) in his named selections. Although the flowers of *C. mirabile* are similar to those of *C. cilicium*, *C. mirabile* can often be distinguished by the presence of jagged teeth at the apex of the petals and by a fragrance reminiscent of coconut; *C. cilicium* has a sweeter, more honey-like scent (see also 'The scent of cyclamen', page 202).

At one time, *C. mirabile* was considered to be a rather rare species and restricted in its distribution in inland south-western Turkey. Field studies by Turkish botanists and teams from the Cyclamen Society have, however, shown it be more widespread and locally abundant. Nevertheless, the Turkish authorities have understandably included *C. mirabile* among those species banned from commercial collection, although it may occasionally be collected by mistake as its territory overlaps that of the common *C. hederifolium* which is, or has been, collected in huge quantities.

The appearance of *C. mirabile* obviously much impressed the botanist Friedrich Hildebrand who first described it, for it was he who chose the epithet *mirabile* meaning 'wonderful'.

C. mirabile Hildebr., *Beih. Bot. Centralbl.* 19: 370 (1906). Type: based on cultivated specimens that were thought to have originated in the neighbourhood of Izmir (specimen in Berlin, destroyed in WWII).

Description **Tuber** globose or slightly flattened above, often up to 8 cm diameter at maturity, deep brown, with a rough and cork-like surface; roots produced in a ring around the lower part of the tuber, just above its base. **Petioles** suberect to erect. **Leaf** blade usually almost orbicular or very broadly ovate, 1.5–4 cm long, 1.5–4.5 cm wide, the margin scallop-toothed; commonly the pattern on the upper side comprises a central deep green 'fir tree' pattern surrounded by a jagged grey-green or silver-green zone, then the same deep green out to the margins; usually there is an almost metallic sheen to the whole leaf and not infrequently there is a pinkish-carmine tinge over the whole upper surface, especially when the leaves first emerge; the underside is normally a beetroot reddish-purple. **Pedicel** suberect or erect directly from the tuber, after fertilisation coiling from the top downwards. **Flowers** produced in early autumn, normally just before or with the developing leaves, usually with a scent remarkably like that of a coconut; petals 8–26 mm (usually 16–20 mm) long, 2–12 mm (usually 5–8 mm) wide, pale to deep pink with a darker purple basal blotch, twisted lengthways and held in a suberect position, without swollen appendages ('auricles') at the base, usually conspicuously and irregularly toothed at the apex and sometimes markedly acuminate; style apex slightly exserted from the mouth of the corolla. $2n = 30$.

Flowering period Early autumn, usually October or November in the wild.

Tuber of *C. mirabile*; note the thickened roots produced in a ring (BM)

Flower of *C. mirabile* showing fimbriate petals (BM)

The scallop-toothed foliage of *C. mirabile* (BM)

Near Eskiçine, south-western Turkey. *C. mirabile* often grows in regions of mica schist rocks where *Pinus pinea* also thrives (BM)

A hillside where *C. mirabile* grows wild in south-western Turkey near Eskiçine in sparse *Pinus pinea* and olives (TW)

Cyclamen mirabile can survive in land that is not too intensively cultivated, such as here in an olive grove between Çine and Akçaova, south-western Turkey (BM)

A cemetery where *C. mirabile* has been left untouched and is now the dominant ground vegetation. Near Nazilli, south-western Turkey (TW)

Variation in leaves of *C. mirabile* growing at Wisley (BM)

Distribution *Cyclamen mirabile* is endemic to south-western Turkey, recorded in the vilayets (provinces) of Aydin, Denizli, Isparta, Izmir, Manisa and Muğla.

Habitat *Cyclamen mirabile* is often found growing on mica schist formations but sometimes also on conglomerate rocks or limestone, from about 160 m to 1,150 m. It usually grows in partial shade at the edge of scrub of evergreen oak (*Quercus coccifera*) and sometimes also among deciduous oaks, along with other Mediterranean species such as *Cistus*, *Pistacia*, *Pinus*, *Crataegus*, *Styrax* and *Olea* (wild and cultivated).

Variations In the wild, *C. mirabile* is remarkably uniform in its characteristics, although a search among hundreds of plants will often reveal a range of leaf markings within the general description given above. Perhaps the most striking are those in which the leaf is covered with an overlay of pink or carmine. The flower colour too is variable in wild populations, in shades of pale to mid pink, almost always with a dark stain at the mouth. Occasionally, this stain is almost absent and very rarely white forms occur with or without the purple blotch at the mouth. A pure white form without a purple blotch at the mouth has been named and described as forma *niveum* Grey-Wilson & J. White (1997). The degree of toothing of the petals is very variable and it is not unusual to see forms in wild populations that have acuminate apices. The length/width ratio of the petal and degree of twist also varies enormously,

Cyclamen mirabile flowering leafless, in dry ground between Çine and Yenipazar, south-western Turkey (BM)

Flower of *C. mirabile* forma *niveum* 'Tilebarn Jan', a pure white selection from Peter Moore (MD)

Pure white *C. mirabile* on west side of Lake Eğirdir, Turkey: forma *niveum* which has no coloured blotch at the mouth (BM)

producing flowers of very different appearances. Although the coconut like scent is usually very noticeable in cultivated plants, it is often difficult to detect in wild plants, possibly because plants in cultivation are normally in the confined space of a glasshouse in which there is little air movement.

Named cultivars

'Tilebarn Anne': bright silvery leaves with attractively etched veins; the leaves are flushed with pink at first; flowers pink; selected by Peter Moore of Tile Barn Nursery.

'Tilebarn Jan': the leaves are banded as in most *C. mirabile* and are attractively pink-flushed as they unfurl; flowers white, but not consistently pure white, some seedlings having flowers with a pink mouth; selected by Peter Moore of Tile Barn Nursery.

'Tilebarn Nicholas': leaves bright silver with a dark green central 'Christmas-tree' pattern, the silver areas flushed pink in the young stages, flowers pink; selected by Peter Moore of Tile Barn Nursery.

A form of *C. mirabile* with long, elegantly twisted petals, seen near Uluderbent, Manisa province, Turkey (TW)

A close-up of the leaves of *C. mirabile* 'Tilebarn Nicholas', raised by Peter Moore (MD)

Cyclamen mirabile 'Tilebarn Nicholas', a fine selection with foliage having a pink-carmine overlay when young (MD)

Further observations on *Cyclamen mirabile*

Habitat of *C. mirabile* In the autumn of 2003 and 2004, the Cyclamen Society undertook field studies of *C. mirabile* to determine the extent of its distribution and other details such as its habitat preference. The following is an edited extract from the Society's Journal *Cyclamen*, Vol. 28 (2004):

"Several observations could be made concerning the habitat of *C. mirabile*. Most populations were growing in soil overlying mica schist formations — in fact this often appeared on the surface as rounded outcrops. Occasionally they were on a type of conglomerate and in the case of the eastern-most (Eğirdir) population on limestone. On the other hand, *C. alpinum*, which occupies a similar region of south-western Turkey, seems to be usually on limestone formations. This probably explains why their distributions meet but apparently do not actually merge. To date, as far as we know, no mixed populations have been found. There was quite a spread of altitude range from 161 m to 1,146 m, the higher altitude populations tending to be towards the eastern end of the distribution area. The pH range recorded was a surprisingly high 7.2–8.4, with little difference between the surface layers of 'organic' soil and the subsoil. It is always interesting to record the associated species and one can soon build up a habitat profile using a combination of rock type and companion plants so that is often possible to spot likely and unlikely sites from a distance. A frequent associate at (nearly) all sites, was *Quercus coccifera*, sometimes also deciduous oaks. Other commonly seen species were *Cistus* spp., *Pistacia*, *Pinus*, *Crataegus*, *Styrax* and *Olea* (wild and cultivated). These are mostly species indicative of the Mediterranean climatic zone but in the case of some of the eastern, higher altitude populations there was a marked tendency for some of the central Anatolian steppe plants to be in evidence. In spite of the latter, *Cyclamen mirabile* is undoubtedly more of a 'Mediterranean species' than 'Irano-Turanian' or 'Euro-Siberian' (these constitute the three phytogeographical regions identified in the *Flora of Turkey*)."

Variation in *C. mirabile* The Cyclamen Society's fieldwork referred to above was partly aimed at noting the degree of variation in wild populations and it is useful to record the observations here:

"With regard to the leaves the most common form of leaf marking was a central dark green 'Christmas-tree' pattern of variable size surrounded by a light green or silvery zone. Leaves with a pink tinge on the upper surface were to be found at most locations, although the majority of these were not as attractive as the Tilebarn selections and on the whole there was not nearly as much variation as can be found in cultivated material at a Cyclamen Society show. One striking plant had almost wholly silver leaves with a narrow green margin and a green spot in the centre. Nearly all those seen were of a broadly ovate shape, about as wide as long, but one specimen had unusual very narrowly ovate leaves with a c. 2:1 length:width ratio and sometimes they were approximately orbicular (*coum* shape) and very occasionally they were wider than long so as to appear more reniform (kidney-shaped) in outline. The leaf margin was fairly consistently and conspicuously toothed in the characteristic manner of this species but a curious example at one site had inward-facing notches on the margin instead of outward-facing teeth.

Flowers The flowers, on the other hand, were noted as remarkably variable. The flower colour was checked using the RHS Colour Chart and this showed that the range was Purple Group 76A–C and predominantly 76B in most populations. Darker pink-purple forms were quite uncommon but very pale ones were plentiful. In fact these were often so pale as to appear almost white although nearly always with a dark marking around the mouth of the corolla; only one pure white form was found, with no trace of colour. At one site there were several plants with barely any marking at all around the mouth of the flower which gave them the appearance of *C. intaminatum* flowers. In fact the blotch at the mouth was very variable, both in shape (one was jagged like a *C. libanoticum* blotch), size and position (sometimes situated as much as 2 mm above the base of the petal).

The variation in overall appearance of the flower was greater than expected, brought about by the variable length/width ratio of the petals. The actual sizes of the measured flowers (a random selection from most of the sites) were as follows: petal length: whole range 8–26 mm long but most frequently 16–20 mm long; petal width: whole range 2–12 mm wide but most frequently 5-8 mm wide. Those with wide petals resulted in a much more substantial-looking flower; for example one specimen had petals only 15 mm long but they were 10 mm wide (whereas the length:width ratio was most commonly between 2:1 and 3:1) and one small 'chubby' form had a petal size of 9 × 7 mm. Some of the longer-petalled ones were more reminiscent of the *C. repandum* flower shape while one particularly long flower had broad petals (25 mm long, 11 mm wide) so this had a very substantial appearance. The degree of petal twist varied from less than 45 to 180 degrees and in one remarkable specimen with particularly long narrow petals (23 × 3 mm) a 360 degree twist was noted, producing a very elegantly shaped flower. A few flowers had flat petals with no twist, resulting in a more shuttlecock shape.

The toothing or fimbriation at the tips of the petals, regarded as a characteristic of the species, was sometimes very slight and occasionally nearly absent. In fact the basic shape of the petal tips varied considerably from rounded to acuminate and not infrequently they were bifid (i.e. notched) at the apex. The distinctive coconut aroma of *C. mirabile*, such a distinctive feature in cultivation, was seldom very noticeable in the wild.

Aberrations noted included flowers with seven petals and one 'stargazer' type with the corolla mouth facing upwards."

2.2e The Cyprium group

(= Subgenus *Corticata*)

Comprising *C. cyprium* and *C. libanoticum*

Two species comprise subgenus *Corticata*. Superficially, they are not very alike but they do hybridise and produce fertile offspring (*C. × wellensiekii*). The fragrant flowers are white or pale pink and have contrasting jagged markings at the mouth of the corolla, which has swellings ('auricles') in *C. cyprium* but not in *C. libanoticum*. The stamens are not exserted from the corolla but the style can be shortly exserted. In fruit, the pedicels start to coil downwards from the apex. Chromosome number 2n = 30.

Cyclamen cyprium

FLOWERING AUTUMN TO EARLY WINTER

Although an attractive autumnal species with very fragrant flowers, the endemic cyclamen of Cyprus, *C. cyprium*, lacks 'flower power' in cultivation as the modestly sized flowers are often produced a few at a time over a long period. This extended flowering might be regarded as a bonus, although for those who display or show cyclamen, it is perhaps a less desirable asset. *Cyclamen cyprium* is widespread on Cyprus but has never been a highly popular or common species in cultivation. It has been known since 1862, when the intrepid nineteenth century botanist and explorer Theodore Kotschy first introduced specimens and tubers from Cyprus. He cultivated these specimens in Vienna before describing and naming the species after the island in 1865.

Cyclamen cyprium has now been adopted as the National Flower of Cyprus. John and Maureen Leach from Agios Georgios noted in their local paper, *The Cyprus Weekly*, that in February 2006 "the Council of Ministers took a break from the ponderous affairs of state to proclaim the Cypriot cyclamen and latzia (golden oak, *Quercus alnifolia*) as the national flower and tree, respectively". Whatever *C. cyprium* may lack in showiness it has certainly made up for it in status as the species also has a classic car meeting named after it, 'The Cyclamen Rally of Cyprus' (see page 528).

In bloom, *C. cyprium* is one of the most distinctive cyclamen species having small, white, strongly fragrant flowers with conspicuous auricles around the mouth of the corolla, where there are also clearly defined deep carmine markings.

A well-flowered plant of *C. cyprium* in Cyprus. It seldom flowers so freely in cultivation, usually producing a few flowers at a time over a long period (BW)

Cyclamen cyprium. Painted by Pandora Sellars (reproduced at 125% of original size)

Rocky woodland with mixed trees and shrubs including *Arbutus andrachne* and pines, a habitat for *C. cyprium* (BW)

Theodore Kotschy, nineteenth century explorer and botanist who first collected and named *C. cyprium*

C. *cyprium* Kotschy in Unger & Kotschy, *Die Inseln Cypern* 295 (1865). Type: specimen from tubers collected in Cyprus at Galata, 1862, and cultivated in Vienna, *Kotschy* (holotype W!).

Description **Tuber** depressed-globose or subglobose, up to 7 cm diameter at maturity but usually much smaller, greyish, rough or sometimes cork-like in texture, roots produced from the underside in a compact cluster but often off-centre. **Petioles** prostrate in the lower part below ground, then ascending to erect above ground. **Leaf** blade broadly cordate, 3–14 cm long, 2–11 cm wide, marbled or zoned pale green or silvery-green around a central deep green 'fir-tree' pattern; the zoning sometimes breaks up into pale spots and blotches on a green ground towards the margins; margin varying from almost entire to coarsely toothed or slightly lobed; the underside normally reddish-purple. **Pedicels** horizontal in the lower part, turning upwards as they emerge above ground, after anthesis coiling from the top downwards. **Flowers** strongly fragrant, produced in autumn to early winter, normally slightly before or with the developing leaves; petals white or very pale pink with an M-shaped (has been likened to a bird in flight)

Tuber of *C. cyprium* (BM)

Flower of *C. cyprium*. Note the prominent auricles and jagged markings at the mouth (BM)

Range of leaves of *C. cyprium* from Martyn Denney (MD)

Cyclamen cyprium in a pine wood habitat. In the Pentadaktylos/Beşparmak range west of Bahçeli, 755 m (MD)

magenta basal blotch, twisted lengthways but not strongly so, with markedly swollen appendages ('auricles') at the base, 1.0–2.5 cm long; style slightly exserted from the corolla. 2n = 30.

Flowering period September to January.

Distribution *Cyclamen cyprium* is endemic to Cyprus, especially the northern Kyrenia (Pentadaktylos) mountains, the central Troodos range and west to the peaks of the Akamas Peninsula; it is absent from low altitudes in the south of the island.

Habitat Variable in habitat, on limestone or igneous formations in the shade of trees and shrubs on banks, in shady, rocky places, old vineyards and sometimes under pines, 100 m to c. 1,950 m.

Variations *Cyclamen cyprium* has little variation in flower size and colour: although normally white, the flowers may be pale pink and a decidedly pink-flowered plant was exhibited by Ian Robertson at a Cyclamen Society show in October 2009. As for most cyclamen, the leaves come with a wide range of markings and are sometimes just plain dull green with no zoning at all. The horticulturally more exciting forms have a conspicuous silver or pewter-coloured zone,

Some striking leaves of plants cultivated by Marco ten Hoope (MH)

A fine leaf form of *C. cyprium*. Ridge track in the Pentadaktylos/Beşparmak range west of Bahçeli towards the forest station at Alevkaya, 561 m, November 2011 (MD)

Further observations on *Cyclamen cyprium*

***Cyclamen cyprium*: some personal observations.** For **Brian and Jo Walker**, Cyprus is a very special island. They have observed populations of *C. cyprium* in several locations on their frequent visits to Cyprus and are experienced in its cultivation. Brian writes:

"This Cyprus endemic can be found from low altitudes, approximately 100 m, right up almost to the top of Mount Olympus (1,952 m). Its habitat is in shaded limestone or igneous rocky places generally beneath trees and shrubs, stream banks and sometimes on screes and in vineyards and olive groves. I have even seen it in the forks of branches and holes in tree trunks. The distribution of this endemic species is western and northern Cyprus, often locally abundant in the Troodos range where it flowers from late September to early January. The high altitude populations from the Cedar Valley area up into the Troodos Range have the dullest, rather lobed or angled leaf forms of two shades of olive green and almost always rich purple or crimson-purple below. This leaf type also grows in the areas of Platres, Stavros Tis Poskas, Ayia, between Kato Pyrgos and Pano Pyrgos, also between Pano Pyrgos and Kambos, Galat, Kalopanayiotis, Evrykhou, Toxeftera, Perapedhi, Vroisha, Lagoudhera and Nikos. At the lower altitudes, the variants from Ayios Neophytos and Kato Kivides, which grow along the banks of seasonal streams in great profusion, have the most attractive leaf forms which are beautifully marked with silver. In the Polis area where they grow in rocky ground under shrubs and pine trees the marking is

Holes in tree trunks provide an alternative habitat for *C. cyprium* (BW)

whereas others are variably splashed with silver or pewter on a darker ground. The leaf shape too varies: it is usually coarsely toothed but may be shallowly lobed or, at the other extreme, almost entire.

Named cultivars

'ES': this refers to Elizabeth Strangman of Washfield Nursery who made the plants available to Peter Moore of Tile Barn Nursery. The original plants were selected by Hilda Davenport Jones in the 1960s; they have deep green leaves that are boldly marked with splashes of contrasting silver.

'Galaxy': leaves with a deep sombre grey-green background, heavily spattered over the entire surface with silver flecks and spots (like the stars of a distant galaxy against a night sky); flowers typical for the species, selected by John Fielding, distributed by Ashwood Nurseries.

Best grown as a pot plant, *C. cyprium* flowers late in the year and is deliciously fragrant (MD)

not quite so distinct. In northern Cyprus it grows in the region of Kyrenia, St Hilarion, Ayios Amvrosios, Karmi, Vasilia, Bellapais, Kazan, Sisklipos, Pentadaktylos and the northern flank of the Kyrenia Range (Mt Kourtella)."

On the matter of its value as a horticultural subject, Brian Walker notes that:

"*Cyclamen cyprium* makes a good pot plant for a cool greenhouse and can even be grown outside in a favourably sheltered position. A mature well-grown plant will produce many flowers which are sweetly fragrant. When the plants die down in the spring they should be given a complete summer rest by withholding all water. However if the weather is very hot, and the pots are not plunged in sand so as to keep them cool and prevent drying out, a little water should be given occasionally. After this summer dormancy the plants should be given only a little water until they are in full growth, then they can be given copious amounts."

Habitat and variation of *Cyclamen cyprium* in the wild. Martyn Denney has also made some very pertinent observations about this Cypriot endemic, noting that:

"This is essentially a woodland plant, and like *C. creticum* except at altitude it seeks the darkest and most shaded sites where the most moisture can be found … . The most extensive, dense populations where it particularly thrived were under *Platanus orientalis*, these (often streamside) areas being significantly damper than the majority of locations under *Pinus* … . There is no doubt that *C. cyprium* is a widespread plant which can be locally abundant and where conditions particularly suit it can be at densities of 20–80 plants per square metre … . Like *C. creticum*, at higher altitudes — above 1,100 m in the Troodos, or 700 m in the Pendadahtilos (Pentadaktylos, Kyrenia) Range — it grows happily in stable scree and rock crevices and here the plant density is significantly lower."

He goes on to describe the many variations in flower colour (white or pink) and in leaf markings that are liable to be encountered, showing that although this species is an island endemic it is nevertheless extremely variable. In particular, he found the tuber characteristics particularly interesting as this species often has roots that are produced in a markedly off-centre position on the underside of the tuber. Denney found that:

"Whilst many tubers rooted entirely from one side of the base, a significant proportion rooted from the centre. The positioning of the roots did not appear to be linked to the size of the tuber [Hildebrand suggested that off-centre rooting was caused by the tuber growing more on one side than the other], but it was noticeable that where plants were growing on a slope with the tuber near the surface, the roots often projected either horizontally or at an angle of up to 45° from the side of the tuber into the slope, thus easily retaining the tuber in an upright position."

Taking the recorded sites for *C. cyprium* and checking on at least some of them provided some useful data on their status with regard to threats and survival. For the lower altitude populations, Denney notes that these "depend for their survival upon good deep shade to provide a relatively cool and damp environment during the heat of the summer. Suitable habitats are relatively rare and are particularly vulnerable to the changing land use linked to a constantly growing tourism industry…". In the Avakas Gorge, he reports a fairly dense population growing in stable scree in deep shade, but the population of pink-flowered plants recorded between Kato Pyrgos and Pano Pyrgos almost certainly no longer survives because the area around the latter village has been almost cleared of woodland for the charcoal industry. The higher-altitude populations fare better. For instance, the Paphos Forest — a large area on the western slopes of the Troodos consisting of a network of valleys, descending from north-east to south-west, with a mixed canopy of plane and pine — has good sites for *C. cyprium*. Many of these valleys contain permanent streams and it is in the damper bottoms of these valleys that *C. cyprium* occurs, usually in small groups but with larger, more dense populations in places where the decaying plane leaves collect. Denney also confirms that the plane- and pine-shaded valleys of Dhiarizos and Piatys, between the villages of Komititzi and Kefelos, are rich in *C. cyprium*.

On the negative side, Denney notes that formerly recorded populations in the Marathasa Valley on the northern side of the Troodos may have been lost as the result of agricultural use of the valley floor and road and dam development.

Nevertheless, *C. cyprium* appears to be far from endangered on the island and can be seen without difficulty in the autumn months. Nevertheless, changing climate probably represents a much greater threat to these narrow endemics than to more widespread plants.

***Cyclamen cyprium* in western Cyprus** On frequent visits to Cyprus, **Chris Clennett** has made various observations of both the native species (*C. persicum* is also found on the island). Writing about *C. cyprium* in the Cyclamen Society's journal, he notes that despite gradual and continuing encroachment of building, the Akamas peninsula, an area with a rich natural flora, retains a great deal of undisturbed habitat, including forests, woodland, scrub and shoreline. Many of the trees in this area are among the oldest and tallest in the country, and protecting these venerable specimens is foremost among the concerns of foresters and conservationists on the island. *Cyclamen cyprium* is very restricted in this part of the island, being found in the Paphos Forest which forms the western foothills of the Troodos Range, and in isolated pockets on the highest hills of the Akamas peninsula. One of the reasons for this seems to this species' preference for serpentine or other igneous or metamorphic rock. This rock type is found extensively in the Troodos, but also forms a small 'cap' on the largely limestone hills of the Akamas. Throughout the Paphos Forest, *C. cyprium* can be found under plane trees and at the margins of dense conifer woodland. It does not form extensive colonies and the population as a whole is scattered thinly over larger areas. Even in cultivation where 'rainfall' can be introduced artificially, *C. cyprium* is one of the last of the autumn-flowering group to begin growth; in its native haunts, the best months to find the species in flower are November and December.

In March 2008, Clennett and his wife went again to Cyprus to visit a known site for *C. cyprium*, the Avakas Gorge, a spectacular feature consisting of limestone rock. Sure enough, a small colony of *C. cyprium* was found in the shady gorge, and a much larger colony was also located on steep north-facing limestone cliffs in a smaller gorge north-west of Neo Chorio. So, the apparent preference of this species for serpentine or other igneous or metamorphic rocks is not without exception. It was noticeable, however, that of the several forms of limestone found in the Akamas peninsula, only the hard form that would leach relatively little lime into the surrounding soil supported *C. cyprium*.

***Cyclamen cyprium* in the Avakas Gorge** The habitat of this species also interested **Doug Thompson**, who provides the following comments:

> "The plants were found only above the narrows in the Avakas Gorge at about 150 m above sea level … ... many plants were seen in rock crevices and in deep leaf litter in the shade of rocks, oleander and *Arundo donax* (giant reed), exclusively facing north and west. All the plants I saw were, without exception, located in a narrow band about 2 metres above the stream. This was still quite active at the end of April and is understood to be prone to flooding. None were seen in south- and east-facing areas of the gorge. The quite extensive colonies were all in deep shade in narrow and humid parts of the gorge. There seemed to be a quite specific requirement for shade and moisture, and conditions suggested a liking for more moisture than cyclamen species previously seen elsewhere. This is however somewhat at odds with the current cultural recommendations of complete summer dryness (Grey-Wilson, 2002). Perhaps a moister atmosphere is required (but not moisture at the roots) to maintain a low activity level in the plants during dormancy." [see also cultivation comments above, from Brian Walker –Ed.].

Cyclamen libanoticum

Flowering in spring

This beautiful species is one of the most distinct of all cyclamen having large, broad-petalled flowers in a shade of clear pink, rather than the more frequent carmine or purple-red coloured flowers of other species. As well as being wide, the petals, unlike those of most cyclamen, are almost flat or are twisted only slightly, and this too helps to produce the impression of a flower of great substance. *Cyclamen libanoticum* also has a distinctive scent, again unlike that of any other cyclamen species. Another point of note is that this species is considered to have one of the most restricted distributions, being confined to a small region in the Lebanese mountains. However, an interesting and valuable account by the late Richard Haber and his wife Myrna Semaan (page 307) suggests that there is no immediate threat to its wild populations. From the plants seen in cultivation, *C. libanoticum* would seem to be one of the least variable cyclamen species, and this might well be expected for a species of limited distribution, but again, Haber and Semaan's report provides evidence to the contrary. Their observations show that there is considerable variation in the flower form and colour and in the leaf markings.

Cyclamen libanoticum. Painted by Pandora Sellars (reproduced at original size)

Cyclamen libanoticum is a superb species for an alpine house or frame. The wide petals are often not as markedly twisted as those of most species (BM)

Surprisingly, the nearest relative to *C. libanoticum*, botanically speaking, is now considered to be *C. cyprium*, although superficially the two appear to have little in common. It is not unusual, however, for modern analyses based on studies at the molecular level to reveal unexpected surprises. This relationship is supported in practice by the fact that if the two species are cross-pollinated (pollen must be saved as their flowering periods do not overlap), they hybridise to produce intermediate offspring. These hybrids are fertile and are collectively known as *C.* × *wellensiekii*.

In cultivation, *C. libanoticum* is rarely seen growing outside in the garden, but is certainly a choice addition to an alpine house or frame where there is a little protection from the elements. There is no real reason, however, why it should not thrive in a sheltered position in semi-shade.

C. libanoticum Hildebr., *Engl. Bot. Jahrb.* 25: 477 (1898). Type: Lebanon, 40 miles north-east of Beirut, 800–1,400 m, December 1895, *E. Hartmann*; material cultivated at Freiburg, *Hildebrand*.

The Lebanese mountains, home of *C. libanoticum* in the Ibrahim River valley. The woody flora includes oak (*Quercus*), hop hornbeam (*Ostrya*), hawthorn (*Crataegus*), Judas tree (*Cercis*) and bay (*Laurus*) (MS)

Description **Tuber** subglobose or depressed-globose, often with a depression in the centre of the upper surface, up to c. 4.5 cm diameter at maturity and up to 2 cm thick, dark brown, smooth and velvety-pubescent at first but becoming more rough and corky on the surface in old specimens; roots slender, produced from the centre of the underside and some from the sides, mainly concentrated on one side. **Floral trunk** often present (at least in wild specimens). **Petioles** densely adpressed-pilose with brown hairs, prostrate in the lower part below ground, then ascending to erect above ground. **Leaves** developing from autumn onwards, well developed by flowering time, rather fleshy in texture. Leaf blade ovate, broadly ovate or sometimes suborbicular, 4–14 cm long, 4–12 cm wide, dull, slightly glaucous green variably marbled grey-green and sometimes with irregular darker patches, usually without clearly defined zones but sometimes with a small green triangular pattern in the centre, slightly pilose on both sides especially on the veins; margin often with shallow angles, usually smooth but sometimes with small, well-spaced teeth; the underside normally suffused purplish-red. **Pedicels** usually horizontal in the

Tuber of *C. libanoticum* (BM)

Flower of *C. libanoticum*. Note the substantial petals and the wide mouth with conspicuous, jagged carmine markings (BM)

The rather thick, fleshy leaves of *C. libanoticum* are well developed at flowering time and are attractively mottled, although not as dramatically as most species (RW)

lower part, turning upwards as they emerge above ground, after anthesis coiling from the top downwards. **Flowers** produced in spring, with a distinctive 'chemical' smell and a pepper-like component; petals broad and overlapping, obovate with subobtuse to subacute tips, c. (1.5–)2.5–3.2 cm long, usually rather weakly twisted lengthways, without or with not very conspicuous swollen appendages ('auricles') at the base, swept upwards to an erect position, near white in the lower part shading to pale pink in the upper half, with rich magenta markings at the base forming a jagged ring around the mouth of the corolla, the central mark reaching the mouth and extending into it; style held within the mouth of the corolla or exserted from it. 2n = 30.

Flowering period Late February to early April.

Distribution *Cyclamen libanoticum* is endemic to Lebanon, in the valley of the Ibrahim River. The *Flora of Turkey* (Meikle, 1978) records the occurrence of *C. libanoticum* from Belen in the Amanus Mountains. This matter has been addressed in some detail by C. Clennett (see 'Further observations').

Habitat Mixed deciduous and evergreen oak woodland, 600–950 m.

Variations Apart from the variations recorded in the general description of the species above, particularly of the leaves, there appears to be limited variation in the *C. libanoticum* plants in cultivation compared with those of some other cyclamen species. In cultivated material, the flowers are a fairly uniform clear pink, but there is a little variation in the shade of pink and in the shape of the blotch around the mouth of the corolla; often, this blotch is more or less in the shape of a letter 'M', sometimes it is a zone of jagged markings without much of a defined shape. In wild populations, however, there is a marked and interesting degree of variation, recorded in the detailed article by Richard Haber and Myrna Semaan (page 307).

Further observations on *Cyclamen libanoticum*

Cyclamen libanoticum in Turkey? The great Cyclamen researcher botanist Otto Schwarz published a record for *C. libanoticum* based on a herbarium specimen collected in southern Turkey's Amanus mountains at Belen on 27 February 1865. The specimen was sent to the herbarium in Jena, Germany, labelled *C. coum* but was later identified by Schwarz as *C. libanoticum*. On the face of it, this identification is unlikely and has since been taken as a mistake, with the plant being thought in reality to represent one of the local species of the region, *C. coum* or *C. pseudibericum*. **Chris Clennett** decided to investigate the matter in detail (see *Journal of the Cyclamen Society*, December 1999) and requested a loan of cyclamen specimens from Jena. However, the herbarium sheet in question did not provide a clear answer:

> "It is immediately apparent … that this specimen is neither of these two species [*C. coum* and *C. libanoticum*] … It does not fully conform to the accepted description of *C. pseudibericum* either, having no visible basal blotch [on the petals] and a strongly exserted style, the latter ironically a feature of *C. libanoticum*. The leaf shape is also unusual, with obcordate leaves showing very strongly developed marginal teeth … ."

Chris Clennett goes on to explore various possibilities: the specimen has become badly disfigured over the years (unlikely, as this doesn't normally happen, certainly when we are

C. libanoticum in the wild in Lebanon among mossy limestone rocks with trailing ivy (*Hedera*) (MS)

considering characters such as leaf shape and an exserted style); or maybe it could be a mixed collection of flowers from one species and the leaves of another, but the samples would need to have been taken from an aberrant plant to possess these various features. Another more exciting possibility posed by Chris Clennett is that the specimen represents an as yet unknown species. This too is unlikely, given that the region around Belen is botanically quite well explored.

The matter may have been left there, a puzzle for which there seemed to be no answer, but Christopher Grey-Wilson (2002) also considered the possibilities and, with a bit of lateral thinking, came up with a plausible alternative suggestion. He took the view that the whole episode could be a mistake resulting from a curatorial mix-up of specimens or their labels — as happens from time to time in herbaria, no-one is perfect! If this were the case, and if the pressed plant did not originate from Belen in Turkey, which species could it represent? In Grey-Wilson's opinion it is one of the *C. repandum* alliance: the toothed leaves and narrow-petalled flowers showing an exserted style certainly support this view. One of the (detached) leaves on the herbarium sheet could be that of *C. coum* and this might be the only part of the original Belen specimen, as this species does grow in the area. Grey-Wilson even suggests that somewhere in the Jena Herbarium there might be a specimen of a Turkish cyclamen with Greek field data!

A journey in Lebanon In 1997, the inveterate travellers and cyclamen enthusiasts **Erna and Ronald Frank**, in the company of Manfred Koenen, made a journey to Lebanon in the hope of seeing among other exciting plants, *Cyclamen libanoticum*. In view of their long association with the Cyclamen Society and the astute observations made on their numerous travels in cyclamen country, we here repeat a small section of their unpublished journal relating to this Lebanese species:

> "24 April 1997. Mairouba — Nahr Ibraihim. We continued through a confusion of roads back towards the north and the Adonis river on yet another stiflingly hot day. We were searching for *C. libanoticum* along the indifferent road near El Mechete … . We found just a few, hiding in the shadow of rocks and concealed by undergrowth. A short distance outside the village, Manfred noticed a rocky, shaded [over] hang some way off from the road, at the top of a steep bank rising from the far side of a stream … . There were plenty of *C. libanoticum* peering out between the rocks. A notable find! Throughout this day, *C. persicum* was also in great evidence in many shades. Manfred found one which was almost completely white, with the slightest pink tinge to the mouth of the flower."

2.2f The Persicum group

(= Subgenus *Persicum*)

Comprising: *Cyclamen persicum*, *C. persicum* var. *autumnale* and *C. somalense*

A group of only two species flowering from autumn through winter to early spring. The heart-shaped leaves are held on erect pedicels from the tuber, not trailing underground before emerging as happens with many cyclamen species. The fragrant flowers usually have long, elegant petals and a contrasting darker stain at the mouth of the corolla which has no swellings ('auricles'). The stamens are not exserted from the corolla but the style can be slightly so. In fruit, the pedicels do not coil tightly but arch over to touch the ground. *Cyclamen persicum* 2n = 48 (72, 96, 136 in some of the 'florists' cultivars); *C. somalense* 2n = c. 96.

Cyclamen persicum var. *persicum* and var. *autumnale*

FLOWERING IN WINTER, EARLY SPRING OR AUTUMN

For most people, mention of cyclamen will conjure up images of the large-flowered pot plants that are so popular and readily available from florists and garden centres. No hybridisation is known to have taken place in the breeding of the myriad of *C. persicum* cultivars, but the results of selection from within this one species are truly amazing. A comprehensive account of the history of the florists' cyclamen by Diony de Bont and John Grimshaw can be found on pages 373–414.

The wild forms of *C. persicum* are excellent plants for an unheated glasshouse or conservatory, free flowering and strongly fragrant (BM)

Cyclamen persicum. Painted by Pandora Sellars (reproduced at 80% of original size)

In this chapter, we concentrate on the wild species, named and described long ago by Philip Miller (1691–1771) in his remarkable eighteenth-century work *The Gardeners' Dictionary*. Miller was the gardener-botanist in charge of the Chelsea Physic Garden, where he cultivated a remarkable range of plants over a period of nearly 50 years. His *Dictionary* first appeared in 1731 and ran to eight editions, the last of which (1768) is the most important in botanical terms as it contains plant names in the form of binomials following the then new system of Linnaeus; in some cases, these are the earliest (i.e. post-1753) validly published names. Examples of names of *Cyclamen* species first published by Miller are *C. persicum*, *C. coum* and *C. purpurascens*. Miller says of *C. persicum*: *Cyclamen with sawed heart-shaped leaves*; *or Persian Cyclamen*. Although rather scant, the text goes on to say that the flowers are white with a purple bottom and are produced in March and April. Unfortunately, the wild origins of *C. persicum* were not known with any certainty and, for some reason, it was assumed to have arrived in Europe from Persia (Iran). Despite much botanical work in that country, however, there have been no positive sightings of the species in Iran and its presence there must be in considerable doubt. It may of course have been cultivated in historic gardens in the Near and Middle East. As our illustration on page 484 shows, *C. persicum* was certainly known to the horticulturally sophisticated Ottoman Turks in the early seventeenth century, and might well also have been known to Persian gardeners. There is one report, in the *Journal of the Cyclamen Society* of December 2000, by Erna Frank that a German friend of hers, Peter Zeininger, had found *C. persicum* on the north-east slopes of the Elburz mountains in Iran. Unlikely as this seems, there are relic Mediterranean elements to be found in the Caspian region of Iran (for example, *Ophrys* orchids and *Paliurus spina christi*) but there have been no other sightings that we can trace.

Cyclamen persicum is usually found at sea level or comparatively low altitudes, mostly around the eastern end of the Mediterranean. It is known from the Aegean Islands, Cyprus, southern Turkey, Syria, Lebanon, Israel and also further west in North Africa, in Tunisia and Algeria. Thus it is very much a Mediterranean species and unlikely to occur far inland in the more continental climate of Iran. *Cyclamen persicum* is one of the more tender species and best suited to cultivation in a frost-free environment, although a succession of mild winters in Britain has led to a proliferation of small-flowered cultivars being used in window boxes and hanging baskets, particularly in the larger towns with their even milder microclimates. The wild forms of *C. persicum* flower naturally in winter or early spring, when there is still plenty of moisture in the soil of the species' summer-dry habitats. An autumn-flowering version has been located in Israel, now named var. *autumnale* by Christopher Grey-Wilson, of which there is an account in the 'Further observations' section by Oron Peri (page 120).

Cyclamen persicum is mainly a low altitude plant occurring primarily near the Mediterranean coast. Here, tucked under a bush at Loryma archaeological site, south-western Turkey (BM)

Cyclamen persicum is distinct from all other species except for the related but little-known *C. somalense*. Its overall appearance makes it a rather distinct species, even without looking closely at the botanical detail. Most cyclamen species have leaf and flower stems that travel outwards from the crown of the tuber for a short distance before turning upwards to emerge from the soil, but those of *C. persicum* stand erect from the tuber. This is one of the features that make *C. persicum* such a useful and attractive pot plant, the result being a compact plant with a 'bunch' of leaves and flowers. The natural wild form also has elegant long flowers with twisted petals and a rich perfume, sadly characters that have been

In *C. persicum*, the fruiting pedicels do not coil tightly as in other species but arch over to touch the ground (BM)

disregarded in the selection process in the avid search for larger and larger flowers. However, plant breeders have now returned to favouring more compact plants that have smaller flowers, and these often have a pleasant, perceptible fragrance. Another unique characteristic of this species and of its ally *C. somalense* is the lack of coiling of the fruiting stalks; they stay more or less straight, perhaps slightly arched, and bend over to touch the ground. In all other cyclamen species, the pedicels commence coiling soon after the fertilised flower has fallen and end up like watch springs. This is probably a useful feature in view of the usually rocky habitat of *C. persicum*, which often grows in rock crevices. The seed capsule stands a better chance of landing in an adjacent crevice rather than in the same one as the parent plant, which might lead to great competition in a limited space.

There is no doubt that it is this species that has made the genus *Cyclamen* so well known and popular, even with the non-gardening public.

C. persicum Miller var. ***persicum***, *Gardeners' Dictionary* ed. 8, n. 3 (1768). Type: specimen cultivated in Chelsea Physic Garden, deposited at the Natural History Museum, London (BM).

Synonym

C. latifolium Sm. in Sibth. & Sm., *Fl. Graec.* 2: 71, t. 185 (1813).

C. persicum var. ***autumnale*** Grey-Wilson, *Cyclamen*: 174 (1997). Type: specimen prepared from a plant cultivated by *P. Moore*, Kent, England (holotype K).

Description Tuber rounded at first, becoming more depressed-globose with age, up to c. 15 cm diameter at maturity, brown, rough or even cork-like in old age, rooting from the underside, the roots thick but not markedly fleshy. **Petioles** erect from the tuber. **Leaves** rather fleshy, well developed before flowering time. Leaf blade broadly ovate, 2.5–18.6 cm long, 2.5–15.3 cm wide, green with very variable marbled pale green or silvery-green patterns and zones, sometimes predominantly pewter-coloured; margin thickened with cartilaginous teeth; the underside normally purplish suffused. **Pedicels** rising erect from the tuber, after anthesis not coiling but arching over to touch the ground. **Flowers** produced

Tuber of *C. persicum*; note the thick fleshy roots (BM)

The flower of *C. persicum* has long elegant petals with no auricles at the mouth and and usually a dark stain (BM)

Some striking leaves of *C. persicum* at Akamas, Cyprus (MD)

in winter through to early spring or more rarely autumn, fragrant; petals swept upwards to an erect position, white to medium purplish-pink, usually with a darker pink or magenta basal stain around the mouth of the corolla, oblanceolate-elliptic with acute or subacute tips, 2–4.5 cm long, 0.7–2 cm wide, strongly twisted lengthways, without swollen appendages ('auricles') at the base; style slightly exserted from the mouth of the corolla. 2n = 48 ('florists' selections are often polyploid).

Flowering period December to May (var. *persicum*), October to January (var. *autumnale*).

Distribution *Cyclamen persicum* has a wide distribution in the eastern Mediterranean region and North Africa. It has been recorded in northern Tunisia and Algeria, Crete (?native), the Aegean Islands of Rhodes, Symi, Kárpathos and Chios, the Athos Peninsula in northern mainland Greece, Cyprus, southern Turkey, western Syria, Jordan, Lebanon and Israel. In some of these localities, it is found on ancient sites and in graveyards, so this raises the question as to whether it is truly native or has been introduced for decorative purposes. Var. *autumnale* is widespread in Israel, also in adjacent Jordan, Palestine and the Golan Heights region of Syria.

Habitat Stony or rocky places, often in crevices, in open rocky maquis and garigue, often on limestone formations but not necessarily so, sea level to 1,200 m.

Variations In the wild, *C. persicum* is noticeably variable in leaf characters, although perhaps not so much in leaf shape (the leaf shape being predominantly broadly ovate) as in the leaf markings, which range from rather dull deep greens to beautifully silver-zoned forms or those marked with creamy veins in a netted pattern on a dark ground. Flower colour too varies considerably

Cyclamen persicum in Rhodes, probably the most frequently seen colour form, white with a carmine stain at the mouth (MH)

in the wild forms, although white or pale pink with a carmine 'nose' are the variants most frequently encountered; pure white forms without the pinkish colour around the mouth of the corolla can also be found. In the wild, scent is a noticeable characteristic, although this tends to be lost in the highly bred forms. As in all cyclamen species, the flower shape varies considerably, from those with long, elegantly twisted petals to forms with shorter, more 'tubby' flowers. An interesting variant is the Israeli autumn-flowering form described by Grey-Wilson in 1997 and given the botanical status of varietas as var. *autumnale*. This dictates

Cyclamen persicum in Cyprus; it is nearly always seen growing on rocky (limestone) hillsides or in crevices (BW)

A pink form of *C. persicum*, between Loryma and Bozburun, south-western Turkey, photographed in April (TW)

that there is automatically a var. *persicum*, the two varieties being distinguished solely by flowering time: December to May in var. *persicum*, October to January in var. *autumnale*.

Names for colour variants of var. *persicum* are available as follows: forma ***persicum*** — flowers white or pink with a carmine mark at the corolla mouth; forma ***albidum*** (Jord.) Grey-Wilson — flowers wholly white; and forma ***puniceum*** (Glasau) Grey-Wilson — flowers deep carmine or reddish.

Named cultivars of var. *persicum*

'Album': flowers pure white; known in the past and now re-introduced from Israel by the Cyclamen Society; referable to forma *albidum*.

'Silver Leaf': originally collected south of Tripoli in the Lebanon by Jim Archibald (*JCA* 1070); an exceptional silver-leaf form was selected out by Melvyn Jope.

'Tilebarn Kárpathos': leaves unremarkable for the species; flowers deep cerise with notably twisted petals.

Further observations on *Cyclamen persicum*

***Cyclamen persicum* on the Akamas peninsula of Cyprus**
This species is not common or widespread in Cyprus, so the following first-hand account by **Chris Clennett** (*Cyclamen*, December 2001) is useful:

"The geology of Cyprus is key to the distribution of *Cyclamen* species. Unlike many of the smaller Mediterranean islands, it consists of an igneous mass of rock, overlaid at lower altitude by a soft type of sedimentary chalk or limestone known as 'kafkalla'. This fundamentally alters the soil conditions for the flora .... and increases considerably the range of specialised endemics that result. Where *Cyclamen cyprium* is almost always found on or near the igneous rocks, *C. persicum* is found at lower altitudes on the kafkalla. It is possible that this species has been introduced to ancient sites such as the Tombs of the Kings near Paphos, but on the Akamas it is found at considerable distance from any settlements and would appear to be a genuine native.

On the Akamas, *Cyclamen persicum* is widespread, forming dispersed colonies of plants through low-growing woodland of *Quercus coccifera* and *Q. infectoria*, and more exposed slopes of scrub. The plants are decidedly more common on the north side of the peninsula, presumably because of the improved conditions for both shade and moisture. Along the nature trails that lead from the Baths of Aphrodite, the species can be found in numerous places, growing particularly at the base of trees or large rocks, where rain water is naturally channelled. As is common elsewhere for this species, it is also to be found growing directly on large boulders in cracks and crevices, where, presumably, the seed was lodged many years before. Where it is found away from the rocks, the large plants more than hold their own amongst the verdant spring growth of wild flowers. A feature of spring in this area is the proliferation of coarse annual grasses, and it seems somehow incongruous to see cyclamen in what looks for all the world like a long grass meadow. Of course, the grasses are short lived, from germination to seeding taking little more than ten weeks. This grass growth does provide *Cyclamen persicum* with assistance, however, in shading the plants from the strong sun as the temperatures rise in April and May. Other wild flowers found in close association with the *C. persicum* in this part of Cyprus include *Asphodelus albus*, *Gynandriris* [*Moraea*] *sisyrinchium*, *Ranunculus asiaticus* and *Ranunculus ficaria*.

As the season progresses, the lush vegetation surrounds the cyclamen as their seed capsules ripen, and seed is set at around the time most of the annual plants begin to die back through higher temperatures and lack of moisture. The cyclamen themselves then become dormant around the early part of June, and remain so until the autumn rains return. The time of rain can vary enormously in the area, and despite the fact that it is the wetter end of the island, even allowing commercial banana cultivation near the coast, extended drought is not uncommon. First rains can trigger immediate growth in mid-September if heavy enough, but in recent years rain has been light and growth does not commence until early winter. It was interesting to see *Cyclamen persicum* in fruit in mid May under the shade of carob trees (*Ceratonia siliqua*), immediately above the paths to the Baths of Aphrodite and literally within a stone's throw of thousands of tourists who remained completely oblivious to the presence of this (or any other) plant.

Flower colour here is much as it is seen throughout the range of this species, varying from mid-pink to white and always shading to the deep nose. Neither our friends nor I have seen a pure white-flowered plant in this area, but that's not to say they do not exist since we have not scoured the whole peninsula."

***Cyclamen* on Cyprus** Further valuable information with regard to the distribution of *C. persicum* on Cyprus is provided by **Martyn Denney** in the Cyclamen Society's journal of December 2001, some of the information being gleaned from other publications such as Meikle's outstanding *Flora of Cyprus*:

"In the south of Cyprus *Cyclamen persicum* is found around the ancient sites of Amathus, Kourion, Paphos Tombs of the

Kings, Paphos Petra tou Dighenis, Limassol old city and the Baths of Aphrodite near Polis. It is also found on the north coast of the Akamas Peninsula around Agios Georgios. The emergence of *C. persicum* leaves is almost entirely dependent on the autumnal rains, and where October and November are dry, the leaves are particularly late. Though I did not see this myself, local enquiries confirmed that it still thrives on the limestone bluffs above Amathus, Kourion and the Tombs of the Kings, and that in the latter two cases it grows in rock crevices around or on the sites themselves. At Amathus the gate keeper has transplanted a plant from the bluff above to a spot by his hut, and with regular watering, this was in full flower on 2 December 2000. Around the Baths of Aphrodite *C. persicum* grows on a north-facing escarpment in grass around the base of small shrubs and rocks about 75 m (250 ft) above the sea. Here, possibly due to local ground-water, the plants appear to be less dependent on rainfall and were in full leaf at the time of my visit, with flower buds just beginning to appear.

In the north of Cyprus *C. persicum* is evidently more widespread along the Pendadahtilos (Pentadaktylos, Kyrenia) Range and the Karpas Peninsula to the east. Meikle reports it as growing in Pendadahtilos from Kornos to Kantara, whilst Fielding and Moore did not find it growing west of Kyrenia (Girne), but reported that to the east it was common, growing at low altitudes including amongst rocks along the shore. Mutch (*Bulletin of Alpine Garden Society* 23: 164–168 (1955)) says that it grows as high as 215 m (705 ft) on the north side of the range, but to 550 m (1,800 ft) on the south side. The *persicum* growing near or above Bellapais Abbey and Kantara Castle is again associated with ancient sites, but here the connection must be insignificant in light of the distribution of the plant throughout these hills. It is reported by Meikle as being locally abundant on the Karpas Peninsula, specifically around Koma tou Yialou and Eleousa Forest."

***Cyclamen persicum* on Crete** In 1995, a team from the Cyclamen Society visited Crete to study *C. creticum*. Needless to say, they also took the opportunity to look at any other cyclamen that 'got in the way', including *C. persicum*. **Trevor Wiltshire** notes that they did find *C. persicum* in full flower near a small church past Toplou Monastery, growing with *Pistacia* spp., *Salvia trifoliata*, *Cistus* spp., *Geranium robertianum* and grazed grasses. The population covered a maximum of 40 m^2 (50 sq yds). They searched for over a kilometre around the site and found no other populations, despite the presence of similar habitats. The team were of the opinion that the population was a result of an escape from cultivation, which perhaps occurred a great many years ago.

***Cyclamen persicum* on Rhodes** Writing for the Cyclamen Society's journal of December 1994, **Martyn Denney** describes observations made during a visit to Rhodes:

"In the cemetery of the Murad Reis Mosque, *C. persicum* had all but finished flowering on and around the graves, but it is pleasing to be able to report an excellent seed set, which promises to help the regeneration of the colony. Some seed pods were quite large, being over 15 mm in diameter. The leaves of the plants showed very little variation and were large, giving a 'cabbagey' look. A foray in Rodini Park produced sightings of a small number of plants of *C. persicum*, in leaf only and with virtually no seed set. These

A Moslem cemetery in Rhodes town, *C. persicum* thriving in the crevices of tombstones (BM)

Kárpathos Island, an attractive pink form of *C. persicum*. The species usually grows at sites with a north- or west-facing aspect (JF)

plants were principally on almost bare banks above the path, around the base of oak trees … . On Mount Filerimos, to the south-west of Rhodes Town, *C. persicum* was still in flower on the slopes immediately below the monastery. Most plants had the familiar whitish flowers, but there were occasional examples with a good pink flush. Although large quantities of seed had been set, this was noticeably less mature than that found on the plants in the old Turkish cemetery. The leaves were generally large, and again had no distinctive features. Further down Filerimos, *C. persicum* gave way to *C. repandum rhodense*."

***Cyclamen persicum* on Kárpathos** When the Cyclamen Society was planning the itinerary for fieldwork on Kárpathos, they had no precise locations for *C. persicum* on this island. Enquiries revealed, however, that it could be observed near Kárpathos town and that it clearly occurred in some quantity as the local people gathered its flowers for decoration. In April 1991, the team found this to be true but the populations near the town and by the cemetery were at risk from developments. Among these plants were some interesting variants, notably unusually dark-flowered ones that were more strongly coloured than most wild plants, almost as dark as some of the florists' selections. Other variations included plants that had extra petals, formed by mutation of stamens. The range suggested to the team an exceptional degree of genetic diversity. The site was described as "a series of abandoned terraced fields with dilapidated walls and small, steep rock faces" with the plants growing in sharply drained screes, in walls and in crevices. The team questioned whether this occurrence on Kárpathos is natural as the plants were growing so close to the town and harbour, although they noted that this site is well within the known natural range of *C. persicum*. The sites on Rhodes too are associated with a cemetery, a public garden and a monastery, so it is possible that such an attractive plant has had its natural distribution widened by human activity.

***Cyclamen persicum* on the East Aegean island of Simi** In 1998, the Cyclamen Society undertook a survey of cyclamen in the islands of the Dodecanese, primarily aiming to study *C. rhodium* but including any other cyclamen species in the

C. persicum is also found on Symi Island in the Dodecanese (BC)

region. Their report, written up in the Society's journal of June 1999, includes some useful data on *C. persicum* on Simi. The team recorded the species as being "all over the island" and made detailed observations at nine sites. Curiously, there appeared to be two variants: the 'standard' form, growing around Simi town and the coast towards Emborios, and a diminutive form with often only a single leaf, growing throughout the rest of the island. The tubers of this diminutive form were often very large, out of all proportion to the size of leaf surface. This form also had uninteresting leaf patterns, whereas the standard form had a great range of leaf patterns with some good silver-centred foliage. On the whole, the flowers were typical of the species, white with a pink stain at the mouth.

***Cyclamen persicum* in Syria** The occurrence of *C. persicum* in Syria is not well recorded, so the report by **Lynn and Michael Almond** in the Cyclamen Society's journal for June 1996 provides valuable information:

> "We had seen *C. persicum* in the wild in western Turkey, in Rhodes and, in greater quantity, in the south of Turkey. Nothing, however, had prepared us for the vast quantity we saw in north-west Syria in spring 1995. And not only was there a great quantity, but both the flower and (particularly) the leaf forms were very variable.
>
> The road which leads north-eastwards from Latakia on the Mediterranean coast through the coastal mountains towards the Orontes river was draped on either side for miles on end with curtains of cyclamen. Most of this we saw only from the moving bus, but when we stopped for half an hour's refreshment at a small settlement along the way, we were able to explore the rocky outcrops all around and take dozens of photographs of cyclamen of all shapes and shades of pink. At Saladin's Castle (Qal'at Sahyun), further south, although the cyclamen were not in quite such abundance, the plants growing on the sides of the ravine to the north of the castle demonstrated a remarkable variety of leaf forms. At Qal'at Marqab, above the Mediterranean coast north of Tartus, the terraced hillsides below the northern side of the castle were again cascades of pink *C. persicum*, and there was more inside the citadel itself.
>
> In the first two places we saw *C. persicum* in Syria, it was growing in conditions much more lush than those in which we have seen it in western and southern Turkey (where it was either on bare cliffs or under sparse, thorny scrub in dry conditions). In both these two places the rocky outcrops on which it was growing were among trees and tall, thick shrubs and other dense vegetation. Even in the other places, where it was growing in more open conditions, the surrounding vegetation was greener and more dense than it had been where we had seen it in Turkey. Later we went on to Jordan. We saw a little *C. persicum* in flower at Ajlun castle and some evidence of it gone over in the ruins of Jerash."

Cyclamen persicum in north-west Syria at Qal'at Sahyun, 30 March 1995 (LA)

High-altitude *Cyclamen persicum* in Israel In 1990, some fieldwork was carried out by the Cyclamen Society in Israel with a view to studying *C. coum* in its most southerly locations. The opportunity was also taken to make observations about *C. persicum*, a relatively common species in the region. During this visit, a good silver-leaved form and a form with dark flowers were collected, but it was on one foray to the mountains to study *C. coum* that perhaps the most interesting observations were made. **Melvyn Jope** writes:

> "*C. persicum* was found in several locations in the Golan [Heights], but we were particularly interested in the plants found growing with *C. coum* at the summit of Mt Meron. It was very cold with rain and hail … and there were the last remnants of earlier snow falls. There were many *C. persicum* growing at the summit and as well as *C. coum* the companion plants were *Ornithogalum lanceolatum*, *Iris histrio*, *Sternbergia clusiana*, *Hyacinthus orientalis*, *Gagea* and *Colchicum* spp."

Two of the *C. persicum* tubers were collected at an altitude of 1,200 m with a view to introducing hardiness into this otherwise frost-tender species.

Habitat site for *C. persicum* in Israel (MELJ)

A dense population of *C. persicum* in Israel, south of Bet Shemesh at 300 m altitude (MELJ)

Autumn-flowering *Cyclamen persicum* in Israel. Oron Peri has studied the autumnal variant of *C. persicum* and has provided for us the following informative account.

"*Cyclamen persicum* var. *autumnale* was thought to be endemic to Israel, but in recent years it has also been found growing in the Golan Heights (Syrian Territory) and in Palestine. In autumn 2011, I found populations in Wadi Shoayb in Jordan, growing in semi-desert conditions. It seems that it grows in many of the wadis (seasonal rivers) that reach the Jordan valley from both sides of the Great Rift. It is probably also in southern Lebanon.

In Israel, var. *autumnale* is found growing at varying altitudes throughout the entire Galilee region, the Golan Heights, Mt Carmel, the Jordan valley, Judea and the Samaria mountains, and the Judean and Samarian deserts, at altitudes from around –250 m up to about 400 m. It grows in shady, stony areas, mostly on north-facing rocky slopes and walls; it is seldom found in more open rocky ground and then always in crevices or just at the base of large rocks. It seems that var. *autumnale* has become well adapted (like *C. graecum*) to growing in very small spaces such as holes and crevices in rocks and walls; occasionally only one or two leaves with one or two flowers can be seen popping out of a tiny hole. Most of the colonies grow on *terra rossa* or rendzina soils derived from hard limestones but the variety sometimes occurs on volcanic soils (e.g. in the Golan Heights). Unlike var. *persicum*, var. *autumnale* does not grow on sandy soils.

Colony size can vary from a few dozen to many hundreds of plants, but colonies of var. *autumnale* are never as big and widespread as those of the common variety. These plants do not grow in the vicinity of rocks by chance, it seems that dew that forms on rocks provides a source of humidity as a result of the changes between day and night temperatures; in autumn, the nights start to get cooler and consequently drops of water form on the surface of the rock and trickle down into the crevices where the cyclamen grow. This might also provide the explanation for the fact that the flowering time of this variety is so precise, the first flowers are present by 15 October, whereas the flowering of var. *persicum* depends entirely on the first rains of the winter.

Cyclamen persicum var. *autumnale* blooms from mid October to January, depending upon the area in which it grows. Colonies in dryer, hotter areas will end their flowering by mid December, and thus there is no overlap between the flowering times of the two varieties and less probability of hybridisation in these areas. When var. *persicum* begins to flower, var. *autumnale* already has fruit capsules in an advanced state. Interestingly, the capsules of the two varieties mature at the same time (April or May), which means that var. *autumnale* fruit are slower to mature by two more months. Furthermore, var. *autumnale* is very shy to set seed, only a small percentage of the flowers are fertilised and it is not unusual to find a mature plant that

A paler form of *C. persicum* var. *autumnale*. Note the lack of leaves at flowering time (OP)

The deep pink form of *C. persicum* var. *autumnale* growing wild in Israel. It flowers between October and January (OP)

The white and pink bicoloured form of *C. persicum* var. *autumnale* in Israel (OP)

has not set seed at all. The origin of this phenomenon is not clear: it might result from a lack of pollinators in the autumn or from a genetic disorder. This poor seed-set might explain why var. *autumnale* colonies are quite small and not very abundant.

With regard to variation, the flowers of var. *autumnale* normally appear before the foliage, but in more humid areas, the leaves are sometimes partially developed by the time of flowering. Usually, rain does not fall in the localities in which *C. persicum* grows before mid November, but occasionally there are early rains that can cause the appearance of leaves at the same time as the flowers; for example, in the autumn of 2006, the rains started at the end of September and so by the end of October, the plants were in full bloom and had fully developed leaves. There are no colour or size differences between the two varieties, but I have noticed that the purple or pink eye of var. *autumnale* flowers generally stretches up to about one third of the flower length, in some cases even further so that it often blends with the basic colour of the flower to form a monochromic flower. This might be another reason for poor seed set: the lack of a clearly marked entrance to the flowers might impair the activity of pollinators.

Variation in leaf colour and pattern are definitely less interesting in var. *autumnale* than in var. *persicum*, silver-leaf forms being very rare. Unlike those of var. *persicum*, colonies of var. *autumnale* are typically quite distant one from the other and this results in characteristic differences between some of the colonies; for example, some of the colonies on Mt Carmel have smaller leaves, much smaller tubers, and markedly short pedicels; others in the Galilee region and south Golan, are more robust plants having larger leaves and tubers and larger flowers on much longer pedicels; whereas colonies in the desert have thicker leaves and short pedicels. This phenomenon of short pedicels exists only in var. *autumnale*: having poorly developed leaves at flowering time means that the flowers (unlike those of var. *persicum*) do not need long pedicels to raise them above the foliage. I have found plants with pedicels of just 4 cm in length, whereas in var. *persicum* pedicels are seldom less than 15 cm and often longer than 25 cm.

In cultivation, var. *autumnale* is a fine and interesting plant, but the tubers are evidently smaller than those of wild plants and take a much longer time to grow; I have some 8-year-old plants with tubers that are only 5–7 cm in size. When growing from seed, I find that var. *autumnale* is not only shy to set seeds, but also that its germination is slow and unreliable, and that the resulting plants take from three to five or more seasons to produce their first flowers.

In my opinion, away from its country of origin, this plant should be treated in a similar way to *C. rohlfsianum* so as to encourage it to behave normally and produce its flowers before the leaves. I suggest growing it in small pots so that the soil dries out quicker, but the type of soil is relatively unimportant as long as it drains well. Water should be withheld completely from May until the plants have mature flowers, but check the tubers once a month to see that they do not get soft or start to shrink, in which case, add a bit of water. When the flowers are mature, the plants can be watered regularly. Although var. *autumnale* grows in shady areas in its natural habitat, I recommend growing it in full sun so that it thrives. Hand pollination is necessary to secure fertilisation."

Fortunately, the future of this variety seems secure as most of its colonies are found in protected areas, and some are inside nature reserves. Nevertheless, two of the largest colonies were completely destroyed a few years ago when new neighbourhoods were created, the first in Jerusalem (Beit-Hakerem) just above the University Botanic Gardens, the second in Haifa (Vardiya).

Development of *Cyclamen persicum* as a pot plant In Chapter 5 (page 373), **Diony de Bont** and **John Grimshaw** recount the history of how *C. persicum* has been developed to produce the remarkable range of florists' pot plants we have available for indoor decoration today. Even more remarkably, this huge diversity has resulted from selection from within just one species.

Cultivars of *C. persicum* at an RHS Show, indicating the extent of breeding in this species (see Chapter 5) (BM)

Cyclamen somalense

AUTUMN- OR WINTER-FLOWERING

When the discovery of a new *Cyclamen* species was announced in 1986, it was greeted by cyclamen enthusiasts in a variety of ways ranging from astonishment, delight and scepticism ("probably just an outlying *Cyclamen persicum*") to total disbelief ("I just don't believe in it!"). However, those of the botanical community with a knowledge of the Mediterranean and African floras certainly had no doubts, first because the Swedish botanist Mats Thulin and his Somali colleague Ahmed Mumin Warfa were the botanists reporting this discovery — they would not get it wrong — and second because such a link between these two regions is a known and proven fact. In former times, the area with a Mediterranean-type climate was much more extensive than it is at present, and the cyclamen found in a small region of north-eastern Somalia represent an enclave of survivors from the distant past. It is very likely that cyclamen were once much more widely represented throughout the region, and it is even possible that there have been other species, now extinct. It is interesting to speculate that a mountain block, such as Tibesti in Chad, might once have had a flora consisting of Mediterranean plants, perhaps even a cyclamen, as also might parts of the Arabian peninsula.

Mats Thulin, the botanist who first set eyes on *C. somalense*, has provided a detailed and interesting account of its discovery and of the behaviour of the few tubers that were introduced into cultivation, together with more information about its habitat and the climate *C. somalense* experiences in this fascinating region of eastern Africa (page 124). Also, and rather importantly, Mats Thulin has now for the first time determined the chromosome number of this exciting discovery (see page 128).

Plant of *C. somalense* flowering in Uppsala, Sweden. Thanks to the cultivation skills of Bo Willmer of the Botanical Garden of Uppsala University, the living specimen of *C. somalense* is still in cultivation after 25 years in his care. Photograph by Magnus Lidén, November 2005 (ML)

Coastal plain near Qandala with spiny *Euphorbia prona* in the foreground and Cal Miskaat rising up in the background. Photograph M. Thulin, January 1997 (MTh)

C. somalense Thulin & Warfa, *Pl. Syst. Evol.* 166 (3–4): 249 (1989). Type: Somalia, Al Miskat Mts, SW of Qandala, 26 November 1986, *Thulin & Warfa* 6084 (holotype UPS).

DESCRIPTION **Tuber** subglobose to elongated or flattened, becoming irregular with age with protuberances, 3–4 cm diameter, brown, rough or cork-like, rooting from the underside. **Petioles** erect from the tuber. **Leaves** developing at flowering time. Leaf blade cordate-reniform, slightly angled, 3–10.5 cm long, 4–12 cm wide, dull green with a silvery zone and silvery veining; margin irregularly dentate; the underside purplish suffused. **Pedicels** 7–12.5 cm long rising erect from the tuber, elongating after anthesis, not coiling but arching over to touch the ground. **Flowers** produced in autumn or early winter, ?fragrant; petals swept upwards to an erect position, pale purplish-pink, usually with a darker pink basal stain around the mouth of the corolla, narrowly oblanceolate-elliptic, 1.5–1.7 cm long, 0.5–0.7 cm wide, twisted lengthways, without swollen appendages ('auricles') at the base; style slightly exserted from the mouth of the corolla. 2n = c. 96 (M. Thulin see below, page 128).

FLOWERING PERIOD November.

Tuber of *C. somalense* in the Botanical Garden of Uppsala University, November 2005 (ML)

Flower of *C. somalense* in cultivation in the Botanical Garden of Uppsala University, November 2005 (ML)

Leaf of *C. somalense* in hole of limestone rock near the type locality, 4 January 1997 (MTh)

Leaves of *C. somalense* in crevices of limestone rocks near the type locality, 4 January 1997 (MTh)

Distribution *Cyclamen somalense* is endemic to north-eastern Somalia.

Habitat Holes and crevices of north-exposed limestone rocks in the mist zone at 1,200–1,600 m.

Variations From further observations made in the wild, it is now possible to say that the tuber can be very variable in shape, from almost globose to rather flattened, The leaves, like those of nearly all cyclamen, vary considerably in their silvery markings as can be seen in the accompanying photographs.

Cyclamen somalense, an extraordinary discovery

Mats Thulin

On the 14 November 1986, I woke up at 3 o'clock in the morning and made my way out into the dark streets of Mogadishu. With my Somali collaborator, Ahmed M. Warfa, I had to be at the airport just south of the town before 4 o'clock to catch the weekly domestic flight to north-eastern Somalia. We carried equipment for plant collecting, and it took some negotiation before we were allowed to bring our plant presses and camping gear on board the Somali Airlines Focker plane heading for Alula in the north-easternmost tip of the country. The plane took off just after sunrise and, after a stop in Bosaso on the way, we reached the small dusty airfield of Alula at 11 a.m.

My aim for the journey was general plant collecting for the Somali Flora project, which we were hoping to get started, while Ahmed Warfa's main aim was to collect material of *Cordia* for his PhD project, which I was supervising at Uppsala University. We had made arrangements with people at the French Integrated Date Palm Project to meet us in Alula, and we were happy to see a car with our friend Abdirizak waiting for us. He took us over the sandy plains to the coastal village of Geesaley, some 40 km to the south-west. I had travelled extensively in Somalia before but this remote corner was new to me, and in the late afternoon I made a first walk on the plains to get acquainted with the local flora.

Going inland from the coastal plain in northern Somalia, there is a hilly subcoastal zone of varying width, then an extensive block of steep limestone escarpment reaching above 2,000 m in places, and finally an uplifted and often more or less dissected plateau lying to the south. The climate is strongly influenced by the monsoon winds. The erratic rainfall in the coastal and subcoastal zone is less than 100 mm per year, whilst the upper

Mats Thulin discovered *C. somalense* while studying the flora of Somalia. Here he is seen working in the Herbarium at Kew (BM)

part of the escarpment can receive a mean annual rainfall of over 700 mm in favourable places. A major portion of the rain falls during the winter months, during the north-east monsoon, when mists are also frequent. The summers are hot and dry.

Our main target during this three-week fieldtrip was the limestone escarpments and we wanted to study them in as many places as possible. In the main, the only way to do this is was by foot and we started off on the 15 November by exploring the escarpment south of Bandar Murcaayo. We went up to about 500 m above sea level that day and saw numerous frankincense trees along the way. A dwarf species of *Commiphora*, a relative of the myrrh tree, that we found on the rocks turned out to be a distinctive new species, *C. spinulosa*, still only known from this locality.

The next day, we decided to go further up with a guide, and brought sleeping bags so that we could stay overnight. We struggled in the heat of the merciless sun along a path that would take us up to the top of the escarpment. At 1,000 m, we stopped for the night and just slept on the ground underneath the starry sky. The next day, we explored the vertical cliff faces at 1,100–1,150 m. Here, the charming *Anemone somaliensis*, a relative of the Mediterranean *A. coronaria*, was in flower in holes and crevices of the limestone. *Andrachne fragilis* (Euphorbiaceae), also growing in holes in the limestone cliffs, was one of the new species here, still only known from the type collection. When we got down again in the evening, we were loaded with plants, very tired, and perhaps above all, very thirsty, the small amount of water we had carried with us having been long since finished: the cool water in the ancient well Tayeega near the coast tasted wonderful.

Our days were then filled with fieldwork in the area. One day, for example, we visited Cape Guardafui (or Raas Caseyr) the very tip of the Horn separating the Gulf of Aden and the Indian Ocean, and with a lighthouse no longer in use. To reach the Cape, we walked for three hours, all the time in a drizzling rain, an unusual but rather pleasant experience in this desert environment. In Geesaley, we heard that 40 mm of rain had fallen that day, an amount that could well be the ration for the whole year and enough to make the coastal plains impassable for cars for some time.

Anemone somaliensis in crevice of limestone rock, November 1986 (MTh)

Harbour of Qandala, December 1985 (MTh)

Leaves of *C. somalense* in crevices of limestone rocks at the type locality, 26 November 1986 (MTh)

Plant of *C. somalense* in flower at the type locality, 26 November 1986 (MTh)

One area I particularly wanted to get to in north-eastern Somalia was Cal Miskaat, a large mountain block between Qandala and Bosaso that is entirely without roads. At previous visits, I had nagged at the edges of it, but this time I hoped for a closer encounter with this little-known area. With a car equipped with a radio transmitter and borrowed from the Integrated Date Palm Project, we proceeded on very rough roads to Qandala, another small coastal town in north-eastern Somalia, with a history going back at least to the times of the visits by Greek seafarers 2,000 years ago when the frankincense trade was flourishing. In 1986, there was still some frankincense trade, but the main export now was living sheep and goats taken in small boats across to the Arabian peninsula.

From here, accompanied by a local guide, we started a three-day walk on the morning of the 25 November to find out more about the secrets of Cal Miskaat. This time we carried a tent, as well as reasonable amounts of spaghetti, dates and water. We spent the first night at about 800 m above sea level in a place I had visited before, but the second day we got into new territory. Again we met *Anemone somaliensis* and several other plants giving a Mediterranean impression, such as *Olea europaea* subsp. *cuspidata*, *Pistacia falcata*, *Brassica somalensis*, *Salvia somalensis* and *Cheilanthes catanensis*. Here were also a number of plants I had never seen before, several of which I later described as new species, such as *Salvia bariensis*, *Teucrium eburneum* and *Scutellaria somalensis* to take three examples of members of the Lamiaceae that are still only known from Cal Miskaat. Now a light rain started and we got quite wet, but it did not matter so much as it was still warm, the main issue was to keep the plant presses dry.

In the late afternoon of the 26 November, the first *Cyclamen* leaves turned up in holes and crevices of the limestone rocks. At least they looked exactly like *Cyclamen* leaves, dark green with a white marmoration above and purple beneath, but I could not quite believe that we had found a member of this genus in Somalia. After some searching, however, I found a few plants with a single flower each, as well as some young fruits, and now there was no doubt about it, this was a *Cyclamen*! I quickly took photographs of the plants and made a small collection, enough to be able to describe this species, which I immediately thought must be new to science. I also kept three tubers for cultivation, but no ripe fruits were seen. As we continued to climb, I found the plant to be confined to elevations between 1,250 and 1,600 m, and that it was restricted to more or less north-facing limestone rocks in the mist zone. We spent the second night at about 1,600 m and now it was really good to have a tent. There was more rain during the night and the temperature dropped quite a bit.

No people live permanently in the area, but nomads come with their goats and sheep to graze after the rains. We also saw a few camels that did not look quite comfortable in this rocky environment. The nomads live in small huts or in the natural caves found in some places, and we were invited to share tea and fermented goat milk with them. People must, however, have lived here long ago as witnessed by the pre-islamic graves that we saw here and that are found in many places in northern Somalia.

The *Cyclamen* was described as *C. somalense* Thulin & Warfa in a paper published in *Plant Systematics and Evolution* 166: 249–252 (1989). We compared the new species with the cyclamen species closest to it, *C. persicum*, the main distinguishing features being the small subglobose to elongate tubers, the leaves that are generally wider than long, with a somewhat angled outline and coarsely toothed margin, the small flowers on stalks hardly exceeding the leaves, and the narrow corolla-lobes. Among the hundreds of new species I have had the privilege to find in Somalia, this is the one that has attracted by far the most attention. There was a period when I almost needed a special secretary to deal only with the *Cyclamen* correspondence.

The tubers were planted in the Botanical Garden of Uppsala University and they still survive there. They have produced leaves regularly, and until recently they have produced flowers only three times, in 1999, 2001 and 2005, always in November and each time with a single flower only. However, in 2011 one plant produced its best display of flowers ever in the garden, exactly 25 years after it was planted. Six flowers appeared in December, unfortunately all from the same clone. To celebrate the 25th anniversary the garden had a small ceremony with me and Bo Willmer, the gardener who has looked after it all the time, as special guests. At the time of the first flowering, the plant figured in Swedish newspapers as well as on Swedish television. The botanical artist Bo Mossberg also made a painting of the first flower of this species to appear in cultivation. The flowers of the cultivated plants agree well with those in nature, but the flower stalks well exceed the leaves, perhaps because of shortage of light. Of course we hope to get seeds from the cultivated plants, but this has not yet been possible. Self pollination has been tried, but it seems impossible to get viable seeds this way.

The most flowers produced by the plant of *C. somalense* in cultivation were in November 2011. Photographed at Uppsala Botanical Garden (ML)

I have had the chance to revisit the *Cyclamen* locality once, in January 1997, accompanied by a group of Somalis, including the guide from 1986, and my wife Gunilla. This time we spent four days (2–5 January) in Cal Miskaat and hired a camel to carry our things. On this occasion, I also found plants at slightly lower elevations, down to 1,200 m, and estimated the total

Pre-islamic grave in Cal Miskaat surrounded by rounded bushes of *Euphorbia balsamifera*, January 1997 (MTh)

Our guide Id Said Aden loading the camel in Cal Miskaat, January 1997 (MTh)

Fruit of *Cyclamen somalense* near the type locality showing arched pedicel unlike those of most species which are coiled tightly but similar to that of *C. persicum*, 4 January 1997 (MTh)

number to be several thousands within an area of perhaps a square kilometre, but no new localities were found. The plants were in full leaf but no longer in flower, and after checking hundreds of them, I found only a handful that had flowered, all of them again with just a single flower on each plant. One relatively well-developed fruit was collected, and I thought that now we would have viable seeds, but unfortunately they turned out to be too young to germinate. In any case, *C. somalense* seems to be a species that has very limited flower production. Another thing noted in 1997 was that all of the tubers were not more or less elongate, as is the normal condition in the species, but some were distinctly flattened and approached the shape found in *C. persicum*.

Temporarily inhabited cave in Cal Miskaat. Two women preparing food outside and one above the cave collecting wood, January 1997 (MTh)

The nomads in the area know the *Cyclamen*, although it seemed to be of no particular importance to them. The plants also seemed to be largely untouched by grazing animals. Under the present circumstances, the species is under no obvious threat, but with the limited size of the known population, this situation could of course quickly change.

Cyclamen somalense belongs to a disjunct Mediterranean element in the mountain flora of northern Somalia that comprises a fairly large number of species, and most of these are quite distinct from their Mediterranean relatives. The evergreen bushland, found in favourable places along the escarpments, includes species of *Olea*, *Buxus*, *Pistacia* and *Ceratonia* and gives the impression of a Mediterranean macchia vegetation. In the *Cyclamen* site in Cal Miskaat and its vicinity, apart from species already mentioned, are also *Coris*, another member of Primulaceae in a wide sense, which is otherwise confined to the Mediterranean region, as well as two species of *Helianthemum* (Cistaceae). This Mediterranean element may be, at least partly, a relic of late Tertiary origin that has survived in the Mediterranean-like climate, with dry hot summers and winter rains, that prevails in this area.

Thulin & Warfa (1989) originally postulated the nearest relative of *C. somalense* to be *C. persicum*, and Grey-Wilson (1997) even placed these two species together in his subgenus *Persicum*. A phylogenetic analysis based on morphological data by Anderberg (1994) placed *C. somalense* as sister to *C. rohlfsianum*, but in another analysis (Anderberg *et al.*, 2000), based on both morphological data and nuclear ITS sequence data, *C. somalense* indeed comes out as sister to *C. persicum* with moderate support.

C. persicum and *C. somalense*, along with *C. rohlfsianum* and *C. graecum*, were placed in subgenus *Eucosme*, a group characterised by its basic chromosome number $x = 12$, by Anderberg *et al.* (2000). *Cyclamen persicum* has $2n = 48$ in wild populations, *C. graecum* $2n = 84$ and *C. rohlfsianum* $2n = 96$. The chromosome number of *C. somalense* has been unknown up to now, but I take this opportunity to report the count of $2n =$ c. 96 that I have made in the root tips of one of the plants cultivated in the Botanical Garden of Uppsala University. This chromosome number confirms the position of *C. somalense* in subgenus *Eucosme*.

2.2g The Hederifolium group

(= Subgenus *Cyclamen*, Series 1)

Comprising *Cyclamen hederifolium*, *C. hederifolium* subsp. *crassifolium*, *C. confusum* and *C. africanum*

Three species that are very similar in appearance form this group. They are autumn-flowering with, most typically, large, lobed leaves and large pale to deep pink flowers that have very conspicuous swellings ('auricles') around the mouth. The stamens are not exserted from the corolla but the style can be shortly so. In fruit, the pedicels start to coil downwards from the apex. These species have a range of chromosome numbers based on n = 17; 2n = 34, 68, 102.

Cyclamen hederifolium subsp. *hederifolium* and subsp. *crassifolium*

Flowering in autumn

Apart from the florists' selections of *C. persicum*, this, the ivy-leaved cyclamen, is by far the most likely species to be encountered in gardens and nurseries and is one of the delights of autumn. Following its floral display, the attractively marked foliage provides further interest throughout the winter until late spring and is an excellent foil for early bulbs such as snowdrops, winter aconites and *Anemone blanda*. In Britain, *C. hederifolium* had certainly been introduced into cultivation by the sixteenth century, possibly much earlier than this. It is now well established in naturalised populations in some areas, sometimes even in built-up areas where self-sown plants grow on roadside verges. This is rather surprising for a Mediterranean plant that has a widespread distribution at relatively low altitudes from France and Italy through Greece to western Turkey, but it is undoubtedly a very hardy species. Its tubers will usually survive heavy frosts even when exposed. It is also an amazing plant in terms of its capacity for variation, its characteristics are forever changing as new seedlings arise. This great propensity for variation is a delight for enthusiasts, and it is well worth a visit to the shows of the specialist and alpine societies, such as the Cyclamen Society and Alpine Garden Society, to look through the exhibits in the classes devoted to plants that are grown for their foliage effect. Although these numerous variants are unclassifiable in the botanical sense, they have great horticultural significance and so some have been given the status of cultivar, for example 'Ruby Glow', 'Silver Cloud' and 'Album'. This course is justified only if seedlings from the original parent plant are relatively uniform, a state that can be achieved by repetitive selection over several generations as Peter Moore has shown with this and several other species.

As to botanical variants, there is one subspecies, *crassifolium*. It was Friedrich Hildebrand, Professor of Botany at Freiburg University who first became aware of this southern Greek plant,

Cyclamen hederifolium is the most widely cultivated species, hardy and ideal for semi-shaded positions (BM)

Cyclamen hederifolium. Painted by Pandora Sellars (reproduced at 110% of original size)

A pine wood habitat for *C. hederifolium* at Olympia, Peloponnese (BM)

which he regarded as sufficiently distinct from *C. hederifolium* to merit description as a species: *C. crassifolium* (Hildebrand, 1907). For nearly a century, this name was disregarded or overlooked, and even if mentioned, it was either dismissed as an unidentifiable obscure name or incorporated into *C. hederifolium* without comment. Following some recent research by Cyclamen Society members in collaboration with Reading University, this Peloponnesian variant now has recognition as *C. hederifolium* subsp. *crassifolium* (Culham, Denney & Moore, 2009). As its name informs us, this inhabitant of the Peloponnese has noticeably thicker, fleshier leaves than subsp. *hederifolium*. This subspecies differs from the typical form not only in appearance, it has also been shown to be genetically distinct in having a chromosome number of 2n = 68 (subsp. *hederifolium* 2n = 34).

On the island of Crete, there is another relative of *C. hederifolium* that has been the subject of much investigation during the past decade. It was first described by Christopher Grey-Wilson (1997) as *C. hederifolium* var. *confusum*, then raised by him to the rank of subspecies (2002). Now, following further fieldwork by members of the Cyclamen Society working with botanists from Reading University, it has been upgraded further to the status of species: *C. confusum* (Culham, Jope & Moore, 2009). Accordingly, it is dealt with in this book as a separate species (page 145). Although clearly part of the *C. hederifolium* alliance, this species too can be distinguished cytologically as it has a chromosome count of 2n = 102. Further studies are necessary to establish the exact distributions of *C. hederifolium* subsp. *crassifolium* and *C. confusum*, and to determine more convincingly their distinguishing features as they both have noticeably fleshier foliage than *C. hederifolium* subsp. *hederifolium*, thicker flower stalks and fragrant flowers. The Reading University–Cyclamen Society team have also ascertained that DNA sequencing adds weight to the view that *C. confusum* should be regarded as a distinct species. The paper explaining the background and reasons for this decision is repeated under that species on page 151.

C. hederifolium Aiton, *Hort. Kew.*, ed. 1: 196 (1789) subsp. ***hederifolium***. Type: lectotype [designated by J. Compton in *Cyclamen* 32, 1: 22 (2008)], the illustration LXX in Regnault, *La Botanique* 70 (1774).

Synonyms

C. linearifolium DC., *Fl. France* 3: 433 (1805).
C. neapolitanum Ten., *Prodr. Fl. Nap.* 66 (1813).

C. hederifolium subsp. ***crassifolium*** (Hildebr.) Culham, Denney & Moore, *Journal of the Cyclamen Society* 33 (1): 14 (2009). Type: [neotype designated by Culham, Denney & Moore *loc. cit.*] Greece, Peloponnese, Pentelikon, *Cyclamen Society Expedition* CSE01006G, 28 April 2001, specimen cult. *P. Moore* Tile Barn Nursery, Kent 2009 (neotype RNG).

Synonym

C. crassifolium Hildebr., *Beih. Bot. Centralbl.* 22 (2): 195, t. 6 (1907).

Description (both subsp.) **Tuber** depressed-globose, up to 25 cm diameter at maturity, rough and cork-like on the surface, rooting mainly from the sides and base. **Petioles** horizontal below ground before turning up to emerge from the soil. **Leaves** developing just before or with the flowering. Leaf blade extremely variable in shape, often broadly ovate-triangular with a few shallow angular lobes but sometimes lanceolate, 3–15 cm long, 2–13 cm wide, often with a central ovate-triangular green pattern surrounded by a pale green, greyish or silver marbled zone, but sometimes the central zone silver; the variations are almost endless from plain green to wholly

Tuber of *C. hederifolium*, with roots emerging around the circumference (BM)

Flower of *C. hederifolium*; note the very conspicuous auricles at the mouth (BM)

Range of leaves showing the remarkable variation in *C. hederifolium* (BM)

silvery-grey; margin subentire to irregularly dentate; the upper surface usually not glossy; the underside green or purplish suffused. **Pedicels** horizontal before turning upright, after anthesis coiling from the top downwards. **Flowers** produced in autumn, unscented or with a distinct fragrance; petals swept upwards to an erect position, in varying shades of pink, stained reddish-purple at the base around the mouth of the corolla, sometimes white (commonly so in cultivation but not in the wild), obovate-elliptic with obtuse or subacute tips, c. 1.5–2.3 cm long, 0.7–1 cm wide, twisted lengthways, with very prominent swollen appendages ('auricles') at the base; style held within the mouth of the corolla or slightly exserted. 2n = 34 (subsp. *hederifolium*), 68 (subsp. *crassifolium*).

Flowering period August to October.

Distribution *Cyclamen hederifolium* is a widespread species with an extensive area of distribution in southern France, southern Switzerland, Italy, Corsica, Sardinia, Sicily, Croatia, Montenegro, Bosnia, Serbia, Macedonia, Albania, Bulgaria, Greece, Crete, Aegean Islands, and western and south-western Turkey; subsp. *hederifolium* — as for the species with the possible exception of the southern Peloponnese; subsp. *crassifolium* — Southern Greece (Peloponnese), the exact distribution for subsp. *crassifolium* has not yet been determined but may include southern Sicily and the Greek islands of Corfu, Cephalonia, Cos, Kythera, Lefkada, Zakinthos and Poros. The Cyclamen Society is carrying out a programme of fieldwork, particularly on the Greek islands, to obtain as much information as possible, and hopefully the collaborative studies with the University of Reading will shed some light.

Habitat Light woodland, usually under deciduous trees but sometimes pines, or in rocky situations in scrub but always favouring shady positions, sea level to 1,300 m.

Variations It is almost superfluous to stress the extreme variability of *C. hederifolium*, the fact is so well known and enjoyed by gardeners and there is almost no end to the possibilities for those who like collecting different forms. Whenever a new batch of seedlings is raised, something new and unexpected might appear. Suffice to say that the leaves can vary from the standard ivy shape to narrowly arrow-shaped at one extreme and to broadly angular, sometimes with coarse lobes, at the other. Plants with lanceolate or sagittate leaves were given the name *C. neapoletanum* Ten. var. *poli* by Delle Chiaie in 1824 and *C. neapolitanum* var. *hastatum* by Tenore in 1830. In an article in 2002, Angelo Porcelli makes the combination *C. hederifolium* var. *poli* and notes that this variant is scattered throughout several provinces of central and southern Italy.

The patterns on the upper surface often take the form of a central green triangular or jagged 'Christmas-tree' shape, surrounded by pale green or silvery zones, but sometimes it is the central area that is silvery to nearly white; almost any combination is possible, from wholly pewter-coloured or grey to wholly deep green, although the latter is not commonly seen — maybe it is considered less desirable and therefore not so frequently selected as a parent from which to collect seeds. A curious form with primrose yellow leaves was reported by Mike Bayliss, but presumably the lack of chlorophyll would result in a lack of vigour.

The flower colour too can be anything from pale pink to deep purple-red and the white forms either pure white or white with a pink stain around the mouth of the corolla. The overall corolla size and the size and shape of the individual corolla lobes or petals are also remarkably variable. Some forms are scented, although there does seem to be a tendency for Italian plants to be unscented; fragrance has been noted in Greece and in Turkey where the plants usually have a pleasant fragrance, reminiscent

Cyclamen hederifolium in the Kardamili Gorge. Populations in the Peloponnese are often referable to subsp. *crassifolium* but more research is needed (SB)

Variation in leaf markings from the collection of Marco ten Hoope (MH)

Wide variation in flower form from a population of *C. hederifolium* in the Peloponnese (BM)

of that of *C. purpurascens*. Records of fragrance are, however, few and in any case are very subjective unless tested and assessed scientifically (see Chapter 3, pages 202–206).

Some variants have been given specific names in the past but there is little evidence to suggest that these have any worthwhile botanical standing. For practical purposes, however, it is useful to have an all-embracing name for the white-flowered forms, as explained below. Some particularly dark-flowered forms collected by Melvyn Jope on the island of Corfu in 2007 have been given the name *C. hederifolium* Argirades Group. He has also recorded specimens at Leonidi in the Peloponnese that are very distinct with silver leaves that have a pink overtone; these may be referable to subsp. *crassifolium*.

Although *C. hederifolium* is immensely variable, and the variation is scattered rather randomly throughout its area of distribution, there are certain characteristics that do have some correlation with geography. This allows for the recognition of two subspecific taxa: subsp. *hederifolium* and subsp. *crassifolium*. The latter tends to have broad, coarsely lobed, rather fleshy leaves that are shiny on the upper surface, thicker pedicels and scented flowers; it has a chromosome number of $2n = 68$ ($2n = 34$ in subsp. *hederifolium*). Research into subsp. *crassifolium* is ongoing and its status may well change as its genetic and morphological distinctiveness is further investigated. Two variants of subsp. *hederifolium* at the rank of forma are recognised by C. Grey-Wilson (1997), f. *hederifolium* with pink flowers and f. *albiflorum* for those with white flowers that lack any pink coloration. The latter was based on the illustration in Jordan & Fourret's *Icones Floram Europae* 3: plate 417 (1903), first published as a species, *C. albiflorum*.

A fine plant of a pale form of *C. hederifolium*, probably subsp. *crassifolium*, at Harouda in the Peloponnese (SB)

With a more restricted distribution than subsp. *hederifolium* less variation has been noted for subsp. *crassifolium*. Nevertheless, a quite considerable range of foliage markings have already been selected for this subspecies and no doubt many more will follow as it becomes more widely cultivated. Some have an overall shiny-silvery almost metallic appearance; others have an almost white zone right in the centre of an otherwise dark leaf. Flower size and colour are in much the same range as for subsp. *hederifolium*, so when in flower there is little chance of distinguishing the two subspecies. One particularly large form of subsp. *crassifolium* was reported by Hilary Luker in *Cyclamen* 27 (1): 7 (2003). This was seen in October 2002 on the Peloponnese, growing amid ordinary-sized plants, and had flowers that were 5 cm long from the base of the auricles to the tips. Although the flower colour of subsp. *crassifolium* is normally pale with a darker pink-purple mouth, this too varies and a white-flowered form has been found. This has been described as *C. hederifolium* subsp. *crassifolium* forma *virgineum* B. Mathew in *Journal of Cyclamen Society* 33 (1): 14 (2009).

Named cultivars of *hederifolium* forma *hederifolium*

With such a variable species it has been very tempting for gardeners to select out favourites and give them cultivar names, a course that is acceptable only if the offspring turn out to be reasonably uniform. The following are some of the more notable:

'Antiochus': leaves attractively marbled; flowers deep pinkish-red.

'Apollo' ('Bowles's Apollo', Bowles's Apollo Group): the original plant named by Gerard Parker was found by him in the garden of E. A. Bowles at Myddelton House; the leaves were remarkable for the large shield-shaped area of silver in the centre and another around the margin, separated by a green zone; they also have an overlying flush of pink, although this is variable; flowers pink.

Cyclamen hederifolium 'Bowles' Apollo' with pink-tinged leaves (MD)

Cyclamen hederifolium 'Fairy Rings', unusual shaped leaves with striking markings (JB)

Cyclamen hederifolium 'Lysander' with its remarkably lobed leaves (JB)

Cyclamen hederifolium 'Rosenteppich' has one of the deepest colours of all cultivars (MD)

'Coquette': leaves green suffused pewter with an arrowhead green mark in the upper half of the blade; flowers white with a pinkish mouth and flushed lavender at the petal margins. The unfurled white buds have a delicate tracery of lavender which makes this cultivar easy to recognise; best regarded as a pale picotee type.

'Elsie Thomas': leaves attractively marked; flowers with broad petals, pink richly stained deeper at the mouth and spreading up the petals but with the margins almost white.

'Fairy Rings': leaves with pale green centre surrounded by a dark zone, then a pale band margined by a further dark green zone; the pattern is said to take five years to reveal itself fully; flowers mid-pink.

'Green Elf': leaves green, less than 3 cm long, with a central pale green hastate pattern; flowers pale pink, carried on slender stalks.

'Highfield': distinctively glossy foliage; pink-flowered; a selection made by Dave Hoskins of Ringwood.

'Joy's Choice': leaves two-toned silver and pewter, seven-angled; flowers pink, of normal form and size.

'Linnett Stargazer': a variant with flowers facing upwards rather than in the usual pendent position.

'Lysander': originated in the Peloponnese near Sparta; variable, with coarsely lobed leaves; flowers pink.

'Monstrosum': flowers very large with petals 28 mm long from mouth to tip and 25 mm wide, carried on very thick stalks (around 4 mm in diameter), looking almost as if they were fasciated; noted in Greece in October 1972 by Herbert and Molly Crook; possibly referable to subsp. *crassifolium*?

'Peter Moore': leaves silver-zoned in centre surrounded by dark green; flowers pink. Named by Jan Bravenboer.

'Rosenteppich': leaves rather oval in shape, less lobed than most forms of *C. hederifolium*; flowers variable, at best rich deep pink to reddish-purple, German origin.

'Ruby Glow': leaves deep green, rather sombre with a hint of black; they are not particularly well marked; flowers deep rich magenta or magenta-purple, particularly intense around the mouth, but suffused over much of the petals except for a rather paler margin.

'San Marino Silver': a selection with silvery leaves made by Ray Johnstone, presumed to be derived from plants or seeds originally collected near San Marino.

'Silver Cloud': leaves silvery, somewhat variable in shape from distinctly lobed to heart-shaped almost without lobes; the intensity of the silver or pewter is also variable; flowers pink.

'Silver Shield': leaves deep green with a pewter or paler green shield pattern in the centre; flowers pink.

'Stargazer': leaves unremarkable; flowers upward-facing, pale pink with a deeper mouth.

'Tilebarn Greville': leaves narrowly ovate, very dark green with a silver centre; flowers pink.

'Tilebarn Silver Arrow': leaves arrow-shaped, silvery-green; flowers pink with a darker purple-pink mouth.

Cyclamen hederifolium 'Silver Shield' (MD)

A fine pot of *C. hederifolium* forma *albiflorum* at a Cyclamen Society show (MD)

Cyclamen hederifolium forma *albiflorum* 'Tilebarn Shirley' (MD)

of *C. hederifolium* forma *albiflorum*

'Album': in Christopher Grey-Wilson's opinion, this name should be attached only to pure white-flowered forms, lacking any pink tinge at the mouth.

'Artemis': silver or pewter-coloured leaves; distinctive and attractive leaf markings with a double shield pattern in grey or pewter rather than silver; flowers white.

'Cotswold White': selected white-flowered plants from seeds originally brought from southern Greece by Bob Brown; leaves rather narrow, metallic green with an attractive pattern in mid-green.

'Daley Thompson': leaves unremarkable; flowers large, white, with the petals measuring up to 4 cm long.

'Discovery': ivy-like foliage in shades of green and grey-green; flowers pure white to 3.5 cm long; the first flowers to open are the largest, the later ones more normally sized.

'Ellen Corker': leaves rather average for the species; flowers white, suffused with salmon-pink around the mouth; as they age this pink coloration takes on a purplish hue; selected by Chris Clennett.

'Nettleton Silver': leaves silver or pewter-coloured, with a rounded, slightly toothed but not lobed margin; flowers pure white.

'Perlenteppich': dainty white flowers that can have pink in the throat of the corolla.

'Tilebarn Helena': leaves arrow-shaped, silvery-green; flowers white; selected by Helena Wiesner, named by Tile Barn Nursery.

'Tilebarn Shirley': leaves narrowly ovate, dark green with a central bright silver shield-shaped zone; flowers white.

'White Bowles's Apollo': silver or pewter-coloured leaves; flowers white.

'White Cloud': silver or pewter-coloured leaves varying in shape from almost entire and heart-shaped to markedly lobed; flowers white; selected by Phil Cornish.

Cyclamen hederifolium forma *albiflorum* 'Tilebarn Helena' (MD)

Further observations on *Cyclamen hederifolium*

With such a huge distribution, it is impossible to present a thorough survey of *C. hederifolium* in all its countries of origin, but we hope that the following selection of articles will provide interesting reading and a good overall view of the species. They are mostly extracted from contributions by enthusiasts and specialists to the Cyclamen Society's journal. Some of these writings ably demonstrate the valuable contribution to knowledge that observant people can make whilst enjoying themselves on holiday!

***Cyclamen hederifolium* on Corfu** This species grows in abundance on Corfu and it has been reported that subsp. *confusum* [i.e. now *C. confusum* — ed.] grows there, but the plants seen in October 2005 by **Richard Appleton** when he visited the island were, he considered, better assigned to *C. hederifolium* subsp. *hederifolium*. In the Cyclamen Society's journal of June 2006, he made the following observations:

> "We explored the hilly west coast of the island from Paleokastritsa in the north to Lake Korrission in the south and found *Cyclamen hederifolium* throughout this area, growing at all altitudes down to a few metres above sea level. The plants grow in a variety of situations including mixed woodland and scrub, roadside verges and olive groves. A favoured location is along the top of retaining walls in terraced vineyards, in which situation superb and venerable specimens can be seen right on the outskirts of Corfu town.
>
> Some plants grow in full sun but generally they favour areas of dappled shade, frequently in association with *Arisarum vulgare*. They were growing in a variety of substrates including terra rossa, leaf soil and moss. We found some growing out of holes in limestone rocks in the manner more usually associated with *C. graecum*. In most situations the number of seedling plants present indicated an encouraging level of regeneration. The flowers were in the main pale to mid pink but a small proportion was noticeably darker, approaching the colour of the cultivated 'Ruby Group'. Although we must have observed thousands of plants we did not see a single albino. All the flowers we sampled were scented. In exposed situations the pedicels rose more or less vertically presenting the flowers in a showy group. However in woodland where the tubers are deeply buried, the pedicels travel some distance horizontally before emerging.
>
> Aside from their sheer abundance, the most notable feature was the wide variety of leaf size, shape and markings observed. The majority of leaves are lobed, some deeply, and some quite prominently toothed. Colour and markings varied from plain deep green at one extreme to some strikingly banded patterns. The leaf surface is dull rather than shiny and the reverse generally showed a limited degree of purple colouration … . On a day trip to nearby Albania we saw cyclamen from the bus window but were unable to examine them more closely. It would be interesting to see if these plants show characteristics similar to the species on Corfu and to hear whether there are substantiated records for the occurrence of *C. hederifolium* subsp. *confusum* [now *C. confusum* — ed.] in this area."

More on Corfu In November 2007, a Cyclamen Society team consisting of **Melvyn Jope**, **Peter Moore** and **Pat Nicholls** visited Corfu with the intention of adding more information about the occurrence of *C. hederifolium* and/or *C. confusum* on the island, a question which is currently being addressed by researchers. Their account of the visit appears in the Society's journal of December 2007. During the time they were on Corfu, they encountered hundreds of thousands of plants and remarked on the lack of interesting leaf forms and flower colour. They did, however, find some of a unique red shade in one area and these are of course horticulturally interesting. Although populations that had larger leaves than usual for *C. hederifolium* (a characteristic of *C. confusum*) were seen in the south-west of the island, the team did not consider them to be morphologically similar to the Cretan plants of *C. confusum*. Nevertheless, specimens were brought back for cytological and molecular assessment by the University of Reading, so this is an ongoing study.

***Cyclamen hederifolium* on Zakinthos** On a family holiday to the Greek island of Zakinthos, **Pat Nicholls** saw some cyclamen in the ruins of the Venetian fort above Zakinthos town that he considered to be reminiscent of *C. confusum*. This would be consistent with the record of this species' presence on the island in Christopher Grey-Wilson's writings. The plants were growing under pines in quite dense shade, and although they were dormant, some idea of the size of the leaves was clear from the dry remains; they were much larger than normal, as indeed were the seed pods. Pat concedes that this may have been due to the habitat, but the plants were similar to those of *C. confusum* he has seen on Crete and others that he grows himself. It seems most likely, however, that these plants are referable to *C. hederifolium* subsp. *crassifolium*, of which a more detailed study is considered highly desirable, as stated by Alastair Culham and his colleagues (*Cyclamen* 33, 1: 14 (2009)):

> "The taxon we recognise at subspecific level (*Cyclamen hederifolium* subsp. *crassifolium*) is distinct predominantly for chromosome number and morphology. Although the extent and variety of the morphological differences [i.e. from *C. hederifolium* subsp. *hederifolium*] are not yet clear and the geographic boundaries of the distribution have not yet been defined, for practical purposes it is necessary to distinguish these plants for the benefit of botanists, nurseries and horticulturists. The overall characteristics of these plants coincide with Hildebrand's 1907 description of *C. crassifolium* but are not, at least at this stage, considered sufficiently distinct from those of *C. hederifolium* Aiton to warrant specific status."

***Cyclamen hederifolium* fieldwork on Zakinthos** With the knowledge gained by both **Pat Nicholls** and **Melvyn Jope** on separate family holidays to Zakinthos, the Cyclamen Society carried out a field study on the island in November 2008. In view of their previous experience, the team was made up of Pat and Melvyn, plus Chris Gifford.

The aim, as with the visit to Corfu the previous year, was to ascertain which of the *C. hederifolium* group is present on the island: possibly subsp. *crassifolium*. Interestingly, in contrast to the cyclamen plants seen on Corfu, the plants on Zakinthos showed great variability:

> "We saw a vast array of leaf shapes, sizes and patterns throughout the island and the variation surprised us all. Arrowhead leaves were present in good numbers and there were some extreme shapes amongst them. There were plants with very small leaves and plants with extremely large leaves. Some of the larger ones resembled *C. hederifolium* subsp. *crassifolium* and were collected for research, including DNA analysis at Reading University." [Note: the Society always has the appropriate documentation for such work – ed.]

During the course of the fieldwork, the team made voucher herbarium specimens and took numerous measurements of leaves and flowers for statistical analysis; they also took soil samples and made habitat notes. The Cyclamen Society plans to publish a full report on the series of field studies when they have been completed, hopefully clarifying the relationships between *C. hederifolium* subsp. *hederifolium*, subsp. *crassifolium* and *C. confusum*.

Cyclamen hederifolium is sometimes found in more open rocky places although usually on cool slopes. Here on Zakinthos Island, Greece, they are probably referable to subsp. *crassifolium* (MELJ)

Remarkably floriferous plants of *C. hederifolium* on Zakinthos Island (MELJ)

A striking leaf form of *C. hederifolium* on Zakinthos Island, probably referable to subsp. *crassifolium* (MELJ)

Another extraordinary form of *C. hederifolium* on Zakinthos Island, from Site 08/09G (Plant number CSE08020G) at Anafrontiria, 310 m (MELJ)

***Cyclamen hederifolium* in Crete** The occurrence of *Cyclamen hederifolium* on the island of Crete, although not surprising, is a relatively recent discovery. The stated distribution of the species in Christopher Grey-Wilson's *The Genus Cyclamen* (1988) did not include Crete, but by the time of his next monograph (1997), the plant we now know as *C. confusum* had been found and was described in the latter work as *C. hederifolium* var. *confusum*. **Melvyn Jope** has studied the populations in Crete and shown that *C. hederifolium* also occurs there. He writes (in *Cyclamen*, June 2000) that he observed some plants a few kilometres south of Topólia that had leaves which were very different from those of the *C.* [*hederifolium* var.] *confusum* growing at Topólia. This he called the Panorama site, after the taverna that is just below the site. The distinctive feature of the plants was that some of them were just in flower in late February, while others had well-formed seed pods. Also, the leaves lacked the distinctive substance and glossy sheen [of *C. confusum*], and the leaf pattern and appearance was more akin to *C. hederifolium* than to the plants at Topólia and Polirinía (both are sites for *C. confusum*). Melvyn has since confirmed this identification and provided a little more information:

"The plants that I found at the Panorama site are *C. hederifolium* ssp. *hederifolium*. I recall very clearly sitting in my car looking at the site expecting to see *C. creticum* having seen it there before. I then noticed what was clearly *C. hederifolium* and getting out of the car found several plants. I had never seen *C. hederifolium* ssp. *hederifolium* on Crete before (and I think that no-one else had recorded it there). On a subsequent visit Peter [Moore] and I found that the population grew over a much larger area than originally thought."

Further comments about *C. hederifolium* in Crete can be found in the 'Further observations' section under *C. confusum* as the two species occur in the same region and occasionally together, as described by David and Helen Sissons on page 150.

Cyclamen on the island of Naxos Most of the almost countless Greek islands no doubt have *Cyclamen hederifolium* on their flora lists. Naxos is among the larger ones in the Cyclades group, about 50 km long and with mountains up to 1,000 m altitude. In October 2005, **Palle Christensen** visited the island and made the following useful notes, which were first presented in the Cyclamen Society's journal of June 2006:

"The centre of the island is fertile and used for agriculture. The west coast is dry, in many places garigue-like, while the eastern side is more humid with shrubs and woods .... In all we found cyclamen flowering in seven places. At Mount Zas and by the Kouros of Apollonas we saw them immediately. In the roadside near Coronidas, cyclamen were growing right up to the bitumen layer. On the northwest coast, the main road runs halfway up the mountainside, and numerous ravines cross below the road. We knew where to look, and at four stops at such places we easily spotted small colonies of cyclamen. I was most inclined to believe that all these plants were *Cyclamen graecum*, but I had to revise this assumption. First of all, all seven places had fairly damp soil, and they were often overshadowed with light shrubbery, which is not the most beloved habitat of *C. graecum*. All the findings were higher up than 200 m (660 ft). There were still relatively few well developed leaves, so only the flowers were available for identification. The flowers, however, had blotches and lines above the auricles very typical of *C. hederifolium* subsp. *hederifolium*, although the pale bluish-pink flowers on short stems were very different from those of the *C. hederifolium* in my home garden in Denmark. In spite of careful search in dry, sunny places we did not identify any *C. graecum*, and any member of the *repandum* group would hardly be visible at the time of our visit."

***Cyclamen hederifolium* on Samos**. Writing in Cyclamen June 1996, **John Reeve** extols the virtues of Samos as a holiday island, describing it as green and picturesque, and for cyclamaniacs a chance to find plenty of the plants growing 'at home'. He writes:

"I spent two weeks in October 1995 on the Greek island of Samos, just off the Turkish coast. The walk from Vathi to the monastery of Ag. Zoni was particularly rewarding in cyclamen terms, with many *C. hederifolium* in flower on the steep banks and retaining walls by the road. They covered the whole spectrum of pinks from nearly white to really dark. I found some particularly good forms growing in a deep litter of pine needles not far from the monastery. At the monastery itself, lots of cyclamen grew in the grounds around the church, one huge tuber erupting through the cobbles of the courtyard with dozens of blooms. There were many plants too in coastal pinewoods lining the Samos–Kokkarion road. The grounds of a long-demolished house, marked by its gate piers, steps and wrought-iron gate complete with rusted chain and padlock, were particularly rewarding, with numerous plants in flower coming through the pine needle litter and rubble. I found no cyclamen around Platanos, where there is a good covering of mixed woodland, until the village of Ydrusa: here they were abundant in field boundary walls and path verges, often growing in great numbers in bare earth at the foot of walls. The only place between Mytilini and Chora where I found cyclamen in bloom was by a small chapel with one lone pine tree. In the extensive pine woods between Samos, Kamara and Zoodochu Pigis I found no cyclamen apart from some fine specimens at the monastery of Zoodochu Pigis itself."

***Cyclamen hederifolium* on Lesbos** The following year **John Reeve** visited Lesbos, the third largest island of Greece, and found *C. hederifolium* near the village of Skala Sikaminea very near to the beach:

"The flowers covered a wide range of pinks and of petal shapes, and were scented … . I found one white-flowered plant — not a pure album as it had a cerise nose, but a joy all the same. Another site for *C. hederifolium* was just outside the seaside resort of Petra … . Here most of the plants were on the sides of the steep-sided dried-up water courses that dominate the area — again a wide variety of colours and of petal forms, and again scented."

***Cyclamen hederifolium* on Thassos** From this report by **Pattie Peck**, extracted from *Cyclamen* June 1998, it is clear that *C. hederifolium* occurs on Thassos, the most northerly of the inhabited Greek islands in the Aegean with mountains up to 1,200 m altitude. She describes Thassos as being a very beautiful island about 93 km around the coast, with spectacular scenery and hairpin bends with hidden coves and wooded denes. She writes:

"I had hoped to see cyclamen but the burnt hillsides were full of typical garigue with white and pink cistus among sage, marjoram and thyme. The smell was lovely. Many of the roadsides were covered in the lovely blue flowers of chicory and yellow *Chrysanthemum coronarium*. One day, we drove down to an empty beach we had seen on our travels, with a deserted and abandoned hotel, its garden overgrown and neglected. It was surrounded by pine woods, and a small grassy area bordered the fine sandy beach … . Where the bay curved round, the trees came right down to some huge rocks, and I wondered if there might be some cyclamen growing among them. I looked up for handholds as I scrambled off the beach, and I found myself surrounded by cyclamen foliage and deep pine litter with brambles and broken branches all knitted in together … . The leaves were quite a sight as they were of every shape, size and pattern — apparently *C. hederifolium*. They varied from round to spear-like, and ivy-shaped to one plant with frilly edges; size varied from 4 cm to 13 cm. They were beginning to go yellow [this was May 1997]. There was little evidence of flowering. A few plants had at the most two seed pods, not yet quite ripe, and just one or two had babies above them. I would estimate the tubers to be around 30 cm (12 in) below the surface of the leaf litter."

Leaf studies of *Cyclamen hederifolium* in Sicily In 1996, **Martyn Denney** visited Sicily as part of the Cyclamen Society's programme to study the ivy-leaved cyclamen and its variants and relatives (*hederifolium*, *confusum* and *crassifolium*), a project also involving the University of Reading. There have been several such field forays mainly to Greece, Crete and other Greek islands, and to date it has been established that the variant from Crete known as *confusum* is worthy of recognition as a separate species (see pages 145–152).

Encouraged by the fact that there seems to be little detail known about *C. hederifolium* in Sicily, Denney decided to visit in late November when the plants would be in full leaf, as the morphology separating the various taxa is most obvious at that stage rather than at peak flowering time. The following is an extract from his full report which can be seen in the Cyclamen Society's journal of December 2006.

"Along the northern, Tyrrhenian coast of Sicily a series of limestone mountains form an almost unbroken chain from Monte Peloritani near Messina in the east, through the Nebrodi Mountains to the Madonie Mountains in the west towards Palermo. First, the area around Floresta in the extreme east of the Nebrodi to the north of Randazzo and around the headwaters of the Fiume Alcantara was explored. Here, the upper slopes of the hills are clothed in mixed oak, beech and chestnut woodland but in spite of some promising habitats there were no cyclamen to be seen and perhaps the area was too dry. A few kilometres west of Randazzo on the other side of Monte Colla is the valley of another small river, the Fiume Flascio. An area of the valley is enclosed as the 'Demanio Belle Caronie' but, following the track around to the east of the park, one passes through an area of mixed oak and juniper woodland; *C. hederifolium* subsp. *hederifolium* is to be found in many patches where there is reasonable shade. Here again there were a number of plants with acuminate leaves, but no flowers. The Portella Femmina Morta, at 1,524 m, is the highest road pass in the Nebrodi and has four particular advantages. First, nearby is the Villa Miraglia, a mountain refuge-hotel with excellent food; second, there is a small forest road that climbs still higher to the summit of Monte Soro; third, it is at the start of a wonderful walk through beech woods to the Lago Biviere di Cesaro, and fourth, the surrounding woods are full of *C. hederifolium*. The cyclamen grows principally around the base of the multi-stemmed beech trees. In the early morning with the sun just beginning to thaw the overnight frost, even standard-looking leaves appeared shiny. However as they dried, it was apparent that in many the shine was a natural feature whether the leaf itself was of arrowhead shape or resembled the leaves of *confusum* [or subsp. *crassifolium*?] as found in Greece. As the road climbs, the beech gives way to oak, and the cyclamen appear less frequently until, as the oak becomes more open towards the summit, the plants disappear.

A little way along the Monte Soro road, a forest track diverges to the north-east and drops down behind Monte Soro through beech and oak woods to the Lago Maulazzo. The path then continues east and passes through tall beech forest before arriving at the Lago Biviere di Cesaro, a natural lake with Etna in the background to the southeast. The beech forest with occasional holly bushes is occasionally carpeted with cyclamen and in other places they surround the base of the trees and holly bushes. From the Lago Biviere the path continues to the ancient beech wood of 'Bosco di Mangalaviti'.

To the west a series of mountains form two ridges separating the Nebrodi from the Madonie Mountains. Notable amongst these is Monte Sambughetti in the 'Riserva Naturale Orientata Sambughetti — Campanito'. A forest road on its northern slopes winds through oak forest and has extensive populations of *C. hederifolium* which even in late November were still in flower. There were fewer remarkable leaves here, though a scattering were acuminate.

The Madonie Mountains cover a smaller area than the Nebrodi, but contain Sicily's highest non-volcanic peak, the 1,980 m Pizzo Carbonara. *Cyclamen hederifolium* is found essentially throughout the Madonie. On the western side it can be found from 780 m just north of Piano Torre, to Piano Battaglia in the central Madonie at 1,580 m, around the peak of P. Carbonara and south again at 1,000 m near Polizzi Generosa. It is found in various habitats consisting of either pure beech forest or mixed oak, *Sorbus*, *Pinus*, *Crataegus* and *Ilex aquifolium*. Populations here include acuminate leaf forms with both darker patterns and more silvered leaves but it is my impression that the proportion of these is lower. The Madonie Mountains have spectacular scenery and autumn colours, are excellent for walking with well-marked and mapped paths (Alaimo, 2002) which pass through protected habitats containing rare plants such as *Abies nebrodensis*, and have mountain refuges which enable one to stay amongst the peaks.

Cyclamen hederifolium in Sicily with oak leaf mulch. Site 06/08 on Monte Sambughetti, 1,229 m, 21 November (MD)

Deciduous woodland in Sicily on Monte Sambughetti habitat for a large population of *C. hederifolium* (MD)

Near the north coast resort of Cefalu, but still within the Regional Park of the Madonie, is the Pizza Sant' Angelo. This 1,080 m peak stands almost on its own overlooking the Tyrrhenian Sea. A forest road meanders up the mountain through oak woodland to a geophysical observatory and here I found good populations of *C. hederifolium* still in flower. The leaves here were of a more standard subsp. *hederifolium* type yet many still exhibited the finely dentate margin and shiny lamina surface associated with *confusum* and *crassifolium*.

For comparison purposes I spent a day looking for *C. hederifolium* on Monte Iblei in the extreme southeast of Sicily. This is a region of spectacular limestone canyons and a flora to match. Even in late November there were healthy populations of autumn-flowering crocus and *Iris planifolia* flowering by the roadside. Here, at an altitude of 670 m *C. hederifolium* subsp. *hederifolium* was still in flower. Most of the leaves were matt on the upper surface and the flowers (based on a relatively small sample) were more rounded with corolla lobes which were broader and less twisted than in the north of the island.

So where does this leave us? In my view there is no doubt that the plants growing on the northern slopes of Mount Etna, in the eastern Nebrodi and at the site on Mount Iblei in the extreme southeast of the island are subsp. *hederifolium* but elsewhere the situation is more interesting. On Sicily we have plants that superficially have many of the characteristics of *confusum* (or Hildebrand's *crassifolium*) but without the same thickness or fleshiness of the leaf lamina. A consideration of the sepals shows that their shape fits Hildebrand's prescription. Also they have entire margins, a character used by both Grey-Wilson, when describing *C. hederifolium* subsp. *confusum*, and Hildebrand, in discussing his *crassifolium*. This feature was used to distinguish their taxa from subsp. *hederifolium*.

The collections made, consisting of leaves in silica for DNA extraction, living plants for root tip collection for chromosome counts and voucher herbarium specimens, will provide material for further study that may help to answer some of these questions, or pose further new ones."

***Cyclamen hederifolium* in the Abruzzi** Italy abounds in *Cyclamen hederifolium* and this is most likely the country of origin of the bulk of the plants in gardens: the very old cultivated and naturalised stock that was known for a long time as *C. neapolitanum*. Doubtless a whole book could be written about *C. hederifolium* in Italy. However, we have chosen just one article on the subject, written by the late **Graham Simpson** who for many years headed the 'panel of experts' of the Cyclamen Society. Writing in *Cyclamen*, December 1998, he describes a journey made in the Abruzzi region:

"The Abruzzi mountains, about 75 km west, or slightly north-west, of Rome, form the central part of the Apennines which stretch north to south as a sort of backbone to the Italian peninsula … . It soon became apparent that the area has abundant supplies of *Cyclamen hederifolium* which are almost everywhere. In early October, they were at the peak of their flowering period. The leaf patterns were very variable but the flowers were all of a good rich pink colour … . In the Abruzzi, the sites for *C. hederifolium* are somewhat unusual in that, to accommodate themselves in the situations in which they find themselves, they have had to get accustomed to growing under deciduous trees, but on long and quite narrow terraces below vertical cliffs.

At Lake Scanno, around the junction with the side road that leads to Sulmona, we found a good colony only about 10 m away from the car park to the two lake-side restaurants. We photographed two plants, each with over 100 flowers and with exposed tubers which revealed themselves as having a diameter of over 25 cm (10 in).

A largish colony of cyclamen was spotted by the roadside about 10 km north of Scanno, on the road to Anversa degli Abruzzi … . After the Abruzzi, we moved to the Sorrento Peninsula, to the village of Massalubrense. All around here *C. hederifolium* was present, as I had expected from a previous, inopportune, visit in 1944. We even saw it on the Island of Capri … in the garden of the famous Villa San Michele."

Cyclamen hunting in Bulgaria In eastern Europe, *C. hederifolium* appears to reach its most north-eastern localities in Bulgaria, where there is still a considerable element of the Mediterranean flora, at least in the more southern and eastern parts of the country. **József Szendefi**, writing in *Cyclamen* June 2002 describes how he travelled from Hungary through Romania and Bulgaria in early March 2001, searching for cyclamen. His comments about *C. coum* are to be found under the 'Further observations' section of that species (page 56) but he also found *C. hederifolium* in Bulgaria:

"I travelled first to the Danube delta then followed southwards along the Black Sea to the Turkish border, then towards Rodope mountains, then northwards crossing Stara Zagora, Veliko Tarnovo and Ruse. At Ruse I crossed the River Danube continuing my way to Bucharest, then Craiova, Turnu Severin, Timisoara, Arad and Hungary …. The first cyclamen I found was southwards from Burgas [Bulgaria] some kilometres from the border of the National Park. There, finally, I found a population of *C. coum* ssp. *coum* which must be similar to the ones found in Turkey … . Following my way near the Greek border I was stopped by police, and after an hour waiting was informed I was not allowed to continue on my way … and I had to change direction. Fortunately, this allowed me to find a *Cyclamen hederifolium* site with very interesting narrow-leafed and almost completely triangular-leafed plants. (Until then I had found wide-shape leafed *C. hederifolium* plants only.) The wood was an oak forest with sparsely grown trees, and on the other side of the road only a few trees but oak bushes, perhaps because of previous fire or tree-cutting. Sheltered under the bushes I could see older plants with large tubers. The next place where I found *C. hederifolium* was on the north side of Stara Zagora along the road. If you cross the railway and the brook and go up on the hillside you can find a *C. hederifolium* habitat which I would recommend to everybody who is passionate about this species. A great number of plants can be seen with an enormous variety in leaf shape and pattern. From that point on I did not find more cyclamen plants, but it is probably due to flatland … . Arriving in Romania, I visited Bucharest and some other cities, stopping at different times in woody parts, but I found no more cyclamen plants until I reached my home garden."

A site for *C. hederifolium* in western Turkey: between Bademli and Biçakçi (BM)

***Cyclamen hederifolium* in Turkey** During a Cyclamen Society survey of *C. mirabile* in Turkey carried out in the autumns of 2003 and 2004, *C. hederifolium* was inevitably encountered on several occasions as the two species have an overlapping distribution, although they were not observed growing together. The *C. hederifolium* populations seen in western Turkey were fairly 'ordinary' in leaf characteristics, if it is possible to use that word in the context of this species, except that the flowers were mostly scented and on the whole pale, with strikingly contrasting carmine-purple stripes at the mouth — 'pyjama-striped' was the comment from one of the team members. The following observations relating to *C. hederifolium* have been compiled by Brian Mathew from notes made during the *C. mirabile* fieldwork. A more extensive account of these two excursions is to be found under *C. mirabile* on pages 286–295.

"In 2003 the journey to '*C. mirabile* country' started with the ferry from Istanbul across the Sea of Marmara to Bandirma, then to Gönen, Edremit and Izmir. On the way, populations of *C. hederifolium* were observed, some of these known to be in sites where commercial collecting takes place; although this practice is decades old, the cyclamen were still common and regenerating. They occurred in varied habitats, sometimes in humus-rich deciduous oak woodland (*Quercus infectoria*?) with *Primula vulgaris* and *Helleborus orientalis*, sometimes in mature beech (*Fagus orientalis*) forest.

Cyclamen hederifolium between Bademli and Bicakci, October. The plants from western Turkey have deliciously fragrant flowers (BM)

These species are indicators of the Euro-Siberian (i.e. Black Sea coast) element whereas further south, heading more towards the Izmir area and the Aegean, the *C. hederifolium* habitat became Mediterranean in character with *Pinus brutia*, *Platanus orientalis*, evergreen oak (*Q. coccifera*), *Pistacia lentiscus* and *Cistus* spp. All the populations seen had the appearance of being fairly unremarkable *C. hederifolium*: quite variable but with nothing particularly of note except for the fragrance.

The following year, at the end of October and in early November, starting out from Izmir, *C. hederifolium* was observed again in a surprisingly dry habitat east of Tire, growing in the boles of evergreen oaks. At a contrasting site, they were found under large plane trees in rich leafmould by a small stream. In the mountains north of Aydin, the habitats in which *C. hederifolium* was observed were highly variable: at one site, this cyclamen was found in hot, dry *Quercus coccifera* scrub, whereas in another, it was thriving at the foot of a tufa cliff with water constantly seeping down; here there were fine drifts of maidenhair fern, *Adiantum*. The habitat in which *C. hederifolium* was found most consistently, however, was mixed rather dry scrub, over an altitude range of 135–795 m. In the valley west of Aydin, north of Inçirlova, a dam was under construction so the populations above the site are inevitably doomed. Nevertheless, *C. hederifolium* is a plentiful species in this whole region and on the whole is not considered to be seriously threatened. Further exploration south-east of Aydin and south-west of Nazilli, on the south side of the Büyük Menderes river, added more populations of both *C. hederifolium* and *C. mirabile* (although never together). Here, the flowers of *C. hederifolium* were again nicely fragrant and conspicuously bicoloured with a pink-purple striped nose on a pale ground."

Arbutus forest on the west coast of the Datça Peninsula, south-western Turkey, a habitat for unusual populations of *C. hederifolium* (BM)

Cyclamen hederifolium on Turkey's Marmaris-Datça Peninsula

The extreme south-west of Anatolia and its peninsulas extending out towards the Aegean Islands were visited by a Cyclamen Society team in the spring months of 1997–1999, with the task of mapping the distribution of and studying *C. alpinum* (for an account of this fieldwork see under that species, pages 258–269). *Cyclamen hederifolium* was also encountered and the following notes by Brian Mathew indicate that this would be an area worthy of critical study in the autumn:

"Comments have been made about *C. hederifolium* in western Turkey, where the plants look fairly typical of the species, if it possible to make such a sweeping statement about such a variable plant! There is one area, however, where the plants of this species appear rather distinct: the long, narrow peninsula south-west of Marmaris. Here, *C. hederifolium* occurs at low altitudes (most populations seen were below c. 400 m, down to near sea level) in mixed woods and scrub with *Pinus brutia*, *Pistacia*, *Styrax*, *Myrtus* and *Smilax*, sometimes beneath *Liquidambar orientalis* trees. In one area on the northern coast of the Datça peninsula, it was plentiful in thickly wooded hillsides where old trees of *Arbutus unedo* and *A. andrachne* formed a dense canopy. The overall impression of the *C. hederifolium* plants in this region is that they have large, fleshy somewhat shiny leaves that are conspicuously 3–5 lobed, thus morphologically rather more like those of the Cretan *C. confusum* or Peloponnesian *C. hederifolium* subsp. *crassifolium*. Unfortunately, the plants from this area have not been studied in detail and there are no chromosome counts or DNA analyses available for them, so it is impossible to make any meaningful statement about the relationship with Cretan or southern Greek plants. Needless to say, such information is badly needed in order to make further progress in understanding this widespread and variable species."

Cyclamen hederifolium on the Datça Peninsula has thick glossy leaves, reminiscent of *C. confusum* (BM)

Cyclamen hederifolium near Marmaris, south-western Turkey (BM)

Cyclamen confusum

Flowering in autumn

Cyclamen confusum on a mossy bank near Kasteli (DS)

In 2009, cyclamen enthusiasts must have been excited to hear the news that a new species from Crete had been identified, until perhaps they heard that it was one they had known about for some time under another name: *C. hederifolium*! Nevertheless, the discovery of this plant was a noteworthy development and the result of some careful collaborative work by members of the Cyclamen Society and researchers from the University of Reading.

Superficially, *C. confusum* — an apt name — is rather like *C. hederifolium* in overall appearance, although there are some notable differences and more detailed investigations have shown that this Cretan plant is genetically distinct from the more widespread species. Christopher Grey-Wilson first distinguished this plant as a variety of *C. hederifolium* in 1997 but then elevated it to the rank of subspecies in 2002. Further field studies of the wild populations on Crete, carried out over the past ten years particularly by Melvyn Jope and Peter Moore, have provided much information. The Greek authorities allowed living samples to be collected for research at the University of Reading and the resulting molecular data indicate that further raising of the taxonomic status to that of species is justified. An extract from the 2009 paper by Culham, Denney, Jope and Moore, in which the case for recognising *C. confusum* is presented, appears below. In the same paper, Hildebrand's *C. crassifolium* is established as a subspecies of *C. hederifolium*.

In outward appearance, the leaves of *C. confusum* are often a brighter green than those of *C. hederifolium*, rather shiny, thicker in texture and more succulent. Although there is variability in leaf shape, the leaves of *C. confusum* tend to be larger than those of *C. hederifolium*, often as wide as they are long, and have several obvious angular lobes on the margin. The leaf and flower stalks

Cyclamen confusum. Painted by Christabel King (reproduced at 90% of original size)

Flowers of *C. confusum* and *C. hederifolium*; the former often has a wider corolla mouth than *C. hederifolium* (BM)

(petioles and pedicels) of *C. confusum* are also inclined to be rather thick. The flowers of *C. confusum*, although obviously similar to those of *C. hederifolium* in overall appearance, have a noticeably wide mouth to the corolla with very conspicuous swellings at the mouth (auricles) and they are fragrant.

Melvyn Jope notes that when this species is encountered *en masse* in Crete, the fragrance is one of the pleasures of cyclamen hunting on the island in the autumn. It has also been observed that *C. confusum* has a tendency to flower later than the general run of 'hederifoliums', which is a useful attribute from the garden viewpoint; it can sometimes flower into the new year. Thus, the two species can be distinguished on the basis of their leaf and flower characteristics; furthermore, the two are geographically separate, with *C. confusum* occurring in discrete populations on the island of Crete, although they do meet in some locations. The two species also have different chromosome numbers: *C. confusum* 2n = 102, *C. hederifolium* 2n = 34 or 68.

C. confusum (Grey-Wilson) Culham, Jope & Moore, *Journal of Cyclamen Society* 33 (1): 14 (2009). Type: Western Crete, Platanos to Polirinía, south of Kíssamos, *Grey-Wilson* s.n. (holotype K).

Synonyms

C. hederifolium var. *confusum* Grey-Wilson, *Cyclamen*: 174 (1997).

C. hederifolium subsp. *confusum* (Grey-Wilson) Grey-Wilson, *Cyclamen* (ed. 2): 207 (2002).

Description Similar to *C. hederifolium* but having thick, fleshy **leaves**, shiny on the upper surface and often prominently lobed; petioles often thicker. **Flowers** scented; corolla usually with wider mouth and with the swollen appendages ('auricles') even more conspicuous; **pedicels** usually thicker than those of *C. hederifolium*. 2n = 102.

Distribution Western Crete; the exact distribution of *C. confusum* has not yet been delimited but could also include extreme south-western Turkey (Marmaris and Datça Peninsulas). Reports of its occurrence on various Greek islands (for example Corfu, Cephalonia, Kos, Lefkada, Zakinthos and Poros) probably refer to *C. hederifolium* subsp. *crassifolium*, which is somewhat similar in outward appearance.

Flowering period October and November (and sometimes into the early part of the year).

Habitat Mixed coniferous (mostly *Pinus brutia*) and deciduous woodland, sometimes in rocky situations and olive groves, c. 250 m.

Variations As yet, nothing like the great array of variation in foliage shape and markings recorded in *C. hederifolium* has been recorded for *C. confusum*, but no doubt the range will increase through selection as the species becomes more widely cultivated.

Cyclamen confusum in full leaf showing the lush, thick-textured foliage of this species. Near Kastéli, Crete (DS)

Flowers of *C. confusum* from the site at Topólia (MELJ)

Cyclamen confusum near Topólia, Crete, under *Pinus brutia*. In a dry autumn the flowers will appear before the leaves (BM)

The flower colour is normally pale pink but darker forms are available in the nursery trade; these are derived from specimens that originated from Topólia or Polirinía in Crete but searches to date have failed to reveal the other extreme, a pure white form. Melvyn Jope notes that of the many plants in Crete that have been checked none has yet been found without a fragrance. Two plants collected during some Cyclamen Society fieldwork just south of Topólia have a particularly pronounced scent. A form with very large flowers has been selected by Ashwood Nurseries and will no doubt become available if a true-breeding stock can be raised.

Further observations on *Cyclamen confusum*

***Cyclamen confusum* in Crete** The few populations of *C. hederifolium* and *C. confusum* in a limited area of western Crete have been visited and studied regularly over the past two decades by several enthusiasts, including Melvyn Jope and Peter Moore and (in 2009) David and Helen Sissons. These visits have resulted in the accumulation of valuable information about the somewhat enigmatic *C. confusum*. The following interesting comments have been published before, but for convenience it is useful to bring them together here.

In the Cyclamen Society's journal of December 1998, Melvyn Jope writes:

"Two populations [of *C. confusum*] are known, both in the district of Kíssamos, one just south of the village of Topólia and the other south-west of the village of Polirinía. The site at Topólia has been known for some 25 years and has been regularly visited; the one at Polirinía rather less perhaps because it is just a little less accessible. During a visit to Crete in October 1998, Peter Moore and I decided to take a more detailed look at both colonies and to see if we could find any evidence of a plant population between these two villages which, as the crow flies, are only 5 km (3 miles) apart. Considerable time was spent in the area of the villages of Kalathenae and Sirikári but no further evidence of the plant was found.

The most striking impact of the Topólia colony on the warm October day when we visited the site was the fragrance — perhaps enhanced a little by the number of flowers of *Narcissus serotinus* in flower at the same time, a lovely visual combination when joined by large numbers of *Scilla autumnalis*. Quite different from the smell that greets you if you visit the site in May — *Dracunculus vulgaris* is established in a large clump on the other side of the road and the foetid smell of its flowers pervades the whole lower part of the site. Most of the plants at Topólia have flowers which are undoubtedly larger than regular *C. hederifolium* and flowers with really long petals are not uncommon. Growing conditions — mainly under *Pinus brutia* — are good and there is a lot of evidence of regeneration.

A deep-coloured *C. confusum* with long elegant petals from the populations at Topólia (MELJ)

The population at Polirinía had been thought to be rather small, with plants growing along the side of a ditch running south-west, and a few spreading out into the olive groves. Further investigation revealed that the colony extends into a wooded area on the north of the gully, and we concluded that the populations at both sides are about the same size. However, there was less evidence of regeneration at Polirinía. When looking at the range of depth of colour of petals on both sites, we found several plants with what would be regarded as darker flowers, as well as one or two with white petals (although not an albino form). It is quite obvious that the locals at Topólia enjoy the plant as much as we do because on several occasions we have been joined by Greeks coming to pick bunches. The fine display of flowers this year is evidence that the regular picking of flowers has had no adverse impact on the site."

Crete in February. Melvyn Jope, reporting in *Cyclamen* 24 (1):3 (2000) on a visit to western Crete from 20–27 February, noted that perhaps the most interesting discovery was that of thousands of *C. hederifolium*, including 20 or 30 plants still in flower and even in bud, a few kilometres south of Topólia [this new site was referred to as Panorama], among a mass of flowering *C. creticum*. To him, the plants seemed very different from and horticulturally more attractive than the glossy-leaved and plainish-green *C. confusum* growing at Topólia. This was of course very late for *C. hederifolium* to be flowering, but flowering may have been delayed by a wet and cold winter. Melvyn also noted that *C. confusum* was regenerating in countless thousands at Topólia and at Polirinía.

Substantial information was added in a subsequent article by **Peter Moore**, **Melvyn Jope** and **Nick Moore**, published in *Cyclamen* June 2001, where more sites for *C. confusum* on Crete were recorded as a result of visits made in autumn 1999 (MJ and PM), spring 2000 (MJ) and October 2000 (PM and NM). Three sites were visited: Topólia, Polirinía and Panorama.

"The plants were in flower but the leaves were not yet visible. The flowers at Panorama [i.e. *C. hederifolium*] were generally a very pale pink, whereas the plants at the other two sites [*C. confusum*] ranged from pale to very deep pink. The flower sizes of 20 randomly selected plants at Topólia and 20 plants at Panorama were nearly identical. The flowers at the Panorama site were less evident and there were many surface tubers with no signs of growth. Some 2 km further south of Panorama, there is a new bridge over the Tiflis. Below the bridge running north, a bulldozer has been at work clearing debris and rocks from the bed of the river and carving out a track along the west bank. *Cyclamen confusum* was to be found growing all along the track as far as the cave church of Agia Sophia. This is as far as the bulldozer has got so far; if they continue down the gorge, it could reveal an extension of the site. At the moment, the bottom of the gorge beyond this point is pretty well impassable. Melvyn also visited in late October and investigated the site further travelling south, and it became apparent that the site extends by several kilometres. The most southerly plants were by the roadside on the way to the mill just before the sign reading 'Welcome to Mithimna'.

We then decided to look for further sites between Topólia and Polirinía. This proved easier said than done, as we seemed to find a plethora of 'roads' which our map makers obviously had not! A lot of the destinations ended up at the bottom of gorges with no way through, or the top of mountains with no way over. Eventually we were successful and found a new site just above the village of Sinenianá. The plants were in flower but there were no signs of leaves as yet. The flowers were similar to the plants at Topólia and Polirinía, and an added bonus was a profusion of *Sternbergia sicula* growing amongst the cyclamen. The terrain and roads are not easy but we were convinced that the area still contains further sites yet to be found. The habitat was right, and it is just a question of finding the locations."

Melvyn wrote a further update to these observations in the journal *Cyclamen* in December 2003 as a result of a visit to Crete in the first week of October 2003:

"Anyone interested in cyclamen and able to visit Crete is likely to have visited the site at Topólia. I was very disappointed to learn from Vic Aspland that the area had been devastated by fire, which raised the question: was it still worth going? The answer is that even though the area covered by cyclamen is reduced to about a tenth of its former area it is emphatically worth visiting. The area to the right of the church steps seems to have survived the heat better than other areas and the few surviving plants to the left of the church seem to have had a good seed set from last year. Whether or not they will cope with the change in habitat, with a mallow species now being the dominant plant, is perhaps less certain. Young fir trees are also much in evidence so in 20 years or so maybe the site will be very different again. My next visit was to the site now known by those who have visited it as the Panorama site. The owner has turned the terraces near the entrance of the site into an enclosure for goats. This has produced a barren area that on the day I visited was like a dust bowl — I had to close my eyes at the slightest breeze. Having walked the length of the site, I found just six flowers, it just goes to show that with a week to spare you cannot guarantee getting the season right. Those catching the last charter into Chania at the end of October would probably fare better. My plan for the next day was to visit a site mentioned in correspondence by Peter Bielby. Following the Number 13 walk in the Sunflower Landscapes Western Crete book, Peter had noticed leaves of *C.* [*hederifolium* var.] *confusum* below the church at Sirikári. Peter Moore and I had not been aware of this site on our previous visits. The day started well, watching baby turtles going into the sea for the first time, aided by volunteers for the Hellenic Society for the Protection of Sea Turtles, seeing large numbers of what I take to be trout in the river at Gerani and watching kingfishers patrolling up and down the river. As on the other days of my visit the temperature was about 28°C (80°F) by early afternoon and quite warm for walking.

During previous trips Peter Moore and I had tried to find a link between the sites at Topólia and Polirinía without success, but a separate visit by Peter and Nick Moore resulted in *C.* [*hederifolium* var.] *confusum* being found at Sineniană. Even though Sirikári is not that far away it would help to fit another part of the jigsaw puzzle. I started on the downward walk immediately opposite the church gate and had descended for only about 10 minutes when I came across the plant in the walls of olive tree terraces, a variation on the habitat at the other two main sites. While looking at the plants I was suddenly surrounded by a large flock of sheep that seemed completely unperturbed by my presence. Once they had gone I found about six more plants; two of the tubers had very neat oblong pieces bitten out of them by the sheep — perhaps looking for anything with a reasonable moisture content. The gorge at Sirika goes to Kastéli but walking its entire length would have to wait for another day. Always good to have an excuse for a return visit.

The following day called for a visit to Polirinía. Those of us who have been to Crete many times can be very prone to saying it is not like it was — well Polirinía isn't. A beautiful old building on the road to the site which was full of holes for pigeons to nest in has been tastefully restored with most of the lovely stonework covered with cement render. A tarmacadam road also goes all the way to the packhorse bridge. Evidence of local cultivation of tomatoes etc. is much more widespread than in the past but not to the point where the main population of *C.* [*hederifolium* var.] *confusum* would be threatened. As for flowering plants, too early once again — just a dozen or so plants in flower."

More Cretan *Cyclamen confusum* Inspired by a previous autumnal visit to western Crete in 2007, **David and Helen Sissons** decided to venture back between the 8th and the 21st of November 2009, and as a result have added further to the knowledge of *C. confusum* in its natural locations:

"The aims of our return were to define the extent of known populations of *C. confusum* and *C. hederifolium*, to seek new populations, and maybe to discover a white-flowered *C. confusum*. We were soon to realise the benefits to be gained from setting off on our travels two weeks later than we had in 2007: not only were more cyclamen plants in leaf, populations were at or near their full extent of growth, enabling accurate observation and identification.

We based ourselves in Kastéli, which was the ideal place to stay as it is very close to known cyclamen sites in Kíssamos province. However, having picked up a Citroen C3 at Chaniá airport, we still managed to add several hundred kilometres to its odometer over the following fortnight.

One of the well-known *C. confusum* sites is the gorge which runs in a NNE direction from Sirikári to Polirinía. It is known as the Sirikári (or Tsichlianó) gorge. We decided

Site of *C. confusum* population in a gorge adjacent to the Sirikári Gorge, Crete (DS)

to explore the gorge upstream to the SSW from Sirikári towards Marouliană, but found no cyclamen beyond the junction with the road to Sineniană.

There is, immediately to the east of the Sirikári gorge, a large unnamed gorge which runs in a long bow from the village of Sirikári towards Polirinía. As far as we were aware, no one had looked for cyclamen here — with good reason. It is a deep gorge with uninvitingly precipitous walls, and the only way in was to make our way upstream along the river-bed from the packhorse bridge at its northern extremity. We managed to get about 1 km SSE along the gorge by climbing over boulders and squeezing under plane trees. It was well worth the effort. There were more *C. confusum* plants here than we had ever seen. Many had deeply coloured flowers and due to the low light levels, very large leaves. As far as we could tell, this population continued far into the gorge.

The Tsichlianó river then flows through a gorge to the east of Polirinía. We can confirm that *C. confusum* populations continue north through this gorge and into the Dháfni valley, following the river. Here we found it in the grounds of a monastery (signposted 'ΙΔΙΟΚΤΗΣΙΑ ΙΕΡΑΣ ΜΟΝΗΣ ΠΑΡΘΕΝΩ'), as well as near a small church signposted 'ΙΕΡΟCΝΑΟC ΧΡΙCΤΟΥ' (Church of Christ), but most significantly at a site just 700 m south of the junction of the road to Dháfni with the main coast road. This means that we can report that *C. confusum* grows as close as 1.5 km to the sea at Kastéli!

The grounds of the Church of the Panaghía at the southern end of the village of Topólia is another well-known site for *C. confusum*. We wanted to follow this

Cyclamen confusum growing wild in Crete, near Kastéli (DS)

population down to the floor of the Topólia gorge and to find its northern and southern limits. We found plants just north of the village of Kapsanianá, but no further north. We found that *C. confusum* extends south through the Topólia gorge for about half of its length, then stops. There is then a gap of some 450 m before *C. hederifolium* plants start to appear on the sides of the gorge. This population of *C. hederifolium* continues south along the southern third of the Topólia gorge to Katsomatádos. From here it grows along an eastern tributary into the hills towards Sásalos. We followed it south until it seemed to disappear just short of the village of Míli, some 2 km south of Katsomatádos.

During the fortnight, we examined many other gorges in Kíssamos province in search of new sites of *C. confusum* and *C. hederifolium*, but found none. These include the Mesavlia gorge, the Magaghístra gorge, the Milonés gorge and the gorges to the north and south of Sásalos. We also looked closely at the area around Plátanos in the far west. Here we found no evidence of *C. confusum* or *C. hederifolium*, nor did we find any gorges or river valleys that would provide typical growing conditions for such plants, as it seemed too dry with insufficient shade.

On the last day of your visit, when your intention is to potter and take some photographs, but your wife suggests something a little more strenuous — ignore her at your peril! After our visit to western Crete in 2007 we reported finding a new site of *C. hederifolium* in the Keramarianó gorge. We had not intended to re-visit this site, but having been disappointed in not finding cyclamen in several other gorges, I managed to persuade David to go back there — "Just to see how the *C. hederifolium* are doing." And they were indeed thriving. As we made our way up the gorge admiring the many familiar scented flowers and well-marked leaves of *C. hederifolium*, I noticed a somewhat darker flower growing on a large boulder to one side of the path. Closer inspection with binoculars revealed a plant with thick glossy leaves growing next to a *C. hederifolium* plant. On the north side of the boulder, in a damp hollow we found numerous *C. confusum* plants, all with typical thick glossy leaves, which were duly photographed. David also photographed two plants growing in rock pockets about 50 cm apart, one a *C. confusum* and the other a *C. hederifolium*. We understand that these two species have never until now been reported as growing in close proximity.

So, did we manage to achieve our aims? Yes and no. We have found several significant new sites for *C. confusum*, one of these demonstrating clear overlap with a site of *C. hederifolium*. We have defined the extent of known sites, but no, we have not found a white-flowered *C. confusum* — not yet!

We would like to thank Melvyn Jope for his valuable assistance in our preparation for this trip; Yiánnis Bikákis and his mother Artemis at the Family Bikákis hotel, Kastéli for making us feel so welcome; and Olga Korakáki and her mother Aretí at the Aretoúsa taverna, for keeping us well fed."

A new species of *Cyclamen* from Crete Extract from the paper by **Alastair Culham**, **Martyn Denney**, **Melvyn Jope** and **Peter Moore** in which the case for recognising *C. confusum* as a species distinct from *C. hederifolium* is presented. The full paper is published in the *Journal of the Cyclamen Society*, June 2009.

"*Morphology — the outward appearance* Within *C. hederifolium* there is variation in leaf thickness, petiole and pedicel thickness and other more cryptic variation that takes many measurements and statistical analysis to pick out. Our new species (from western Crete) is separated from typical *C. hederifolium* by generally thicker dark-green leaves with a glossy upper surface, a thicker petiole and a more robust and usually scented inflorescence.

Phenology — seasonal behaviour This *Cyclamen* taxon flowers slightly later than typical *C. hederifolium* Aiton continuing to flower further into the autumn. The flowers usually appear before the leaves, which at flowering time may be absent or very immature.

Chromosome number and breeding behaviour Cytological work carried out on samples from a very limited number of locations by Bennett and Grimshaw in 1991 and by Johnson and Boyd in 2001 showed that most *C. hederifolium* Aiton is diploid (2n=34) but suggests that plants from the Peloponnese, on Pentalikon and on various Mediterranean islands including Kythera and Zakynthos are tetraploid

(2n=68); samples of the new species from Crete have been found to have a higher chromosome number still (they are hexaploid, 2n=102). However, as Bennett and Grimshaw point out, Greek mainland populations were poorly represented in their study and a full survey of Greek plants would be very desirable. It is similarly desirable that the same plants be subject to a complementary study of their DNA sequence data. These differences in chromosome number immediately establish a breeding barrier.

DNA sequence data The use of DNA sequence data has begun to uncover what are often called 'cryptic' species in many groups of plants and animals. A preliminary survey shows the populations of our new species to have a small number of unique DNA sequence markers (consistently two DNA bases in one chloroplast DNA region). By contrast, the tetraploid variants do not appear to have unique DNA sequence in the regions we have sampled.

Distribution While some combinations of the characters given above are found in plants throughout the southern range of *Cyclamen hederifolium* Aiton the full combination including the DNA sequence markers are presently found only in western Crete, while the tetraploid plants are found in the other southern areas. *Cyclamen hederifolium* subsp. *hederifolium* also grows adjacent to, but not overlapping the population of the new species in western Crete and no hybrids have been found.

Discussion There are no hard and fast rules governing the definition of a species. However, the rules that are applied include any, from recognition by differing appearance, by breeding barriers, by geography and by DNA sequence. The taxon from western Crete which we consider deserves recognition at the species level (named *Cyclamen confusum*) is distinct for all of these factors."

Cyclamen africanum is a fine plant for a frost-free glasshouse or conservatory (BM)

Cyclamen africanum

FLOWERING IN AUTUMN

There has been much debate over the years as to how to tell the difference between *C. africanum* and its obviously close relative *C. hederifolium*, the most practical of which seems to be to plant them out in a frost-prone position: the survivors in spring will be *C. hederifolium*! Taxonomically this is not very helpful, so we must look for other pointers, one of which is the behaviour of the pedicels (flower stalks). On the whole, the pedicels of *C. hederifolium* are markedly inclined to grow horizontally away from the centre of the tuber and then turn upwards to emerge from the soil, whereas those of *C. africanum* usually grow more or less upwards directly from the tuber. The leaf stalks (petioles) of *C. hederifolium* also do this, and very markedly so, but in *C. africanum*, this is not such a distinct characteristic: in some plants the petioles are more or less horizontal, in others rather more erect; the superb illustration from Cosson's *Illustrationes Florae Atlanticae* shows this very well, as does Walter Hood Fitch's watercolour from *Curtis's Botanical Magazine*, although Fitch had apparently not dug up the tuber as the roots are shown coming from the centre of the underside rather than from all over its surface. Unfortunately, *C. africanum* and *C. hederifolium* hybridise, clouding any distinctions between them, so observations can really be used only with reference to plants that are derived from a reliable source. It can be said that *C. africanum* is usually a larger, more leafy-looking plant than *C. hederifolium*, but it is questionable as to whether anything along these lines could be quantified.

Cyclamen africanum. Painted by Christabel King (reproduced at 140% of original size)

Cyclamen africanum from Ernest Cosson's *Illustrationes Florae Atlanticae* VI: tab. 147 (1893). By permission of the Trustees of the Natural History Museum (London)

In his 2002 edition of *Cyclamen*, Christopher Grey-Wilson provides a lengthy discussion in which he explores various other characters that might be used as distinguishing features. Of note is the tuber shape: hollowed on the upper side in *C. africanum* but not remarkably so in *C. hederifolium*. The position of the roots is also considered worthy of comment: emerging from anywhere on the surface in *C. africanum* but confined to the sides and top in *C. hederifolium*. The flowers of the two species are very similar and it is doubtful that they could be used for identification, although Grey-Wilson comments that the anthers are yellow in *C. africanum* and reddish or purplish in *C. hederifolium*. If consistent, this would be a very neat way of checking the identity of a particular flowering plant, although there is the potential problem mentioned above of possible hybridisation.

C. africanum is not the most popular or sought-after of species, no doubt in part because of its lack of hardiness. Nevertheless, it is as attractive as *C. hederifolium*, although not exhibiting as much interesting and exciting variation. This might result simply from the fact that it is not as frequently cultivated, so there is a much more limited amount of material from which to select; or it may be that with its more restricted distribution in its wild state in Algeria and Tunisia, there is less genetic variation than in its much more widespread relative. In spite of this, *C. africanum* is certainly a worthwhile plant to cultivate in a frost-free glasshouse or conservatory. Growers should be encouraged to obtain a 'true' plant from a reputable source and to keep it separated from plants of *C. hederifolium* so that any offspring will not be contaminated.

Plate 5758 from *Curtis's Botanical Magazine* Vol. 95 (1869): *C. africanum* by W. H. Fitch (MD)

C. africanum Boiss. & Reut., *Pugill. Pl. Afr. Bot. Hisp.*: 75 (1852). Type: Algeria.

Synonyms

C. numidicum Glasau, *Planta* 30: 528 (1939).
C. ambiguum Schwarz, *Feddes Repert. Sp. Nov.* 58: 275 (1955).
C. commutatum Schwarz & Lepper, *Feddes Repert. Sp. Nov.* 69: 91 (1964).

Description Tuber depressed-globose and often with a concave depression on the upper surface, dark brown, up to 14 cm diameter at maturity, rough and cork-like on the surface, rooting from almost anywhere on the surface. **Petioles** erect or semi-erect from the tuber. **Leaves** developing at or just before flowering time. Leaf blade fleshy in texture, broadly ovate, 5–16 cm long, 4.5–18 cm wide, usually with a central ovate green pattern surrounded by a pale green or greyish marbled zone that may be continuous or broken into separate patches, outside of which is a green border around the edges; margin dentate, sometimes quite finely so; the underside normally green but may be purplish suffused. **Pedicels** arising more or less erect from the tuber, after anthesis coiling from the top downwards. **Flowers** produced in autumn, with a slight scent (?similar to that of violets); petals swept upwards to an erect position, usually pale to medium pink, each with a dark crimson-purple V-shaped basal stain forming a jagged ring around the mouth of the corolla, obovate to narrowly elliptic-oblanceolate with subacute or obtuse tips, c. 2–3.5 cm long, 0.7–1.2 cm wide, twisted lengthways, with prominent swollen appendages ('auricles') at the base; style held within the mouth of the corolla or slightly exserted. $2n = 34, 68$.

Flowering period September to November.

Distribution *Cyclamen africanum* occurs in North Africa, in northern Algeria and north-western Tunisia.

Habitat In scrub, rocky places and crevices on limestone formations, usually in light shade; sometimes in cedar forests.

Variations The leaf variation in *C. africanum* is not nearly as great as that of *C. hederifolium*, although there is considerable similarity between the two species; it would be quite easy to take a leaf of the former and match it almost exactly with one of the latter. There is normally a dark green roughly triangular area in the centre of the leaf, variable in size, and this extends along the main veins for a variable distance and sometimes out to the margins forming a 'Christmas-tree' pattern. The remainder of the leaf is usually paler green with little tendency to the silvery-greens so common in *C. hederifolium*. Usually, the leaves have numerous small cartilaginous teeth on the margin but this varies, certainly in cultivated plants. In flower colour, there is some variation in depth from pale pink to a deeper rose and there is a slightly variable darker reddish-purple V-shaped stain running back from the mouth of the corolla.

Tuber of *C. africanum*; note that the roots appear almost all over (BM)

Flower of *C. africanum* with fruiting pedicel starting to coil (BM)

Leaf variation in cultivated specimens of *C. africanum* (BM)

Further observations on *C. africanum*

Few published observations on *C. africanum* have been traced; maybe North Africa is not such a popular destination for plant enthusiasts in the autumn months, when they are often lured away by the delights of bulbous abundance in Greece and Turkey. **John Whitehead**, who has written about his search in 2007 for the other North African endemic cyclamen, *C. rohlfsianum* (page 192), had 25 years earlier encountered *C. africanum* in the Kabylie mountains of northern Algeria, "growing amongst rocky outcrops on open grassy slopes in the cedar forests".

The origin of *C. africanum* in North Africa no doubt accounts for its lack of hardiness in cultivation in colder climates and this is a point that most writers, when discussing its garden value, have commented upon. Its attractiveness was clearly in no doubt for the RHS committee who judged a plant shown by W. E. Th. Ingwersen Ltd. on 19 September 1967: they bestowed upon it the RHS Award of Merit. The knowledgeable cyclamen enthusiast Doris E. Saunders, writing an account of the species in 1968 in the *Bulletin of the Alpine Garden Society* following its award, suggests that it "might, perhaps, be tried in a very warm sheltered position where the soil never becomes waterlogged." Her astute observations on the characteristics of the species are also worth quoting:

> "The leaves are usually larger than those of *C. neapolitanum* [= *C. hederifolium*] with a leathery texture and polished surface and are very frequently rounded or kidney-shaped and dark sombre green. Brighter lobed leaves also occur and there is a very attractive two-tones grey form which has only recently arrived from Algeria. The margins of the leaves are edged with coarse horny teeth and the under surface is generally light green. The violet-scented flowers up to 3.5 cm long are large and handsome and vary in colour from pale to deep rose-pink, but there seems to be no record of a white form. At the base of each petal there is a wide crimson patch with a forked tip, and the rim of the corolla tube bears conspicuous auricles. The anthers are yellow with a purple groove on the back, not rust-coloured as in *C. neapolitanum*, and the stigma equals the corolla tube."

2.2h The Purpurascens group

(= Subgenus *Cyclamen*, Series 2)

Comprising: *Cyclamen purpurascens* and *C. colchicum*

The two late summer- or autumn-flowering species comprising this group are very similar in characteristics. They have leathery, rounded leaves without lobes on the margins and are more or less evergreen, the old leaves being replaced by new without a long dormancy period. The fragrant flowers have no or only a slight contrasting darker stain at the mouth of the corolla, which has no, or poorly developed, swellings ('auricles'). The stamens are not exserted from the corolla mouth but the style usually protrudes slightly. In fruit, the pedicels start to coil downwards from the apex. 2n = 34.

Cyclamen purpurascens

Flowering mid summer to autumn

Apart from being an attractive species and having an alluring fragrance, *C. purpurascens* is particularly valuable in that it helps to bridge the gap between the late spring-flowering and the early autumn-flowering cyclamen. In fact, some growers claim that they can have cyclamen in flower every month of the year and, if this is indeed so, it is this species that fills the void during the 'off' period when most cyclamen are at rest. Being a species of the European mountain woodlands (although it has also been found as low as sea level, see page 165), it is not subject to the excessive heat and drought of the Mediterranean region, so it does not have such a marked period of dormancy; in fact, it is more or less evergreen with the new leaves of autumn sometimes overlapping with those of the previous year. Only its close relative *C. colchicum* and the

Cyclamen purpurascens, a near-evergreen species flowering in mid-late summer (JožeB)

Cyclamen purpurascens. Painted by Christabel King (reproduced at 120% of original size)

Mt Triglav in Slovenia, a splendid backdrop for *C. purpurascens* (BM)

Turkish *C. parviflorum*, which is truly an alpine species, also behave in this way. Curiously, these three are not the easiest of species to cultivate in Britain despite being very hardy, and all of them are seen in collections much less frequently than popular species such as *C. hederifolium*, *C. coum*, *C. repandum* and *C. cilicium*.

Although *C. colchicum* is very similar at first sight, *C. purpurascens* is an easily recognised species, not only because of its early flowering time but also because the flowers are generally a warm rosy-pink or reddish-purple, a softer colour than the rather vibrant carmine shades of many of the other species. More significant is the fact that the flowers have no, or only very slight, swellings ('auricles') at the mouth of the corolla. A species that flowers in summer might be expected to produce ripe seeds earlier than other species but this is not the case; Melvyn Jope has noted that the capsules of plants that flowered in July were not ready for harvest until April or May the following year.

Not surprisingly, being European, *C. purpurascens* featured prominently in the early Herbals and has been popular with artists ever since; surely it must be one of the most frequently depicted of all the cyclamen species. Our example illustration by E. D. Smith is taken from Sweet's *British Flower Garden* (1826).

The Caucasian *C. colchicum* and *C. purpurascens* differ most noticeably in their leaves, those of the former having a thickened, finely toothed margin, which in close-up takes on the appearance of a row of minute beads. Further comments about the differences can be found on page 169.

Cyclamen purpurascens has been illustrated countless times over the centuries. This is from Robert Sweet's, *The British Flower Garden* t. 176 (1826); a copper plate engraving by Edwin Dalton Smith after a painting by Weddel. Two other cylamen were illustrated in this work: *C. vernum* (plate 9) and *C. repandum* (plate 117) (MD)

C. purpurascens Mill., *Gardeners' Dictionary* ed. 8, n. 2 (1768). Type: Based on specimens in the *Burser* and *Clifford* herbaria at the Natural History Museum, London (BM).

A dramatic silver-leaved form of *C. purpurascens* (MD)

Synonyms

C. europaeum L., *Species Plantarum* 1: 145 (1753) partly, an ambiguous name combining three species.
C. clusii Lindl., *Bot. Reg.* 12, t. 1013 (1826).
C. litorale Lindl., *Bot. Reg.* 32, t. 56 (1846).
C. europaeum var. *immaculatum* Hrab. in *Česk., Bot. Listy*. 3: 35 (1950).
C. purpurascens subsp. *immaculatum* (Hrab.) Halda & Soják in *Fol. Geobot. Phytotax.* 6(3): 322 (1971).
C. fatrense Halda & Soják, *Cas. Nar. Muz. (Prague)*, 140 (1–2): 64 (1971).

Description Tuber globose in the younger stages but becoming irregularly shaped and knobbly later, up to c. 4.5 cm diameter at maturity, brown, smooth at first, then corky; roots appearing from almost anywhere on its surface. **Petioles** usually more or less erect from the tuber but in the case of deep-seated tubers sometimes growing sideways a little before emerging. **Leaves** developing before flowering time, the previous season's can also still be present as the new ones appear. Leaf blade suborbicular or broadly ovate, 3–8 cm long, 2.5–8 cm wide, usually dull green variously marbled with pale green to silvery green zones, but enormously variable from plain green to an overall pewter colour; margin almost entire to shallowly toothed; the underside normally suffused deep reddish-purple. **Pedicels** arising more or less erect from the tuber, after anthesis coiling from the top downwards. **Flowers** produced in mid-late summer through to late autumn, with a strong fragrance; petals swept upwards to an erect position, usually pale to medium rosy pink or reddish-purple and darker towards the base at the mouth of the corolla but not with a well-defined blotch, obovate-elliptic with obtuse to subacute tips, c. 1.5–2.5 cm long, 0.5–1 cm wide, twisted lengthways, without conspicuously swollen appendages ('auricles') at base; style exserted from the mouth of the corolla. 2n = 34.

Flowering period June to December.

Distribution *Cyclamen purpurascens* occurs in eastern France, Switzerland, Austria, southern Germany, the Czech Republic, the Slovak Republic, northern Italy, Slovenia, Croatia, southern Poland and western Hungary.

Tuber of *C. purpurascens* with roots forming almost anywhere on the surface (BM)

Flower of *C. purpurascens* which has no, or only poorly developed, auricles at the mouth (BM)

Cyclamen purpurascens in woodland near Mt Triglav (BM)

Population of *C. purpurascens* on slope near Vrhnika, central Slovenia, beneath *Tilia platyphyllos* (lime) and *Carpinus betulus* (hornbeam) (JožeB)

Habitat Usually in woodland of deciduous or mixed deciduous and evergreen trees, occasionally in the open, mostly found on limestone formations with a humus top-dressing of decaying leaves, from sea level to 350 m (Niels Sørensen, in the *Journal of the Cyclamen Society* 22 (1): 11 (1998), notes an unusual habitat in Austria: 2 m below the River Danube's high-water mark).

Variations It is superfluous to say that *C. purpurascens* is variable in leaf characteristics, as most species are. The photographs by Jan Bravenboer reproduced here demonstrate this attribute more successfully than any number of words. Nevertheless, this species does have an exceptional range of leaf variants and some of them are very attractive indeed. In flower colour, there is also quite a range from pale to deep pinks and rich reddish-plum. White-flowered forms have also been found and named, so we now have three groups of *C. purpurascens* described at the rank of forma: f. *purpurascens* for all the pink or reddish ones, f. *album* for pure white forms and f. *carmineolineatum*, described by Paul Hendrikx. The flowers of this last form are also white but with a carmine zone at the mouth, which extends a short way along the veins of the petals. Unlike f. *album*, which seems to occur only very rarely as single individuals among pink-flowered plants, the plant described by Hendrikx as f. *carmineolineatum* was one of an entire population of about 400 plants. It was located by Manfred Koenen in 1983 near the Switzerland–France border, between Geneva and Albertville, but more recently it has also been found in Slovenia by Jože Bavcon.

Named cultivars

There are very few named cultivars of *C. purpurascens* in commerce, perhaps because the species is one of the less easily cultivated and therefore not raised in large quantities, unlike the very popular species such as *C. coum* and *C. hederifolium* which have many named selections.

Leaf variation in *C. purpurascens*, from the collection of Jan Bravenboer (JB)

Cyclamen purpurascens f. *carmineolineatum* growing in The Netherlands in the garden of Paul Hendrikx, who described and named the form in 2001 (PH)

This form of *C. purpurascens* collected by Jože Bavcon, white with a pink mouth, is referable to forma *carmineolineatum* (JožeB)

'Album': should be referred to as forma *album*.

'Garibaldi': leaves silvery with a green margin; flowers short and broad; selected by Jan Bravenboer

'Green Ice': leaves with a broad green margin and a silver Christmas-tree pattern in the centre; selected by Jan Bravenboer.

'Green Lake' is similar to 'Green Ice' but with a light green zone in the centre of the silver zone; selected by Jan Bravenboer.

'Pico Rosso': leaves with a pink overlay; selected by Jan Bravenboer.

'Verdume': leaves plain green; selected by Jan Bravenboer.

Fatra Form: plain green leaves that can be matt or somewhat shiny; this is treated as a distinct taxon by some Czech and Slovak botanists, *C. fatrense* or *C. purpurascens* subsp. *immaculatum* (see synonymy).

The rare pure white variant *C. purpurascens* f. *album* at a Cyclamen Society show (MD)

Cyclamen purpurascens 'Garibaldi', a Jan Bravenboer selection (JB)

Jan Bravenboer's *C. purpurascens* 'Green Ice' (JB)

Cyclamen purpurascens 'Picco Rosso', another selection by Jan Bravenboer (JB)

Cyclamen purpurascens 'Verdume' a plain green-leaved form from Jan Bravenboer (JB)

A fine silver-leaf form of *C. purpurascens* (MD)

Silver Leaf Group: leaves with a silvery overlay.

Lake Bled Form: leaves beautifully silvered with a netted pattern of veins towards the margin.

'Lake Garda': leaves silvery with a thin green margin and central arrowhead zone.

Limone Form: leaf upper surface almost entirely silvered with narrow green bands around the margin and a faint grey hastate pattern in the centre; flowers deep pink.

Further observations on *Cyclamen purpurascens*

As a much-loved species growing in mountain regions that are popular for hiking and holidaying in the late summer, it is not surprising that much has been written about *C. purpurascens* in books and periodicals, especially the Cyclamen Society's journals. Much is to be learned from enthusiasts and specialists who travel to see their favourite plants, so below, we repeat selected passages from some of these articles and notes, in particular those of the late **Graham Simpson**, a *C. purpurascens* aficionado, and **Jan Bravenboer**, who has made *C. purpurascens* a particular focus of his attentions. The following two modified extracts are from articles written in the *Journal of the Cyclamen Society*.

Cyclamen purpurascens **around Lake Garda, Italy. Graham Simpson** (in the *Journal of the Cyclamen Society* 22 (1): 28, June 1998) has some useful comments to make about the distribution and variation of *C. purpurascens* in the area:

"This species is already well recorded from the west coast of Lake Garda, at Limone, and further north, in the Val do Ledro. What seems to be less well known, even now, is the extent to which the species grows on the east side of the Lake.

In mid-June 1996, an earlier time than I would normally choose to look for cyclamen in the Central Alps, I located a site above Malcesine at a height of about 560 m (1,840 ft). The plants were indeed difficult to find, largely because new growth was slow after an abnormal period of three to four weeks of complete drought.

On the first of three visits, only a few new leaves were visible above the surface, as in this drought the ground had been covered by an extra layer some 8–10 cm (3–4 in) thick of leaves fallen from the overhanging deciduous trees. However, a little gentle prodding revealed large quantities of ripe seed pods. There was little change on my second visit a few days later, but by the next week (24 June) 10 or 12 plants were in flower. These were spread over the site as a whole, rather than being bunched together in one spot.

The plants were identical in appearance, though somewhat unusual in two respects. The flowers had long, narrow petals giving them the appearance of a typical *C. repandum*. The flower colour was an all-over deep carmine with no gradation in colour between the tips of the petals and the mouth.

One would like to think that these seemingly unusual plants were typical of those on the site as a whole. However, it could instead be that all the plants which shared these two unusual (and desirable) characteristics also shared a third — early flowering — and that later flowering plants in this population were more typical *C. purpurascens*.

Habitat of *C. purpurascens* on Monte Baldo (MD)

The extensive network of footpaths shown by a local 'walkers' map' proved much neglected, with most ending in an impenetrable tangle of undergrowth. Despite this, during the 14 days I spent in the area I did search for other sites. Although I found no others, there must surely be further sites on these lower slopes of the Monte Baldo range."

***Cyclamen purpurascens* in the Lake Garda area. Jan Bravenboer** writes:

"Searching for *Cyclamen purpurascens* in northern Italy is a great pleasure and you will find them everywhere, but the exciting silver-leaved forms can be found only in the Lake Garda area. This is a great place to be, particularly the western part, for although in summer there are many tourists and it can be very hot, you can always find a quiet and cooler place by leaving the lake side and walking in the forests. *Cyclamen purpurascens* can be found in all the mountain valleys and slopes of the western part of Lake Garda. However, locating silver-leaved plants is more difficult and you may search a whole day without finding a single one. Once, I walked for five hours and found only thousands of the 'normal' plants but suddenly found a group of beautiful silver-leaved plants. Why in that specific place? Some plants had totally silver leaves, some had a narrow to broad green margin, others were pewter-coloured and some had a 'Christmas-tree' pattern.

Most of the plants grow in shaded positions but some can be found in full sunshine and these have very thick, leathery leaves compared with the 'normal' ones in the shade. The species prefers shaded places in mixed forests, often accompanied by *Hepatica nobilis* and *Helleborus niger*. Sometimes the mixed forest changes to pure beech or pine forest, in which case *C. purpurascens* is often no longer to be found.

The first plants start flowering at the end of June, sometimes earlier, but you'll find flowering plants at Lake Garda as late as the end of September. Many plants have dark pink flowers but there are paler pink ones as well and if you are very lucky you may even find a white form.

In cultivation, the silver-leaved plants mostly give 10–30% seedlings true to type but there are exceptions and one plant gives me more than 90% good silver-leaved seedlings."

Jan also has a few useful tips for those visiting the area in search of cyclamen:

"Lake Garda looks great on a picture postcard, but walking in the mountains around the lake is not child's play. An Italian mountain path never rises slowly so you need good conditions and good mountain shoes because most places are very steep. The loamy soil can be dry and hard in summer, but very slippery and dangerous after rainfall. The Lake Garda mountain gorges are always dangerous, even after weeks without rainfall so it is always better not to walk alone and to have a good map and compass."

In a later article, Jan Bravenboer describes the area and habitat of the cyclamen and associated plants in greater detail:

"*Cyclamen purpurascens* likes to grow in mixed forests …… in which you will find beech, red-deal, larch, silver-fir, Scots pine, hornbeam, ash, oaks, elms, maples, laburnum, hazel trees and wild fruits like the walnut. But they don't grow in forests of pure pine or beech. *Cyclamen purpurascens* likes clay and limestone… . The tubers and roots always grow in pure clay and above this there are several centimetres of humus. Many plants have floral trunks, but they also grow on the surface. In many places, 'naked' tubers hang on the rocks, waiting for some water. The best and most beautiful silver-leaved plants grow in the very steep and dark gorges where the humidity is high … . Along the path, where sunshine gives more light, you will find fewer but the flowering period starts much earlier. Most cyclamen are on north-facing aspects but you will find some south-, east- and west-facing as well. On average, the Lake Garda area is coloured pink from the beginning of June till the end of September and sometimes until the first night frosts in November or December."

***Cyclamen purpurascens* in Switzerland** Although widely distributed in northern Italy and Slovenia, *C. purpurascens* is rather restricted in Switzerland. In view of this, it is worth repeating a short note taken from the *Journal of the Cyclamen Society* 25 (1): 3 (June 2001). **Joy Cuany**, writing from Switzerland, describes an interesting place for cyclamen enthusiasts to visit:

"For those en route in the latter part of August, I would recommend a stop over in the southern tip of Switzerland — the Monte San Georgio area. On the southern slopes of this hill at that time of year, the subtle perfume invades the beech woods, reminiscent of Bond Street in the 1950s! Whether sunshine or rain, the great pools of all shades of pink *C. purpurascens* are truly worth the detour."

***Cyclamen purpurascens* in the French Jura** Writing in the Cyclamen Society's journal of June 1996, **Martyn Denney** gives some valuable first-hand information about the species in the Jura mountains. "Referring to *C. purpurascens*, Grey-Wilson mentions eastern France in its distribution, but his sketch maps suggest this to be in the south-east of the country close to Italy. Saunders describes it as coming "... from limestone districts as far west as Haute Savoie and Southern Switzerland ... ". By contrast, the Michelin green guide for Jura Franche-

Comte when describing the local flora, says "... alors que dans quelques gorges calcaires et buissonneuses s'est refugie le bulbeux cyclamen d'Europe aux fleurs odorantes." Michelin gives two sites, Cascade de la Billaude and Gorges du Flumen. The Cascade de la Billaude lies at the bottom of a narrow gorge surrounded by mixed coniferous and beech woodland. The sides of the gorge are steep (1:1 to 1:0.5) and no cyclamen were seen in this area. At the rim of the gorge behind the Chalet des Cyclamens Bar, three plants of *C. purpurascens* were found, one with a single flower and another with two flower buds. The woods above the gorge and beyond the road on the western side are mixed, but within this were distinct patches containing either beech or conifers with gorse and grasses. *Cyclamen purpurascens* was found as isolated plants or sparse populations at a density of one plant per square metre or less. Plants tended to be found along the edges of the beech and in the more open areas. Tubers were 7–10 cm (2¾–4 in) deep in decomposed leaf litter with a high proportion of calcareous rock fragments typically 5–7 cm × 1–2 cm (about 2 in × ½ in) thick. Typically, 2–3 cm (1 in) floral trunks were in evidence. Leaves were nearly orbicular, 3–4 cm (1¼–1½ in) in diameter, with plain grass-green upper surfaces and reddish-purple undersurface. The leaves were reminiscent of the form of *C. purpurascens* often referred to as *fatrense*. A very small number of plants had faint grey markings on the upper surfaces of their leaves. No flowers or seed capsules were seen and there was no significant evidence of regeneration. Enquiries at the bar elicited the information that flowering in 1995 had been comparatively poor and was considered finished at the time of my visit on 18 August. The bar had half a dozen flowers in a wine glass. Although not fresh, these gave off a strong scent, and were fairly squat with petals no longer than about 1 cm (½ in). They were not a clear pink, and although one may attribute this to deterioration since picking, they were very similar to the single flower seen on the rim of the gorge… . The site at the Gorges du Flumen received only a cursory investigation. No obvious habitats were found and no plants sighted, although this does not necessarily indicate that the report in Michelin is false."

***Cyclamen purpurascens* in the Haute Savoie** There are some other interesting notes about the species in France, this time the Haute Savoie, in an article by **Richard Manuel** (*The Journal of the Cyclamen Society*, June 1998) who writes that:

"*Cyclamen purpurascens* is very common in the woods around Outredière, above Sallanches, Haute Savoie. Outredière is a small and very exclusive village with stunning views to Mont Blanc, hence its exclusivity — it is not even signposted from below. The area is on a very steep east-facing limestone cliff/scree slope, and the village is about 850 m (2,800 ft) up. The woods above and below the village are of mixed pine and beech with some ash, with a very deep litter layer. The cyclamen are everywhere, even in the darkest areas of otherwise bare leaf litter; mostly just leaves, of course, but even here the occasional flower occurs. In more open parts — tree falls, along footpaths, etc — a few plants with multiple flowers were found, but these were relatively uncommon compared to non-flowering plants. Most of the plants had slightly mottled leaves but some were plain. We did not see any outstanding variants in either leaf or flower form".

Richard Manuel was in the area from 6 to 9 July, and there were plenty of flowers to come while none had gone over, but it was a late year.

***Cyclamen purpurascens* on the Adriatic coast** For those who associate *C. purpurascens* with mountain habitats, it may come as a surprise to find that it descends to much lower altitudes, in fact almost to sea level, in the Adriatic region. **Joseph Rupreht** has studied it in the area and his article in the *Journal of the Cyclamen Society* 25 (1): 18 (June 2001) contains valuable information:

"In April 2000, I ascended Mount Učka above the Croatian Adriatic resort of Opatija in search of *Cyclamen hederifolium* and *C. repandum*. What I found looked like *C. purpurascens* with leaves and tubers quite distinct from the other two species. To be quite sure I planted the tubers, still in leaf, and waited until the end of July, when the typical purple scented flowers appeared. The plants thrive in the Dutch climate with sturdy leaves of varying pattern surrounding a multitude of flowers. One plant has remarkably heart-shaped leaves. Several tubers were much bigger than is usually found in the inland district of sub-alpine Slovenia.

These *C. purpurascens* were found in mixed *Carpinus*, *Fagus*, *Ostrya* forest, at an altitude of 400 m (1,300 ft), less than 1.5 km from the Adriatic. The forest undergrowth included typical Mediterranean *Ruscus aculeatus* interspersed with the cyclamen. The location was on a shady slope on the mountain, and cyclamen grew also on calcareous rocks where they were embedded in moss cushions. The area evidently receives frequent showers and fog forms there during the day. Another small colony of *C. purpurascens* was found further inland, on the Istrian peninsula, also on calcareous ground. Wherever terra rossa changed to impenetrable flysch, cyclamen and hellebores disappeared. On this trip no *C. repandum* or *C. hederifolium* were spotted".

Dr Rupreht goes on to say that the distribution map for the species in Hegi's *Illustrierte Flora der MittelEuropa* Vol. 5 (1931) shows *C. purpurascens* reaching close to the Adriatic coast. In the

Cyclamen purpurascens is a very hardy species. Here, not far inland in Croatia it sometimes inhabits these wooded hollows (limestone dolines) (BM)

same work, it is stated that *C. purpurascens* reaches the shores of the Kvarner Bay on the Adriatic coast at Rijeka and Opatija. Dr Rupreht also notes that it has been seen at Crikvenica growing under the maquis only metres from the Adriatic and that he had heard of reports of it growing on some of the Adriatic islands, in which case, there might be a somewhat unexpected overlap of the distributions of *C. purpurascens*, *C. repandum* and *C. hederifolium*.

In fact, the occurrence of *C. purpurascens* and *C. repandum* together on the Adriatic islands was confirmed in a separate article in the same 2001 issue of the journal by **József Szendefi** from Hungary. He writes:

"In mid January 2001, I visited the region of Split in southern Croatia, and then in early February Krk Island in the north of the country. From the overall distribution of cyclamen species, it should at least be possible to find *C. repandum* in both places. On Krk Island, on 4 February, I found *C. repandum* all around the island with the exception of the area around the bridge from Rijeka at the northern tip of the island. However, especially on the northern part of the island I did find very different *C. purpurascens* plants. There were some silver-leaved forms, similar to those from Limone del Garda, Italy, but a completely plain silver leaved plant was not found, only one with a very narrow dark green border, and others with twice as much silver as dark green. I found a place with both *C. repandum* and *C. purpurascens* growing together under one tree, which was the first time I have seen both species together. One interesting thing I noticed was that many *C. purpurascens* plants are found on the mainland coast around the Senj to Rijeka road almost at sea level, being only 10–20 m (33–66 ft) in altitude. With only very little possible shade (a few trees in summer and full sun in winter time) they appeared very healthy — for example, I found a 'real salad' with a tuber size of around 10 cm (4 in) at 10 km from Rijeka. There are a lot of *C. purpurascens* plants in the woods in the mountains above the sea, and I found a place where there were some plain dark green leaved plants, similar in appearance to those of the Fatra form, although without flowers of course."

In the same issue of the Cyclamen Society's journal (Vol 25 (1): 9, June 2001), **József Szendefi** also noted some interesting differences in the way in which flowers of *C. purpurascens* from various places open:

> "Flowers on plants from Slovenia, south-west Zala and the Bakony mountains start to open at the bottom of the bud, gradually peeling back from its tip. By contrast, flowers from other places, including the Fatra forms from Slovakia, start opening more from the top of the bud, near the calyx, with the petals bulging out there nearer the calyx than the tip, to form a pear shape, and then bursting open. It would be interesting to know if this is a constant characteristic and, if so, the geographical limits of the two variants."

***Cyclamen purpurascens* in the Slovak Republic. Barbara Boize**, writing in *Cyclamen* 24 (1) (2000), provides some useful first-hand observations based on a visit to Zilina in the Slovak Republic in August and September 1998. She found *C. purpurascens* in two areas: in the Velká Fatra south-east of Martin and Demänovská dolina and in the Low Tatras south of Liptovský Mikoláš. The museum in Martin had displays of the local flora with a photograph of '*Cyclamen fatrense*' and a helpful local provided valuable information, reporting that cyclamen were to be found only in three valleys in the Velká Fatra: Belianska dolina, Necpalská dolina and Lubochnianska dolina. Of these, Necpaly, 9 km south-east of Martin, was visited on 31 August and *C. purpurascens* was found at an altitude of around 520 m in the woods. Individual cyclamen plants were widely scattered in the woods, which consisted of Norway spruce that had been selectively felled over time so that there were gaps in the canopy. The herb layer was dominated by *Asarum europaeum* and wood sorrel, *Oxalis acetosella* with *Euphorbia amygdaloides*, violets, ferns and grasses; there were few shrubs, mostly regenerating beech. The underlying rock was limestone with the dryish loamy soil having a pelleted structure and being covered by a thin layer of needles and beech litter. The cyclamen were growing on a steep north-facing slope in a very scattered population with an estimated density of about 0.1 plants per m^2. Six individual plants were examined and these all had plain leaves with red undersides; they varied in size from 3 cm wide × 3.5 cm from apex to sinus to 5.5 × 5 cm. Only three plants had flowers and then only one each. Two were a mid-pink colour with a purple mouth, and the other was deep pink with a red mouth; all had a strong sweet scent. Further on there were more, all with plain leaves but no signs of flowers or fruits and the density here was higher, estimated at 3–4 plants per m^2.

The Demänovská dolina south of Liptovský Mikoláš in the Low Tatras was visited in early September and cyclamen were found growing at about 700 m, scattered across an open slope. All had plain leaves with a dirty purple underside and no signs of flowers or fruits. The leaves were smaller than those seen in the Velká Fatra, a typical measurement being 2 cm wide × 1.5 cm from apex to sinus. Associated plants were *Asarum europaeum*, *Oxalis acetosella*, *Geranium robertianum*, moss, ferns and grasses. The slope was west-facing with thin dry soil over limestone and scattered litter. Cyclamen were not growing where the herb layer was thick nor on the limestone outcrops. Lower down, other denser populations were seen in association with mosses, wood sorrel, grass and orchids under sparse Norway spruce; the density of these was estimated to be 15–20 plants per m^2. All had plain leaves with deep purple undersides, but further on, some of the colonies had leaves with lighter green rays radiating from the sinus.

***Cyclamen purpurascens* in Slovenia** Working from Ljubljana University Botanic Garden, **Jože Bavcon** has been studying wild populations of *C. purpurascens* in Ljubljana. His observations

Jože Bavcon has studied the natural variation of *C. purpurascens* in Slovenia. This is part of his collection in Ljubljana (JožeB)

The collection of *C. purpurascens* in the University Botanic Garden, Ljubljana, Slovenia, where Jože Bavcon studies the plants and their cultivation (JožeB)

and experiments have now been published under the title of *Common Cyclamen* (Cyclamen purpurascens) *and its Diversity in Slovenia*, a book of 163 pages devoted to just one species! The author's aim was to determine how variable the species is in its wild state in Slovenia and to collect samples for cultivation in the Ljubljana University Botanic Garden. The idea of the survey came about after the author saw a dense colony of *C. purpurascens* plants, noting that he was "overwhelmed by the diversity of their leaves. I took a closer look at this patch and found no less than twelve different types". He has separated out individuals of particular merit and given them cultivar names, some of them striking and desirable whereas others have extraordinary shapes. Still others have been selected for flower colour, for example 'Belkasti' has white flowers stained pink at the mouth and would be referable to forma *carmineolineatum*. There are some interesting observations in the book, showing just how thorough one can be when living in a region where a cyclamen species occurs naturally.

A fine specimen from Jože Bavcon's collection of *C. purpurascens*. His book about the species (see Chapter 9) describes selected cultivars (JožeB)

Cyclamen colchicum. Painted by Pandora Sellars (reproduced at 125% of original size)

Plant of *C. colchicum* showing the distinctive leaf margin; exhibited on 18 September 2011 at a Cyclamen Society show, cultivated by Ian Robertson (JA)

Cyclamen colchicum

Flowering late summer to late autumn

This could be taken, as indeed it first was by the Russian botanist Nicolai Michailovich Alboff (Albov) in 1895, as an eastern variety of the familiar *C. purpurascens* (then known as *C. europaeum*). However, this cyclamen has distinct features and these prompted Alboff to reassess the status of the plant a few years later, when he recognised it as a species in its own right. Most recently, Christopher Grey-Wilson in his thorough assessment of the genus published in 2002 also accepts *C. colchicum* as distinct, but in the preceding century this plant has been regarded variously as a variety or subspecies, sometimes under its synonym *ponticum*. The epithets *colchicum* and *ponticum* both give us an immediate clue to its origin, Colchis being one of the classical regions of the south-eastern Black Sea (Pontus Euxinus) and Pontus a province a little further to the south-west, overlapping into north-eastern Turkey. The known locations, including the type locality, are in a restricted area in the Okum Gorge near the village of Okumi (42°42' 48" N, 41°44' 11" E) in Abkhazia in the western Caucasus. Although this is not that far from the Turkish border, *C. colchicum* has not been recorded in Turkey although there is a claim (in *Turkiye Bitkileri* by Hikmet Birand, 1952) that '*C. europaeum*' (i.e. *C. purpurascens*) has been collected in Giresun province at Kayalik in 'rocky places'; this might be just a case of misidentification as only *C. coum* is known in northern Turkey. Little seems to have been published about the status of *C. colchicum* in the wild, so a survey of its natural populations would be desirable.

The points of distinction between *C. colchicum* and *C. purpurascens* that were stressed by Grey-Wilson lie mainly in the leaf characters. Perhaps the most noticeable features of *C. colchicum* are the thick leathery texture of the leaves and the manifestly toothed leaf margin. There is possibly also a slight difference in leaf shape, although this is so variable in *C. purpurascens* that it probably cannot be used as a definitive characteristic: the leaves of *C. colchicum* are considered to be generally broadly ovate, those of *C. purpurascens* usually more rounded or broadly reniform. The flowers of the two species are almost identical in appearance and any observations about their comparative sizes do not seem to have much foundation. In the case of the plant depicted here in Pandora Sellars' beautiful illustration, the flowers of *C. colchicum* are certainly just as large and substantial as those of any good form of *C. purpurascens*. The flowers of both species have a powerful fragrance.

As with *C. purpurascens*, this is an almost evergreen species having no real period of dormancy. It has not to date proved to be an easy garden plant, although there are some successful growers. It is likely to remain a specialist plant for the enthusiast as, horticulturally speaking, it has no great advantages over its much more frequently seen relative.

C. colchicum (Albov) Correvon, *The Garden* 53: 193 (1897). Type: Abkhasia, Defile d'Okoum (Okum), 300 m, 1 September 1893, *Albov* 247 (G, LE,); *Albov* 525, 14 October 1893 (K, syntype).

Synonyms

C. europaeum var. *ponticum* Albov, *Bull. Herb. Boiss.* 2: 254 (1854).

C. europaeum var. *colchicum* Albov, *Trudy Tifl. Bot. Gart.* 1: 166 (1895).

C. europaeum subsp. *ponticum* (Albov) Schwarz, *Gartenflora N.F.* 1: 16 (1938).

C. ponticum (Albov) Pobed., *Journ. Bot. URSS* 33: 226 (1948).

Description Tuber globose in the younger stages but becoming more irregular later, up to c. 5.5 cm diameter at maturity, brown, rather smooth; roots appearing from the sides but sometimes from other places on the surface. **Petioles** usually more or less erect from the tuber. **Leaves** very thick in texture, developing before flowering time but the previous season's might also be present as the new ones appear. Leaf blade broadly ovate, sometimes suborbicular, 2.5–7 cm long, 2.5–8 cm wide, plain rather dull or shiny green or marked with a jagged silvery-green zone towards the margin; margin thickened with conspicuous teeth giving a bluntly serrated appearance; the underside is normally suffused deep reddish-purple. **Pedicels** arising more or less erect from the tuber, after anthesis coiling from the top downwards. **Flowers** produced in late summer to late autumn, strongly fragrant; petals swept back to an upright position, rosy pink or reddish-purple, not or only slightly stained darker at the mouth,

Tuber of *C. colchicum*; leaves are present for most of the year (BM)

Flower of *C. colchicum*, which is very similar to that of *C. purpurascens* (BM)

Variation in leaf markings in *C. colchicum*, from Jan Bravenboer's collection (JB)

oblanceolate-elliptic with obtuse to subacute tips, c. 10–22 mm long, 5–10 mm wide, twisted lengthways, without conspicuously swollen basal appendages ('auricles'); style exserted from the mouth of the corolla. 2n = 34 (cf. Grey-Wilson, *Curtis's Botanical Magazine* 15 (3): 181 (1998)).

Flowering period July to October.

Distribution Endemic to western Caucasia, particularly in the Okum Gorge region, south-east and inland from the port of Sukhumi; there are also unconfirmed reports that it occurs to the north of Sukhumi. The exact distribution seems to be open to question as the *Flora URSS* (1952) reports it as being fairly widely distributed in the western Caucasus. This is contrary to the information published by Otto Schwarz who indicated that repeated attempts to find it outside the known (Okum Gorge) area had failed. Recently, however, another population has been found in the West Georgia and an account of the discovery by Clare Trivedi is reported below.

Habitat On limestone formations on cliffs and in the shade of trees and shrubs, 300–1,400 m.

Variations In leaf pattern, this eastern species is much less variable than its hugely variable western counterpart, which has an astonishing range of patterns from almost plain green through variously silvery-patterned forms, and sometimes with an overall pewter suffusion. In *C. colchicum*, plain green, or nearly so, seems to be the norm but there can be silvery-green zoning, as shown in the accompanying photograph. As noted above, the flowers vary considerably in size, as they do in all cyclamen species, and to some extent this is determined by the availability of moisture and nutrients.

Named cultivars

'Deirdre': a completely plain green-leaved selection from Jan Bravenboer.

C. colchicum 'Deirdre', a plain-leaved selection by Jan Bravenboer (JB)

Further observations on *Cyclamen colchicum*

Few people have published observations about *C. colchicum* in the wild and in cultivation, so first-hand reports are scarce. Nevertheless, the following comments contain some valuable information about this interesting species. An important recent discovery is the recording of this species outside the Okum Gorge area. **Clare Trivedi**, International Projects Co-ordinator, Millennium Seed Bank Partnership (MSBP), Kew, and **Tsira Mikatadze-Pantsulaia**, Head of the Department of Plant Conservation, National Botanical Garden of Georgia (NBGG), have contributed the following account of this exciting work. **David Kikodze**, Deputy-director of the Georgian Institute of Botany and **Tinatin Barblishvili**, Research Scientist in the Department of Conservation at the NBGG, have provided the accompanying photographs.

Discovering new populations of *Cyclamen colchicum* in Georgia Since 2007, Georgian botanists have been working with Kew's MSBP to collect and conserve the amazing flora of their country. The programme, 'MSBP: Saving the Flora of the Caucasus', has already conserved seed from almost 1,000 species and has enabled the Georgian team to build a much better picture of the state of their wild populations. In October 2010, a particularly exciting discovery was made when local naturalists alerted the team to the presence of cyclamen in a district in West Georgia.

As soon as the team saw a photo, they suspected the plant could be *Cyclamen colchicum* (Albov) Albov, although the species had never been recorded for that district before.

The genus *Cyclamen* is represented in Georgia by five species, among which the colchic cyclamen is characterised by a very narrow and patchy distribution on limestone substrates. The medicinal properties of the species are documented in ancient texts and it is thought that the plant is still extensively collected for medicinal use, as well as for its ornamental value. According to available data, the wild populations of *C. colchicum* are declining.

The MSBP team comprises staff from the National Botanical Garden of Georgia (NBGG) and the Institute of Botany. The local naturalists told them about a small population of what they called *kochiota* in the Megrelian language. Fortunately,

The river gorge in West Georgia where a previously unrecorded population of *C. colchicum* was discovered (DK)

A marbled-leaved *Cyclamen colchicum* in the newly discovered population in western Georgia (TB)

A plain-leaved form of *C. colchicum* in the new population (TB)

A fine population of *C. colchicum* on a vertical cliff at the new site in Georgia, recorded by Roman Tolordava (RT)

the team still had funds left from a fieldwork grant provided by the Rufford Foundation, and were able to visit the field immediately to catch the population in flower. The population was in a remote river gorge, on almost vertical dolomitic limestone rocks, and was definitely *C. colchicum*. Two more populations were also found nearby.

In June 2011, a small seed collection was made from this new population, and the team at NBGG and Kew have established trials to determine the germination and propagation requirements of this species. There is an urgent need to survey potential sites for further populations and to collect more seed for *ex situ* conservation and further research. Historically, this species was more widespread and there could be further populations, although each population is likely to be of a small abundance.

Conservation matters Paul Hendrikx (*Journal of the Cyclamen Society* 24 (1): 3, June 2000) reports that 'in Georgia' the flowers of *C. colchicum* are picked for decoration and the tubers are dug up for medicinal purposes, so the species is becoming increasingly rare. It is good to know, however, that there are efforts to cultivate *C. colchicum* in Tbilisi Botanical Garden in an effort to conserve it.

Experiences of growing *C. colchicum* from seed. Meg Baker writing in the *Journal of the Cyclamen Society* 22 (2): 56–57 (December 1998), quotes Prof. Schwarz's opinion that *C. colchicum* is "the most slow-growing species in the whole genus, never reaching flowering size before the fifth year". Her experience does not support this. Of the three seedlings that germinated in June or July 1996, one was ahead from the start. This plant had very thick leaves, which were completely plain dark green and very shiny like waxed cardboard. It proved to have very *hederifolium*-like flowers (almost with auricles) but of a colour typical of *C. purpurascens* and the plant was compact and rather coum-like. Baker imagined therefore that in the wild and out of flower, it could be mistaken for a plain-leafed *C. coum* (this may account for some of the records for *C. colchicum* which suggest that it is widespread in the Caucasus). This was the first plant to flower, on 3 August 1998. The other two took longer but it is worth quoting Meg Baker's comments on the variation:

> "Another of the plants was always much smaller with small round, very dished (concave) matt leaves with a suspicion of a pattern. The leaf shape this season is altering markedly and is now rather similar to Plant 3 although much smaller. It also has a smaller flower, somewhat dumpy (petals 1.8 cm long by 0.8 cm wide in maturity, also well twisted) opening as a bicolour with white edges to pale pink petals, and still no more than a rather transparent pale pink in maturity. Its nose marking is the same deep magenta as Plant 1 but only colours the rim, whereas in Plant 1 it suffuses a third of the petal. The third plant hasn't yet flowered, but has a crowd of thick leathery tooth-edged and beaded leaves (average 6 cm each way) that have a heavily patterned grey band around the edge — rather like joined-up finger-mark blotches. The leaves are heart shaped, between matt and shiny, dark green (duller and larger than those illustrated in Christopher Grey-Wilson's book, with a more solid pattern)."

Some further notes on *C. colchicum* appeared in the *Cyclamen* Society's journal of December 1998, in which there was a review of various previous papers about this interesting species, notably those of **Prof. Otto Schwarz**. Schwarz had noted that in the alpine house at Jena Botanic Garden, he found *C. colchicum* to be the leafiest and most floriferous of all the *Cyclamen* species, preferring an unshaded position but away from direct sunlight, and never being allowed to dry out; it was grown in a free-draining compost of leaf mould with limestone chippings. In his experience, the species needed cross-pollination to set fertile seed.

In an earlier issue of the journal (June 1981, pp. 17–19) there is a drawing of a leaf cross-section showing the cellular differences between *C. purpurascens* and *C. colchicum* in detail. It is these cellular characters that give the foliage of *C. colchicum* such a distinct texture. In correspondence with the then Editor of the journal, James Mars, Prof. Schwarz had added that, when growing in the open at Jena, these thick leaves were much more sensitive to cold than those of *C. purpurascens*, and were readily injured by frost.

Niels Sørensen has made some interesting comments based on his experience of growing *C. colchicum* in Denmark. His best plant was very floriferous, putting out a mass of scented flowers all at once in summer while the previous year's seed capsules were still ripening, so that the plant became congested with new and old capsules. (This tardiness in maturing was noted by Prof. Schwarz as well.) He also made an observation about the evergreen habit of *C. colchicum*: that at the beginning of August as the old leaves turn yellow many new ones appear. On the question of flower size, Niels Sørensen noted that his *C. colchicum* could have relatively large flowers (petals 22 mm long), within the size range of those of *C. purpurascens*, contrasting with those observed at Kew by Christopher Grey-Wilson (1997: 52–54), which had rather small flowers.

Roger Poulett reported seeing a plant at Munich Botanic Garden that had originated from Professor Schwarz, and this too had small 'dumpy' flowers; so it seems that, as with *C. purpurascens*, there is considerable variation in flower size. However, Sørensen agrees with Grey-Wilson's view that the seed capsules are small and says the seeds are as well.

2.2i The Graecum group
subsp. *candicum* and subsp. *anatolicum*

(= Subgenus *Cyclamen*, Series 3)

Comprising: *Cyclamen graecum*, *C. graecum* subsp. *candicum* and *C. graecum* subsp. *anatolicum*

Botanically, *C. graecum* is a rather isolated and very distinct autumn-flowering species. The leaves are heart-shaped, usually not markedly lobed and with a satin-like overlay. The flowers have a contrasting darker stain at the mouth, which has obvious swellings ('auricles'). In fruit, the pedicels start coiling upwards from the mid-point or basal half. The stamens are not exserted from the corolla mouth but the style often does protrude. 2n = 84 (85, 86, 87 also recorded).

Cyclamen graecum

Flowering in autumn

C. graecum is one of the floral treasures that can be seen by autumn visitors to central and southern Greece. It is particularly prevalent in the Peloponnese, where it often flowers in great abundance in open sunny areas amid the grey limestone. In the shadier sites, perhaps where pine trees have taken over, there may be just as many plants but these never have the dense domes of flower seen out in exposed spaces. Although common in the Peloponnese and southern mainland Greece, *C. graecum* is much more widespread, extending to Crete, the eastern Aegean Islands, Cyprus and southern Turkey.

Cyclamen graecum flowers at the end of hot, dry summers, often before the leaves appear if there has been no autumn rain (SB)

Cyclamen graecum subsp. *graecum*. Painted by Christabel King (reproduced at 110% of original size)

Peter Moore and Melvyn Jope of the Cyclamen Society examine a population of *C. graecum*. Peloponnese: between Hora and Hatzis, November (BM)

A fine specimen of *C. graecum* subsp. *graecum* derived from a plant collected by the Cyclamen Society (CSE96342) on the island of Hydra. Cultivated by Ian Robertson (MD)

Cyclamen graecum subsp. *candicum*, an award-winning plant exhibited by Ian Robertson at a Cyclamen Society show. The original introduction was from the Askifou Plain, Crete (CSE96466) (MD)

An amazing specimen of *C. graecum* subsp. *anatolicum* exhibited in September 2010 by Ian Robertson at a Cyclamen Society show, grown from seeds from Monte Smith, Rhodes (MD)

Although the plants are obviously at their most attractive at flowering time, the leaves of *C. graecum* are also an outstanding feature and hence it is among the most appealing of all cyclamen in cultivation. Well-flowered selections of any of the three subspecies of *C. graecum* are one of the great sights of the Cyclamen Society's autumn shows, but these plants can be almost as attractive in the foliage classes of a competition. The leaves are quite leathery in texture with an almost cartilaginous margin. A satin-like or sometimes velvety sheen covers the whole upper surface, but it is the markings that are so fascinating, with an apparently endless range of possibilities. The leaves of *C. graecum* plants are very different from the usually larger, softer leaves of *C. hederifolium*, so when the plants are in leaf, there is never any problem of identification. The flowers of these two species are, however, quite similar, although the flower and leaf stalks of *C. graecum* rise directly upright from the tuber whereas those of *C. hederifolium* are horizontal initially. There is even more of a difference in the roots: *C. hederifolium* tubers produce thin fibrous roots from almost anywhere on their rough surface (though by and large the rounded bottom of the tuber remains relatively free of roots) whereas the fissured, cork-like tubers of *C. graecum* have a tuft of thick cord-like roots emerging from the centre of their underside. This *C. graecum* root system has probably evolved for the purpose of delving into cracks in hot, dry rocky ground, the habitat in which this species is most often found. *C. graecum* is distinct even in the fruiting stage: it is the only *Cyclamen* species in which the flower stalk commences to coil not from the top as in most species but from a point lower down its length, varying from near its base up to about half way. *C. graecum* and *C. hederifolium* tend not to form mixed populations as they have different habitat preferences: full sun for *C. graecum*, semi-shade for *C. hederifolium*. Nevertheless, they can be found growing alongside each other in some situations, for example where scrub or pine trees have taken over a colony of *C. graecum* and now provide the shade necessary for *C. hederifolium* to flourish; ultimately *C. graecum* would probably die out in this situation because of a lack of flowering and hence regeneration, but the longevity of the individual tubers is such that this would take a long time.

J. H. Ietswaart (1990) studied *C. graecum* in some detail and detected differences between specimens from mainland Greece, Crete and Turkey, resulting in the proposal that at least some of the populations on Crete were worthy of recognition as *C. graecum* subsp. *candicum* and the Turkish ones as *C. graecum* subsp. *anatolicum*. Christopher Grey-Wilson further developed this concept and three subspecies, albeit overlapping in their characteristics, are currently accepted. Such overlap is not unusual at this level of taxonomic hierarchy, and the subspecies level remains a useful status indicating that a degree of distinctiveness is present in combination with discrete distribution patterns or perhaps habitat preferences. The geographical epithets above refer to the 'Kingdom of Candia', one of the historic names for the island of Crete, and Anatolia, the term for the whole of the Asiatic part of Turkey.

Cyclamen graecum is one of the earlier species known to botanists, having been described in 1834 by Johann Heinrich Link on the basis of specimens collected by Franz Berger at Nauplion, Attica. This site on the Greek mainland is therefore the type locality of the species, and so populations from mainland Greece and the Peloponnese automatically take the name *C. graecum* subsp. *graecum*.

It is not surprising in view of the extreme morphological variability of *C. graecum* that, over the years, it has acquired several synonyms. Botanists, and especially the cyclamen specialist Hildebrand, found it difficult to accept such a spread of variation within one species, so we have a proliferation of names — *C. mindleri*, *C. maritimum*, *C. aegineticum*, *C. pseudomaritimum*, *C. pseudograecum* and *C. cyprograecum* — none of which has any great significance and all of them are now regarded as synonyms of one or other of the three subspecies.

The subspecies of *C. graecum*

Three subspecies are currently recognised and they are distinguished by flower and leaf characteristics as shown in the following key. Note that these features are to some extent overlapping and that knowledge of the origin is helpful when attempting to confirm the identity of a particular plant.

Key to the subspecies (after J. H. Ietswaart (1990); C. Grey-Wilson (2002))

1a. Corolla whitish or very pale pink, the mouth 9–14 mm diameter; flowers not sweetly scented; leaves generally narrowly ovate, pointed and with dark hues . . . . subsp. *candicum*

1b. Corolla mid to deep pink, the mouth 5–10 mm diameter; flowers scented or not; leaves usually broadly ovate and rather blunt, often paler and brighter green 2

2a. Leaves mostly less than 8.5 cm long with a beaded but not lobed margin; flowers scented or unscented, the corolla with well-developed auricles; basal darker stain of each corolla lobe confined to and extending along the veins, especially the central one . . . . subsp. *graecum*

2b. Leaves often more than 8.5 cm long with a beaded and usually also lobed margin; flowers sweetly scented; corolla usually with poorly developed auricles; basal stain of corolla lobes more compact, not extending appreciably along the veins and the central vein is thus poorly defined . . . subsp. *anatolicum*

C. graecum Link subsp. ***graecum***, *Linnaea* 9: 573 (1834). Type: Greece, 'ad rupes circa Naupliam [Nauplion] frequens flore-bat Septembri' *Berger* (M).

Synonyms

C. aegineticum Hildebr., *Gartenflora* 57: 296 (1908). Type: Greece, Island of Aegina, specimen cultivated in Freiburg, *Hildebrand*.

C. mindleri Heldr., *Bull. Herb. Boiss.* 6: 386 (1898). Type: Greece: Island of Aegina, *Mindler* (P).

C. persicum subsp. *mindleri* (Heldr.) Knuth, *Pflanzenr.* 4 (237): 248 (1905).

C. graecum subsp. *mindleri* (Heldr.) A. P. Davis & Govaerts, *CITES Bulb Checklist*: 131 (1999).

C. graecum subsp. ***candicum*** Ietsw., *Journal of the Cyclamen Society* 14 (2): 50 (1990). Type: Crete, between Pervolitza and Rhizokloko, c. 80 m, November 1987, *Ietswaart* 1031 (holotype L).

Synonym

C. pseudograecum Hildeb., *Gartenflora* 60: 629 (1911).

C. graecum subsp. ***anatolicum*** Ietsw., *Journal of the Cyclamen Society* 14 (2): 51 (1990). Type: Turkey, Antalya province, c. 3 km west of Kemer, c. 100 m, October 1988, *Ietswaart* 1167 (holotype L).

Cyclamen graecum subsp. *candicum*. Painted by Christabel King (reproduced at 110% of original size)

A secure home for *C. graecum* subsp. *candicum*, wedged tightly in a limestone pocket (BM)

Great variation in leaf markings of *C. graecum* subsp. *anatolicum* near Silifke, southern Turkey (MD)

Synonyms

C. cyprograecum E. & N. Mutch, *Bull. Alpine Garden Society Gr. Brit.* 23: 164 (1954) [illegitimate name].

C. maritimum Hildebr., *Gartenflora* 57: 293 (1908).

C. pseudomaritimum Hildebr., *Gartenflora* 57: 293 (1908). Type: based on material from Mersin, Turkey, cultivated in Freiburg from specimens collected by *Siehe*.

Description (all subspecies) **Tuber** globose or subglobose, up to 20 cm diameter at maturity, the surface brown or greyish-brown, very rough and cork-like usually with cracks running lengthways, rooting from the centre of the underside; roots thick and fleshy, rather long and straight. **Petioles** usually arising more or less erect directly from the tuber. Leaf blade ovate, often broadly so, 5–15 cm long, 3–13 cm wide, the upper surface with a satin-like sheen, very variably marked with silvery or grey-green zones on a green ground, sometimes dull green with creamy veining, sometimes with a central green or silvery area surrounded by continuous or broken grey-green or silvery markings, sometimes wholly silvery or pewter coloured; margin thickened, with distinct but small teeth ('beaded'), sometimes lobed; the underside normally purple. **Pedicels** erect from the tuber, after anthesis starting to coil in the lower half. **Flowers** produced in autumn, normally just before or with the developing leaves, with no scent or a slight scent; petals whitish or pale pink to deep pinkish-purple with a darker stain at the mouth of the corolla, often in the form of three deep magenta veins, each petal twisted lengthways, oblanceolate, 15–25 mm long, provided with distinct but often not markedly swollen appendages ('auricles') at the base; style apex not reaching beyond the mouth of the corolla. 2n = 84 (85, 86, 87)

Tuber of *C. graecum* subsp. *graecum*; note the cork-like surface and thick fleshy roots (BM)

Flower of *C. graecum* subsp. *graecum* (BM)

A range of leaf forms of *C. graecum* subsp. *graecum* from cultivated plants at Wisley (BM)

Flowering period September to November (occasionally to January).

Distribution whole species — *C. graecum* is a native of central to southern mainland Greece and the Peloponnese, Crete, Cyprus, East Aegean Islands, south-western and southern Turkey; subsp. *graecum* — eastern and southern mainland Greece, Athos peninsula, Peloponnese, Cyclades, north-western Crete (Korikos and Rodopou peninsulas), Aegean Islands (excluding Rhodes); subsp. *candicum* — western and central Crete; subsp. *anatolicum*: southern and south-western Turkey in the provinces of Antalya, Muğla and Içel, Rhodes and possibly some of the other East Aegean Islands, northern Cyprus.

Habitat Mostly on limestone formations in *terra rossa*, on rocky slopes, in rock crevices, in scrub and in lightly cultivated or old uncultivated olive groves; sometimes in light woodland but then in sites that have probably been overgrown by, for example, pines; found from sea level to about 800 m.

The limestone hills and mountains of Crete, home to *C. graecum* subsp. *candicum* (BM)

A typical situation for *C. graecum* subsp. *graecum* wedged in rock crevices. Here seen in a sun-drenched cliff on Crete (SB)

A splendid plant of *C. graecum* forma *album* at a Cyclamen Society show, cultivated by Ian Robertson, September 2010 (MD)

Variations *Cyclamen graecum* is markedly variable in flower colour from pale to deep pinkish-purple with diverse darker markings at the mouth, although these are normally in the form of dark streaks and veins rather than a discrete blotch. The Cretan subsp. *candicum* is particularly striking in usually having pale flowers with a very strongly contrasting purple stain around the corolla mouth. White forms of *C. graecum* subsp. *graecum* can be found in wild populations; in cultivation, these have resulted in almost pure-breeding white-flowered strains, although with variable leaf markings. The botanical name encompassing all of these pure white forms (i.e. forms that lack a purplish 'eye' at the mouth of the corolla) is *C. graecum* subsp. *graecum* forma *album* R. & E. Frank. This name was based on a plant collected by Ronald and Erna Frank in the Peloponnese in 1980 but white forms occur sporadically throughout the distribution and several have now been reported, including examples from Crete.

It is in its leaf colouring that *C. graecum* is at its most variable, from almost plain green forms through examples with almost every conceivable type of net-veining, patterning and zoning to wholly pewter-coloured forms; the last are now quite commonly represented in cultivation by a form derived from a plant introduced from near Athens in 1971 by Brian and Margaret Mathew. This form breeds fairly true from seed and its offspring are consistent enough to qualify for the cultivar name 'Glyfada', after the village near the collection site. In fact there are many similar plants to be found in this area of Attica, as reported below by Martyn Denney. Some forms of *C. graecum* have a very dark background colour to the leaf, perhaps over-enthusiastically described as blackish but certainly heading in that direction, again with variable markings. Almost always, however, the foliage of *C. graecum* has a satin-like or velvety overlay, which gives the species such a distinctive appearance. The leaf characters have been used

Flower of *C. graecum* subsp. *candicum*. The contrasting mark at the mouth is often very pronounced in this subspecies (BM)

Variation in leaves of *C. graecum* subsp. *graecum* from the Peloponnese (JF)

The occasional white form occurs amid normal pink. *C. graecum* forma *album* on Crete (SB)

Leaf variation in *C. graecum* subsp. *candicum* (JF)

Cyclamen graecum 'Elizabeth Drew' has white flowers and grey leaves (MD)

to some extent, along with other features, to distinguish the three subspecies. The leaves of subsp. *candicum* are often more markedly narrow-pointed than those of subsp. *graecum*, whereas the foliage of subsp. *anatolicum* has a tendency to be shallowly lobed or jagged-edged, with fine cartilaginous teeth (a feature sometimes referred to as 'beading'). Exactly how significant these 'differences' are requires very thorough investigation over the whole range of the species.

Named cultivars

subsp. *graecum*

'Elizabeth Drew': leaves pewter coloured with thin green margin; flowers white; a seedling from Glyfada raised by Ray Drew and named after his daughter.

'Glyfada': leaves of an overall pewter colour; flowers large, pale pinkish-purple; collected by Brian and Margaret Mathew near Glyfada, Attica in March 1971.

'Rhodopou': leaves silver or pewter with narrow dark green border; flowers pink.

Leaf variation in Turkish *C. graecum* subsp. *anatolicum*, at Alanya (BM)

Cyclamen graecum 'Glyfada', with wholly pewter-grey leaves (BM)

Further observations on *C. graecum*

In many places in central and southern Greece, Crete and western Turkey, *C. graecum* grows in profusion. Clearly, we cannot give a thorough account of the species over its whole range so there follows a small but informative selection of articles taken from the pages of the Cyclamen Society's journal.

***Cyclamen graecum* on the Greek mainland. Martyn Denney**, in the Cyclamen Society's journal of June 1998, writes that:

"Late January may not immediately seem the most obvious time to study *C. graecum*. In fact it provides an opportunity to look at the variations in leaf form at a time when they have matured and are no longer obscured by the more obvious flowers. Athens is surrounded by three mountains which limit the size of the city… and provide the plant enthusiast who has limited time with an ideal introduction to the flora of the area. Of course, January is not exactly peak month for very much, except for *Crocus sieberi* ssp. *atticus*, flowering in the snow beneath ice-shrouded pine trees on top of Mount Parnitha, and *Muscari* at lower levels. To the north of Athens, Parnitha rises to 1,413 m and here *C. graecum* is in its element. It grows in profusion from the lowest levels almost to the top where *C. hederifolium* appears to take over. As you pass up the switchbacks of the escarpment by car you almost drive over the plants which have seeded themselves in the roadside grit. Although there are very attractive leaf forms to be found, it has to be said that on Parnitha generally you get quantity rather than quality. On the other side of the city Mount Hymettos is a curved ridge which starts to the east, rising above the city to 1,026 m and running first south-southwest, then south-southeast, finally ending in low hills which run into the Saronic Gulf between Vouliagmeni and Varkiza. A very good asphalt road starts near the University and passes the ruins of the Kessariani Monastery on its way up the west face to serve the radar and radio stations on the summit. *Cyclamen graecum* is to be found at all levels, in woodland, phrygana and the scree beneath the summit where it can be found growing in association with *Sternbergia lutea*. There is a wide variety of leaf forms in this area, from almost plain green through a profusion of silvery-grey hastate patterns to a minority of almost completely pewter-leaf forms. It is interesting to note that the leaves of plants in woodland tend to be 'plain' green, while the forms with more defined patterns or silver-grey markings tend to grow in more open and rocky situations.

As the ridge is followed southwards, the proportion of silver and pewter-leaf forms increases. Behind Voula and Vouliagmeni the hills are lower and closer to the sea. The surface is covered in weathered limestone outcrops with the plants growing principally under the edges of the rocks, and chasmophytically in deep fissures and pockets. This area is very close to where Brian and Margaret Mathew found the parent of the *C. graecum* 'Glyfada' strain many years ago, and there are large numbers of plants to be found here with similar characteristics, ranging from pewter leaves to a more definite clear silver.

On the sun-drenched hillsides south of Athens, *C. graecum* subsp. *graecum* shows great variation in leaf characters from attractively zoned to almost wholly silver or pewter (BM)

Where the hills drop into the sea as a promontory between Vouliagmeni and Varkiza, the limestone forms a pavement covered in fissures and small holes. Within these, *C. graecum* flourishes, covering the pavement with a multitude of silvery and pewter-coloured leaves, for it is here that the population with the greatest density of silver-leaved plants grows. Leaves vary from plain silver, plain pewter or silver with a fine dark green dentate margin, to dark green with a silver hastate pattern or silver centre. On a conservation note, it is important to point out that the plants are not only protected by international law, but would be impossible to dig up or extract from the rock as they are very mature with their tubers effectively encased in solid rock.

The north-eastern side of Hymettos is more remote, and indeed is to a large extent sheer and unwelcoming. Even so, it is possible to gain access at such points as the Kuttuki Cave above Peania; *C. graecum* grows here also with *Clematis cirrhosa* scrambling through the rocks, but it keeps to more stable areas of the limestone scree around bushes of *Quercus coccifera*, crevices of rock outcrops or in pine needle litter under the occasional rare pine tree. Leaf forms in the scree and crevices are generally interesting, and there is the occasional silver-leafed plant. In the shade of the pine trees the leaves tend to be a plain dark green.

On Hymettos there is a clear trend in leaf patterns from a predominance of 'plain' green and hastate patterns in the north-west, through pewter and grey, to a predominance of silver leaves in the south-west.

There is evidence to suggest an association between environmental factors and leaf characteristics:

Altitude: irrespective of the factors which follow, the incidence of silvering within leaf patterns tends to increase as altitude diminishes.

Light: plants in more shaded locations tend to have less silvering in the leaves. This could be explained by natural selection, since presumably the higher chlorophyll levels in the greener leaves are better able to synthesise light.

Soil: where plants grow in pine-needle litter, and pH values are (almost certainly) lower, leaf silvering is non-existent, or minimal, irrespective of light levels.

Limestone: as the association with limestone outcrops becomes closer, the predominance and intensity of leaf silvering increases. Where plants grow around outcrops in essentially terra rossa soil with rock chippings, the proportion of plants with all-over silvering becomes significant, but with an intensity of pewter or grey-green. Where plants are growing in rock crevices or chasmophytically, the intensity reaches silver or silver-grey, often with a glossy appearance.

Perhaps the most interesting discovery along the Hymettos massif was a colony of *C. graecum* in full flower in the last week of January, in stable limestone scree within terra rossa. These mature plants exhibited fresh foliage in a condition one would expect to see in early October. There was evidence that they flower regularly in the spring, as there were almost mature seed capsules which were clearly from the previous season's production, when compared with the less mature pods within autumn-flowering colonies. There was good regeneration, with the young seedlings of a maturity which would suggest winter/spring, rather than autumn germination. Following the precedent suggested by Grey-Wilson (1997) with his naming of *C. persicum* var. *autumnale*, if this colony proves to be a consistent, stable spring-flowering form, it may be appropriate to designate it *C. graecum* subsp. *graecum* var. *vernalis*. During a return visit on 10 March, two of the plants were still flowering."

***Cyclamen graecum* on the Greek island of Angistri** In 1999, **John Reeve** visited the island of Angistri in the Saronic Gulf near the more popular Aegina. In *Cyclamen* 24 (2000), he describes how in autumn it is possible to have the best of both worlds, a peaceful holiday combined with the joy of looking at cyclamen on this pine-covered island:

"Before my visit during the last week in September and first in October 1999, they had had a considerable amount of rain, which probably helped to bring the countless *C. graecum* that cover the island into flower, in some places creating vast areas of pink. One area around Aghia Barbara was particularly wonderful in its array of plants. The area had been cleared at some time for a vineyard and the stones piled up to form walls all over the landscape. These were like ribbons of pink with the plants clearly growing well in this undisturbed environment. They covered the spectrum of pink from the palest to deep cerise. Most had long, sinuous petals somewhat reminiscent of *C. repandum*. Other plants were growing in the open glades in the pine woods.

The island is criss-crossed by a large number of tracks and on one such track above Metoxi, at its highest point (at around 187 m (600 ft)) is the most wonderful array of plants any cyclamen enthusiast could want to see. Imagine an English bluebell wood, but in pink and that is what it looked like. Some of the cyclamen were growing among rocks protruding through the pine-needle litter, creating a completely natural rock garden. One large rock covered in leaf-mould had six plants each with a quite distinct leaf pattern (though none was in flower) — a rock garden in miniature. Those with a particular interest in the varied leaves of *C. graecum* would see some wonderful patterns here … . Plants in the more sheltered, shady areas of the woods seemed, by and large, to have more rounded petals than those in the open. I found lovely plants just yards from the sea at Milos. One particularly large plant was growing on its own in what was once a builder's yard, near Scala, the main town. It seemed quite happy among the sand, gravel, spilt cement and other building material. It would certainly have filled a 25 cm (10 in) pan comfortably and must have had a couple of hundred blooms. There was a distinct perfume with so many plants so I presume that most of them were scented. To my delight and excitement, I found two pure 'album' plants, one growing in a walled garden area near the ferry port in Scala, the other above Milos."

***Cyclamen graecum* on Rhodes** On the island of Rhodes, *C. graecum* [the Rhodes variant is regarded as subsp. *anatolicum*] is very restricted in its distribution, apparently confined to just one small area. **Peter Moore** visited the site on Monte Smith and made the following notes about its features both there and in cultivation:

"The site is an exposed windswept cliff with a certain amount of shrub cover. The shrubs grow horizontal to the ground, a sure indication of the power of the wind. The windy conditions have created the most compact cushion form of C. *graecum* often with leaves as broad as they are long. This characteristic feature is passed on to the next generation and maintained in cultivation. Only when completely sheltered from the wind by shrubs do the plants grow in the more usual habit of the species. The leaf patterns are excellent; frequently banding patterns are found, rather than the more common blotch pattern. A good percentage of silver markings is also present."

Martyn Denney also made a study of the species on Rhodes and provides a detailed account, extracted from *Cyclamen*, December 1994:

"Monte Smith rises some 150 m (490 ft) behind Rhodes Town and boasts the Rhodes Acropolis, ancient stadium and theatre, and a host of radio antennae and satellite dishes. It has a flat top, and an escarpment to the north-west which falls steeply to the sea and the main road between Rhodes Town and the airport. *Cyclamen graecum* grows most of the way along this escarpment in very large numbers, from the road at the top, down past the main road, to within a couple of metres of the sea. Its distribution has been described by Richard Bailey as '...abundant, though only in a very limited area', however I found the population spread most of the way along the escarpment at all altitudes.

At the top of the slope, the plant grew in a hard calcareous conglomerate, in places with the tubers up to 20 cm (8 in) in diameter, fully exposed. It often grew very densely, covering much of the ground. In the less accessible areas half way down, *C. graecum* grew in great abundance in very deep (30 cm or more) undisturbed pine needle litter, in very deep shade. These plants had very long floral trunks which in some cases ran parallel to the surface for some distance — usually down the slope. A great many seedlings were in evidence. In places, *C. graecum* was found growing in caves under rock outcrops in very low light levels. Although there was a generally high level of seed set throughout the site, these 'cave' plants invariably had none. Richard Bailey has reported that *C. graecum* is found in both full sun and deep shade and dismisses the theory that those found in shade had started life in an open situation and become accustomed to shade as trees grew, on the basis that seed was set, seedlings were in evidence, and the plants were therefore regenerating. I would support this view: *C. graecum* flourishes either where it grows in the shade of trees or rocks or where the herbaceous vegetation shades the tuber from the sun. I would speculate that this is due to a need for moisture and a cool root run.

Further down the slope, *C. graecum* can be found in large numbers between the carriageways of the main road, and sparsely on the seaward side of the road to within a couple of metres of the waves. Ole Sønderhousen reported having found *C. graecum* near the beach on the Rhodes Town side of the row of Turkish houses known as Kritika. While I did not find *C. graecum* near the beach so far south, this report is entirely consistent with the isolated plants that I did see near the beach a few hundred metres further north.

The wide range of interesting leaf patterns found on Monte Smith has been reported before. While the variations occur throughout the site, I noticed that the most extreme variations in both shape and pattern occurred in the middle altitudes in half shade. Some plants had leaves with distinctly scalloped edges. Richard Bailey stated in 1992 that Monte Smith is the only site for *C. graecum* on Rhodes. However, I can report that *C. graecum* can be found in the suburbs of Rhodes Town in a street known as Pavlou Mela. It grows along the top, and seeded into the base of a 2 metre high wall, over a distance of some 100 metres. From the leaf markings, these plants would appear to be indistinguishable from those on the Monte Smith escarpment. However, the number and age of the plants would indicate that if they have been introduced, this must have occurred some long time ago."

***Cyclamen graecum* on Cyprus** On Cyprus, *C. graecum* [regarded as subsp. *anatolicum*] is a very localised species. **Martyn Denney** reviewed the background to its discovery there and visited the site. In the Cyclamen Society's journal for December 2001 he writes:

Cyclamen graecum subsp. *anatolicum* in Cyprus (MD)

"*Cyclamen graecum* has been known to exist on Cyprus only since 1948 or 1949 according to Saunders (1979), and one may infer from Mutch (1955) that it was discovered by a Mrs C. C. Grove. It grows in a single relatively confined location either side of the road just to the southeast of Liveras (Sadrazamkoy) near Cape Kormakiti (Koruom Burnu). The site is formed of low *Pistacia lentiscus* scrub growing in terra rossa and brown soil with limestone boulders and outcrops at an altitude of about 75 m, sloping to the south-east and north-east and within about 1 km of the sea. The plants are generally growing under the edges of the *Pistacia*, around the rocky outcrops and in areas surrounded by other vegetation where there is some shelter. The site is not especially botanically rich, but does have some other interesting bulbous/tuberous species including *Narcissus serotinus* and *Spiranthes spiralis*.

Mutch (1955) considered the plants to be very similar to the Cretan form of *C. graecum* which at that time was often referred to as *C. pseudograecum*, but he preferred to call it *C. cyprograecum*, though he never validly published it as such. When Ietswaart (1990) revised the nomenclature of the species, he considered that the plants on Cyprus should be grouped along with those along the Turkish Mediterranean coast and the island of Rhodes, as *C. graecum* ssp. *anatolicum*, but it is not clear whether he included plants from Cyprus amongst those specimens investigated. However, he goes on to say: 'I have not found the Cypriot population to be different from the Turkish mainland populations, excepting that it shows quite uniform, pale flowers. The large leaves found in the Cypriot population which might be differential (Grey-Wilson, 1988) are not maintained under cultivation. It is quite possible that these natural large leaves are due to some fertilisation by cattle dung'.

Ietswaart's reference to pale flowers supports Mutch's assertion that the '... corolla lobes were very pale indeed — an off-white pink ... at the base was a bright red patch, but the characteristic feature was the delicate coloured veining extending from the mouth almost to the upper ends of the petals', and that of Meikle, that the corolla is '... white or less often pale pink ... with a conspicuous, crescent-shaped magenta blotch (at the base) ...'

The population on Cyprus does contain plants with particularly large leaves which are reminiscent of some forms of *C. persicum*, and in December coiled seed capsules serve as reassuring confirmation that the plants are *C. graecum*. It should also be noted that in shaded parts of Monte Smith on Rhodes, *C. graecum* ssp. *anatolicum* also exhibits similar large rather fleshy leaves. Though not a botanist, I would agree with Ietswaart that the Cyprus plants essentially resemble those from Rhodes and Turkey, and if splitting into subspecies is to be supported then his diagnosis is correct.

A range of leaf variation in *C. graecum* subsp. *anatolicum* seen at Liveras, Cyprus (MD)

Meikle and Grey-Wilson (1997) consider the leaves of the Cyprus plants to be rather duller and less attractively marbled than typical *C. graecum*, but I agree with Moore (1996) that this is not the case, as I found significant numbers of plants with interesting bright silver or cream markings in a variety of attractive shapes. The underside of the leaves range from paleish green to purple, with many examples of green with a pale purple overlay."

***Cyclamen graecum* in Turkey. Erna Frank**, writing in the *Journal of the Cyclamen Society* (December 1996), provides some useful comments about *C. graecum* in Turkey (i.e. subsp. *anatolicum*):

"Whereas the eastern limit of *C. graecum* in Turkey seems to be well defined, in the west it probably merges with the eastern limit of the Greek distribution. Many small islands are sprinkled along the west coast of Turkey ... ... it might be an interesting study to paddle from island to island and investigate *C. graecum* on all of them. On the mainland of Turkey *C. graecum* follows the southern coastline, accompanying the main road for long stretches. Between Fethiye and Kaş there are sizeable populations along the roadside under pines, with strikingly large and deep pink flowers, the petals slender and slightly twisted. In some locations, notably on Baba Dağ and in the Bey Dağ west of Kemer, *C. graecum* rubs shoulders with *C. trochopteranthum* [*C. alpinum*]. About 30 km inland

A pine wood habitat for *C. graecum* subsp. *anatolicum* in south-western Turkey, near Çıralı, November 2011 (IR)

from Kalkan *Crocus mathewii* is scattered among *C. graecum*, which here are not very floriferous, probably because the tree cover is too dense. Between Finike and Elmalı at about 1,000 m (3,280 ft) altitude we found a population of C. *graecum* with long narrow leaves.

In the Cilician Taurus *C. graecum* meets another cyclamen species, namely C. *cilicium*. I have never seen them actually intermingled but have seen them less than a stone's throw apart. Just north of Alanya, at Mahmutseydi, for instance, there was only a narrow path between the two species. Another spectacular population is to be found east of Alanya, where in a very rocky and quite shady spot *C. graecum* has settled in rock crevices, displaying silver leaves of positively artistic designs. And so along for well over 200 km to west of Taşucu to a location which is thought to be the eastern outpost of *C. graecum*. Here the plants grow in a dried-up, very stony river bed. They seem fairly uniform, the flowers of a clear medium pink with barely twisted petals and small auricles, and a tendency for small basal blotches which diffuse a little way into the petals rather than sending up the typical prongs. There seems to be an enormous variety of forms of C. *graecum* in Turkey, often within one population, although I cannot base my observations on detailed studies… ."

Safely tucked into limestone crevices, *C. graecum* subsp. *anatolicum* at its most easterly point in southern Turkey near Silifke (MD)

A rocky, wooded habitat for *C. graecum* subsp. *anatolicum* in Turkey, Bey Dağ west of Antalya (BM)

***Cyclamen graecum* subsp. *candicum* in Crete** One of the most informative accounts of this subspecies, along with a superb selection of photographs, appears in the magnificent volume *Flowers of Crete* by John Fielding and Nicholas Turland (2005) and we can do no better than to quote directly from this:

"*Cyclamen graecum* subsp. *candicum* is endemic to Crete and fortunately fairly common. In the north-west it grows at Platanias where there is a dwindling colony on the now heavily built-on bluff above the town; this is within sight of a population of subsp. *graecum* on the Rodopou peninsula. It is unlikely that the two subspecies are incompatible but they appear not to have interbred. Subspecies *candicum* is common in hilly areas through to central Crete and then appears to terminate in the hills south of Malia about two thirds of the way towards the east of the island. It occurs from sea level to over 1,100 m in the Lefka Ori, where it grows in full sun with *C. creticum* on the E4 route above Askifou and also just north of the Omalos plain. The flowers, compared with those of subsp. *graecum*, are generally short-petalled and compact, usually almost white with a dark magenta sinus marking. Darker pink flowers do occur throughout the range but seem uncommon. No doubt there would be occasional pure white forms without the dark sinus but they would be hard to spot. Often in the wild the plants seem small, especially at high altitudes, with only one or two flowers, or none at all. At lower elevations one very occasionally sees large plants with 15 or more flowers. The leaves are by far the darkest of the subspecies of *C. graecum* and, even when poorly marked, are especially striking in contrast with the pale flowers. Indeed, the darkest ones appear almost black, though in reality are very dark green. When a leaf is heavily marked with a silvery pattern in contrast to the 'black' ground colour it is fabulous and looks as if made of velvet."

2.2j The Rohlfsianum group

(= Subgenus *Cyclamen*, Series 4)

Comprising: *Cyclamen rohlfsianum*

This group has a sole representative that is unlike any other cyclamen. It is an autumn-flowering species with large, coarsely lobed leaves and large flowers with darker staining and swellings ('auricles') around the mouth. Both the stamens and style are exserted, a unique characteristic in cyclamen. In fruit, the pedicels start to coil upwards from at or near the base. 2n = 96.

Cyclamen rohlfsianum

AUTUMN FLOWERING

Undoubtedly one of the most tender of all the *Cyclamen* we cultivate, this rather local endemic of the Gebel Akhdar region of Cyrenaica is also one of the more uncommon species in cultivation, perhaps for this very reason of lack of hardiness. This dictates that the species has to be grown under glass in frost-free conditions but it is well worth the effort as a well-flowered specimen can be very striking. It is probably true to say that there have been few introductions of the species to cultivation, and it is said to have been introduced around 1929. In an article in the Alpine Garden Society's Bulletin for March 1945, W. E. Th. Ingwersen comments on the "almost mythical" *C. rohlfsianum* as thriving in the cliff house of Dwight Ripley in Sussex; this was a wall of rock within a glasshouse, the crevices all planted with plants that were unlikely to survive unprotected outside. The species was certainly established in specialist collections in the 1950s and 1960s when the alpine house at Wisley displayed some fine large and obviously old specimens. These may well have originated from the

Cyclamen graecum subsp. *candicum* in Crete growing in dissolution holes in limestone, a typical habitat; the flowers of this subspecies are often pale and the leaves among the most striking of all cyclamen (SB)

Well-grown plant of *C. rohlfsianum*. The flowers and leaves may appear together depending on the watering regime (SB)

Cyclamen rohlfsianum. Painted by Pandora Sellars (reproduced at 120% of original size)

plantsman Eliot Hodgkin, who travelled widely and had a valuable circle of contacts. Peter Davis and Jim Archibald also collected tubers of *C. rohlfsianum* in Libya in the mid-twentieth century and no doubt some of the plants in cultivation today are derived from their collections.

Cyclamen rohlfsianum is one of the more distinct species of cyclamen, unique in that its cone of anthers protrude out of the mouth of the flower, whereas they are held well within the corolla tube in all other species. It also has a distinctive tuber that, with age, becomes very large and knobbly with several growing points, eventually facilitating division, although this is not a particularly practical method of propagation. Such vegetative propagation might, however, be useful in the case of a selected variant, such as the white-flowered form described below, if seed propagation proves unsuccessful. The foliage too is unlike that of any other cyclamen, the leaves being very coarsely lobed and toothed at the margin and up to 16 cm across in vigorous specimens. Another point worthy of comment is that *C. rohlfsianum* has the highest chromosome number (2n = 96) of any cyclamen, except for *C. somalense* and a few of the florists' *C. persicum* cultivars. The species with the next highest count is *C. graecum* (2n = 84). It is interesting that this and *C. rohlfsianum* share a feature that is unusual in the genus, that of coiling their flower/fruit stems from the lower part upwards, unlike all other *Cyclamen* species which begin to coil from the top downwards after fertilisation.

There is no doubt that the area from which *C. rohlfsianum* originates, Gebel Akdar (Jabel Akhdar) in Cyrenaica, has been isolated for an exceedingly long time, having been cut off during the chaotic geological history of the Mediterranean region. Thus, its native plants have evolved along paths separate from the rest of their relatives. There are other local endemics in the area including *Crocus boulosii*, named in 1968, which is related to species in Cyprus (*C. veneris*), Syria (*C. aleppicus*) and Crete (*C. laevigatus*) indicating past floristic links between these now physically separated regions.

The unique combination of the characteristics of *C. rohlfsianum* means that it is difficult to know where to place it in the overall botanical classification of *Cyclamen*, so the obvious course for researchers has been to place it in a series of its own, series *Rohlfsianum* as proposed by Christopher Grey-Wilson, but allied to series *Graecum*. Very few species of cyclamen have gone through life since their discovery and formal description without acquiring a string of synonyms but this Libyan 'one-off' has achieved that distinction.

As to the epithet, one might ask who was Rohlfs? And how did it come about that in 1857 Paul Friedrich Ascherson commemorated him in a cyclamen? His name is certainly not a familiar one in plant collecting or botanical circles. The same question puzzled John Grimshaw and the interesting findings of his research are repeated below.

C. rohlfsianum Aschers., *Bull. Herb. Boiss.* 5: 528 (1897). Type: Libya, Benghazi, November 1879, *Rohlfs & Stecker.*

DESCRIPTION **Tuber** regular in the young stages but becoming increasingly misshapen and knobbly, up to 20 cm diameter, brown, rough and cork-like, roots slender, tending to be produced from the sides of the tuber but sometimes also from the base. **Petioles** prostrate in the lower part below ground, then ascending to erect above ground. **Leaf** blade suborbicular to wider than long (almost reniform), often dark lustrous green with a silvery-green zone near the margin, 3.5–12 cm long, 5–16 cm wide; margin coarsely and irregularly lobed and toothed; the underside stained reddish-purple. **Pedicels** papillose-hairy, horizontal in the lower part, turning upwards as they emerge above ground, after anthesis coiling in the lower part first. **Flowers** produced in autumn, normally just

Tuber of *C. rohlfsianum* (BM)

Flower of *C. rohlfsianum*. Note the exserted stamens, a unique character in cyclamen (BM)

An old, knobbly tuber of *C. rohlfsianum* (MH)

Leaves of *C. rohlfsianum* (BM)

before or with the developing leaves, fragrant; petals usually pale to mid pink with a deeper magenta stain at the base around the mouth, 10–23 mm long, 6–9 mm wide, twisted lengthways, with distinct swollen appendages ('auricles') at the base; stamens and style clearly exserted well beyond the mouth of the corolla.

Flowering period September to November.

Distribution endemic to Cyrenaica, Libya, recorded only in the Gebel Akhdar, the mountain range lying north-east of Benghazi.

Habitat In scrub in limestone areas among shady rocks, in crevices and gullies, sea level to 450 m in typical Mediterranean vegetation comprising woody species such as *Ceratonia siliqua*, *Cupressus sempervirens*, *Laurus nobilis*, *Olea europaea*, *Pistacia lentiscus* and evergreen *Quercus* spp.

Habitat of *C. rohlfsianum* in the Gebel Akhdar, Libya, 6 November (JW)

Libya, the Wadi Kuf in the Gebel Akhdar mountains of Cyreniaca, locality for *C. rohlfsianum* (JW)

Range of leaf markings in *C. rohlfsianum* from Peter Moore (BM)

Variations For a rather localised endemic species, *C. rohlfsianum* is surprisingly variable in leaf characters although, whatever the markings, the foliage has an unmistakeable appearance and is unlikely to be confused with that of any other species. The flowers do not seem to be very variable in colour, and again, immediate recognition of this species' flowers is not difficult because of the consistently protruding anther cone. A white-flowered variant has occurred in cultivation in the collection of Susan and Michael Ayling of Kingsley, Hampshire, arising as one of a batch of seedlings from long-cultivated stock. This has completely white flowers with no trace of colour around the mouth and was described as *C. rohlfsianum* forma *album* by Brian Mathew in the June 2002 issue of the Cyclamen Society's journal.

Cyclamen rohlfsianum forma *album*, a pure white-flowered form which occurred in cultivation (MD)

Further observations on *C. rohlfsianum*

A visit to Libya Few enthusiasts have seen *C. rohlfsianum* in the wild and there are even fewer written observations, so we are indebted to **John Whitehead**, an inveterate traveller-plantsman for contributing the following observations about the species and its habitat:

"My first encounter with North African cyclamen was in 1983 when I saw many *Cyclamen africanum* growing amongst rocky outcrops on open grassy slopes in the cedar forests of the Kabylie mountains of northern Algeria. The second time it was to see *Cyclamen rohlfsianum* in Libya many years later in early November 2007, when entry visas were available but travel still limited to groups of tourists for visiting the many famous historical sites. The opportunities for botanising native plants in the wild was difficult, compounded by other problems of long distances to travel and a tight itinerary run by a local guide who was an expert on historic Libya but failed to appreciate the significance of interesting special endemic plants. Also the lack of autumn rains resulted in a sporadic flowering period.

The isolated Jabel Akhdar limestone mountain range is bordered by the Mediterranean Sea to the north and the Sahara Desert to the south, resulting in an interesting endemic flora. Plant specialities include the Libyan strawberry tree *Arbutus pavarii*, *Crocus boulosii* and the Libyan endemic, *Cyclamen rohlfsianum*; these are all recorded as occurring in natural habitats throughout the mountain range in the ancient area of Cyrenica.

During the long journey travelling by mini bus from Benghazi along the Jabel Akhdar mountains, eastwards to Cyrene, are landscapes of many wooded slopes and valleys, dominated by typical Mediterranean vegetation which could be tantalizingly viewed along the route. We stopped south of Cyrene to see the world's highest newly built suspension bridge, spanning the spectacular Wadi Kuf which is grand enough to be called a canyon. Surprisingly, the guide announced that the group could walk across the impressive bridge. This was our first stop in a wild location and by the road we were surrounded on the upper slopes with natural vegetation. Time was limited and luckily on an upper wooded slope I spotted a clump of attractive reddish stems and orange bunches of fruit of a strawberry tree. I asked the guide if I could go to see the tree and with a favourable reply and against the clock, Brenda and I climbed over the roadside barrier and at last we were able to botanise the limestone pebbled slope. Immediately we found plants in flower on the first open area, finding a white *Narcissus serotinus*, yellow *Ranunculus bullatus* and hooray, a pink cyclamen but without foliage! Frantically, I tilted the flower upwards and again hooray, it had the protruding stamens unique to *C. rohlfsianum*. Further up the slope we found many cyclamen in the shade and within thickets of ground level branches under shrubby trees. Due to the recent drought period most of the cyclamen plants produced only early single flowers and a few leaves. Two days later we were able to make a short visit to the bottom of the Wadi Kuf and along the riverside were forests of towering cypress trees. On a vertical shaded roadside bank were cyclamen plants with interesting variable leaf markings and one with a single flower and splendid luxuriant foliage. I noticed white spots on the foliage and when turning over a leaf to inspect the purple colouring underneath, a small cloud of whitefly took flight. During the week in Libya we had acquired plenty of pictures of good ruins but were also satisfied botanically that we had a close, albeit brief, encounter of the cyclamen kind."

A plant of *C. rohlfsianum* with remarkably lobed leaves. Jebel Akhdar, Libya (JW)

Cyclamen rohlfsianum in the Jebel Akhdar, Libya, showing clearly the projecting stamens and style, 6 November (JW)

Friedrich Gerhard Rohlfs, 1831–1896, and *Cyclamen rohlfsianum* While reading what he describes as "a ponderous tome" about the history of the Tanganyika Railways, **John Grimshaw** came across a mention of a German Consul-General in Zanzibar in 1885 by the name of Rohlfs. The book mentioned that, before he became a diplomat, he had been an explorer with a journey from Tripoli to Lagos to his particular credit. The probability seemed strong that this was the Rohlfs of *C. rohlfsianum* and Grimshaw resolved to find out more about him. Grimshaw quotes *Encyclopaedia Britannica*, which sums up Rohlfs's career thus: "Explorer renowned for his dramatic journeys across the deserts of North Africa." It goes on to say that he was "more adventurer than scientific explorer" but graciously concedes that he did manage to compile some valuable geographical information. From this and other sources, Grimshaw was able to patch together an outline of his career and writes:

"Rohlfs was born at Vegesack near Bremen, in Germany, on 14 April 1831. Presumably at quite an early age he ran away to sea because by 1850 he was back ashore, fighting against the Danes in Schleswig-Holstein. After this he joined the Austrian army, which he deserted in favour of the French Foreign Legion in 1855; from this, however, he was discharged. He must have put his time with the Legion to good effect for he learnt Arabic sufficiently well to undertake a journey in Morocco in 1862 disguised as an Arab. His disguise was less complete than that of his contemporary Sir Richard Burton who, a few years previously, had penetrated in disguise even into the Kaaba in Mecca. Rohlfs was less fortunate and he was robbed, wounded and left for dead by brigands. Undaunted, in 1864 he explored the Fezzan region of Libya. His greatest journey took place in 1865–1866 when he became the first European to cross Africa from the Mediterranean at Tripoli to the Gulf of Guinea at Lagos, following the river Niger in the latter part of the journey. Following this adventure it was back to the wars, accompanying General Napier's Magdala expedition against the Emperor Theodore of Abyssinia in 1868. The mad emperor was holding the British consul and some missionaries hostage; to rescue them Disraeli's government sent 5,000 men — including Rohlfs — with artillery and a baggage train of Indian elephants. The expedition returned home after Theodore shot himself with a revolver given to him, presumably on an earlier occasion, by Queen Victoria.

The accounts of Rohlfs's life after this are unfortunately rather hazy. His journey in 1874 from Tripoli to Egypt is hastily passed over as rather an anticlimax and for the years between then and his appointment to the Zanzibar consulate in 1884 no information has come to light. This is unfortunate, for the type specimen of *C. rohlfsianum* was collected (by Rohlfs and Stecker), according to Grey-Wilson (1988), in 1879. If this date is correct, Rohlfs could not have collected it on his Tripoli–Egypt journey but must have returned to Libya in 1879 [Rohlfs did indeed return to Libya, see below — Editor]. The appointment to Zanzibar was undoubtedly influenced by Rohlfs's noted imperialist ambitions for the Fatherland. It was the period of the notorious scramble for Africa and it was during Rohlfs's brief term of office that Germany's most notorious scrambler, Karl Peters, set off into the interior on the expedition that resulted in the annexation of what is now Tanzania by Germany. Rohlfs's stay in Zanzibar lasted only seven months and from then, July 1885, to his death at Bad Godesberg in 1896 no information about his activities has come to light.

As *Encyclopaedia Britannica* indicates, Rohlfs was an adventurer and his scientific achievements seem limited. No birds or mammals bear his name, which argues strongly in favour of the conclusion that he was a less than avid collector but he does have a few plants bearing his name: a grass, a sedge, an orchid and of course he has the distinction of being the only person to be commemorated by a cyclamen species."

Friedrich Gerhard Rohlfs' Libyan Expedition In 1879, a scientific expedition was sent by Kaiser Wilhelm I to Libya, the members of which were Gerhard Rohlfs, Anton Stecker, Franz Eckhart and Karl Hubner. There is a detailed account of this in *Reise von Tripolis nach der oase Kufra* (Leipzig, 1881). The expedition carried splendid presents of gold and silver, no doubt with the intention of smoothing the path for the explorers as they attempted to penetrate southwards far into the desert to study the oases of the Kufara area. There are many references to Rohlfs and his hazardous journey in the very readable *The Secret of the Sahara: Kufara* by Rosita Forbes who repeated the journey some 40 years later and was only the second European to reach this wild region. Rohlfs apparently arrived back in Benghazi in October and clearly, in spite of all the hardships of his journey, was still interested enough in his surroundings to collect the cyclamen that is now named after him.

Rohlfs was one of the great explorers of the age in Africa and his work was acknowledged by Charles Darwin. In a letter dated 8 June 1871 Darwin "thanks GR, who made one of the most wonderful expeditions on record" and goes on to say that he will consider his [Rohlfs's] notes if he ever has to prepare a new edition [?of *Origin of Species* — Editor].

2.3 *Cyclamen* hybrids

On the whole, cyclamen do not hybridise freely, unlike plants of many genera where, after a few years of cultivation, it is difficult to identify whether one has 'true' plants or a 'soup' of hybrids. There are hybrid cyclamen but it is probably true to say that these are more of passing botanical interest, demonstrating that such hybrids are possible, rather than of any great horticultural merit. With almost endless variation available within most of the species, the huge range of superb house plants provided by *C. persicum*, and a nearly year-round flowering time for the genus as a whole, it is difficult to see how much 'improvement' could be made by hybridisation.

Perhaps the most desirable hybrids from the point of view of the cyclamen enthusiast are *C.* × *wellensiekii* (*C. libanoticum* × *C. cyprium*) and *C.* × *schwarzii* (*C. libanoticum* × *C. pseudibericum*).

Cyclamen × schwarzii, (*C. libanoticum × C. pseudibericum*). Cultivated by Ian Robertson (JA)

Cyclamen × perpuris 'Odorella', a hybrid raised in Germany by a team from Erfurt (JB)

Named hybrids are:

C. × atkinsii — described 1853 by T. Moore; purported to be *coum × persicum*.

C. × drydeniae — described 1997 by C. Grey-Wilson; *coum* subsp. *coum × alpinum*.

C. × hildebrandii — described 1955 by O. Schwarz; *africanum × hederifolium*.

C. × marxii — described by J. Halda in 1973; *colchicum × purpurascens*.

C. × meiklei — described 1997 by C. Grey-Wilson; *creticum × repandum*.

Cyclamen × wellensiekii (*C. libanoticum × C. cyprium*) (TB)

C. × saundersiae — described 1997 by C. Grey-Wilson; *balearicum × rhodium* ssp. *peloponnesiacum*.

C. × schwarzii — described 1997 by C. Grey-Wilson; *libanoticum × pseudibericum*; Grey-Wilson considers some of the plants that are cultivated as pale or whitish forms of *C. pseudibericum* to be of hybrid origin.

C. × wellensiekii — described 1997 by Ietswaart; *libanoticum × cyprium*.

C. × whiteae: described 1997 by C. Grey-Wilson; *graecum × hederifolium*.

C. hederifolium × C. persicum — a hybrid made in 1994 by Hiroshi Ishizaka (Japan) between *C. persicum* 'Sonja' and *C. hederifolium*. The resulting plants were sterile but after doubling the chromosome number the amphidiploid plants were fertile. The same cross was made soon after this by the Dutch firm Sahin, who obtained a hybrid plant by embryo rescue; the parents were a diploid *C. persicum* and a *C. hederifolium* with relatively large lilac flowers; it is said to flower almost continuously.

C. persicum × C. purpurascens — in 1995 Hiroshi Ishizaka obtained sterile hybrids from crosses between these two species; the result is known as 'Kaori No Mai' and is propagated by tissue culture.

The same interspecific cross has been made in Germany by Dr. A. Ewald from Erfurt and also has to be propagated vegetatively. It has a good scent and there are four colour variants: white, pink, rose and pale lilac. One clone is named *C. × perpuris* 'Odorella'.

Quercus coccifera scrub provides a home for *C. alpinum* although even this does not survive expanding human activity (BM)

2.4 Conservation

To some extent, cyclamen species are under no immediate threat of extinction as many of them grow in places, such as in extreme rocky environments, where there is unlikely to be housing development or conversion for agricultural use. Rapid climate change will, however, be a very serious issue for cyclamen and there are many other threats to individual populations. These are matters for concern as they can lead to a much-reduced and fragmented distribution, which results in a lack of gene exchange outside the remaining small, local populations.

Cyclamen species are included in the Convention on International Trade in Endangered Species (CITES) Appendix II. This regulates the movement of plants from one country to another; not only when being traded commercially but by any form

Agricultural development threatens many wild plants. Here, next to a newly planted olive grove, *C. mirabile* grows in the remaining scrubland which will inevitably be ploughed up (BM)

Changing land use can lead to complete destruction or fragmentation of wild populations. For the moment, *C. hederifolium* remains safe in the enclaves of scrub on steep hillsides. Western Turkey (BM)

Continuing construction of dams will threaten individual populations. Here in an area where *C. mirabile* occurs a valley will be flooded, probably destroying some populations. Fortunately the altitude range of the species means that it will survive above high water level (TW)

Although in the past populations of cyclamen have been under great threat from commercial collection for the horticultural trade, this practice has now greatly diminished in favour of propagation. Here in Turkey, *C. cilicium* is being produced in large numbers for the international market (MD)

Collection of cyclamen tubers for the 'bulb trade' at one time reached staggering proportions. These are *C. mirabile* tubers, a by-product of cultivation for a new olive grove (BM)

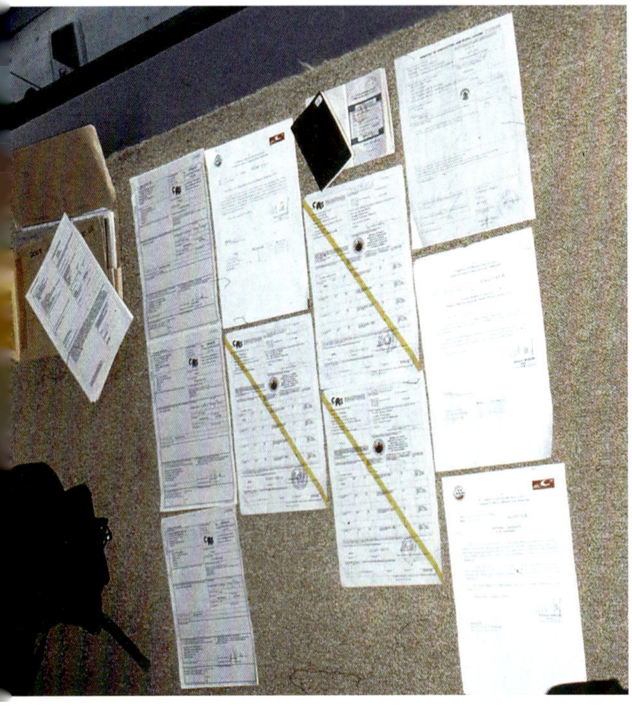

Official papers accumulated ready for Customs inspection at the airport! (BM)

The Cyclamen Society's field studies have done much to advance knowledge of the distribution of *Cyclamen* species and their status in the wild. For studies in Turkey the Society received an award from DHKD (Society for the Protection of Nature) (VAA)

of transportation, whether it involves money or not. Appropriate CITES documents are required both for exporting from one country and for importing into another. In some countries permits must also be granted to anyone carrying out any research in the field. In the case of cyclamen, documents are therefore required for CITES (export and import) as well as for research; acquiring these can involve much work before a collecting trip and result in a considerable sheaf of papers!

The interesting contribution from Max Debussche and John D. Thompson below, although restricted to the species belonging to subgenus *Psilanthum* (*C. repandum* and its allies), outlines various threats to cyclamen. Many of the observations are applicable to the genus in general.

What are the threats to natural populations of cyclamen?

Max Debussche and *John D. Thompson*

Eats flowers, tubers and leaves The scientific ecological literature contains many examples of how herbivory can impact on population growth and distribution in many plant species, particularly rare ones. During our field studies, we have often observed the effects of invertebrate and vertebrate herbivores on the different cyclamen taxa of subgenus *Psilanthum*. First, snails and slugs, during rainy periods, consume leaves, flowers and fruits. On Mallorca and in southern France, we have documented an impact of these herbivores just prior to ripening of fruits of Balearic cyclamen (*C. balearicum*), which can result in damage to or destruction of three-quarters of all fruits in a population (Affre *et al.*, 1995). Second, caterpillars of various Noctuidae and Geometridae consume leaves and flowers (see also Jope, 2006). We have observed up to 25% of leaf surface area destroyed, very probably eaten by an alliance of gastropods and caterpillars in a population of Rhodian cyclamen (*C. rhodium*). However, these observations could be made on many species in the wild and the levels of damage are unlikely to impact on the long term viability of cyclamen populations due to the longevity of plants and their reliance on tuber-stored reserves (see e.g. Andrieu *et al.*, 2011).

Wild boars, on the other hand, do serious damage to natural cyclamen populations in some regions. They are by far the most important consumers of tubers and often leave many to dry up on the soil surface after a night of frenzied digging and scraping on the forest floor. In this way, they can cause very visible and sometimes quite spectacular damage to Balearic cyclamen (*C. balearicum*) populations in southern France and to Tyrrhenian cyclamen (*C. repandum*) as well as to ivy-leaved cyclamen (*C. hederifolium*) on Corsica. After observing such effects, one can easily understand how cyclamen have become known as 'sowbread' in many countries (see Rupreht, 1995). Wild boar numbers have increased dramatically in several European countries over the past few decades due to concomitant causes, mainly because of forest spread that has followed rural depopulation and land abandonment and the introduction for hunting purposes of animals from very fertile stocks (e.g. Saez-Royuela & Telleria, 1986). But once again, boars are unlikely to have a significant long-term impact on cyclamen populations.

Land-use changes are currently the most important factor impacting on Mediterranean biodiversity, and are likely to remain so through the twenty-first century, despite the growing attention paid to climate change, which is predicted to be of secondary importance relative to land use changes over coming years in the Mediterranean region (Sala *et al.*, 2000). Having said that, contemporary changes in land use are not dramatically threatening to the taxa of subgenus *Psilanthum*. The exception is in North Africa, where the Kabylian cyclamen (*C. repandum* var. *baborense*) is reported to be seriously endangered by systematic forest destruction and regular and huge fires. Where fires remain moderate and irregular, however, they do not generally have dramatic direct effects because they occur in the summer, the dormant period of the annual cyclamen life-cycle, and often result in rapid resprouting of the burnt vegetation. By contrast, where evergreen woodlands and shrublands become very dense and closed because of the absence of any human activity or because few clearings are made by the falling of old remnant trees, flowering is reduced and the populations of both cyclamen and other Mediterranean geophytes are decreasing (Andrieu *et al.*, 2007). Except for the Kabylian cyclamen, it is mostly cyclamen populations that are near the Mediterranean coasts where tourism and human density are expected to increase dramatically in the coming decades that are under serious threat; such populations include those of Tyrrhenian cyclamen in the Maures massif in southern France and of Balearic cyclamen near St Florent on Corsica.

Climate change The 4th Assessment Report on Climate Change (IPCC, 2007) predicts both a global mean increase of 3°C during the twenty-first century and a 20% decline in rainfall in the Mediterranean region. Faced with such change, there are two possible ways for plants, like any other organism, to survive: migration or adaptation. For cyclamen taxa, and many other plants and animals, migration is difficult. First, as their seeds are primarily dispersed by ants, the migration of cyclamen taxa is a slowly progressive process (Hildebrand, 1898, Affre *et al.*, 1995). Second, in subgenus *Psilanthum*, the spread of several taxa is impeded by major geographic barriers, primarily the Mediterranean Sea for those occurring on islands, or high mountain ranges and cultivated and urbanised areas. Nevertheless, many plants do migrate by following river valleys and gorges or by shifting their elevation limits upwards. This would be nothing new to many species, which have undergone previous episodes of migration during the repeated glacial-interglacial periods of the Quaternary. What will be new is the need to migrate more rapidly than in the past and through a highly disturbed and fragmented human-dominated landscape. In this context, plants that are capable of long-distance dispersal will obviously have the best chance of survival, and cyclamen taxa will not be amongst such plants.

The presence of most cyclamen taxa over a range of altitudes, and thus of climatic variants, suggests that these plants have a capacity to respond locally to climate change in the different geographic regions where they occur. Other than Rhodian cyclamen, the various taxa in subgenus *Psilanthum* have an altitudinal amplitude that is of the order of 1,200–1,400 m. Translated into climatic variability (a 0.6°C decrease per 100 m elevation), this means that the local range of climatic tolerance (9.5°C and 7.5°C for Tyrrhenian and Balearic cyclamen, respectively) largely exceeds the 3°C warming (i.e. 500 m variation in altitude) predicted for the twenty-first century. A similar pattern is evident for mean annual rainfall. So, in relation to changes in average temperatures, most taxa of subgenus *Psilanthum* should be able to persist within the current geographical distribution by modifying their presence along elevation gradients, or more simply by remaining on the same site. In this context, Rhodian cyclamen is certainly the taxon most at risk, but also under threat are the insular and/or isolated populations of Balearic cyclamen on Cabrera, Ibiza, and Corsica, Tyrrhenian cyclamen on mainland France, Cretan cyclamen on Kárpathos, and any isolated population of whatever taxon occurring at low altitude without a corridor of favourable habitats to facilitate migration to higher elevations.

Despite this, a simulation study of the consequences of global warming by Yesson and Culham (2006) resulted in pessimistic predictions (for the year 2050): reductions in the distribution area of 71–97% for Balearic cyclamen, 50–100% for Cretan cyclamen and 44–87% for the group including Tyrrhenian, Peloponnese and Rhodian cyclamen. Our field data provide no support for these wide-ranging predictions. We feel that the interplay of climate change and habitat fragmentation will be more important than changing climate alone. Habitat fragmentation will impede migration in a changing climate and will bring many isolated coastal and low-altitude populations to the rim of extinction, but it will not necessarily cause the extinction of whole taxa.

CHAPTER THREE

Scientific studies and fieldwork

3.1 INTRODUCTION

This chapter contains the results of botanical studies by various researchers and of teams working in the field. Some of this work, such as the floral scent survey, has encompassed the whole genus; other investigations are confined to a single group — for example the subgenus *Psilanthum*, which provides perhaps the greatest challenge to taxonomists — or even just one species. The Cyclamen Society has been particularly active in this last respect, carrying out field surveys on several cyclamen species, notably in Greece and Turkey. Another project has been carried out in recent years by Trevor Wiltshire, working on behalf of the Society. Wiltshire has been advising on the cultivation and propagation of cyclamen at the Başkonuş research station near Kahramanmaraş in southern Turkey. A trial ground was set up at the DHKD (Society for the Protection of Nature) Conservation Training Centre with Evrim Tabur as the project officer, working in collaboration with Prof. Neriman Özhatay of Istanbul University. The work is supported by the Baku-Tiflis-Ceyhan Company, which is constructing an oil pipeline through this region of Turkey.

Cyclamen propagation beds at Baskonuș research station, Turkey (TW)

3.2 THE SCENT OF CYCLAMEN

Robin Clery, Trevor Wiltshire & Dominique Lelièvre

Introduction

This is a summary of the work carried out at Wisley and Quest in 2002 on the scent of a number of cyclamen species. That cyclamen have a scent is well known, but the variety of odour types represented and the similarities and differences between them have not been studied in any detail before.

Equipment for 'headspace sampling' for scent analysis (TW)

We present here a list of odour descriptions of the type used in perfumery for each of the species that have been assessed. Several of these have also been investigated by 'headspace analysis', a procedure that captures the fragrance from the blooms so that its chemical composition can be determined in the laboratory. From these analyses and descriptions, it has been possible to recognise three distinct fragrance groups among the different species, each with a characteristic scent: *C. purpurascens* has a rosey, muguet fragrance; *C. mirabile* a strong coconut odour; and *C. rohlfsianum* a waxy, dry "old hymn books" smell. *Cyclamen graecum* and several variants of *C. hederifolium* are example of plants with intermediate scents that fall within the range of the characteristic odour-types.

Odour descriptions

The following list is a compilation of odour descriptions on various different occasions in the format:

Species, plant location and/or number if known [day/month/year]
Description.

C. purpurascens W945445 [18/9/1998]
Muguet (citronellol), rosey (geraniol), slightly lemony, slightly spicy cinnamon or cinammic.

C. purpurascens [16/10/2002]
Clean floral, muguet or rose (citronellol), citrus, fresh slightly lime.

C. mirabile [18/9/1998]
Coconut, undecalactone, gamma decalactone, "pech pure".

C. mirabile W895822 [5/10/1999]
Coconut, typical undecalactone note.

C. mirabile Various individuals, all in full bloom [16/10/2002]
Strong clean, coconut or undecalactone note.

C. hederifolium (Zante) W1997 1964A [18/9/1998]
Citronellol, muguet, cyclamen aldehyde, spicy or cinnamic. (Less rosy than *C. purpurascens*.)

C. hederifolium (Zante) [18/9/1998]
Very similar to above, but slightly spicy or sharper, with notes of lilac or terpineol, methyl salicylate.

C. hederifolium (Zante) [16/10/2002]
Muguet or rose (citronellol) with aldehydic or apple blossom notes.

C. cilicium [18/9/1998]
Sweet vanilla or chocolate, "heather honey", indolic, bergamot or linalyl acetate.

C. cilicium [5/10/1999]
Faint honey, aldehydic scent; also coconut, slightly floral-rosey.

C. cilicium Various individuals, all in full bloom [16/10/2002]
Honey (phenyl acetaldehyde), hyacinth (green aldehydic, 9-decenal-like) slightly coconutty (lactones: undecalactone and so on), slightly muguet or rose (citronellol) and also fresh green terpenic notes, salicylates.

C. cyprium [18/9/1998]
Citronellol, like *C. purpurascens*, rosey, cyclamen aldehyde.

C. rohlfsianum [18/9/1998]
Very weak scent, musty or dusty, "old hymn books", earthy, aldehydic.

C. rohlfsianum Single plant with one flower [16/10/2002]
Dry aldehydic, wax-jacket, "old hymn books", slight geranium/metallic 9-decenal note and slight dry spicy woody notes.

C. graecum album [18/9/1998]
Very faint scent, cyclamen aldehyde, slightly sour-fruity, butyric, aldehydic.

C. graecum album [16/10/2002]
Weak, coconutty, slightly fruity — "apple blossom" note; also aldehydic, similar to *C. rohlfsianum* but weaker.

C. graecum 895889 (pink/purple flowers) [16/10/2002]
Honey, muguet or rosey (citronellol), also some sweet estery notes and apple-blossom or aldehydic notes.

C. intaminatum W953598A [5/10/1999]
Slightly coconutty, softer than *C. mirabile*.

C. intaminatum Various individuals, all in full bloom [16/10/2002]
Honey and coconutty (lactones: undecalactone and so on) notes predominant; also some iso-butyrate or angelate notes like camomile oil.

Odour mapping

To represent the odour types graphically, we have used a triangular odour map. Each corner represents a characteristic odour type: rosey, coconutty and dry waxy. Those species with intermediate scents are positioned on the map according to the relative strength of the three odour-types in their scent. Some species have other components to their scent, such as honey, medicinal or spicy notes, but most could easily be described in terms of these three characteristics.

Cyclamen scent map

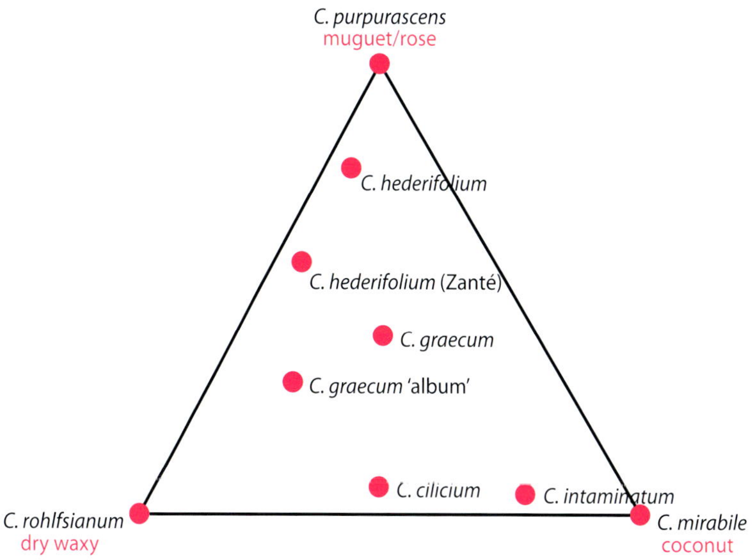

The chemical composition of the scents of six species has been investigated in more detail by trapping the volatiles and recovering them for analysis by gas-chromatography/mass-spectrometry (GCMS) see 'Appendix'. *Cyclamen purpurascens* and *C. mirabile* have the strongest scent, which could be analysed to provide greater detail. The key volatile chemicals responsible for the characteristic scent are citronellol for *C. purpurascens*, a series of gamma lactones for *C. mirabile*, and undecan-2-one for *C. rohlfsianum*.

Odour groups and associated scent chemicals

To determine whether specific fragrant chemicals could be associated with the distribution of scent types emitted by each species, the relative proportions of citronellol, gamma-undecalactone and undecan-2-one emitted by several species were used to plot these species on a triangular graph.

Rosey or floral	Coconutty or fruity	Waxy, dry or aldehydic
Citronellol Geraniol Linalool Phenylethyl alcohol	All lactones C5–C8 esters	C8–C12 aliphatic aldehydes Undecan-2-one Phenylacetic aldehyde

For example, the proportion of 'rosey' odour was calculated as:

$$\text{Rosey}/(\text{Rosey} + \text{Coconutty} + \text{Waxy}) \times 100\%.$$

As all three scent components are not emitted by all of the species tested, this tends to plot the species along the sides of the triangle, but nonetheless close to their positions on the odour-map.

A better distribution was achieved by selecting other key fragrant chemicals from the analyses and by using the totals for

Proportions of odour types in the scents of each species

Species	Plant no.	Proportion of compounds in each odour group (%)			Normalised to 100%		
		Rosey	Lactones	Aldehydes	Rosey%	Lactone%	Aldehyde%
C. purpurascens		47.5	0.04	0.06	99.8	0.1	0.1
C. mirabile	895822	0	57	1	0.0	98.3	1.7
C. rohlfsianum		0	0	4.6	0.0	0.0	100.0
C. cilicium	[1998]	0	0.2	5	0.0	3.8	96.2
C. cilicium	[1999]	1.2	0.2	13	8.3	1.4	90.3
C. intaminatum	953958A	0.5	17.5	29	1.1	37.2	61.7
C. hederifolium	W1997, 1964A	17.43	0.1	3.5	82.9	0.5	16.6

the three odour groups. These relative proportions of each odour group were calculated by sub-totalling the relative peak areas from the GC analysis for compounds in the three odour groups, and then expressing this sub-total as a percentage of the total for all three groups (see Table above).This distribution shows that the scents of different *C. hederifolium* plants lie in a range between the rosey and aldehydic (dry waxy) characters, whereas the scents of *C. intaminatum* and *C. cilicium* lie in the range of aldehydic (dry waxy) and coconutty types.

Even using a carefully chosen selection of scent chemicals cannot account for the full variation in odour types as each individual chemical has a different intensity of odour; thus a tiny amount of octalactone, which has a powerful coconutty and peach note, will have a very significant effect on the odour. Such differences in odour intensity explains why *C. cilicium* plots closer to the 'lactone' corner on the odour description graph than on the chemical composition graph. The scent plot based on the chemical analysis of a few of the key ingredients does not match exactly the plot of odour descriptions, but it is clear that the two diagrams are similar.

Conclusions

On the basis of the plants assessed so far, cyclamen can be categorised as having rosey, waxy or coconutty scents. Within these broad groupings, the scents are modified with green, citrussy, aldehydic, honey and fruity notes. This categorisation is supported by the chemical analysis of the scents of six species. The key odour compounds in the rosey group are citronellol, geraniol and phenylethyl alcohol, the key compounds in the coconutty group are gamma lactones and the key compounds of the waxy group are undecan-2-one and aldehydes.

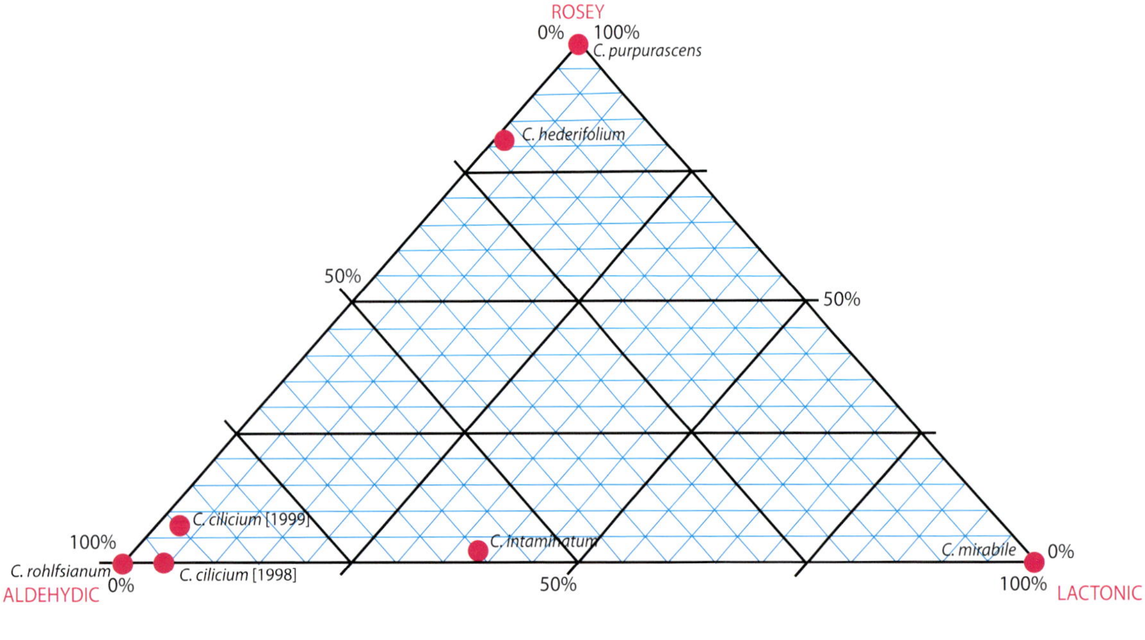

Triangular plot of fragrance components

APPENDIX

Names in CAPITALS are confirmed identifications. Names in lower case are tentative or generic identifications.

tr = trace, i.e. <0.01% relative peak area (RPA)

***C. purpurascens* W945445**

Identification	%RPA
ALPHA-PINENE	0.06
SABINENE	tr
BETA-PINENE	0.08
6-METHYLHEPT-5-EN-2-ONE	1.1
MYRCENE	0.01
PARA-CYMENE	0.02
LIMONENE	0.37
1,8-CINEOLE	0.01
TERPINOLENE	0.01
METHYL BENZOATE	0.27
LINALOL	0.87
NONANAL	0.05
CIS-ROSE OXIDE	0.1
METHYL OCTANOATE	0.07
TRANS-ROSE OXIDE	0.04
CITRONELLAL	0.95
BENZYL ACETATE	0.02
INONYL ACETATE	0.03
ALPHA-TERPINEOL	0.01
METHYL SALICYLATE	tr
DECANAL	0.01
CITRONELLOL	43.01
GERANIOL	0.69
METHYL CITRONELLATE	2.64
GERANIAL	0.02
CITRONELLYL FORMATE	tr
CINNAMIC ALDEHYDE	0.13
CITRONELLYL ACETATE	0.05
METHYL ANISATE	0.03
BETA-ELEMENE	1.39
ALPHA-CEDRENE	0.01
CARYOPHYLLENE	0.24
GERANYL ACETONE	1.47
farnesene isomer	0.03
HUMULENE	0.01
GAMMA-DECALACTONE	0.04
GERMACRENE D	0.08
ELEMOL	1.24
NEROLIDOL isomer	0.29
SESQUI ALCOHOL (farnesol type)	2.05
10-EPI-GAMMA-EUDESMOL	0.37
10,11-DIHYDROFARNESOL	21.95
farnesol isomer	1.58
farnesol isomer	0.46
SESQUI ALCOHOL (farnesol type)	2.76
BENZYL BENZOATE	0.11
CITRONELLYL BENZOATE	0.05
Total	**84.78**

***C. hederifolium* W1997 1964A**

Identification	%RPA
BUTYL ACETATE	1.22
CIS-HEX-3-ENOL	tr
HEPTAN-2-ONE	tr
NONANE	0.45
HEPTANAL	0.71
ALPHA-PINENE	6.56
CAMPHENE	0.19
GAMMA-VALEROLACTONE	tr
BENZALDEHYDE	0.1
SABINENE	tr
BETA-PINENE	tr
6-METHYLHEPT-5-EN-2-ONE	7.3
MYRCENE	tr
DECANE	1.32
OCTANAL	0.53
3-CARENE	2.65
P-CYMENE	0.51
LIMONENE	1.59
1,8-CINEOLE	0.32
BENZYL ALCOHOL	tr
PHENYLACETALDEHYDE	tr
ISO-AMYL BUTYRATE	tr
OCTANOL	0.17
CIS-LINALOL OXIDE	0.48
GUAIACOL	0.18
LINALOL	4.25
NONANAL	1.33
CIS ROSE OXIDE	0.26
ORTHO-DIMETHOXYBENZENE	1.04
CAMPHOR	0.1
CITRONELLAL	0.75
METHYL SALICYLATE	tr
DODECANE	0.5
DECANAL	1.28
CITRONELLOL	9.85
GERANIOL	3.33
GERANIAL	1
CITRONELLYL ACETATE	2.83
GERANYL ACETATE	3.45
TETRADECANE	0.39
DODECANAL	tr
GERANYL ACETONE	4.04
DODECANOL	0.83
CITRONELLYL ISOVALERATE	0.14
GERANYL ISOVALERATE	0.22
Total	**59.87**

C. *mirabile* W895822

Identification	%RPA
ALPHA-PINENE	0.7
GAMMA-VALEROLACTONE	tr
SABINENE	tr
BETA-PINENE	0.1
6-METHYLHEPT-5-EN-2-ONE	0.51
OCTANAL	0.18
PHELLANDRENE-ALPHA	0.25
LIMONENE	0.31
PHENYLACETALDEHYDE	0.26
GAMMA-HEXALACTONE	0.12
NONANAL	0.49
a gamma lactone	tr
BENZYL CYANIDE	0.26
GAMMA-HEPTALACTONE	0.05
NONANOL	0.05
DECANAL	0.7
a gamma lactone	0.33
a gamma lactone	0.08
PHENYLACETALDEHYDE OXIME 1	0.33
GAMMA-OCTALACTONE	8.89
PHENYLACETALDEHYDE OXIME 2	0.3
2-PHENYL NITROETHANE	0.26
GAMMA-LACTONE	23.5
a gamma lactone	1.0
a gamma lactone	10.96
4-PHENYL-2-HYDROXY-3-BUTANONE	0.75
a gamma lactone	12.47
ISOLEDENE	0.53
a gamma lactone	0.01
a gamma lactone	0.01
ALPHA-HIMACHALENE	13.17
GAMMA-DECALACTONE	0.09
GAMMA-MUUROLENE	0.92
BETA-SESQUIPHELLANDRENE	1.13
10-EPI-BETA-EUDESMOL	0.29
Total	**79.0**

C. *rohlfsianum*

Identification	%RPA
ALPHA-PINENE	0.655
BENZALDEHYDE	0.541
PHENOL	2.236
6-METHYLHEPT-5-EN-2-ONE	0.689
DECANE	0.948
PARA-CYMENE	0.14
LIMONENE	0.64
BENZYL ALCOHOL	0.53
poss. 7-methyl octanal	1.04
UNDECANE	4.732
NONANAL	0.898
DODECANE	1.49
DECANAL	0.05
UNDECAN-2-ONE	2.04
ALPHA-COPAENE	0.2
TETRADECANE	17.84
CARYOPHYLLENE	0.361
TRIDECAN-2-ONE	0.1
PENTADECANE	13.19
Total	**48.32**

C. *cilicium*

Identification	%RPA
BUTYL ACETATE	tr
ISO-AMYL ACETATE	tr
ALPHA-PINENE	0.1
6-METHYLHEPT-5-EN-2-ONE	2.67
DECANE	0.5
OCTANAL	1.72
BENZYL ALCOHOL	0.5
PHENYLACETALDEHYDE	1.89
ACETOPHENONE	49.35
NONANAL	2.42
PHENYLETHYL ALCOHOL	1.25
NONANOL	2
ETHYL BENZOATE	2
DECANAL	7.78
GAMMA OCTALACTONE	0.1
a gamma lactone	0.1
GERANYL ACETONE	1.6
Total	**73.98**

C. *intaminatum* W953598A

Identification	%RPA
2-methyl-2-propenyl butanoate	12.28
6-METHYLHEPT-5-EN-2-ONE	6.34
DECANE	1.32
OCTANAL	4.13
UNDECANE	0.66
NONANAL	4.99
PHENYLETHYL ALCOHOL	0.5
DECANAL	17.21
a gamma lactone	2.28
UNDECANAL	1.61
a gamma lactone	12.59
a gamma lactone	1.71
a gamma lactone	2.89
DODECENAL	1.37
GERANYL ACETONE	2.85
Total	**72.73**

3.3 COMMENTS ON THE TAXONOMIC STATUS OF SPECIES IN *CYCLAMEN* SUBGENUS *PSILANTHUM* SCHWARZ

Alastair Culham & Pedro Bota

Without doubt, the most controversial part of the classification of *Cyclamen* is that of taxa within *Cyclamen* subgenus *Psilanthum* Schwarz. When first described, this subgenus included only three species: *C. repandum*, *C. creticum* and *C. balearicum* (Schwarz, 1964). A year earlier, the name *C. rhodium* was first used to describe the *Cyclamen* on Rhodes, adding a fourth entity to the subgenus (Gorer, 1963), but this name was not recorded by Schwarz as it was not validly published until 1975 (Schwarz & Lepper, 1975). On the face of it, *Psilanthum* is a small and simple subgenus, but this disguises the complexity of morphology and distribution of *C. repandum*, which could in fact be an aggregate of species or lower taxa.

Grey-Wilson has varied his treatment of the subgenus in his various books (Grey-Wilson, 1988, 2002). In their review of the nomenclature within *Psilanthum*, Compton & Culham (2003) reported that the use of names within the group had been in flux for many decades. We use this observation as our basis for sampling, and thus have examined the following taxa within the section:

Genus	Species	Infraspecies taxon	Authority	Date	Brief name
Cyclamen	*repandum*		Smith	1806	*C. repandum*
Cyclamen	*balearicum*		Willk.	1875	*C. balearicum*
Cyclamen	*creticum*		(Döerfl.) Hildebr.	1906	*C. creticum*
Cyclamen	*rhodium*		R. Gorer ex O. Schwarz & Lepper	1975	*C. rhodium*
Cyclamen	*repandum*	forma *album*	Grey-Wilson	1988	
Cyclamen	*repandum*	var. *baborense*	Debussche & Quezel	1997	
Cyclamen	*repandum*	var. *repandum*			
Cyclamen	*rhodium*	subsp. *peloponnesiacum*	(Grey-Wilson) J. Compton & Culham	2003	*C. peloponnesiacum*
Cyclamen	*rhodium*	forma *albiflorum*	(B. Mathew) J. Compton & Culham	2003	
Cyclamen	*rhodium*	subsp. *vividum*	(Grey-Wilson) J. Compton & Culham	2003	*C. vividum*

To avoid undue use of long names, we used brief names for each of the six taxa studied by Debussche and Thompson (2002) and subsequently by us. Debussche and Thompson (2002) present an excellent starting point for this review. They studied *C. balearicum*, *C. repandum*, *C. peloponnesiacum*, *C. vividum*, *C. rhodium* and *C. creticum* using six floral traits and five leaf traits. Their predominantly field-based study included almost 800 samples from these six taxa and revealed significant identifiable differences between all six taxa. In addition, their paper reported the occurrence of *C. balearicum* on Corsica (St Florent), where it appears to hybridise with *C. repandum*. Debussche and Thompson concluded that it would be appropriate to treat *C. balearicum* as a subspecies of *C. repandum*, given that the differences between these two entities were at a similar level to those within other subspecies of *C. repandum* recognised at that time. This added much needed standardisation of the way in which names are applied within this group.

Debussche and Thompson used a large number of samples but only measured a small number of vegetative characteristics. With this in mind, we expanded our study to include 17 vegetative traits, investigating whether additional vegetative characteristics are useful in distinguishing the species. We also added data from DNA sequencing to determine whether the underlying genetics tie in with the morphology (Bota, 2006).

Cyclamen subgen. *Psilanthum* occurs mainly on the northern edge of the Mediterranean but *C. repandum* var. *baborense* (Debussche & Quézel, 1997) grows on the north coast of Africa. Unfortunately, we were unable to include this variety in our study as insufficient material was available to us.

Leaf characteristics were assessed by a mixture of direct measurements (lengths and angles) and coded characters, such as leaf margin toothing. Some of the characters that were measured are illustrated here (blue = leaf apex angle, yellow = angle from tip to widest point, green = distance between lower lobe tips, red thick line = angle of sinus, red thin line = angle from leaf stalk between tip and second major lobe. Other characters that were recorded included the shape of marginal teeth, overall leaf shape (based on Cyclamen Society codes) and various leaf-patterning characters.

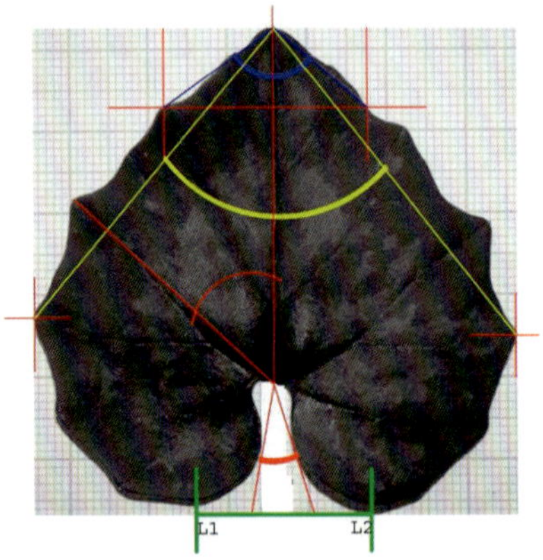

Principal component analysis of the 17 recorded variables showed that the six taxa formed two distinct groups: 1) *C. balearicum*, *C. creticum* and *C. rhodium*, which occur on islands, and 2) *C. repandum*, *C. peloponnesiacum* and *C. vividum*, which are mainland species. This separation is broadly in agreement with the findings of Debussche and Thompson. The separate groupings of island and mainland taxa could be indicative of climatic differences between islands and continental areas in the Mediterranean: the island species might have adapted to island life in a climate that is generally milder and damper than that on the mainland.

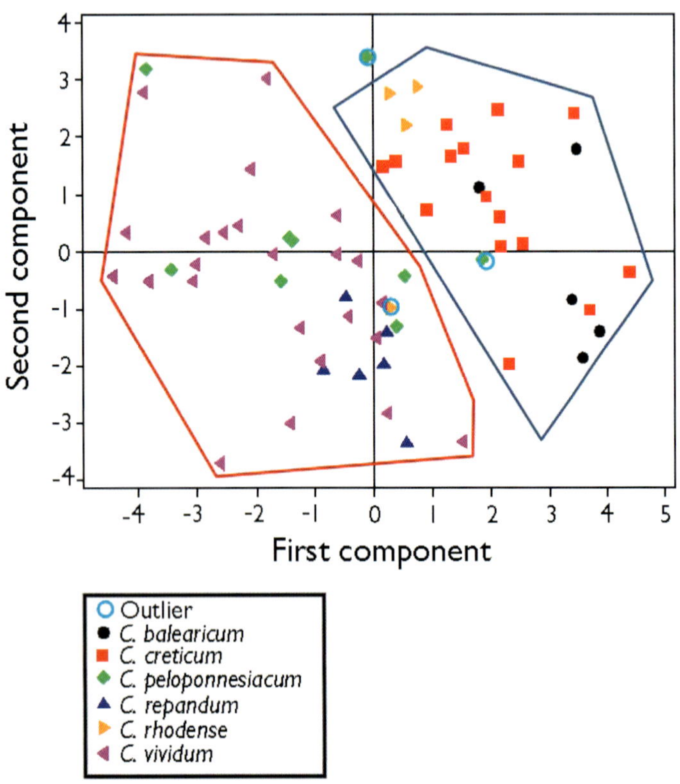

Analysis of the individual characters shows that each of the six taxa can be distinguished by at least four of the assessed variables; some species combinations show significant differences in up to 9 of the 17 characters that were recorded. This indicates that these six taxa can be identified by comparing a combination of average leaf characters for each species.

Numbers of the 17 assessed leaf characters that differed for each species pair

Cyclamen	*balearicum*	*creticum*	*peloponnesiacum*	*repandum*	*rhodense*
creticum	7				
peloponnesiacum	6	5			
repandum	8	8	4		
rhodense	7	6	8	6	
vividum	8	7	6	7	9

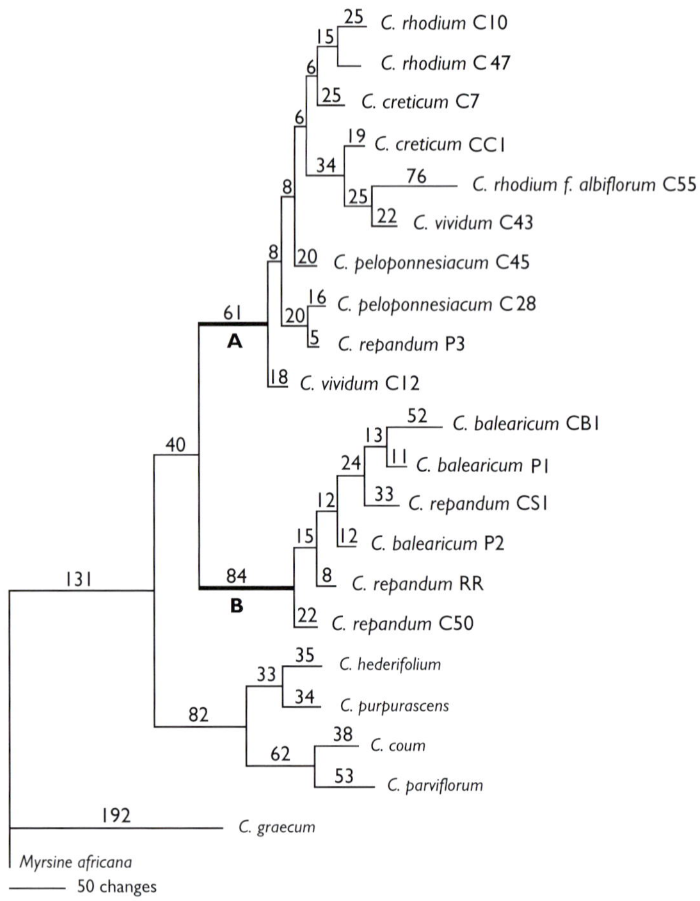

To augment the morphological data, DNA sequences in two chloroplast DNA regions and one nuclear DNA region were compared for 31 plants. DNA length is measured in base pairs; each base is a piece of information carried on the DNA strand and is represented by the 'rung' of the ladder in the conventional spiral (double-helix) diagrams of DNA structure. Some 2,522 bases of DNA were sequenced for each sample (giving a combined dataset containing almost 80,000 pieces of information in total). Just over 400 bases showed variability among samples, the uniformity among plants seen in the remaining bases essentially confirms that they were all *Cyclamen*.

Analysis of the DNA sequences shows them to fall into two distinct groups: A) *C. creticum*, *C. rhodium*, *C. peloponnesiacum* and *C. vividum*, and B) *C. balearicum* and *C. repandum*. Multiple samples sometimes had identical DNA sequences, in which case only a single representative sample is shown. The results from DNA data make geographic sense; furthermore, they tie in with known periods of population separation (Yesson *et al.*, 2008) and patterns of possible hybridisation and gene flow (Affre & Thompson, 1997a; Affre *et al.*, 1995, 1997; Gielly *et al.*, 2001).

In this initial survey, at least, the DNA data support the grouping of these taxa into two species corresponding with groups 'A' and 'B', with infraspecific taxa being recognised within them. Thus, we would have: A) *C. creticum* subsp. *creticum*, *C. creticum* subsp. *rhodense*, *C. creticum* subsp. *peloponnesiacum* and *C. creticum* subsp. *vividum*, and B) *C. repandum* subsp. *balearicum* and *C. repandum* subsp. *repandum*. We do not propose to make these combinations here. The level of variation within these groups suggests that further sampling and more sophisticated analysis would be useful in unravelling them, and that research in this area is still needed before nomenclatural changes should be made on the basis of DNA sequence data.

We are therefore left in the curious position of having taxa that are demonstrably identifiable through their appearance (i.e. by morphology) throughout most of their range and for most individuals, but that have a level of DNA sequence variation that indicates that the taxa are not yet fully separated. This leaves several options: 1) we could study the problem further using more DNA markers and possibly more morphological markers, 2) we could wait for evolution to combine or separate the lineages more fully (this could take some time), or 3) we could balance the current evidence and make a classification that is likely to be changed again in the near future.

Further resources

For those who are interested, a number of websites now archive old literature, so it is possible to see, free of individual charge, the descriptions for some of these *Cyclamen*:

Cyclamen repandum was first described by Smith in *Floræ Græcæ Prodromus* (Sibthorp & Smith, 1806; viewable free of charge at http://biodiversitylibrary.org/page/6114456 although a much fuller description and illustration appeared later in *Flora Graeca* (Sibthorp & Smith, 1836; http://www.bodley.ox.ac.uk/users/millsr/isbes/FG/FGE1/target354.html and http://www.bodley.ox.ac.uk/users/millsr/isbes/FG/FGE1/target355.html).

Cyclamen balearicum Willkomm (1875; http://biodiversitylibrary.org/page/9603730)is an old name of long standing for cyclamen plants in Majorca.

3.4 ECOLOGY AND EVOLUTION OF *CYCLAMEN* SUBGENUS *PSILANTHUM*

Max Debussche & John D. Thompson

Distribution and differentiation of subgenus Psilanthum across the Mediterranean

A dream voyage

Imagine the scene: it is spring, in successive stages you are moving eastwards across the Mediterranean from the Balearic Islands off the coast of Catalonia to the south-eastern islands of the Aegean Sea. On the island of Mallorca, your excursion will take you through rocky shrubland of evergreen lentisc (*Pistacia lentiscus*), with unexpected clumps of dwarf palms (*Chamaerops humilis*) on the rim of cliffs plunging down in the blue sea. In the south of France, you will have the opportunity to scramble up a narrow river bed, jumping from one limestone block to another, along the edge of a dense evergreen holm oak (*Quercus ilex*) coppice. Across the sea again, this time to Corsica, for a hike through ancient chestnut (*Castanea sativa*) orchards with carpets of eagle fern (*Pteridium aquilinum*) spotted with Corsican hellebore (*Helleborus argutifolius*) and littered with granite boulders. Then, across the rolling hills of Tuscany, the landscape will turn to woodlands of deciduous sessile and Hungarian oaks (*Quercus petraea* and *Quercus frainetto*) spattered with the white blossoms of flowering ash (*Fraxinus ornus*). In the Peloponnese peninsula, you will picnic next to a mossy spring in the shade of majestic oriental plane trees (*Platanus orientalis*), visit monasteries and climb through open woodland of Greek firs (*Abies cephalonica*) and Syrian junipers (*Juniperus drupacea*). Then on to the islands again, to Crete, where you will experience the famous limestone gorges where cypress trees (*Cupressus sempervirens*) cling to the cliffs, and finally to Rhodes, where bent double in dry river

Cap Formentor, Mallorca (DT)

A drift of Tyrrhenian cyclamen in Gargano mountains southern Italy (TH)

beds, you will crawl through dwarf carob trees (*Ceratonia siliqua*) and oleanders (*Nerium oleander*). Throughout your trip, each local excursion will bring a common pleasure, carpets of spring-flowering cyclamen, sometimes white, sometimes pink, sometimes with flowers that combine the two.

A fragmented geographic distribution

Cyclamen subgenus *Psilanthum* occurs across Mediterranean Europe eastwards to the Asian islands of Rhodes and Kos and southwards to the African Kabylia, and Crete. Mostly European in extent, this distribution has two main characteristics.

First, the distribution of the subgenus is not continuous from west to east. In fact, the fragmented distribution contains three main segments, within each of which the different taxa themselves have disjunct distributions, often on different islands and continental areas. The western-most segment has two parts: the Languedoc–Roussillon region west of the Rhône Valley in southern France and the Balearic Islands (Ibiza, Mallorca, Cabrera and Menorca). The geographically central segment occurs around the Tyrrhenian Sea, where it covers the Italian peninsula, the large islands of Corsica, Sardinia and Sicily to the west, the western coast of the Balkans to the east and two small isolated areas, one in Provence, France and one in Kabylia, Algeria. The eastern segment runs in an arc around the southern Aegean, from the Peloponnese peninsula, down across the islands of Crete, Kárpathos and Rhodes and up to Kos.

The second characteristic of the distribution of subgenus *Psilanthum* is that it is entirely included in the Mediterranean climate zone. Because this distribution covers 2,400 km from west to east and 1,100 km from north to south, varying in altitude from coastal areas up to mountain ranges, the different taxa experience almost all the variants of the Mediterranean climate. Thus, although summer drought is always severe across this distribution, below-zero winter temperatures and snow can be either non-existent or very common, and the major peak of rain can be in either autumn or winter and can be very variable in amount.

The underlying rock can be sedimentary (limestone, dolomite and sandstone), metamorphic (schist, gneiss and serpentine), intrusive (granite), or volcanic (basalt and rhyolite). Thus, soils can be basic, neutral or acidic with a range of different textures and chemical properties. As is typical for the Mediterranean (see Thompson, 2005), climatic and edaphic heterogeneity, combined with natural fragmentation of favourable habitats as a result of climate-induced changes in sea level and the striking relief of the terrain, have greatly contributed to the isolation of populations of *Cyclamen*. In fact, these factors both influence patterns of gene flow and provide the template for genetic differentiation and taxonomic divergence. How these factors have acted to create contemporary variation in subgenus *Psilanthum* is a theme that we develop at the end of this chapter. First, we describe the variability of morphological traits, ecological characteristics and genetic markers across the different taxa, illustrating their similarities and differences.

Spring flowering, lack of swellings (auricles) at the corolla mouth and roughly triangular leaves add up to a member of subgenus *Psilanthum*: Tyrrhenian cyclamen in Gargano, southern Italy (TH)

Unity and diversity in subgenus *Psilanthum*

The taxa in *Cyclamen* subgenus *Psilanthum* share a number of common traits that, together, allow us to distinguish this group from all other *Cyclamen* species (see Schwarz, 1964; Anderberg, 1994; Grey-Wilson, 1997; Clennett, 2002; Debussche *et al.*, 2004):

- flowering occurs in spring when the leaves are in full development;
- the flowers have elongated corolla lobes with no auricle at the mouth of the corolla;
- after flowering, the pedicel retracts in a corkscrew fashion and brings the fruit to ground level;
- the seeds are very small;
- the leaves have a thin, roughly triangular, limb with a patchy pattern that can be silvery, and, when underground, a thread-like petiole;
- the tuber is small, rounded and often flattened, with roots attached to the central part of the underside.

In addition, the taxa of this subgenus can be distinguished from other *Cyclamen* on the basis of genetic constitution (Anderberg *et al.*, 2000; Compton *et al.*, 2004). All taxa in subgenus *Psilanthum* have the same chromosome number (2n = 20), with the exception of some plants on Crete that have 2n = 22. This low chromosome number suggests an ancient origin within the genus. Moreover, attempts to hybridise plants from Crete, Corsica and the Balearic Islands show that these plants are fully cross-fertile (J. D. Thompson & L. Affre, unpublished data), another feature that underlines the unity of this subgenus and the close phylogenetic relationships of the component taxa.

Within the subgenus, plants do of course show a number of characteristic differences that enable us to distinguish the different taxa. These differences include variation in flower colour, floral design and leaf morphology.

Variation in flower characters

Flower colour has immediate visual impact, and in some cases, the basal ring of the corolla also has a distinctive colour. The flowers form a range of distinct colour variants, from the pure-white forms observed on Crete to the vivid carmine forms of the eastern part of the Peloponnese peninsula. There are plants with white flowers with a very thin light pink edging of the corolla lobes on the Balearic Islands and southern France or with a pink basal ring on Rhodes. In the Peloponnese, there are plants with contrasting pale pink upper and carmine lower parts of the corolla lobes, and there are specimens with pink-carmine flowers on Corsica. In a given population, however, it is often difficult to detect marked differences in flower colour except on the Peloponnese, where some populations contain both vivid carmine-coloured flowers and contrasting bi-coloured flowers. As a consequence of this rule of colour homogeneity, we were totally surprised when we discovered plants with a very large range of colours growing side by side in three very special populations on Corsica.

Flower size and the degree of stigma–anther separation also show much variation among different geographic locations. In the Balearic Islands and southern France, the white flowers are less than 2 cm long and have a stigma positioned within the corolla close to the anthers. By contrast, everywhere else, the flowers often measure 2.5 cm or more in length and all (or most in Crete) have a style that protrudes from the corolla with the stigma well away from the anthers, which themselves are placed within the corolla.

Variation in leaf characters

The leaf lamina is very variable within the subgenus in terms of colour and patterning, even within a population. Having said this, in some regions, the leaves are very distinctive. In the Balearic Islands and southern France, the lamina is slightly toothed, sometimes untoothed, blunt or rounded at the tip, with most of the upper surface patterned with grey or silvery markings. Across the eastern segment of the distribution, the leaf lamina of many plants not only have grey or silvery patches but are speckled in the same colour. On Crete, the lamina is often elongated and acutely pointed, whereas in the Peloponnese, the leaves can be as wide as they are long.

Questions and studies

Such variation in flower and leaf morphology raises a number of questions. What is its biological significance? How has it evolved and how is it maintained? Can we use such variation to help classify the genus into distinct taxa? How does it compare to the analysis of molecular variation? Ultimately, the question becomes whether or not we can identify distinct species within the subgenus on the basis of such variability. Our research on the ecology and genetics of this subgenus over the past 15 years has been centred on these questions, which we have addressed in relation to the major historical, ecological and evolutionary processes that are known to shape the evolution of plant biodiversity in the Mediterranean region (Thompson, 2005). Its geographic distribution and patterns of morphological variation make subgenus *Psilanthum* an ideal study system in this respect.

Our approach to studying variation within subgenus *Psilanthum* has been multi-faceted, beginning with detailed, extensive sampling of wild populations, which gave us a representative and well-balanced sample of all of the variation encompassed by this group of plants. We then carried out multivariate statistical analysis to detect which ecological features and morphological traits were able to discriminate different groups. Finally, we collaborated with molecular geneticists to produce a phylogenetic analysis of resemblance and differentiation within the group, which we compared with the morphological analysis. We have also made painstaking experimental studies of reproduction and dispersal, to examine both the functioning of plants in the wild and the causes of the patterns of genetic variation present within and among natural populations. When brought together, the findings from these different approaches reveal a fascinating story that has implications for our understanding of the evolution of plant biodiversity in the Mediterranean. They also provide the basic knowledge necessary for the application of conservation management plans for endemic species.

Classifications of taxa in subgenus *Psilanthum*

You only have to read the recent taxonomic studies of subgenus *Psilanthum* by Debussche and Thompson (2002), Compton and Culham (2003), Compton *et al.* (2004) and Clennett (2004) to realise that the classification of this group is difficult. Despite the alternative suggestions made recently, we stick to the position that we published in 2004 (Debussche & Thompson, 2004), which was based on morphology (Debussche & Thompson, 2002), reproductive biology (Affre & Thompson, 1999) and molecular phylogenetic analysis (Gielly *et al.*, 2001; Thompson *et al.*, 2009). Quite simply, we recognise two different species in the subgenus: *C. balearicum* Willk., which has homogeneous morphology across its disjunct endemic distribution in the west, and *C. repandum* Sm., which contains a range of subspecific variants comprising four geographically delimited subspecies, two of which have local varieties. This is in contrast to the 'four species' classification proposed by Compton and Culham (2003), although this contains the same seven taxa that are included in our classification, albeit at different levels. In order to overcome the intricacies involved in continuously moving nomenclature, we name these taxa after their main geographic distribution; [1] refers to the names and combinations adopted by Debussche and Thompson (2002), [2] refers to those adopted by Compton and Culham (2003).

Balearic cyclamen = *C. balearicum* Willk.[1,2]

Tyrrhenian cyclamen = *C. repandum* Sm. subsp. *repandum*[1,2]

Kabylian cyclamen = *C. repandum* subsp. *repandum* var. *baborense* Debussche & Quézel[1,2]

Peloponnese pale-flowered cyclamen = *C. repandum* subsp. *peloponnesiacum* Grey-Wilson var. *peloponnesiacum*[1], *C. rhodium* subsp. *peloponnesiacum* (Grey-Wilson) J. Compton & Culham[2]

Peloponnese vivid-flowered cyclamen = *C. repandum* subsp. *peloponnesiacum* var. *vividum* (Grey-Wilson) Grey-Wilson[1], *C. rhodium* subsp. *vividum* (Grey-Wilson) J. Compton & Culham[2]

Cretan cyclamen = *C. repandum* subsp. *creticum* (Dörfl.) Debussche & Thompson[1], *C. creticum* (Dörfl.) Hildebr.[2]

Rhodian cyclamen = *C. repandum* subsp. *rhodense* (Meikle) Grey-Wilson[1], *C. rhodium* R. Gorer ex O. Schwarz & Lepper[2]

The Balearic cyclamen, often has beautifully silvered leaves. Flowering and reproducing plants growing in the Hérault gorges in southern France, a site long known by botanists, such as Magnol, since the seventeenth century (GD)

Balearic cyclamen

The morphology, reproductive biology and genetic variability of this species makes it the most clearly identifiable taxon in the subgenus.

Distribution

This is the most western of all *Cyclamen* species, with a narrow endemic and disjunct geographic distribution, being present in the Balearic Islands and southern France. On the Balearic Islands, it is widespread and often abundant on Mallorca, present in more localised populations on Menorca, and much rarer on Ibiza and Cabrera (Rosselló *et al.*, 1996; Debussche *et al.*, 1997). In southern France, where its indigenous status is in no doubt, this cyclamen occurs in the back country west of the Rhône Valley in the Languedoc–Roussillon region (Debussche *et al.*, 1995), where it is a legally protected species. It was first recorded here as early as the beginning of the seventeenth century (Strobelberger, 1620) growing "abundantly" in sites that were difficult to access. It is now known to occur over a distance of 220 km from the gorges of the Gardon River, just before this river flows into the Rhône, westwards to the Galamus gorges close to the Pyrenees. Along this arc, it occurs in five well-delimited areas. It is particularly abundant in several small valleys (where at least several hundreds of thousands of flowering plants occur) in the lower part of the Cévennes hills, where it occurs over about 30 km^2. The four other areas contain fewer populations and much smaller numbers of plants.

Phenology and reproductive biology

These plants bear 2–6 leaves that appear gradually from late-autumn into winter, becoming fully developed by flowering time, and then die back in late-spring. The triangular cordate leaves are up to 6 cm long, slightly toothed or un-toothed with a rounded apex, and have their upper surface marked with grey or silvery patches that cover most of this surface. Leaf colour shows little variation in southern France, but it can vary from almost completely silver to almost completely green on Menorca. The undersides of the leaves are mostly burgundy coloured. The petiole is thread-like in the soil, attached to the tuber via a 2–3 mm wide and up to 3 cm long appendix or 'floral trunk'. The round tubercule is 2–4 cm in diameter, often with a flattened appearance.

Flowering occurs from February and March in the mildest sites on the Balearic Islands to April and May in the Cévennes hills in southern France. Plants bear 1–6 very discretely scented flowers (of 12–20 mm in length), which often have, when one takes a careful look, light pink markings at the edge and thin light pink tracings on their corolla lobes. In one population in the southern part of the island of Menorca, we have seen plants with a pale-pink basal ring on their corolla, which are slightly reminiscent of flowers on Rhodes and Kos (Rhodian cyclamen), a long way across at the eastern limits of the distribution of subgenus *Psilanthum*. Only rarely are flowers visited by insects that could effect pollination, particularly in southern France (Affre *et al.*, 1995). The close proximity of the stigma and anthers (which are only about 0.5 mm apart) means that the self-compatible flowers can easily self-pollinate in the absence of any insect pollination (Affre *et al.*, 1995; Affre & Thompson, 1999). In this species, self-pollination produces as many seeds per flower as does outcrossing. Hence, this species can reproduce easily when pollinators are rare or when plants are separated by large distances or the barrier of a dense forest.

A study of genetic variability in this species on the Balearic Islands and in France showed that the individual populations there contain very little variability, providing strong evidence that in the

The Balearic cyclamen in the Cévennes region, southern France (MaxD)

wild this species reproduces primarily by self-fertilisation (Affre *et al.*, 1997). Genetic differentiation was, however, marked among populations, a finding that surprisingly is more apparent in southern France than on the Balearic Islands. In fact, pairs of populations from different parts of the range in southern France differ much more than pairs of populations on different Balearic Islands. We hypothesise that this is the consequence of historically smaller populations and greater isolation among sites in southern France, resulting from the negative combination of successive glaciations and human destruction of the forests since the Holocene. On the Balearic Islands, these two factors probably had less impact on the sizes and connectivity of populations of Balearic cyclamen, giving rise to lower levels of genetic variation among populations that might have been historically more inter-connected and bigger than those in southern France.

A fruit of Balearic cyclamen produces 20–40 small seeds as the leaves dry back. The seeds (each about 2–4 mg) are not as small as those of *C. intaminatum*, but are smaller than those of any other *Cyclamen* species. Seed dispersal is primarily assured by ants such as *Crematogaster scutellaris*, which typically live in the woody vegetation where Balearic cyclamen occurs. These ants move the seeds and consume the substance surrounding them as the fruit dehisces (Affre *et al.*, 1995). Ants often only disperse seeds a few decimetres, although we have observed them carrying seeds up to 5.5 metres from the maternal plant. Otherwise, water run-off during heavy rain may contribute to local dispersal. This limited dispersal further contributes to the marked patterns of genetic differentiation observed among populations of Balearic cyclamen. Seed germination occurs during damp mild conditions in late winter and early spring, and results in the growth of one very small cotyledon. Seeds can remain dormant for one or two years if the conditions for germination remain unfavourable.

About four years after germination, plants begin to flower in the wild. Plant longevity can no doubt reach several decades: we have plants in garden conditions that have lived 25 years after being collected as flowering plants in the wild, and yearly annual mortality of 15–20-year-old plants in pots in our experimental gardens at Montpellier is currently low (< 2 %).

Ecology

On the Balearic Islands, *C. balearicum* can be seen from almost sea level (e.g. 5 m above sea level on Menorca) to almost the summit of the Puig Mayor on Mallorca (1,445 m above sea level) (Debussche *et al.*, 1997). This large altitudinal gradient means that the species occurs in a range of climatic conditions on the islands: from 300 mm of rain per year and a mean monthly minimum in the coldest month of +8°C on Cabrera to 1,300 mm of rain per year and a mean monthly minimum of the coldest month of +1°C at altitude on Mallorca (frost is usual on the summits of the Sierra de Tramuntana). At the warmest and driest extreme of this gradient, freezing temperatures have never been observed; there, the plants have small flowers, and might remain vegetative or even fail to produce leaves in especially dry years. The 300 mm rain isohyet seems to be the climatic limit for any *Cyclamen* as this limit is shared by Balearic cyclamen, *C. persicum* and *C. rohlfsianum*. The climates in which populations in southern France occur resemble those in which the high-altitude populations occur on the Balearic Islands. The altitudinal range of Balearic cyclamen in France is smaller than that on the Balearic Islands, 70–480 m, (Debussche *et al.*, 1996), with an annual rainfall that is generally much higher, ranging from 650 mm per year up to 1,800 mm per year in the Cévennes hills (three times the annual rainfall of Paris or London!). Most of this rain falls in autumn and is rapidly absorbed by the stony well-drained soils and limestone substrate. At all French sites, the mean monthly minimum of the coldest month is close to 0°C. In the Cévennes, winter freezing is regular and heavy snow is not unusual; the leaves survive nights as cold as –7°C at soil level as long as a daytime thaw occurs (Debussche, 2003).

Throughout its distribution, the Balearic cyclamen is mostly found on north-facing (north-east to north-west) slopes, in relatively sheltered situations (e.g. at the bottom of dry-river valleys or at the foot of steep slopes and cliffs) under woody evergreen vegetation from shrublands to forests (Debussche *et al.*, 1996, 1997). Balearic cyclamen mostly occurs on hard limestone, occasionally on dolomitic limestone and, exceptionally and only on the island of Menorca, on sandstone and schist. The shallow, stony and well-drained soils in which this cyclamen grows are neutral to basic, even on sandstone and schist (pH 6.9–8.0).

On the Balearic Islands, the vegetation in which this cyclamen grows falls into one of three main types. The most common vegetation type is dense holm oak or Aleppo pine woodlands and tall evergreen shrublands (mostly lentisc), covering the bottoms of valleys up to mid slopes. Second, on Mallorca and Menorca, the plants grow among the same trees and shrubs but in more open and lower vegetation at the cliff foot, on ledges and on rocky outcrops (where we have counted densities reaching 650 leaves of more than 3 cm in length in 60 m^2). Third, in the mountains of Mallorca, the plants grow in open evergreen scrublands on limestone screes among numerous endemic species such as *Crocus cambessedesii*, *Digitalis dubia*, *Hypericum balearicum*, *Pastinaca lucida* and *Teucrium asiaticum*. The herbaceous layer often includes the tall grass *Ampelodesmos mauritanicus* and the widespread *Brachypodium retusum*. Only once, on Menorca, have we seen plants growing on schistose rock under heathland-type vegetation, dominated by *Erica arborea*; these plants showed a pale-pink basal ring on their corolla.

In southern France, Balearic cyclamen again occurs in vegetation that is evergreen and variable in height. There are two main habitat types: (a) holm oak mossy woodlands with an understorey dominated by evergreen box (*Buxus sempervirens*), on dolomitic limestone covering valley bottoms and lower slopes (typical in the Cévennes region); and (b) in the same holm oak woodland but in more open situations and at lower altitudes on hard limestone on ledges and shady gorges (e.g. in the Hérault, Gardon and Galamus gorges). In such scenic sites, leaf densities rival those that occur in the Balearic islands, reaching 635 leaves of more than 3 cm long in 60 m^2, and there are other endemic plants to be admired: *Antirrhinum latifolium*, *Fritillaria nigra*, *Galium maritimum* and *Narcissus dubius*. The evergreen large-leaved mock privet (*Phillyrea latifolia*) and butcher's broom (*Ruscus aculeatus*) are common in many sites. Common herbaceous plants are grasses such as *Brachypodium retusum* and *Bromus erectus*.

The more humid and cooler conditions in southern France are associated with four main vegetation differences when compared with the Balearic Islands: (a) the woody vegetation is denser, (b) only one study site (out of 40) has a vegetation dominated by Aleppo pine, (c) though not dominant, deciduous woody species (e.g. downy oak (*Quercus humilis*), Montpellier maple (*Acer monspessulanum*), serviceberry (*Amelanchier ovalis*), mahaleb cherry (*Prunus mahaleb*) and scorpion senna (*Hippocrepis emerus*)) are much more common, and (d) mosses and ferns are more abundant. Balearic cyclamen never occur on acidic soils in southern France, despite their contiguousness to basic soils, and their ecological amplitude in southern France is more specialised than on the Balearic Islands (Debussche & Thompson, 2003).

Tyrrhenian cyclamen (including Kabylian cyclamen)

The Tyrrhenian cyclamen has the widest distribution of any taxon in subgenus *Psilanthum*. We know a fair amount about the biology of this taxon but less about its ecology, except for that of populations on Corsica and of its one and only population in continental France, where this taxon is protected by French law. The Kabylian cyclamen, a variety of the Tyrrhenian cyclamen, is a poorly known restricted endemic of the mountains of Kabylia (Algeria) and the sole taxon of subgenus *Psilanthum* that occurs on the African continent.

Distribution

The geographic distribution of Tyrrhenian cyclamen covers the central part of the range of subgenus *Psilanthum*, stretching for maximum distances of around 1,000 km both east-to-west and north-to-south (Debussche *et al.*, 2000). Centred on the Tyrrhenian Sea, the fragmented distribution of Tyrrhenian cyclamen covers the Italian peninsula northwards to Carrara, the large islands of Corsica, Sardinia and Sicily, the small islands of the Tuscan archipelago, the Adriatic coast of the Balkans (from northern Croatia into Bosnia and Montenegro) and several islands of the Adriatic. One outlying small population exists to the north near St Tropez in the Maures Massif of southern France. The other outlying population, to the south, is the Kabylian cyclamen. Most of the distribution of the Tyrrhenian cyclamen thus covers an ancient micro-plate that migrated and fragmented in the late Miocene (Thompson, 2005).

The indigenous status of Kabylian cyclamen, which was first discovered in 1880, is not in doubt (Debussche & Quézel, 1997). In the 1950s, this plant was observed in six small areas that are difficult to access along a 100 km stretch between

The Tyrrhenian cyclamen in Corsica, an island where this species thrives in millions (GD)

Bejaia and Jijel, no more than 25 km inland, at 400–1,800 m elevation (Quézel, 1956). In the 1980s, it appeared that at least two of these sites have shown marked regression in population size as a result of forest destruction and regular fires (L. Zeraïa, personal communication). Since then, we have had no reports on the status of this plant. Other plants that have a eurasiatic distribution and that are absent from the rest of Africa have been reported in this area; these include field maple (*Acer campestre*), dog's mercury (*Mercurialis perennis*), bird's-nest orchid (*Neottia nidus-avis*) and aspen (*Populus tremula*) (Quézel, 1956). The native status of the population of Tyrrhenian cyclamen in the Maures Massif in mainland France has been questioned as this plant was first reported to be growing there in the mid-nineteenth century, and in a rather easily accessible habitat close to areas disturbed by humans (Debussche *et al.*, 2000). Nevertheless, its distribution pattern, encompassing the Maures Massif (with or without the associated small islands) and Corsica, is shared by three restricted endemics: the annual buttercup *Ranunculus revelierei*, the tall thistle *Ptilostemon casabonae* and cat thyme *Teucrium marum* (Gamisans & Marzocchi, 1996), strengthening the claim that the Tyrrhenian cyclamen is indigenous at this site in mainland France.

Life-history and phenology

Tyrrhenian cyclamen produce 2–8 leaves, almost simultaneously from the end of winter or early spring onwards. They die back at the start of the summer drought. This taxon has the shortest leaf life-span of any *Cyclamen* and, as a consequence, its leaves have the highest rate of resource assimilation by photosynthesis (Debussche *et al.*, 2004). The foliage, which is totally developed at peak flowering, lives fast and dies young. The triangular cordate leaves are toothed, up to 12 cm long, marked with a hastate patch pattern, more or less silvery, sometimes yellowish green. Some plants have completely green leaves; others have almost completely silver leaves, with just a green leaf margin. The petioles are attached to the round, flattened tubercle (2–6 cm in diameter) by a floral trunk.

Flowering occurs from March to June, depending on altitude, with up to 20 generally sweetly perfumed flowers per plant. The flowers have a morphology and function that are very different from those of Balearic cyclamen (Affre & Thompson, 1998, 1999). The long (16–30 mm in the wild) corolla is pink-carmine in colour, darkening towards the base; pale or very occasionally albino variants sometimes appear in the wild. The stigma protrudes from the corolla and is well separated from the anthers (by about 1.8 mm on average), which remain enclosed in the flower. Although this taxon is, like Balearic cyclamen, self-compatible, the spatial separation of stigmas and anthers means that simple autonomous mechanical self-pollination rarely occurs, and hence it produces few seeds in the absence of pollinators. From what we have seen on Corsica, most pollination is carried out by bumble-bees which touch the stigmas with the underside of their abdomen when they alight on the flower. Bumble-bees hang upside down and grip the base of the corolla and cause the pollen to drop from the poricidal anthers onto their body by vibration of their wing muscles, which causes a high frequency buzzing. This type of pollen removal is known as buzz-pollination and occurs in other groups with pendulous flowers and poricidal anthers (Ericaceae, Solanaceae, and Primulaceae such as *Dodecatheon*).

The fruits of Tyrrhenian cyclamen are mature at the end of the spring and the seeds each weigh 3–5 mg. Seed germination and seedling emergence occurs primarily in early spring. We have little information on plant longevity, although the mortality rate in pots at our experimental gardens in Montpellier suggests that the longevity of Tyrrhenian cyclamen might be shorter than that of Balearic cyclamen. This would fit recent suggestions, based on the ecology and fecundity, that Mediterranean restricted endemic species have enhanced longevity and local persistence relative to their widespread congeners (Lavergne *et al.*, 2004; Thompson, 2005). Tyrrhenian cyclamen is a fairly decorative species, hence several naturalised populations have been reported in Europe, mostly in the Mediterranean climate regions but also under an oceanic climate in sites as far north as Devon (England) (Aspland, 2005). Naturalised populations of this species do not, however, compare to those of the autumn-flowering ivy-leaved cyclamen (*C. hederifolium*).

The few known plants of Kabylian cyclamen in gardens and herbaria tend to have larger flowers and leaves than Tyrrhenian cyclamen, hence their status as a variety in the botanical literature (Debussche & Quézel, 1997).

Ecology

Our knowledge of the ecology of Tyrrhenian cyclamen comes primarily from studies of this taxon on Corsica (Debussche *et al.*, 2000). On this island, this plant is a common feature of woodland floors from sea level to 600 m elevation, and it can reach up to 1,300 m. Annual precipitation therefore differs greatly among sites, ranging from 500 mm to 1,500 mm per year. Likewise, the mean monthly minimum temperature of the coldest month varies from +6°C near the coast, where freezing temperatures are not reached, to −1.5°C, in the colder mountain sites where snow and freezing temperatures are a normal part of a winter. In the dryer lowland and coastal areas, this taxon occurs almost only in shady talwegs; even so, during dry springs, precocious wilting of leaves and shorter flowering periods can occur. In the mountains, foliage development is delayed and spring frost hardly injures leaves, which can resist temperatures of −6°C for a few hours (Debussche, 2003). The altitudinal limits reach by this plant are probably set by a combination of winter cold and summer drought, which together limit the annual period available for plant development.

The habitats of this *Cyclamen* have three clear similarities to those of Balearic cyclamen (Debussche & Thompson, 2003): they are north-facing, have woody vegetation cover and highly variable height (1–40 m) of this cover. The similarities begin and end there; the two taxa differ markedly in the ecological characteristics of their habitats, with Tyrrhenian cyclamen occurring in a wider range of habitats (in terms of soils, topography and vegetation) on Corsica than does Balearic cyclamen on the Balearic Islands and in southern France. Tyrrhenian cyclamen is most commonly found in tall shrublands and deciduous or evergreen woodlands, where it is particularly abundant in clearings and along woodland edges. It occurs primarily on intrusive and metamorphic underlying rocks (granites, rhyolithes, diorites, gneiss, schists and so on), which are the dominant rocks of the island, but also grows on most of the rare limestone outcrops. The soils in which Tyrrhenian cyclamen grows are thus highly variable in their chemical and physical characteristics, being very stony and shallow with a pH of up to 7.7 in some limestone sites to less stony and deeper with a pH that can be as low as 5.0 on granite. Thus, Tyrrhenian cyclamen occur in a diverse range of vegetation communities: in tall *Erica arborea* heathlands, in evergreen holm oak coppices, in wild olive tree (*Olea europaea* var. *sylvestris*) thickets on dry river banks, in moist riparian forests of alder (*Alnus glutinosa*) and flowering ash (*Fraxinus ornus*), in chestnut orchards, in Corsican pine (*Pinus nigra* subsp. *laricio*) forest where trees reach 40 m in height, and in grazed open shrublands with eagle fern, bramble (*Rubus ulmifolius*) and the foliage odorant Montpellier rock rose (*Cistus monspeliensis*). We have recorded Tyrrhenian cyclamen on Corsica alone under as many as 12 different dominant tree species, both deciduous and evergreen, whereas Balearic cyclamen is associated with only two main tree species, holm oak and Aleppo pine, throughout its entire distribution. We recorded the highest density of Tyrrhenian cyclamen leaves (960 leaves of more than 3 cm in 60 m^2) in an open Corsican pine woodland on granite at 1,050 m elevation. The ecological amplitude of Tyrrhenian cyclamen makes it impossible to describe a typical habitat for this taxon on Corsica. The population that occurs in the Maures massif of mainland France fits into the range of the above descriptions.

Other than the Corsican hellebore, which is very common in different habitats of the Corsican backcountry, the narrow endemic species that grow with Tyrrhenian cyclamen mostly occur in rocky habitats (see Gamisans & Marzocchi, 1996). Some of these endemics have a disjunct distribution that extends to the Balearic Islands. These include the Balearic sandwort (*Arenaria balearica*), two lords-and-ladies species, one autumn-flowering and cow-dung-smelling (*Arum pictum*) and the other spectacular spring-flowering and carrion-smelling (*Dracunculus muscivorus*), a small daisy (*Bellium bellidioides*), a dwarf hyacinth (*Brimeura fastigiata*), a dwarf toadflax (*Cymbalaria aequitriloba*), and a thyme (*Thymus herba-barona*). On Corsica, Tyrrhenian cyclamen frequently occurs intermixed with the autumn-flowering ivy-leaved cyclamen.

Elsewhere in its distribution, the Tyrrhenian cyclamen probably has fairly similar ecological amplitude. It has been observed in Tuscany (Italy) close to sea level and up to 600–700 m, primarily in shady tall shrublands and holm oak woodlands (J. D. Thompson, unpublished observation; F. Selvi, unpublished observation; Selvi & Stefanini, 2005) mixed with sclerophyllous trees (e.g. cork oak (*Quercus suber*), strawberry tree (*Arbutus unedo*) and large-leaved mock privet) and shrubs (e.g. *Viburnum tinus*, *Rhamnus alaternus* and *Erica arborea*) and/or with deciduous trees (e.g. Montpellier maple, service tree (*Sorbus domestica*), flowering ash and downy oak). As one moves south, the species' elevation limits rise to about 1,200 m in Calabria or Sicily (F. Selvi, personal communication).

In Algeria, the mountains of Kabylia form an abrupt barrier to sea breezes, resulting in a range of precipitation of 800–2,000 mm per year in the areas where Kabylian cyclamen is likely to occur, slightly more than that experienced by Balearic cyclamen in Southern France. Forest inventories from the 1950s provide all we know about the habitat of Kabylian cyclamen (Quézel, 1956). This taxon occurs on limestone and sandstone, that is, on a range of pHs. It appears to be a forest species, primarily growing with deciduous Algerian oak (*Quercus canariensis*) and its hybrid derivative (*Quercus afares*) from crosses with cork oak. At high elevation, freezing temperatures and snow are common. The occupied habitats are thus likely to be in the range of those of the European Tyrrhenian cyclamen. Kabylian cyclamen commonly occurs alongside the autumn-flowering African cyclamen (*C. africanum*), a narrow endemic cyclamen of north-east Algeria and northern Tunisia (which is the sister species of ivy-leaved cyclamen); in the same manner, Tyrrhenian cyclamen occurs with ivy-leaved cyclamen in Europe.

Relicts of Balearic cyclamen on Corsica

Unexpected populations of white and bicoloured flowers

In *La Flore de Corse* produced from 1927 to 1938, Briquet and de Litardière indicate that white flowers can be observed amongst typical pink cyclamen flowers on Corsica, and they specifically mention five sites. In fact, white-flowered plants are more abundant than these authors suspected. On Corsica, one can find white flowers in almost one-third of the sites we have studied, but at a very low frequency (less than 2%) (Debussche & Thompson, 2000; see also Denney, 2003). These flowers are essentially albino variants; other than their colour, they have a morphology and breeding system that is the same as that of pink flowers in the same population (Thompson *et al.*, 2009).

John Thompson and Max Debussche at the site of a hybrid population Balearic cyclamen × Tyrrhenian cyclamen at St Florent, Corsica (GD)

At one of the five sites mentioned in "*La Flore de Corse*", which Briquet and de Litardière visited in 1936, white flowers were described as quite abundant in limestone shrubland (a rare habitat type on this granite and schist island!). Our curiosity thus brought us near the village of St Florent in the north of the island at the base of the western side of the Cap Corse in the late 1990s. We immediately realised that this site has little in common with other sites where Tyrrhenian cyclamen occurs on Corsica (Debussche & Thompson, 2000, 2003). It is less than 2 km from the sea, on a limestone outcrop with steep north-west-facing slopes and limestone cliffs that dominate a half-dry rivulet shaded by oleanders and wild olive-trees. The plant community resembles a typical limestone garrigue, dominated by Montpellier rock rose, Corsican rock rose (*Cistus creticus* subsp. *corsicus*), and holm oaks that do not exceed 3 m in height. Moreover, we discovered that each of the two neighbouring limestone outcrops also harbour one population that has white-flowered plants. In these three populations, the frequency of white-flowered plants ranges from 10–16%, although there are patches where they make up a third of the plants present. In all three of these populations, we found flowers that are identical to those of Balearic cyclamen and others that resemble the albino variants of Tyrrhenian cyclamen that we have frequently seen elsewhere. In addition, there were a number of bicoloured flowers with whitish or pale pink corollas and a darker pink basal ring of highly variable diameter. These closely resemble the flowers of hybrid plants that we have obtained from crosses among Balearic and Tyrrhenian cyclamen in cultivation. The leaves of some plants also bear a resemblance to those of Balearic cyclamen. Most importantly, the diversity of floral and leaf form, size and colouration is higher than that we have seen in any *Cyclamen* population (Debussche & Thompson, 2002) and, in fact, encompasses a large spectrum of the variability that we have observed across the distribution of the whole subgenus.

Does Balearic cyclamen occur on Corsica?

The high variability of leaf and floral traits in the three limestone sites near St Florent could have two origins (Debussche & Thompson, 2000). First, the historical distribution of Balearic cyclamen could have encompassed the Tyrrhenian islands. Under this scenario, the three sites might contain previously undescribed relict populations of Balearic cyclamen, which have undergone hybridisation and introgression with Tyrrhenian cyclamen. Second, the sites might contain a recently evolved white variant of Tyrrhenian cyclamen that resembles Balearic cyclamen but is in fact something very new. We compared the genetic and morphological characteristics of plants in the three populations and the ecological characteristics of the three sites with those for Tyrrhenian cyclamen elsewhere on Corsica and in mainland France and for Balearic cyclamen on the Balearic Islands and in mainland France. These comparisons suggest that Balearic cyclamen is indeed indigenous to Corsica. The arguments for this claim can be summarised as follows.

1. The limestone habitats near St Florent on Corsica have ecological characteristics that fit within the range of habitats occupied by Balearic cyclamen across its distribution, but that are distinct from those of the habitats of Tyrrhenian cyclamen elsewhere on Corsica and in mainland France (Debussche & Thompson, 2003).
2. The floral morphology of many white-flowered plants in the St Florent populations is very distinct from those of white- and pink-flowered Tyrrhenian cyclamen in other populations on Corsica and of pink-flowered individuals within the same St

A unique hybrid plant with both typical leaves of Balearic cyclamen and pink flower bud of Tyrrhenian cyclamen at St Florent, Corsica (GD)

Florent sites. At the other sites on Corsica, pink- and white-flowered individuals have flowers that are morphologically the same in terms of size and of stigma and anther characteristics (white-flowered plants are simply albino plants). In addition, a large number of the white-flowered plants in the St Florent populations have flowers that are morphologically similar to those of Balearic cyclamen in terms of their small size, marked proximity of stigmas and anthers, and amount of pollen produced per flower (Thompson *et al.*, 2009).

3. Some traits that are never seen in typical Tyrrhenian cyclamen and that are specific to Balearic cyclamen are present in some plants within the St Florent populations, namely a pink tinge to the margin of the otherwise white corolla lobes, few small and smooth teeth on the leaf lamina, and a whitish colour of leaf patches (Debussche & Thompson, 2002).
4. Gene sequences for a maternally inherited chloroplast gene show that the typical Tyrrhenian cyclamen sequence and the typical Balearic cyclamen sequence are both present in white- and pink-flowered plants that occur at the St Florent sites (Gielly *et al.*, 2001). The Balearic cyclamen sequence has been reported in one other population of Tyrrhenian cyclamen on western-central Corsica, but there are no white-flowered plants in this population.
5. Plants that have bicoloured flowers can have a size and floral design similar to either of the two parental taxa, although their leaf traits are similar to those of Tyrrhenian cyclamen.
6. Last and certainly not least, we conducted molecular genetic analyses of plants from the three St Florent sites (Thompson *et al.*, 2010), which we distributed among four classes on the basis of their flower morphology: a) small white flowers with a stigma inside the flowers close to the anthers, akin to those of Balearic cyclamen, b) white flowers with an exserted stigma, c) pink flowers with an exserted stigma, akin to those of Tyrrhenian cyclamen, and d) bicoloured flowers. These four classes of plants were compared with plants of Tyrrhenian cyclamen from elsewhere on Corsica and with plants of Balearic cyclamen from the Balearic Islands and France. The results were very clear: the plants of all four morphological classes of St Florent cyclamen were intermediate between Tyrrhenian cyclamen and Balearic cyclamen and contained greater genetic variability than the two typical taxa in a way that indicates that they combine the genetic constitution of these two taxa.

A hybrid plant with intermediate leaf pattern and small white flowers similar to that of Balearic cyclamen at St Florent, Corsica (GD)

Points 1–4 suggest that both Balearic cyclamen and Tyrrhenian cyclamen coexist at the St Florent sites. Point 6 is strong evidence for hybridisation between these two taxa in the St Florent sites. The occurrence of morphological intermediates and the fact that Balearic and Tyrrhenian cyclamen are cross-compatible in cultivation and produce fertile hybrid offspring, some of which bear bicoloured flowers that resemble plants in the St Florent site, further indicate that the dramatic floral diversity at this site has a hybrid origin.

A sample of flowers showing the high diversity of colour and size within a single hybrid population at St Florent, Corsica (GD)

So Balearic cyclamen, though essentially as hybrid and introgressed plants, can be added to the list of species that have a disjunct endemic distribution in the Balearic and Tyrrhenian Islands, corroborating the historical geological links between the two regions. Other populations that have a high frequency of white flowers were identified on Sardinia at the end of the nineteenth century (Hildebrand, 1898); some of these still occur (M. Tonioli, personal communication), and are thus sitting and waiting to be counted, measured and analysed in order to reveal whether Balearic cyclamen is also a Sardinian species.

Peloponnese pale- and vivid-flowered cyclamen

Distribution

A picnic in the Langada gorges near Sparta or a short walk after visiting the Malevis monastery in the Parnon mountains allows you to discover the pale-flowered and vivid-flowered varieties, respectively, of Peloponnese cyclamen. This taxon has a narrow endemic distribution in the Peloponnese peninsula; all reports of this taxon's occurrence north of the gulf of Corinth and on islands are probably mistakes, often resulting from their being confused with ivy-leaved cyclamen (see Debussche *et al.*, 2006). In its pale-flowered variety, this taxon can be seen near the Gulf of Corinth on Mont Helmos (Aroania Mountains), and more abundantly in the south of the Peloponnese (Taygetos Mountains and Mani peninsula). The vivid-flowered variety occupies an eastern-most part of the peninsula, particularly in the Parnon and Madara Mountains (Johnstone, 1992; Moore, 1993, 1996; Grey-Wilson, 1997). Nevertheless, the two geographic distributions have an overlap area where we have observed mixed populations of the two varieties that have a whole range of intermediate floral colours (see also Moore 1993, 1996).

Flower of Peloponnese pale-flowered cyclamen (BM)

Flower of Peloponnese vivid-flowered cyclamen (BM)

Life-history and phenology

Peloponnese cyclamen leaf out earlier than Tyrrhenian cyclamen, but later than Balearic cyclamen, when grown in pots at our experimental gardens in Montpellier. The leaf of Peloponnese cyclamen has a triangular-cordate shape, sometimes rounded and wider than long, and is only slightly lobed. Patch pattern can be hastate or fuzzy, greenish grey to silvery, and with or without speckles of the same colour. This speckling is a common trait that is shared with the other eastern taxa of the subgenus from Crete, Kárpathos, Rhodes and Kos. The underside of the leaf is redder in plants of the vivid-flowered variety.

The large flowers (20–42 mm in wild populations) have a stigma that exceeds the corolla, indicating that they rely on insects for pollination. The pale-flowered variety has a characteristic three-tone pink corolla, with a deep pink well-developed ring at the base, and light pink lobes with a deeper hue at their extremity. Occasionally, the ring at the base of the corolla is pale pink, rather like that of Rhodian cyclamen (Ten Hoope, 2004), or absent and the flowers are almost white (Mathew, 1999). In striking contrast, vivid-flowered Peloponnese cyclamen, as its name suggests, has a deep pink to carmine coloured corolla, in a lighter hue at the extremity of the lobes, similar to, albeit often more vivid than, that of the flowers of Tyrrhenian cyclamen.

Ecology

Overall, Peloponnese cyclamen has an ecology reminiscent of that of Tyrrhenian cyclamen: it occurs mostly on north-facing slopes, along a wide altitudinal and climatic range, in different soil pHs, and under an important woody cover, either deciduous or evergreen, including conifers (M. & G. Debussche, unpublished data; Cyclamen Society, unpublished data).

Peloponnese cyclamen is observed from almost sea level to 1,400 m elevation (Cyclamen Society, unpublished data). It can thus tolerate freezing and snow in some of its populations. It grows in tall shrublands and in woodlands (with at least 50% woody cover, of 2–15 m in height), with a scarce herbaceous layer, primarily with a north-east to north-west aspect, on limestone, schists and mica schists, and thus on a range of soil pH (6.0–7.3). The woody vegetation is dominated by a number of deciduous trees such as oriental plane tree, Hungarian oak, chestnut or terebinth (*Pistacia terebinthus*), or by evergreen trees such as Kermes oak (*Quercus coccifera*), large-leaved mock privet, Greek fir, Crimea pine (*Pinus nigra* subsp. *pallasiana*). The habitats of this taxon are often extensively grazed by sheep and goats.

Our own unpublished data and those of Moore (1993, 1996) suggest that pale-flowered and vivid-flowered varieties Peloponnese cyclamen have some differences in habitat. Pale-flowered plants appear to be mostly associated with fairly dense humid woodlands of deciduous tree species, whereas vivid-flowered plants are more commonly found in more open evergreen, including coniferous, formations on rocky soils. The two varieties can, however, be found together in the same habitat, so both the ecological distinction and the taxonomic separation are far from being clear cut.

The densest spread of this taxon that we have observed (550 leaves of more than 3 cm long and 220 flowers in 60 m^2) was in a population of vivid-flowered plants growing under an open coppice of Kermes oak, on limestone on a north-east slope, 760 m above sea level. Two other cyclamen (autumn-flowering) species can co-occur with Peloponnese cyclamen: Greek cyclamen (*C. graecum*), which is found in the most open and drier sites, and ivy-leaved cyclamen, which is found in humid forest habitats. Anyone who comes across the habitats of this taxon will be delighted by numerous blooming plant species, such as a red anemone (*Anemone pavonina*), the purple rockcress (*Aubrieta deltoidea*) which is widely cultivated in rock gardens, and several *Ophrys* and bulbous species.

Cretan cyclamen

Distribution

Now to the most southern location of your voyage across the Mediterranean (excepting Kabylia, which we haven't been able to visit for security reasons only): the islands of Crete and Kárpathos. The white-flowered Cretan cyclamen has a narrow endemic distribution limited to these two adjacent islands (Grey-Wilson, 1988; Bailey, 1992). It is widespread on Crete, with populations occurring across the whole island from west to east and north to south (Aspland, 1994; Wiltshire & Dale, 1995; Denney, 1996). By contrast, only two sites are known to harbour this taxon on the eastern coast of the much smaller island of Kárpathos (Bailey, 1992).

Life-history and phenology

In pots at our experimental gardens in Montpellier, a few leaves of Cretan cyclamen appear in late autumn, at the same time as those of Balearic cyclamen (Debussche *et al.*, 2004). The leaf is triangular in outline, often with a pointed apex and sharp lobing, but the lobes can also be blunt or even non-existent. The shape and colour of the patch pattern of the limb are very variable: they can be hastate and greenish grey without speckles, or fuzzy and silvery, or deep green with a dense splash of silver speckles. The leaves can thus resemble either Tyrrhenian cyclamen or Rhodian cyclamen.

At our experimental gardens in Montpellier, this species flowers later than the other eastern taxa of the subgenus, albeit earlier than Balearic or Tyrrhenian cyclamen (Debussche *et al.*, 2004). It bears large white flowers (17–28 mm in cultivation, up to 38 mm in the wild) that have a slight fragrance and a style which is slightly

Cretan cyclamen on Crete, between Apostoli and Ag. Fotini (JF)

The pink *C. creticum* forma *pallide-roseum* of Cretan cyclamen is much less common than the white-flowered variant (JF)

exserted from the corolla, and sometimes included. Its intermediate stigma anther separation, which is less than that of Tyrrhenian and Peloponnese cyclamen but greater than that of Balearic cyclamen (Debussche & Thompson, 2002), appears to be correlated with intermediate incidence of selfing and genetic variability in natural populations (Affre & Thompson, 1998, 1999). Occasional very pale pink flowers (forma *pallide-roseum*) of this species have been observed in a small number of populations on Crete, where they occur at frequencies of less than 10% (J. Thompson, unpublished observations). This species shows variability in its chromosome complement among populations on Crete; in addition to populations that have the typical 2n = 20 chromosome number, there are populations with 2n = 22 chromosomes, some of which have marked differences in chromosome arrangement when compared to 2n = 20 populations (Greilhuber, 1989).

Ecology

The range of habitats in which Cretan cyclamen occurs bears a striking resemblance to those of Peloponnese and Tyrrhenian cyclamen. Cretan cyclamen has been observed from 60 m to 1,250 m elevation on Crete (Denney, 1996), so some populations must withstand snow and freezing temperatures in winter (see Fry, 1992). In a Mediterranean garden, leaves of this species tolerate freezing down to −5°C, as long as such events last no more than a few hours (Debussche, 2003). This taxon mostly occurs on north-east- to north-west-facing slopes in woodlands and shrublands, in gullies and at the foot of cliffs; it is less associated with shady and north-facing sites with increasing elevation (Aspland, 1994; Wiltshire & Dale, 1995; Denney, 1996; Grey-Wilson, 1997; Thompson, 2006). It occurs under both evergreen trees, such as Kermes oak, Turkish pine, cypress tree, lentisc, carob tree, and deciduous trees, such as oriental plane tree and chestnut. Although it occurs mostly on limestone, Cretan cyclamen can also be found on sandstone and schist, that is to say, it occurs on both basic and acidic soils. It occasionally occurs with Greek cyclamen, in low-altitude open sites with a range of orchids, anemones and arums. On Kárpathos, Cretan cyclamen has been found at 400 m above sea level on limestone, mostly on north-facing slopes, under lentisc and Turkish pine, and in rock crevices on less shady spots (Bailey, 1992).

Rhodian cyclamen

Distribution

The last stage of your voyage across the Mediterranean from the Balearic Islands almost to the shores of Turkey brings you to Rhodes. Near Kattavia on the southern tip of the island, extends a dry landscape of scrublands grazed by goats, and crossed by a rocky oued shadowed by a dense thicket of carob trees and oleanders. Among the roots and litter of the thicket, you discover the white flowers of the Rhodian cyclamen. This narrow endemic taxon is limited to Rhodes and also Kos, where it was first discovered only in 1980, and correctly identified in 1996 (Grey-Wilson, 1997; Denney, 1999). On Rhodes, its distribution covers 70 km from north to south over a belt of about 20 km wide at its maximum (Bailey, 1991). This taxon has the most limited spatial distribution of all the taxa in subgenus *Psilanthum*.

Life-history and phenology

In our experimental gardens in Montpellier, the leaf and flowering phenologies of Rhodian cyclamen are similar to those of Peloponnese cyclamen. The limb pattern and coloration of the leaves of Rhodian cyclamen are also quite similar to those of Peloponnese cyclamen, although in shape, the leaves are more triangular and longer. The flowers are large (18–32 mm in the wild) with a style that comes out of the corolla, which is indicative of the role of insects in assuring pollination. The corolla is white or very pale pink, with a narrow deep-pink ring around the gorge. In several populations, a small number of plants (no more than 5%) have a pinkish colour at the apex of the corolla lobes, a colour pattern similar to, albeit much less marked than, that observed in flowers of the pale-flowered Peloponnese cyclamen (see also Bailey, 1991). On Rhodes, we have very occasionally observed albino variants, which might be more frequent on Kos (Denney, 1999).

Ecology

Rhodian cyclamen is found from 15–780 m elevation on Rhodes (Cyclamen Society, unpublished data) and thus does not occur where freezing is likely to occur, unlike the other eastern members of the subgenus. In a Mediterranean garden, leaves of this species tolerate freezing down to –5°C, as long as such events last only a few hours (Debussche, 2003). Mostly limited to north-west to north-east facing slopes, Rhodian cyclamen is always found under woody vegetation (more than 60% cover) of variable height (2–20 m), in a rather scarce herbaceous layer. It occurs on limestone, dolomite, marl, conglomerate and sandstone, and as a consequence on a large range of pH (5.6–7.7) (M. & G. Debussche, unpublished data; Cyclamen Society, unpublished data). Its habitats are often extensively grazed and it can often be seen on roadsides, even piercing the macadam.

Population of Rhodian cyclamen growing on Rhodes (MH)

The Rhodian cyclamen occurs exclusively under evergreen trees and shrubs, primarily Turkish pine and less often cypress, lentisc, carob tree and Greek strawberry tree (*Arbutus andrachne*). In the south of Rhodes, and at low altitudes elsewhere, it primarily occurs in shady talwegs and on slope bottoms. Its most dense occurrence was found at 100 m elevation in Turkish pine woodland on limestone (560 leaves of more than 3 cm long and 110 flowers in 60 m^2). Rhodian cyclamen can occur, for example on Mt Profitis Ilias, with Persian cyclamen (*Cyclamen persicum*), a dwarf pale-blue anemone (*Anemone blanda*) and the magnificent white-flowered endemic Rhodian peony (*Paeonia rhodia*) (see also Denney, 1994). On Kos, Rhodian cyclamen occurs from 35 to 260 m in a highly localised area in the south-west of the island. Its habitats are like those on Rhodes: north-facing slopes in Turkish pine woods or in gullies under Kermes oak, lentisc and Greek strawberry tree, on limestone and igneous rocks (pH 6.4–7.7) (Denney, 1999). Rhodian cyclamen thus exhibits two main ecological differences from Tyrrhenian, Peloponnese and Cretan cyclamens: its lower maximum elevation (780 m), which means that it is not exposed to freezing events, and its occurrence only under evergreen trees.

Patterns of variation, processes of evolution

Phylogeographic relationships in the subgenus

What first attracted our attention to this group of spring-flowering *Cyclamen* is the distribution pattern of the different taxa, particularly their disjunct distribution. For several years, we have tried to unravel where and how such divergence has occurred, whether the evolution of new endemic taxa is simply the result of divergence in already geographically isolated populations, or whether some divergence evolves before geographic separation in geographically peripheral or ecologically marginal populations. The combination of molecular, morphological and ecological studies that have contributed to the description of the different taxonomic and geographic entities within subgenus *Psilanthum*, also provides insights into the evolution of the different taxa in relation to the geological and climatic history of the Mediterranean basin (Thompson, 2005).

The gene sequence analyses of samples obtained from across the entire distribution (except Kabylian cyclamen and Cretan cyclamen from Kárpathos) of the different taxa of the subgenus by Gielly *et al.* (2001) produced a phylogenetic tree composed of five haplotypes — which we named "repandum", "balearicum", "peloponnesiacum", "vividum" and "creticum" — re-grouped into two main clades.

The first clade contains all the samples of Tyrrhenian, Balearic and Rhodian cyclamen, with two haplotypes, namely "repandum" and "balearicum". The "repandum" type was found in all the samples of Tyrrhenian cyclamen, except one from Corsica, and all the samples of Rhodian cyclamen. The "balearicum" type was found in all the samples of Balearic cyclamen and in one sample of Tyrrhenian cyclamen from Corsica. The samples collected from the St Florent populations of Corsica contained both "repandum" and "balearicum" types.

The second clade contains all the samples of Peloponnese cyclamen (both varieties) and Cretan cyclamen, with three haplotypes, namely "peloponnesiacum", "vividum" and "creticum". The "peloponnesiacum" type was found in all the samples of both varieties of Peloponnese cyclamen, except for one sample of the vivid-flowered variety in which was found the "vividum" type. The "peloponnesiacum" type was also found in one sample of Cretan cyclamen, from eastern Crete, whereas the "creticum" type was found in the samples from western Crete.

These two clades suggest that there has been an important phylogeographic separation of taxa from the Peloponnese peninsula and Crete from taxa occurring elsewhere in the distribution of the subgenus (i.e. taxa from the western Mediterranean basin and eastwards with the Aegean Islands of Rhodes and Kos). In the first clade, the divergent position of different samples of Tyrrhenian cyclamen, which either occurred together with Balearic cyclamen sample and showed the "balearicum" type or occurred together with Rhodian cyclamen samples and showed the "repandum" type, suggests that this taxon has diverged at the geographic limits of its range to produce Rhodian cyclamen to the east and Balearic cyclamen to the west. Hence, Rhodian cyclamen does not appear to have evolved at the end of a land chain from the Peloponnese peninsula across Crete and Kárpathos to Rhodes, but more probably as a result of geographic isolation following the loss of land bridge connections across the Cyclades, perhaps during the Pliocene (see Thompson, 2005).

Taxonomic differences in the botanical groups found on the islands off the Turkish coast and those on the more westward but neighbouring islands of the Aegean Sea are often observed and support this hypothesis. Rhodian cyclamen is indeed more closely related to Tyrrhenian cyclamen than it is to Cretan and Peloponnese cyclamens. The fact that $2n = 20$ populations of Cretan cyclamen show clear differences in karyotype organisation from populations of Rhodian cyclamen provides further support for this idea (Grielhuber, 1989). The structure of the second clade suggests that Cretan cyclamen has diverged in allopatry from Peloponnese cyclamen and that the two (so-called) varieties of Peloponnese cyclamen are probably extremes of a continuous range of morphological variation that is fixed in some populations.

Evolutionary relationships among Balearic and Tyrrhenian cyclamens

The distribution and genetic affinities of Balearic and Tyrrhenian cyclamen suggest that they are very closely related. However, Balearic and Tyrrhenian cyclamen show a distinct difference in their overall ecology. The link between species disjunction, reproduction and ecology, and the possibility of local speciation in the subgenus *Psilanthum* has become a distinct possibility in the light of the discovery of Balearic cyclamen on a limestone massif on Corsica, outside of its previously documented range. The cyclamen present in these Corsican populations have DNA types that are characteristic of both species and are introgressed plants created by crosses among the two taxa, although many have a floral morphology that is identical to that of Balearic cyclamen. The maintenance of a floral morphology akin to Balearic cyclamen, in the face of strong genetic introgression from Tyrrhenian cyclamen, strongly suggests that the floral form of Balearic cyclamen might be advantageous in some conditions.

Without wishing to be too speculative, we can suggest that temporal variability in the pollination environment, perhaps resulting from the absence of pollinators or years with a spring drought that create a need for rapid flowering and fruiting, could favour the persistence of a highly selfing strategy in these sites.

Without some selective advantage, it is hard to understand what maintains the floral phenotype of Balearic cyclamen in these sites given the levels of introgression observed. The ecology of this site is typical of Balearic cyclamen habitats and is distinct from that of other Tyrrhenian cyclamen populations on Corsica. Balearic cyclamen is a highly inbred species and although Tyrrhenian cyclamen is more reliant on outcrossing, some populations can show high levels of inbreeding, which might cause populations to diverge and show gene frequencies more typical of Balearic cyclamen (Thompson, 2005).

All in all, we would be inclined to conclude that the discovery of Balearic cyclamen on a limestone outcrop located on an otherwise "schist and granite island" implies that initial divergence occurred in geographically peripheral and ecologically marginal populations on an unusual soil type. The cyclamen populations in this zone of Corsica are very small, the plants barely exceed several hundred in number and occur in a small area when compared to those of populations of Tyrrhenian cyclamen elsewhere on Corsica. Differentiation may have been facilitated by developmental modifications in an inbreeding population. We can note here that the newly opened flowers of Tyrrhenian cyclamen are similar in size and floral design to Balearic cyclamen (Thompson, 2005). An abrupt developmental change or some form of instability in development could thus have triggered the evolution of the floral phenotype of Balearic cyclamen. Thus, a combination of ecological specialisation and inbreeding might have been involved in the initial divergence of Balearic cyclamen in peripheral populations of Tyrrhenian cyclamen. If this interpretation is true, then initial divergence must have been ancient, pre-dating the tectonic activity that isolated Corsica from the Balearic Islands and southern France (Thompson, 2005).

3.5 THE CYCLAMEN SOCIETY'S FIELDWORK

Introduction

In 1985 the late Fred Buglass, the Cyclamen Society's president, put forward the idea of field studies to look in depth at *Cyclamen* species as they occur in the wild: to study their precise distributions, habitats and variability, and how these factors might affect their classification. The "Research & Exploration Sub-Committee" was duly formed, with the aim of planning a series of field trips, most of which are reported below.

In addition to summarising the Society's fieldwork, we have included an account of the studies of the rare local endemic of Lebanon, *C. libanoticum*, made by the late Dr Richard Haber and his wife Dr Myrna Semaan (page 307).

The first of the Society's field studies took place in 1987 and since that time many more have been undertaken. They are presented roughly in chronological order of the first study undertaken for each species:

- **3.5a** *C. coum*: 1987, 1988, 1990
- **3.5b** *C. creticum*: 1991, 1994, 1995
- **3.5c** *C. rhodium* subsp. *rhodium*, subsp. *peloponnesiacum* and subsp. *vividum*: 1991, 1992, 1993, 2004
- **3.5d** *C. alpinum*: 1997, 1998, 1999
- **3.5e** *C. intaminatum*: 2001, 2004
- **3.5f** *C. pseudibericum*: 2002
- **3.5g** *C. repandum*: 2003
- **3.5h** *C. mirabile*: 2003, 2004
- **3.5i** *C. cilicium*: 2008, 2009
- **3.5j** *C. libanoticum*: Richard Haber and Myrna Semaan's study

There has also been a succession of visits to study the most widespread species, *C. hederifolium*, mainly in Greece and the Greek islands. This is on-going fieldwork not yet completed and there remains much to be done before a comprehensive assessment can be presented. The study is being carried out in collaboration with Alastair Culham at the University of Reading and involves cytological and molecular investigations as well as detailed field observations. One of the benefits to date has been the clarification of the status of *C. confusum* in Crete and its recognition as a species rather than as a variant of *C. hederifolium*. An explanation of the science behind this decision can be found in Chapter 2 under *C. confusum* (page 151). The Society's programme of fieldwork on *C. hederifolium* has, to date, involved visits to the Mediterranean in 2001, 2002, 2003, 2004, 2005, 2006, 2007 and 2008, with further forays planned. Short extracts and comments from some of the reports (e.g. Sicily 2006) can also be found in Chapter 2 under "Further observations on *C. hederifolium*" (page 136).

3.5a *CYCLAMEN COUM*

Jeremy Wood

Introduction

Following the resolution to launch a series of field studies, it was proposed that the first field trip should be to Turkey to study *C. coum* and its close associates. It was envisaged that this work would probably be spread over 4–5 years, after which the studies could be extended to include other cyclamen species.

Cyclamen coum was a good starting point as it was known to be a very variable species, particularly with regard to leaf shape, leaf markings, flower size and flower colour. Desmond Meikle, writing in the 1978 *Flora of Turkey* Vol. 16, had given a large number of locations for this species in Turkey, a country that did not appear to pose any problems with regard to access. The broad aim was to study variations in the field, and to see whether they could be explained in terms of geographical and environmental factors.

Meikle had in fact published his view of the species in Turkey, dividing it into two varieties, which he termed var. *coum* and var. *caucasicum*. Using his own words, "in var. *coum* the leaves are orbicular or suborbicular, entire or subentire; the corolla lobes are short, obtuse or rounded, usually less than 10 mm; whereas in var. *caucasicum* the leaves are broadly ovate-cordate, often acute or subacute and toothed; the corolla lobes are acute or subacute, 10–15(–17) mm". He goes on to say that the species exhibits a clinal pattern of variation: plants from north-western Turkey, around Istanbul, tend to have broad, rounded leaves and small flowers with short, blunt corolla lobes; further east, the leaves gradually become more acute and cordate, and the flowers larger with more acute corolla lobes. Grey-Wilson, in his monograph of 1988, accepted Meikle's findings, except that he upgraded the two taxa from variety to subspecies. Meikle's findings were based entirely on examination of herbarium specimens, and he never visited Turkey. Nevertheless his statements, quoted above, seemed so precise that a major aim of the first field trip to Turkey, and possibly also of subsequent field trips, would be to determine whether Meikle's view of the species could be substantiated by examining populations in the wild. Nevertheless, the fieldwork would be broad enough to cover other aspects of the plants, and additional factors, both geographical and environmental, that were not considered by Meikle.

At the eastern end of the range of *C. coum* in Turkey plants often have longer petals and ovate leaves (MD)

An example of *C. coum* from north-west Turkey, with short petals and round leaves (BM)

Planning and execution of the fieldwork

1987 Turkey Team: Richard Bailey, Jill White, Trevor Wiltshire and Jeremy Wood

This first study visit to Turkey took place from 20 March to 9 April 1987. With Meikle's suggestion of a west to east cline in mind, the trip ranged from Istanbul in the west and travelled as far east as possible without entering the USSR, stopping at intervals to make a detailed examination of the cyclamen. This gave an overall west–east distance of about 1,060 km. The area covered lay along the northern part of Turkey bordering on the Black Sea, the greatest distance inland being about 60 km. A proforma had been designed in order to record the leaf and petal measurements (to the nearest mm), leaf shape, petal shape and leaf markings, each judged by comparison with a set of standards, and flower colour by comparison with the RHS Colour Chart. Details of each site that were recorded included its latitude and longitude, altitude, aspect, slope, degree of shade and accompanying plant species. In addition, herbarium specimens of leaves and flowers were taken, and approximately 50 live plants were collected, paying particular attention to some of those that had striking leaf patterns.

The Cyclamen Society's team at work in 1987, Belgrad Forest in European Turkey (JWhi)

As well as recording scientific data the expedition looked for plants of horticultural merit, such as leaf variation (TW)

A superb silver leaf form of *C. coum* in cultivation (no. CSE 87055T) selected from a population between Ordu and Ulubey, northern Turkey (MD)

Cyclamen coum at its lower altitudes is frequently found growing in mixed woodland situations on the Black Sea coast, such as can be seen inland from Trabzon (BM)

The managed, coppiced hazel groves support a good ground flora including *C. coum* and primroses (MD)

Helleborus orientalis often accompanies *C. coum*, a natural association worth replicating in gardens (MD)

With little previous experience of Turkey, the team members were naturally concerned about their ability to find populations of *C. coum* that were suitable for detailed study. But, upon travelling along the coast road to the east of Samsun, it soon became apparent that the species was very common and could be found wherever the habitat was favourable. Wooded areas were the most likely places to find cyclamen, and also worth a search were hazel groves, where the cyclamen were sometimes accompanied by hellebores (*Helleborus orientalis*) and primroses, sometimes the pink form *Primula vulgaris* subsp. *sibthorpii*. In stark contrast, when crossing the Anatolian plateau between Ankara and Samsun, small groups of cyclamen could be seen in open, rocky terrain, usually emerging from crevices between the rocks. Much further east, at one site just above the coast road, cyclamen were found on a rock face with their exposed tubers anchored in wet moss. Several excursions were made up the side roads heading inland, up into the foothills of the Pontic Alps, where cyclamen were found up to 1,370 m. Weather conditions were apt to be uncomfortable away from the coast, and on one occasion it was snowing right down to 500 m. Unexpectedly, to the east of Ankara, the cyclamen plants at any given site exhibited extreme variation in leaf shape, petal shape and leaf markings, and there was no such thing as a 'typical' plant in a population. A big surprise was finding *C. parviflorum* at intervals along the road to the Sumela Monastery, south of Trabzon, and up the steep footpath to the monastery itself. The altitude range at which this species was noted ranged from 730 m to 1,370 m and most populations were growing in the shade of trees. This

With year-round rainfall on the Black Sea coast, *C. coum* can grow on steep rocky banks with minimal soil, here seen with *Primula vulgaris* subsp. *sibthorpii* (MD)

Plant hunting is not always fun! The Cyclamen Society's team in northern Turkey on 4 April 1987 (JWhi)

was a very different habitat from the oft-quoted site at the top of the Zigana Pass, where *C. parviflorum* is found in open, alpine conditions at over 2,000 m. In some places on this route, *C. coum* and *C. parviflorum* were found growing side by side, but the two species were always distinct with no sign of hybridisation.

The team continued to record data over the next two weeks with some limited collection of plants (for this field trip, despite the best efforts of Turkish colleagues, all of the permits necessary for collection had not been forthcoming). The trip finished in the Belgrad Forest just outside Istanbul. Unfortunately, the cyclamen plants here had all finished flowering, so it was only possible to record details of the leaves. But at the beginning of the trip, the whole area around Istanbul had been under deep snow, making fieldwork impossible. A final visit was paid to Dr Özhatay at Istanbul University to hand over some of the herbarium specimens. On arrival back at Heathrow, the collected plants were taken by Peter Moore for cultivation at his nursery on behalf of the Society.

Here in the woods below the Sumela Monastery, *C. coum* and *C. parviflorum* can be found growing in close proximity (MD)

1988 Turkey Team: Richard Bailey, Ross Lilywhite and Peter Moore

The second field trip to northern Turkey took place the following year, 11–30 March 1988. The broad aim was to complement the work of the 1987 field trip, revisiting some of the most promising sites, but also extending the search to cover some new locations. The methodology used for recording cyclamen populations was the same as that used in 1987, but on the advice of Dr Basil Smith, a sample of 50 plants was examined at each site wherever possible, using objective methods to select the plants. It was decided to go first to the Belgrad Forest near Istanbul, but again most of the cyclamen had finished flowering there. Nevertheless, sufficient remained in flower to allow worthwhile information to be gathered. The team then proceeded to the north-eastern part of Turkey, from Ünye to Hopa and Artvin, the areas that had been of greatest interest the previous year. Preliminary findings from 1987 suggested that altitude, as well as longitude, was affecting the leaf and flower characteristics that distinguish C. *coum* subsp.

At higher altitudes, *C. coum* grows in more open places. In 1987, the field work investigated the hypothesis that leaf and flower characters were linked to altitude (MD)

The mixed deciduous woodland of the Belgrad Forest (BM)

coum from subsp. *caucasicum*. To obtain further information on this point, sites were chosen in pairs along the coast, one at a high and the other at a low altitude, each at approximately equal longitude. Unfortunately, the cyclamen had nearly finished flowering at some of the chosen low-altitude sites, and it was not always possible to find anything approaching 50 plants in flower. The collection of plants for horticultural purposes (the necessary permits having been obtained) focused on selection of the following characters: white flowers, especially dark flowers, bicoloured flowers, absence of a white eye, scented flowers, and fine leaf forms. No forms of *C. coum* with white flowers and a purple basal blotch were found, but a plant with white flowers and no basal blotch was collected: this form has now been named forma *albissimum*. Another collected plant had white flowers with the faintest tinge of pink replacing the normal basal blotch. Dark flowers and really attractive leaves were seldom, if ever, found on the same plant, but some very fine silver-marbled plants were collected above Artvin. One of the most unusual plants, collected near Trabzon, had pink marbled leaves. Plants of *C. parviflorum* were also collected, covering a wide range of flower colours, including one with flowers so pale as to be almost white. Despite extensive searches, not a single plant of *C. parviflorum* with marbled leaves could be found. In all, nearly 150 cyclamen plants were brought back to Britain. Herbarium specimens were gathered at many of the sites and these were divided between the Royal Botanic Gardens, Kew and Gazi University, Ankara.

The pure white variant of *C. coum* forma *albissimum* (CSE88020A) in cultivation, originally introduced from near Sumela (JB)

An example of the wonderful leaf forms of *C. coum* introduced in 1988 (CSE88397) from near Artvin, north-eastern Turkey (JB)

1990 Israel Team: Richard Bailey, Melvyn Jope and Peter Moore

After a gap of a year, the studies on *C. coum* resumed in 1990 when two groups took to the field, the first visiting Israel, the second, north-western and south-eastern Turkey. These visits clearly extended the area of search considerably, particularly in a southerly direction, so that the four field trips undertaken by the end of 1990 spanned a total distance of 990 km from north to south. This was sufficient to allow any possible effect of latitude on plant characteristics to be investigated.

The decision to visit Israel was made following receipt of a letter from the Hebrew University, Jerusalem, pointing out that *C. coum* occurred in their country but was not recorded by Grey-

C. coum in the most southerly part of its distribution in Israel, growing in a basalt scree between Mas'ada and Quneitra at about 1,050 m (MELJ)

Wilson (1988). The Israel study visit took place on 4–18 February 1990. Correspondence with botanists at the Hebrew University had identified three sites for *C. coum*, all in the north of the country. Of these, the one on Mount Hermon, on the border with Lebanon, lay in a military zone and proved impossible to visit. The other two sites, on the Golan Heights and Mount Meron, proved to be accessible and both were studied. These populations represent the most southerly locations for *C. coum* known at present. Among the large population on the Golan Heights, most plants were fairly uniform in flower colour, with hardly any very dark or very pale flowers, the exception being a small group of *C. coum* forma *albissimum* similar to those found in Turkey in 1988. The cultivar 'Golan Heights' is derived from this collection. Leaf pattern and shape were also fairly constant, most leaves having a faint hastate pattern, although a few completely plain leaves were found. On Mount Meron, two very small populations were located, totalling some 60 plants. Snow was still lying and few plants were in flower. Again the cyclamen were fairly uniform, and no plants that extended the range of variation noted on the Golan Heights were found.

The field team had also been asked to study *C. persicum* which is, in Israel, at the south-eastern edge of its range. This species is very widespread in Israel, and it was studied over the whole extent of its range from Tel Maresha in the south to the Syrian and Lebanese borders in the north. Its flower colour varied from deep pink to white with a purple basal blotch, and even pure white with no basal blotch. Of special interest were plants growing on Mount Meron at an altitude of 1,200 m, close to *C. coum*, and flowering in the snow. Some of these plants were collected in the hope that they might prove hardier than strains of *C. persicum* already in cultivation. Of even more interest was a report from Israeli botanists of autumn-flowering populations to the north-east of Jerusalem. These were duly found and some plants collected.

Cyclamen coum forma *albissimum* 'Golan Heights' in cultivation (MD)

Leaf samples were taken throughout the three seasons of field work to determine the range of shapes, sizes and patterns. Those in Israel were dull in comparison with plants from north-eastern Turkey (MELJ)

The white-flowered, plain-leaved selection of *C. coum* seen here in the wild, now known in cultivation as 'Golan Heights' (MELJ)

Cyclamen coum on Mt Meron, Israel, where higher altitude plants at 1,200 m were recorded (MELJ)

1990 Turkey Team: Chris Clennett, Erna Frank and Jeremy Wood

The fieldwork in Turkey was carried out from 23 February to 9 March 1990. This year no collecting was permitted by the Turkish authorities, so the emphasis was on recording the details of cyclamen populations and gathering herbarium specimens. The team first journeyed north-west from Istanbul to Kirklareli near the border with Bulgaria, and there they found three sites where *C. coum* was in flower. These were some 90 km further west than the Belgrad Forest site near Istanbul, and it was hoped that extreme forms of *C. coum* subsp. *coum*, with orbicular leaves and small, rounded petals, would be abundant. But although most of the leaves were orbicular, only one or two plants had strictly rounded petals. The team then flew on to Adana, near the eastern end of Turkey's Mediterranean coast, and explored an area in the Amanus Mountains and south towards the Syrian border. The cyclamen there presented a very different appearance from those near Kirklareli, and with only one exception, the populations were characterised by leaves that were ovate, some markedly so. At some sites, the leaf margins were distinctly toothed, a character hardly ever noted in northern Turkey. The hastate pattern was usually rather indistinct, and silver leaves were never seen. The flowers tended to be small, particularly at high altitude. Several populations had unusually pale flowers, a few being white with a dark basal blotch; a common characteristic was for the eye to be much smaller than normal, and plants with a pink eye were not uncommon.

In the Amanus mountains of southern Turkey, *C. coum* and *C. pseudibericum* can occasionally be found together. *C. coum* flowers (centre) with leaves of *C. pseudibericum* (right) (JW)

While driving along the coast road towards the Syrian border, patches of *C. persicum* could be seen, and at one village a little inland, these plants were flowering abundantly. In woodland at the foot of the Amanus Mountains, *C. pseudibericum* was coming into flower, some of the plants being adjacent to those of *C. coum*. There was no obvious sign of hybridisation between the two, but it is interesting to report that when Desmond Meikle was shown herbarium specimens of *C. coum* leaves from that site, they were so prominently toothed that he at first mistook them for leaves of *C. pseudibericum*.

Summary of the scientific findings

One of the aims of the four field trips to study *C. coum* was to seek evidence to support Meikle's conclusions (1978) regarding a gradual change in leaf shape, petal length and petal shape along a progression from west to east along the distribution range of *C. coum*. From visual observations alone, no such change was evident during the first field trip, although the members were rather expecting it to be obvious. They did, after all, travel about 1,000 km from west to east, making many intermediate stops to observe and record details of the cyclamen. It was not until the team had returned to Britain, and the results were analysed, that certain trends began to emerge. In conformity with Meikle's view, the tentative conclusions were that, from west to east, the leaves became more ovate, and the petals became both longer and more pointed. But another, unexpected, finding was that altitude was affecting some of these same characters; at high altitude the leaves became more rounded and the petals became shorter. No such effect of altitude had ever been reported before. It is probable that one reason for the difficulty in observing the west-east trends in the field was that they were partially obscured by differences in altitude. But another reason was the wide range of variation among the cyclamen at a given site, with round leaves and ovate leaves on adjacent plants, and with small flowers next to larger flowers. This presented a very confused picture to anyone trying to judge differences between one population and another during the course of the field trip.

At this stage, Basil Smith, a professional statistician, was asked to look critically at the results of not only the first field trip but of each subsequent field trip as they became available. He was able eventually to combine the findings from all four field trips for purposes of analysis, a procedure that he considered valid as the same methodology was used in the field for all four trips. In addition to the effects of longitude (distance from west to east) and altitude, he also included the effect of latitude (distance from north to south). His conclusions can be summarised as follows:

Leaf shape: round leaves are more abundant to the north, to the west and at high altitude (although the effect of altitude is not statistically significant); ovate, subacute leaves are more abundant to the south and to the east.

Plants of *C. coum* at the higher altitudes tend to have small rounded leaves and short 'dumpy' flowers (MD)

Petal length: short petals are more abundant to the south and to the west (again this altitude effect is not statistically significant); long petals are more abundant to the north, to the east and at low altitude.

Petal shape: obtuse petals are possibly more abundant to the north and to the west, though this is straining the statistics; acute petals are not correlated with any variable.

This analysis confirms the tentative conclusions from the first field trip in respect of leaf shape and petal length, based on a much larger sample of plants. It also shows that latitude affects these two characters, the leaves becoming more ovate and the petals shorter as one travels south. Petal shape, however, varies in a somewhat inconsistent way. To give some idea of the magnitude of the differences in petal length, the only one of the three characters to be actually measured, in northern Turkey the average length was 12.1 mm in the west and 14.9 mm in the east, based on total counts of 170 plants and 305 plants, respectively. The difference between the two, although statistically significant, is not all that large, and is probably another reason for the difficulty in spotting trends visually.

In north-eastern Turkey and at lower altitudes, *C. coum* often has longer flowers and the leaves are more ovate, although the distinctions are not clear-cut (MD)

However, the findings summarised above do not tell the whole story. Two other facts emerged from Basil Smith's analysis. The first was that differences that were associated with longitude, latitude and altitude only explained a relatively small proportion of the variance in the three plant characters being considered. This means that other factors must be influencing leaf shape, petal length and petal shape. At each site where a detailed study was made, the aspect, slope and degree of shade were recorded, but these factors did not appear to be having any influence. Deciding what additional factors might be worth including presents something of a challenge to future researchers.

The second fact was that variations in one of the three measured characters were not correlated with variations in either of the other two: all three were behaving independently of one another. Although not specifically stated, it seems to be implicit in Meikle's key to the two subspecies (1978) in which plants with short petals tend to have orbicular leaves and round petals, and in which plants with long petals tend to have ovate leaves and acute or subacute petals. The field trip findings suggest that this is not the case, certainly not in those areas visited from 1987 to 1990. In individual plants, orbicular leaves can be found in conjunction with short round petals, with short acute petals, with long round petals or with long acute petals; similarly, ovate leaves can found in conjunction with short round petals, with short acute petals, with long round petals or with long acute petals. It is tempting to postulate that these three plant characters are inherited independently of one another. Nevertheless, it has to be remembered that the distribution of *C. coum* extends a long way to the east of Turkey, and we hope very much that the Cyclamen Society will undertake a study of the species in Georgia. Populations that far east might well be more uniform than those in north-eastern Turkey, and the characters that typify subsp. *caucasicum* more obvious. In southern Turkey, and to a lesser extent in Israel, a majority of plants were found to have ovate leaves and short petals, a combination that does not readily fit either subspecies. It seems possible that these southern populations have evolved in comparative isolation over a considerable period, and that in the process they have developed the particular characteristics that can be seen today.

Other leaf and flower characters that play no part in distinguishing taxa were included in the studies as they are important in other respects. The hastate pattern on the leaves, sometimes referred to as marbling, is a character to which growers pay a great deal of attention, and is often the first feature to catch one's eye in the field. In north-western Turkey, 10% of the sample had totally plain leaves; in other areas, this proportion was 4% or less. Marbled leaves were particularly in evidence at low altitude in central-northern and in north-eastern Turkey and the effect of altitude on the proportion of marbled leaves was statistically significant.

Throughout the field trips, the length and width of the leaves were measured, and variations in leaf size were particularly obvious. In northern Turkey, 80% of all leaves measured had a width of 50 mm or less, but some populations had leaves measuring 90 mm or more in width, one giant leaf having a width of 147 mm. No convincing relationships could be found involving average leaf size. Nevertheless, the incidence of large leaves, with a width of 50 mm or more, increased from west to east and was greater in dense shade than in light shade. The effect of shade is not unexpected, and confirms previous suspicions based on visual observation.

Flower colour was determined in the field by matching each flower, as far as possible, to the nearest sheet in the RHS Colour Chart. This proved surprisingly difficult as many of the flowers seemed not to match any of the available colours in the chart, the intervals in which are uncomfortably large over the range covered by cyclamen flower colours. Another problem is that two observers might have different ideas as to the closest match, so that the recorded colour depends critically on the particular observer making the judgment. The arrangement of the chart makes it impossible to treat the results mathematically, so no statistical analysis could be performed. What can be said is that the most common flower colour on all four field trips was a purple or red-purple of medium intensity (saturation). Flowers that were more intense than the average were often found, particularly in some populations recorded in 1988. Very pale flowers were infrequent, except in three populations in southern Turkey where they formed 50% or more of the sample. These three populations, uniquely among all those recorded on the four field trips, contained a few white flowers with a dark basal blotch, a form frequently seen in cultivation. Two plants found in Israel and one in north-eastern Turkey were complete albinos with no basal blotch, whereas two more found in north-eastern Turkey were white with the faintest pink blotch. Bicoloured flowers were present in some populations; at one site near Artvin, 40% of the sample was judged to have bicoloured flowers. Apart from a slight tendency for bicoloured flowers to be found at high altitude, inspection of the records failed to find any link between flower colour and location. The colour of the eye is reported by Grey-Wilson (1988) to differ between the two subspecies, the eye being white in subsp. *coum* and pink in subsp. *caucasicum*. Based on this, one would expect to find more plants with a pink eye as one travels eastward in Turkey. The records from 1987 and 1988 show that a pink eye is rare in Turkey, but is in fact more prevalent in the north-west and the south than in the north-east, a fact that does not really fit in with Grey-Wilson's statement.

The Cyclamen Society's Gold Medal exhibit featuring the *C. coum* fieldwork. RHS Show, January 1990 (MD)

Horticultural aspects

One of the most striking features of the plants seen during the first field trip was the marbling of the leaves. From Unye eastwards, some really fine leaf forms were found, one of the choicest examples consisting of a dark green 'Christmas-tree' shape surrounded by a contrasting area of bright silver, in some cases with a narrow dark rim. On this field trip, leaf pattern was the chief factor that determined which plants were collected. Nowhere else in Turkey or in Israel were such leaves seen, the intensity as well as the extent of the marbling setting them apart from the ordinary. The 1988 field trip concentrated largely on this same area of north-eastern Turkey, but the collecting was widened to include other features of the plants, and a greater proportion of the time in the field was spent searching for choice or unusual plants. As already mentioned, one plant with a pink marbling was collected, but unfortunately neither the original plant nor any of its offspring have shown the pink marking in cultivation. Presumably the environmental conditions must have played a part in determining this most unusual pattern. In general, however, the collected plants maintained their characteristics, particularly their flower colour and leaf marbling characters, in cultivation.

In 1990, the Society showed a selection of the plants from the first two field trips carried out to study *C. coum* at the RHS flower show in Vincent Square. The plants were accompanied by photographs, herbarium specimens, maps, explanatory text and some of Trevor Wiltshire's watercolour paintings. The display was awarded a gold medal. Through the seed exchange operated by the Cyclamen Society, the progeny from the collected plants has been made available to a large number of members, and it is fair to say that the range of variation in cultivated *C. coum* today is far greater than it was before 1987. As one example, the pure white form, now known as forma *albissimum*, was brought into cultivation and is now offered by specialist nurseries.

Cyclamen coum forma *albissimum*, perhaps one of the best introductions made during the Cyclamen Society's field work. This is a plant of CSE88020B, grown by Ian Robertson (MD)

3.5b *CYCLAMEN CRETICUM*

Until the 1980s, *C. creticum* was thought to be a Cretan endemic, but it was then found to be present on Kárpathos, an island midway between Crete and Rhodes. In view of the meagre detailed information that was available about the cyclamen of these islands, the Cyclamen Society decided to send a group to Rhodes and Kárpathos in 1991. Their aim was primarily to study *C. rhodium* but they would also take the opportunity to make detailed observations and distribution records for *C. creticum*, *C. persicum* and *C. graecum*. In 1994, 1995 and 1996, teams carried out field studies of *C. creticum* on Crete. The following is a compilation, extracted from the Society's Journals, of the information on *C. creticum* submitted by the various groups.

***Cyclamen creticum* on Kárpathos 1991** Team: Richard Bailey, John Fielding and Peter Moore

Richard Bailey

Habitat and distribution

Although a large number of seemingly suitable sites on Kárpathos were investigated, *C. creticum* was found in only two places, within a few kilometres of each other, on the east coast of the island. Each site was at an altitude of approximately 400 m. Both sites were steeply sloping, with an overall slope of approximately 1:1 to 1:2. The soil was variable in depth, a somewhat clayey loam with a surface layer of leafmould, over fine-grained limestone. It is reasonable to suggest that it would become extremely dry in summer. The tubers, however, were often deeply buried, up to 23 cm, at which depth they are unlikely to dry out completely even in summer. In neither location was the species abundant. Groups of plants were scattered sparsely over the whole area. Young seedlings were found at both sites, however, and the species appeared to be at least maintaining its population.

Cyclamen creticum with seedlings, an indication of a healthy population on the island of Kárpathos (JF)

At the first site, *C. creticum* was growing over a very limited area in dappled shade below a canopy of *Pistacia lentiscus*, very often in moist hollows or where there were signs of water seepage below boulders. The slope had an overall westerly aspect, but *C. creticum* appeared to grow best at the foot of north- or north-west-facing boulders. A few plants were, however, found on the south-facing slopes of the undulating hillside, always well buried in rock crevices or under overhangs of the waterworn limestone. The surrounding vegetation was not abundant. Large areas were sparsely covered by grasses or had bare soil. *Pistacia lentiscus* was the most abundant shrub but there were also specimens of *Ceratonia siliquosa*, *Cistus salvifolius*, *Cistus incanus*, *Clematis* sp., *Lithodora hispidula*, *Phlomis* sp., *Salvia triloba* and *Sarcopoterium spinosum*.

At the second site, *C. creticum* was growing in deep stony soil (with very sharp drainage) in light shade below sparse *Pinus*. The aspect was, overall, north-north-east and plants were widely scattered over a large area not seeming to need the extra 'protection' of north-facing boulders. Like so many places on Kárpathos, this site had been extensively burnt, probably within twelve months of our visit. There was very little growth of shrubs, herbaceous plants or annuals, but in view of the recent burning, it was difficult to decide whether this was the normal state. Several shrubs, notably *Arbutus andrachne*, *Lithodora hispidula*, *Phlomis* sp. and *Pistacia lentiscus*, were regenerating from rootstocks, *Prasium majus* appeared to have survived the fire unscathed (or regenerated quickly) and there were a number of *Cistus* seedlings.

Variation

Flowers: *Cyclamen creticum* was not very variable on Kárpathos. The flowers were almost all pure white, although a very small number had a pink flush so pale that it had to be categorised as white when using the RHS Colour Chart. All the flowers examined were slightly scented. Petal length varied from 26 mm to 14 mm and width from 7.5 mm to 4 mm.

Leaves: leaf shape, pattern and size were studied. All leaves had pointed tips, 29 were longer than wide, 20 approximately as long as wide and 22 wider than long. The pattern of the markings was rather more uniform, some with speckled leaves similar to those of *C. rhodium* subsp. *peloponnesiacum*. Out of the 79 plants examined, the marbling of nine was silver rather than green. The largest leaf examined was 81 × 69 mm and the smallest 27 × 23 mm.

Cyclamen creticum on Kárpathos seemed to be a very uniform population. It certainly added little, if anything, to the range of variation already known in cultivation and in the literature. The nature of the two sites at which the species was found gave no hint that the plant might have been introduced; it would appear to be a truly native plant that could well have been more widely distributed in the past. Its limited distribution, the intensity of grazing and the acknowledged susceptibility of the island's woodlands to fire make its future a matter for concern.

Cyclamen creticum on Crete

Although *C. creticum* has been described as being very abundant in Crete, other reports have suggested that it had a relatively restricted distribution with the majority of populations being concentrated at the western end, only limited populations in the east, and two isolated populations in the centre of the island. The Cyclamen Society made three visits: in 1994, to the western end of Crete, making detailed observations at 20 sites and noting many more populations; in 1995, to the drier eastern end of Crete where *C. creticum* was found to be far more extensively distributed than had been known (29 sites were recorded); and in 1996, to the central section, where again, many more sites were added to the distribution.

Western Crete, 1994 Team: Vic and Janet Aspland, John Fielding and Melvyn Jope

Vic Aspland

Crete is the largest and most southerly of the Greek islands, 257 km from east to west and in width varying from 57 km to 15 km. It is dominated by a backbone of limestone mountains with four main massifs: Lefká Óri (White Mountains) rising to 2,452 m, Oros Ídi (2,456 m), Oros Dikti (2,148 m) and Ori Sitias (1,476 m). The compactness of the island makes it possible to pass through many vegetation zones in a very short time. A journey of as little as 25 km can begin at sea level among banana, olive, orange and lemon trees, then climb through vineyards, pass wheat fields, and end with *Crocus sieberi* flowering through melting snow. Because most of the terrain is underlain by limestone, Crete has no rivers or streams of any substance, and most of the water flow is underground. This is not to say that the whole island is parched from spring to autumn; many gorges, even quite small ones, have sufficient ground moisture to support a population of deciduous trees and shorter plants in their shade.

Cyclamen creticum on Crete, here seen on the east side of the Lefka Ori at 680 m (JF)

Cyclamen creticum was found growing at altitudes varying from 90 to 1,250 m and was confined (with one exception mentioned below) to a recognisable ecological niche with three key components: shade, moisture and north-to-west aspect. These three factors are subtly blended to create the conditions needed. At lower elevations, *C. creticum* was often to be found in shallow gorges with good tree cover giving 90–100% shade. Even in these localities though, only the gorge bottom and the north- to west-facing side were occupied. The type of tree supplying the shade seemed unimportant: *C. creticum* was found beneath *Platanus orientalis*, *Acer cretica*, *Ficus carica*, *Quercus coccifera*, *Laurus nobilis* and *Cupressus sempervirens*. At higher altitudes, where temperatures were lower, a gorge or watercourse was less necessary, but even here, the same aspect seemed to be obligatory. On two of the passes leading to the Omalos Plain (at 1,140 m in each case), *C. creticum* was found growing beneath *Cupressus sempervirens*. Although these habitats had a generally dry look, some moisture was present in the soil below the trees; it is quite likely that ground moisture was maintained by a combination of the dense shade and rainfall, or at least overnight mists. On the western pass, a colony of *C. creticum* was found on the south-facing slope (also at 1,140 m), but none was seen beyond the dense shade of the cypresses under which they sheltered. In this locality also, but higher up the slope, was the only habitat in which the species was found growing in full sun. Here, altitude is presumed to have tipped the balance in its favour.

One site of particular scientific interest was visited. It had been reported that the population in a sweet chestnut grove near Agia Irini had a chromosome count of 2n = 20, the same as *C. repandum* and *C. balearicum*, whereas previously the recorded count for the species was 2n = 22. This site was particularly interesting for other reasons: the lower reaches contained sweetly scented yellow primroses and all of the *C. creticum* flowers examined were also scented; most had pale pink 'noses'. A further oddity of this site was the presence of running water, with one *C. creticum* growing in saturated soil by the rill, and one being below water level.

We have already referred to the three key aspects of a 'typical' *C. creticum* habitat, and it is remarkable that the species seems to be indifferent to most other conditions. *Cyclamen creticum* was found as a chasmophyte occupying apparently soil-free limestone crevices, or in soils that varied from *terra rossa* through clay-like,

Here on the north side of the Omalos Plain at 1,140 m, a population of *C. creticum* in association with lichens under *Acer creticum* and *Quercus coccifera* scrub (JF)

sandy, schisty and organic-rich, to pure leafmould. It also seemed indifferent to soil pH. On-site tests indicated that it would tolerate values ranging from 2.8 (which would be regarded as probably too acid for many rhododendrons) to 10 (which is more alkaline than many chalky soils).

The previous experience of the field study members had indicated that the degree of variation found in *C. creticum* in the wild was far wider than is suggested by the plants in commerce, a conclusion that was supported by the current work. Leaf margins varied from entire to deeply jagged, with the patterning being anything from a simple hastate mark in silver or green to overall silver splashing, like that found in *C. rhodium* subsp. *peloponnesiacum*. In western Crete, the future looks bright for *C. creticum*. Individual habitats are comparatively small but mostly unsuitable for cultivation, and hence unlikely to be disturbed. Populations were often large, with much regeneration by seed. In addition, the species was able to creep back into man-made habitats where cultivation was not too intense, as demonstrated by the chestnut-grove population and another stand that inhabited the edge of a concrete channel carrying an irrigation pipe.

Eastern Crete, 1995 Team: Heather Dale, John Fielding and Trevor Wiltshire

Trevor Wiltshire & Heather Dale

The 1995 visit to eastern Crete, commencing on 18 March, aimed primarily to record populations of *C. creticum*, but also to note any other cyclamen species that were found. It was particularly important to visit a reported site for *C. persicum* (see page 117). The cyclamen were in full flower everywhere except at high altitude on the Lassithi plateau. Detailed records were made at 29 sites and *C. creticum* was seen in 94 different locations. It was much more common than previously stated in the literature, occurring in all areas where the habitat was appropriate. Generally, this was a north-facing gully often densely shaded by mature trees or heavy scrub. The fact that *C. creticum* was found on wooded hillsides, on un-wooded north-facing cliffs and on north-facing slopes of low dry scrub suggests, however, that the important factor is shade from the sun for the majority of the day as opposed to any specific conditions associated with either trees or cliffs. This is supported by the variation, both in altitude (from sea level to 850 m) and in the soil conditions found at the sites. The leaf litter layer on the surface varied from none to several centimetres. The litter was usually decaying *Quercus coccifera*, although at two sites, the litter was composed of pine needles. The soil beneath deep litter had a high organic matter content, which improved the very clayey soil; where there was no litter, the soil was hard-packed clay, usually red-brown except in the Lassithi plateau where it was grey-brown. In the Zakros gorge, *C. creticum* were growing in a very different soil that was sandy and rocky with little organic matter. This wide range of soil types suggests that environmental factors that are not related to soil determine the distribution of *C. creticum*.

Despite the previous summer's serious fires, which removed the canopy of trees and thus exposed the cyclamen to high light levels, the plants looked very healthy having many flowers and, often brightly patterned, dark green leaves that tightly hugged the ground. Sufficient seed had escaped the fire to provide evidence of good regeneration. In fact, regeneration was good at most sites, so at present at least, the distribution of *C. creticum* on Crete is not declining further. It was our impression that in the past, when a greater area of the island had natural tree cover, the distribution of *C. creticum* would have been more widespread. It is probably since the deforestation of large areas for timber and farming that the species has been restricted to gullies and other areas that are impossible to farm.

The populations in eastern Crete were variable in leaf pattern and although they were generally fairly dull in colour and patterning, there was a higher incidence of well-marked and brightly coloured leaves than was seen in the west of Crete.

Central Crete, 1996 and overall assessment of three years of fieldwork Team: Melvyn Jope, John Fielding and Martyn Denney

Martyn Denney

The 1996 expedition was to cover the remaining central part of Crete. This area extended from Lefka Ori (White Mountains) eastwards past Psiloritis (Mount Ida) to a point almost due south of the city of Iraklion, and represented some 50% of the land mass of the island. It was clear from the early planning stage that such an area could only be "sampled" rather than covered thoroughly and in detail. A number of other potential sites had been identified, but the work of mapping the distribution of *C. creticum* required that time be spent investigating areas where there had been no previous sightings as it was as important to establish where it did not grow, as where it did grow. In two weeks, 33 sites for *C. creticum* and *C. graecum* were studied, starting on 12 April. First, the team explored the Kouloukinas mountains, a low range on the northern coast of the island between Rethymnon and Iraklion. Two sites for both *C. creticum* and *C. graecum* were recorded at between 300 and 600 m; at the higher of the two sites, large numbers of *C. creticum* were found amongst dense pistachio scrub adjacent to a large stand of *Arum concinnatum*. The next area to be covered was Psiloritis, which we approached from the north along the road that leaves Anogia and crosses the northern flanks of the mountain to the Nidha Plateau and the Idaien Cave. *Cyclamen creticum* was found at 850 m growing with the dwarf yellow and white *Orchis pauciflora*. Other interesting plants on the Nidha Plateau included *Corydalis rutifolia* subsp. *uniflora* and *Chionodoxa nana*.

Here, on the road south from Iraklion to Pirgos at 575 m, *C. creticum* is growing in the open but with shade protection from seasonal annuals and perennials such as the prickly *Notobasis syriaca* (JF)

A markedly speckled-leaved plant of *C. creticum* north of Ligortinos (JF)

The attractive pink-flowered forms of *C. creticum* can be found in various populations on Crete (JF)

Close-up of *C. creticum* growing on Mt Psiloritis (JF)

Cyclamen have never been reported in the Asteroussia Mountains on the coast of Crete directly south of Iraklion, but these mountains had been clearly identified as an area that should be explored. The trip south across the island was uneventful other than the discovery of a site for *C. creticum* high above the road where some excellent pink flowered forms were found. The hills around Pirgos were green and lush with many species of *Ophrys* and *Serapias* but no cyclamen. The Asteroussia Mountains stretching west towards Matala had a number of potential sites on both their northern and southern flanks but these felt wrong, were too dry and were trampled by goats, and nothing was found except for some good colour forms of *Moraea sisyrinchium* and extensive stands of pale yellow *Arum creticum*, in both its green- and maroon-stemmed forms. Whether *C. creticum* existed here before the mountains were so extensively grazed is pure speculation.

Under a dense canopy of *Platanus orientalis* with a leafy mulch, *C. creticum* is thriving along a river bank on the south-west side of Mt Psiloritis (JF)

The team then returned to Rethymnon along the southern flank of Psiloritis. Both *C. creticum* and *C. graecum* were found more or less as expected at various locations. From Rethymnon, we visited the Gipari Gorge. Uniquely, this runs roughly east to west. It is in the centre of the island and ends at the mountain village of Asigonia; *C. creticum* was found growing in very dense populations in the wooded river valley under *Quercus coccifera* and *Platanus orientalis* with robust plants of *Dracunculus vulgaris* just coming into flower. Amongst the *C. creticum* were some quite exceptional leaf forms and flowers with very long petals, some of which were collected: CSE96252, petals 38 mm long with 360° twist; 96256, acuminate leaves 98 mm × 62 mm; 96258, very bright silver hastate pattern; 96259, silver leaves with a few flecks of green; 96260, silver leaves. There was also an interesting plant (CSE96261) that had foliate calyx and corolla.

From there, the team headed south across the island towards the mountain village of Spili, finding further sites for both *C. creticum* and *C. graecum* on the way. Then, from a base on the south coast at Agia Galini, the next few days were spent exploring the gorges and gullies that dissect the coast between there and where the Lefká Óri meet the sea in Sfakia to the west of Chora Sfakion. The southern coast of Crete is noticeably warmer and drier than the north of the island and here *C. creticum* hides deep in damp shady places or at the bottom of the deep gorges, though even here it remains on the north side of rocks and cliffs. Some of the sites are well documented and the team confirmed the existence of these populations in all but one case: the Imbros Gorge.

A dense population of *C. creticum* in the Gipari Gorge near Asigonia in central Crete (JF)

The large population in the Gipari Gorge contained some interesting variation in leaf patterns and flower shape (JF)

Undisturbed spots such as the hollow boles of trees provide ideal sites for *C. creticum*. Gipari Gorge, central Crete (JF)

The Aradhena Gorge, Sfakia, where *C. creticum* grows in dense shade in moss on rocks (JF)

Two sites were of particular interest. Site 96/23 between Plakias and Kato Rodakino had plants with leaves that were apparently thicker with a more leathery texture than is usual. The plants collected here will require further investigation to establish whether they exhibit any morphological or chromosome variation. Site 95/25, the Aradhena Gorge, had populations of *C. creticum* in very dark crevices, growing as chasmophytes in limestone rock and epiphytically in moss, which had a significant proportion of flowers in varying shades of pale pink (69D, 76B by RHS Colour Chart, or paler).

We had particularly wanted to find populations of *C. creticum* at high altitude. The 1994 expedition had found it at 1,250 m between Agia Irini and the Omalos Plain in Lefka Ori; the 1995 expedition had found it at 1,050 m above the Lassithi Plain in Dikti; in 1996, we had found it no higher than 850 m in Psiloritis. We were now fortunate enough to find a site at 1,000 m on Mount Akones, which is an outlier to the east of Lefka Ori (Site 96/27). This site was very exposed and comprised a north–north-west facing gully, with a few stunted trees of *Acer sempervirens* growing with mountain grasses, which supported a rich flora. The cyclamen were found growing only in the gully, where they sheltered under rocks that would probably be submerged in the stream in early spring, and around the stunted trees where they could make the most of the scant shade. Few plants were in flower.

After surveying a site for both *C. creticum* and *C. graecum* on the Askifou Plain, we headed north-west towards Chania via Platanias. On the way, an excellent population of *C. creticum* was surveyed near the village of Agiou Pantes. The plants ranged around the edge of ancient *Cupressus sempervirens* woodland and included some excellent silvered leaf forms. Three plants were selected for collection; two with very distinct silver hastate patterns (CSE96494 and 96496) and one with very heavy silver spotting (CSE96497).

Returning along the north coast of the island to Rethymnon, some limestone gullies to the west of the town were explored but, although these carried a fairly rich flora, there was no sign of any cyclamen. The last foray was to explore the north-west foothills of Psiloritis near the Arkadi Monastery at about 450 m. Here, *C. creticum* was growing in a limestone ravine that still had pools of water left from the torrent which flowed in winter and in drifts on the wooded hillsides, where it was accompanied by large numbers of *Arum concinnatum*.

Assessment of results

The field trips to Crete have yet to observe anything that indicates the presence of members of subgenus *Psilanthum* other than *C. creticum* (i.e. *C. repandum* or *C. rhodium*) in any of their forms on Crete. Thus, there is no evidence that the 'pink' forms of *C. creticum* have arisen as a result of a cross between *C. creticum* and another pink-flowered species.

Cyclamen repandum and *C. balearicum* have a chromosome count of 2n = 20, whereas *C. creticum* has a count of 2n = 20 and 22. Collections made by the 1994 expedition near Agia Irini (Site 94/17) west of the Omalos Plain and at Agios Stefanos (Site 94/06) just south of the Rodopou Peninsula, have been confirmed as having a chromosome count of 2n = 20. The collection from Agia Irini confirms a previous report by Greilhuber but, as far as we are aware, 2n = 20 chromosome counts have not previously been recorded from Agios Stephanos.

The analysis that follows is based on field data collected during the three field studies on Crete and at the one site on Kárpathos in 1991. It must be borne in mind that it brings together statistics collected over four years and therefore in differing seasons, which may have affected the data collected and thus the conclusions that have been drawn. The data themselves have been recorded by different personnel each year and therefore the subjective evaluation of such factors of leaf shape and leaf pattern might have affected what was recorded. Still, it is doubtful that such caveats would invalidate the results.

At each site that was suitable for analysis, depending upon the size of the population, either 25 or 50 pairs of a leaf and a flower are collected from random plants. The following data were recorded: leaf length, leaf width, leaf shape, leaf pattern, flower colour, petal length, petal width, degree of petal twist, and colour of leaf reverse. Leaf shape and pattern were determined by matching collected leaves against photographs of designated leaf standards.

For simplicity, leaf reverse was recorded as 'red' or 'green' rather than by separating out 'pale red' and so on. Flower colour was treated as 'white' or 'pink'. In this context, pink included all flowers that were not entirely white. This category included petals whose colours were clearly identifiable as either 69D or 76B according to the RHS Colour Chart, petals that were so pale pink that they did not to register as pink on this chart and petals with a pink nose or a pink flush.

Growing conditions

The field study group found *C. creticum* growing at almost every altitude from 60 m to 1,250 m. There were four basic habitat types: phrygana (mixed dwarf shrubs, often spiny and aromatic), generally on the northern side of the island; mid- to high-altitude grassland, with the shelter of rocks; deciduous or *Cupressus sempervirens* woodland at mid to low altitude in east-west inland valleys; and southern gorges and gullies.

The majority of populations were growing with an aspect in the northern half of the compass. Where the aspect was basically southerly, the plants were growing in deep shade, usually within a gully, often with running water in summer. The degree of shade varied from 0–90%, to some extent with altitude: the higher populations were typically in more open habitats, possibly in the shelter of rocks. There was no correlation between the percentage of shade and any other factor, although a subjective view is that the larger leaves occurred in the deepest shade. Soil was generally a clay, usually terra rossa but sometimes black or brown, sometimes overlaying limestone and usually with a humus layer of decaying organic matter. The soil at the Agia Irini site where plants with a chromosome count of 2n = 20 were found differed in that it was based on a schist. At most sites, there was an identifiable amount of moisture in the soil; at certain times of the year, some of the plants in gully locations would have been submerged in water.

Variation

Leaves: Leaves varied dramatically in size, up to 162 mm long by 148 mm wide. There was a direct correlation between leaf length and leaf width. Nine sites had a ratio of leaf length to width of 1.1:1, 30 sites had a ratio of 1.2:1, and three sites had a ratio of 1.3:1. There was a very slight increase in leaf size as longitude increased, i.e. upon going from west to east; this did not, however, carry through to Kárpathos, where the average leaf size was amongst the five smallest found.

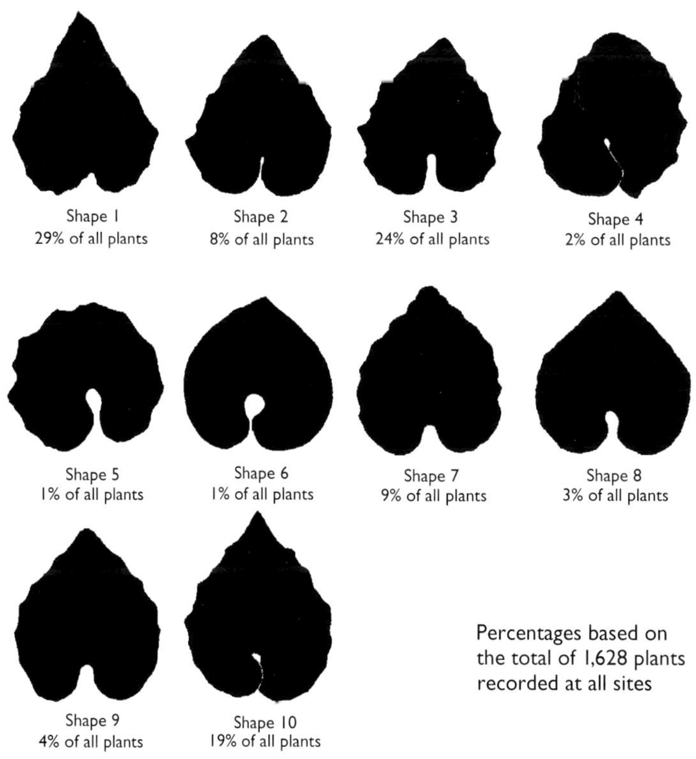

Leaf shape standards (MD)

There was little correlation between longitude and leaf shape, except that west of Dikti, there was a significant increase in the percentage of plants of 'Leaf Shape 1'. Generally, there was no correlation between aspect and leaf shape (as categorised by leaf shape standards), but at sites with a southerly aspect, 'Leaf Shape 3' became predominant at the expense of 'Leaf Shape 1'. Except for a slight increase in the predominance of 'Leaf shape 1' to the north, there was no significant correlation between latitude and leaf shape.

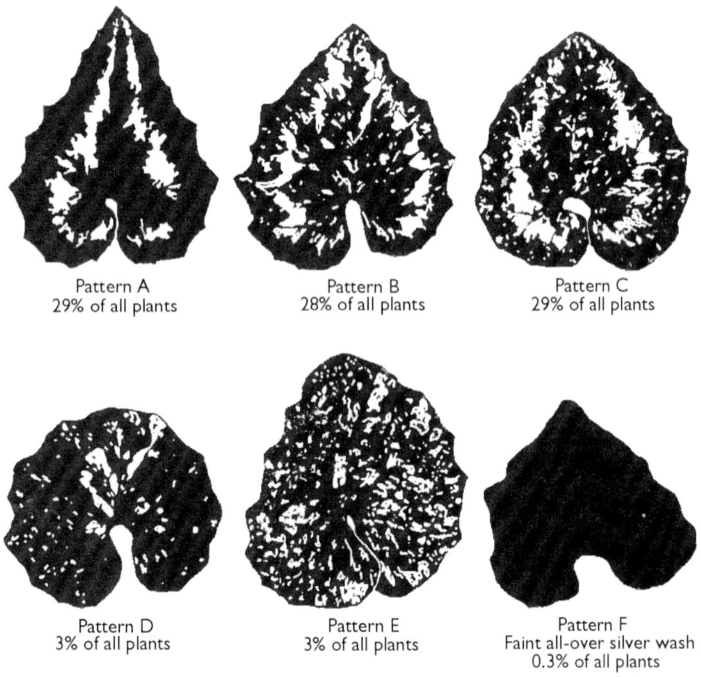

Plain green: 8% of all plants

Percentages based on the total of 1628 recorded at all sites

Leaf pattern standards for *C. creticum*. The photographs used as Leaf Patterns came from: A, *repandum album*; B, *rhodium* subsp. *rhodium*; C–F, *rhodium* subsp. *peloponnesiacum* (MD)

'Leaf Pattern A' predominated at the western end of the island, 'Leaf Pattern C' in the centre of the island, and 'Leaf Pattern B' in the eastern end of the island.

The predominance within *C. creticum* of a *C. rhodium* subsp. *rhodium*-like leaf pattern in the east supports a gradation along the ancient Aegean land mass from subsp. *rhodium* in the east to subsp. *peloponnesiacum* in the west. It is the only factor that supports this gradient. Contrary to this, at the site in Kárpathos, 87% of the population recorded had 'Leaf Pattern C'.

A percentage of plants had plain green leaves with no marbling pattern, but these all occurred to the east of longitude 24° 21' E (around Episkopi, half way between Lefka Ori and Rethymnon). At two sites at the extreme eastern end of the island, over 40% of the population had plain leaves.

Leaves with a red reverse were found at all sites and formed the vast majority of plants recorded. Green-backed leaves were found at only 12 sites, but at one site (94/02) 93% of the plants had such leaves. Sites where plants with green-backed leaves were found generally had an aspect in the northern half of the compass, that is from east-facing, through north-east-, north- and north-west-facing. No green-backed leaves were found at altitudes exceeding 680 m, and no green-backed leaves were recorded at the site in Kárpathos.

Leaf shape (across all sites, based on a total of 1,628 plants)

Leaf shape standard	1	2	3	4	5	6	7	8	9	10
% Proportion of plants	28.5	7.9	23.8	1.9	1.0	1.4	8.5	3.4	4.2	19.2

Leaf patterns (across all sites, based on a total of 1,628 plants)

Leaf pattern standard	A	B	C	D	E	F	Plain green
% Proportion of plants	28.6	28.4	28.6	3.1	3.2	0.3	7.8

Flowers: The longest petals found were 38 mm long; the widest 8 mm wide. The average length of petals varied from 16.8 mm to 28.4 mm. Average width varied from 3.8 mm to 6.8 mm. The degree of twist varied from 0° to 360°. There was no correlation between petal length and width of petal, degree of twist or flower colour. The largest flowers recorded were all plain white, but this is probably not significant as white is the predominant flower colour. The ratio of average petal length to average petal width varied between 3.5:1 and 5:1, except at the Kárpathos site where this ratio was 3:1 (with petals averaging 17.6 mm long by 5.9 mm wide). At over 50% of sites, the average ratio of petal length to width was 4:1.

The degree of petal twist correlated with a number of other factors: all petals with a twist greater than 180° occurred at less than 400 m altitude, with the exception of one site at 625 m. There is a significant correlation between latitude and petal twist. No plants with a petal twist greater than 180° were found north of 35° 16' 975" N, and the further south the forays went, the greater was the proportion of highly twisted petals. There is also a very distinct correlation between longitude and the degree of petal twist. With the exception of a small number of plants (2%) in the Gipari Gorge, and 15% of plants at a site north-west of Dikti, all of the plants recorded with a twist greater than 180° were at the eastern end of the island, i.e. east of longitude 25° 39' E (or east of Dikti). There, most sites had a significant proportion of plants with a petal twist greater than 180° and most had some plants with a twist of 360°. The site on Kárpathos had 6% of plants with a petal twist of 270°. All sites where plants were found with a petal twist greater than 180° had an aspect between due east, through north to north-west. All sites with a southerly or westerly aspect had plants with a petal twist of 180° or less.

White flowers were found at all sites, whereas flowers in varying shades of pink were found at 85% of sites in proportions varying from 2% to 74%. There was no correlation between the proportion of pink flowers and any other factor. The site at Agia Irini had 60% pink flowers, but at the site on Kárpathos, all of the flowers were white.

Petal twist and colour (across all sites, based on a total of 1,628 plants)

Petal twist	0°	45°	90°	135°	180°	225°	270°	315°	360°
%	25.7	17.9	40.1	4.1	8.9	0	1.1	0.3	1.7
Petal colour	white	pink							
%	79.6	20.4							

3.5c *CYCLAMEN RHODIUM*

The following accounts of fieldwork carried out by members of the Cyclamen Society were first published in the Society's journal. These versions have been edited for the purpose of this book, but the valuable observations and findings are presented unchanged. The name of each taxon has been amended to coincide with the nomenclature used in Chapter 2.

Field studies of *C. rhodium* subsp. *rhodium*

From 1991 to 1993 three studies that were funded by the the Cyclamen Society were undertaken in Greece. A further field trip, to the Dodecanese, took place in 1998. The overall intention was to fill in as much detail as possible about the distribution, habitat preferences and natural variation of three cyclamen taxa, one on Rhodes and two in the Peloponnese. The 1998 foray provided additional information, including observations of the species on the island of Kos. In addition to the compilation of data, selected living plants were to be collected under licence to enhance the range of material available in cultivation. Accounts of the separate visits are presented below. In all cases, a great deal of data was gathered and has much improved the overall fund of knowledge relating to these taxa.

Rhodes, 1991 Team: Richard Bailey, John Fielding and Peter Moore

Richard Bailey

Habitat and distribution

During this fieldwork, *Cyclamen rhodium* subsp. *rhodium* was confirmed in all its major and well known sites, principally the area around Salakos, the neighbouring Mt Profitis Ilias (probably the most widely known site) and Mt Filerimos, but also at Epta Piges (Seven Springs), Embonas and elsewhere on Mt Ataviros and around Petaloudes (Butterfly Valley). At all these sites, subsp. *rhodium* was abundant in suitable habitats. It was also found in suitable habitats throughout Rhodes.

The density of plants varied very considerably, from isolated specimens widely scattered over a woodland floor or in scrub, to densely packed masses in which leaves and flowers covered up to 80% of the soil surface. At one site, in *Pinus brutia* forest on the

Peter Moore sampling a *C. rhodium* subsp. *rhodium* population on Mt Profitis Ilias (JF)

A thriving population of *C. rhodium* subsp. *rhodium* on Mt Profitis Ilias, Site 91/02, in pine woods at 450 m (JF)

north-facing slope of Mt Profitis Ilias at an altitude of 820 m, one population covered an area of some 30 × 6 m, with an estimated average of 55 adult (i.e. flowering) plants per sq. m. Seedlings were also present, at an estimated density of 4 per sq. m. This adds up to a population total of over 10,500 plants. The site where this population grew was typical of many sites studied at higher (over 500 m) altitudes: a woodland floor below *Pinus brutia* with isolated groups of *C. rhodium* of varying sizes but sparsely populated by other plants. Associated shrubs were varied. *Pistacia lentiscus* was undoubtedly the most constant but other shrubs that were commonly present included *Salvia triloba*, *Prasium majus*, *Lithodora hispidula*, the spiny horrors *Genista acanthoclada* and *Sarcopoterium spinosum*, and, more rarely, *Ceratonia siliqua*, *Myrtus communis* and *Styrax officinalis*. Occasionally, most often in the south, the three strawberry trees, *Arbutus andrachne*, *A. unedo* and *A.* × *andrachnoides*, were found associated with *C. rhodium*. The herbaceous and bulbous plants were varied and included several species of *Ophrys* and other orchids (of which the giant *Barlia robertianum* was the most spectacular), *Muscari* and *Bellevallia* species, *Anemone blanda*, *Aristolochia clematitis* and the serenely beautiful endemic *Paeonia rhodia*. At several sites, small-leafed forms of *Iris unguicularis* were noted, including one with almost grass-like leaves on Mt Profitis Ilias.

At lower altitudes (the lowest site where *C. rhodium* was found was a mere 20 m above sea level), the surrounding vegetation was more varied. The lower sites were all shaded and, at least in spring, moist. A typical low-altitude site would be on the steep, north-west to east-facing slope of a valley (or gully) through which a stream was flowing. Plants of *C. rhodium* would most likely be in woodland or scrub, often with fairly vigorous competition from grasses and other flowering plants. The most abundant shrub was usually *Pistacia lentiscus*, but any of the shrubs noted above could also be found. Such sites must have been common and widespread all over Rhodes, at both high and low altitudes, before fires, development and agricultural encroachment caused dramatic changes in the natural and semi-natural vegetation of the island. At present, *C. rhodium* is found, for all practical purposes, all over the island where such habitats still survive. The southern-most site noted was 2 km north of Kattavia in dappled shade below *Arbutus*, the most northerly at Mt Filerimos in shade below *Pinus*. The species was noted on the east coast at Vliha Bay in *Pistacia* scrub, and to the west on the western slopes of Akramitis in *Pinus* woodland and mixed scrub.

At most sites, at both high and low altitudes, recently germinated seedlings were present. The greatest density observed was 17 per sq. m. There appears, therefore, little doubt that *C. rhodium* is maintaining its population adequately where suitable habitat conditions still exist. That the species can persist in extremely adverse circumstances was amply demonstrated by its occurrence in an area of former *Pinus* forest that was very recently (and thoroughly) burnt near Profilia. Here, *C. rhodium* survived in clumps of regenerating *Myrtus communis* and in the angle of the bank along a winter water course. A few young seedlings were also present. In another, almost equally adverse, situation, a few plants were found in an abandoned olive and citrus grove, at an altitude of only 20 m and within 50 m of new, actively expanding, developments close to the airport at Paradision. Higher up the hillside, flourishing populations were present. In virtually all sites, *C. rhodium* was growing in a highly humic soil that formed a relatively shallow layer (10–55 cm) over a mineral subsoil, often a stiff, gritty, clay or coarse sand. In this situation, the tubers usually occurred just above the base of the humic layer and had floral trunks of varying lengths. In all cases, the situation was such as to suggest that it would be dry, but not bone dry, over the summer.

Pinus brutia woodland on the way from Embonas to Mt Profitis Ilias, site of a large population of *C. rhodium* subsp. *rhodium* with abundant seedlings (JF)

A typical form of *C. rhodium* subsp. *rhodium* on Rhodes, between Embonas and Amartos, 500 m, Site 91/04 (JF)

Variation

Flowers: Flower colour was evaluated using the RHS Colour Chart. More variation in colour was observed than is usually found in cultivated plants of *C. rhodium*, but the overwhelming majority of flowers were, apart from the purple basal blotch, of the typical very pale pink: too pale to register as other than as white on the RHS Colour Chart. A few plants of colours 75A, 75C and 75D, all a pale pink, were found, and also a number whose petals were pure white above the basal blotch. No true albino form was found.

The depth (i.e. extent) of the basal blotch was also noted as this is one character in which subsp. *rhodium* differs from subsp. *peloponnesiacum*. This varied little, the usual depth being 0.5–1 mm, although a few plants were noted with blotches 2 or 3 mm deep.

Petal size was more varied than petal colour. Both length and width were recorded, the longest being 32 mm and the shortest 11 mm, while the widest was 10 mm and the narrowest 4 mm. The degree of twist in the petals varied from those that lay completely flat to those that achieved three complete 360° turns along their length.

When flower characters alone were considered, some plants were similar to those of subsp. *peloponnesiacum*. For example, some flowers with a good pink colour and a deep (3 mm) basal blotch would not appear out of place in the Peloponnese.

Leaves: Leaf size, shape and pattern were studied using comparison charts based on the range of material cultivated at Tile Barn Nursery.

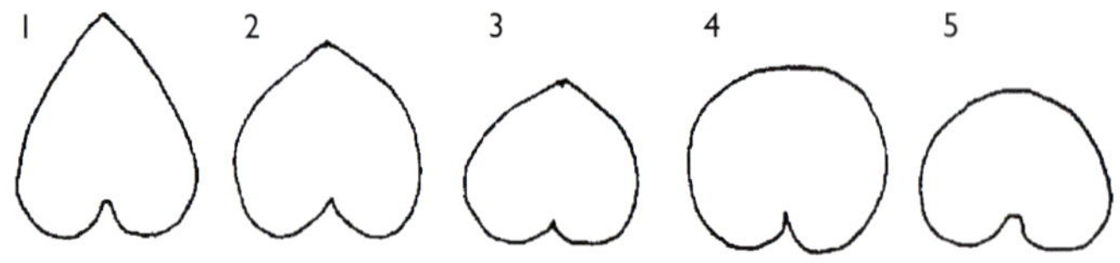

Leaf shape standards

Leaf pattern standards (see p. 246) ranged from an entirely green leaf, through to those with a hastate central pattern but with none of the speckling frequently seen in subsp. *peloponnesiacum*, through to those that were more or less uniformly speckled, and on to those that were entirely silver.

The largest leaf recorded was 152 × 148 mm and the smallest 22 × 18 mm, but the majority of leaves fell within the range of 50–70 × 45–60 mm. The most common leaf shapes were overwhelmingly the pointed and heart-shaped leaves of 'Leaf Shape 1' and 'Leaf Shape 2', with 66% and 27% of leaves, respectively, conforming to these leaf shape standards. 'Leaf Shape 3' described 6% of leaves examined, with only two leaves having 'Leaf Shape 4' and no leaves of 'Leaf Shape 5' being recorded. Any variation in leaf shape that was found was uniformly distributed over the island. This was not the case, however, for leaf pattern. No completely silver leaves were recorded, but some 4% of leaves were entirely plain and around 8% of leaves had the speckled pattern typical of subsp. *peloponnesiacum*. While not confined to the area, the majority of these occurred west of a line from Embonas to Kattavia. No great statistical or taxonomic significance can be attached to this, as only 16 plain-leaved and 35 speckled-leaved plants were recorded out of a total of over 450. The latter would, if examined away from their flowers, be acceptable as leaves of subsp. *peloponnesiacum*.

Discussion

There is little doubt that, historically, *C. rhodium* was distributed throughout Rhodes. Even now, after substantial environmental change, it is very much more widespread than the existing published records suggest, being found in suitable habitats all over the island from north to south and from east to west. Nonetheless, away from the mountainous spine along the western edge of the island, suitable habitats are under continuing threat, and *C. rhodium* could, very soon, become endangered. A careful search in suitable habitats on Kárpathos failed to reveal any plants of subsp. *rhodium*.

The range of variation discovered in the flowers and leaves was such that some flowers could be taken for those of subsp.

peloponnesiacum, as could some leaves. However, no single plant combined the leaf and flower characters to the extent that it could be placed confidently in subsp. *peloponnesiacum* as that subspecies is currently understood, although several came very close.

Nonetheless, the variation observed suggests that *C. rhodium* subsp. *rhodium* is a more varied taxon than has hitherto been thought to be the case from studies of herbarium and cultivated material. It must also be emphasised that the variation noted is well within the range one would expect and accept in what is, after all, a relatively small and isolated population. The taxonomic problems presented by a variable species complex such as the *C. repandum* group will always stimulate discussion, even controversy. Considering the results of this study, it is possible to suggest that *C. rhodium* subsp. *rhodium* is as distinct from the other subspecies as it is from *C. repandum*, *C. balearicum* or *C. creticum*.

Kos, 1998 Team: Chris Clennett, Brian Cresswell and Martyn Denney

Martyn Denney

Cyclamen rhodium subsp. *rhodium* was found at eight sites at the south-west end of Kos just south of Kefalos, largely between Mt Latra and Mt Zini, and at one site on the north side of Mt Zini. These nine sites, all basically north-facing, contained four different habitats:

- sparse or dense *Pinus brutia* woodland with phrygana on a brown sandy loam (Sites 98/02D–98/05D);
- a steep wash-out gully on the north side of Mt Zini, just above the high-water mark, on a sandy loam (Site 98/07D);
- *Quercus coccifera* and *Pistacia lentiscus* phrygana with isolated groups of *Pinus brutia* on *terra rossa* (Site 98/08D); or
- in and around a steep north-west facing gully under mature *Pinus brutia*, *Arbutus andrachne* and *Pistacia lentiscus*, on a brown clay loam (Site 98/09D).

The differences between these habitats are significant, but two sites stood out from the others. Site 98/07D was on the north side of Mt Zini in a steep gully running down to the sea, with cyclamen plants growing predominantly under *Quercus coccifera* phrygana in a thin sandy loam. Although the plants were generally not found in the gully bottom, a small colony was growing in an isolated group of more mature *Q. coccifera* at the top of the beach, probably as a result of a landslip. We explored this gully upwards until it became too steep to proceed, and at this point the cyclamen plants were thinning out.

The gully was not accessible from the summit of Mt Zini, but there were no cyclamen growing on the silica sand in the area above the gulley that makes up the summit. We concluded, therefore, that the cyclamen were confined to the area in which the sand gave way to a topsoil with a degree of loam content.

The island of Kos, Mt Zini and the gully where *C. rhodium* subsp. *rhodium* was found (BC)

Cyclamen rhodium subsp. *rhodium* on Kos. Leaf patterns showed far less variation than on Rhodes (MD)

It was interesting that on both sides of Mt Zini, the cyclamen populations were clearly limited by the well-defined change in the soil from a brown sandy loam to a silica sand containing black crystals that were possibly a form of quartzite.

Site 98/09D was on the north-west flank of Mt Latra. Here, *C. rhodium* was clearly performing at its best, growing in a gully in open shade beneath mature trees. This was reflected in the vigour and density of the plants.

At the other sites, *C. rhodium* was growing with the tubers 15–30 cm beneath the surface, generally below a thick semi-decayed humus layer that was principally comprised of both new and decaying *Pinus brutia* needle litter.

In total, 295 plants were surveyed, at five sites. Unfortunately, because of an extremely mild winter and a hot period in February, the season was very advanced in 1998. Although isolated plants were in flower, allowing the identity of the plants to be proved and providing herbarium specimens, there were insufficient flowers to allow the team to carry out surveys of petal characteristics and dimensions. Of the plants seen in flower, a significant proportion, greater than that recorded in the 1991 surveys on Rhodes, had an entirely white corolla (98059D, 98060D, 98061D, 98178D and 98181D). One of these albino plants (98059D) had an upward-facing inflorescence of the 'stargazer' type, with quite narrow petals. A number of plants (98002D, 98003D and 98057D) with pale pink petals (RHS Colour Chart 76B, 76C and 76D) and a darker pink nose were found at a couple of sites.

On 1 May, the team visited populations of *C. rhodium* on Rhodes, on Mt Profitis Ilias and at Filerimos. Their aim was to compare the apparent early nature of the season on Kos with that on Rhodes, in order to establish whether the plants at each location were at a different stage in their growth cycle and, if so, whether this resulted from meteorological or other factors. We concluded that the season on Rhodes was advanced in a similar way to that on Kos, and that at similar altitudes the condition of the plants (i.e. the lack of flowers) was also similar. We found significant numbers of *C. rhodium* in flower only above 580 m, and these were starting to go over and had many flaccid petals.

Preliminary survey results

The following are initial brief results from the analysis of the five surveys of *C. rhodium* on Kos, and comparing them with a similar analysis of the surveys carried out on the island of Rhodes in 1991. Flower statistics are not given, as insufficient flowers were found on Kos for these to be measured.

	Kos 1998	**Rhodes 1991**
Altitude range	35–260 m	120–500 m
Shade	38–83%	0–100%
Slope	Level to 1 in 0.5	1:3 to 1:1
Aspect	NW-N-NE	NW-N-E
Soil pH	6.4–7.7	n/a
Average leaf length	71.5 mm	70.5 mm
Average leaf width	61 mm	60.5 mm
Average leaf length:width ratio	1.17:1	1.16:1

Leaf pattern standard		A	B	C	D	E	F	Plain green
% Proportion of plants	1998	36	57	5				2
	1991	33	31	23	6	3		4

Leaf shape standard		1&2	3	4	5	6	7	8	9	10
% Proportion of plants	1998	30	37	5	1	4				23
	1991	93	6.5	0.5						

Note: The same leaf standards have been used by all Greek expeditions between 1991 and 1998 (see pages 245 (shape) and 246 (pattern)).

Although individual leaves varied considerably in size, the overall average was very similar for the two islands. Nevertheless, we had the subjective impression that the leaves on Rhodes had much wider variation in leaf pattern, and that those patterns were much more clearly defined than those on Kos. This is borne out to some extent by the leaf pattern statistics.

Field studies of *C. rhodium* subsp. *peloponnesiacum* and subsp. *vividum*

Apart from the obvious difference in the flower colour the two subspecies of *C. rhodium*, the cyclamen in the Peloponnese also appear to have subtly different distributions and habitats. The two research projects of 1992 and 1993 did much to enhance our knowledge of this and also made some successful introductions to cultivation, such as the collection CSE92155, which was exhibited at a Cyclamen Society show in 2007 by expert cultivator Ian Robertson.

A fine specimen of *C. rhodium* subsp. *peloponnesiacum* at a Cyclamen Society show. CSE92155 introduced from Site 92/10. Peloponnese, on road from Varvitsa to Megali Brisi, 1,100 m (MD)

Cyclamen rhodium subsp. *peloponnesiacum* showing the irregular silvery splashing of the leaves and bi-coloured pink and red flowers, characteristics of this subspecies (MD)

Peloponnese, April 1992 Team: Chris Clennett, Ray Johnstone and Melvyn Jope

Ray Johnstone

The two variants of *C. rhodium* were studied: *C. rhodium* subsp. *peloponnesiacum* has pale pink flowers bearing a deeper pink nose, and hastate-patterned leaves with varying amounts of 'splashing', generally with a green reverse. It is found in the central and south Peloponnese, mainly in the Taigetos mountains and Mani peninsula; *C. rhodium* subsp. *vividum* has uniformly deep reddish flowers and plainer leaves, generally showing a red reverse. The distribution of this variant is in the Parnon mountains and its peninsula to the south.

No 'rhodiums' were found until we were about 24 km south of Tripolis. A roadside war memorial marked the entrance to a beautiful wooded dene, which had a small clear stream babbling its way along the bottom. Celandines, *Trollius*, wood anemones and deep-coloured violets with sheets of primroses gave the impression that we were back in the British countryside; but the presence of *Anemone coronaria*, *Arum italicum*, euphorbias and *Ophrys* reminded us of where we were. A large sheet of primroses lured me to take a photograph and as I focused on the primroses

Cyclamen rhodium subsp. *vividum* is easy to distinguish with its glowing reddish flowers. Site 93/12, between Leonidi and Vaskina, an exposed habitat at 525 m (JF)

a splash of pink caught my eye — we had found cyclamen at last! One or two plants were growing among the primroses. The rest grew mainly on the steep wooded slopes above a rock face, but not in any great numbers.

From a base in Sparta, we searched the Parori gorge 5 km out of Sparta and here we found lots of flowers including more *C. rhodium* subsp. *peloponnesiacum*. The cyclamen were distributed sparsely at first but became more numerous as we continued up the gorge. We carried out our first survey at a suitable point, measuring and making notes on the leaves and flowers from a random selection of 25 plants. A few hundred metres up the gorge, another survey of 25 plants was carried out near a little church.

Along the Kalamata road, at a wooded area near Tripi, there were masses of *peloponnesiacum* displaying some very good leaf forms. Next, we headed north and east into the Parnon range and a stop near Karies proved successful; a few kilometres further on, cyclamen were present along the banks of a stream and a nearby ditch in large enough numbers to support a survey. Soon afterwards, cyclamen turned up in large numbers in a pine wood, but the leaves at this site tended to be fairly plain.

The next focus was the Taigetos mountains. From Paleopanagea, a track heads west into the hills and the flora gets better as the track gets higher and rougher: subsp. *peloponnesiacum* was present most of the way up. Eventually, there is a picnic area that seems to be the base from which the Taigetos refuge could be reached on foot in 1½ hours and the peak of Profitias Ilias (2,470 m) in 4 hr 15 min. This was a marvellous area for plants, probably the best of the entire trip: *C. rhodium* subsp. *peloponnesiacum*, *C. hederifolium* and *Galanthus reginae-olgae* carpeted the light woodland along the river banks. Deep red *Anemone coronaria*, a stunning deep blue form of *A. blanda*, violas, primroses and colchicums were also present. Cyclamen and snowdrops could even be found growing in the moss-covered trunks of mature plane trees. A most unusual *C. hederifolium*, with the leaves wavy-edged and lobed and with a reticulate appearance, was collected at this site.

From the next base in Githio, we explored part of the Mani peninsula. *Cyclamen rhodium* subsp. *peloponnesiacum*, together with many *Ophrys*, was present in quantity in oak-wooded hillsides near Passavas. After the survey, our customary look around for good plants began and before long a shout from up the hill proclaimed the finding of "Jope's Joy", an albino *peloponnesiacum* [now named forma *albiflorum* — Ed.]. *Cyclamen hederifolium* in some good leaf forms and *C. graecum* were also present at this site. We then headed east from Githio, intending to go as far down the Parnon Peninsula as we could; 8 km out of Githio subsp. *peloponnesiacum* was found in small quantities, and *C. hederifolium* and *C. graecum* were also present until we reached an agricultural area on the plains around Skala.

Some 2 km out of the village of Geraki, the team found the first specimens of subsp. *vividum*. This site was different from others as it was on open scrubby hillside, the cyclamen growing in the shelter of kermes oaks (*Quercus coccifera*), which helped them to survive the depredations of the flocks of goats in this area. After carrying out a site survey, there followed the usual hunt around for good specimens and this time we found "Johnstone's Frill", a form with a strikingly different leaf, in the centre of an oak bush. The flower belonging to this plant was also notably different, the

Flower of *C. rhodium* subsp. *peloponnesiacum* forma *albiflorum* (BM)

Flowers from Ag. Petros, Mt Parnon, showing *C. rhodium* subsp. *peloponnesiacum* and subsp. *vividum* together (JF)

untwisted petals 14 mm wide and stubby, with the petal tip being replaced by a fimbriated edge. We checked the distribution of the cyclamen in this area and found that they were present from the village of Geraki up to the summit of the road some 6 km away. Returning to a base in Sparta, we then set out east for Goritsa. An impressive gorge contained subsp. *vividum*, although not in great quantities, but upon heading for Kallithea, the cyclamen were present in sufficient quantities to allow a survey to be carried out on a rocky hillside in *Quercus* scrub. Heading out of Sparta on the return journey, we took the road that crosses the Parnon range via Vourvoura and Aghios Petros to Astros. Subspecies *peloponnesiacum*, alongside very good forms of *Iris unguicularis* and some colour forms of *Anemone blanda*, grew in quantity on steep slopes above the road 2 km west of Vourvoura. We found subsp. *vividum* again, in sight of Aghios Petros growing under chestnut trees on the steep slopes of two gullies. Subspecies *vividum* was also present below the village and we concluded that this area was an overlap zone for the two subspecies. We then headed back to Galatas and crossed to Poros, marking the end of our *rhodium* studies; in the next two days, we saw only *C. hederifolium* and *C. graecum* in the environs of Poros and Galatas.

A rather open, rocky site for *C. rhodium* subsp. *vividum*: Site 93/12 near Vaskina, April 1993 (JF)

Peloponnese, April 1993 Team: Melvyn Jope, John Fielding and Peter Moore

Peter Moore

The purpose of the Cyclamen Society's second field trip to the Peloponnese in spring 1993 was to determine the extent of the area in which *C. rhodium* subsp. *peloponnesiacum* and subsp. *vividum* were to be found. After an overnight stop on Poros, where there were masses of *C. graecum* and *C. hederifolium* and some fritillaries, the team took the ferry to Galatas on the mainland. We headed west, through Nafplio to Astros, following the route of the 1992 trip. Thereafter, on a previously unexplored road between Aghios Andreas and Aghios Pandeleimon, we found and surveyed our first site of *C. rhodium* subsp. *vividum* (referred to hereafter as subsp. *vividum*) growing in profusion among *Fritillaria graeca*: a memorable sight for our first site. The typical Greek scrub was also host to a sea of *Anemone blanda*, *A. coronaria* and fritillaries. Next day, we decided to follow the same road to its highest point and drove above the village of Prastos to a height of 1,500 m. The forest was still partially covered in snow and although we found two species of crocus, *Scilla biflora* and a tulip species, there were no signs of cyclamen. We dropped down 100 m and found subsp. *vividum*, not in flower, but thinking that plants growing at this altitude would be hardy, we collected a few. Recording sites down the road towards Aghios Andreas, we were rewarded by the sight of *Iris unguicularis* flowering among the cyclamen.

Next day, we took the coastal road towards Leonidi and collected and surveyed cyclamen at only 25 m altitude. Here, we found a colony of silver-leafed *C. hederifolium* quite different from those in cultivation, the deeply lobed leaves mainly with an overall silver wash. Upon driving from a base in Leonidi on the road towards Vaskina, we encountered a sight never to be forgotten: Site 12, a mass of subsp. *vividum* growing in very old derelict terraces as far as the eye could see. Add to that *Fritillaria graeca*, *Ophrys tenthredinifera*, *Orchis quadripunctata*, *Crocus cancellatus*, *Cyclamen hederifolium* and *C. graecum*, and our cup was full to the brim; it overflowed on finding a white subsp. *vividum* with a pure white 'nose' to the flower and the faintest pink flush to the tips of the petals!

A good stand of the vivid reddish-plum *C. rhodium* subsp. *vividum* at Site 93/12 (JF)

An unusual form of *C. rhodium* subsp. *vividum* with heavily silver-splashed leaves. Site 93/14 near Kosmas (JF)

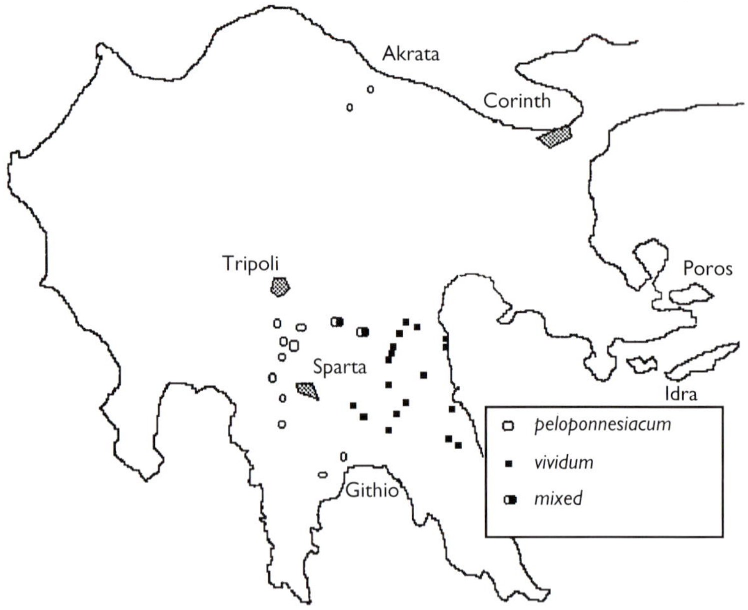

The next site was on the way to Githeo, where we made a collection at 1,050 m near Kosmas and surveyed a site at 500 m. Beyond Kosmas, we saw the first signs of *C. rhodium* subsp. *peloponnesiacum* (referred to hereafter as subsp. *peloponnesiacum*), mixed with subsp. *vividum*. From there, we crossed the Yarnon down into subsp. *peloponnesiacum* country, an area already covered by the 1992 expedition. Then, we set out from Githeo for Monemvasia, heading into the Madara mountains and into subsp. *vividum* territory. Two sites were surveyed near Kremasti in the Madara mountains to the south of the Parnon mountains. The last sighting we had of subsp. *vividum* was in a windy tundra area, before the road towards Monemvasia descended into flat farming country where there was no further sign of cyclamen. From Monemvasia, we headed south to the Lakonia peninsula where there was a noticeable lack of cyclamen, although we did find a *C. hederifolium* with a leaf that measured 18 cm across. North of Githeo, we surveyed a site in an area of thousands of acres of burnt scrub where subsp. *peloponnesiacum* was thriving.

From the next base in Sparta, the team headed for Aghios Petros, in search of a site visited by the 1992 expedition where subsp. *peloponnesiacum* and subsp. *vividum* meet. At the first stop, the colony consisted of just subsp. *peloponnesiacum*, but at the next site, where a survey was made, a small quantity of subsp. *vividum* was recorded among the subsp. *peloponnesiacum* and amid masses of violet-scented primroses. A mile further on, the terrain had changed to an exposed area where subsp. *vividum* had taken over: the cyclamen population comprising roughly 80% subsp. *vividum* and 20% subsp. *peloponnesiacum*. We then headed north through Tripoli, to visit the mountain village of Vitina in the Menalo range. Although the Menalo mountains appeared the most obvious area north of the Parnon mountains in which cyclamen of this group might be found, and search hard though we did, we could not find any cyclamen here. The terrain seemed correct and the usual associated plants were evident. We were prevented from covering some of the area as roads had been destroyed by avalanches, but we did cover a large proportion of the mountain range. At Stemnitsa, *Crocus sieberi* was flowering just below the snow line, and lower down, sheets of the violet-scented primroses and *Anemone blanda* were in flower. Further north in Kalavrita, a mountain village at the base of Mount Helmos, no cyclamen were found but we were encouraged by reports from Jim Archibald that subsp. *peloponnesiacum* was to be found on Mount Helmos. The vast plateau below the snow-covered slopes was a sea of purple *Crocus sieberi* and a spring-flowering colchicum, both interspersed with a yellow-flowering crocus but not a subsp. *peloponnesiacum* to be seen. Following a telephone call to Jim to get precise location details, we set off again along the northern flank of the mountain. Before we even got to Jim's location, we found subsp. *peloponnesiacum* just below the monastery at Mega Spileo and again further north and west.

Our initial impression gained in the field was that subsp. *vividum* is confined to the Parnon and Madara mountains and that it mingles with subsp. *peloponnesiacum* in a limited area. The map shows sites for the two taxa found in the 1992 and 1993 expeditions. It would appear that the conditions in which subsp. *vividum* grows — open exposed sites similar to those occupied by *C. graecum* — are very different from those inhabited by subsp. *peloponnesiacum*; subsp. *vividum* requires less shade and protection from shrubs and rocks. By contrast, subsp. *peloponnesiacum* is found only in moist woody conditions, often close to streams.

Individual flowers of subsp. *vividum* are very uniform in colour and differ from those of subsp. *peloponnesiacum* in the extent to which the colour at the base of the petal extends to the tip of the petal. The colour at the base of subsp. *vividum* flowers is a very uniform 64B on the RHS Colour Chart, and there is no sign

Leaves of *C. rhodium* subsp. *peloponnesiacum* on Mt Helmos, one of the most northerly sites in the Peloponnese. Site 93/21 (JF)

of the distinct colour zone that is evident at the base of subsp. *peloponnesiacum* flowers. In subsp. *vividum*, the colour at the base of the flower merges up the flower and, in the best forms, the colour at the base extends to the tip of the flower. The leaves of subsp. *vividum* are generally dull in comparison with those of subsp. *peloponnesiacum*, much more uniform and usually red underneath, whereas the leaves of subsp. *peloponnesiacum* are more often green beneath. It is not easy to distinguish the two subspecies by studying the leaves alone, but their flowers are quite distinctive.

The range of subsp. *peloponnesiacum* from the south to the north of the Peloponnese is very extensive, and it would be very interesting to know whether it occurs further north on the Greek mainland. Both subsp. *vividum* and subsp. *peloponnesiacum* seem to be thriving, and there was strong evidence of regeneration. There is no suggestion that either of the subspecies is endangered.

Leaf variation in *C. rhodium* subsp. *vividum*, at Site 93/12 near Vaskina, Peloponnese (JF)

Leaf variation in *C. rhodium* subsp. *peloponnesiacum* from cultivated specimens (BM)

Kythira 1997 Team: Melvyn Jope, Peter Moore and Hilary Temple

The purpose of the Cyclamen Society's visit to Kythira, which lies between the Peloponnese and Crete, was to establish whether a member of subgenus *Psilanthum* occurs there; the presence of such plants on Kythira could be logically assumed as there is known to be strong floristic links between the Peloponnese and Crete. Extensive searches were made and many cyclamen — *C. graecum* and *C. hederifolium* — were observed but no spring-flowering species were found. Of course, when trying to establish distributions for plant species, recording an absence can be almost as valuable as noting the positive sightings: but rather more difficult to prove. The team then moved on to the Peloponnese and studied populations of *C. rhodium* subsp. *peloponnesiacum* at various sites including Petrina, Sparta, Aremissia, Saïdona and between Karavouni and Milia. A single plant was also found between Areopoli and Githio, thus extending the known distribution to the east and south of the previous records.

Cyclamen rhodium was subsequently found to be present on Kythira when Vic and Janet Aspland located it on a personal visit to the island in 2009. The account of their visit appears on page 44 under the species entry for *C. rhodium*.

C. rhodium subsp. *peloponnesiacum* seen in 1997, showing the damp shady conditions preferred by this subspecies (MELJ)

3.5d *CYCLAMEN ALPINUM*

Brian Mathew & Neriman Özhatay

Teams 1997/8/9 comprising: Vic Aspland, Brian Mathew, Trevor Wiltshire (Cyclamen Society), Sema Atay (DHKD, the Turkish Society for the Protection of Nature), Andy Byfield (Fauna and Flora International), Başak Kocak and Neriman Özhatay (Herbarium, Istanbul University, ISTE).

L to R: Neriman Özhatay, Brian Mathew, Andy Byfield, Sema Atay, Vic Aspland in discussion over herbarium specimens in Istanbul Herbarium (ISTE). (TW)

Introduction

In the late 1990s, there was a suggestion at a meeting of the Royal Horticultural Society's Rock Garden Plant Committee that there might be two variants of *C. trochopteranthum* as it was then known, a high- and a low-altitude 'form', with some identifiable distinguishing characteristics. The only way to ascertain whether this were true, would be through a programme of fieldwork, studying the various known populations and at the same time providing new information about the distribution of what was then a comparatively poorly known species in the wild. The Cyclamen Society proposed to undertake the survey in Turkey over a period of three successive springs, from 1997 to 1999. A great deal of information was gathered about this species, which was relatively poorly known at the time, although there was no convincing outcome as to the validity of the proposed 'high' and 'low' variants. On the whole, the really large-leafed plants were found at low altitudes, whereas very small-leafed ones could be found at high altitudes, but there was much overlap. Interestingly, the large and small leaf characteristics tend to be maintained in cultivation, so the variation is at least to some extent genetic rather than purely environmental.

A large-leaved form of *C. alpinum*, often although not always associated with low altitudes (BM)

A high altitude small-leaved example of *C. alpinum*, near Arpacik, south-western Turkey, at 1,220 m (BM)

Random assessment of populations is often not easy in such terrain. *Cyclamen alpinum* country near Kasimler, Site 97/22T (BM)

The following is an account of the fieldwork, which took place over six weeks spread over three seasons. Three separate visits (early, mid and late spring) were necessary because of the widely divergent flowering times (early February to late April) of populations at sea level and those at the highest elevations, and because of the very considerable extent of the known distribution. Many new populations were located and these, together with already known populations, were found within an area covering over 20,000 square kilometres.

To gather data on the overall size of *C. alpinum*, which was one of the aims of the survey, leaf length and width and the diameter of the flowers were measured, taking as random a sample as possible. The actual method for doing this had to be worked out on the spot because the populations differed so much in their constitution, and on only one occasion was it possible to find the edges of a discrete population and take anything like a random line across it. In most instances, the populations were either so sparse that nearly all individuals were measured in order to acquire 50 samples of leaf and flower or were in terrain that made it impossible to achieve a straight, random line for sampling. On several occasions, the populations were so broken into an irregular pattern of clusters of 'mini-populations' coupled with scattered individuals that, rather ironically, a 'random' line was in fact chosen in order to make sure of encountering enough specimens for the survey.

In addition to the actual measurements, notes were made on leaf-pattern type, flower colour (using the RHS Colour Chart), the width of the petals, the degree of petal twist and the colour of the darker stain at the base of the petals. Supplementary notes included observations on fragrance, fimbriated petals, and any particularly striking or abnormal characteristics. The localities were determined precisely using a GPS instrument, altitudes recorded, and pH readings taken. Habitat features that were noted included aspect and steepness of slopes, amount of shade, overall soil type, presence of rock formations, dominant vegetation type and any plant species that were closely associated with the cyclamen. A photographic record was made of the habitats, plant associations, populations of cyclamen and individual plants of *C. alpinum* that demonstrated natural variation. Herbarium specimens were collected, and a set of these was deposited in Istanbul Herbarium (ISTE).

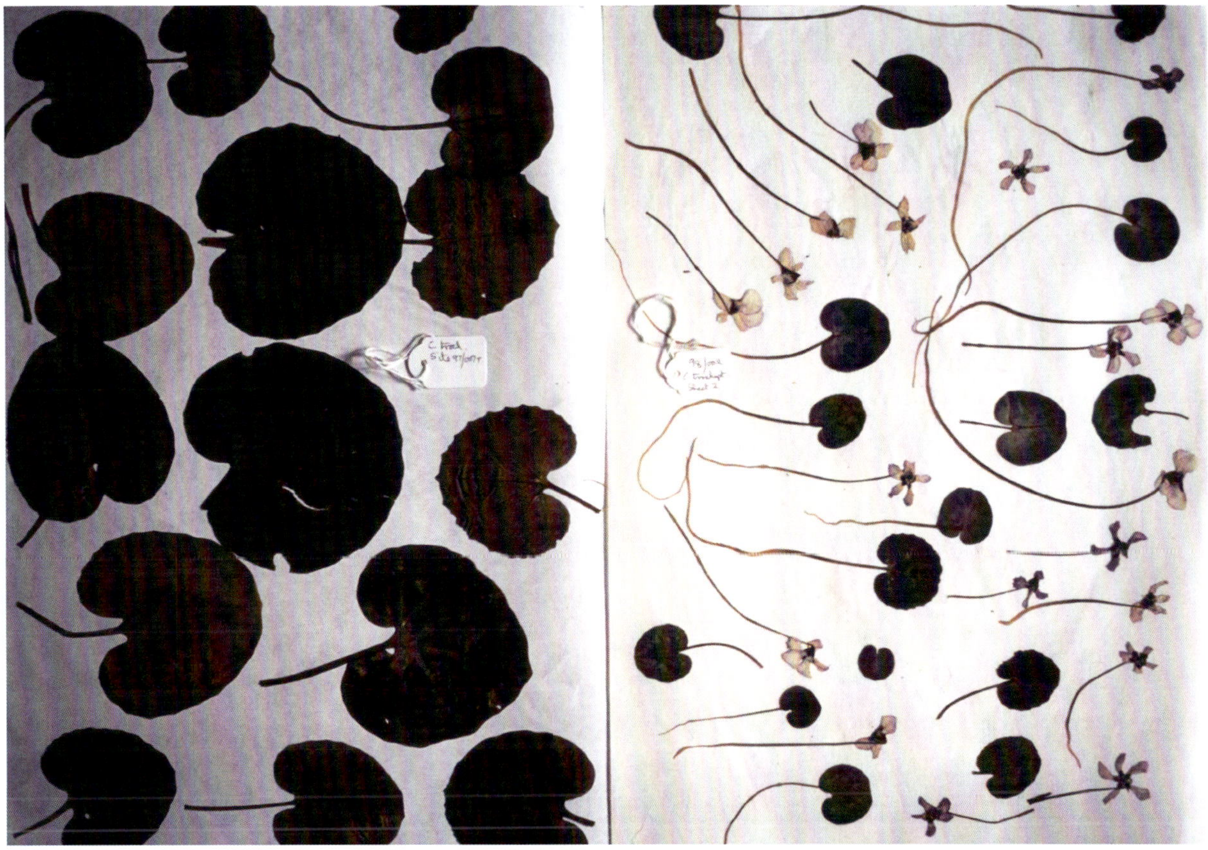

Cyclamen alpinum herbarium specimens. One aim of the fieldwork was to establish if the effect of altitude had taxonomic significance. Site 97/07T near Marmaris at 75 m and Site 98/01T in the Bey Dağ at 850 m (BM)

The team obtained permission to collect live plants with a view to comparing the behaviour of plants from low and high altitudes when brought together under standard conditions in cultivation. In Turkey, it is against the national laws to collect any plants (including seeds) or to undertake any form of research unless the necessary permissions have been obtained in advance. In addition, as cyclamen are listed by CITES (see p. 197), two further documents are required if living specimens are to be collected, one from Turkey to export the tubers and one from the UK to import them into the EC region. Colleagues in Turkey, notably from Istanbul University and DHKD, were extremely helpful in this respect.

In 1997, our fieldwork concentrated mainly on the south-west corner of Turkey at lower altitudes as there had been some late snowfall, and later moved inland to higher regions. The start date was 16 April and the itinerary outline was as follows: Büyükonak, Akçaova, Marmaris to Datça, Sandras Dağ, Mendos Dağ (Fethiye), Seki to Elmalı, Camkyüsü, the Sinekçi Pass, the Kuraova Pass, Lake Kovada, the Lake Eğırdır area, and finally the region from Antalya to Kumluca. Over 20 populations of *C. alpinum* were surveyed in a wide range of habitats from near sea level to above 1,600 m and in scrub or woodland to more open scree-like conditions. Woodland populations were found under a wide variety of species, both coniferous (*Cedrus*, *Pinus* and *Abies*) and deciduous (*Platanus*). Horticultural interests were satisfied by observing some unusual forms of the species.

In evergreen forested habitats, *C. alpinum* tends to occur in the more open rocky areas (TW)

An attractive leaf form of *C. alpinum*, in a damp situation by a river beneath *Platanus* trees, south of Eğırdır, Site 97/16T (BM)

In the second year (1998), two main areas were visited: the mountains to the north-west and west of Antalya, based in the small market town of Elmalı, and the north-east of Antalya in the mountains around Lake Eğırdır, which had also been visited in the previous year. The reasons for visiting the former region were two-fold: first it appeared, from the previous season's work and from existing records (or rather the lack of them) that there is a gap in the distribution of *C. alpinum* to the north and north-west of Antalya; second, there were known to be good populations south of Elmalı. The region around Eğırdır was selected in order to ascertain just how far to the east and north-east *C. alpinum* is distributed.

Upon heading from Antalya to the north-west, then turning in an arc southwards towards Elmalı, it soon becomes apparent that the highlands here are closely allied to the inland plateau area of Turkey — the 'Irano-Turanian' floristic region — a harsh, cold climate with considerable snowfall in winter and long dry summers. There are few trees and the vegetation is of the steppe type with spiny cushion plants and not at all like cyclamen country, hence the reason for the apparent (and now thought real) gap in distribution. Rainfall maps for the region confirm that this region is dryer than the area to the south of Elmalı.

Elmalı itself is situated on a rich plain that has been turned over to agriculture, so little of the original vegetation remains around the town. Heading southwards out of Elmalı, a change of vegetation type soon becomes apparent. Within a few kilometres, bushes and trees (*Quercus*, *Juniperus* and *Prunus*) appear and it became clear that we were moving into a higher rainfall area; we found the first population of *C. alpinum* just 17 km south of the town at 1,100 m, growing among bushes and limestone rocks in humus-rich terra-rossa. Several white-flowered plants were seen here among the many 'standard' ones, whose flowers fall into the red-purple group of colours in the RHS Chart (mostly Group 74B, 74C or 74D); one plant had flowers that were more 'coum-like', with reflexed rather than spreading petals.

Surveying another population a little further on, on the Avlanbeli pass, revealed just how inclement the weather can be at the end of March. At only 1,100 m, there had been snow overnight with torrential rain and bitingly cold winds following on; the plants growing there are certainly hardy. On the route southwards from Elmalı, travelling towards the Mediterranean coast, *C. alpinum* is replaced by *C. graecum* in the region with altitudes below 500 m about 50 km south of Elmalı. There, the populations include some very attractive silver-leaved plants. As would be expected the flora changes as the altitude declines, with lower areas featuring typical Mediterranean plants such as *Euphorbia characias*, *Ophrys* spp., *Iris unguicularis*, *Colutea*, *Poterium*, *Arbutus*, *Pinus brutia* and *Daphne gnidioides*. Other routes to the south and south-west of Elmalı also have good populations of *C. alpinum*.

Cedar forest and limestone scree formations at Çamkyüsü, 1,650 m. Trevor Wiltshire photographing a population of *C. alpinum* at Site 97/12T (BM)

The mountains by Lake Eğırdır, home to thriving populations of *C. alpinum* (BM)

Interesting variants seen in the region included many plants with truncate leaves (i.e. as if chopped off straight across the top, not rounded or pointed), plants with flowers that were flat (the petals spreading out horizontally) and whirlygig-shaped, large-flowered forms with wide, substantial petals, variants with fimbriate (fringed) petals, forms with interesting leaf patterns (e.g. silver-flecked) and so on.

Many of the higher-altitude populations in this region grow in open conifer woods of *Cedrus*, *Juniperus* and *Pinus*, but slightly lower down, cyclamen populations are often in thorny scrub of *Quercus coccifera*, *Crataegus* and *Prunus*. The final site visited in this general area, at 1,400 m on the Kuraova pass, had the former type of habitat and was bitterly cold, to the extent that the tubers were actually frozen into the soil. All this suggests that *C. alpinum* might be a lot hardier than it has been given credit for in the past – or at least that some of these higher-altitude forms could prove to be so.

In the second phase of the 1998 work, the team headed towards Lake Eğırdır, taking a diversion inland to Beşkonak on the way in an effort to locate the eastern-most edge of the distribution of *C. alpinum* in this region. However, the vegetation type in the Beşkonak area is typically Mediterranean (featuring *Arbutus*, *Myrtus*, *Cistus* and so on) and the cyclamen is *C. cilicium*, here in one of its western-most localities. From the observations made on these and later visits, it appears that the eastern extent of the distribution of *C. alpinum* is to the west of the Dedegöl Dağlari, a mountain range that runs approximately north to south (roughly to the south of Lake Beyşehir); *C. mirabile* also appears to be found only to the west of this range, whereas *C. cilicium* occurs mainly to the east of it. There are thriving populations of *C. alpinum* in the hills between Lakes Eğırdır and Beyşehir and some good leaf forms were noted there. Several other populations were located and surveyed in the area to the south of Lake Eğırdır, but heading north along the eastern side of the lake, round the top and down the western side indicated that the species is unlikely to extend any further in those directions. The terrain there becomes dryer, colder and more steppe-like with many bulbous plants and spiny cushions, and looks very unsuitable for cyclamen. The presence of an *Iris* of the 'Juno' group, possibly *I. galatica*, is a pointer to having entered the Irano-Turanian floristic region once more.

Cyclamen alpinum east of Lake Eğırdır on the road to Sariidris (Site 98/11T) (BM)

Interesting variations in the leaves at Site 98/11T (BM)

A short foray was also made west from Antalya through steppe country in order to look at *C. alpinum* in the more north-westerly parts of its known distribution. After some 100 km of this very unsuitable harsh territory, some populations were found in rather greener habitat leading up to the Dirmil pass on the way to Altinyayla, growing in a very spiky mix of *Berberis*, *Crataegus* and *Quercus coccifera*. However, *C. alpinum* appears not to continue onto the pass itself, which is essentially a serpentine ridge noted for several choice endemic plants such as *Fritillaria serpenticola*, *Muscari mirum* and *Crocus baytopiorum*. A further excursion was made along the road leading south-west from Antalya between two mountain ranges (the Bey and Tahtalı ranges) to Altinyaka and Kumluca. We saw excellent populations of *C. alpinum* here and one of these contained forms with totally unmarked deep green leaves, not beautiful but interesting. From the 'spot' records made along the road, it seemed quite likely that *C. alpinum* occurs throughout the beautiful cedar forest that runs almost continuously along the western flanks of these ranges.

For the 1999 fieldwork, earlier dates were chosen in an effort to find the really low-altitude populations, anything between sea level and at most 800 m, in flower; we started on 24 February. A short foray into the mountains to the west of Antalya (into the Bey Dağlari) proved to be valuable as one of the best populations recorded during the three years was located. This population was 'best' in terms of density of plants: in the thickest part of the colony there were about 40 plants per square metre. Although physically separated from other populations, this population had considerable

The beautiful Bey Dağ range, with rich cedar forests (BM)

The foothills of the Bey Dağ have excellent populations of *C. alpinum*. At Site 99/01T there is a wide range of variation, this is one of the paler shades (BM)

variation in depth of flower colour, some plants had flowers with exserted styles and some had 'coum-shaped' flowers, whereas others were of the more familiar flat-propeller shape; the leaves too varied quite a lot. Our next move was along the coast towards Fethiye and the Marmaris-Datça peninsula, where low-altitude populations had been seen in 1997. These were also at the western extremity of the known distribution, so would also be of interest from this viewpoint. The only population of *C. alpinum* seen in the area along the way was reported by Erna Frank and by Bob and Rannveig Wallis, in a most inauspicious scrubland site on the main road opposite a petrol station! This was an 'island' site, situated in a sea of low-altitude Mediterranean spiny plants and maintained by a water seepage that provided a shady, humid habitat. In fact, the plants were very fine with large-flowers, mostly in the range of 30–36 mm diameter and one of them a staggering 44 mm across.

Plants with drooping petals were apparent at several populations. A plant from this Site 99/04T has maintained the characteristic in cultivation (BM)

The colour was a fairly uniform bright pink-purple but the shape varied greatly from 'propeller' or 'windmill' to 'coum-like' and 'elegans-like'; several plants had droopy petals that hung down below the calyx (not with age or wilting), a feature that was also observed in several other colonies further to the west. On the whole, the leaves were rather dull in pattern and quite large (up to 59 mm wide) for this altitude of 520 m; *C. graecum* was present in the more open spots nearby.

A short diversion inland from Fethiye into a very unsuitable-looking dry pine forest turned up another *C. alpinum* population in a small depression where a spring issued forth out of the rocks, providing a microclimate of humidity; in fact, at this time of year, it was so wet that some of the cyclamen leaves were under water. This site was at a high altitude, 920 m, and the leaves were mostly smallish, about 12–36 mm wide, although in shadier spots they did exceed this size.

At only 50 m altitude near Marmaris *C. alpinum* was still in flower on a river bank in a moist mild valley with plane trees, oleanders and myrtle, a very different habitat from that of the snowy, frozen passes around Elmalı where we had seen it in the previous year. As at most other sites, this population was not large (the site measured 50 m long and only 10 m wide) but there were plenty of interesting variations, including the 'droopy' flowers referred to above in which the petals hung down below the level of the calyx. They were perfectly fresh and the plants were certainly not dry at the root. The plants at this site had noticeably large leaves, ranging from 40–90 cm across but mainly 50–70 cm.

Just south of Marmaris another pine-wood colony of *C. alpinum* was of interest in that it was clearly not on limestone (and probably on metamorphic rock) and the associated plants were *Erica manipuliflora*, an ericoid *Hypericum*, *Muscari macrocarpum* and

Between Marmaris and Datça, *C. alpinum* inhabits scrubby woodland with low altitude Mediterranean elements such as *Arbutus andrachne* (Site 99/06T) (BM)

Fritillaria forbesii. Heading out on this long peninsula towards Datça, a population was encountered on a small pass at 350 m, clearly on serpentine formations. Here, the dominant plants that formed the scrub were *Arbutus andrachne*, *Phillyrea latifolia*, *Cistus* spp., *Quercus coccifera*, *Ruscus* and *Euphorbia acanthothamnos*. The variation in leaves (which were fairly large here) and flowers was much the same as had been seen on several occasions, including 'droopy' and fimbriate petals; an attractive and frequent characteristic at this site was a dark red-purple stain on the leaf stalks that ran out on to the veins of the leaf. Further attempts at locating *C. alpinum* on this peninsula failed, suggesting that this population is the most south-westerly, but there are many places that would take a long time to reach on foot, so one can never state categorically that this is the case. Out towards the end of the Datça peninsula, there are some magnificent forests of *Arbutus*, including mixtures of both *A. andrachne* and *A. unedo*, that form a dense canopy with flourishing populations of *C. hederifolium* beneath. These cyclamen have very large, fleshy and curiously 3- to 5-lobed leaves, so it is possible that they are referable to *C. confusum*.

One of the highlights of the 1999 season came about by pure chance. A convenient overnight stopping place was Dalyan, where the team chatted to some young men who were helping out at the pension. They turned out to be students from the local agricultural and horticultural college, and one of them said that he was studying *Cyclamen* and had a collection growing at the college. Furthermore, he knew a site nearby where '*C. coum*' grew and that it was still in flower. This was all too tempting, so the next day saw a large group trekking across a large flat marshy area towards a spur of mountains that terminate in some cliffs. Sure enough, at less than 10 m above sea level (the water of the estuary was lapping alongside), there was a flourishing population of *Cyclamen* — not *C. coum* of course, but *C. alpinum*. Although this is generally a hot area in summer, the site was moist and cool in a rock slide at the foot of vertical cliffs. As with other low-level populations

Although at less than 10 m above sea level here, south-east of Dalyan, *C. alpinum* found cool, moist scree conditions at the base of a cliff (Site 99/12T) (BM)

In the mountains north of Fethiye, rich populations of *C. alpinum* occurred under deciduous scrub at 1,220 m (Site 99/16T) (BM)

in the west, there were plants that had very large leaves (106 mm long was the largest!), plants with 'droopy' flowers, variants with fimbriate petals and so on. The student also showed us his *Cyclamen* collection and we were able to dispense some hints on cultivation.

This encounter with the lowest possible population would have made a fitting end to the fieldwork, but there were plenty of excellent sites to come. Just inland from the coast road back to Fethiye, at 100 m, was a very large colony with a density in places of 25 plants per square metre; we could not determine the extent of it as the plants just went on and on up the very steep slopes into the distance. Then, north of Fethiye, and climbing back into the mountains, there was (at 450 m) another very large population. Here, there were up to 30 plants per square metre, all in full flower and colouring the ground beneath the bushes, some plants with leaves in the 'pewter' category. Still higher, at 1,050 m in open deciduous oak and pine woodland, *C. alpinum* occupied a range of habitats from almost fully exposed to deep shade in the boles of the trees; as might be expected, the leaves were on the whole smaller than those lower down but varied greatly in size according to exposure. A bonus found near here was a splendid drift of a white form of *Crocus antalyensis*, which is normally lilac. The final population of *C. alpinum* seen was at 1,220 m, still north of Fethiye and on a sunny hillside. Many of these plants were in situations that are much more open than is usual or under still-leafless deciduous bushes and were flowering profusely, perhaps due to the greater exposure to light.

As a result of this fieldwork, *C. alpinum* is now one of the better known species in terms of precise distribution, range of altitude, habitat (including rock and soil type, pH and plant associations) and morphological variation. Having observed a very large number of specimens of *C. alpinum*, the team concluded that there is no convincing evidence that low- and high-altitude variants should be distinguished as separate taxa. There is a little correlation between overall size (particularly of leaf) and altitude, and the size characteristics appear to be reasonably stable when plants are brought into cultivation. It can be said, however, that the populations at the higher altitudes were often small-leaved, and that some of the populations seen at low altitudes had plants with huge leaves, larger than any previously seen in cultivated specimens, but there was much variation at all sites and often considerable overlap

A particularly well-flowered specimen at Site 99/16T (BM)

in measurements between populations. Extreme examples can be quoted: a population on the Marmaris Peninsula at 75 m altitude possessed leaves of 47–106 mm long and 55–104 mm wide; on the other hand, in a population at Çamkyüsü at 1,650 m, the leaves were 8–38 mm long and 10–38 mm wide.

Observations on the habitat preference of *C. alpinum* were perhaps more interesting. Turkey can be divided, for phytogeographical purposes, into three broad vegetation types: the moist Euro-Siberian region in the north, along the Black Sea and its adjacent mountains; the central plateau area (the Anatolian Plateau), with its extremely cold winters and steppe-like vegetation (known as the Irano-Turanian region); and the Mediterranean region, characterised by autumn, winter or spring precipitation and warm, dry summers. This is a fairly crude but useful view of the phytogeography; there are, for example small 'enclaves' of Mediterranean vegetation on the Black Sea coast and small areas of Euro-Siberian vegetation further south in what is otherwise a Mediterranean zone. Botanists who are familiar with the flora of Turkey are able, by and large, to place individual species into these phytogeographical groups. It soon became clear, by looking at the plant associations, that *C. alpinum* is a 'Mediterranean' species; on the other hand, *C. coum*, which is generally considered to be its nearest relative, is clearly a 'Euro-Siberian' species. When in *C. alpinum* country, and this was particularly noticeable in the mountains to the south of Lake Eğırdır, a traveller could be amid various plants of the Mediterranean element, only to turn a corner and be confronted by a piece of 'Euro-Siberia', with a quite different range of plants including violets, primroses, coltsfoot and *Eranthis*, all familiar European species. Even the atmosphere in these 'Euro-Siberia' areas had a different feel, colder and damper, and the soil was a heavy sticky greyish clay; *C. alpinum* was never encountered in these habitats. Around another corner, there might be a reversion to the warmer Mediterranean climate with a change of plants again; it was soon found possible to predict where *C. alpinum* would appear, and, similarly, where it would not. Unfortunately, *C. coum* was not found in the 'Euro-Siberian' areas, although that is not to say it is not there. The rock types and soil pH in which *C. alpinum* was found were strikingly variable. Although the species was usually found on limestone formations, it was also recorded on serpentine and sandstone, and in soils with pH readings from 6.8 to 10.1.

The survey also extended the known range of the species in a north-easterly direction, into the area between Lake Eğırdır and Lake Beyşehir. In fact, it appears that the eastern-most boundary for *C. alpinum* is the mountain range between these two lakes: *C. alpinum* was found on the western slopes of this range, which runs roughly in a north to south direction, but on the eastern slopes it was replaced by *C. cilicium*.

Another useful aspect of the visit was that it proved to be a 'tidying up' operation for some of the records. For example, an outlying locality cited for *C. alpinum* between Bilecik and Bozüyük, well to the north of the general area for the species, was visited and (not unexpectedly) was found to contain *C. coum*. Other records on the outer fringes of the distribution appear to represent, in the east, *C. cilicium* and, in the west, *C. mirabile*; such records must have been based on non-flowering dried material in herbaria. In addition to the information gathered, some horticulturally desirable plants were collected under licence and have added to the range of variants in cultivation.

A plant of *C. alpinum* at a Cyclamen Society show, originally collected on Mendos Dağ under cedars at 1,440 m (CSE97006, from Site 99/16T) (MD)

3.5e *CYCLAMEN INTAMINATUM*

Richard Bailey

Cyclamen intaminatum between Inönü and Kütahya, 1,020 m in scrubby woodland of *Quercus cerris*, Site 04/01T (MD)

In 2001 and 2004, the Cyclamen Society, together with Turkish botanists, undertook fieldwork in Turkey to observe and record details of wild populations of *C. intaminatum*. Richard Bailey the leader of the British contingent, provided detailed accounts of their interesting findings in the Society's journal, from which we have extracted the following information. In both years, the group received invaluable assistance in the planning and research application stages from Prof. Dr Neriman Özhatay (Istanbul University) and Sema Atay (DHKD: Turkish Society for the Protection of Nature).

Cyclamen intaminatum, 2001 and 2004

Teams: 2001: Richard Bailey, Jim Novis, Ian Robertson (Cyclamen Society) and Narin Sadıkoğlu (University of Istanbul)

2004: Richard Bailey, Martyn Denney (Cyclamen Society) and Sırrı Yüzbaşıoğlu (University of Istanbul)

The first fieldwork was carried out from 14–23 October 2001 and covered a small area in a triangle of western Anatolia, roughly defined by the towns of Bursa, Inegöl and Eskişehir to the north and east and Kütahya in the south. The central parts of this area are mountainous, often steep (even precipitous) and densely wooded, whereas the fringes of the triangle and some valleys in the centre are often intensively farmed. The primary aims of the study were to determine the extent of the distribution of the species, to achieve a greater understanding of the species' precise habitat and as a consequence its requirements in cultivation, to study natural variation within the species and to introduce a few selected plants into cultivation to increase the range available to enthusiasts. Any living material collected would also be used for on-going DNA studies.

Richard Bailey and Sırrı Yüzbaşıoğlu assess the habitat of *C. intaminatum* at Oylat, Site 04/07T (MD)

Living material was introduced into cultivation for DNA studies and horticulturally desirable variants were selected, such as this robust delicate pink with fimbriate petals (CSE04058 from Site 04/06T near Inayet, Bursa province) (MD)

Near Oylat (Site 04/07T), *C. intaminatum* and *C. coum* grew intermixed (MD)

Research suggested that the most likely habitat for *C. intaminatum* in the wild would be steep, well-drained slopes, of any aspect, on limestone formations below scrub or light, mixed, broad-leaved woodland, at altitudes of 700–1,200 m. Nonetheless, generalisations, while useful, always have their exceptions. Initially, this seemed a reasonable summary of the *C. intaminatum* habitats that the team encountered; at one site near Oylat, where a limestone precipice plunged into acid sandstone, *C. intaminatum* grew on the limestone, with its tubers often completely filling water-worn holes and crevices. It also flourished in woodland on limestone scree below the precipice, but was not found on adjacent acid soils. Later, however, *C. intaminatum* was found amongst acid rocks under *Pinus* spp. on the road between Boğazova and Çayyaka.

Although the most likely situation for *C. intaminatum* was on steep slopes, or on the summit of narrow ridges, the plant was also found on more level ground. This ground, however, was usually very rocky and with good drainage. The soils at sites where *C. intaminatum* was found were remarkably consistent, all being clay, often a *terra rossa* type, often sandy or gritty and with free limestone. In all sites with a stable soil, the mineral soil was topped with varying depths of leaf mould, although the *C. intaminatum* tubers were normally growing in the mineral soil. The characteristics of the majority of the sites can be summarised as follows: altitude range, 670–1,565 m; slope, 1-in-20 to 1-in-0.5; aspect, north-west to south; soil type, fine clay to gritty clay; soil pH, 6.4–7.8. Nevertheless, conditions at one site fell outside the norm in having a sandy soil with a pH of 5.3. At wooded sites, the canopy was usually mixed, often including *Fagus orientalis*, *Fraxinus excelsior*, *Juniperus oxycedrus* and species of *Quercus*. Scrubland sites mostly featured coppiced *Quercus cerris* and other oaks. *Cyclamen intaminatum* was not found amongst dense herbaceous vegetation, although it did occur beneath completely closed tree cover. In some sites, it grew amongst unshaded limestone rubble with very sparse vegetation, much of which consisted of mosses and lichens. It was most frequently associated with herbaceous plants and bulbs, mostly out of flower. Of special interest was the occurrence of *C. intaminatum* and *C. coum* intermixed, amongst unshaded limestone boulders and low vegetation at Oylat; *C. coum* was also found, but without *C. intaminatum*, in dense beech (*Fagus orientalis*) woodland near Domaniç.

One aspect of the study was to record the natural variability in wild populations so the variation in leaf size, shape and pattern, and in flower colour and size, was recorded at all the sites studied. Flower colour and petal size (length and width) were recorded and an estimate was made of the degree of petal 'twist'. Flower colour ranged from the white with grey (colourless) veins most common in cultivated plants to deep pinks. A few bicoloured flowers were noted. Conscious of the subjective nature of any colour measurement and possible change in colour after picking, the colours were recorded using the RHS Horticultural Colour Chart in daylight, by the same person throughout and actually on the site. Petal size was extremely variable: length 12.1–28.8 mm (average 17.3 mm), width 4.2–8.9 mm (average 6.6 mm). The longest petals were not necessarily the widest, for example: the longest petal was 5.6 mm wide, the shortest 6.2 mm wide; the narrowest was 16.9 mm long and the widest 19.4 mm long. Although usually regarded as one of the smallest-flowered cyclamen, the largest *C. intaminatum* flowers fall within the size and colour range of the closely related *C. cilicium* and *C. mirabile*. Petal twist is difficult to estimate and ranged from 145° to 45°, but in the majority of flowers, it was between 90° and 100°.

A range of flower characteristics was noted to determine natural variation. Bilecik Province, south of Bozüyük (MD)

At each site samples of one leaf per plant were taken to note shape, size and markings. Bilecik Province, south of Bozüyük (MD)

With regard to the distribution, considerable time was spent seeking the plant outside the area in which it was known to occur; this search was not successful, although a number of new sites were found in the western part of its known distribution. It now seems reasonably certain, however, that *C. intaminatum* does not extend north of the main road from Eskişehir to Inegöl (the main Bursa to Ankara road) or significantly east of the road from Eskişehir to Kütahya. In both directions, the wooded, mountainous, hard rock landscape and geology of the plant's central area of distribution is largely replaced by lower, gently rolling landscapes, mostly on softer rocks, and by arable agriculture. Nonetheless, *C. intaminatum* did occur in patches of uncultivated scrub along the road between Eskişehir and Kütahya, suggesting that it might be found as scattered relict populations elsewhere in that direction where intensive arable agriculture is not dominant. Time did not allow investigation of areas south west of a line between Kütahya and Bursa.

On the 2004 study (from 15 to 26 October), the first sighting of *C. intaminatum* was on the road from Inönü to Kütahya. This was the eastern-most site recorded and was followed by sites in the north between Söğüt and Çalti, in the west close to Bursa, and to the south in woodlands at a higher altitude south-east of Domaniç. Within the area circumscribed by these four sites, work was concentrated in two areas: one in the Ulu Dağ range and the other in the more intensively farmed area towards Bilecik. Several new sites were noted in the Ulu Dağ, extending the range slightly both to the south and to the west. Most importantly, *C. intaminatum* sites were found for the first time on the southern slopes of these mountainous sites. Overall, *C. intaminatum* was found at 14 sites and, as during the previous visit, detailed observations were made. Herbarium specimens and leaves for DNA studies were also collected.

The habitat studies confirmed *C. intaminatum* to be a plant of scrub and woodland situations, but it appears to be tolerant of a wide range of conditions, growing on acidic or calcareous rocks and under broad-leaved or coniferous trees and shrubs. Most often, it was found under broad-leaved, deciduous trees and shrubs such as *Cornus mas*, *Corylus maximus*, *Crataegus*, *Fagus orientalis*, *Quercus cerris* and *Thelycrania sanguinea*. In such sites, the underlying rock was calcareous, usually limestone of the carboniferous type, with a soil pH ranging from 6.0 to 8.0. The soil itself was normally humus rich, and varied from almost pure clay to a gritty clay-loam. In such situations, *C. intaminatum* was often accompanied by *C. coum*, *Helleborus orientalis*, *Osyris alba*, *Poterium sanguisorba*, *Ruscus aculeatus* and *R. hypoglossum*. In the Ulu Dağ range, however, it was growing under conifers, mostly *Pinus* spp., although a wide range of broad-leaved trees and shrubs were often also present. Here, the rocks were siliceous, supporting acid-loving lichens such as *Parmelia conspersa*, *P. glabratula*, *P. saxatilis* and *Rhizocarpon geographicum* agg.. The soil pH varied from 5.1 to 7.3 and was usually a humus-rich gritty or sandy loam. *Crocus pulchellus*, *C. speciosus*, *Euphorbia amygdaloides* and *Digitalis ferruginea* were often present at these more acidic sites. Plants that are commonly associated with cyclamen in the Mediterranean flora, such as *Cistus* and *Pistacia lentiscus* were recorded, but rarely. [The above plant associations suggest that *C. intaminatum* is part of the Euro-Siberian floristic region, whereas *C. mirabile* and *C. cilicium* fall more within the Mediterranean element — Ed.].

Cyclamen intaminatum is tolerant of a wide range of habitat types, here growing beneath evergreen oaks near Bozüyük (MD)

A fine specimen of *C. intaminatum* at Site 04/08T (MD)

Near Çayyaka, at Site 04/08T, *C. intaminatum* occurred on wooded slopes in a layer of pine needle/bracken at 1,270 m (MD)

Leaves from Site 04/08T near Çayyaka showing variation from almost plain green through to a good silvery 'Christmas-tree' pattern (MD)

In looking for potential sites, the team soon learned to seek north-facing slopes above 700 m altitude and most of the sites in which *C. intaminatum* was found satisfied both of these criteria. The lowest site was at 704 m and the highest 1,796 m, and most were indeed on north-facing slopes. On the southern side of the Ulu Dağ, however, *C. intaminatum* grew down a south-eastern slope to an altitude of 1,125 m, the lowest site at which the plant was recorded on the southern slopes of the Ulu Dağ.

As in 2001, variability in both leaves and flowers was studied at most sites using material taken from randomly selected plants, but no greater variation was found in 2004 than in the previous studies. Leaf pattern was remarkably uniform; the majority of plants had a dark green, 'Christmas-tree' marking, usually outlined by broken yellow-green markings, although the pattern was very occasionally silver rather than green. A few plants had plain green leaves. We saw no leaves with a pink-flushed upper surface like those noted by Grey-Wilson (2002), but many plants had red-backed leaves (we noted 52 with red backs and 95 with green backs). The proportion of red-backed to green-backed leaves did vary significantly from population to population from over 90% red to all green. Most leaves were reniform, slightly wider than long when measured at maximum width and length.

Flower colour was checked again using the third edition of the RHS Horticultural Colour Chart (1995); whenever possible, colour was assessed in the field rather than under artificial light. Virtually all of the flowers were in the RHS Purple Group of colours, which in most peoples' eyes would be thought of as pale- to mid-pink. Pure white flowers, thought for so long to be typical of *C. intaminatum*, were relatively uncommon, although they were found more frequently than in 2001. As before, flower size and shape was defined and measured by petal size and degree of petal 'twist'. Petal lengths were 10.2–23.1 mm (most commonly 13–18 mm) and widths 4.0–9.0 mm (most commonly 5–7 mm).

The 'twist' in the petals, estimated from base to apex, ranged from 45° to 145°, but by far the most common angle of twisting was 90°. Some flowers with long, thin petals, of over 19 mm long and less than 5 mm wide, and with a high degree of twist produced very elegant flowers, often similar to those of *C. repandum*. These flowers were, however, relatively unusual; most had petals around twice as long as wide with 90° of twist.

Although it seems likely that *C. intaminatum* has the most scattered and restricted distribution of all Turkish cyclamen, the difficulty of reaching sites in the mountainous, central area of its distribution means that the Cyclamen Society's studies were largely confined to the fringes. Nevertheless, a great deal of information has been added to our knowledge of this species in the wild.

A delicate pink form seen at Site 04/12T, Inegöl to Pazargeri, 23 October (MD)

3.5f *CYCLAMEN PSEUDIBERICUM*

Vic Aspland

Team: Vic Aspland, Harry Gill, Pat Nicholls and Dr Murat Ekiçi (Gazi University, Ankara)

In 2002, the Cyclamen Society undertook a field study in Turkey of *C. pseudibericum*, a project considered to be well worthwhile as this is one of the more restricted species and there had been no previous detailed survey of it in its native territory.

The aim of this Cyclamen Society field trip, undertaken from 10 to 24 March 2002, was to increase the knowledge of *C. pseudibericum* in the wild. With the advance of modern scientific methods, a further purpose was added: to collect specimen plants from which the Society can supply material for molecular studies.

Three bases were planned, in Adana, Osmaniye and Kahramanmaraş; from these three centres, radial forays would be conducted into the surrounding arc of mountains. First, the team set off from Adana towards the north-west, seeking *C. pseudibericum* near the western limits of its distribution. Heading northwards along the valley of the mighty Seyhan River, stops at the roadside revealed *Anemone blanda* in a range of forms, many with a contrasting white centre, and specimens of *Daphne sericea*. The first sighting of *C. pseudibericum* turned out to be the most variable population found on the whole trip. A north-facing bank was liberally scattered with cyclamen, so work began to document the finds. This work included measuring and recording the leaves and flowers from 50 randomly selected plants, noting observations and taking photographs to describe the site, and photographing individual plants. A representative selection of the specimens was transferred to the herbarium press for later use by the Cyclamen Society and by Istanbul University Herbarium. This proved to be the lowest site found for *C. pseudibericum*, at 240 m.

Joining up with Dr Murat Ekiçi, the team set off towards the fruitful north-west. On an un-named pass to the south of Aladag, a quite different habitat was encountered. The road travels almost due north and rises to the pass quite gently. Low ridges of dissected limestone rise to either side. An inspection of the natural shelves in the slope to the west of the road (i.e. in the east-facing slope) revealed *C. cilicium* growing with a *Scilla* somewhat similar to *S. sibirica*. The slope had little tree-cover other than a few scattered *Pinus halepensis*, so was likely to be quite dry in summer. This accorded well with observations of *C. cilicium* made during the

Site 02/01T, where *C. pseudibericum* grows in scrub of Nerium, *Daphne sericea*, *Pinus halepensis* and *Quercus coccifera*. Seyhan River valley, March 2002 (VAA)

Striking leaves are a feature of *C. pseudibericum*, this one at Site 02/01T with particularly bold markings. Adana Province, in the Seyhan River valley, 240 m (VAA)

A random sample of leaves and flowers from Site 02/01T, to be measured, pressed and deposited in herbaria in Turkey and the UK (VAA)

Cyclamen pseudibericum and a *Muscari* sp. in a natural association between Kozan and Feke, Adana Province, 770 m (Site 02/03T) (VAA)

Cyclamen Society's fieldwork on *C. alpinum*; it is a plant of rather dry habitats. In the other direction, the ridge was set back further from the road, but a trek of a few hundred metres brought us to its foot. It was impressive! Imagine a deeply dissected limestone pavement with razor-sharp edges. Then tilt it to an angle of perhaps 30–35 degrees. Within clefts, and under overhangs, nestled *C. pseudibericum*. The only other plant that grew in any number was the immensely prickly *Quercus coccifera* (Kermes oak). In general, *C. pseudibericum* occupies habitats that are more moist than those typical for *C. cilicium*. We may theorise that in this specialised habitat, more moisture is available to plants growing within the clefts and under the rocks, and the altitude (1,070 m) and westerly aspect may also keep it cooler.

Moving east, the next site found was north of Feke at an altitude of 770 m, and this offered a completely different habitat. Considerable shade was cast on the west-facing slope by woods of *Pinus halepensis* with occasional *Quercus cerris* and *Platanus orientalis*. The ground flora included a *Muscari*, *Anemone blanda*, *Viola ?odorata*, *Bellis perennis* and *Cistus salvifolius*. Although the location was moist at the time of the visit, it would undoubtedly become quite dry later in the year. The heavy shade would, however, moderate the soil temperatures. The rock here was highly metamorphosed to an irregular schist type.

Following this, the focus moved on to Kahramanmaraş from where we could strike north to revisit a site known to Murat from eight years previously. In the intervening period, his recollection of the precise access to the valley in question had become a little hazy, so a stop was made at a small wayside establishment. With considerable help from locals, the team made its way — somewhat perilously — to a wide valley surrounded by snow-covered hills. The views of the limestone landscape were wonderful, with a waterfall, and streams disappearing into a sinkhole. Here, in addition to *C. pseudibericum*, a perfect flower of *Iris persica* was poking from the hard-packed terra-rossa soil. At an altitude of 1,420 m, this was the highest sighting of *C. pseudibericum* recorded.

The next target area for exploration was the Amanus Mountains. *Cyclamen pseudibericum* had been recorded by a Cyclamen Society team during earlier field studies of *C. coum* on the north side of the Belen Pass, which crosses the central Amanus. The two populations recorded in this area were found, the higher of which was a mixed population of the two species, in which the plants were growing mostly in a fair amount of shade in woods, consisting mainly of *Pinus halepensis*.

Eastwards over the pass, on the eastern flank of the Amanus, the soil type changed abruptly to a very fine pale khaki, in which the pines continued but the ground flora was almost completely absent, presumably because of lack of nutrients. In addition, this was clearly a rain-shadow area: certainly not *C. pseudibericum* country. A later search, to the south of the Belen Pass, also produced no sightings of *C. pseudibericum* as the land was either intensively cultivated or too hot and dry.

Following this, the centre of operations moved to Osmaniye to allow exploration of the northern Amanus, where an apparent 'pseudibericum gap' was discovered. Most of the published distribution maps show *C. pseudibericum* covering an arc through the eastern Taurus, anti-Taurus and Amanus Mountains. Despite searches in the Nur Dağ Pass area, where *C. coum* was confirmed, and several penetrations from the north and north-west, no *C. pseudibericum* were found. This absence seemed to be related largely to the nature of the soils, which varied from the impoverished

The Amanus mountains where John Watson and colleagues found the pink form of *C. pseudibericum* (BM)

At 1,190 m between Düziçi and Gümüs, *C. pseudibericum* growing with a *Corydalis* in oak and beech leaf litter (Site 02/05T) (VAA)

khaki type through to highly metamorphosed types with, in places, what was very obviously serpentine with its high content of toxic heavy metals. Thus, as far as we were able to establish, there is a large break in the distribution of *C. pseudibericum* in the northern Amanus. The species had been previously recorded north of Osmaniye, east of the village of Düziçi and, with local help, the team reached a north-northwest-facing slope clothed with *Fagus orientalis*, *Platanus orientalis* and occasional *Quercus cerris*. The ground below was carpeted with *C. pseudibericum*, growing in some areas with *C. coum*. Descending a hill from this site gave a view into a declivity splashed with patches of colour, and investigation revealed a very large area in which *C. pseudibericum* grew in company with various perennials such as *Primula vulgaris* and a *Corydalis*. The whole area was moist, with a healthy stream at the bottom. The plants grew in the open, with the cyclamen spreading up the hillsides under *Fagus orientalis* scrub, which was bare of leaves at this time of year, and reaching as far as we were able to walk. The main ridge extended away to the south-east for many kilometres, much of it covered by *Fagus* scrub, so it seems likely that the cyclamen habitat is quite extensive. Lower down, a sharp eye spotted three splendid cyclamen growing at the top of a roadside bank, fairly obviously planted by the local villagers.

In most of the populations observed, the variation between plants was relatively small, with the majority looking just like an average plant as seen in cultivation. The most variation had been noted at the first site so a return visit was made. An interesting discovery here included three plants with white flowers, one of them with only a very faint basal blotch, and a plant with vivid glowing flowers for which Harry Gill coined the description 'cherry pink'. After considerable searching, a mate was eventually found for it, so it is hoped that seed will set in cultivation. The main purpose of this fieldwork was, however, to gather facts about *C. pseudibericum* in the wild, which are presented below.

Habitat

Cyclamen pseudibericum was found at altitudes between 240 and 1,420 m. In the lower half of the altitude range, the habitats faced from west, through north, to east, and were clothed by tree cover of varying density, from quite dense woodland to rather sparsely scattered trees. The principal trees here were *Pinus halepensis*, deciduous *Quercus* spp. and the occasional *Platanus orientalis* and *Fagus orientalis*. At higher altitudes, the dominant covering tree became the deciduous *Fagus orientalis*. Only at altitudes above 850 m was the cyclamen able to occupy more open situations. Here, the habitat invariably comprised deeply dissected limestone pavement where *Quercus coccifera*, often either stunted or heavily goat-browsed, was not uncommon.

A pale-flowered form of *C. pseudibericum* in the wild in the Seyhan River valley (Site 02/01T) where populations with the most variation were observed (VAA)

Cyclamen pseudibericum often grows in woodland, in this instance pines (*Pinus halepensis*), *Quercus cerris* and *Platanus orientalis* (VAA)

Substrate

The underlying rock types at *C. pseudibericum* sites included sandstone, highly metamorphosed schist and limestone. Where it was possible to examine the structure of the soils (at the lower sites with tree cover), the cyclamen tubers were almost invariably positioned on the top surface of a purely mineral soil, or lay just within its upper boundary. Above and often around the tubers lay a much more organic-rich soil, which was open in texture and moisture-retentive. In the higher, limestone habitats, the soils were rich and heavy, with minimal organic content. The soil pH was determined at each site studied. The range of variation was modest, between 6.8 and 7.8. Note that even the limestone-derived soils were not highly alkaline. It could be expected that these would experience considerable leaching, either by winter rains or by spring snowmelt, which would reduce the pH.

Range of variation

Cyclamen pseudibericum is a remarkably uniform plant in the wild. A total of 226 plants were fully recorded, of which 55% had flower colours that fell within the RHS Colour Chart group 74 (red-purple) and 38% were group 78 (purple). Excluding the three

A strikingly patterened leaf form at Site 02/01T, north of Adana (VAA)

white-flowered plants and the two with cherry pink flowers described above, the remainder departed from these shades to only a minor extent.

Leaf shapes and patterns were classified using pre-determined scales. The shape scale ran from 1 (reniform, like classical *C. coum*) to 8 (very pointed, like *C. repandum*). The pattern scale ran from A (very narrow hastate pattern, rest of leaf silvery) to G (hastate pattern very large, indistinct, little other markings or flecking). Leaf patterns were quite uniform, with 91% of all leaves graded being in grade C or D. This is the type with which we are all familiar in cultivation: a moderately sized, dark green, hastate pattern at the centre of the leaf surrounded by a more-or-less continuous silver band, and the outer margin of the leaf also dark green, often with some silver flecking. Leaf shape was slightly more variable, but nevertheless 54% of plants were of types 5, 6 or 7: a rather broad heart shape.

Distribution

Sites visited ranged from the Seyhan River valley 60 km north of Adana, along the Taurus and anti-Taurus mountain chain eastwards to a point 35 km north-north-west of Kahramanmaraş, then southwards into the Amanus as far as the northern side of the Belen Pass. The distribution of *C. pseudibericum* is constrained by a number of features. The plain surrounding Adana is intensively cultivated, with no ecological niches available to the species. I infer, however, that even in the absence of cultivation, the higher temperatures at this low altitude would probably render it unsuitable. As far as we could determine, *C. pseudibericum* was absent from a 30 km stretch of the northern Amanus, running south from the main Osmaniye–Gaziantep road. On the western and northern sides, this absence was attributable to the nature of the rocks, some of which appeared to be highly nutrient-deficient siltstones, and others highly metamorphosed types including serpentines with their burden of phytotoxic heavy metals. On the eastern flank, the influence of the former rock type was supplemented by the rain-shadow effect, the climate being far too dry for *C. pseudibericum*. This rain-shadow effect extended across the valley to the east and on into the lower range of hills beyond this, an area with a distinctly Mediterranean feel. In both of these areas, no cyclamen were found.

Conservation status

One of the higher limestone pavement habitats was adjacent to a village near Honaz Daği, and horizontal areas there had clearly been intensively grazed, but the cyclamen were quite safe in their crevices. This ridge extended for several kilometres, and others could be seen in the distance across and down the valley. There was recent tree planting activity in the area (the trees carefully wrapped in bundles of brushwood to protect them from harsh weather), so the general habitat conditions might be expected to improve in the future. One other limestone area seen was the subject of minor terracing, but the massive nature of the rock meant that substantial limestone faces remained for colonisation by the cyclamen. Other higher-altitude limestone environments were remote from obvious human habitation. Although some evidence of grazing was apparent, the expected impact on the cyclamen would have been minimal. Lower-altitude sites appeared to be unsuitable for agriculture because of their generally smaller extent and slope, especially given the proximity of the plain, so appeared to be reasonably secure. To the east of Düziçi, the picture was even more positive. The Forestry Service was actively conserving the habitat, and villagers further down the valley were clearly aware of the attractiveness of the plants. This habitat was very extensive indeed; the team could easily have spent two weeks determining its extent. I am aware of no record in the UK of wild-collected tubers of C. *pseudibericum* having been offered for sale. From an albeit brief sampling of the area, the team concluded that the future of C. *pseudibericum* in the area is secure for the present.

Implications for cultivators

Cyclamen pseudibericum is clearly not a plant of the familiar Mediterranean climate, which has comparatively moist winters with little or no frost, and very hot, dry summers during which geophytes are dormant, and which frequently provides a home to many unfriendly spiny plants. Although the habitats where *C. pseudibericum* was found might well be *fairly* dry in summer, the tubers would not be subjected to the very hot conditions occurring at lower altitudes nearer to the Mediterranean. Temperatures would be ameliorated by west through north to east aspects, in places supplemented by shade from trees. In limestone pavement areas, the location of the tubers in clefts or under shelves in the native rock would similarly modify their environment. Most of the higher habitats were snow-covered as late as February and in the absence of snow cover the conditions would be cold. The association with primroses suggests that given good drainage, *C. pseudibericum* should be an excellent plant for the open garden in Britain. In my own experience, this has proved to be the case. I have now had it growing outside for four years, and in 2001, the first self-sown seedlings appeared. It occupies a west-facing site beneath *Euonymus alatus*, in a sandy humus-rich soil and, to date, the lowest temperatures experienced have fallen to −7°C. Readers in North America might like to note that the climate in my area of England corresponds to USDA Zone 5, which is also shared by most of France, northern Italy, Belgium, The Netherlands, and much of north-central and north-eastern Spain.

The tireless efforts of Prof. Dr Neriman Özhatay during and after our time in Turkey cannot be overemphasised. Dr Murat Ekiçi, our Turkish companion in the field not only was very pleasant company but also unobtrusively smoothed our path. Bob Wallis, an accomplished observer and cultivator of cyclamen supplied invaluable information and advice.

3.5g *CYCLAMEN REPANDUM*

Martyn Denney

Cyclamen repandum is widespread in Corsica; this population is on the north-western side of the Col de Eustache (MD)

***Cyclamen repandum* in Corsica 2003** Team: Richard Bailey and Martyn Denney

Background

Corsica, often justifiably described as the 'L'Ile de Beauté' is the fourth largest island in the Mediterranean after Sicily, Sardinia and Cyprus. It lies 170 km south of the French coast, 83 km west of Italy and just 12 km north of Sardinia. Most of the island is a crystalline massif cut by river valleys and gorges, but there are a small number of isolated limestone outcrops. Although the highest mountain, Monte Cinto, rises to 2,706 m and its peak is covered by snow until June, Corsica is far more than simply a mountain rising from the sea. On the western side, the mountains approach the sea closely and the coastline is very dramatic with rugged cliffs and interesting rock formations. The east coast, however, is flat enough to have a wide agricultural plain backing long sandy beaches. The inland valleys often contain perpetual streams and the hillsides are clothed in dense forest of holm oak (*Quercus ilex*), cork oak (*Quercus suber*), strawberry tree (*Arbutus unedo*), manna ash (*Fraxinus ornus*), sweet chestnut (*Castanea sativa*), or magnificent stands of the Laricio pine (*Pinus nigra* subsp. *laricio*). Corsica is a very green island.

The fieldwork was facilitated by our friend and colleague Max Debussche from Montpellier, who assisted us in obtaining permission to collect plant material and by providing notes on known populations of cyclamen. The purpose of the trip was to extend our knowledge of *C. repandum* by mapping its distribution and by collecting statistics to stand alongside those collected in earlier years and those relating to other taxa in subgenus *Psilanthum*. In addition, plant material was to be collected in the form of herbarium specimens and living tubers that would yield material for the studies taking place at the University of Reading. As a secondary matter, the plants would also provide seed for the Society's seed distribution.

Fieldwork

The starting point, on 13 April 2003, was in San Martino di Lotta, which is on the eastern side of Cap Corse in dense woodland about 400 m above the coast. The woods below San Martino yielded not only *C. repandum* but also considerable populations of *C. hederifolium*. There was a striking uniformity among the great majority of plants, both in flower form and leaf pattern, which was reminiscent of the typical *C. repandum* we are all used to seeing in cultivation. This was generally true throughout the island, though fortunately there were odd plants that had unusual characteristics: often either flowers with no dark nose or leaves with a particularly

A cool, shady habitat for *C. repandum* near San Martino di Lotta (Site 03/01C), 14 April 2003 (MD)

prominent leaf pattern. Significant variation was found at a very small number of sites, but these were certainly special cases. The general lack of variation in *C. repandum* could be attributable to a relatively high level of inbreeding within the species, resulting from the method of pollination and the morphological characteristics of the sexual organs (Affre & Thompson, 1997). It was below San Martino that we saw our first *C. repandum* f. *album* growing in soil collected on a rock outcrop at the edge of woodland. This is by no means a common plant, and most of the sites that we visited had no white-flowered *C. repandum*. Where f. *album* did occur, it was generally only an odd plant; we encountered significant populations at about half a dozen places, but even there, white-flowered forms were never predominant.

The west coast of Cap Corse is rugged and remote with isolated villages often perched high above the sea, but with cliffs that in spring are a riot of yellow with the scented flowers of brooms and gorses. *Cyclamen repandum* grows here too in almost pure sand, in areas at the edge of the maquis where there is some shelter but less competition from more robust plants. It was along this west coast of Cap Corse that we encountered the effects of wild boar for the first time. The pigs trash whole hillsides of woodland, churning up the soil as they root around for anything edible. In some areas, the cyclamen seem to survive this attention, whereas in others, known populations have disappeared.

Near St Florent, there are limestone outcrops that support a flora that is more diverse and far richer than that found on the metamorphic rocks that occur throughout the island. There has been a detailed analysis of the populations of *C. repandum* that grow there and it has been hypothesised that these are a result of hybridisation between *C. repandum* and relict populations of *C. balearicum* that existed on these outcrops in the past (Debussche & Thompson, 2002). These resultant populations include a significant number of plants with white flowers, sometimes with a reduced stature that is somewhat reminiscent of *C. balearicum*, but there are also flowers that have a pink nose and petals that are either white or the palest shade of pink and thus resemble *C. rhodium* subsp. *rhodium*. It has to be said, however, that we encountered isolated *rhodium*-like flowers at a number of other sites on underlying metamorphic rock and, at some of these sites, the proportion of white flowers was also higher than normal.

As we drove south along the north-west coast of Corsica from St Florent to the Fango Valley, we started to encounter significant stands of *Helleborus argutifolius* for the first time. This, the most robust of all hellebores, grew to as much as 1.2–1.5 m where the conditions were particularly suitable. A large number of plants were commonly encountered where *C. repandum* grew but a small number were frequent companions, though it should be stressed that these species did not appear everywhere or in all types of habitat. Along with the hellebore, there was *Erica arborea* (the tree heath), *Viola alba* subsp. *dehnodtii* (with violet- or lavender-coloured flowers, in spite of its name) and *Smilax aspera*, an extremely vicious, very prickly plant that scrambles like a vine for up to 15 m through maquis and scrub, hanging down in an often impenetrable and lethal curtain. *Asphodelus aestivus* also grew at many sites.

The Valle di Fango, for practical purposes, runs from below the north wall of the 1,329 m Bocca di Capronale, west-north-west to the coast just north of the small town of Galeria. There is a tarmac road from Galeria as far as the village of Barghiana at 212 m.

From Barghiana, a forest track is driveable as far as a bridge over the Capronale stream at 509 m. *Cyclamen repandum* grows all along the valley wherever there is suitable habitat, though there are only isolated populations between Pont de Tuvarelli-Chiorna at 95 m and the coast. Above this point, the banks at the side of the road and track were carpeted with cyclamen, growing in places with magnificent plants of *Helleborus argutifolius* and *Erica arborea*. We sampled five populations at 15 m, 70 m, 215 m (in a side valley), 635 m and 725 m. At the bottom of this range, the plants were growing in a soil comprising sand with humus in deep shade of *Pistacia lentiscus* and *Erica arborea* scrub, and were probably coming to the end of their flowering period. At 725 m, the highest point at which cyclamen were seen, the plants were growing in gritty clay in more open scree-like conditions and were only in bud. Throughout the valley, there was very little variation in flower colour and f. *album* was seen only at the 70 m site. Similarly, the leaves were unremarkable, except at the 635 m site where at the woodland edge some well-defined silver hastate patterns were encountered — ironically, in an area that had been trashed by pigs. To Richard's delight, at 595 m by the side of the path, we found a cyclamen growing epiphytically part way up the trunk of a tree.

From Fango, we moved south once again through a particularly remote and rugged area with very narrow roads to Porto, where we were to turn inland once more, through the Gorge de Spelunca and the Forêt d'Aïtone, over the Col de Vergio and down into the central valley to the University town of Corte. Spelunca is both a dramatic and delightfully pretty gorge, where *C. repandum* can be found both in the wooded slopes above the gorge sides and down in the bottom by the river. The D124 road crosses the river at the lower entrance to the gorge near Ota and, from here, a good path follows the river through light *Alnus* and *Fraxinus ornus* woodland. For me, the highlight of the gorge was finding *Osmunda regalis* (the royal fern) growing by the river with its feet in the water.

The Forêt d'Aïtone has magnificent stands of mature *Pinus nigra* subsp. *laricio* clothing the valley sides from about 900 m to 1,400 m. Both *C. repandum* and *C. hederifolium* grow here, and we found both growing together up to about 980 m, though in the case of *C. repandum* not yet in flower. The Col de Vergio at 1,477 m is the highest road pass on Corsica and on 20 April there was still snow by the roadside and *Crocus corsicus*, in both its lilac and white forms, was flowering in profusion.

We drove down the eastern side of the col in torrential rain, pausing at any likely patch of forest to search for the highest population of cyclamen. Eventually, by this time rather damp, we found both *C. repandum* and *C. hederifolium*, quite unremarkable, growing in gritty sand around the base of rock outcrops under *Pinus nigra* subsp. *laricio* at 1,030 m. The drive from there down to the central valley, though through dramatic scenery, held little of botanical interest except for our only encounter with the aroid *Helicodiceros muscivorus* in the Scala di Ste Regina.

We had been asked to investigate a limestone outcrop north of Corte to establish if the cyclamen there exhibited the same characteristics as those on the limestone outcrops near St Florent — they don't! The climb up this limestone hill in the rain was interesting for the view and the clumps of *Pancratium illyricum*, but apart from a few plants of *C. hederifolium* with attractively silvered leaves and some pleasant but ordinary *C. repandum* there was little else. As we reached the bottom of the hill, we encountered the most extensive and dense population of cyclamen of the entire trip. On abandoned terraces under young holm oak (*Quercus ilex*) and manna ash (*Fraxinus ornus*), *C. repandum* was growing as you might imagine the English bluebell carpeting a woodland floor. Generally the plants were unremarkable — it was their density and the extent of the population that were astounding. In one small area, however, we found plants that were the highlight of the entire trip. These plants of *C. repandum* had normal pink flowers but their leaves were bright metallic silver all over but with a narrow dark green edge. Five plants were collected, with the hope that the characteristic could be stabilised and a named cultivar introduced. As of 2011, however, they remain shy of flowering and there has been little progress.

The Solenzara Valley is one of Corsica's most scenic places. The road follows the river inland to the south-west, climbing initially through pasture and then *Pinus nigra* forest, until at Ponte de Fiumicelli it leaves the river, passing high above it, first over the Bocca di Laronu and eventually over the 1,218 m Col de Bavella. Soon after leaving the coast, we stopped at about 70 m by the

A wonderful leaf form amid more typical foliage of *C. repandum*. Limestone outcrop north of Corte (Site 03/21C, Plant number CSE03179C) (MD)

Cyclamen repandum in the Solenzara Valley, 630 m (Site 03/24C) (MD)

roadside to visit another of Max Debussche's sites for *C. repandum*. Above the low cliff on the south-east side of the road, there was a dense scrub of *Erica arborea*, *Quercus ilex*, *Pistacia lentiscus*, and a *Rhamnus* sp. festooned with *Smilax aspera*. *Cyclamen repandum* was growing along and through the edge of the scrub, down the cliff and along a roadside drainage ditch. Among the rather uniform plants, we found and collected a good f. *album* and a plant with pale shell-pink flowers that had no dark nose. Back on the road, from Boca di Laronu the mist hung low over the pine trees. When we stopped at about 630 m on the climb up to the Col de Bavella to carry out another leaf and petal survey of a limited but fine population of *C. repandum* by the roadside, we were in low cloud and the muffled silence was punctuated by a constant splatter as large drops of moisture rolled off the leaves of the sweet chestnut trees. Later, we moved on down to the village of Zonza for the night. After Zonza, surveys were made at a further four sites in the south of the island before moving north along the east coast of Corsica to Castagniccia.

Castagniccia is a remote part of Corsica named after the sweet chestnut (*Castanea sativa*) trees that clothe the sides of the three valleys that make up the area and which once provided a living for the myriad of small villages that can be seen dotted around amongst the trees. In this region, *C. repandum* grows everywhere, especially on the often vertical roadside banks and on almost any piece of woodland not thoroughly trashed by pigs or swamped by bracken. The pigs are a serious problem and effectively destroy much of the local habitat. However, we found two particularly good populations of *C. repandum* growing under *Castanea* and *Corylus* among moss-covered boulders.

View from Piedicroce towards Carcheto, north-eastern Corsica (MD)

Both sites had relatively large groups of f. *album*, some with quite elegant flowers, but the leaves were very uniform. At one of these sites, we found two *C. repandum* with bicoloured flowers — a plum-coloured nose extending half way along the length of the petals with the remainder white — most attractive and unusual plants. Another two plants had white flowers, but with the palest pink or lavender streaking up the edges of the petals and around the mouth of the corolla. Elsewhere in Castagniccia, we found a number of good populations of *C. hederifolium* mixed with *C. repandum* but with only leaves to assess, they were not especially interesting. One particularly robust *C. hederifolium* with very large leaves with a bold hastate pattern caught my eye as a potentially wonderful garden plant and this was collected, also adding to our living DNA bank for later scientific research.

Conclusions

Cyclamen repandum is undoubtedly one of the more common plants on Corsica and grows wherever there is suitable habitat. For the horticulturalist, the following observations about the conditions required might be of interest:

- a lack of overwhelming competition from *Pteridium aquilinum* (bracken), *Asphodelus aestivus*, or grasses;
- a lack of severe grazing or trashing by pigs;
- at least some shade from the sun, either due to aspect or from trees or shrubby plants; plants were encountered in unusually sunny locations but they were inevitably severely stressed and the leaves were flaccid; no cyclamen were found in the very densest shade;
- slope immaterial — from vertical to horizontal;
- aspect from east-north-east through north to west north west;
- altitude 15–1,030 m;
- soil from almost pure sand, through brown loam and gritty clay, to almost pure humus;
- soil pH anything from 4.8 to 7.8.

In spite of the very broad range of habitat in which *C. repandum* grows, it can safely be classified as a woodland plant; it is clearly happiest in cool, moist, but well-drained conditions, with shade from the sun at least during the hottest part of the day.

The members of the team, and the Cyclamen Society, would like to thank the following for their assistance, without which the field study could not have taken place: Max Debussche of the Centre d'Ecologie Fonctionnelle et Evolutive, Centre National de la Recherche Scientifique, Montpellier, France; Bernard Rond, Directeur Régional de l'Environnement de Corse, Ajaccio; and Mlle Laetitia Hugot of the Conservatoire Botanique National Méditerranéen de Porquerolles, Corte.

A pale variant of *C. repandum* but retaining the darker mouth. Castagniccia, Corsica, at Site 03/32C (Plant number CSE03365C) (MD)

A near white *C. repandum* at Castagniccia, Corsica, lacking the dark mouth marking; Site 03/32C (Plant number CSE03366C) (MD)

3.5h *CYCLAMEN MIRABILE*

Brian Mathew & Sema Atay

Teams 2003 and 2004 comprising: Harry Gill, Brian Mathew, Tim Murphy, Trevor Wiltshire (representing the Cyclamen Society), Sema Atay (DHKD: the Turkish Society for the Protection of Nature) and Neriman Özhatay (Herbarium, Istanbul University, ISTE)

During autumn 2003 and again in 2004, the Cyclamen Society undertook extensive field studies of *C. mirabile*. Being one of the more restricted species in Turkey, *C. mirabile* was thought to be a prime candidate for study, although little was required in the way of taxonomic observations as this taxon is morphologically distinct from all other species. The flowers are superficially similar to those of *C. cilicium* but the two are distinct in many other ways. Nevertheless, with almost any species, there is always plenty of scope for mapping exact distributions, filling in the finer details of habitat, noting plant associations and recording the natural variation within populations. In this case, assessments of the species' current conservation and of any perceived threats to its survival status were also thought to be desirable. Some years ago, *C. mirabile* appeared regularly in consignments of wild-collected cyclamen, and was sometimes traded under the name 'purpurascens', so the extent of its distribution and status were of some concern. Apart from the observations and dried specimens gathered (one set of specimens was deposited in the Istanbul Herbarium ISTE), living specimens were collected under licence from the Turkish government for a range of reasons: some were for DNA studies, some to establish new site records, some were representatives of the highest-altitude populations (with potential for hardiness in the UK), whereas others were selected for their interesting foliage, flower shapes, sizes and colours (i.e. for their horticultural potential). These collections would introduce new stock into the cultivated gene pool and hopefully extend the range available to Society members who had provided the finance for this fieldwork. In the two seasons, nearly 60 populations of *C. mirabile* were recorded, giving a good indication of the extent of distribution and the conditions under which *C. mirabile* grows. Apart from the actual (GPS) location, the following features of the habitat were noted: altitude, degree of shade, aspect and angle of slope, soil pH and type, plant associations and (when possible) the cyclamen population size. At sites where there were plenty of flowering plants, one flower was taken from each plant (up to about 50 plants) for more detailed observations: measurements of petals, degree of petal twist, colour and scent. It was intended to record which insects were attracted to the unusual coconut scent of this species but this aspect of the survey was disappointing. During the whole trip, only one bee was observed visiting a cyclamen; nevertheless, it was clear that fertilisation had taken place as capsules were already forming.

Trevor Wiltshire (standing) and Harry Gill, recording a population (Site 04/35T) during the Cyclamen Society's survey of *C. mirabile* (BM)

An award-winning plant of *C. mirabile* (CSE03017T) at a Cyclamen Society show, exhibited by Tim Murphy who was a member of the 2003 team in Turkey. From north-east of Nazilli (MD)

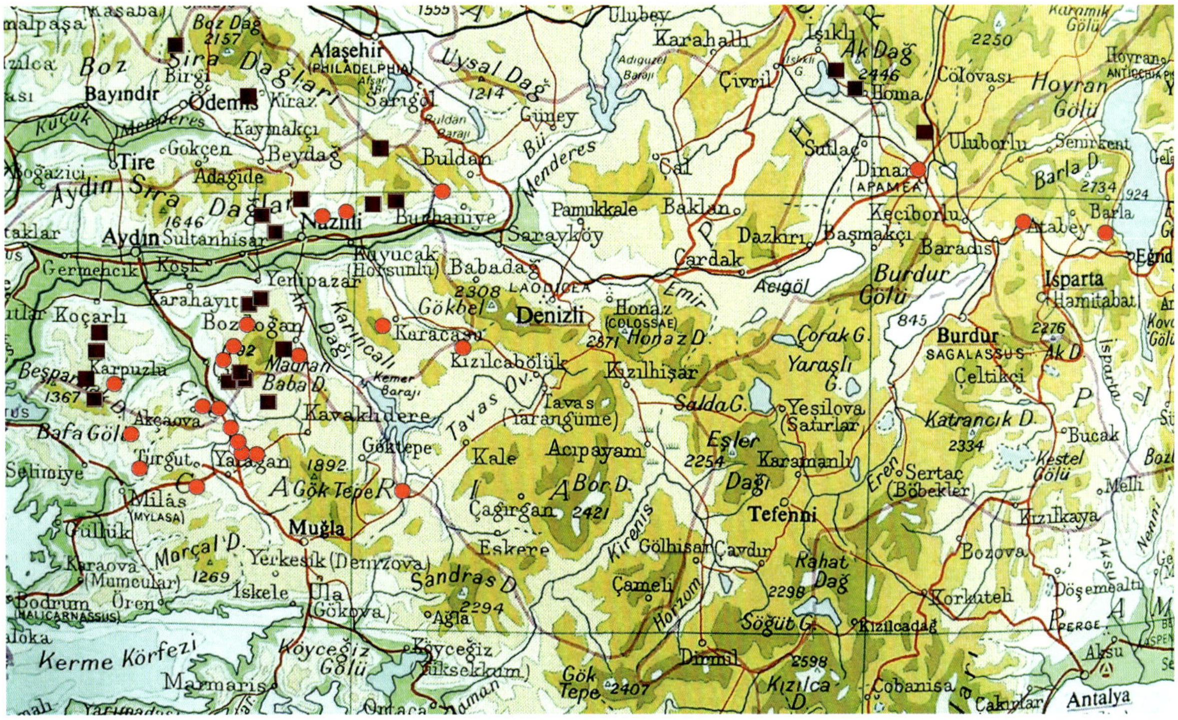

Map showing extent of the recorded distribution of *C. mirabile*. Red dots are from 2003, black squares 2004. © Collins Bartholemew Ltd 1988. Reproduced by kind permission of HarperCollins Publishers

Study area

Using information gleaned from literature and herbarium specimens, a period in late October and early November was chosen for the fieldwork, and this proved to be just about the correct timing. During the two seasons, the extent of the whole region visited in search of *C. mirabile* was approximately an area defined by Aydin and Milas in the west, Muğla in the south, Nazilli in the north and extending east to Denizli, Dinar and Eğırdır. The primary aim was to check on areas outside the known range of the species to establish whether it had a wider distribution than that previously recorded, particularly to the north and east of the known range. There was known to be a population near Lake Eğırdır and it was thought that there might be a break in distribution between Denizli and Dinar, separating the eastern and western populations. Although in 2003 more localities for *C. mirabile* were added in the east and to the north, no populations could be found in this 'Denizli to Dinar gap' in spite of much meandering and diligent searching; so this gap in distribution did seem to be a reality.

To extend the search further to the north, the 2004 survey headed eastwards from Izmir into the Boz Dağ range. Although there was no suggestion that *C. mirabile* occurred in the Boz Dağ massif itself, the idea was to work southwards from there until the first populations were located, thus determining the most northerly extent of the species. In the pinewoods around Gölçük, there were no cyclamen, but heading southwards towards Birgi on a hillside of scattered *Quercus coccifera*, the first plants of *C. mirabile* were located, some 70 km north of any previously known location (Site 04/15T). So ended another theory, that *C. mirabile* did not extend into the mountain ranges north of the Büyük Menderes river valley! Continuing southwards, no further cyclamen were seen

The 2004 fieldwork located populations well to the north of previously known sites. Site 04/15T near Birgi, Izmir Province (BM)

A frequent associate of *C. mirabile* is the spiky-leaved *Quercus coccifera* which provides shade, humus and protection from grazing animals (BM)

until Bademli, east of Tire, where the team found *C. hederifolium*, nicely fragrant and with prominent pink pyjama stripes on a nearly white ground colour. These were in very dry country, growing in the boles of the evergreen oaks, whereas at another site, a little further on, they were under large plane trees in rich leaf mould by a small stream, although a local inhabitant said that there had been no rain all summer. Further south towards Atça, *C. mirabile* appeared again (Site 04/18T) on a very dry hillside, used mainly for growing figs. It does not appear to be represented west of this in the valleys to the north of Inçirlova, where *C. hederifolium* is common in a range of habitats that are very different from those preferred by *C. mirabile*.

The next region for inspection was to the south-west of Aydin, in the mountainous region of the Batı Mentese Dağlari (range); *C. mirabile* had been recorded the previous year (Site 03/18T) near Karpuzlu, so further searching north of this would provide useful information, whether registering the absence or the presence of the species. In fact, the team were excited to record the presence of *C. mirabile* in this region, thereby extending its known range in this north-westerly direction; six separate populations were observed over a limited altitude range of 502–665 m, all in a silty-sandy soil derived from mica-schist rock and always in association with *Quercus coccifera*. Other bonuses for the team were the striking *Crocus nerimaniae*, a recently described autumnal species with blue flowers and black stamens, named after our colleague Neriman Özhatay, and a cemetery full of *Sternbergia lutea*. Further work, carried out slightly to the east in the Doğu Mentese Dağlari (south-east of Aydin and south-west of Nazilli), on the south side of the Büyük Menderes river, added more populations of both *C. hederifolium* and *C. mirabile*, although never together. The flowers of *C. hederifolium* were nicely fragrant and conspicuously bicoloured with a pink-purple striped nose on a pale ground. On the north side of the great river valley, just 10 km north of Nazilli, a fine population of *C. mirabile* was located in a little-cultivated olive grove at the very low altitude (for this species) of 165 m. This was a flourishing colony with as many as eight cyclamen plants per square metre in the densest part. Nevertheless, it would take only a change in agricultural practice to more intensive cultivation methods, such as those often used in olive groves, to wipe out or severely damage such populations. Another site (04/36T) was found rather higher up at 375 m further to the north, so this was a useful link with the first new northerly location at Birgi. This too was an old olive grove, overgrown with various annuals, herbs

Prof. Dr Neriman Özhatay preparing a herbarium specimen of the *Crocus* named after her: *C. nerimaniae* (TW)

One of the northern-most localities for *C. mirabile*, north of the Büyük Menderes river valley (Site 04/36T) (BM)

and grasses with a few deciduous and evergreen oaks. Here, the cyclamen were flowering very profusely in a small but healthy population, only about 30 × 10 m in overall size. Further forays into the area north-east of Nazilli proved interesting and several populations were seen, the most northerly in this direction being near the village of Uluderbent (marked as Derbent on some maps) at Site 04/39T. *Cyclamen mirabile* was flourishing at this site on a hillside of coppiced oaks and this population was of particular interest in that it encompassed more variation in leaf and flower than any other population seen during this and the previous season; at the densest part, there were about 15 cyclamen per square metre. Some interesting forms were noted at this site. All of the populations seen to date were on similar mica schist formations which gave rise to sandy, gritty or silty soil, depending upon the degree of decomposition.

A form of *C. mirabile* with long, very twisted petals at Site 04/39T, between Karacaali and Uluderbent (plant number CSE04210T) (BM)

Travelling eastwards from this rich 'western' region for *C. mirabile*, there is the relatively uninteresting stretch of country (i.e. in cyclamen terms) corresponding to the above-mentioned 'Denizli-Dinar gap'. To the east of Denizli is the Acigöl salt lake and harsh, dry steppe country with scattered junipers, not looking at all like 'Cyclamen territory'. North of the Denizli-Dinar road, tortuous routes were followed through the hills and mountains but the team was not rewarded by finding cyclamen, although some habitats looked not too uninviting. It was worth noting that in this area, a different *Cistus* was commonly seen (*C. laurifolius*), one we have not recorded growing with *C. mirabile*, and in places there were very fine forests of *Pinus nigra*. The outcrops of this region were of limestone; although it is tempting to say that *C. mirabile* does not usually grow on limestone, the more easterly populations usually seem to be confined to soils derived from such rock. This perceived gap in distribution corresponds approximately with a rainfall pattern, which indicates that much of the range of *C. mirabile* falls within the 600–1,000 mm annual rainfall belt. To the east of Denizli and north-east towards Dinar, the rainfall drops significantly to just 400–500 mm but then increases again into the 600+ mm level towards Lake Eğirdir, where *C. mirabile* is again represented.

With the knowledge that *C. mirabile* occurs at Dinar (where it was observed in 2003) and eastwards to Lake Eğirdir, the team decided to search once more to the north-east of Denizli but this time further on, in the mountains to the north-west of Dinar. Cyclamen have been recorded (in literature) on one of these mountains, although the plants were identified as *C. coum*; this

The very dry area in the region of the salt lake Acıgöl is unsuitable for cyclamen, hence there is apparently a natural break in the distribution of *C. mirabile* (BM)

is possible of course, but equally, this location could be another site for *C. alpinum*. The team was therefore surprised to find *C. mirabile* growing on an overgrazed crumbling limestone hillside between Çivril and Dinar. Apart from the ubiquitous *Quercus coccifera*, there were a few other species here typical of limestone country: *Euphorbia rigida* and the awful spiny bush *Paliurus spina-christi*, which is so often seen along the Mediterranean coast. A passing villager reckoned there were spring cyclamen growing in the mountains, underlining the literature report of *C. coum*. The team's find of another site for *C. mirabile*, 16 km on towards Dinar, suggests that there are others scattered around in this extensive area of mountains that would take years to survey properly. A short probe from Dinar in the direction of Afyon proved that the species is there as well, growing at a higher altitude of 1,145 m in an open, almost alpine-looking habitat with *Acantholimon* and *Astragalus* cushions and a low scrub of *Berberis* and *Crataegus*. Large limestone outcrops gave shelter for the cyclamen, which had already finished flowering at this higher, cooler level. Leaving this eastern area and working back westwards (west of Denizli and north to north-east of Muğla), it was an interesting contrast to be back in country with rounded mica-schist outcrops and the associated gritty-sandy soils. Sometimes the landscapes were enhanced by the stone pine, *Pinus pinea*; the cyclamen here are relatively safe from cultivation as the trees are preserved for their supply of pine kernels, although there is a tendency to cultivate beneath the trees.

Observations

Habitat

Most *C. mirabile* populations were growing in soil overlying mica schist formations; in fact, this often appeared on the surface as rounded outcrops. Occasionally, they were on a type of conglomerate and, in the case of the eastern-most (Eğirdır) populations, on limestone. On the other hand, *C. alpinum*, which occupies a roughly similar region of south-western Turkey, is usually to be found on limestone formations. This probably explains why the distributions of these two species meet but apparently do not actually merge. To date, as far as is known, no mixed populations have been found. A considerable spread of

One of the higher altitude sites for *C. mirabile* in the extreme north-east of its known distribution area (BM)

altitude range was noted for *C. mirabile*, from 161 m to 1,146 m, the higher-altitude populations tending to be towards the eastern end of the distribution area. The pH range recorded was a surprisingly high 7.2–8.4, with little difference between the surface layers of 'organic' soil and the subsoil. It is always interesting to record the associated species and a habitat profile, using a combination of rock type and companion plants, can be built quickly, so that is often possible to spot likely and unlikely sites from a distance. The evergreen *Quercus coccifera*, and sometimes also deciduous oaks, was a frequent associate at (nearly) all sites. Other commonly seen species were *Cistus* spp., *Pistacia*, *Pinus*, *Crataegus*, *Styrax* and *Olea* (wild and cultivated). These are mostly species indicative of the Mediterranean climatic zone, but at the sites of some of the eastern, higher-altitude populations, there was a marked tendency for some of the central Anatolian steppe plants to be in evidence. Nevertheless, when considering the three phytogeographical regions identified in the *Flora of Turkey*, *C. mirabile* is undoubtedly more of a 'Mediterranean species' than an 'Irano-Turanian' or 'Euro-Siberian' one.

In the west of its range *C. mirabile* usually occurs on mica schist rock which weathers to rounded outcrops and a rather gritty-sandy soil (BM)

A broad-petalled form of *C. mirabile* at Site 04/33T south-west of Nazilli. The soil is a gritty sand from crumbling mica schist (TW)

Typical scenery at Site 04/33T where *C. mirabile* was in abundance (BM)

Variations

There was a lot of variation among the *C. mirabile* plants seen in the course of the fieldwork, but not as much as one hoped. It is probably true to say that there is more variation in cyclamen species in cultivation than in their wild populations. This is perhaps not surprising. In cultivation, we raise large numbers of plants from seed and select those that we regard as being improvements: those with darker or albino flowers, beautiful or unusual leaf patterns and so on. There is no such aesthetic selection process in the wild, the only criteria being related to the plant's survival; characters that are selected here might affect crucial matters such as hardiness, suitability to the habitat and attraction of pollinating insects, rather than the strong flower colours and leaf patterns that the human eye finds appealing. Furthermore, in the wild, only a small percentage of the offspring reaches maturity, so the chances of unusual variants cropping up are much reduced.

A deep pink form of *C. mirabile* at Site 04/33T in Aydin Province, south-west of Nazilli (BM)

To record flower colour, the RHS Colour Chart (3rd Edition, 1995) was used and this showed the range to be in the Purple

Group 76A–C, and predominantly 76B in most populations. The range was similar at most sites. Darker pink forms were quite uncommon but very pale ones were plentiful. In fact, some flowers were often so pale as to appear nearly white (particularly in bright sunshine) and it was often necessary to bend down to check the 'nose' of the flower to see whether it had a purple blotch or was truly a white form. Certainly some of these flowers could be described as white with a dark nose, but only one undeniably white-flowering plant, a complete albino with no trace of colour, was found, in 2003 on the west side of Lake Eğirdir. The variation in the overall appearance of the flowers was greater than expected, brought about by the variable length/width ratio of the petals. The actual sizes of the measured flowers (a random selection from most of the sites) were as follows: petal length, whole range 8–25 mm long but most frequently 16–20 mm long; petal width, whole range 2–12 mm wide but most frequently 5–8 mm wide.

Wide petals resulted in a much more substantial-looking flower; for example, one specimen had petals only 15 mm long but 10 mm wide (whereas the length:width ratio was most commonly between 2:1 and 3:1). Some of the longer-petalled flowers were of a shape that was reminiscent of the flowers of *C. repandum*.

The toothing or fimbriation at the tips of the petals, regarded as a characteristic of *C. mirabile*, was sometimes very slight and occasionally nearly absent. In fact, the basic shape of the petal tips varied considerably from rounded to acuminate, and not infrequently, the petal tips were bifid (i.e. notched).

Flower scent was disappointing: the distinctive coconut aroma is such a distinctive feature of *C. mirabile* in cultivation, but in the wild it was seldom very noticeable and often apparently absent altogether. This seemed to have nothing to do with time of day or temperature: the weather during the fieldwork ranged from cool mornings to hot sunny days, both wet and dry.

There was quite a lot of variation in freedom of flowering, although this is probably linked mainly to the habitat and not a genetic trait. The most floriferous specimens were found where the plants were receiving extra light, not under dense vegetation such as stands of evergreen oaks. The most profusely flowered cyclamen (with as many as 50–60 flowers per plant) were mainly in open areas, usually the result of scrub clearance, and on roadside banks where competition was less. It is possible, however, that this might be a temporary phenomenon and that plants in full exposure would not survive in the long term. Cemeteries tended to be particularly favoured habitat, and one of these provided a remarkable sight. There were thousands of plants, each well-flowered and at their peak: the scene even more dramatic because it was a well-tended graveyard where most other plants had been weeded out.

The leaf variation was not remarkable, the most common form being a central dark green 'Christmas-tree' pattern of variable size surrounded by a light green or silvery zone. Furthermore, the degree of variation was similar in most populations. Leaves that

A pure white form of *C. mirabile*, one of a large population in *Quercus coccifera* scrub. West of Lake Eğirdir at Site 03/33T (TW)

Elderesi cemetery, Aydin Province, 4 November 2003 (BM)

Some strikingly marked leaves at Site 03/27T near Kızılgüney, Aydin Province (TW)

Pink-tinged leaf forms were not uncommon in most populations (BM)

had a pink tinge on the upper surface were seen at most locations, scattered through the population; at a few sites, there were some good carmine-tinged leaves. Leaf shape, on the other hand, varied a great deal from approximately orbicular (*coum* shape) to ovate, the latter sometimes with a pointed apex. Usually, the length and width of a leaf were approximately equal, regardless of whether the leaf was orbicular or ovate (in the latter case, measured across the broadest, lower part of the leaf); occasionally, the leaves were wider than long, so as to appear more reniform (kidney-shaped) in outline. If a leaf was longer than wide, the difference was quite small, usually only about 2–4 mm but could rarely be as much as 6 mm. The entire size range of the leaves measured was 11–50 mm long by 7–47 mm wide, but these figures are most probably not very meaningful as, in many cases, the leaves might not have been fully developed. The leaf margin was fairly consistently and conspicuously toothed in the manner we expect of *C. mirabile*, but in one curious example, the margin had inward-facing notches instead of outward-facing teeth.

Note the toothed leaves, a characteristic of *C. mirabile* — perhaps a survival adaptation to resemble the prickly Kermes oak leaves (in foreground) beneath which it often finds shelter? (TW)

North of Milas, the lake formed by the Geyik dam; *C. mirabile* populations have probably been submerged but it occurs just above the water line and higher up the slopes (TW)

Conservation

Overall, it can be said that *C. mirabile* is not in immediate danger as a species. Agricultural activity involving the clearance of scrub-covered country has, however, reduced some populations to very small fragmented 'subpopulations', which seem unlikely to survive in the long term. At some locations, the natural vegetation has been destroyed in order to plant, for example, pines, cedars and *Robinia*. The conifer plantations will probably become too dense to allow the survival of cyclamen and other vegetation beneath. New dam projects, for example that near Çine, will undoubtedly submerge some cyclamen populations but water is a precious commodity in the dryer parts of inland Turkey; hopefully the altitude range of *C. mirabile* is such that many populations will escape. Commercial collecting is not allowed for this species, and in any case, the practice of cyclamen collecting is very much on the decline in favour of propagation. On the positive side, the populations of *C. mirabile* in deciduous oak woodlands are probably fairly safe from development projects, as the oaks are maintained and coppiced as a source of wood, likewise populations that grow in the vicinity of the valuable *Pinus pinea*. The observations made over the course of the fieldwork suggest that grazing, which is often so destructive, can actually be beneficial to *C. mirabile*, maintaining open patches in the ubiquitous *Quercus coccifera* scrub; if left unchecked, this scrub would ultimately become far too dense for most of the small herbs beneath.

Thanks to the work of the Cyclamen Society team and our Turkish colleagues, there is now a much clearer picture of the distribution, habitat and variation of this beautiful and distinct endemic cyclamen.

3.5i *CYCLAMEN CILICIUM*

Martyn Denney & Chris Clennett

Teams 2008 and 2009 comprising: David Bent, Chris Clennett, Martyn Denney (representing the Cyclamen Society) and Evrim Tabur (DHKD, the Turkish Society for the Protection of Nature).

Cyclamen cilicium near the village of Aladag in the eastern Taurus mountains, 927 m on 2 November (MD)

In cultivation, there is little variation in *Cyclamen cilicium* f. *cilicium* (the pink-flowered plant) and it seems less robust than *C. cilicium* f. *album*. Could it be that the reason for the lack of variation in f. *cilicium* is that the villages in Turkey where commercial collection occurs are clustered together in the western part of the plant's range? In 2008 and 2009, the Cyclamen Society carried out field studies to try to answer this question and to map the distribution of the species. In both years, the studies took place during the first two weeks of November; this proved to be the ideal time, with plants in flower and most plants also in leaf. In 2008, the study started in the extreme east of the distribution of *C. cilicium* near the village of Ala Dağ to the north of the city of Adana and proceeded westwards to just west of the town of Silifke, the ancient city of Seleucia. In 2009, the team started in the extreme west of the range on Barla Dağ on the western shore of Lake Eğırdır and worked their way south-eastwards to Silifke.

In the east, *C. cilicium* only grows on the southern side of the main ridge of the Taurus Mountains, whereas to the west of Silifke, it grows on both the northern and southern sides of the mountains. Near the village of Ala Dağ on the southern slopes of the Taurus, at its most easterly station, *C. cilicium* grows near populations of *C. pseudibericum* surveyed by the 2002 field study. It is always exciting to find the first population of a species that one hasn't seen before in the wild, and it is a relief to find that you have 'hit' peak flowering time as this enables the morphological part of the work (measuring leaves and flowers) to be carried out properly. This was an unremarkable population, growing in a small limestone river valley with *Crocus cancellatus*. The leaves were quite plain and the population exhibited very little variation in either leaf or flower. We continued north into the mountains and then west along a forest track that follows the contours through *Pinus halepensis* forest towards the Cilician Gates pass. Over a distance of 50 kilometres or so, we encountered just a few isolated, sparse populations of *C. cilicium*. The surface of the track provided evidence of the fact that the soil here is sandy. The cyclamen were growing in deep pine needle litter, often around small limestone outcrops.

The next day, we planned to explore the north slopes of Ala Dağ to establish whether cyclamen grow there. We explored the rocky slopes around Çamardi, but this is clearly not cyclamen country, even in the valleys where there is tree cover, and we are confident that *C. cilicium* does not extend from the Cilician Gates eastwards along the northern side of the Taurus Mountains.

The Cilician Gates Pass (Gülek Boğazı) just south of the town of Pozanti is a well-documented location for *C. cilicium*, but it is better known as being the site of a Roman victory over the

Site 08/06T at 1,388 m near the Cilician Gates Pass, a well-known location for *C. cilicium*. *Pinus halepensis* provides a pine needle top-dressing to the sandy/clay soil (MD)

View from site 08/09T near Gülek looking northwest to the main ridge of the Taurus, west of the Cilician Gates Pass (MD)

A great diversity of leaf markings at Site 08/10T on the way to Çamliyayla (MD)

Parthians in 39 BC. On a small road just to the east of the pass and in the Elmadı valley, which climbs northwards into the mountains to the west of it, we started to encounter populations that had variations in leaf patterns with well silvered markings. The Elmadı Valley culminates under the peak of Medetsiz Tepe in an amphitheatre used for grazing sheep and goats. Here, we found *C. cilicium* growing in limestone crevices at 1,670 m with just a few plants still in flower.

The area to the west of the Cilician Gates road and north of the cities of Mersin and Tarsus is especially rich in *C. cilicium*. Here, the Lower and Middle Miocene limestone is at the surface and the plants grow wherever there is a suitable habitat. One of the most dense and interesting populations grows in and around a spectacular isolated Cretaceous limestone outlier. The habitat here is perfect for *C. cilicium*, with plants able to grow in rock crevices and in *Quercus coccifera*, *Pinus halepensis* and broad-leaf litter in damp and shaded gaps between the rock outcrops amongst a rich diversity of other herbaceous plants and sub-shrubs. Plants in more open positions benefit from shade from the *Pinus halepensis*. Surprisingly, this site (08/10T) is at an altitude of only 450 m. Although the leaf markings at this site show a good diversity, the flower colour is relatively uniform.

Moving gradually south-westwards, we started encountering some variations in flower colour: first a plant with almost white flowers and a red nose near the village of Gözne, and then two plants with superb deep-coloured flowers south of Findikpinari.

Erna Frank once wrote in the *Bulletin of the Alpine Garden Society* (Vol. 54: 51–53 (1986)) that she and Ronald "… made various forays into the Toros along the very minor roads leading off the coast road, and established that *C. cilicium* grows along most of them". We didn't find this to be the case. Many of the roads heading north from the coast do pass through populations of *C. cilicium* but this is only the case where the geology is correct. This really is a key factor. The plants only grow in limestone areas, that is, either where there are limestone outcrops or where the underlying rock is limestone, and this is reflected in the soil, which

A thriving population of *C. cilicium* near Tepeköy, on 7 November (MD)

A gorge north of Erdemli (Site 08/21T). A knowledge of the geological preferences of the species is useful when hunting for populations: *C. cilicium* grows in the shady crevices of the limestone rocks (MD)

A pleasing form of *C. cilicium* at Site 08/21T, combining good flower colour and leaf markings (Plant number CSE08230T) (MD)

often contains limestone pieces or fragments. Although the Taurus are essentially limestone mountains, the limestones there are heavily folded and thrusted because of the Alpine Orogeny, which is a result of the collision of the European and African tectonic plates. The central Taurus are composed of Permian limestones, which are metamorphic, commonly known as marble. Mesozoic limestones and dolomites occur in layers that are over 1,000 m thick. These are surrounded by various rocks from Cambrian to recent age. The limestone is often surrounded by an ophiolitic melange from the Late Cretaceous, which forms an impermeable base or cover. In the Taurus between Mersin and Erdemli, there are both fingers of Cretaceous serpentine rock at the surface that come south from the main ridge of the mountains and areas of Lower Miocene conglomerates and marls (Ternek, 1957); *C. cilicium* does not grow in these areas. Similarly, the Mut basin behind Silifke is composed of Late Oligocene–Early Miocene lagoonal, fluvial and marine sedimentary deposits, and *C. cilicium* doesn't grow in this area either. Here especially, the change in geology is evident in the vegetation of the area.

At Site 08/29T, between Aydincik and Gülnar, an interesting long-petalled form of *C. cilicium* (Plant number CSE08349T) (MD)

West of the Mut Basin and the Göksu River the Devonian, Permo-carboniferous and Upper Cretaceous limestones take over, and *C. cilicium* is again present where the habitat is suitable. In this area, there is sufficient altitude to provide slightly cooler and moister conditions. As we moved west of Silifke for the first time, we found a number of plants with more elegant flowers, either a narrow flower with especially long petals (over 29 mm) or flowers whose petals had a twist of up to 270°, but there tended to be less variation in flower colour. In general, the leaves were unremarkable.

In the eastern Taurus, *C. cilicium* is found in *Pinus halepensis* forest and also very often grows in association with *Quercus coccifera*. Grazing by sheep and goats is especially heavy in the Taurus Mountains and this has a detrimental effect on much of the flora. In many places, *C. cilicium* only survives where it grows within or around *Q. coccifera* bushes, where it obtains some protection from the holly-like leaves.

The November 2009 field study began with a survey of the extreme western population of *C. cilicium* on Barla Dağ, then worked eastwards to join up with the 2008 research area. In this western part of the range, *C. cilicium* was known to grow on the northern side of the mountains, and one of the team's principal objectives was to assess both the frequency and extent of the species in this region. This area included all the places where *C. cilicium* is collected commercially and the only known site of *C. cilicium* f. *album*.

In the western part of the range of *C. cilicium*, the geology is again key to its distribution. Barla Dağ on the western shore of Lake Eğırdır marks the north-western extremity of a band of Mesozoic series strata that includes limestones and extends in a line south-eastwards. To the northwest of Barla Dağ, the surface geology is fragmented and comprises a Quaternary river valley followed by Eocene rocks, Andesites, Spilites and Porphyrites. It seems likely that these rocks form a natural barrier that limits further distribution of *C. cilicium*. On Barla Dağ, *C. cilicium* is unremarkable in every way. The leaves have the usual silvery hastate pattern (but with no particular definition) and the flowers are an average pink (76A–76D on the RHS colour chart). The area is heavily grazed and even the *Quercus coccifera* bushes, around and under which the cyclamen grow, look as if they had been clipped with a pair of shears.

We followed the line of the limestone mountains, known as the Dedegöl Dağlari, in the direction of Akseki, far away to the southeast. We continued towards the village of Çamlik where commercial collecting still takes place. While the team was surveying a site with a lot of plants, a passing villager told us that collection has been much reduced in recent years and that only one family was now involved because the money was so poor. The plants here are very similar to the mass of *C. cilicium* that has been in cultivation for many years, with a uniform flower colour and attractive but unremarkable leaves. All through this area to the northwest of Akseki, the cyclamen grow beneath deciduous oak trees (*Quercus infectoria* and *Q. pubescens*).

Just to the north of Akseki on the old road over the Irmesan Pass, we found *C. cilicium* growing in warm sunshine on a southeast-facing hillside under *Cedrus libani* and *Abies cilicica*. The plants here had some darker-coloured flowers but were otherwise unremarkable. The road from Akseki to Ibradi and beyond is botanically interesting as it has a lot of bulbs, and though these are mainly spring-flowering, we found many *Crocus pallasii* and *Biarum pyrami* amongst the populations of cyclamen. In limestone rubble, some of the cyclamen had leaves with silver- or cream-coloured markings covering much of their surfaces.

Quercus coccifera often provides some protection for *C. cilicium* from grazing animals. This white form is at Site 08/30T on the Aydincik to Gülnar road (Plant number CSE08354T) (MD)

Between Akseki and Murtiçi, the team visited the only known site of *C. cilicium* f. *album*. The plants were discovered in October 1982 by Ronald and Erna Frank and Manfred Koenen, and described by them as forma *album* the following year in the Alpine Garden Society's *Quarterly Bulletin* (Vol. 51: 150). The plants grow under *Juniperus excelsa*, *J. oxycedrus* and *Quercus coccifera* in a mixed population with *C. cilicium* f. *cilicium*. At the time of our visit, there were just 15 white-flowered plants, but putting our own data with that from elsewhere, we can say that this small population seems to be stable but not expanding. Each plant was photographed and measured.

The area southeast of Akseki towards Gündoğmuş has many populations of *C. cilicium* growing amongst limestone outcrops in the shade of *Juniperus drupacea*, *J. excelsa* and *Pinus halepensis*, often in deep, remarkably dry layers of pine needle litter. To the northeast of Alanya in the Dim Valley, a new reservoir was under construction. The new road was only partly constructed and took us high above the valley before rejoining the old road as the valley narrowed. Here, we found evidence of the previous day's rain with boulders strewn across the road and cracks where there was subsidence. On a large rock, we found in flower *Galanthus elwesii* var. *monostictus* surrounded by a dense population of *C. cilicium* that was very wet and beaten down by the rain. This road continues over the Taurus to Göktepe, Sarivaliler and Ermenek to the north, crossing two high passes. *Cyclamen cilicium* grows all along the road for a few kilometres on the northern slopes of the lower (more southern) of the two passes. Further on, above Göktepe, we found the highest population we were to encounter. The habitat appeared at first unlikely, being a hot southeast-facing slope with little shade. Here, at an altitude of 1,898 m, *C. cilicium* grows with *Crocus cancellatus* amongst grass and sparse low scrub around limestone boulders. As is often the case in such extreme conditions, the plants were unremarkable.

Inland from Gazipaşa, there is a network of small roads that follow the river valleys. We encountered cyclamen in small pockets, but in many places the geology was not suitable for cyclamen. What limestone did exist appeared to be crystalline in nature. In one place in the side of a dry stream gully growing under *Quercus coccifera*, we found a few *C. cilicium* plants with purple markings on the upper surface of their leaves. This site also had large numbers of *Biarum pyrami* and a population of *Allium callimischon*

A wide-petalled white form of *C. cilicium*, at Site 09/18T north of Anamur (MD)

A good pink form of *C. cilicium* at Site 09/22T on the Anamur–Ermenek road at 1,617 m (MD)

The road north from Anamur passes over the Taurus to Ermenek in the Göksu Valley and it was here that we transferred to the northern side of the Taurus. On the watershed, the road passes through several kilometres of dramatic karst landscape. *Cyclamen cilicium* grows here in odd pockets of soil and needle litter from *Juniperus drupacea*, *J. excelsa*, *J. communis*, *Abies cilicica* and in some places *Cedrus libani*. Just over the northern side of the pass, where the limestone forms outcrops rather than bare rock, the populations of cyclamen are larger. In this area, the villagers harvest *Cedrus libani* cones, possibly for an indigenous tree propagation programme.

Following the northeast-facing slopes from Kazanci towards the villages of Sarivadi, Pinarönü and Göktepe, we found significant populations of *C. cilicium* that included some of the most interesting leaf forms that we encountered, both in the extent of the 'silver' in the leaf patterns and in the presence of purple markings on the upper leaf surface. Travelling north to Hadim, after passing through an area where the geology was unsuitable, we found a small population of cyclamen just south of the town. This was the first of a dozen populations found in an arc to the north-west, which ended back at Akseki. To the north north-east of Hadim, there is a fine population under *Quercus infectoria* and *Q. pubescens*. The flowers there were unexceptional, but the leaves included some fine patterns with excellent silvering. The distribution of *C. cilicium* on the northern side of the main ridge of the Taurus between Ermenek and the Alacabel Pass was widespread where the geology was correct and some of the populations were extensive.

At the end of the field study, we visited the village of Hocaköy. This is one of the few remaining villages where *C. cilicium* is collected commercially for export under a quota approved by the Turkish Ministry of Agriculture. The Muhtar, Mehmet Ali Yilmaz, welcomed us into his home and over çay, he and his wife explained about the collections. Until a few years ago, the village collected between 1 and 2 tonnes of tubers each season for export to The Netherlands, but recently the amount collected had declined to as little as 150 kg due both to fewer requests and the low price paid to the village: they received just 0.60 Turkish Lira (approximately £0.30) per kilo of tubers in 2009. The plants are collected in May, often being

Limestone slope at Site 09/21T on the Anamur to Ermenek road, c. 80 km south of Ermenek. Open pine and fir woods providing a deep layer of needle debris for the cylamen (MD)

Large populations were found on the north side of the Taurus watershed. Here at Site 09/26T *C. cilicium* is flowering well in the bare ground after a fire in the cemetery (MD)

At the higher altitudes *C. cilicium* is often found in association with conifers. Here at Site 09/40T between Seydişehir and Akseki, *Pinus nigra*, *Juniperus excelsa* and *Abies cilicica* (MD)

Tubers of *C. cilicium* are often relatively easy to dig up for commercial purposes. The practice of collecting them in millions has now greatly diminished (MD)

found by collectors seeing the ripening seed pods. Only tubers with a diameter of 4 cm or more are collected. All remaining top growth and roots are removed and all soil is cleaned off before they are weighed. The Muhtar's wife had last collected cyclamen in May 2009, and then only 35 kg.

Surveys and statistics

A total of 85 *C. cilicium* sites have been recorded by the Society, 33 in November 2008, 44 in November 2009 and the remainder during other field studies. At 29 of these sites (13 in 2008, 16 in 2009) measurements were taken of representative leaves and flowers.

Habitat

Cyclamen cilicium grows from longitude E 30° 45' 05" in the west to longitude E 35° 23' 53.5" in the east. It typically grows at altitudes of 400–1,700 m but was recorded as low as 174 m and as high as 1,898 m. The mean altitude, in bands of 100 m, was 1,000–1,099 m, with 15% of sites recorded falling into this band.

Sites that were open to all aspects were recorded, but 40% of recorded sites were facing the southern quartile. The plants grow on slopes of from 0–75% with 55% of sites having a slope of 25–50% (22.5°–45°). Although *C. cilicium* tolerates completely open places with no shade and places with total shade, 47% of the sites recorded were in 50% shade. From studying sites in the spring, it is clear that shade provides a longer growing season that allows seed to mature before the heat of summer dries out the soil.

At 79% of sites, the soil was some form of clay (generally a brown sandy clay), 11% of soils were loam based. The clays were fine mineral soils, generally with limestone fragments and sometimes with small fibrous organic matter. Soil pH varied from 6.4–8.5, with 50% of sites having pH between 7.5 and 7.9 and 92% between 7 and 8.4.

Companion plants

At 12% of sites, all in the extreme east, *C. cilicium* was found growing in predominantly *Pinus halepensis* woodland. A wide range of companion plants were recorded, but the following were the most widespread:

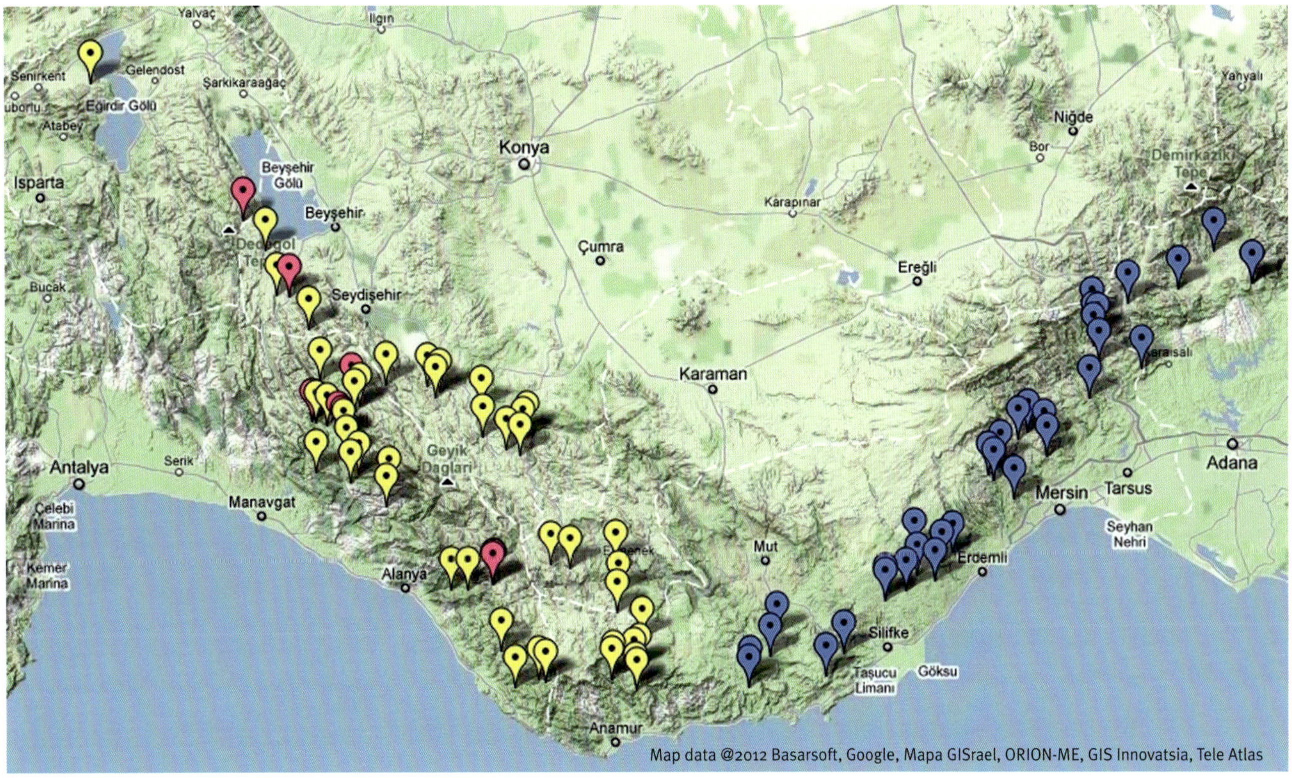

Map of c-s Turkey showing distribution of *C. cilicium*: 2008 (blue), 2009 (yellow), other records (pink) (MD)

Companion plant	% of sites
Quercus coccifera	62%
Styrax officianalis	46%
Pistacia terebinthus	40%
Poterium verrucosum	36%
Pinus halepensis	31%
Juniperus excelsa	30%
Quercus pubescens	28%
Crocus cancellatus	28%
Salvia officinalis	25%
Juniperus communis	23%
Juniperus oxycedrus	20%
Quercus infectoria	18%
Cistus creticus	17%
Daphne sericea	17%
Ceterach officinarum	17%
Juniperis drupacea	12%

Morphological characteristics

At a number of sites spread across the range, where there was a sufficient number of plants in both leaf and flower, we took measurements of representative pairs of flowers and leaves and recorded data that describe leaf shapes and patterns.

Leaves: the maximum leaf lengths recorded at each of the sites varied from 28.5 to 75.8 mm; minimum leaf lengths recorded varied from 14.1 to 33.3 mm. The maximum leaf widths recorded at each of the sites varied from 25.0 to 70.3 mm; minimum leaf widths recorded varied from 10.7 to 29.9 mm. We made a point of selecting mature leaves as far as possible but there remains the possibility that immaturity may be a factor in the measurements of minimum dimensions.

Our leaf shape and pattern standards each represent a series of five variations between two extremes. It is therefore unsurprising that the greatest number of leaves fall into the middle variation.

Records of flower size, shape and colour were made at each of the *C. cilicium* survey sites. This one at Site 09/24T, with long, elegant petals is being visited by a hoverfly (MD)

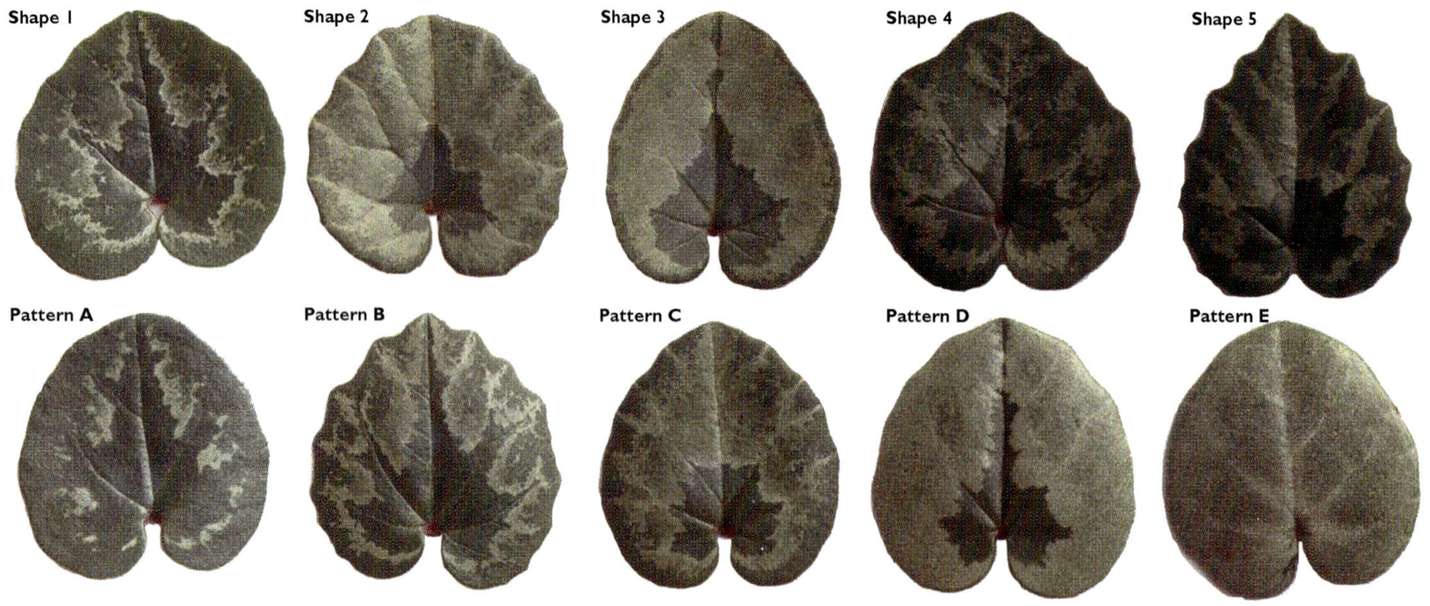

Template for assessing leaf shape and markings of *C. cilicium* (MD)

Leaf shapes		**Leaf patterns**	
Shape 1	3.7%	Pattern A	4.1%
Shape 2	20.4%	Pattern B	41.3%
Shape 3	57.7%	Pattern C	54.1%
Shape 4	6.1%	Pattern D	0.5%
Shape 5	12.1%	Pattern E	0%

The full range of leaf sizes, shapes and patterns was found across the entire distribution of the species, with no graduation either by latitude or longitude. Note: leaf Pattern E was encountered, but not at a surveyed site.

Selected specimens were brought into cultivation under licence from the Turkish authorities. Plant number CSE08196T from Site 08/18T near Tepeköy (MD)

Flowers: the maximum petal lengths recorded varied from 18.1 to 29.2 mm; minimum petal lengths recorded varied from 11.9 to 19.1 mm. The maximum petal widths varied from 6.9 to 11.3 mm; minimum petal widths recorded varied from 3.9 to 7.4 mm. As with the leaves, immature flowers could influence the minimum petal sizes. All flowers had twisted petals and the vast majority had a 90° twist, though most sites surveyed had a few flowers with 180° and two sites had at least one flower with a 270° twist. Just 1.6% of flowers examined had fimbriated petals. 54.9% of flowers were scented.

Flower colour is measured using the Royal Horticultural Society colour chart. Each sheet is numbered and sub-divided further into four sections designated A (palest) to D (darkest). The recorded colours of *C. cilicium* f. *cilicium* petals were as follows:

74A–74D (Red-purple group)	2.8%
75A–75D (Purple group)	14.5%
76A–6D (Purple group)	78.1%
78A–78D (Purple group)	4.6%

There was no geographical trend for flower characteristics generally, but there was a correlation between longitude and flower colour. All flowers matching RHS sheet 74 and almost all matching sheet 75 were in the eastern part of the distribution, whereas those flowers matching RHS sheet 78 were in the western part of the distribution.

The flowers of *C. cilicium* f. *album* had petals varying in length from 14.6 to 22.6 mm and in width from 5.8 to 8.2 mm. All petals had a 90° twist, none were fimbriated; 50% of the plants had scented flowers.

3.5j *CYCLAMEN LIBANOTICUM*

Cyclamen libanoticum is, for cyclamen enthusiasts world-wide, one of the less well-known species, seldom visited in its native land by 'outsiders', unlike the many Greek and Turkish species. We are indebted to colleagues in Lebanon for writing a first-hand account of the species based upon their studies of it in the wild. Sadly, Richard Haber died unexpectedly during the preparation of the following account but his wife Dr Myrna Semaan has completed their interesting study of this beautiful plant. Both being botanists and conservationists, they have provided much information and their paper contains a wealth of detail about the natural habitat and conservation status of the species.

One of the most attractive of all cyclamen species, *C. libanoticum* is also one of the rarest (MS)

Cyclamen libanoticum, a rare and beautiful endemic of Lebanon

Ricardus M. Haber & Myrna Semaan

In a valley of Phoenician myth, the Valley of the Ibrahim River in Lebanon, the forces of evolution have conjured up a unique cyclamen. In that valley, the deity Astarte is said to have mourned her deceased lover, Adonis, killed by the wild boar that he was fighting. She wept on his grave where red flowers grow every year celebrating their eternal love. A botanist would be tempted to associate Astarte's tears with a botanical oddity particular to the valley, thus discerning that the flowers they breathe life into should be coloured pink signed with a purple marking. In 1898, Hildebrand became aware of these flowers and identified them with Lebanon, naming them *Cyclamen libanoticum*.

The Valley of Ibrahim River is carved into the western slopes of the Mount Lebanon Range that runs parallel to the Mediterranean Sea. It starts from sea level, where it opens out into the Mediterranean and rises to 1,300 m, following a serpentine course curving along a series of mountains. At its upper limit, high mountains crown the valley and carry the snow-cap that enriches the river system. The base of the valley is mostly narrow, occupied by the ever-flowing Ibrahim River. Wherever it grows wider, the riverbanks broaden, accumulating sand and soil invaded by the water-loving poplar and plane trees, creating some particular habitats. The hard limestone of the mountain slopes is configured into cliffs, steep slopes, smooth slopes, immense rock escarpments and artistic rock faces. Maritime wind sweeps through the valley bathing the slopes of surrounding mountains with warmth and humidity most of the year. This combination of physical and biological elements has endowed the valley with a rich and highly diverse natural ecosystem. As for human habitation, many villages are nestled within the valley or spread along its crest.

Cyclamen libanoticum is one of three cyclamen species that exist in Lebanon. *Cyclamen persicum* and *C. coum* survive in and around the Valley of the Ibrahim River: *C. coum* inhabits high altitudes on the mountains crowning the valley top and *C. persicum* thrives from the mid altitudes down to the mouth of the valley. Although *C. libanoticum* also occupies the middle mountain zone, it shows an allopatric distribution with *C. persicum* enforced by habitat segregation. In this zone, *C. libanoticum* survives in very distinctive habitats and it is distributed in dispersed populations thriving solely on the southern flank of the valley. Some populations occur close to the riverside, others grow high on the mountain slopes. In addition, *C. libanoticum* populations extend into a subsidiary valley on the same flank of the Valley of the Ibrahim River, surviving under similar conditions.

To see the graceful *C. libanoticum* on the wooded slopes, we have to crawl or slide down under an entangled canopy criss-crossed by the viciously thorny rough smilax (*Smilax aspera* L.). The slopes are of a northern to north-western exposure, and thus are less exposed to the scorching sun, which allows them to maintain moisture for longer periods throughout the year. A thick layer of

The endemic Lebanese *C. libanoticum* occurs either in wooded valleys near the River Ibrahim or higher up on limestone slopes, although still in shade (MS)

leaf litter and mossy rocks signal the proximity of the cyclamen and the anticipation grows. In a stable microhabitat, where litter has accumulated and degraded over a sufficiently long time, the cyclamen grow near the base of deciduous trees between their woody roots and exposed rocks. They also grow in rock outcrops and between boulders overgrown by roots of trees, where pockets of soil are preserved and covered with a mat of fresh green moss. Searching for the cyclamen, their small flowers flash at you in well-shaded areas, where few rays of light infiltrate the net of naked branches in the canopy. Flowering in March, the 'Cyclamen of Lebanon' blooms before the canopy of deciduous trees starts to put on its leaves. Some trees might carry their blossom concomitantly with the cyclamen. One expects the inside of these woods to turn quite dim after the trees sprout their foliage. Once the first cyclamen is discovered, it points the way to the rest of the population. The well-dispersed plants are readily distinguished by their green leaves on the brown carpet of fallen leaves. A variety of deciduous trees contribute to the layer of leaf litter and they are easily identified from the colour and shape of their dry leaves. The arboreal flora includes the deciduous gall oak (*Quercus infectoria*), hop hornbeam (*Ostrya carpinifolia*), terebinth (*Pistacia terebinthus palaestina*), Judas tree (*Cercis siliquastrum*), hawthorn (*Crataegus monogyna*), and the evergreens Kermes oak (*Quercus coccifera*) and sweet bay (*Laurus nobilis*).

Digging by hand through the layer of leaf litter is the best way to reach the underground parts without breaking off the leaves or the flowers of the cyclamen. The top layer of intact dry leaves hides layers of broken leaves; the deeper we dig, the more recognisable becomes the effect of decomposition reaching down to layers of black wet debris suffused with white mycelia. Underneath is a layer of dark soil rich in organic matter. Beneath this leaf mould, the cyclamen tubers lie almost on the surface of the soil or barely buried, anchored by their fine roots. Just a little touch is sometimes sufficient to uproot the tuber and set it loose almost clean of any soil but with some composting debris deposited on top. Cyclamen that live in the mossy areas have their tubers situated on the surface of the soil completely covered by the wet moss.

The habitat of *C. libanoticum* in dense undergrowth in leafmould-rich soil, often covered with moss (MS)

In the populations that exist at the level of the river course, the cyclamen are noticeably found in very shaded areas, such as at the foot of wet rock faces hidden behind tall deciduous trees of oak and plane that completely block direct sunlight. The soil is black, damp and covered with decomposed matter, but the layer of leaf litter is absent in these areas because of the high humidity and rapidity of decomposition.

December to March constitute the wet season when most of the precipitation falls in the habitat of *C. libanoticum*. The flowering of the cyclamen occurs near the end of the wet season, when the microclimate is heavily laden with moisture and the soil saturated with rainwater.

The wild populations of *C. libanoticum* comprise hundreds of specimens, which display marked variability particularly in their leaves and flowers. A tuber might carry one leaf and one flower to three leaves and five flowers. The leaves vary in their shape, size, margin, pattern and colour. They can be ovate, little longer than wide, to broadly ovate, wider toward the apex. The margin can be entire to slightly angular with faint white dots marking vein endings, or strongly angular with very discrete white endings, sometimes even toothed especially at the leaf base. The leaf pattern consists basically of a central deep green blotch that is heart-shaped in the lower half of the leaf and runs narrowly around the midrib in the upper half, mostly not reaching the margin. A contrasting zone borders the heart-shaped blotch and extends into a marginal zone, which is densely maculated with tiny patches of the deeper green. The contrasting zone varies in colour from bright green to yellowish green or with different grades of a whitish tint resulting in a range of silvery sheen. The heart-shaped blotch can also be variegated at its middle with or without the extension around the midrib. When combined, pattern and colour variation produce, at one extreme, leaves of deep and bright green with a fully expressed pattern, and at the other extreme, whitish-silvery leaves with only the outline of the deep-green heart and light well-spaced marginal maculation. In addition, there are different intensities of purple colour on the reverse of leaves.

In spite of a limited distribution, *C. libanoticum* displays considerable variation in leaf pattern and petal shape in the wild (MS)

Variation in the flower encompasses size, shape and colour. Petals vary in width: some look particularly slender when they are elliptic at the base and lanceolate at the apex; wholly elliptical petals generally look broader. A gradation of dark to faint pink is observed in the populations, but a white collar is always present at the base of the petals. The basal mark can be coloured vivid purple to violet, sharply edged but never reaching the margins of petals and rarely reduced to a central pattern.

The regeneration potential of *C. libanoticum* in the wild is promising. In corners of the habitat, tiny single leaves emerging from tiny tubers are found in good numbers in the vicinity of mother plants.

The 'Cyclamen of Lebanon' does not enjoy any form of conservation. Fortunately, its distribution areas are currently distant from the centres of intensive degradation forces. The major disturbance that endangers its habitat is logging for firewood and charcoal production. Every number of years, the woods where the plants grow are thinned out or more intensively cut down. Urban sprawl, development and land reclamation, which constitute extensive threats to the flora of Lebanon, are not rampant in the Valley of the Ibrahim River, but even so, a development project has distorted the habitat of one of the populations and reduced some of its distribution zone.

CHAPTER FOUR

Cyclamen in cultivation

4.1 GROWING CYCLAMEN: CULTIVATION AND PROPAGATION

Alisdair Aird & Peter Moore

Cyclamen are primarily Mediterranean plants. In their natural habitats, almost all are dormant during the warm dry summer, and grow during the wetter autumn to spring months. Some species experience mild or largely frost-free winters. Others, growing in colder-winter areas, generally have the protection of snow cover against any severe frost, and come into growth once the snows are past. Most cyclamen are plants of woodland or grow among sheltering shrubs or rocks, and they shun waterlogged soil. These natural growing conditions give obvious clues as to how best to cultivate cyclamen.

In Britain and other western European countries, summer is normally wetter and cooler than it is in the Mediterranean region. In winter, the weather tends to be much more changeable than around the Mediterranean — periods of mild, grey, wet weather alternate with colder clear spells that have bright days and frosty nights. Nevertheless, the more adaptable cyclamen species grow really well in British gardens as long as the gardener has borne in mind the drier and warmer summer weather and the steadier less "stop-start" winters that they would experience in their natural environments.

A greenhouse or bulb frame provides much more control over growing conditions, allowing gardeners to grow cyclamen species that would be less happy outside in the garden. Almost all cyclamen species take well to pot cultivation, and luckily the species that seem least-suited to long-term pot cultivation under glass are the best species for growing outside in British and other western European gardens.

Cyclamen are ideal for an alpine house and are the one of the mainstays of an autumn display, here seen at the Royal Horticultural Society's garden, Wisley, in October (BM)

The most adaptable of species, *C. hederifolium*, is seen here thriving in rough grass in a churchyard (BM)

Cyclamen in the garden

In *Cyclamen*, as in many other plant genera, it is the species with the widest natural distributions — *C. hederifolium* flowering in autumn and *C. coum* flowering in late winter — that are the easiest to grow in the garden. Just as their greater adaptability has allowed them to thrive in a more varied range of natural environments, so it helps them cope with a wider variety of garden situations and circumstances. Nevertheless, the more care you take to give them conditions that suit them well, the better they will flourish. Providing conditions that are ideal for these two species will also put you well on the road to having an environment that also suits other cyclamen species.

The first requirement is good drainage. A light, free-draining soil, ideally containing plenty of organic matter such as leaf-mould, is perfect for cyclamen. On heavier soils, look for or prepare a sloping site, or improve the soil by digging in plenty of coarse grit. Even then, if water does not drain away quickly when you test it by sprinkling on a good can-full of water (say 10 litres), the soil is still too heavy. The best bet then is to build up a simple raised bed above the normal surface. This can be quite informal and of any size, anything upwards from just a small, low mound such as a tipped-out light barrow-load. Start with a layer of drainage material: builder's rubble is ideal, but any coarse material such as broken crocks or stones will do. For the top planting layer, which should be at least 7 cm and preferably 10 cm deep, a good mix is equal parts of garden soil, coarse grit and some sort of humus, such as composted pine bark. The ideal soil for cyclamen is slightly on the alkaline side of neutral, but in practice, pH is unimportant as most cyclamen will tolerate a wide range of soils with different pHs from peaty to very chalky.

In summary, except for their requirement for good drainage, cyclamen are not at all fussy about soil. They will be happy in any normal growing medium.

A south-facing slope is a good location for growing cyclamen, partly because it will be warmer and drier during the summer dormant period, and partly because it will have better light in the dark days of the winter growing season. Nearby trees or shrubs are helpful in several ways. Most important, their roots help to mop up excess soil moisture — cyclamen's worst enemy. They give shelter in winter from the cold sharp winds that desiccate cyclamen leaves. Even bare deciduous branches and twigs overhead give useful protection against radiation frosts. The fallen leaves can provide a helpful mulch and, in time, work down into the soil to enrich it with humus. But remember that cyclamen are small ground-hugging plants. A blanket of big leaves such as those of chestnut or large-leafed rhododendron can almost suffocate them, so gardeners must be scrupulous about sweeping up such leaves each autumn; this task is considerably easier when growing the late-growing *C. coum* than when growing *C. hederifolium*, which would already be well into growth by leaf-fall. A mulch of leaves around the size of those of oak, beech, birch or apple

On heavy clay soils raised beds are a simple but effective method of improving the drainage, essential for successful cultivation. *C. cilicium* in a low, raised bed over London clay (BM)

is fine. Conifer needles are also particularly good, and as they break down they favour the development of beneficial micro-organisms that seem to aid the establishment of self-sown cyclamen seedlings. There is as yet no scientific evidence that this is the case, but many growers have reported that cyclamen seedlings do proliferate remarkably successfully in pine duff.

One of the best spots for cyclamen is at the base of a Leyland's cypress or other hungry evergreen hedge. Other particularly good places include rock gardens (they look particularly attractive against a rock background), good-sized gaps in informal paving, paths or drive edges, the surrounds of bird baths or other stonework features, thin fine grass under lightly foliaged trees, grassy or stony banks, narrow dry 'problem beds' by walls, fences or house walls, and rose gardens (giving interest and colour at a time of year when the rose bushes themselves are just bare clumps of thorny sticks).

In a rock garden or similar spot, a top dressing of coarse grit or even of small pebbles can show off the cyclamen leaves well, and do something towards deterring snails and slugs. Elsewhere, an annual mulch (applied during the dormant season) of well-rotted garden compost, leaf-mould, fine composted pine bark or other organic material is appreciated but by no means essential. Don't mulch with coarser wood or bark chippings as these take up so much nitrogen as they slowly rot down that cyclamen are starved.

Cyclamen hederifolium well placed at the foot of a large pine tree in a partly shaded but well-drained situation (BM)

Visiting a nursery at flowering time enables one to select the healthiest and most desirable plants. A fine batch of silver-leaved *C. coum* at Tile Barn Nursery near Benenden, Kent (BM)

The best cyclamen to plant out are pot-grown as dried tubers can be hard to coax back into life. Buying plants from a nursery or garden centre will allow you to chose a specific flower form or leaf pattern, and obviously such plants will be in growth at the time of purchase. Planting out such growing plants is a fiddly job as the leaf and flower stems, which are flimsy and thread-like between the tuber and soil surface, can flop about all over the place. You need a lot of patience to reposition them just as they were in the pot, so that the plant goes on growing without a check. Consequently, it is much easier to buy and plant out during the dormant season, when you don't have these recalcitrant stems to cope with. And if you buy from a good source, the plants will have been grown from seed from parent stock plants that are carefully selected for their attractive leaves and flowers.

The actual process of planting is simple. For a few plants, make a hole for each tuber, using your trowel to make sure that the surrounding soil is nicely crumbly and easily broken up. The holes should be around 10 cm apart for *C. coum* (and most other species) and around 15 cm for *C. hederifolium*. They should be arranged irregularly and not in a definite pattern so that the planting looks natural. The planting holes should be wide enough for you to spread the tuber's roots out around it, and deep enough to allow for just 2 cm or so of soil to cover the top of the tuber. This thin cover lets the surface of the tuber warm up during the summer dormant period, which seems to help flower bud initiation. If, however, you expect more than just a few degrees of frost, or if your garden is prone to scratching blackbirds, cats or squirrels, you can play for safety by planting substantially more deeply, up to 15 cm or so. When doing so, you must make sure that drainage is good even at this depth — the tubers will rot if they are sitting in a wet and sticky sump.

If you have a lot of cyclamen to plant, it will be quicker and easier to dig out the entire surface of the planted area, to a depth of 2.5 or 3 cm; then, after forking over the soil, scatter the tubers over it (check each one to make sure that it has landed the right way up), then replace the soil around and on top of them.

It is not necessary to add any fertiliser but some people like to fork in a light dressing of Growmore or slow-release fertiliser before planting, others like to scatter a little bulb fertiliser at the end of summer just before growth starts. If, however, you want the plants to look as compact as they do in the wild, don't fertilise.

Don't mix cyclamen of different species in the same area. cyclamen plantings look better and more natural when they are of a single species. Also, you will find that the stronger-growing or larger species quickly swamp the smaller ones in a mixed patch. The further apart you keep each species, the better so that when ants carry off the seed (the main way that cyclamen spread naturally in gardens), they are then less likely to drop it among plants of a different species.

If you want to grow cyclamen in grass, you have the problem associated with planting snowdrops or daffodils in lawns: you cannot mow until the cyclamen die back in early summer and have to stop mowing before they emerge again. If you are happy with that regime (or patient enough to work round each plant on your hands and knees, cutting the grass with a pair of scissors), the two best ways of establishing a colony in grass are mirror images of each other. One way is to plant the cyclamen into bare soil, and then sow a rye-free mix of fine fescues and bents around them. The other method is to scatter cyclamen seed into existing grass. ("Empty" cyclamen seed pods that you have already harvested the seed from can also be used as there always seem to be one or two seeds hidden away in crannies that ants will sow for you.) *Cyclamen hederifolium* is robust enough to show quite well in even rough grass; in more managed grass, *C. coum*, *C. purpurascens* and *C. repandum* can look lovely. In a lawn setting, *C. purpurascens* does need regular detailed grooming around each plant because, unlike other cyclamen species, it grows and flowers in summer while the grass is growing. You must be a devoted fan to cope with growing *C. purpurascens* in grass.

Good companion plants for cyclamen include small bulbs such as crocuses, snowdrops, sternbergias and dwarf narcissi. Thymes are useful in that they cover the ground after the cyclamen die back in summer but do not swamp the cyclamen. Also, bear in mind that in the wild many cyclamen are found growing in scrub, so they are perfectly happy among small shrubs and sub-shrubs. *Cyclamen coum* is very much at home growing among primroses, as it does naturally in the Black Sea coastal mountains. Open-growing dwarf conifers go well with cyclamen, and rock gardens give good scope for positioning the smaller species in prominent spots that do them justice. On a larger scale, cyclamen look good growing around mature trees, conifers or deciduous,

Cyclamen repandum in grass at the National Trust's garden at Killerton, Devon (BM)

***Cyclamen coum* in a woodland situation with snowdrops and other winter or early spring 'bulbs' (BM)**

Spring bulbs growing together in a delightful setting at the National Trust's garden at Knightshayes, Devon (BM)

A natural plant association: *C. coum* with *Primula vulgaris* near Sumela on the Pontic mountain slopes of the Black Sea. *Colchicum* leaves also herald an autumn display (MD)

and don't suffer from the root competition. Leaf fall from large-leaved trees such as chestnuts or sycamores can, however, swamp cyclamen plants.

As noted above, *C. coum* and *C. hederifolium* are the mainstays for garden planting. *Cyclamen hederifolium* is the more robust of this pair, and normally lives to a great age. *Cyclamen coum* tends to live for only a few years, but this is no matter in the garden as both species tend to seed themselves around, so a patch tends to be self-perpetuating. Both species also withstand even very hard frost (with the ground frozen solid to well below their roots), though at temperatures below about −10°C (14°F) without snow cover, their leaves and flower buds might be burned off.

Cyclamen repandum, *C. purpurascens* and its close relative *C. colchicum* are just as cold-hardy, but need protection from cold east and north-east winds. In British conditions, many people find three other species — the dainty *C. cilicium* and *C. intaminatum*, and the striking bolder *C. pseudibericum* — easier to grow. In sheltered spots in favoured gardens, *C. alpinum*, *C. creticum*, *C. cyprium*, *C. libanoticum*, *C. mirabile* and a high-altitude form of *C. persicum* that has recently come into cultivation can also be good. *Cyclamen graecum* does survive out-of-doors in warm places in southern Britain, but it very rarely flowers and does not look happy.

Cyclamen under glass

Under glass, the most difficult aspect of the cyclamen's natural environment to mimic is the steady and predictable progression of the seasons, with their relatively constant temperatures from day to day but with the slow but very marked change from hot dry summer to cool wet winter and then back to hot dry summer again. A big greenhouse with well-controlled ventilation and shading (and heating for the tender species) can provide even pots standing on benches with a satisfactorily Mediterranean-style regime. In a small greenhouse, however, temperatures tend to fluctuate so wildly that, for a near-comprehensive collection of the different species, cyclamen pots are best plunged quite deeply in sand, so as to stabilise conditions as much as possible.

If grown under glass in the shadow of buildings or trees, cyclamen tend to look etiolated and "leggy", having over-long leaf stems. Also, the marbling patterns that make the leaves of so many forms particularly attractive tend to be duller under these conditions. Furthermore, because such a greenhouse would not have the benefit of much solar warmth, it would need to have closed ventilators more often, which would increase disease risks. Finally, most cyclamen are less likely to flower and set seed well in an overshadowed greenhouse. So the best position for a greenhouse for cyclamen (as for most plants) is an open one, well away from overshadowing trees or buildings.

Pots are best plunged in sand to provide more stable moisture and temperature levels (BM)

This brings us immediately to the opposite problem: too much sun. Even cyclamen species such as *C. graecum*, which in the wild seem to revel in full exposure to the hot southern sun, can quickly show distress under glass if unshaded from even the relatively weak sun of temperate climates. It is in fact not the sunlight itself that damages them, but the heat it generates in their leaves. In the open, this heat dissipates quickly, as even on still days there is plenty of air movement around the leaf surfaces as they warm up. In a greenhouse, though, unless the ventilation is exceptionally good with powerful fan assistance, the air around the leaves is both too still and too warm to give any relief.

So, while it is important to give the plants as much light as possible on dull days at any time of year, and even on brighter days in the dimmest depth of winter, it is essential to shield them from overheating by over-much sun in autumn and spring. In warm weather, they will appreciate as much ventilation as you can give them — it's no exaggeration to say that if you are buying or building a greenhouse for cyclamen, the best thing for them is to install as much ventilation as is structurally possible. In an existing greenhouse, it's unlikely that ventilation alone will keep the temperature down to an acceptable level — say, around 27°C (80°F) — on a very sunny day in April or May, so you will also need shading.

The simplest and cheapest form of "adjustable" shading is a spray-on wash such as Coolglass, which dries to white but which becomes more transparent on damp days, allowing more light in then. At the opposite end of the scale are complex structures made of alloy slats, a bit like venetian blinds, which can be angled either to let maximum light through or to shut out some or all of the sun's rays. Fitted externally (which is best, to keep any absorbed heat outside the greenhouse), these have to be mounted well above the ventilator panels so that the opening of these panels is unimpeded. On commercial glasshouses, the shading is operated automatically, by climate-control computers. Even an amateur manual installation is expensive, although it is possible to use an 'old-fashioned' version of this system consisting of wooden slats linked together by screw-in eyelets, rolled up as a blind, and attached to the greenhouse ridge, which is raised and lowered down the roof by a pulley. This system has the disadvantage that unless it is fixed on rails well above the ventilators, it can only be fitted between them.

Flexible shade netting is an option that falls somewhere between the extremes of spray-on and slat shading. Such netting allows around 50% of the light through and can be draped over a greenhouse, with flaps cut in it to let the ventilation panels open. It has to be tied down to wires stretched along the sides of the greenhouse and to the ridge peaks at each end.

Some cyclamen growers set up specialised frames for their plants. An appropriate frame has a removable glass or hard plastic roof above a sand plunge bed. The plunge bed, held in a timber or masonry box-frame, is deep enough to plunge the cyclamen pots nearly to their rims. The roof is normally fixed up to 35 cm or so above the surface of the plunge bed, leaving all four sides open for air circulation. In bad weather, it can be lowered on to the box-frame, effectively closing it. In fine weather during the growing season, it can be removed altogether. During the dormant season, the box-frame helps the plunge bed to warm up and keep nearly dry, giving the plants a Mediterranean summer. Ingenuity can add various refinements, such as adjustable side screens that can be used as appropriate in variable weather.

Keen cyclamen growers will argue endlessly about whether terracotta or plastic pots provide better growing conditions, quite apart from any aesthetic considerations. Terracotta has the advantage that it makes the cyclamen's great enemy — a waterlogged compost — less likely. But it does have disadvantages: it's more expensive, less easy to keep clean and pathogen-free, and more prone to drying out. For you, the best type of pot for cyclamen is the pot that you are used to growing other plants in.

As with pots themselves, different growers prefer different potting mixes. The common factor is that the recipe — say, one part of coarse grit to one part of soil (or soil equivalent) and one part of a humus-giving material — must provide excellent drainage. Successful potting compost permutations are endless; for example,

Clay versus plastic: a matter of choice but growing conditions must be adapted accordingly (BM)

Perlite or vermiculite can be added or used instead of the grit element; John Innes No 2 can be used as a soil base or a soil-less substitute; and peat, fine composted bark or screened leafmould can be used as the humus element. So long as the drainage is good, use a mix that is similar to whatever you find suits the way you grow other plants.

Cyclamen look most natural when fertiliser levels are kept low; they don't need additional feeding.

In pots, most cyclamen are happy to have the tuber placed on the surface of the compost, with the compost coming half-way up the sides of the tuber. Then they should be covered with a layer of very coarse grit. A pale-coloured grit can make an attractive contrast with their dark leaves. The main exception is *C. repandum* and its allies, which should be planted a little more deeply, with the tuber completely covered by at least its own depth of compost.

Propagation

Cyclamen seed should ideally be sown at the end of summer, in a 15 cm pot. Soak the seeds in water for 24 hours beforehand. Nearly fill the pot with a compost similar to that which you chose for adult plants, and space-sow the seeds on top of the compost. A pot of 15 cm in diameter will comfortably take about 25 to 30 seeds. Then put a layer of coarse grit on top, to a depth of about a finger's thickness (say, up to 1.5 cm). Water well initially, then make sure you don't let the compost dry out completely. An ideal temperature to aim at for germination is 15°C (60°F), and darkness helps. If you don't have suitable space under the staging of a greenhouse, the seed pots will do perfectly well out of doors, perhaps against a north wall (in the Northern Hemisphere), with some overhead protection from the worst weather (and protection from marauding slugs). You should expect to see the first seed leaf in a few weeks' time.

The baby plants will do best shaded from sunlight, and it's well worth trying to keep them growing on into their first summer for as long as possible without their going dormant, by keeping the pot cool and just barely moist. The plants should stay in this pot until they go dormant at the end of their second year of growth, by which time the little tubers should be big enough to pot up singly in 7.5 cm pots.

Seedlings at the end of their second growing season, ready for potting singly or planting out. These were probably sown too thickly but have done well anyway! (BM)

A very well-grown batch of seedlings of *C. purpurascens* in the collection of Jan Bravenboer, Holland (JB)

Bringing cyclamen indoors

Any pot-grown cyclamen species can be brought into the house so that it can be enjoyed while in flower. A north-facing window sill is ideal, and the cooler the room the better; avoid south-facing rooms and nearby radiators. Although they like fresh air, cyclamen dislike draughts. The air in most houses is drier than ideal for cyclamen, so standing the pot on a small tray of damp Hortag or similar moisture-retentive material is helpful. If you have an open porch, putting the cyclamen out there overnight, in a cardboard box open at the top, will refresh it.

Care of individual species

C. hederifolium and its relatives

Cyclamen hederifolium is the real tough guy of the family, and the best species for outdoor planting in temperate climates. It copes well with lighting that ranges from full sun to shade, and with competition from grass. If you want to try other species outdoors, keep *C. hederifolium* separate from them; when planted together with other cyclamen species, it will swamp them. *Cyclamen hederifolium* can cope with dampish ground better than *C. coum*, the other main outdoor species, and will also survive in even drier conditions.

A good way to get a striking effect from a naturalistic planting is simply to scatter seed roughly in the area you want. Ants might harvest much of it, but the seeds they drop on the way to their nests seem to end up in more fruitful planting positions than the most carefully sited human sowings. If you have been gathering your own cyclamen seed, even scattering the pods you think of as empty will probably produce a surprising number of plants. We know of one showy bank-side colony produced solely from pods scattered there by its owner after he had "emptied" them.

The white-flowered form seems to be more vigorous than the pink-flowered forms; in some mixed white-and-pink plantings, the white form can gradually gain ground at the expense of the pinks, but some individual pink plants can be just as strong.

Spring bulbs such as crocuses interplant well with *C. hederifolium*; for a mixed planting, remember that over the years this species' tubers can grow to 25 cm (1 ft) in diameter.

Both *C. confusum* and the closely related *C. africanum* are individually more striking than typical *C. hederifolium*, being bolder, more eye-catching plants. They also flower at least two or three weeks later than *C. hederifolium* (attempts to isolate particular early- or late-flowering strains of *C. hederifolium* itself have to date failed). Although these two species seem to be at least fairly hardy, in the sense that they can withstand a few degrees of frost, they have not been in general cultivation for long enough to have seen a really hard winter in the UK. So their long-term suitability for growing outside in temperate climates remains to be determined; in the wild, their preference for sheltered spots at relatively low

The ivy-leaved cyclamen, *C. hederifolium*, is robust enough to hold its own in a ground cover of ivy (*Hedera helix*) making a delightful combination (BM)

Here in the drier conditions of a scree bed beneath a Daphne bush *C. hederifolium* is equally successful and seeding freely (BM)

A splendid drift of pink and white *C. hederifolium*. It is interesting to see how the balance of pink and white changes over the years (SB)

One of the more tender species, *C. africanum* is an admirable plant when pot-grown in frost-free conditions (SB)

Some species have leaf stalks that spread out underground and when pot-grown can result in a bare centre, as in this specimen of *C. mirabile*. Using a smaller pot would help to keep the foliage together (MD)

Another specimen of *C. mirabile* with a smaller pot to plant ratio (MD)

altitudes, usually not that far from the sea, makes it unlikely that they could ever rival typical *C. hederifolium* for sheer robustness and versatility.

Most people who grow *C. confusum* or *C. africanum* keep them in pots, usually under glass. Their upright habit of growth, with leaf and flower stems rising straight up from the tuber, makes for a neater display in a pot than is typically provided by *C. hederifolium*. But a really large plant of *C. hederifolium* can look magnificent in a big pot. Don't be tempted to make that pot over-large: a pot that just fits a large specimen, almost underpotting it, seems to encourage the foliage to fill to the pot's centre, whereas over-generous pots let the foliage run out towards the rim, leaving the centre less well furnished. This is worth remembering for other species too as several, for example *C. mirabile*, behave in the same way.

As we have said, *C. hederifolium* is the most robust cyclamen species. Its one weakness is that, of all cyclamen, it seems the first to be attacked by aphids. Such attacks are, however, at worst unsightly rather than deadly.

C. coum and similar species

Cyclamen coum can withstand more cold than *C. hederifolium*; one of the nicest sights in a bleak February is to see its cheerful pink flowers thrusting up from a light covering of snow. In open ground, even in full sun, it will also do well in dry summer conditions such as those resulting from competition from the roots of a nearby conifer hedge. Nevertheless, the roots of *C. coum* are more fragile than those of *C. hederifolium*, so it does appreciate at least some occasional moisture in summer. On the other hand, don't keep it too damp; *C. coum* damps off more readily than *C. hederifolium*. The named cultivars of *C. coum* all seem to be just as vigorous as more run-of-the-mill plants, needing no special cosseting. Indeed, *C. coum* subsp. *caucasicum*, from Russia, is not merely bigger than

A welcome sight in early spring, *C. coum* is a tough species! (JB)

"ordinary" *C. coum* but seems to be even more vigorous. The one exception to this general rule is *C. elegans*, which even expert growers regard as difficult; the plants in cultivation derive from only a limited pool of genetic material, and it might be that this stock simply chances to lack vigour.

As the leaves of *C. coum* are nowhere near as big as those of *C. hederifolium*, the named cultivars and other special forms of *C. coum* do deserve individual placing in the open garden, where they will be noticed individually. But in a mass planting of *C. coum*, silver-leafed forms show up well among their plainer-leafed siblings.

The very versatile *C. coum* is equally happy outdoors or under cover, making a very satisfactory plant for the show bench (MD)

The very special white form, *C. coum* f. *albissimum*, is probably best grown under cover where its purity is untouched by adverse weather (MD)

In pots, *C. coum* benefits from fairly frequent re-potting, at least every second year.

Cyclamen alpinum flowers a little later than *C. coum* but can otherwise be treated very much the same; it is a choice plant for growing in a pot.

Although hardy enough to grow outside in many gardens, *C. alpinum* also makes a superb alpine house subject (MD)

C. parviflorum

The natural habitat of this species indicates what a challenge it is to grow well: heavily drained alpine pasture deeply watered by melting snow, then continually drenched by mist and low cloud. Unless your own weather matches this, there's no point in your trying to grow *C. parviflorum* in the open. Furthermore, as almost all other cyclamen are more attractive, this species will probably appeal only to those trying to build up a collection of as many different species as possible, or to those who find the words 'difficult cultivation' an irresistible lure. If you are one of that select band, then you will probably already have realised that the key to success is keeping *C. parviflorum* cool, potted in an open-textured compost (even in orchid chippings), and in attending scrupulously to the watering needs which that entails.

One of the more difficult species to cultivate, *C. parviflorum* needs cool growing conditions and well-drained compost that never dries out completely (MD)

C. cilicium and *C. mirabile*

This pair, like *C. coum*, should not be allowed to go bone-dry in summer. Both tolerate full light in temperate climates but are happier in light shade, and both prefer light or at least well-drained soils. They are difficult to grow out of doors in heavy soils. *Cyclamen cilicium* has fine roots, similar to those of *C. coum*, whereas the roots of *C. mirabile* are slightly more thong-like, almost reminiscent of those of *C. graecum*. Over-watering *C. mirabile* in summer can be fatal, as these roots tend to rot and the rot spreads into the tuber. The tuber of *C. mirabile* has the unnerving habit of splitting its sides as it grows — don't worry, these fissures, which heal themselves, are a normal natural feature.

Both these species are well-suited to pot cultivation, their leaves and flowers show well in pots and often repay the closer inspection that pot growth allows. Like *C. coum*, both *C. cilicium* and *C. mirabile* reward regular re-potting.

Although *C. cilicium* is hardy and can be grown outside, it is also well-suited to pot cultivation. This is the attractive white form, f. *album* (MD)

Cyclamen intaminatum is hardy but it is best appreciated as a pot plant, being one of the smallest species (MD)

If you grow *C. mirabile* from seed, keep only those plants that show the characteristic red or purple colouring in the young leaves; however heartless it seems, dispose of any lacking this hue. Through the generations, this will ensure that a higher proportion of seedlings carry the feature that won this species its name 'marvellous'.

Selected forms of *C. mirabile*, such as Peter Moore's 'Tilebarn Nicholas' can be better appreciated when grown in pots under glass (SB)

C. intaminatum

This species is broadly similar to *C. cilicium* in its needs for growing out in the garden. It's such a small plant that you should choose a prominent spot for it so as to show its dainty form and delicate flowers well. It will repay you by being the first of the autumn cyclamen to flower. *Cyclamen intaminatum* also does well in a pot, and can stay in the same pot without re-potting for many years; don't let it dry out completely in summer.

C. repandum and its allies

Cyclamen repandum is decidedly a woodland plant. Its hair-like roots are very fragile. Appreciating shade or at least semi-shade, it naturalises well in woodland, enjoying fairly damp though never water-logged conditions. It needs reasonably light soil: the ideal is the open-textured leafmould of long-established light deciduous woodland. It is best planted with a hand's width of such soil (say 5 cm) above the tuber.

Partial shade in humus-rich soil suits *C. repandum*. Here, it is growing in a woodland situation at the National Trust garden Knightshayes Court, Devon (BM)

Cyclamen creticum, *C. rhodium* and *C. rhodium* ssp. *peloponnesiacum* all need similar conditions. The last of these is well worth trying out of doors in temperate climates; at least in the south-east of England, *C. rhodium* subsp. *peloponnesiacum* seems happier outside than does *C. repandum* itself.

When grown in pots under glass, all of these species dislike extremes of heat and prefer a cool root run (*C. rhodium* ssp. *peloponnesiacum* is perhaps rather more heat-tolerant than *C. repandum* and the others). So, they are best kept on the floor rather than on an exposed bench while they are growing; while dormant, they do well in pots under the bench.

Cyclamen rhodium subsp. *vividum* is broadly similar in its requirements, but takes more sunlight. *Cyclamen balearicum*, with its silvery leaves, also tolerates more heat and drier conditions. In Mediterranean-climate areas, *C. balearicum* is the easiest of this whole group to grow in open gardens, as opposed to woodland, though it prefers light shade.

C. cyprium and C. libanoticum

Cyclamen cyprium is best treated like *C. repandum*, as it needs some shade and dislikes great heat. It is well worth trying outside in reasonably mild spots. It flowers well into November and December and, being strongly scented, is very useful for that time of year. Under glass, the main task is to keep *C. cyprium* fairly cool on sunny days; it's another species that is best grown down on the ground rather than up on staging.

The late autumn or early winter flowers of *C. cyprium* are much appreciated at this time of year and are very fragrant (MD)

Cyclamen libanoticum, a woodland species, also enjoys much the same shady environment as *C. repandum*. Its spectacular flowers, more open than those of other species, and its distinctive foliage, not showy but giving interesting leaf variations from plant to plant, make it worth taking trouble with. It is not easy to grow well in a pot. The prime requirement is to keep the pot cool in hot weather, so under glass, it is best kept on the ground under the staging, where it will flower well.

A very distinctive and attractive species, *C. libanoticum*, not easy to cultivate but well worth the effort (SB)

C. pseudibericum

This bold plant, with its neat habit and conspicuous flowers, makes a showy subject for a pot. Don't let the pot dry out completely in summer. As this is a greedier species than many, it repays regular re-potting, every second year or even annually. In temperate areas, it is well worth trying out of doors where it prefers some shade and shelter.

The large-flowered *C. pseudibericum* is a satisfactory species for pot cultivation, showy and with a good balance of leaf and flower. This is the *roseum* form (MD)

C. purpurascens

This summer-flowering woodland species grows in climates that are more continental than those in which other cyclamen species occur naturally. It survives deep frost and great heat, but in its natural environment, it gets a good deal of rain in early summer, which keeps the soil reasonably moist through the hot season. Consequently, *C. purpurascens* will not tolerate much heat if growing in dried-out ground. For a good display from this species, leafmould is nearly a must.

Its close relative *C. colchicum*, which needs similar treatment, is actually a tougher plant and can withstand even greater extremes of temperature.

The summer flowering and deliciously fragrant *C. purpurascens* is not the easiest species away from its natural climate. In pots, it should never be dried out completely (MD)

C. graecum

This is the best of all cyclamen species for growing under glass in a pot. It is forgiving and long-lived, some would say almost indestructible. The endless variations in its richly patterned leaves provide a collection of just this one species with plenty of interest, and in full flower, it can be spectacular. *Cyclamen graecum* does need sun to flower well, to give it a baking both during its dormant period and during its early-autumn flowering period; much further north than 52° N, it may feel rather deprived of sun. It needs no shade, and its leaves develop their silvery patterns best in full sun. When dormant, *C. graecum* should be kept from drying out completely, ideally by standing the pot on a substrate such as sand that can be lightly moistened from time to time. Once tiny flower buds show on the surface of the tuber or on its floral trunks (in early August or even late July in the south of England; you can gently tilt the grit topping off the pot to check), start watering and make sure that the compost never dries out completely; if it does, those buds will abort.

The massive thong-like roots of *C. graecum* need deeper pots than those of any of the other cyclamen species and, in time, its tuber grows to a great size, as big as a football. Nevertheless, it seems to flower best if it is not repotted frequently. Young plants may need moving on every couple of years or so, but older ones are happy to stay in the same pot and compost for up to five years or more. Diminished flowering will tell you when it's time to repot.

In temperate climates, *C. graecum* is disappointing as an outdoor plant. It tolerates a little frost but usually barely survives; in the south of England, when planted against a sunny wall, it will flower in a better-than-average summer, but rarely well. By contrast, it is a first-class garden plant in warm Mediterranean climates, needing no shade even there.

C. persicum

This is another splendid species for growing in a pot under glass, holding its elegant and abundant fragrant flowers well clear of the foliage. *Cyclamen persicum* flowers through the late winter into spring, and will tolerate a touch of frost but no more than that. Its main enemy under glass is warm sun, which makes its leaves flop, so as spring warms up, it needs some shade or preferably abundant ventilation. Like other cyclamen species, *C. persicum* prefers a lean potting compost; one too rich promotes excessive leaf growth at the expense of flowers and at the risk of disease. The compost can be dried out almost completely while the plant is dormant. The rare autumn-flowering variety should be treated similarly.

The 'Golan Heights' selection, which originates from high altitudes in a cold-winter area of Israel, does well out of doors in temperate climates, given part shade and a well-drained soil. Unlike most plants of this species, it comes through several degrees of frost with no damage to its leaves.

In warm Mediterranean climates, *C. persicum* is an excellent plant for a fairly shady part of the open garden and needs no watering.

Not a good plant for outdoor cultivation in Britain, *C. graecum* is best in an alpine house where it can be one of the most rewarding of all species (MD)

Although not hardy enough for outdoor cultivation, the wild form of *C. persicum* is a superb pot plant for a frost-free glasshouse, winter-flowering and strongly fragrant (MD)

C. rohlfsianum

This species is the most tender of all cyclamen, and is not worth trying out of doors anywhere in north-west Europe or regions with similar climates. It is not the easiest species to grow in a pot either, liking a minimum winter temperature of 9–10°C (48–50°F) and a warm summer dormant period, and having a quick flush of new growth so that its flowers can open while the days are still warm and with good light.

New material introduced into cultivation in the 1960s by Manfred Koenen of Bonn University Botanic Garden has proved much more amenable than the rather weak original stock. This is now what is generally offered by specialist nurseries and seed exchanges. Nevertheless, this stock still needs expert handling to give its best — good-sized fragrant flowers followed by large and strikingly sculptured glossy leaves.

One expert, Sheila Brown of the UK, set out her recipe for success in the *Journal of the Cyclamen Society* (vol. 28, no. 1 and vol. 29, no. 1). For their first few years, she treats young plants of *C. rohlfsianum* roughly similarly to those of *C. graecum*, placing their pots on a substrate that is kept moist through the summer. But she keeps her mature plants of *C. rohlfsianum* completely dry while they are dormant, giving them a thorough watering (by immersing their clay pots) in mid-August, then provides no further water until the flower buds appear, when she resumes moderate watering. She explains that continued watering brings the leaves on too soon, so that they hide the flowers. Once the leaves begin to expand, however, careful watering should continue through the winter so that they attain their full size.

The tender *C. rohlfsianum* is a challenge, requiring dry heat in summer and careful watering to discourage early leaves from hiding the flowers. This plant grown by Sheila Brown illustrates this (SB)

4.2 PESTS AND DISEASES OF CYCLAMEN

Chris Clennett

Fortunately for cyclamen growers, these plants are prone to a relatively small set of pests and diseases, although those that do attack cyclamen are amongst the most difficult to control. Some years ago, an article appeared in the Cyclamen Society journal (Clennett, 1990) that assembled together all the information available at that time on the pests and diseases of cyclamen. In the intervening years, much has changed, particularly the range of chemical controls available to both amateur and professional growers in the countries of the European Union. No trade names were given in the original article, but the references below include one or more trade names associated with each active chemical. These trade names are those used in the UK, but in many cases they will be the same worldwide as these products are manufactured by multinational agrochemical companies.

The *Pest and Disease Control Handbook* published by the British Crop Protection Council (Scopes & Ledieu, 1983) lists five pests and two diseases of cyclamen. This substantial volume, which is aimed at professional growers, includes only pests and diseases that can cause economically significant losses to nurseries that produce pot-plant cyclamen. Nevertheless, it provides a good starting point, and the pests or diseases encountered by growers of cyclamen that are not included in this book are infrequent, although troublesome when they do appear. Recently, Jan Bravenboer noted a new pest of Cyclamen, the gall mite *Calacarus? trotteri*, and his account of this is found on page 335. This account is followed by a note from Melvyn Jope about a very old well-known pest, tortrix moth caterpillar, which at times has been particularly troublesome in his cyclamen collection.

Pests

Vine weevil (*Otiorhynchus sulcatus*)

The adults of this pest are largely nocturnal beetles of around 1.5 cm in length and dark grey coloration. Damage caused by vine weevil is frequently undetected until the afflicted plants are beyond recovery; the larvae of the beetle feed on the roots and tuber, cutting off the supply of water and nutrients to the top growth and causing an abrupt collapse of the plant. The grubs themselves are white with a noticeable brown head, and are about 0.5 to 1.0 cm in length. They adopt a characteristic sickle or "C" shape when removed from the soil, and if observed closely are seen to be coated in short stiff bristles. If the host plant survives for sufficient time and the infestation remains unnoticed, the grubs will pupate after three months, becoming ghostly white versions of the adult

The adult vine weevil which moves slowly and cannot fly but makes leaves unsightly. It can lay thousands of eggs (BM)

but still underground. The adults then emerge during two main periods, the autumn or the following spring, by burrowing their way up to the soil surface. They are extremely vulnerable at this stage as their external skeleton has yet to harden, but unfortunately they emerge only at night when potential predators (and gardeners!) are unlikely to discover them.

Damage caused by the adult vine weevils is usually unimportant, although disfiguring, and rarely found on cyclamen. The adults cut semi-circular portions from the leaf edges as they feed, most often on broad soft leaves such as those of vines (hence their common name), kangaroo vine, grape ivies (*Cissus* and *Rhoicissus* species), and species and varieties of *Rhododendron*.

Control of vine weevil is difficult and rarely completely successful. Some adults invariably remain to cause a fresh infestation, and as each female can lay up to 1,000 eggs in its four-month life, the potential for serious damage should never be underestimated even though some of the eggs are infertile and will not hatch. No male adults have been found to date, so every beetle encountered is a potential cause of infestation. Furthermore, it is not possible to restrict population growth by releasing irradiated sterile males, a technique that has been used successfully to control several North American pests by causing a serious crash in the population number when no offspring appear for a generation.

Typical leaf damage by vine weevil adults (SB)

For many years, chemical control was achieved with aldrin. This chemical was never available to amateurs because of its severe toxicity, and many professionals refused to use it when it was available. Following the withdrawal of aldrin in the 1970s, growers relied on HCH (or hexa-chlorohexane to use its full name), which is also known as BHC or lindane. This product was applied as a drench to the compost and, in my experience, only ever held

Vine weevil grubs, which cause havoc to tubers and can devastate a collection (BM)

the pest in check. The loss of just one plant can be a disaster to a collector, and so HCH was not a complete solution, particularly for amateur growers. HCH too was withdrawn because of potential health risks in 2000. A particular problem for nurseries growing cyclamen as pot plants was that HCH released a vapour that was harmful to the biological control agent for two-spotted mite (which is often, if wrongly, referred to as red spider mite).

Since the withdrawal of these powerful insecticides, effective replacement products have only gradually appeared in the professional market. For chemical control in ornamental plants (including cyclamen) the organophosphate insecticide chlorpyrifos is currently approved for professional growers. This is incorporated into the soil as slow-release granules (such as Fargro's suSCon GREEN) or poured on as a drench (such as Dursban from Astra Zeneca). This product is not available to amateurs in the UK or Europe and its amateur use is being phased out in the USA because

use of the concentrate without protective clothing is hazardous. Fipronil and imidacloprid can also be used to control pests on non-edible plants growing in containers. Like chlorpyrifos, these chemicals are either incorporated into the compost as slow-release granules (such as Vi-Nil from Certis or Intercept from Scotts) or used as a soil drench (Provado). As with any pesticides, please refer to the manufacturer's instructions for the use of these products. Fipronil is a member of the phenyl-pyrazole group of pesticides and was developed by Rhone Poulenc in the late 1980s. It is marketed under a variety of brand names for the control of animal pests, not only in horticulture.

For UK professionals, Scotts have recently launched Exemptor to replace their Intercept product as a systemic insecticide that can be incorporated into potting composts. Based on the chloronicotinyl pesticide thiacloprid, this insecticide is active against aphids, whitefly and leaf beetles as well as against vine weevil. This is also the active constituent of Provado Vine Weevil Killer 2 for amateurs.

Nematodes can be used as an alternative to chemical insecticides. Also called eelworms, nematodes are small (mostly microscopic) unsegmented worms. Naturally extremely numerous in the soil, the vast majority of nematode species are detritus feeders, but some are plant pests and a few are parasites of insects. These insect-attacking (entomopathogenic) nematodes have been developed as a means of controlling vine weevil. The nematodes attack weevil grubs by burrowing into their skin or by wriggling inside the mouth, anus or breathing pores. Once inside the grub's body, they release a biological weapon — bacteria that poison the grub and cause its death within a few days. The nematodes feed upon the remains of the insect and then reproduce. The cycle is completed when the young nematodes leave the remains of the insect to search out another host.

Entomopathogenic nematodes are sold under various trade names and are all supplied as a clay matrix that contains infective-stage nematodes. This is mixed with water and spread onto the soil, whereupon the nematodes seek out the vine weevil grubs and kill them. Nematodes that are commercially available include *Steinernema carpocapsae*, *S. feltiae*, *S. riobrave*, *Heterorhabditis bacteriophora*, *H. marelatus* and *H. megidis*. They are most commonly used for management of soil insect pests in high-value crops. Because nematodes are living organisms, their successful use is influenced by environmental conditions. Nematodes need adequate moisture, temperatures within the tolerance levels for the specific species, and protection from UV radiation (direct sunlight) during application. Most strains of nematodes are dependent on soil temperatures above 12°C, which should not be a problem in glasshouses but does limit their effectiveness outdoors.

The use of *Heterorhabditis* as a commercial parasite seems to be less developed than that of *Steinernema*. *Heterorhabditis* nematodes are versatile in that they attack a range of pests, including both moth caterpillars and beetle grubs (including weevils). However, *Heterorhabditis* species require relatively high temperatures for full effectiveness: the effectiveness of *H. bacteriophora* falls significantly at temperatures below 21°C (70°F) and this species is ineffective below 10°C (50°F). *Heterorhabditis megidis* requires a moist open compost, such as peat or coir, in the temperature range of 14–20°C (58–68°F). It is supplied in the UK under the trade name Nemasys H. The best time for treatment is August or early September, when the compost will be at the right temperature and before the grubs have grown large enough to cause serious damage.

Several species of *Steinernema* have been used in commercial products. Initially, *S. carpocapsae* and *S. feltiae* (synonym *S. bibionis*) were widely distributed in the UK as Nemasys. Like *Heterorhabditis*, these species are restricted by temperature, being ineffective below 10°C (50°F) and most efficient at temperatures above 21°C (70°F). *Steinernema kraussei* has become available more recently in the UK — to amateurs through Defenders Biological Controls under the trade name Grubsure LT, and from Just Green and other suppliers as Nemasys; and to professionals through Syngenta Bioline as Exhibitline SK or through Avoncrop as Nemasys L. This nematode species was found in Scotland, but is native to many parts of Europe at relatively high elevations, where it functions at temperatures down to 5°C (41°F). Experiments have shown that *S. kraussei* will achieve 75% control of vine weevil even when soil temperatures are as low as 3°C (37°F), making it more useful than other nematodes for controlling pests in cyclamen collections in unheated glasshouses. Once again this product is supplied in sachets for application as a soil or compost drench; it is available in pack sizes to treat 12 m^2 or 50 m^2. It is best applied in spring (March to May) or autumn (August to November) as this coincides with the larval stage of the vine weevil life cycle.

Cultural control remains a principal weapon when dealing with vine weevil. Richard Cowles, writing about damage and control on Rhododendron in the USA (Cowles, 1995, 2003), observed two main sources of infestation that can be prevented by cultural means. In nursery situations, young rhododendrons are frequently grown in close proximity to older plants. Damage to young stock can be caused by small numbers of weevil larvae as the plants have few roots at this stage. The larvae kill the plant and then lose their food source and so starve themselves. Cowles' observations showed that adult weevils were moving from the older plants, which can tolerate infestation, to young ones to lay eggs. This highlights the need to keep larger plants that act as hosts for vine weevil at a safe distance from collections of young or vulnerable plants (including cyclamen). Similarly, cyclamen should not be kept close to hedges or other shrubs that are known to host weevil (particularly rhododendrons).

It is possible to take measures to prevent adult weevils reaching susceptible plants where a collection is housed in a glasshouse or frame. As weevils cannot fly, this can be achieved by growing plants on raised benches and then banding the legs with a treacle-like glue or sticky bands (available from commercial outlets for the control of codling moth in apples). One word of warning though, if you already have weevil damage, then the compost could retain a residual weevil population and so this strategy will only prevent further adults moving into your collection. It would also be worthwhile to trap adult weevils, both for control and for monitoring. As they are nocturnal and hide during the day, traps can be made by laying a damp piece of old carpet or such-like in the vicinity of the crop. Additionally, cardboard tubes (such as those at the centre of toilet or kitchen rolls) can be filled with shredded newspaper and placed among the plants. These should be left for a couple of days and then emptied into a bucket to reveal any hiding weevils.

Because black vine weevils are commonly transported as larvae in nursery stock, exclusion barriers can also be used to quarantine plant materials brought in from other nurseries until it can be determined whether these plants are infested. If the quarantined plants are found to be infested, they can then be treated separately from other plant material.

Cyclamen mite (*Polyphagotarsonemus pallidus*)

This is a minute pest with an enormous name. It is a pest of several genera and the symptoms can often be seen on *Impatiens* (busy Lizzie), where it causes the shoot tips to become blackened and distorted and the meristem or growing point to die rapidly. In cyclamen, the symptoms follow a similar pattern of deformed and discoloured growth. The flowers fail to develop properly with the flower stem becoming twisted and the petals puckered.

The pest itself is a white to brown mite of 0.25 mm in length, too small to see with the unaided eye. (By contrast, a keen-eyed observer should be able to differentiate damage caused by the two-spotted mite as the latter can be seen on the leaves). Under heated glasshouse conditions, cyclamen mite can be found at any time of year with a complete life-cycle taking around two weeks. In cold glasshouses, infestation is only likely during the summer, and in either situation, it is usually infected material brought onto site that causes an outbreak. The mites spread by crawling from one plant to another, and bearing their size in mind, the rate of dispersal is obviously slow!

Control is primarily achieved by thorough investigation of all new plants brought into a collection or nursery, always remembering that other plants other than cyclamen can be infested. Once outbreak has started, remedial action becomes more difficult. The use of biological control has yet to become effective, although it is known that another type of mite (the Typhlodromids) are predatory on cyclamen mite outside. In some countries, professionals can turn to chemical control using endosulfan (Thiodan). This is a highly toxic organochlorine pesticide and its use has been withdrawn in several countries. It retains approval in UK only while existing stocks are being used up. Cyclamen mite can be kept at acceptable levels after infestation is observed by using high volume sprays of pirimiphos-methyl (released by an Actellic or Fumite smoke generator), which has approval in the UK for use against two-spotted and other mites on protected ornamentals. This chemical does not provide total control but will prevent further spread.

Unfortunately, no chemicals that are effective in controlling cyclamen mite are available to the amateur and the only solution I have been able to find is to use hot water treatment. This consists of immersing the entire plant in water at 46°C (115°F) for seven minutes: the temperature and treatment duration are critical. This must be a last resort, however, and rigorous prevention can avoid its becoming necessary.

Broad mite (*Polyphagotarsonemus latus*)

This pest is very similar in appearance to cyclamen mite and only microscopic examination will separate the two. It too attacks several genera of glasshouse plants, causing bronzing of the foliage and distortion similar to that caused by cyclamen mite. In begonias, the symptoms are frequently referred to as "begonia rust" although no fungal disease is involved. Broad mites can be controlled in the same way as cyclamen mite and it is possible, and quite common, to have both broad mite and cyclamen mite infesting the same plant.

Thrips (*Thrips tabaci*, *T. fuscipennis and Heliothrips haemorrhoidalis*)

All thrips species are small elongated insects of about 4 mm in length, commonly referred to as thunder bugs because of their habit of landing on washing and other convenient surfaces during sultry conditions. Inconveniently, the term thrips is the singular as well as plural for these pests, so there is no such thing as a thrip. They are strong flyers, on feathery wings that are unlike those of most other insects, and can spread rapidly as a result. Thrips feed by rasping their mouthparts along the surface of leaf tissue, leaving characteristic silvery dots and patches, and some leaf distortion can occur. The exudation of sap from the resulting damage can lead to the appearance of watery or coloured droplets on the leaves, and to infection by sooty moulds. Larvae and adults feed in the same way and usually select the softest tissue; hence thrips cause particular damage to flowers and buds.

Control of thrips in cyclamen is usually achieved culturally as their attacks are infrequent. Commercial producers of crops such as cucumbers, where thrips damage is economically significant, use a product known as Thrip Stick, an integrated control programme that uses mutually compatible combination of biological and chemical controls. This product contains polybutene, a highly sticky substance that attracts the pests to their death. An equivalent for smaller-scale use are square sticky traps with a blue background that attracts thrips to the sticky surface. These should be suspended at crop height horizontally (not vertically) as the thrips settle on the surface rather than flying into it. Note that thrips are not generally attracted to the yellow sticky traps that are widely available for the control of glasshouse whitefly. Biological control is practicable for glasshouse crops: predatory mites such as *Neoseiulus* or *Hypoaspis* and pirate bugs such as *Orius* are used very effectively.

Thrips damage on cyclamen is uncommon, and consequently the use of biological control is unlikely to be efficient unless other crop plants that are attacked by thrips or other prey suited to the biological control agent are close by. Using sticky traps that are made of blue material is probably the most effective means of control in this situation. In the past, chemical control relied on the pyrethroid group of ingredients such as deltamethrin (Decis or Pearl Micro), but these products no longer have specific approval for use in ornamental crops in the UK. Pyrethroid chemicals have a contact action and would all be applied at high volume (that is to the point of run-off from the leaves). None of the available systemic pesticides appear to provide an effective level of control.

Aphids

Around 500 aphid species have been recorded in the British Isles, but only four of these are likely to be troublesome to cyclamen growers. The main culprits are: peach-potato aphid (*Myzus persicae*), cotton-melon aphid (*Aphis gossypii*), apple-grass aphid (*Rhopalosiphum insertum*) and mottled arum aphid (*Aulacorthum circumflexum*).

Most aphids follow a similar life-cycle and respond to similar control methods and so it is easiest to describe a generalised case and any anomalies that occur. Incidentally, although the common names given above indicate the host plants of each aphid species, these are by no means the only host plants. As can be seen from the common names, these aphids favour plants that are not native to British shores but as aphids are easily distributed in high-altitude winds, many alien species have become naturalised in many countries.

Aphid (greenfly) (RBG Kew)

To complicate matters, there are two types of aphid life-cycle. Autoecious aphids spend their entire life on one host (although different species of plant might be involved), whereas heteroecious aphids have two host plants and migrate from one to the other according to the time of year. Of the examples given above, the peach-potato and apple-grass aphids are heteroecious, and the cotton-melon and mottled-arum aphids are autoecious. From the point of view of control, this difference is largely academic except that growers must be aware of the proximity of the winter host (for these aphids, peach or nectarine and apple, respectively) to the cyclamen collection.

Unless conditions are exceptionally mild, or artificially heated, all aphids overwinter as eggs laid in crevices in the autumn by the last generation of females. In some species, males are produced by the preceding generation and the fertilisation of the eggs causes their dormancy. These eggs will hatch in spring to produce a wingless generation of females, which begin to feed and then to produce live offspring parthenogenetically (i.e. without fertilisation). As the aphid population grows, hormonal changes cause the production of winged females, which migrate to other plants. In heteroecious species, the production of winged females is probably stimulated by changes in day-length and allows migration to the summer host. The aphid population then expands exponentially until food shortage stimulates further dispersals throughout the summer. Heteroecious species will then migrate back to the winter host as day length shortens.

The general appearance of aphid species will be well known to most readers, but differentiating between individual species is not always easy. Of those listed above, the peach-potato and cotton-melon aphids are very variable in colour and are unmarked, so their positive identification is difficult. The cotton-melon aphid is exceptionally sedentary, which can affect its control, but this is not a reliable method of identification! The other two species are distinguishable by markings on their abdomen. The apple-grass aphid is a mid-green with a single dark green stripe running the length of its back, whereas the mottled arum aphid, although of similar ground colour, has a strongly defined black arrow or circumflex on the body (hence the Latin name).

Control of aphids is usually carried out by the use of quick-acting contact insecticides, which are specific in their action, or by systemic insecticides, which provide long-term protection of the plants. Contact sprays that contain rotenone (the active ingredient of derris) or bifenthrin (such as Brigade, Gyro or Talstar for professionals, Bug-gun for amateurs) are effective and bifenthrin has some residual effect. The approval of bifenthrin for use in ornamentals in the UK is for two-spotted mite, but the chemical is also active against aphids and other pests. Pirimicarb (which is available under a variety of trade names, including Aphox, to UK professionals) is an effective carbamate pesticide that is approved for aphid control on protected cyclamen. It is, however, listed under Annex 1 under EU Directive 91/414 as an anti-cholinesterase compound, like the organophosphates, and so should be used with caution.

Systemic pesticides such as imidacloprid (Admire from Bayer or Intercept from Scotts for professionals, Provado for amateurs) or dimethoate (BASF Dimethoate or Rogor) provide several weeks of control. Although aphid damage to cyclamen is seldom severe, the pest can carry several types of virus and hence early control is essential, especially for commercial crops or rare species. Biological control is also a practicable possibility with several predators and a fungus available for use. The fungal control agent is *Verticillium lecanii* (available commercially as Vertalec), which is applied as spores. These germinate and infect the aphid with hyphae, causing its death in about two days. As the dying aphid moves through the colony, it spreads new spores to other individuals. Unfortunately, the fungus requires a minimum temperature of 15°C (59°F) and high humidity to maintain control, so it is only of use on pot-plant cyclamen crops.

More universally successful biological control is feasible using a predatory midge (*Aphidoletes*) or a parasitoid brachonid wasp (*Aphidius matricariae*). The larvae of the midge are voracious predators and will seek out aphids in inaccessible sites, thus giving excellent control. They are supplied as pupal cocoons and must be kept moist until release onto the plants. Emerging adults lay eggs on the treated crop, from which the larvae hatch. These will remain active for two weeks at 15°C (59°F) but for a shorter period as temperature rises. One drawback is that the larvae will not pupate when day length is less than 16 hours, instead they enter diapauses, a kind of insect "suspended animation" that breaks the cycle of control because no further eggs are laid until day length increases. Low-level illumination, such as that provided by a 60W bulb, is sufficient to keep the midges active. This might, however, affect the photoperiod responses of cyclamen plants, which have yet to be studied in detail.

The brachonid wasp *Aphidius* is a UK native insect that is very similar in appearance to the chalcid wasp *Encarsia*, which is used to control glasshouse whitefly. It lays eggs into the body of an aphid, which then gradually changes to a straw colour and enlarges until almost spherical as the wasp larva develops within it. The adult wasp emerges through a characteristic round hole in the abdomen of the now deceased aphid, to search out fresh hosts for the next generation. *Aphidius* is available from a limited number of biological control suppliers, but will frequently arrive spontaneously during summer and can maintain complete control of infestations.

Two-spotted mite (Tetranychus urticae)

This pest is more widely known as red spider mite, and can occasionally prove troublesome to cyclamen. It is a small mite, about 0.5 mm in length, and variable in colour from green to red (it is only red when dormant, hence the inappropriate nature of the old common name). Symptoms of attack are yellow speckling on the leaves and defoliation if the infestation is allowed to persist. If severe infestation is allowed to go unchecked, the pests produce webs that cover the growing tips of affected plants; in cyclamen, webs tend to be found around the margins of leaves. The mite is extremely resistant to chemicals and the nurseryman's principal weapon against two-spotted mite, cyhexatin, was withdrawn in 1988. Some degree of control is possible with bifenthrin, dimethoate or pirimiphos-methyl, and professionals can use fenbutatin oxide (Torq from Fargro). Biological control is possible using the South American predatory mite (*Phytoseiulus persimilis*), which will maintain control so long as the temperature remains above 18°C (64°F). This means it is of little use on cyclamen except under glass during summer, and since the plants are losing their foliage naturally at this time, the use of this predator is of doubtful value even then. Use of the predator to keep two-spotted mite populations in greenhouses down over summer would, however, reduce the chances of infestation when the cyclamen plants come into leaf in the autumn.

Female two-spotted mites (*Tetranychus urticae*) with egg (Giles San Martin)

Glasshouse whitefly (Trialeurodes vaporariorum)

Very occasionally this pest will attack cyclamen, although it more usually attacks softer-leaved plants. The adult is a white-winged insect, resembling a minute moth of about 1 mm in length. Both adults and larvae feed by sucking sap in the same way as aphids, and can similarly cause the appearance of sooty mould resulting from their excretion of honeydew. Only the first-stage larvae and the adults have legs and can move to feeding sites, the other three larval stages being sedentary. Control is achieved by sprays of insecticides such as buprofezin (Applaud from Certis), deltamethrin or imidacloprid. Alternatively, biological control is possible using the parasitoid wasp *Encarsia formosa*. This behaves in a very similar

Glasshouse whitefly (*Trialeurodes vaporariorum*) (Eran Finkle)

Red slug (*Arion rufus*) (RBG Kew)

way to *Aphidius matricariae* on aphids, and will maintain complete control through the summer even in unheated houses. Adults are capable of flying several miles but remain inactive unless they have a supply of nectar available as they emerge from the parasitised whitefly. They will survive throughout the year if temperatures of 12°C (54°F) or more are maintained, but can be reintroduced in spring if they fail to do so. Some degree of pest build-up must be allowed as the wasp attacks only the fourth larval stage.

Cultural control is practicable using yellow sticky traps suspended near the crop. Unlike the traps used for thrips, these can be vertical as the whitefly fly actively towards the yellow colour and become entangled in the sticky surface. This will not eliminate the infestation, but it will curtail growth of the population of adult pests and thus will affect the overall population size.

Slugs and snails (Arion, Deroceras, Helix and Cepaea species)

These pests can be extremely troublesome when they attack cyclamen. Damage is characterised by the appearance of slime trails over the plants as feeding progresses, and areas of leaf and stem are eaten. Slugs, and to a lesser extent snails, are dependent on dampness to remain active so that maintaining dry conditions will help to minimise damage. Control by chemicals is straightforward using baits containing metaldehyde or methiocarb. Of the two, methiocarb is the more effective since it is a stomach poison, whereas metaldehyde only paralyses the victim causing it to dehydrate if exposed. Methiocarb is, however, an anti-cholinesterase carbamate compound and is included under Annex1 of EU Directive 91/414. It has therefore been withdrawn from use by amateurs because of safety concerns. Biological control is technically feasible as there are predatory species that attack slugs. Unfortunately, these have yet to be reared commercially so this remains a hope for the future.

Cultural control remains the main weapon in reducing the chances of attack by slugs and snails. Keeping a frame or greenhouse clear of debris reduces hiding places for these pests, and maintaining a dry atmosphere will deter them. Sharp surfaces, such as crushed egg shells and sharp grit, are popular deterrents but their efficacy will depend largely on a combination of other factors. Sharp grit can certainly reduce damage to hostas in their early growth stages, but this is often because there are "easier meals" elsewhere.

Diseases

Grey mould (Botrytis cinerea)

Grey mould is potentially one of the most serious diseases that can attack glasshouse plants, particularly cyclamen. It is encouraged by damp, still conditions and by low temperatures, and is most often encountered during the autumn and winter. The disease is caused by a pathogenic fungus of the group known as Fungi Imperfecti, meaning that resting sexual spores are never produced. Survival through periods of adverse conditions is achieved by sclerotia, which are resistant bodies that are produced underground. These can survive several years and will germinate to produce new fungal strands or hyphae as soon as moisture is present. The fungus spreads by means of asexual spores produced on branched structures known as conidia, and these spores are thus termed conidiospores. They do not survive for long, which is why the disease stops spreading as soon as conditions become too dry in spring. Conidiospores germinate on contact with moist leaf surfaces and penetrate directly into the plant; they differ from many fungi that can only enter through wounds or leaf stomata. A network of fungal hyphae rapidly develops and secretes cellulolytic and pectolytic enzymes, causing the plant tissue to dissolve into a gelatinous mass. As the plant tissue breaks down, the fungus is stimulated to produce masses of conidia from which the spores are released.

Control is achieved by cultural and chemical means. Using a free-draining compost and maintaining maximum possible ventilation will discourage the disease, and ensuring glasshouse hygiene will prevent the fungus from surviving in debris or the glasshouse soil. *Botrytis* is a facultative parasite that can grow in much the same way as mushrooms when a host plant is unavailable, thus hygiene is of vital importance in its control. Chemical control is not particularly easy for grey mould as it develops resistance to sprays remarkably rapidly. Several chemicals are available, however, and if two or three are used in rotation, resistance is unlikely to build up. The professional grower can select from iprodione (e.g. Rovral from BASF), vinclozolin (Ronilan), chlorothalonil (e.g. Syngenta's Bravo, Scotts' Daconil and Certis' Repulse) or thiram (e.g. Thiosan). One of the principal groups of pesticides used against grey mould used to be the methyl benzimidazole carbamates. These included benomyl (Benlate), carbendazim and thiophante-methyl. These pesticides are also included in EU Annex 1, and their use has been gradually withdrawn. Carbendazim (such as Mascot Systemic from Rigby Taylor or Ringer from Barclay) retains off-label approval for professionals for treating *Botrytis* in protected ornamentals, but this approval is due to expire shortly.

Root rots (Thielaviopsis basicola and Cylindrocarpon destructans)

Root rots are also fungal diseases of the Fungi Imperfecti. They are encouraged by damp conditions and a primary cause of loss of pot-plant cyclamen that are overwatered by their purchasers. These fungi are present in all soils and under suitable conditions will invade plant tissues just below ground level. They usually require a damaged site through which to enter the plant and, for this reason, healthy plants are rarely affected. Once a plant is infected, lesions develop on the roots followed by decay of the outer root tissue, leaving a woody core. Examination of infected plants will show wiry blackened roots whose softer tissues have broken down. As the plant fails to obtain water and nutrients through the damaged roots, top growth begins to wilt and other fungi gain entry, causing the leaves to brown and wither.

Control is best attempted by using sterilised composts and by ensuring adequate drainage. If plants can be kept healthy and growing vigorously, these fungi are unlikely to become a serious problem. Where infection has occurred, however, drenching the compost with carbendazim or captan will effect control. Incorporating etridiazole (such as Scotts' Terrazole) into potting composts will also eliminate the fungi, but this chemical is available only to professionals. Tolcofos-methyl (Basilex or Rizolex) is also approved for use by UK professionals against root rots (and damping-off disease) in ornamentals. However, this is an organophosphorus chemical and so may be withdrawn or subject to additional requirements for operator protection. For professionals, fosetyl-alumnium (such as Certis' Aliette) is also approved for use against root rots in protected ornamentals.

Damping-off disease (Rhizoctonia solani)

This disease attacks only the seedlings of cyclamen, although it can cause other problems such as stem and collar rot in poinsettias and other house plants. The first symptom is the collapse of the petiole in germinating plants, and on investigation the tissues of the petiole will be found to be black. A similar effect is also caused by the fungus *Pythium*, which causes the petiole to turn brown. Secondary rots frequently invade as the stem collapses and these lead to the death of the seedling.

It is extremely difficult to control both of these fungi and, as they are members of different fungal groups, certain identification needs to precede the use of chemicals. *Pythium* and the related genus *Phytophthora* can be kept in check by drenches with propamocarb hydrochloride (including Scotts' Filex or Fargro's Proplant) for which there is approval in UK, whereas *Rhizoctonia* requires the use of a different group of chemicals. In the UK, copper ammonium carbonate (Certis' Croptex Fungex), etridiazole and tolcofos-methyl have approval for professional use against *Rhizoctonia*.

Downy mildew (Peronospora spp.)

Until recently, downy mildew would not have been considered a disease of cyclamen. There are several species of fungus that cause downy mildew of crops, particularly lettuce. Detailed control systems are available commercially because of the high economic value of this plant as a salad crop. It is, however, thanks to Jan Bravenboer's observations in the Netherlands that we can add cyclamen to the list of susceptible genera. The fungal mycelium of *Peronospera* resembles a white web with occasional white grains on the threads. Commencing as a 1–2-cm (0.5–1-inch) spot on the underside of the leaves, it will gradually spread to cover the whole leaf within some weeks. Once the *Peronospera* has attacked the whole plant, the leaves will curl downwards on themselves and the disease becomes visible over the whole leaf, not just the underside. Later, the white mycelium develops a purple hue.

Control of downy mildew is difficult, but in the UK, tolyfluanid (Elvaron Multi from Bayer) has off-label approval for professional use. This fungus is related to *Pythium* and *Phytophthora*, so the same chemicals can sometimes be effective. Jan Bravenboer has found propamocarb hydrochloride (available as Previcur N in The Netherlands) to be effective and this is approved for use against *Pythium* in damping off disease.

Wilt diseases (Fusarium and Colletotrichum)

Wilt fungi grow in the tissues of plants, but their damage is caused when they develop masses of hyphae (strands) within the water transport system of the host plant (the xylem tubes). The mass of hyphae causes a blockage or tylosis in the water system and ultimately cuts off the water supply to the leaves and shoots (or to entire branches in trees or shrubs). The wilt fungi as so named because the first obvious symptom of disease is the wilting of large areas of the plant through lack of water.

In Dutch elm disease, the fungus has developed a relationship with a bark beetle. Fungal spores are released into the feeding burrows of the elm beetle, where the emerging adult beetles pick them up and distribute the fungus to new host elms. This is why Dutch elm disease spreads so rapidly with such devastating effect, but fortunately this method of dispersal is rare. Most wilt fungi are distributed over short distances by surface water, or more frequently by growers moving infected plants from place to place.

On a technical point, the wilt fungi (*Fusarium*, *Colletotrichum* or *Verticillium*) are also Fungi Imperfecti. Like all the fungi in this group, however, when specimens are found to produce sexual spores, they are reclassified according to the type of spores they generate. Many *Fusarium* species have now been reclassified as members of the genus *Haemonectria* (a genus within the Ascomycete fungi). Similarly, several species of *Colletotrichum* have been reclassified as *Gloeosporium* (also Ascomycete fungi). For the sake of simplicity and familiarity, the older names have been retained here.

Fusarium and *Colletotrichum* cause similar disease symptoms in cyclamen. Areas of the leaves, or more frequently the stems, become blackened as they die and wilt symptoms develop above this point. If growing points are attacked, they may recover with swift treatment, but for commercial cyclamen production, any level of damage is often economically unrecoverable. These fungi can also cause anthracnose diseases in various plants, including cyclamen, in which the fungus causes dead patches or lesions to develop. These are usually black and can appear on leaves, stems or fruit. The fungus destroys the tissues it is feeding on, and then generates spores from the surface of the lesions, spreading to further parts of the plant and to neighbouring specimens. Chemical treatment uses the same groups of active ingredients as are used for root rots and damping-off disease. For ornamental plants, UK professionals can also use fosetyl-aluminium (Aliette) against *Phytophthora* wilt disease, and this chemical might have some benefit where other wilt fungi are involved.

Summary

These then are the pests and diseases most likely to be encountered by cyclamen growers. Further problems are always potentially possible, and an exhaustive list would be a book in itself, added to which is the constant threat of new invaders from abroad. Climate change has certainly added to the list of pests able to survive and spread in the UK in recent years as ever milder winters have failed to kill larvae and dormant eggs. Increased international trade in plants, particularly through the open borders within the European Union, are also thought to be adding to the transport of non-indigenous pests and diseases around Europe. The UK now has western flower thrips and tobacco whitefly, economically damaging pests of ornamental crops, to add to the lengthening list of imports and the situation will continue to change.

The legal situation in the UK and the wider European Union is also relevant. Since the introduction of the Food and Environment Protection Act (FEPA), the Control of Pesticides Regulations and the Control of Substances Hazardous to Health Regulations (COSHH) in the UK in the early 1990s, the regulation of pesticides has become ever more stringent. Across the European Union there is now in place a series of legal documents, referred to as the "European Six Pack" covering almost all aspects of health and safety. These documents cover working at heights, risk assessment, provision of workplace equipment, lifting and climbing equipment for arborists, manual handling risks and pesticide usage. In the UK, it is an offence to use any chemical not approved by the Pesticide Safety Directorate (PSD), an agency of the Department of Food and Rural Affairs (DEFRA). This means that only sprays listing "use on ornamentals" may be applied to cyclamen, and this applies to both amateurs and professionals — a stipulation not widely understood. Professionals born after January 1965 must pass a competency test before being allowed to use chemical pest controls, and all professionals using sprays on other people's property must be certificated whatever their age.

The regulations also specify the brand names of each chemical that may be used by professionals, and those for use by private gardeners. For this reason, many established brands that were used by amateurs have disappeared because their manufacturers did not seek re-approval for amateur use from the PSD because of the high costs of registration. A good example of this involves the fungicide benomyl (sold in UK as Benlate). This was withdrawn because the profit from a small market did not justify the cost of registration, rather than because it failed any safety tests. Since its withdrawal, however, the group of pesticides to which it belonged (methyl benzimidazole carbamates, see under grey mould) has been gradually withdrawn from use as regulations have tightened.

The chemical names for all the sprays mentioned here have been quoted deliberately as they must now be printed on all spray containers, no matter what brand they are. Brand names vary across Europe and North America, and so sourcing a pesticide by active ingredient is more likely to be successful than searching for a brand name. Some brand names are international, and I have given some examples of those approved in UK as this information might prove valuable in tracking down a particular pesticide.

The EU continues to introduce more restrictive legislation and there have been proposals designed to reduce or eliminate pesticide usage in areas open to the general public. While these proposals obviously apply to parks, gardens and roadsides, they could be interpreted as also including garden centres and nurseries, potentially restricting further the list of pesticides that can be used by growers or preventing non-professionals from entering the nursery. Amateur sprays are set to become still more diluted as competency testing cannot be extended to include all potential users, and the range of active ingredients available to amateurs is certain to contract. Even for professionals, the horticultural market is small in world terms, so chemical manufacturers will be less willing to carry out research and development when stringent regulations force long delays before commercial launch is possible. Already, several chemicals, including the effective whitefly specific insecticide buprofezin, have had their launch delayed by the time taken by (and backlog of) efficacy testing.

If this all sounds bleak for the control of pests and diseases, it need not be. Research into biological control has only assumed commercial importance with the advent of these new restrictions, and as the work on vine weevil shows, it can yield results in a shorter time than can the development of a new chemical. New and novel strategies come to light regularly in the course of academic research (see, for example, van Tol (2007)), but the funding of "market-related" research has been drastically reduced in the UK and elsewhere in recent years as this is seen by governments as something growers should fund themselves. The small size and profitability of the ornamental horticultural industry suggests that finances may become the limiting factor in developing new techniques in the future. Resistance to parasites and predators cannot develop except over vast numbers of generations, during which time the parasite or predator itself can mutate to maintain its advantage. As regulation and resistance conspire to reduce our reliance on chemicals, it will undoubtedly be this form of pest and disease control that comes to dominate the grower's arsenal.

Further observations on Cyclamen pests and diseases

A new pest of Cyclamen

Jan Bravenboer

For some years, some of our *C. purpurascens* showed what appeared to be a typical disease. The margin of the leaves curled down and the leaves showed a grey colour, like small felt-like hairs growing on the surface. A long search ensued but nobody knew what it was. We thought perhaps a false mildew. I removed all the attacked leaves and eliminated each new leaf but the same problem appeared immediately. However, all the plants survived and the next season they grew well and looked healthy again.

While walking in Italy on holiday, we visited a well known area on the left side of the road as it leaves the tunnel from Riva to the Ledro Valley, close to the road to Pregasina. Seventy five per cent of all the *C. purpurascens* in the whole forest had this same 'disease'. We very much doubted that this was a false mildew and decided to send a sample to a "plant doctor", Hennie van Kesteren, who is a retired employee of the Plant Protection Service station in Wageningen. As a hobby, Hennie attends plant markets with a stand to show people all kinds of diseases and gives them lots of

A relatively new problem, known as 'false mildew' is actually caused by a mite (JB)

information on how to solve these problems. He discovered that the problem affecting *C. purpurascens* was not a fungus as we had previously thought but a kind of mite. He contacted his former colleagues and I took an infected plant to the Plant Protection Service station in Wageningen, who began an official research project. After some days, I got the results from Bert Vierbergen: the mite was identified as *Calacarus? trotteri*, which belongs to the gall mites (Eriophyoidea).

A special group of these gall mites is known as the blister- or felt-mites, so called because their phytotoxic saliva causes raised welts or felted areas on the leaves of their host plant. To date, the name *Calacarus? trotteri* has been applied to the *C. purpurascens* mites as it is difficult to put them into a definitive genus. A new genus will probably be identified for this species. All the information Bert could find was written in two small papers. In the first paper (G. Scalia: *Marcellia* 10: 62–64 (1911)), which contains a description of the mite, it is noted that it was found on *C. hederifolium* on Sicily; the second paper (G. Goidànich: *Informatore fitopatologica* 6: 39 (1956)) describes only the damage to a series of 'cyclamens' from Italy.

This blister mite will probably not survive outside in cold winters and is very likely not to thrive in greenhouses either. It has not been found in the wild outside Italy, so this was something new for the Plant Protection Service, Wageningen. It is present, however, in some cultivated cyclamen collections, so if you see leaves that are bending down and have a hairy felted appearance, it is best to remove them.

Effective treatments might include biological agents such as Spical (*Amblyseius californicus*), a predatory mite from Koppert (www.koppert.com), or the use of acaricides such as Vertimec® from Syngenta or Milbeknock® from Bayer. In some countries, the active ingredient bifenthrin is still allowed, available as Bio Sprayday Greenfly Killer Plus, Doff All in One Garden Pest Killer or Scotts Bug Clear. Bifenthrin can help to solve other problems, such as infestation with the spider mite *Bryobia* sp., which can cause miner-like mottled discolouration in cyclamen leaves. Andrew Halstead, Entomologist from the Wisley Garden, also recommends bifenthrin for controlling mites on cyclamen plants.

Caterpillar damage

Melvyn Jope

I have had isolated problems in the past with the tortrix moth caterpillar, which curls the edge of a leaf and makes a cocoon inside. On this occasion, the species involved is the angle shades moth, *Phlogophora meticuloso*. This species lays its eggs during May to June and then again from August to October. Caterpillars of the second generation over-winter and will feed on cyclamen leaves whenever the night-time temperature is above freezing. The first signs of damage are usually small holes in the leaves then, as the caterpillar develops, it munches along the side of the leaf and can very quickly defoliate the whole plant. It also seems to favour flower petals in preference to leaves, but this year, its taste seems to range from *Ipheion* flowers to the leaves of *Crocus* species.

I am reluctant to use pesticides, so my method of control is to pick off and destroy individual caterpillars. As they are active only at night, this involves going into the greenhouse at night and attempting to find the caterpillars on damaged leaves. They usually feed from the underside of the leaf, so you have to gently turn over any leaf showing signs of damage to see if a caterpillar is there. They are very sensitive to either the light from the torch or vibration in the leaf when it is turned over because they often drop from the surface of the leaf and close up tight like a closed letter 'C' and remain motionless.

The colour of the caterpillar can vary from brownish-yellow to bright green, perhaps influenced by the host plant. They can be very well camouflaged against grit, but once you get used to spotting them, a quicker method of finding them can be to brush lightly over the leaves with your hand and then see if anything has dropped onto the grit. If carrying out night manoeuvres is not possible but you are aware of damage, then try lifting the plant pot and gently tipping off the grit — the caterpillars often hide under it in daylight hours.

One curious factor is that some species of cyclamen are badly damaged, notably *C. graecum*, whereas others, such as *C. confusum* from Crete, are not touched at all.

With all of the recent publicity about dramatic falls in the number of moth species, some being threatened with extinction, it seems a shame to be contributing to the problem. Maybe the solution is to try to prevent the adult moth getting into the greenhouse in the first place.

4.3 CYCLAMEN AROUND THE WORLD

(a) North America

John T. Lonsdale

History of *Cyclamen* cultivation in North America

How long have cyclamen been grown in North America? "*The Gardeners Monthly and Horticultural Advertiser*" dated 1864 and edited by Thomas Meehan, published by W. G. P. Brinckloe in Philadelphia, contains two articles, one of them a page and a half describing the culture of both 'hardy' and 'greenhouse' cyclamen species. However, the information appears to have been replicated from British periodicals and provides no insight into early cyclamen cultivation in North America. Although older books contain references to cyclamen, the cyclamen discussed is invariably *C. persicum*.

Liberty Hyde Bailey's *Cyclopedia of American Horticulture* of 1904 indicates that hardy cyclamen were offered in the United States in the Dutch bulb trade at that time, and six species in addition to *C. persicum* (known at that time as *C. latifolium*) are briefly discussed (under the names *C. Africanum*, *C. Neapolitanum*, *C. Europaeum*, *C. Cilicicum*, *C. Coum* and *C. Ibericum*). Bailey comments that they "are grown only as curiosities in this country". Co-author E. O. Opert comments further "All Cyclamen are very beautiful, and would be much more popular were they hardy in our eastern climate. On the Pacific slope many of them probably would be perfectly at home as outdoor plants…". Twenty years later, the renamed and enlarged 1925 edition of this work discusses seven species other than *C. persicum*, and Opert now comments, "They are little known in outdoor planting in N. Amer., however." The implication that they are not hardy in eastern North America is repeated.

With those discouraging remarks in mind, there will be little surprise at the absence of cyclamen from several wide-ranging lists of garden plants grown in North America during the first three decades of the twentieth century. For instance, Sedgwick's extensive listing (*The Garden Month by Month*, 1907) makes no mention of hardy cyclamen, yet it does mention such still uncommon plants as *Colchicum variegatum*, *Narcissus bulbocodium* of the Monophyllus Group, *Leucocrinum montanum*, and a baker's dozen *Fritillaria*. Nor does Neltje Blanchan's 1909 work *The American Flower Garden* mention them, although it does mention *Crinum* × *powellii*, *Amaryllis belladonna*, *Bessera elegans*, *Milla biflora*, *Crocus imperati*,

934 CYCLAMEN CYCLAMEN

plants at all stages of growth. In the frames the plants may be plunged in tobacco stems, and in the greenhouse they must be fumigated or vaporized with some of the nicotine extracts. Great vigilance must be exercised in growing cyclamens. (E. O. Orpet.)

INDEX.

æstivum, 5.	*floribundum*, 5.	*punicum*, 1.
africanum, 7.	giganteum, 1.	purpureum, 1.
album, 1, 2.	*hederæfolium*, 1, 8.	*purpurascens*, 5.
aleppicum, 1.	ibericum, 3.	*pyrolæfolium*, 1.
Atkinsii, 3.	*indicum*, 1.	roseo-superbum, 1.
atro-rubrum, 1.	*latifolium*, 1.	rubrum, 1.
autumnale, 8.	libanoticum, 4.	*saldense*, 7.
cilicicum, 6.	*macrophyllum*, 1.	sanguineum, 1.
Clusii, 5.	magnificum, 1.	splendens, 1.
cordifolium, 5.	neapolitanum, 8.	*subhastatum*, 8.
coum, 2, 3.	*orbiculatum*, 5.	*vernale*, 3, 8.
elegans, 3.	*Papilio*, 1.	*vernum*, 2.
europæum, 5.	persicum, 1.	violaceum, 1.
ficariæfolium, 8.	*punctatum*, 1.	

A. *Plant blooming in spring (or in winter under glass).*

B. *Corolla-lobes not eared.*

1. pérsicum, Mill. (*C. hederæfòlium*, Sibth. & Smith. *C. indicum*, Linn. *C. latifòlium*, Sibth. & Smith. *C. macrophýllum*, Sieb. *C. pùnicum*, Pomel. *C. pyrolæfòlium*, Salisb.). Fig. 1183. The common greenhouse or Persian cyclamen, in many forms: tuber large, flattened endwise, corky on the outside: lvs. appearing with the fls., ovate, crenate-dentate, base deeply cordate, usually marbled or variegated with white: fls. on scapes 6–7 in. high, large, scentless, white, purple-blotched at the mouth, but with rose-colored, purple and spotted forms, the segms. oblong-spatulate in shape, not eared or lobed at the base: pedicel not coiling in fr. Greece to Syria. The large-fld. cult. forms are grouped as follows by Pax & Knuth: Var. **álbum** (*C. alèppicum*, Hort.). Pure white. Var. **gigantèum**. White with red throat; very large. Var. **magníficum** (var. *punctàtum*). White, speckled with red, large. Var. **rùbrum**. Red. Var. **sanguíneum**. Light blood-red, large. Var. **ròsea-supérbum**. Brilliant rose-red, large. Var. **purpùreum**. Purple with bluish markings, large. Var. **violâceum**. Violet-red. Var. **àtro-rùbrum** and var. **spléndens**. Dark red, large. The var. *gigantèum* (*C. gigantèum*, Hort.) is the common large-fld., improved form of the florist's cyclamen. There are also double-fld. forms (R.H. 1880, p. 250); also fimbriate or crested forms, *C. Papílio* (I.H. 43:63. G.F. 5:235. G.C. III. 21:71; 23:173). Other portraits of *C. persicum* are: B.M. 44. I.H. 35:43. Gn. 47:378; 48:182. J.H. III. 34:578. Gt. 44, p. 203; 45, p. 164. F.S. 22:2345. A.G. 14:390-2; 17:261. A.F. 7.521–5; 11:1176–9; 12:499. The species is frequently figured in the trade catalogues.

2. còum, Mill. (*C. vérnum*, Sweet). Tuber smaller than in the last, globose or flattened: lvs. with or preceding the fls., nearly orbicular or round-reniform, entire, firm, not marbled nor variegated: fls. small, scentless, half or less as large as those of the last, purple and spotted in the throat. S. E. Eu. to Persia. B.M. 4. L.B.C. 2:108. F.S. 22:2345.—There is a white-fld. form (*C. álbum*, Hort.).

3. ibéricum, Goldie (*C. còum* var. *ibèricum*, Boiss. *C. élegans*, Boiss. & Buhse. *C. vernàle*, Koch). Dwarf: lvs. appearing before the fls., ovate-orbicular and rounded at the apex, entire or obscurely undulate, more or less zoned with white above: fls. purple with a darker colored throat. Caucasus.—Perhaps a geographical form of *C. coum*. C. **Átkinsii**, Moore, is *C. coum* × *C. ibericum*, Hildeb. Lvs. reniform, apex rounded, more or less shining, deep green, spotted silver-white, the corolla-lobes pale rose or white and usually lined or spotted red. F.S. 23:2425.

BB. *Corolla-lobes eared.*

4. libanóticum, Hildeb. Tuber globose with a cork-like covering: lvs. autumnal, the blade obcordate, sinuate, dentate or crenulate or rarely entire, marked with white above, deep violet or purple beneath: fls. fragrant; calyx-lobes oblong-acuminate, the margin lightly undulate, 5-nerved; corolla-tube somewhat globose-campanulate, the lobes lightly eared at base and broad-ovate, entire, pale or deep rose-color with a deep red T-form mark at the base; style exserted. Lebanon.

AA. *Plant blooming in summer and autumn.*

B. *Corolla-lobes not eared.*

5. europæ̀um, Linn. (*C. Clùsii*, Lindl. *C. æstivum*, Park. *C. cordifòlium*, Stokes. *C. floribúndum*, Salisb. *C. orbiculàtum*, and *C. purpuráscens*, Mill.). Tuber with corky exterior: lvs. ovate-orbicular or reniform, entire or nearly so, with a deep and narrow basal sinus, more or less white-marbled above, purple-tinted beneath: fls. on scapes 4–5 in. high, bright red and very fragrant, not spotted; the corolla-segms. oblong-spatulate (3/4in. or less long); calyx glabrous: pedicel spirally coiling in fr. Cent. and S. Eu. B.R. 1013.—Lvs. appearing with the fls., more or less evergreen. Variable.

1183. Cyclamen persicum. Showing a flower of perfect form, and the crested variety. (×½)

6. cilícicum, Boiss. & Heldr. Much like *C. europæum*: tuber hairy: fls. pale rose, with purple at the mouth, about twice larger; corolla-lobes elongated; calyx puberulent. Sicily. G.C. III. 23:81.

BB. *Corolla-lobes eared.*

7. africànum, Boiss. & Reut. (*C. saldénse*, Pomel). Large: tuber large: lvs. ovate-cordate to reniform, coarsely toothed, pale beneath, dull and pale green marbled above; calyx pubescent, the lobes broadly ovate-acuminate; corolla nearly white, faintly rose- or purple-tinged, the segms. 1 in. long and deep purple at the base; calyx-lobes lanceolate. Algeria. B.M. 5758. F.S. 8:841.—Little known in this country, but sold by the American agencies of the Dutch bulb houses.

8. neapolitànum, Tenore (*C. autumnàle*, Boos. *C. ficariæfòlium* and *C. subhastàtum*, Reichb. *C. hederifòlium*, Ait. *C. vernàle*, Mill.). Tuber very large, black, thick-rinded: lvs. variable, from hastate to round-reniform, more or less wavy-plaited on the edges, green or somewhat parti-colored: calyx small; corolla pink or rarely white, the segms. short and twisted and the edges raised and white-edged at the base; calyx-lobes triangular to oblong. S. and E. Eu. B.R. 24:49. Gn. 51, p. 37

Liberty Hyde Bailey's excellent *Standard Cyclopedia of Horticulture* gave descriptions of eight species of *Cyclamen* and mentioned many more, although several names are now regarded as synonyms. This is the 2nd edition (1939) (BM)

Sternbergia lutea and *Calochortus* of several species. The eclectic and copious lists put together by Stephen Hamblin in 1927, *Lists of Plant Types for Landscape Planting*, make no mention of hardy cyclamen. Louise Beebe Wilder, in her *Adventures in My Garden and Rock Garden* of 1927, makes no mention of them, even in the lists of plants grown from seed. Several of her works from the 1930s do, however, mention hardy cyclamen. Elizabeth Lawrence wrote about her experiences growing cyclamen in the south, beginning in 1946. She authored three books in which cyclamen are discussed: *Gardens in Winter*, *The Little Bulbs*, and *A Southern Garden*. She was widely read and respected both in North America and in England.

Considering the frequency with which *C. persicum* was raised throughout North America as a florists' crop, and what that implies not only about the level of cyclamen cultural skill then prevalent but also about the public awareness of cyclamen, the lack of strong domestic production of hardy cyclamen seems surprising. There were crazes in the early twentieth century for toy dogs and toy fishes, but no one then seems to have realised the potential for "toy cyclamen".

Historically, at least three factors seem to have operated to depress the development of domestic hardy cyclamen production. During the 1920s and 1930s, the US Department of Agriculture (USDA) made substantial efforts to establish domestic production of bulb crops, including hardy cyclamen, but nothing of a permanent nature seems to have come of the efforts with respect to hardy cyclamen. These efforts are documented in a series of Circulars published by the USDA during that time. In Circular No. 311, "*Bulbs from Seed*", of June 1934, David Griffiths writes:

> "Besides *Cyclamen indicum*, which is strictly a hothouse plant, there are a number of other species in the genus, some of which are perfectly hardy and make delightful additions to the wild garden, the perennial border, or the rock garden, although they are almost never seen in this country. The writer has had *C. europaeum* persisting in a wild-flower nook in his garden in the District of Columbia for 10 years. It blossoms regularly in late September and October and goes through the winter in vegetative condition, the beautifully marbled leaves adding a decidedly attractive touch to the border along with the *Helleborus*. The cyclamen corms become very large, but do not divide much; consequently it is necessary to produce them from seed. Late fall planting has proved most successful on Puget Sound. The seed germinates in the spring, and the corms have taken 3 years to reach flowering size. In the District of Columbia, corms of *Cyclamen europaeum* are planted 3 inches deep and receive a covering of leaves in the fall."

The District of Columbia is an old name for Washington, D.C. The cyclamen here called *C. europaeum* was almost certainly *C. hederifolium*. Puget Sound, which is in Washington State (not to be confused with Washington, D.C.), is mentioned because it was the location of the main bulb-raising efforts of the USDA in the early twentieth century.

The widespread cultivation of *C. persicum* as a florists' crop contributed to the continued obscurity of the hardy species. The importation of hardy cyclamen from European sources was evidently seen as a threat in terms of the potential introduction of diseases that could affect the florists' cyclamen crop. Thus, hardy cyclamen were prohibited entry into the United States during the days of the bulb quarantine. The eventual relaxation of the quarantine was soon followed by the Second World War with its disruption of the international bulb trade.

After the Second World War, a new wave of small, specialist nurseries began to spring up in the Pacific Northwest and these started to pick up some of the momentum for bulb culture that had built up in the years before the war. Although hardy cyclamen were grown or at least offered by some of these nurseries, hardy cyclamen remained at best a niche market produced by a cottage industry. The fact that the market came to be flooded with inexpensive collected plants of dubious provenance and identity also served as a strong disincentive to the development of a strong, healthy home-grown stock. This, as much as anything, no doubt contributed to the slow development of the domestic production of commercial hardy cyclamen in the period after the Second World War.

Recently, there have been more successes in the commercialisation of home-grown hardy cyclamen, albeit on a relatively small scale. Siskiyou Rare Plant Nursery in Oregon began to offer at least half a dozen species in the 1960s, as did Edgar Kline's nursery in Portland, Oregon. Russell Graham, in Salem, Oregon, had a first-class nursery for a number of years that sold a number of seed-raised cyclamen species. Nancy Goodwin's Montrose Nursery offered a nice range of home-grown plants in the last decades of the twentieth century and, during the 1990s, David Fischer in Maine offered numerous species. Ellen Hornig's Seneca Hill Perennials until recently offered a suite of handsome selections of several species and Hansen Nursery on the Oregon coast now specialises in hardy and other cyclamen species. John Lonsdale's Edgewood Gardens currently supplies a wide range of species, including less commonly offered forms. However, small hardy bulbs in general, and cyclamen in particular, remain woefully underestimated, misunderstood and underrepresented in North American gardens.

Climate and centres of cyclamen cultivation in North America

As someone who spent the first 35 years of his life in the UK, it was easy to become conditioned to thinking that if a plant was described as 'hardy', then it must be hardy pretty much anywhere, within reason. In the UK, 'hardiness' is generally taken to refer to a plant's ability to withstand cold periods during a normal English winter and is used as the major indicator of a plant's ability to survive outdoors. My move to the USA in 1995 thus provided me with a horticultural culture shock; not only could I not find the familiar John Innes compost, but there appeared to be significantly more climatic contributors to 'hardiness' than I had previously contemplated. It is certainly impossible to simply answer the question "how well do cyclamen grow outdoors in the USA?"!

A simple comparison of the maps of the British Isles and the USA below shows that the vast majority of the British Isles falls within the USDA hardiness zones 8 and 9, with minimum winter temperatures not usually below –12°C (10°F), although recent experience might have challenged that. Minimum temperatures across the USA range from –46°C (–50°F) to above 5°C (40°F). Thus, while there are areas that are clearly too cold to permit the growth of any cyclamen outdoors, many regions have winter temperature minima that appear to be ideal for growing cyclamen. Unfortunately many of these areas, especially in eastern and south-eastern USA, also have consistently very high summer temperatures (well over 32°C (90°F)) in combination with high humidity, which promotes the rotting of tubers, even in the absence of rain. The occurrence of frequent summer thunderstorms compounds the problem. On the other hand, consistent snow cover might permit the growing of species that would normally not stand a chance of survival in areas with winter minimum temperatures as low as –26°C (–15°F). When it comes to the cultivation of cyclamen outdoors, it is thus difficult even to generalise about the suitability of particular regions; altitude and climate vary significantly within individual states, especially in the West.

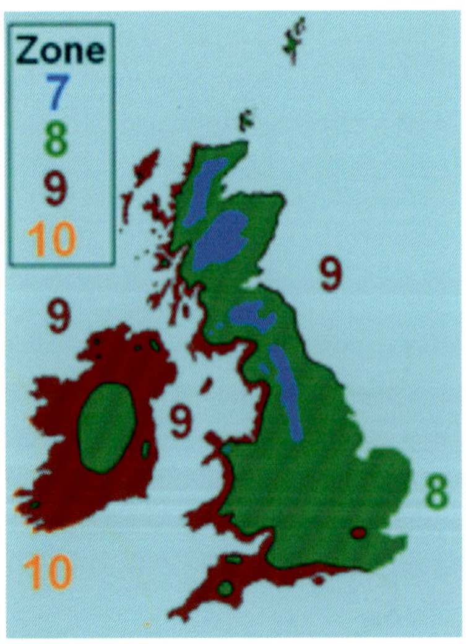

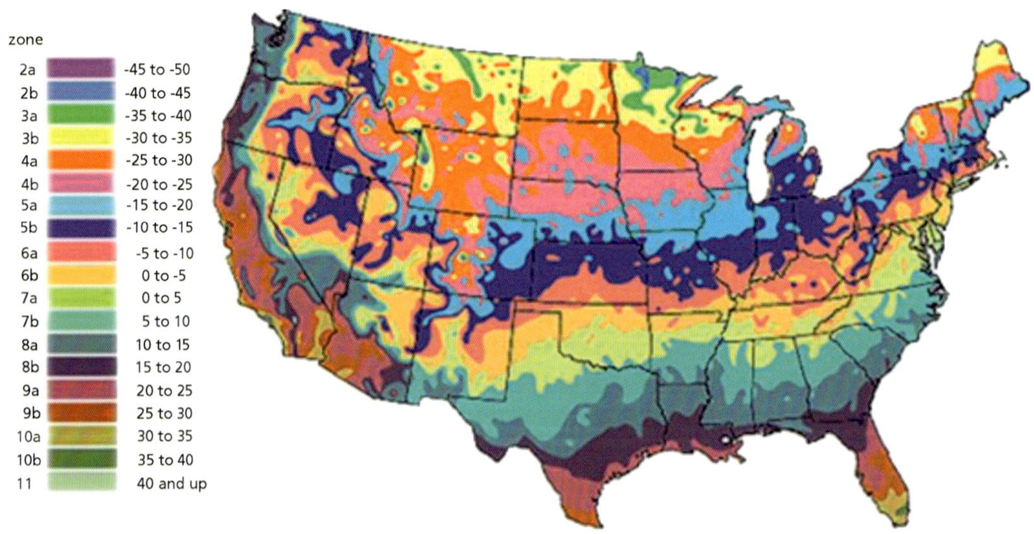

If one was to identify one region as most approximate to the Mediterranean climate experienced by many cyclamen in habitat, it would probably be the maritime Pacific Northwest (PNW), comprising northern California, Oregon and Washington west of the high cordillera. Even in the PNW, however, there are issues caused by excessive winter and spring rainfall and the fauna that these conditions favour! Certain parts of some south-western states, including Colorado, Arizona and New Mexico, are also conducive to the cultivation of at least some cyclamen species outdoors, although the exact areas are very dependent upon factors such as altitude. Denver, Colorado, for example, has a climate that is favourable for the cultivation of many alpine plants and also cyclamen, and the excellent botanic garden there continues to increase enthusiasm for such plants.

Probably the largest concentration of growers of cyclamen in gardens is, however, centred in the Mid-Atlantic states, which include Connecticut, Delaware, Maryland, New Jersey, Pennsylvania, Virginia, New York, North Carolina and Washington D.C..

Popularity of cyclamen in North America

Florists' cyclamen (derived from *C. persicum*) are just as widespread in the USA as elsewhere in the world, appearing as house plants across the country. They are typically grown as 'disposable' plants — once they have bloomed and show signs of wear and tear, it is assumed that they are finished and they are thrown in the trash. The exception is in the milder parts of California, where one often sees florists' cyclamen grown perennially. There is no understanding that, with appropriate treatment, these plants can be brought through dormancy and will provide many years of blooms.

Other species of cyclamen, 'hardy' or otherwise, are almost unknown, even in garden centres. In the maritime PNW, almost any garden with pretensions to horticultural interest includes *C. hederifolium* and *C. coum* but, even within the membership of the North American Rock Garden Society, there is relatively little appreciation of the potential presented by the many species of cyclamen to play a major role in rock and woodland gardens. Across the USA, there are probably fewer than 25–50 individuals who have incorporated cyclamen into significant garden plantings or who grow a reasonable selection in pots under cover. The physical distance between enthusiasts is not conducive to regular get-togethers. The greenhouses that are ubiquitous in English gardens are few and far between in the gardens of mainstream America. Apart from the lack of a history of gardening along English lines in the USA, greenhouses are made impractical by the cost of heating in many regions with very cold winters, together with the need for more robust structures with twin- or triple-wall polycarbonate or acrylic glazing that will withstand major snowfalls as well as falling trees and limbs, hurricanes, ice storms and occasional tornadoes! The number of 'critters' that would love nothing better than to snack on cyclamen tubers or leaves and flowers is also markedly higher in the USA than in the British Isles, more of which later. Americans generally have far fewer vacation days than their European counterparts and tend to prefer 'maintenance free' gardens to sweating in the yard in their spare time. Many suburban homeowners also utilise landscaping companies to carry out grass cutting and garden pest control — such companies are not known for their appreciation of the finer points of gardening.

Flower shows, such as those held in the UK by the Alpine Garden Society, the Scottish Rock Garden Club, the Royal Horticultural Society and local groups affiliated with these societies, are few and far between in North America. In the USA, there are no national shows that feature cyclamen, although small annual shows are held by many chapters of the North American Rock Garden Society. Western Canada, centred on Vancouver, British Columbia, has a more robust history of flower shows, such as those held by the Alpine Garden Club of British Columbia. The Pacific Bulb Society focuses in part on Mediterranean-climate bulbs and draws much of its membership from North America.

All is not lost, however, and the popularity of dwarf bulbs, including cyclamen, is slowly rising. It does however, have a very long way to go before it begins to approach the levels seen in Europe in particular.

North American cyclamen growers

The selection that follows is by no means all-inclusive but contains representatives from all of the main geographic regions that provide favourable conditions for growing cyclamen.

Gerald Firak, Park Ridge, Illinois

Gerald Firak is particularly known for the superb plantings of *C. purpurascens* in his Midwest garden, in which winter temperatures can reach lows of –29°C (–20°F). Based on Gerald's experiences and

Cyclamen purpurascens thriving and seeding freely in Illinois, in the garden of Gerald Firak (GF)

those of others in bitterly cold regions, *C. purpurascens* is favoured as the hardiest of all the cyclamen species. *Cyclamen purpurascens* flowers from early August until the snows of November and is seeding extensively in Gerald's garden. Gerald noted that *C. hederifolium* is not hardy in snowless periods in winter and his display of *C. coum* is ruined by the alternating false springs and low temperatures experienced in late February and early March. *Cyclamen colchicum* is a complete failure in this climate.

Nancy Goodwin, Montrose Garden, North Carolina

Nancy Goodwin pioneered the growing of cyclamen species in the USA at her Montrose garden and nursery in Hillsborough, North Carolina. Her garden is legendary and she has been extremely influential in bringing cyclamen to the attention of thousands of gardeners and in exemplifying the beauty and utility of cyclamen in rock and woodland gardens.

Hillsborough, North Carolina, is in USDA Zone 7a with minimum temperatures between −18 to −15°C (0°F and 5°F) and very hot summers with frequent droughts. In 2010, more than 90 days with temperatures over 32°C (90°F) were recorded. There are normally one or two snowfalls during the winter, but this snow does not persist.

Nancy fell in love with cyclamen when she first saw plants in England in 1963 and started growing them a year after she moved to Montrose in 1977, first in the old rock garden and then through the woods and other areas. She determined to grow as many species as possible and the descendents of these early plantings became the nucleus of Montrose Nursery, which was founded in 1984 specifically to make the cyclamen species grown from her seed available to the public. Cyclamen were the largest selling item throughout the 10 years that the nursery was open and, even now, they are the most popular plants sold at Montrose Garden open days.

Nancy joined the Cyclamen Society shortly after it was formed and has obtained many seeds both from the society's seed exchanges and from the late Jim Archibald and others with whom he put her in touch. She grows virtually all of the species and selections that are in cultivation. *Cyclamen creticum*, *C. balearicum*, *C. africanum*, *C. mirabile*, *C. rohlfsianum* and *C. persicum* are grown in a cold greenhouse, together with special selections of other species for hand pollination. *Cyclamen cyprium* grows only in the greenhouse; it was tried once in the rock garden, it bloomed but didn't survive the winter. Large areas in the woods at Montrose are filled with *C. coum*, *C. hederifolium*, *C. confusum*, *C. cilicium*, *C. graecum*, *C. pseudibericum*, *C. repandum* and *C. rhodium*, and there are smaller areas with *C. mirabile*, *C. purpurascens*, *C. colchicum* and *C. intaminatum*.

The soil at Montrose is moderately acid, pH 5.8 or so, and the cyclamen grow beneath pines, oaks, metasequoias, sweet gums, beeches and maples. *Cyclamen graecum* is planted in rubble; some cyclamen are in sun and others in shade. *Cyclamen alpinum*, in both pink and white forms, has been grown since the mid-1980s on a rocky slope but not so successfully on level soil. Perhaps Nancy's most prized plant in the garden is a specimen of *C. elegans*, which

Nancy Goodwin from Montrose, North Carolina has done much to introduce cyclamen species into the USA. Many others besides the *C. coum* shown here are grown in the open garden (NG)

came from Hans Simon and which is a direct descendant of the plant found by him and Jim Archibald in Iran. Somehow, it persists with no protection. For a number of years, *C. balearicum* grew and flowered in the rock garden and *C. libanoticum* grew but never flowered near an outbuilding. *Cyclamen parviflorum* enjoyed the winters but not the summers and is not currently grown.

Predation of cyclamen flowers (but not leaves or tubers) by deer has been a perennial problem that was solved by the installation of a deer fence three years ago. Since then, there has been a population explosion and cyclamen seedlings have appeared in many new areas. They have naturalised in the lawns near several shady gardens as well as in full sun in adjacent areas.

Seedlings, one to a pot, are grown on shelves in an unheated outbuilding, and transplanted in late summer of the following year either into rose pots for sale or into the garden. Plantone fertiliser together with a little lime is used when planting; plants in the garden are not fertilised further. Greenhouse plants are fertilised every second week from autumn through early spring with a quarter-strength solution.

Nancy's experiences with cyclamen are described in her two books, *Montrose: Life in a Garden* and *A Year in our Gardens*.

Robin Hansen, North Bend, Oregon

Robin Hansen has been producing cyclamen for over 25 years and has grown all of the available cyclamen species in pots in polyhouses at her zone 9 nursery, which rarely receives snow and generally has dry summers with high temperatures up to 32°C (90°F). The polyhouses have a permanent covering of 55–63% shade material, which is augmented between late October and April with 6 mm polythene to protect against winter and spring rainfall.

Ellen Hornig, Oswego, New York

Cyclamen were one of the specialties of the recently closed Seneca Hill Perennials Nursery, which was developed and run by Ellen Hornig. Winter temperatures at Oswego can drop as low as −29°C (−20°F), although there is consistent snow cover between December and March. Oswego is well known for its massive amounts of lake-effect snow caused by its proximity to the south-eastern shores of Lake Ontario. The snow proved to be a significant problem for Ellen's hoop houses, the build-up requiring constant removal and also reducing airflow, sometimes leading to problems with *Botrytis* disease. Ellen introduced *C. coum* subsp. *coum* f. *albissimum* 'Lake Effect', a beautiful seed strain with white flowers, occasionally opening pink if temperatures are low. The blooms of this form are large and full, sometimes with slight serrations on the petals, and the foliage is typically deep matt green, lightly marked in silver. In the garden, Ellen grows *C. coum*, *C. hederifolium* and *C. purpurascens* and reports early indications of success with *C. graecum*, no doubt thanks to the protection afforded by the snow cover. She also has remarkable success with a number of South African plants for similar reasons.

Nina Lambert, Ithaca, New York

Nina Lambert sadly passed away in late 2010, but not before she had created a beautiful and much-toured rock garden at her home in upstate New York. Nina's garden was very well known for her extensive plantings of *C. coum*, which she had grown since the 1960s. For many years, Nina had generously distributed a strain of *C. coum* that she considered to be particularly hardy, tolerating temperatures as low as −29°C (−20°F) without consistent snow cover. *Cyclamen hederifolium* and *C. purpurascens* also thrived in her garden and seeded around.

John Lonsdale, Exton, Pennsylvania

John Lonsdale's aim at Edgewood Gardens in Exton, Pennsylvania, is to grow as many plants, including cyclamen, as possible in the open garden, pushing the traditionally accepted limits of 'hardiness' where possible.

His family's 1.6 acre garden is in USDA zone 6b at an altitude of 200 m, where winter minimum temperatures have reached as low as −20°C (−5°F) and summer maxima are over 38°C (100°F). Summer 2010 had more than 55 days over 32°C (90°F). Humidity is generally low from late autumn to early summer but is very high in July and August, often accompanying temperatures in the 32–38°C (90–100 °F) range. Much of the winter passes without snow cover, a total depth of around 60 cm falling in a normal winter. In the winter of 2009–2010, however, there was over 190 cm. Ice storms can be as frequent as snow storms and bring their own issues for plants. Rainfall is approximately 100 cm per year, being distributed fairly evenly throughout the year; July is the wettest month (11 cm). The percentage of sunshine hours averages around 55%.

At Edgewood Gardens, the soil pH is slightly acid and the soil is very well-drained, quite rocky in places. The climate and perfect drainage in the garden provide excellent conditions for growing 'in-character' plants, especially bulbs, woodland plants and many flowering trees and shrubs. Much of the garden's tree canopy is provided by native deciduous trees, particularly American beech, several oak and hickory species, tulip trees and dogwoods; other natives include *Rhododendron periclymenoides*, *Vaccinium* species, *Arisaema triphyllum* and *Chimaphila maculata*. There are extensive plantings of bulbs and herbaceous plants including *Cyclamen*, *Trillium*, *Helleborus*, *Cypripedium*, *Hepatica*, *Crocus*, *Iris*, *Corydalis* and *Erythronium*. John has found the cold-hardiness of many bulbs, including cyclamen, to be greater than expected, but correct positioning and soil conditions are crucial. Excellent drainage is paramount; it is ice rather than cold that kills, and bulbs and corms in a soil that was only slightly moist when it froze can survive far lower temperatures than those that freeze in a wet medium. This phenomenon is exaggerated when plants are grown in pots. Overnight watering from sprinklers is only provided if conditions become extremely dry in the summer months, as usefully indicated by the collapse of plants of *Asarum canadense*.

The backbone of the autumn display in many a garden, *C. hederifolium* always flowers before the leaves in John Lonsdale's Edgewood Gardens, Exton, Pennsylvania (JL)

The population of white-tailed deer in the eastern USA has exploded in recent years and they are a major impediment to successful gardening. Few plants are truly unappetising to these deer, especially in periods of severe winter weather, although hellebores and peonies have not yet been touched. The vast majority of the garden is protected with a continuous 2.5 m high polypropylene deer fence, with a 4 m gate across the driveway. The black 4 cm square weather-resistant mesh fence is unnoticeable from more than 6 m away. Damage by deer has been effectively eliminated, as long as the integrity of the fence is maintained.

Cyclamen are a major feature of the garden, providing interest from early August through to May. *Cyclamen hederifolium* forms a backbone planting, having self-sown so extensively that there are now more than 10,000 plants throughout the garden. It doesn't seem to favour any particular areas, although growth and re-seeding is most extensive on a south-facing slope on the woodland edge, many parts of which offer relatively little shade. *Cyclamen hederifolium* always flowers well ahead of leaf formation, presumably because of the relatively high temperatures and reduced moisture available at flowering time. It is interesting to note that the few precocious flowers that form in August are often burned off at the base of the stem by the heat of the gravel top-dressing that covers some of the plantings. Also of note is the fact that silver-leaved plants seem more resistant to collapse at temperatures below freezing; leaves look perfectly fine at –3°C (26°F), when those of plants with green leaves have already been reduced to tissue paper.

Cyclamen coum has also self-sown extensively, favouring similar but slightly less hot and dry places to *C. hederifolium*. Flowering does not generally begin until early March. *Cyclamen coum* subsp. *coum* f. *albissimum* 'Golan Heights' has proven winter hardy but is susceptible to rotting, the other pure white *C. coum* 'George Bisson' and 'Lake Effect' are more robust. While *C. coum* tubers are impervious to winter damage, the leaves are not. By the end of the winter, they can be quite burned if there have been a number of snow and ice storms causing many freeze–thaw cycles. On the other hand, leaves that have spent much of the winter under snow emerge in perfect condition. Flowers generally don't appear until early-mid March and are unaffected by the weather.

Cyclamen elegans survives most winters but definitely prefers areas that retain snow cover the longest and therefore are most protected. *Cyclamen purpurascens* will only succeed in areas that do not receive any direct sun and that never dry out. In such conditions, it flowers well and self-sows modestly; when mistakenly tried in more sunny and drier places, the leaves quickly burnt off in early summer. *Cyclamen parviflorum* enjoys similar conditions to *C. purpurascens*. *Cyclamen alpinum* grows well in similar situations to *C. coum* but does not seed around. *Cyclamen cilicium*, *C. intaminatum* and *C. mirabile* have all proved to be reliable outdoors in semi-shaded positions; seed is set on all three species. *Cyclamen mirabile* seems especially susceptible to moisture in summer and needs sites that ensure complete dryness. Some versions of *C. mirabile* 'Tilebarn Anne' are susceptible to severe leaf-curling in early fall when temperatures are still high; the leaves do not unfurl after this has happened. The bright raspberry-red colour of leaves of 'Tilebarn Anne' and 'Tilebarn Nicholas' is especially intense in the heat and high light intensity at this time. *Cyclamen pseudibericum* makes a superb garden subject, not flowering until late March and early April, the time at which the leaves of various 'forms' of *C. rhodium* appear above ground. These are the last cyclamen to flower in Exton, their flowering not starting until mid-April. Species not mentioned above have been tried extensively but do not succeed in the open garden, although volunteer seedlings of *C. graecum* appear from time to time and last for a couple of years.

The remaining species do, however, make superb subjects for pot cultivation in two greenhouses that are heated merely to prevent freezing and that have automatic forced ventilation when temperatures are above 7°C (45°F). Both greenhouses have 70% Aluminet shading from late February to mid September. Cultivation in pots, including seed pots, uses a 50:50 mix of Biocomp BC5 compost (composted peanut shells and pine bark) and perlite, with a 1 cm top dressing of the grower-grade granite grit sold for poultry feeding. Feeding with each watering using MiracleGro is achieved using an-line dosing system. Water is withheld from all pots in late June and watering is re-started in mid-September, but only if temperatures are generally below 27°C (80°F). Trays of seedlings, pots and mature plants other than *C. graecum* and *C. rohlfsianum* are covered with sheets of 1 cm

At Edgewood Gardens, Exton, Pennsylvania, *C. coum* thrives in slightly cooler places than *C. hederifolium*. Its leaves can be damaged in winter if not protected by snow cover (JL)

A white form of *C. coum* at Edgewood Gardens, with *Iris* 'Katharine Hodgkin' (JL)

The striking *C. pseudibericum* is particularly successful at Edgewood Gardens, flowering late March or April (JL)

thick Styrofoam house insulation to prevent excessive heating and dryness, which can cause the desiccation of tubers; *C. coum*, *C. hederifolium* and *C. repandum* seem to be the most susceptible. Pots of seedlings can be lightly watered towards the end of summer if dryness is a problem. Seed is sown without prior soaking in August each year in pots of the above-described compost, which are immediately watered and placed on the floor of the greenhouse beneath the benches. Germination begins in October and is generally complete by late November, with the exception of *C. purpurascens* which always germinates in May.

Jane McGary, Portland, Oregon

Until very recently Jane McGary, one of North America's foremost growers of dwarf bulbs, gardened in Estacada, Oregon, where she grew an amazing collection of dwarf bulbs in unheated cold frames and the garden. The average winter low temperature in her garden is approximately –11°C (12°F), the typical summer high 35°C (95°F) and the average rainfall 115 cm, virtually all between October and mid-June. In the garden, Jane was successful with *C. cilicium*, *C. coum*, *C. hederifolium*, *C. mirabile*, *C. pseudibericum* and *C. repandum*, but in her unheated cold frames, in which the temperature could fall to –7°C (20°F), Jane was also able to grow *C. africanum*, *C. alpinum*, *C. confusum*, *C. graecum* subsp. *graecum*, *C. intaminatum* and *C. persicum*. Now relocated to a Portland suburb where the bulb collection resides in a large, screen-sided alpine house, Jane has moved most of the *C. graecum* into the open rock garden, where it overwinters with little or no damage. Jane urges much greater use of this handsome species, especially in California. In Oregon's wet winters, *C. mirabile* is susceptible to *Botrytis* mould under cover but, surprisingly, remains healthier in the open. Most of the species grown in both gardens are in afternoon shade or in full sun but on an east- or north-facing slope. *Cyclamen purpurascens* is unsatisfactory in the maritime PNW, probably because it does not get cold enough winter dormancy; it will survive for many years but rarely flowers.

Pests and diseases

'Critters' that like to eat cyclamen seem to be far more abundant and variable in the USA than they are in the UK. Constant vigilance must be employed outdoors to minimise feasting on flowers, leaves and tubers by deer, mice, moles, voles, chipmunks, groundhogs, rabbits and squirrels. Some protection can be afforded by burying wire mesh above the tubers or by mulching with at least 2.5 cm of gravel. Mice can also cause damage in greenhouses, and are fond of eating the late winter

Cyclamen pseudibericum in Pennsylvania, looking so at home it could be easily mistaken for the natural habitat in Turkey, except for the background Trillium! (JL)

flowers of *C. coum*. Slugs and snails can be a major problem in the PNW, but generally not in the Mid-Atlantic states.

Greenfly can be a problem at any time, particularly on new fall foliage and vine weevil can be an issue in some places; treatment with imidacloprid is effective against both. The fungal disease *Botrytis* can be a problem anywhere, but especially so in the PNW, which has a climate similar in many ways to that of the UK, with greatly increased humidity during the fall and winter. In the Mid-Atlantic and south-western states, *Botrytis* is much less of an issue, rarely causing problems during extended periods of cool damp weather if spent flowers and stems are left on the plants. Damping off of seedlings can be a problem in late winter in similar conditions, and a prophylactic treatment with Banrot is effective in preventing this problem.

Companion plants for cyclamen in the garden

Cyclamen form an excellent groundcover under trees and shrubs, especially deciduous ones, though they also survive under dense conifers. Planting the more demanding species under conifers, particularly pines, can be especially beneficial in regions that have high winter rainfall. Cyclamen grow well with hellebores, enjoying the same cultural conditions, and other excellent companion plants include *Anemonella*, *Asarum*, *Corydalis*, *Crocus*, *Erythronium*, *Epimedium*, *Hepatica*, dwarf *Iris* and *Phlox*, and *Trillium*. In the PNW, a pleasing combination is *C. hederifolium* among the 'black mondo grass', *Ophiopogon planiscapus* 'Nigrescens'. When planning combinations, remember that most cyclamen species are dormant in mid to late summer, and other herbaceous plants can be used to trail over the resulting bare spots. Planting as many cyclamen species as will thrive in a given garden provides an almost year-round cycle of flowering, but *C. hederifolium* is better isolated from smaller species lest it out-compete them. Given a dry-summer climate, *C. hederifolium* will grow in almost any site from deep woodland shade to vertical cliffs, and once established in a garden, it often seeds into the lawn.

Acknowledgements

Many thanks to Nancy Goodwin, Jane McGary and Jim McKenney for their very helpful input into the preparation of this section.

North American nurseries specialising in cyclamen

Edgewood Gardens, Exton, Pennsylvania, USA (http://www.edgewoodgardens.net)

Hansen Nursery, P. O. Box 1228, North Bend, Oregon 97459, USA

(b) Australia

Jill Weatherhead

Miniature species cyclamen or rockery cyclamen are relatively uncommon but slowly becoming better-known in Australia, while the large florist's cyclamen, selections of *C. persicum*, are very popular as house plants in the winter months. Species cyclamen are hardy and easy to grow in many parts of the country, particularly in the south-east coastal strip. The Adelaide Hills in South Australia, the Great Dividing Range (including the Blue Mountains in New South Wales, and the Dandenong Ranges and Mount Macedon in Victoria) and much of Tasmania have the most suitable conditions, and there are also some growers in Melbourne, Sydney and Adelaide.

History of cultivation of cyclamen in Australia

The earliest reference to cyclamen indicates that *C. coum* and *C. persicum* were grown in 1845 in New South Wales. *Cyclamen persicum* is listed in Alexander Mackay's notebook of plants received at Elizabeth Bay, NSW, in 1845, sent from a Mr Backhouse. In the same year, William Macarthur (1800–1882) cultivated both species at Camden Park, south-west of Sydney. One of the finest of the nation's early nineteenth century country homesteads (the estate comprised 11,300 hectares (28,000 acres) at one stage), Camden Park is now the largest and most intact Australian early colonial garden.

The Camden Park Nursery published catalogues in 1843, 1845, 1850 and 1857. William Macarthur, who was later knighted, corresponded widely and exchanged plants and bulbs with botanical gardens, collectors and botanists in England, Europe and Asia. He purchased many plants from London nurseries (usually Veitch and Loddiges) and was involved in plant breeding. He distributed grape vines, plants and bulbs to other colonies in Australia. Under Macarthur's care, Camden Park became of such horticultural interest that John Gould Veitch wrote "The collection of plants and fruits at Camden is by far the best that I have seen in the colony. No means have been spared to obtain the best varieties in each case." *Cyclamen coum* and *C. persicum* were listed in the 1845 Camden Park Nursery catalogue; by 1850, *C. neapolitanum* (now *C. hederifolium*) was listed too and all three were available in 1857.

James Dickinson sold *C. persicum* (as "Persian Cyclamen") in 1845 from Hobart Town, Van Dieman's Land (now Tasmania). He was transported to the penal colony in 1838 and was released in 1842. By 1845, he had established a florists' business and 10 years later published '*The Wreath: Gardeners Manual, arranged for the Climate of Tasmania*'. Dickinson collected indigenous seeds and by 1870 was exchanging plants with botanist Baron von Mueller, who was curator of the Melbourne Botanic Gardens. Thomas Shepard (1779–1835) left England and arrived in Sydney in 1827. Governor Darling encouraged him to establish the first commercial nursery garden in Australia, which was named for him, the Darling Nursery. Stock came from the Sydney Botanic Garden, Elizabeth Bay House and Camden Park. Shepard published Australia's first gardening book, entitled '*Lectures on Landscape Gardening in Australia*', in 1836. His son T. W. Shepard (1824–1884) published the 1851 nursery catalogue, which listed *C. coum*, *C. europaeum* (now *C. purpurascens*), *C. persicum*, *C. neapolitanum* (now *C. hederifolium*) and *C. vernum* (now *C. coum*). As R. Polya notes in '*Nineteenth Century Plant Nursery Catalogues of Southeast Australia*': "Some 40 years after settlement Thomas Shepard complained that it was difficult to make his nursery economically viable because colonists were able to obtain supplies of fruit trees and other species from the Government Gardens". The Darling Nursery continued until 1900.

Another gardener, Michael Guilfoyle (1809–1884), arrived in Australia from England in 1849. By 1851, he had opened the Exotic Nursery in Double Bay, Sydney, on 1.4 ha (3.5 acres). The first catalogue for this nursery, published in 1851, listed about 1,500 plants, including *C. persicum*, *C. coum* and *C. europaeum*.
His son, William Robert Guilfoyle, is famous for designing the picturesque-style landscape of the Royal Melbourne Botanic Gardens, a fine example of the style and well-loved to this day. John Rule (1831–1887), who was of Cornish decent, went to Australia in 1840 and began one of the first nurseries in Victoria, the Victoria Nursery, in 1850 in Richmond, Melbourne. His 1855 catalogue, known for trees and conifers, listed *C. persicum* ('Persian sowbread') and *C. hederifolium* ('ivy-leafed sowbread'), each priced at 10 shillings and sixpence, but by 1857 the price was reduced to 2 shillings and sixpence or 5 shillings for larger plants. However, the prices fluctuate markedly in other catalogues from that time. Later, Rule also sold *Fritillaria*, *Galanthus* and nine *Crocus* species, and in 1857 he listed 70 roses. He published '*Rule's Economical Gardening for Cottagers; Being Concise Directions for Cultivating Useful Culinary Vegetables*', written by Thomas McMillan, in 1859. By now, the Victoria Nursery was also managed by John's brother, James; in 1856, John Sinclair wrote that it was "richly stocked with flowering plants, fruit trees, native and foreign, with greenhouses and stoves for their culture".

By 1857, *C. coum* and *C. persicum* were available from the Royal Society's Gardens in Queen's Park, Hobart. These gardens went on to list *C. africanus* (sic), *C. atkinsii* (*C. coum* subsp. *coum* or possibly the alleged cross between *C. coum* and *C. persicum* described by Atkins in 1861), *C. coum*, *C. europaeum* (now *C. purpurascens*), *C. neapolitanum* (now *C. hederifolium*), *C. persicum* and *C. vernum* in 1865. From the Cremorne Nursery, William Law and Sumner sold *C. hederifolium*, *C. persicum* "and others" in 1858. In their 1886 catalogue, they describe *C. persicum* as having "flowers white, white and pink or various shades of red". In 1867, their '*Garden Catalogue of Seeds etc*' listed "*C. persicum*, choice mixed, 6 inches". They published a '*Handbook to the Garden and the Farm — a Guide to Australian Cultivation*' in 1889.

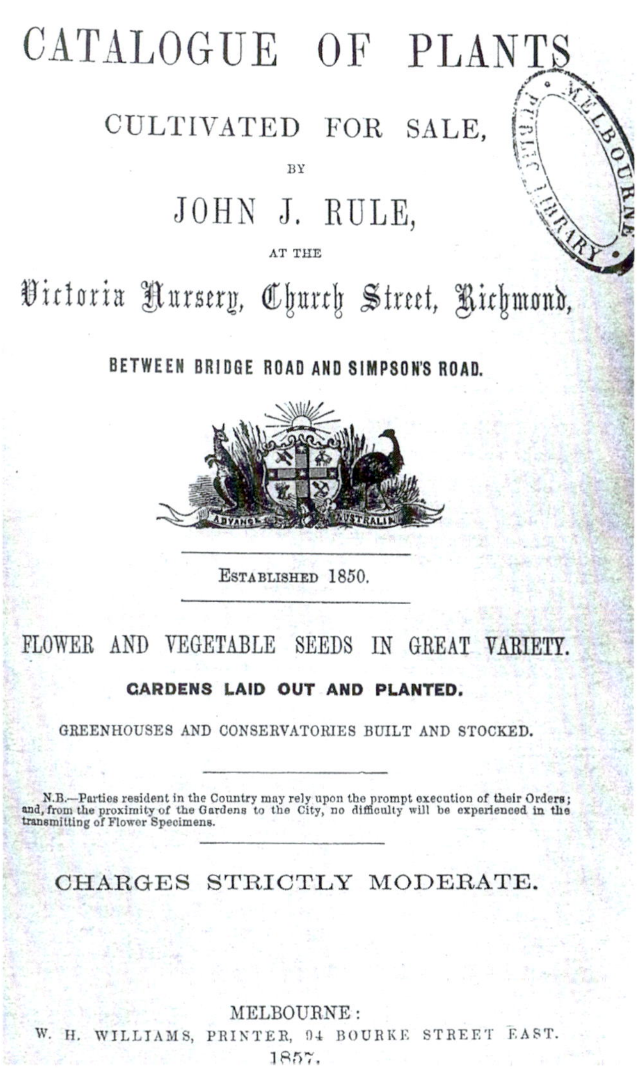

CATALOGUE OF PLANTS

CULTIVATED FOR SALE,

BY

JOHN J. RULE,

AT THE

Victoria Nursery, Church Street, Richmond,

BETWEEN BRIDGE ROAD AND SIMPSON'S ROAD.

ESTABLISHED 1850.

FLOWER AND VEGETABLE SEEDS IN GREAT VARIETY.

GARDENS LAID OUT AND PLANTED.

GREENHOUSES AND CONSERVATORIES BUILT AND STOCKED.

N.B.—Parties resident in the Country may rely upon the prompt execution of their Orders; and, from the proximity of the Gardens to the City, no difficulty will be experienced in the transmitting of Flower Specimens.

CHARGES STRICTLY MODERATE.

MELBOURNE:
W. H. WILLIAMS, PRINTER, 94 BOURKE STREET EAST.
1857.

Cover title from John J. Rule (1857)

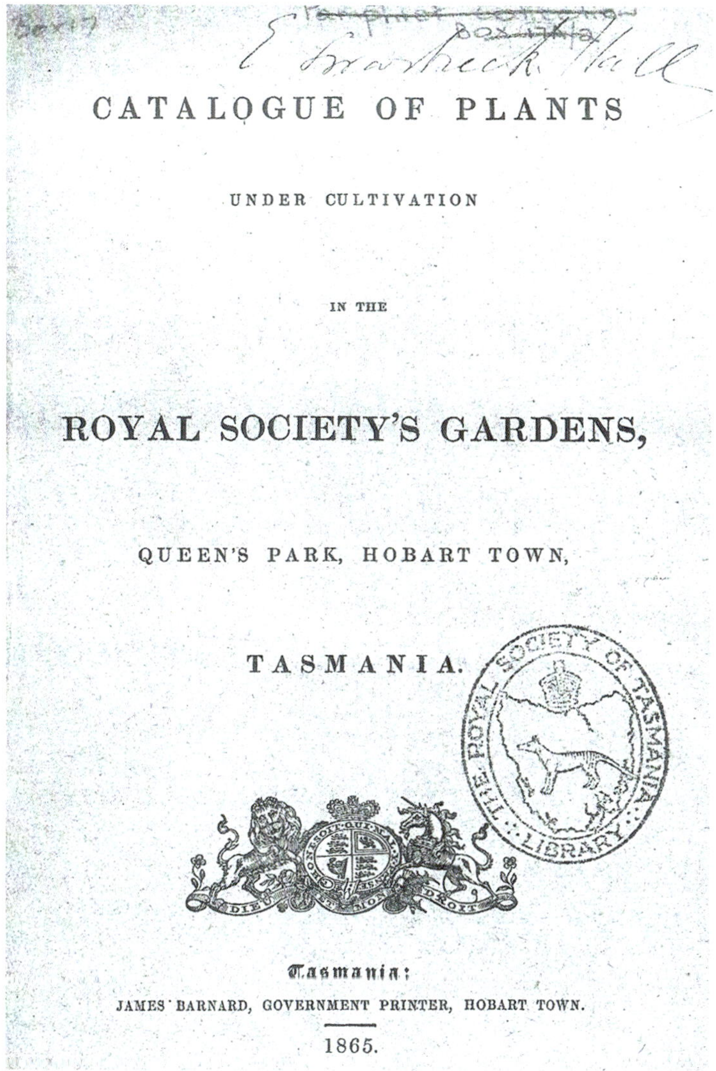

CATALOGUE OF PLANTS

UNDER CULTIVATION

IN THE

ROYAL SOCIETY'S GARDENS,

QUEEN'S PARK, HOBART TOWN,

TASMANIA.

Tasmania:
JAMES BARNARD, GOVERNMENT PRINTER, HOBART TOWN.
1865.

Catalogue of plants at The Royal Society's Queens' Park, Hobart

John T. Baptist (1804–1873) arrived in Australia in 1829. By 1839, he was selling vegetables, plants and flowers in Sydney from his Surrey Hills market garden. He began collecting ornamental plants, buying plants from Camden Park and the Darling Nursery and importing bulbs from England, in 1842. He named his nursery 'The Garden', and his 1861 and 1862 catalogues list *C. persicum*, *C. coum* and *C. neapolitanum* (now *C. hederifolium)*. Baptist imported plant material from England, Germany, France and occasionally Japan. He also collected seed and plants from the Australian bush. Later, the nursery became wholesale, and then in 1885, John Baptist sold all his stock to the Darling Nursery.

Handasyde, McMillan and Co. was established by 1864 in Melbourne. Gilbert Handasyde and Thomas McMillan were British immigrants, both nurserymen and seedsmen. They listed *C. purpurascens* in 1864, adding many more cyclamen to their catalogue in the following year: *C. coum*, *C. graecum*, *C. hederifolium*, *C. hederifolium album*, *C. macrophyllum*, *C. persicum grandiflorum*, *C. persicum album* and *C. persicum odoratum*. So in 1865, we see that both a great range of species cyclamen and also some large florists' cyclamen were available.

Thomas Lang arrived in Ballarat, Victoria, in 1855 from Scotland, setting up a small nursery by 1857. He grew fruit trees, roses and bulbs and he had an "extraordinary range of stock". In 1868, Lang, now partnered by William Elliott, sold *C. coum*, *C. hederifolium* ("rosy-pink") and *C. persicum* (both red and white), for 3 shillings and sixpence each, and *C. europaeum* ("red") for 4 shillings. He was also involved in the Ballarat Horticultural Society, which was formed in 1859. Charles Wyatt trained in horticulture at the Frogmore Royal Gardens at Windsor, and later migrated to Melbourne. He established the Frogmore Nursery at Geelong, Victoria, and sold "Cyclamen of sorts" for 2 shillings in circa 1873.

By 1880, seed merchant William Anderson listed *C. persicum giganteum* in his Melbourne-based nursery catalogue. With F Smith, he published '*The Colonial Gardener*' and their business, later named G. Brunning, has traded continuously from 1852 until the present.

James Scott, an English immigrant, established his nursery, later called the Royal Nursery, in 1853; this was one of the first nurseries in Melbourne. In 1880, his nursery listed 400 ferns, and by 1889, it also listed *C. persicum*.

In 1889, Law and Somner were selling *C. persicum* for between 1 shilling and sixpence and 2 shillings and sixpence. Their 1893 seed catalogue lists *C. persicum* and its varieties *C. persicum giganteum*, *C. persicum giganteum* 'William's Collection' and *C. persicum giganteum album*; and their 1896 list has *C. persicum giganteum* "raised from the finest imported seed". By 1905, Law and Somner also listed *C. persicum giganteum* 'Bush Hill Pioneer' and "new fimbriated cyclamen 'Papillio' ". It is hard to know whether cyclamen arrived in Australia as tubers or seed, but the shipment of tubers would require the use of Wardian Cases (like miniature glasshouses), which were expensive, whereas growing cyclamen plants from seed is very easy. Certainly the State Library of Victoria contains seed catalogues such as John Cattell's English list, which includes white and crimson *C. persicum*, priced sixpence per packet; this was German seed sent via Kent. I think it is safe to assume that most

Back cover of a Law, Somner & Co. Catalogue (1893)

cyclamen plants were grown from seed in the nineteenth century, as outside a Wardian case, tubers would dehydrate on the long trip to Australia by ship. R. Polya (1981) notes: "While seeds are the most easily transportable form of plant life there are inherent problems associated with them. They often deteriorated on the long voyage from England … . Seed merchants continued to stock both imported and local seed."

The cyclamen trail grows a little cool until 1932, when the Chandler family's Como Nursery, in the foothills of the Dandenong Ranges east of Melbourne, listed Swiss cyclamen (*C. purpurascens*) for 1 shilling and sixpence; *C. neapolitanum* (now *C. hederifolium*), *C. repandum* and 15 shades of *C. giganteum persicum* (sic) for 1 shilling; and frilled varieties of *C. persicum* for 1 shilling and sixpence. Interestingly, they also sold "rosy claret" *C. neapolitanum* (now *C. hederifolium*) for 1 shilling or six for 4 shillings and sixpence. Known more for its trees and shrubs, the nursery, by now trading as Chandler's Nursery, closed its doors in 1988.

Phillip Montague had established his nursery in Melbourne by the 1930s when he sold "rosy crimson" *C. neapolitanum* (now *C. hederifolium*) for 2 shillings, separate colours of *C. persicum* for 2 shillings, and mixed colours for 1 shilling and sixpence or 15 shillings for a dozen tubers. Montague moved from Frankston to Murrumbeena and his 1960 catalogue includes *C. repandum*, *C. cilicium*, *C. europaeum* (light and dark pink) and *C. neapolitan* (sic)

CYCLAMEN.
Persicum Giganteum.

Persicum varieties — Raised from the finest imported seed; flowers white, white and pink, and various shades of red. They make lovely border plants, and delight in rich, light soil. The tubers should be planted with the crowns level with the service, and after flowering they should be well mulched. They make beautiful pot plants, and are easily cultivated. When the plants have done flowering gradually ripen them off by reducing the water, but never let them get quite dry; place them in a cool, shady place during the summer months, and re-pot end of January. Price 1s. 6d. to 2s. 6d. each.

Detail from the catalogue of Law, Somner & Co. (1893)

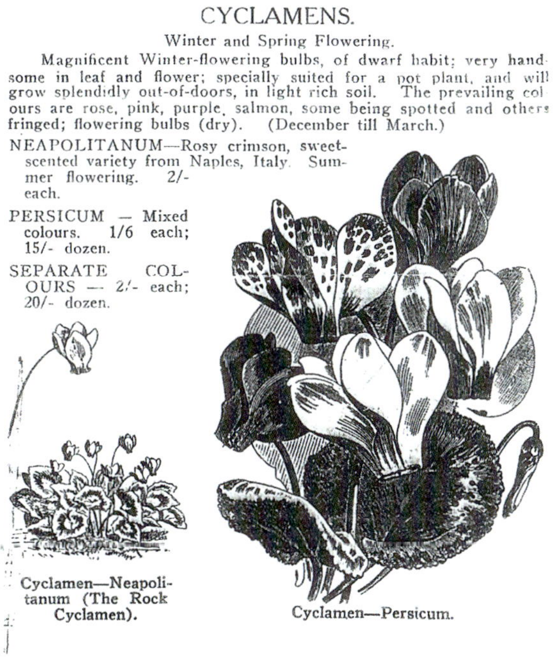

CYCLAMENS.

Winter and Spring Flowering.

Magnificent Winter-flowering bulbs, of dwarf habit; very handsome in leaf and flower; specially suited for a pot plant, and will grow splendidly out-of-doors, in light rich soil. The prevailing colours are rose, pink, purple, salmon, some being spotted and others fringed; flowering bulbs (dry). (December till March.)

NEAPOLITANUM—Rosy crimson, sweet-scented variety from Naples, Italy. Summer flowering. 2/- each.

PERSICUM — Mixed colours. 1/6 each; 15/- dozen.

SEPARATE COLOURS — 2/- each; 20/- dozen.

Cyclamen—Neapolitanum (The Rock Cyclamen).

Cyclamen—Persicum.

Detail from a catalogue of bulbs and tubers, Philip Montague, 1930

(pale and dark pink), all at 3 shillings, or giant tubers for 5 shillings. The early 1960s also saw Jensen's Nursery in Exeter, New South Wales, sell "rose and white" *C. neapolitanum*, *C. coum* and "crimson" *C. repandum* for 2 shillings and sixpence each. I believe it was owned by the Jensen brothers and ran for some decades.

In 1958, Ruth Tindale began to plant *C. hederifolium* in her Dandenong Ranges garden east of Melbourne, now the George Tindale Garden (open daily), which boasts huge sweeps of this hardy gem. This garden promotes cyclamen in Victoria, particularly the easily grown *C. hederifolium* which flowers profusely in autumn.

Bruce Muir began his life-long love of cyclamen in the early 1970s, growing little else bar some *Iris* and *Dierama*. Bruce grew many cyclamen from seed collected during the expeditions of the Cyclamen Society, selling cyclamen at Melbourne markets, until his death in 2003. Meanwhile, Joan Law-Smith created 'Bolobek', one of Victoria's finest twentieth-century gardens, where she created woodland areas populated by *Anemone nemorosa* and *C. hederifolium*. Open for many years was the Garden of St Erth near Blackwood in the ranges north-west of Melbourne, where Tommy Garnett grew a fine range of cyclamen. He wrote about the genus in *The Age* newspaper in 1961 or 1962, inspiring gardeners to try growing *C. hederifolium*. Nearby, Rene and Bill Coffield in Creswick sold *C. cilicium* and *C. hederifolium* in the 1980s and 1990s, and by 2001, Woodbank Nursery in Tasmania listed these, *C. coum* and *C. persicum*. Sadly, both the Creswick and Woodbank Nurseries have now closed. During the 1990s, I listed many cyclamen species through Possum Creek Perennials, including *C. cilicium*, *C. coum*, *C. coum* 'Silver Leaf', *C. intaminatum*, *C. libanoticum*, *C. persicum*, *C. persicum album*, *C. purpurascens*, *C. purpurascens* 'Silver Leaf', *C. repandum* and *C. repandum album*. Glenbrook Bulb Farm listed 14 varieties in 1992, but the premier grower is Marcus Harvey in Hobart, and he lists an astonishing 30 species and varieties.

Cyclamen hederifolium at the George Tindale Garden (LB)

Cultivation of cyclamen in Australia

Australia is a large country but a small continent, with climates ranging from alpine regions and wet rainforests through to the tropical regions, drier forests and deserts.

The populated south eastern coastal strip has a Mediterranean-like climate so, unsurprisingly, most cyclamen perform quite well and flower well in Adelaide, Melbourne and Sydney; but generally grow better, and set seed, in cooler Hobart and winter-cold Canberra, and in cooler, higher rainfall areas: the Adelaide Hills, the Dandenong Ranges east of Melbourne, the Blue Mountains west of Sydney and the Great Dividing Range.

One of Victoria's most beautiful gardens, 'Bolobek', was created by Joan Law-Smith, who writes in '*The Garden Within*' (1991) that *C. hederifolium* grows in "sheets of pink and white … we always know when autumn is in the air by the appearance of the first flowers. It is only later that the leaves appear, having been underground since November. They are a perfect example of nature's artistry in matching leaf to flower, and their foliage will remain long after the flowers have finished. The leaves are basically dark green with an irregular mottling of silver, and no two are alike in size, shape or colour."

Hailing from the Mediterranean (Italy to western Turkey), *C. hederifolium* is the most common species of *Cyclamen* in Australia and very beautiful. It is probably the easiest species to cultivate in Melbourne, Hobart, Adelaide and Sydney, where it is happiest in semi-shade under deciduous trees or shrubs. In the higher rainfall areas, it can grow under evergreen trees. As James Hitchmough notes in '*Garden Bulbs for Australia and New Zealand*' (1989), *C. hederifolium* will grow well throughout the southern half of Australia given a well-drained soil that is moist during their growing season. Perth, however, might be too hot for even this easy species. In the cooler wetter areas, *C. hederifolium* self-sows freely and creates large naturalised drifts, with seeds carried by ants extending the area that they occupy. The pink form is most common but the slightly larger white form is well worth seeking. The plants tolerate dry shade well and relish leaf mould, growing naturally amongst fallen leaves. *Cyclamen hederifolium* flowers prolifically during autumn and can, in some areas, begin flowering before Christmas and last until June. *Cyclamen hederifolium* did not self-sow for botanist Hilary Weatherhead in the sandy soils of Bentleigh, Melbourne, but was successful in the heavy soil of the nearby hills. Once *C. hederifolium* is happy in your garden, other species can be grown. It looks particularly fine with *Colchicum speciosum* flowering behind in the George Tindale Garden in the Dandenong Ranges.

Suzanne Price, writing in '*The Urban Woodland*' (1986), gave the influential advice: "In areas where trees do not meet overhead or where an open area merges with woodland, there is shade either dense or dappled for part of the day and sun for the rest of the day. These situations require ground plants which are equally happy in either sun or shade … the dwarf hardy cyclamen are most useful in these areas as in fact they are anywhere in an urban woodland. If you purchase *C. hederifolium* for autumn flowers and *C. coum* for winter, and … *C. persicum* for spring and *C. purpurascens* for summer, you will have masses of these sweet dumpy flowers all year round." As I already knew and loved *C. hederifolium* and *C. coum*, this book inspired me to try different species, especially *C. purpurascens* for summer flowers. This book and the large sweep of *C. hederifolium* at the George Tindale Garden in the Dandenong Ranges have influenced many gardeners to try growing these hardy gems.

Other autumn-flowering cyclamen include *C. cilicium*, which grows easily and flowers profusely outdoors under a nectarine in my own garden in the foothills of the Dandenong Ranges. From southern Turkey, it is similar to *C. hederifolium* at first glance, but the flowers are often honey scented and have no auricles. The cultivation of *C. cilicium* is similar to that of *C. hederifolium*: both require moist soil in winter but well-drained, dryish soil with leaf mould added in summer. The situation under deciduous trees is ideal, and I find that *C. cilicium* tolerates some summer moisture well. Autumn also sees the flowering of *C. graecum*, which has attractive flowers over usually distinctive, well-patterned thick leaves. This species requires a deep soil in semi-shade or full sun in the higher gardens and dislikes being moved. I have grown *C. graecum* subsp. *anatolicum* from seed from the Cyclamen Society expeditions and it has handsome silver-patterned leaves; it seems happy in a deepish pot. *Cyclamen graecum* is probably the most sun tolerant of the genus; Don Barrett grows it in Adelaide and says it does well there due to summer baking. Nearby, in the Adelaide Hills, Chris Steele-Scott grows it in a hot sunny spot, while Trevor Nottle grows *C. graecum* with the tuber half-exposed. In the Dandenong's, Hilary Weatherhead (at 320 m) grew *C. graecum* in semi-shade, whereas Otto Fauser found it too cold at Olinda (at 540 m) for this species to flower — unsurprising for a lowland bulb from Greece, the Aegean islands, Turkey and Cyprus.

Cyclamen africanum is similar to *C. hederifolium* in appearance but has slightly larger leaves and flowers, and the two hybridise easily. This species is one to grow in Melbourne and Adelaide, whereas in higher or cooler areas, the tubers require good drainage and some protection from frost.

Cyclamen cyprium has elegant, slim, fragrant white flowers with crimson markings on the noses in autumn. Hailing from mountain woodland in Cyprus, it grows well in the Dandenong Ranges but might require some shelter in the coldest winters. Forms with pewter-splashed leaves are very handsome and worth seeking out, particularly the form 'ES' (standing for Elizabeth Strangman), which is boldly splashed with silver. Complete drought in summer is probably best for this species.

The tiny flowers and leaves of *C. intaminatum* have a particular charm. This gem can be lost in the garden, so pot-culture is best; I grow it scattered through a shady trough. Miniature creamy white

Cyclamen persicum in Otto Fauser's garden at Olinda (JW)

flowers appear from December to June over tiny rounded leaves of green or green stippled with cream, and growers say that this species is never really dormant. The pale pink form is desirable also. My plants tolerate summer moisture well; they receive only morning sun and flower sporadically, perhaps because my seedlings are still too young to flower reliably.

Winter sees the chubby flowers of *C. coum*, which are carmine, lilac-pink or white with a pink nose, rarely pure white. This species hails from Bulgaria, Turkey and the Caucasus and is very hardy in south-eastern Australia. Its leaves are kidney-shaped and deep green or green with silvery patterns like Christmas trees — superb varieties are sometimes sold as *C. coum atkinsii*. Some plants have yellow and green leaves; rather anaemic looking, these may be considered unattractive. Forms with fully silver leaves with varying shades of pink flowers are available and very beautiful. Renamed forms include *C. elegans* (syn. *C. coum* subsp. *elegans*), which has attractive larger flowers and silvery variegation to the leaves. This form is grown by Viv Condon, an alpine enthusiast, in her cool glasshouse in Olinda, raised from seed collected in Iran by Jim and Jenny Archibald (number 362.001), which was sown in September 2001 and first flowered in spring 2006.

Cyclamen coum is an invaluable species as it flowers right through winter in south-eastern Australia, ideally positioned in full sun in winter and semi-shade thereafter. I grow many varieties of it in pots, which I place by my front door when the plants are in flower, but it is also very happy in the ground. Lydia Bartlett was a keen grower who kept *C. coum* in pots in Melbourne, and plantsman Otto Fauser grows it in the open ground in cold Olinda, at the top of the Dandenong Ranges. A carpet of this species makes a wonderful picture, particularly when growing near pink *Daphne* or *Correa* 'Dusky Bells' or under the burgundy bells of *Clematis cirrhosa* 'Langsdowne Gem'.

Winter- and spring-flowering *C. persicum* hails from Turkey, Lebanon and Cyprus. This species has fragrant white flowers with a rosy stain around the nose; pink forms are less common. The toothed leaves are often beautifully marbled and patterned in grey on green. I have not tried this species in the hurly-burly of the garden, but it grows easily in pots in semi-shade, requiring a dry summer dormancy period, which can be provided by the uptake of water by the roots of nearby deciduous trees or shrubs. *Cyclamen persicum* does not appear to thrive in Melbourne or Adelaide but prefers higher or cooler gardens.

Cyclamen pseudibericum in the garden of Otto Fauser (JW)

In early spring, the violet-scented, magenta flowers of *C. pseudibericum* appear over the toothed grey-green leaves. From the mountains of Turkey, this species grows well in pots or in the ground in the cooler mountains of south-eastern Australia. Fairly shade tolerant, particularly in low-altitude gardens, this species is very pleasing under crabapple blossom. Next, we see the beautiful soft shell-pink flowers of *C. libanoticum*, which blooms from August to October. From Lebanon, this species seems unhappy in pots but grows well in the ground in Melbourne, Hobart and the surrounding ranges. James Hitchmough notes that this species should perform well throughout south-eastern and south-western Australia, but I have only heard of it growing in south-eastern Australia where it is easily lost in wetter winters while growing in pots. *Cyclamen libanoticum* self-seeds in rich soil on a steep slope in Otto Fauser's garden. I think the key here is to buy tubers in summer and to plant immediately into rich soil with excellent drainage, or alternatively, to grow this species from seed while taking care to avoid over-watering.

Cyclamen repandum, from southern France, Italy, the Adriatic coast of the Balkans, Greece and the Aegean Islands, is a most elegant species that has slim cerise flowers from August to November. The ivy-shaped leaves can be green or stippled white; the form 'Pelops' (*C. repandum* subsp. *peloponnesiacum*) has very white splashed leaves, and is a beautiful plant, well worth seeking. A white form of *C. repandum* is available and seems just as hardy as the cerise forms. This is the most shade-tolerant of all the cyclamen species and grew well in the dense shade of a Himalayan Cedar in Hilary Weatherhead's garden in the Dandenong Ranges. A shady

Cyclamen libanoticum in the garden of Otto Fauser (JW)

Painting of *Cyclamen rhodium* subsp. *peloponnesiacum* by Kay Craig, 2006

Cyclamen rhodium subsp. *peloponnesiacum* in the garden of Otto Fauser (JW)

spot is probably required for *C. repandum* to succeed in Hobart, Melbourne and Adelaide; I grow it in semi-shade where it thrives on neglect both in pots and in the ground, while it has naturalised into a large area in the Adelaide Hills garden of Trevor Nottle. *Cyclamen repandum* is the last species to appear each year with tardy leaves emerging in winter.

Cyclamen rohlfsianum is rarely grown in Australia, and when it is, rarely flowers. I grow it in a sunny spot but it seems shy in all areas except in our driest city Adelaide, where cyclamen connoisseur Don Barrett coaxes his plants to flower but not to produce seed. Again, rich soil is probably best and it should be moist but well drained, particularly in summer. In winter-cold Canberra, Paul Tyerman generally grows *C. rohlfsianum* near the shelter of the house to protect the tubers from frost, in 2011, he discovered that some seedlings tolerated frost to –9 °C in an unheated plastic shade house; they also flowered more profusely than in previous years after he moved them to a position with more sunlight.

Delicate silver-worked heart-shaped leaves make small *C. balearicum* very special and it sports sweetly scented white flowers in spring. This grows well in pots in the garden kept fairly dry in summer and it thrives in both warm south-eastern Australia and the higher gardens where rainfall is greater.

As Suzanne Price writes, *C. purpurascens* (syn. *C. europaeum*) flowers in summer, and it is almost evergreen in Melbourne. Forms with silvery and patterned leaves are available; I have a handsome, fully silver-leaved form that I grew from English seed from Linnett Farm. I grow this species under the veranda facing south-east, where the plants thrive although they receive no direct sunlight after 11 am, with temperatures varying from 2 to 30 °C. Out of 12 plants, I generally have nine flowering at any time in summer while for Paul Tyerman in Canberra, *C. purpurascens* thrives and flowers "virtually year-round". The first decade of the

A silver-marked form of *C. purpurascens* flowering at Christmas in Australia (JW)

Cyclamen purpurascens is, in general, easier to grow in pots (JW)

new century saw severe droughts in southern Australia and in the hottest seasons *C. purpurascens* tubers started to flower in October. I have been advised to keep this species dry in summer, but the plants should not be bone dry: those that I grow under cover do need a fortnightly splash from a watering can as this species comes from an area of central Europe that receives year-round rainfall; *C. purpurascens* can do well in a cool garden in shade but needs excellent drainage as it might be lost in wet seasons. Interestingly, if allowed to dry out, the leaves become leathery, and new leaves are reluctant to follow.

Less common species include *C. colchicum*, *C. creticum*, *C. mirabile*, *C. parviflorum* and *C. alpinum* (syn. *C. trochopteranthum*). These species are rarely available but all are handsome; several collectors grow them in Australia and they generally do well. I recommend trying them in pots first, but raised beds in semi-shade should be safe for most gardens in south-eastern Australia. *Cyclamen mirabile* and *C. colchicum* flower in autumn, whereas *C. creticum*, *C. parviflorum* and *C. alpinum* flower in spring. As far as I know, *C. somalense* is not yet available or grown in Australia: it is proving difficult to acquire seeds.

The florists' cyclamen, selections of *C. persicum*, produce relatively large flowers in winter. These forms are widely available as pot plants; they provide winter cheer inside or outside and can be kept for many years. Popular myth advises drying the bulb by turning the pot on its side in summer, but this can dehydrate the tuber into oblivion. Certainly these plants flower less enthusiastically in their second year after purchase, but thereafter produce some floral charm each winter. I have seen these planted out in the garden under deciduous trees in the Dandenong Ranges, and they can be impressive in a large garden. In a smaller garden, species cyclamen are more in scale and charming. Florists' cyclamen were sold by Chandler's Nursery in 1932, were certainly widely available from nurseries by the 1960s and are very popular today. For profuse flowering, buy new tubers each year; otherwise place the tuber in semi-shade in summer in a fairly dry position, then move the plants inside in winter unless the house is over heated.

Many cyclamen species are available as tubers but generally, to grow an impressive collection, seed-raising is necessary. Fresh seeds germinate best but if the seeds are not fresh, they probably need soaking in warm water overnight prior to sowing. I sow in autumn in good potting mix and mainly trust to rain for irrigation, while in drier areas the pots might need weekly watering. Seedlings appear quickly and I leave these for a year and then pot on into 7.5 cm (3 inch) tubes, adding osmocote fertiliser. Particularly good forms go into attractive pots or pans, and if I have plenty of tubers, of *C. hederifolium* or *C. coum* for example, I plant them out in large drifts in the garden.

Pests and diseases

For most cyclamen, pests and diseases are rare in Australia. Occasional soft rot of large tubers occurs in humid Sydney, but growers there say that by the time the plants begin to rot, they have many seedlings and so they do not lose that species. Chris Steele-Scott in Mt George, South Australia, finds that *C. purpurascens* has occasional mildew. Cyclamen mite (*Phytonemus pallidus*) is occasionally a problem in Tasmania, but the worst pest seems to be blackbirds, which dig up the tubers, and, worse still, the labels.

In the Dandenong Ranges, wallabies occasionally visit to munch both flowers and leaves, remembering and re-grazing cyclamen. Solutions may include wallaby-deterring fences and frequent dusting of cyclamen leaves with ground paprika. More rarely, lyre birds disturb hillside gardens with their powerful scratching; pots of precious plants are best situated near the house.

Australian growers and their techniques

There are a few cyclamen connoisseurs or plantsmen who grow this genus in Australia. In the mid 1950s, plantsman Otto Fauser became interested in cyclamen and grew *C. hederifolium*, *C. purpurascens* and *C. repandum* in Olinda, at an altitude of 540 m in the Dandenong Ranges east of Melbourne. He grew more cyclamen from seed obtained from the English Royal Horticultural Society in 1959, from the Scottish Rock Garden Club in 1961, and later from The Cyclamen Society. Otto grows some cyclamen, including *C. colchicum*, in pots in his cool glasshouse, but has *C.*

Cyclamen species are uncommon in gardens but there are some very successful specialist growers. Here we see *C. hederifolium* in the garden of Viv Condon (VC)

hederifolium, *C. coum*, *C. libanoticum* and *C. repandum* naturalising in the garden. His seed-raising mix comprises sifted leaf mould, coarse sand and a little perlite. Otto soaks seeds for 24 hours in warm water prior to sowing, gives a light dusting of dolomite, then keeps the pot moist in semi-shade. After about a year, he pots on the seedlings into a commercial potting mix with added grit or coarse sand and leaf mould. He grows all the species except for *C. somalense*.

Siblings Barbara and Phillip Gordon began growing cyclamen at the same time and moved from Balwyn, Melbourne up to Olinda, where they established a fine collection; they recommended collecting seed and sowing it while fresh. With Otto Fauser, Ken Gillanders, Olive Craig and Grace and Bill Maxwell, the Gordons formed the Victorian Chapter of the Alpine Garden Society in 1977; this society encouraged swapping of seedlings, including those of cyclamen.

Ken and Lesley Gillanders have grown *Cyclamen* for 50 years, first in The Basin, Victoria then in Claremont, Tasmania at 300 m altitude. They grew them from seed sourced from the Alpine Garden Society and the Scottish Rock Garden Club. Lately, they have grown hybrids from the Cyclamen Society, although these are yet to flower. Ken and Lesley also obtained some seed, including some *C. purpurascens*, from a grower in Austria. *Cyclamen hederifolium* and *C. coum* naturalise for them but *C. graecum* and *C. persicum* require a hot spot, with *C. graecum* needing a summer baking. Their soil is acidic, so they find *C. purpurascens* difficult. For seed-raising, they use a general potting mix of old pine bark and very coarse granitic sand, adding dolomite and gypsum, blood and bone and osmocote fertiliser. They sold cyclamen plants through Woodbank Nursery for about three decades, listing four species in their 2001 catalogue.

Near Blackwood, at 650 m altitude in the Great Dividing Range in Victoria, Tommy Garnett's interesting Garden of St Erth contained many cyclamen in a large, partly shaded rock garden. He opened his garden daily until 1996 and his gardening articles in *The Age* newspaper (published in Melbourne) encouraged gardeners to try growing this genus. Garnett's article about cyclamen published in the early 1960s inspired botanist Hilary Weatherhead to start her collection by buying one large tuber of *C. hederifolium* for a shilling from Phillip Montague's Nursery in 1962; the tubers of some of her early seedlings are now over 20 cm in diameter. In 1982, Hilary moved from Melbourne to cooler Emerald, at an altitude of 320 m with the Dandenong Ranges to the east. There she had a large sweep of *C. hederifolium*, an impressive sight in autumn; ants had moved the seeds 3 m, resulting in cyclamen popping up in the nearby lawn. She grew carmine *C. coum* amongst a prostrate grey juniper forming a beautiful picture in winter. Hilary grew cyclamen from seed in a mix of commercial potting mix, rich sifted soil and lime, collecting her own fresh seed or using seed from the Royal Horticultural Society, Chilterns Seeds or Linnett Farm. She grew many species and several were grown in pots and then added to the garden over time.

From the late 1960s until very recently, Lydia Bartlett grew cyclamen in Melbourne. She obtained seed from the Cyclamen Society and sowed it in potting mix with gravel and leaf mould added, under glass. She then planted out in the garden "as soon as the plants were large enough", at 2 or 3 years old.

Don Barrett, Bruce Muir and Trevor Nottle began collecting cyclamen in the 1970s. Don grows most species in Adelaide with great care, mostly using seed from Witton Lane Seeds, the Cyclamen Society, the Alpine Garden Society and the North American Rock Garden Society. He grows most species in pots under deciduous trees and finds that *C. africanum* and *C. graecum* grow readily in the warm dry climate of Adelaide. Until 2003, Bruce Muir grew many cyclamen species, excluding all other genera with the exception of *Iris* and *Dierama*, at his Glen Iris, Melbourne home. His back garden was entirely filled with pots, many containing cyclamen plants grown from seed collected on Cyclamen Society expeditions. Bruce loved the *C. mirabile* tribe, particularly, 'Tilebarn Nicholas' and 'Tilebarn Jan'. He treated cyclamen seeds by soaking them in petri dishes and then sowing into polystyrene boxes.

In the Adelaide Hills, Trevor Nottle has been growing cyclamen for 35 years. He grew many species from seed from the Cyclamen Society and from the Archibalds in Wales, using a standard potting mix. He grows most of his cyclamen in the garden, where *C. repandum*, *C. coum* and *C. hederifolium* have naturalised with plants numbering in the hundreds, and he has several *C. libanoticum* self-sown in a dry-stone wall.

By the 1980s, Rene and Bill Coffield were growing and selling cyclamen in Creswick, at about 450 m altitude in the Great

A handsome cyclamen in the collection of Bruce Muir, a form of *C. rhodium* subsp. *peloponnesiacum* (JW)

Dividing Range in Victoria. Their raised beds under gum trees contained many treasures and they issued regular catalogues until 1999. Plantsman Stephen Ryan started growing cyclamen in 1980 in Mount Macedon, north-west of Melbourne, and he sells many interesting species in his nursery, Dicksonia Rare Plants. He soaks the seed for 2 days in hot water containing dishwashing liquid, then sows into a premium potting mix with a quarter-part shell grit. After a year, the pots are filled with tiny bulbs, which he pots on. Interestingly, Stephen does not remember seeing cyclamen in Mount Macedon gardens before 1980.

Meanwhile, Don Schofield and Mat Murray began growing cyclamen in Sydney and the Blue Mountains to the west. Don has been growing cyclamen from seed from the New Zealand Alpine Garden Society and the Royal Horticultural Society for 30 years. He uses half (basalt) soil and half coarse sand as a seed-raising mix and for potting on, then plants out into his Blue Mountains garden. Don finds the wet summers make many species difficult and now he grows "the tried and tested few": *C. coum*, *C. repandum* and *C. hederifolium* in particular. Mat Murray has grown cyclamen for over 25 years, obtaining seed from friends, the Cyclamen Society and the Alpine Garden Society. He first grew cyclamen in Sydney, before moving west to Mount Tomah in the upper Blue

Cyclamen coum with *Crocus tommasinianus* in Mat Murray's garden (MM)

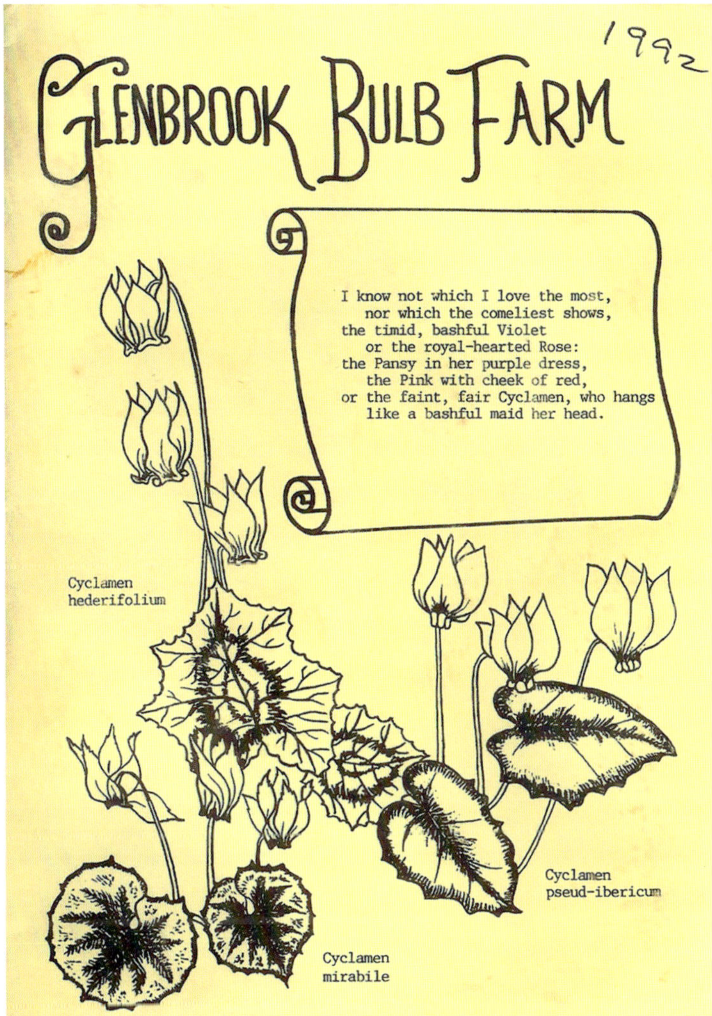

Front cover of the Glenbook Bulb Farm catalogue, 1992

Mountains at an altitude of 1,000 m. Mat grows many species in general potting mix with coarse sand added, and mulches with diatomite (cat litter) to add lime and nutrients, adding this to raised bulbs beds also. Mat sometimes adds crushed egg shells which he says improves the flower colour of *C. libanoticum*. He keeps *C. graecum* warm and dry in summer while keeping *C. persicum* (wild form) in pots as it is a little tender, and he finds that *C. libanoticum* grows particularly well under *Cedrus atlantica* 'Glauca'.

Rod Barwick has been growing and selling *Narcissus* for about 20 years, and by 1992, Glenbrook Bulb Farm in Claremont, Tasmania was growing and selling 14 cyclamen species and hybrids.

By the late 1980s, after starting with *C. hederifolium* and *C. coum*, I began to collect various cyclamen in an attempt to extend the flowering season. I obtained seed from England: from the Cyclamen Society, the Alpine Garden Society, the Royal Horticultural Society and Linnett Farm. I sow seed in premium potting mix in deep 10 cm 'maxi-pots', and pot on in the same mix, adding only osmocote fertiliser, after a year or so. From my mother, Hilary Weatherhead, I received some good forms of *C. coum* with marvellous variegation, along with a *C. hederifolium* with completely silver leaves; I find it grows 80% true from seed. Recently, I bought a form of *C. coum* that has completely silver leaves from Hill View Rare Plants and its flowers, which vary in shades of pink and deep pink, are delightful. In the 1990s, I owned Possum Creek Perennials and over several years sold more than 15 cyclamen species and forms. *Cyclamen hederifolium* naturalises in an area of dry shade in my Selby garden, which is in the foothills of the Dandenong Ranges at 200 m altitude. The white form, *C. hederifolium* f. *albiflorum*, grows happily under a white wisteria tree (*Robinia* sp.) where it marries well with diminutive *Iris cristata alba*. I have added *C. hederifolium*, *C. repandum*, *C. coum*, *C. cilicium* and *C. rohlfsianum* to the garden where they are mulched with fine bark, but I grow other species and varieties in handsome pots or pans. I grow *C. repandum* under a dwarf mock orange (*Philadelphus* sp.), which flowers, then arches over and shades the cyclamen when they are finished. All except *C. rohlfsianum* and *C. parviflorum* flower well.

Chris Steele-Scott began growing cyclamen in the late 1980s too, first in suburban Adelaide and then for the past decade at Mount George in the Adelaide Hills, at 450 m altitude. This area has twice the rainfall of Adelaide (800 mm) and, consequently, Chris is able to grow in the ground many plants either bought from Hill View Rare Plants or grown from seed from the Cyclamen Society. She sows seed into a mix of 2 parts good-quality potting mix, 2 parts coarse sand, 1 part bone meal and 1 part shell grit; she then puts a layer of fine gravel over the seeds and pots on after 2–3 years into a similar mix. In the garden, Chris uses pine needles as a mulch and grows *C. graecum* in a hot northerly spot under a pine tree, where it likes baking. She finds *C. rhodium* subsp. *peloponnesiacum* successful under oak trees and *C. intaminatum* floriferous in

The very successful *Cyclamen hederifolium* forma *albiflorum*, here seen at the Tindale Garden (LB)

pots; *C. purpurascens* flowers well through the summer until *C. hederifolium* reappears in early autumn. Chris collects her own seed but finds that some spreading of individual forms and subsequent hybridisation has resulted from the movement of seeds by ants.

Nearby, Dianne Kaines sold cyclamen and other bulbs and perennials for some years through her Sheringa Nursery (now on-line) and has *C. libanoticum*, *C. hederifolium* and *C. coum* naturalising in the garden. She finds that *C. cilicium*, *C. cilicium album* and *C. repandum* f. *peloponnesiacum* increase slowly in the garden, whereas *C. pseudibericum*, *C. intaminatum* and *C. balearicum* rarely set seed.

In Hobart, Marcus Harvey began growing cyclamen in 1990 and soon began selling an impressive range of species and forms, including eight strains of *C. coum*, from his nursery, Hill View Rare Plants. He first grew some plants from Ken Gillanders and the Goodwin brothers in Bagdad and then began growing from seed from the Cyclamen Society and from Kath Dryden and Roger Poulett in the UK, later collecting species in the wild. When sowing seeds, he first uses a fungicide powder then he uses a mix of 7 parts hammer-milled bark to 3 parts coarse sand in polystyrene boxes, topped with 12 mm gravel, and he leaves these boxes in a shady spot between trees while the seedlings germinate and grow. Marcus keeps a few stock plants in pots and also grows them out in the garden. He harvests his own seeds and sows before they dry out. His successes include *C. colchicum* and *C. elegans* 'Silver Leaf', but in his experience, *C. rohlfsianum* does not set seed in Hobart. He finds *C. mirabile* slow to flower, taking 4 years. He credits Kath Dryden for supplying seed for his marvellous *C. coum* strain, which has completely silver leaves, and the Cyclamen Society for *C. hederifolium* 'Silver Leaf'.

Paul Tyerman has been growing cyclamen for 12 years in Canberra, where temperatures vary from −8°C in winter to 40°C in summer, and has many species doing well. He finds *C. alpinum* (*C. trochopteranthum*) troublesome and has not yet exposed *C. balearicum* to winter frosts. He grows *C. hederifolium* and *C. coum* in the open ground and in pots and finds the former hardy in full sun to full shade, whereas *C. coum* requires some shade and is quite drought-tolerant; both are naturalising in Paul's garden. He grows *C. rohlfsianum* against the house to give some protection from frosts. Paul uses commercial potting mix, sometimes adding coarse river sand, and adding saturaid when re-potting to aid water penetration.

In Olinda, Craig Wilson has grown many cyclamen species for a decade and they grow well in rich acidic soil in his garden behind Gentiana Nursery.

For gardeners who are impatient to grow several cyclamen species, nurseries in the Adelaide Hills, Mount Macedon and Olinda have a few to offer. Many are available from a couple of mail-order nurseries in Australia. Hardy species to start with are *C. hederifolium* for autumn flowers, *C. coum* for winter-flowering and *C. repandum* for blooms in spring.

(c) Japan

Motomu Nagaoka, translated by Yoko Otsuki

Cyclamen hederifolium thriving in the Musashi Kyuryo National Government Park in Japan, 28 October 2008 (MKNGP)

A natural setting for a mixed colony of *C. coum* under pines in the Musashi Kyuryo National Government Park, 3 March 2010 (MKNGP)

History of Cyclamen cultivation in Japan

Cyclamen persicum was definitely the first cyclamen species to be introduced into Japan. The first known record of the plant growing there was in 1892 at Shinjuku Gyoen. In 1884 Saburō Ōkubo, an assistant professor at Tokyo University Botanical Institute, gave *C. persicum* the Japanese vernacular name 'Buta-no-manjū' (literally pig's bun), a translation of its English name sowbread. So we can assume a Japanese familiarity with the genus, and hence an introduction, that antedates the 1890s — say, during the 1870s.

Japan's commercial cyclamen industry started at the beginning of the twentieth century. In the 1950s, production burgeoned, and by 1965, *C. persicum* had become Japan's most popular pot plant. Other cyclamen species were introduced in the early twentieth century, but they have only recently gained in popularity.

Among the first cyclamen species to be introduced were: *C. hederifolium*, which flowered in 1927 at Konishikawa Botanical Garden; *C. purpurascens* (as *C. europaeum*), introduced about 1910; *C. cilicium*, which was introduced in 1932 by Mr Sakurai; *C. coum*, introduced about 1926; and *C. repandum*, introduced about 1925.

A partly shaded scree bed for *C. cilicium* at the Musashi Kyuryo National Government Park, 17 October 2004 (MKNGP)

Start of commercial enterprise

Japan's commercial cyclamen industry started between 1904 and 1905, around the time of the Russo–Japanese War. Tsunesuke Tsujimura built a 1,000 square metre greenhouse in Odawara and started producing primulas, pelargoniums, gloxinias and cyclamen in 1902. According to 'Saishin kaben engei' (translated as '*New Development in Floriculture*', edited by Ishii Yūgi and published by Seibundo-shinkōsha in 1931) the earliest commercial grower was Mr Tadashi Nagai of Shōka-en in Tokyo, who grew cyclamen from seeds imported from Germany.

Along with Chōzaburō Nakatomi in Kanagawa and Takeo Nakanishi of Hōkōen in Tokyo, Tsujimura developed techniques for propagation and cultivation, most notably for controlling seed set and collection. As a result of this work, production soon became widespread in the Tokyo area.

Meanwhile, Takashige Itō of Ena in the Gifu Prefecture imported seeds from the German company E. Binnewies K.G. and started commercial production in 1923. As his neighbouring growers joined him in the enterprise, a cyclamen industry took root. Ena city is located in a mountainous area and it is cool in summer. It provided the ideal climate in which to establish an enterprise that would meet all of Japan's demands for the new pot plant craze. Equally, it was a climate that would favour the development of new hybrids and the harvesting of seed for sale to growers in other regions.

From 1925 to 1941

Because the Ena region in Gifu Prefecture is at high altitude, it favours the growing of cyclamen in summer and soon developed as a profitable centre for their cultivation. Influenced by Takashige Itō, some farmers in the nearby Nakatsugawa City joined the new horticultural industry. They provided both flowering and half-grown plants: they sold flowering-size specimens to customers through catalogues and half-grown plants to commercial pot-plant growers in other areas. Up until 1930, flowering pot plants were sold in March and April. After that date, however, the peak of production shifted to winter so that cyclamen in bloom could be offered as end-of-year gifts. Once Japan was at war, attitudes to floristry changed. By 1941, cultivation was controlled and national priorities set for production: cyclamen were considered a luxury and discontinued as a commercial crop.

Development and popularisation after the war

Before the war, cyclamen had been known as a luxury pot plant for the wealthy. Then they came to be identified as an end-of-year gift, and their production expanded. Meanwhile Ena's old cyclamen-growing industry re-started. Many growers began buying seeds from the region, and selling the results as pot plants. Meanwhile, some growers in other regions, notably Kazunori Mayuyama in Nagano Prefecture, began collecting their own seeds. The climate in Nagano is also cool and favourable for cyclamen cultivation in summer, and so growers from Nagano soon became strong competitors for those in the Ena region. Growers elsewhere soon joined in and some, such as Furōen in Saitama Prefecture and Iwaki Shokubutsuen in Fukushima Prefecture, began to import seeds of new cyclamen cultivars from all over the world.

By 1960, Japan was producing 1.5 million saleable cyclamen plants annually and they were fast becoming the nation's most popular pot plant. Much of the genus' popularity can be explained by cyclamen's flowering season: the plants were a novel but seemingly ideal decoration for the New Year celebrations that are so dear to the Japanese. Then a song, '*Scent of Cyclamen*' sung by Akira Fuse, became a huge hit and reached the Japanese Record Sales Grand Prix. This too helped to boost the sales of cyclamen plants, which were becoming recognised as a fashion item. Since these beginnings, sales have steadily mounted. In 2002, 23 million flowering-size potted cyclamen were sold in Japan.

Despite that impressive figure, the cyclamen industry was hit by the collapse of the economy and the recession of the 1990s. Demand for gifts diminished and prices dropped. Before the bubble burst, a well-grown cyclamen in an 18–21-cm diameter pot, wrapped for presentation as a gift, could retail at anything between £25 and £75. You might think this a very high price for a cyclamen cultivar, but it is very much in keeping with the high values that the Japanese have traditionally attached to special plants that are given on special occasions. Today, post-recession, prices are more realistic: a flowering plant in a 15-cm diameter pot will sell for between £2.50 and £10.

While the demand for expensive prestige pot plants decreased, there was a gardening boom in Japan toward the end of the twentieth century. This consumed vast quantities of cheap plants, mostly for bedding out. These were seldom the species beloved of traditional Japanese horticulture. They were rather colourful 'fillers', many of them exotics. Paramount among them were *C. persicum* cultivars. For bedding purposes, gardeners wanted cyclamen in 9–10.5 cm diameter soft pots. In 1995, 1.74 million such plants were sold; in 2000, that figure had more than doubled to 4 million. These new-style plants are called 'Garden cyclamen', expressly supplied for flower beds and outdoor containers, and they are the main line of cyclamen production at present. To the Japanese, then, a 'garden cyclamen' is a mini cyclamen (*C. persicum* 'Mini Time') supplied in a soft pot for outdoor display.

Change in varieties cultivated

From before World War II to the mid-1950s

Growers began to select for new cultivars among the plants grown from seeds supplied by the German company E. Binnewies K. G. The trend was for ever-larger and brighter flowers. Cultivars developed in the Ena region before the war were divided into Persicum-type, Rococo-type, Papilio-type, Double-type and Silver-leaved-type.

Persicum-types included 'Dark blue-red', 'Scarlet King', 'Rose Red', 'Snow Queen', 'Giant Victoria', 'Lilac', 'Vermilion', 'Delicate Salmon', 'Rose Salmon', 'Harlequin', 'Purple Salmon' and 'Orange Salmon'. Some 40 Rococo-type cultivars were sold including 'Hermosa' and 'Cattleya'. [Some of these names are, as near as possible, translations of Japanese cultivar names — Ed.] In the 1930s, red-flowered plants became especially popular, so the varieties 'Salmon Scarlet' and 'Bonfire' came to represent about 80% of the whole market. This trend continued until the middle of 1970s.

From the mid-1950s to the mid-1970s

After 1946, the cultivars that had been grown in the pre-war period gradually returned to the market and new seeds were imported from Germany. By the mid-1950s, however, consumers were growing tired of these plants and began to import seeds of new cultivars from the Ball Company in the United States and from the Netherlands. These new cultivars were not necessarily great in variety or number: the colour red remained pre-eminent in the popularity stakes, accounting for as many as 70–80% of Japanese-grown plants. Heat was also a factor in selecting these new plants, many of which simply could not cope with Japan's summers and autumns.

New imported selections offered resistance to such endemic problems as virus and climate. In many cases, the Japanese hybridised this new blood with some of their older, tried and tested plants. Before long, a range of original cultivars developed across Japan, including plants such as 'Shinano-beni' and 'Shinano-pink', which are said to be selections from 'Bonfire'. Some of the best of these were produced largely in the Nagano region and they have many excellent characters: thick petioles and flower stems, a compact growth habit with plentiful leaves and flowers, hardiness and performance reliability. Another group of cultivars of hybrid origin is known as the Wakayama Group. These plants originated in the Wakayama region and soon spread through nurseries both there and in the Mie and Aichi Prefectures. These cultivars too were thought to be descended from 'Bonfire', but what distinguishes them (as might be supposed from the area where they were grown) is a marked tolerance of summer heat, a most desirable characteristic in a Japanese cyclamen. This story repeated itself more or less throughout the nation, with growers developing their own cultivar groups in different areas. As a result, Japanese cyclamen culture developed strongly regional themes and characteristics, not unlike wine production in France or Italy.

In 1968, a mysterious new cultivar arrived in Japan from Argentina. This was a spectacular white-flowered plant with blooms as large as 10 cm across. It was originally imported from the Japanese-Argentine grower Sichirō Koike by Seirō Yoshine in Nagano Prefecture, who offered it as a cut flower. This beautiful new plant soon influenced Japanese container-grown cyclamen when it was crossed with older cultivars. Mr Yoshida at Furōen in Saitama Prefecture in particular produced many hybrids of the Argentina-origin Group and they stole the market. Sadly, cultivars with the Argentine white in their parentage did not fare well as pot plants as they were difficult to keep and tended to lose their shape. So this craze was short-lived and most of these new plants disappeared from the marketplace within a short period, the exception being some large lavender-flowered cultivars.

Although the cultivars grown in this period were named clones or seed races, they tended to be tetraploid and this meant that the offspring were rather variable. Growers compensated for this by producing more plants than were ultimately needed (around one and half times the final quantity) and sorting through the seedlings to achieve a uniform 'product'.

A dwarf *C. persicum* called 'Mini Cyclamen' was a significant introduction during this period and it gradually began to spread throughout the trade.

After 1975

Japanese cyclamen growing changed around 1975. Terracotta pots were replaced by plastic containers, and with them came new composts and feeding regimes. There were also new plants: chiefly F_1 hybrids that were arriving from Europe.

By the late 1970s, the market demanded fewer red- and more pastel-flowered plants. The most popular of these were ultimately the Composer Series. These superseded the popular but unstable Pastel-type cultivars, which included important cultivars such as the delicate pink 'Pierce' and 'Lingerie Pink'.

From the mid-1970s to mid-1980s, the most popular cultivar was 'Victoria'. This plant dates back to around 1930 and was genetically unstable by comparison with the new-generation F_1 hybrids, nonetheless, 'Victoria' succeeded for a while in upstaging the newer Pastel-type cultivars and the traditionally popular red-flowered forms. The latter had by now dropped to occupy some 20% of the market. To produce sufficient quantities of 'Victoria' to meet demand, growers would sow twice the number of seeds needed and then rogue out inferior plants. Their efforts were rewarded — a good plant of 'Victoria' in bloom could realise a high price.

Cyclamen persicum 'Pierce', a very popular cultivar selected in Japan

Cyclamen and the domestic gardening boom

Although cultivars of *C. persicum* can survive mild frosts, their winter flowers and foliage are usually damaged or retarded by such temperature lows. As a result, they tended to be treated as indoor plants. Then, towards the end of the twentieth century, a new race of cyclamen was developed. Derived from mini-cyclamen, these compact, small-flowered plants were capable of blooming freely outdoors in winter in sheltered gardens or if given the minimal protection of an eave or an exterior windowsill. These new plants became enormously popular in Japan for winter bedding and outdoor display. Their success coincided with, or was a result of, a boom in domestic, Western-style gardening that occurred there toward the end of the last century. These new cultivars became known as 'garden cyclamen', a name that has since been applied to all mini-cyclamen, not always with justification. In the regions with a minimum temperature of about −3°C, dwarf *C. persicum* cultivars (including the true, hardy 'garden cyclamen') are today often displayed in outdoor containers placed on verandas and window sills.

Varieties and breeding in Japan

Breeding in Japan

It was not until after 1965 that a range of truly Japanese cyclamen cultivars could be said to have emerged. These were the offspring of imported material, forged and tempered by Japanese tastes and the rigours of the Japanese climate. One famous old cultivar is 'Shinano-kō', selected from 'Bonfire'. In the 1970s, breeding trends changed again as growers came to understand better the issues surrounding polyploidy and the desirability of F_1 hybrids. Consumers also looked for a wider range of colours. The deep scarlet red that had been the Japanese standard for so long fell from grace. Then, in the 1990s, advances in tissue culture made it possible to produce uniform clones on a mass scale. Variability and unpredictability, the old perils of seed-raising, disappeared.

Between 1990 and 2005, some 700 cultivars of florists' cyclamen (i.e. *C. persicum* cultivars for indoor display) were introduced to the Japanese trade. In the same period, some 300 cultivars of mini-cyclamen or garden cyclamen (also *C. persicum*) were launched. In both categories, around half of these cultivars were foreign imports, the other half bred in Japan. Today, Japan has some 20 dedicated cyclamen-breeding companies. These companies produce seedlings or plantlets for sale to nurseries that then bulk them up for wholesale. But mass production is only part of the Japanese cyclamen story. Countless small growers and private individuals also join in the cult.

Popular cultivars, groups and seed races of imported Persicum include the F_1 Concerto Series, F_1 Solid Series, F_1 Zodiac Series, New Wave Series, F_1 Novella Series, F_1 Butterfly Series and F_1 Helios Series.

Japanese-bred Persicum include: (a) typical cultivar groups — the F_1 Aurora Series, F_1 Papilio Series, Akebono Series, Argentina-Mammas Series, Iwaki-Sunshine Series, 'Victoria Strain', Candy Series, Gradation Series (Sakata), Sayaka Series, Sun-Lingerie Series, Shooting Star Series (Kanazawa), Serenade Series (Takii), Pierce strain (Hosogane), Plumage Series (Tanaka) and Lips Series (Kanazawa); (b) yellow varieties — 'Kaguya-hime', 'F_1 Golden Boy' (Kage), 'F_1 Golden Girl' (Kage), 'Candle Light' and 'Gold Illumination'; (c) fragrant varieties — Aroma and the Kaori Series (Sakata); (d) Rococo-fringed types — Shell Series, Starmain, Bianca Series, Pretty Princess Series and Prima Donna Series (Takahashi); and (e) others such as the Star Ring Series (silver-leaved), Platinum Series (silver-leaved), F_1 Banny Series (midi type), Pia Series (midi type) and Composer Series (midi type).

Mini cyclamen types include the Alpine Series, Eve Series, Wink Series (Sakamoto), Winter song Series, Carol Series, Cute Series, Cupid Series, Rococo Series, Sun Kiss Series, Genesis Series, Cherish Series, Chimo Series, Dixie Series, Dressy Series (Sakata), Beaver Series, Piccolino Series, Piccolo Series, Figure Series, Forever Series, Flamenco Series, Furufuru Series, Florett Series, Petit angel Series, Petit candle Series, Helena Series, Polka Series, Mabel Series, Micro star Series, Midori Series (Goldsmith), Mini silver-edge Series, Mini star Series, Mini mate Series, Musical Series, Miracle Series, Métis Series (Morel) and Melody Series.

Popular fashions in breeding

The new cultivars that succeeded to 'Victoria's crown were 'Pierce' (Hosogane) and, some five years later, 'Plumage' (Tanaka). Both 'Pierce' and 'Plumage' were stable clones selected in Japan. The success enjoyed by these Japanese cultivars vastly encouraged breeders in that country. At present, they are producing and marketing new flower types such as fringed-flowered forms, rococo-flowered forms, balloon-flowered forms, and even deformed-flower forms, such as crested or windmill types. Other current breeding trends include fashions for fragrant and yellow-flowered clones.

Inter-specific hybridisation

There are 19 *Cyclamen* species. In breeding terms, they are not compatible, but at the Saitama Agricultural Institution, attempts have been made to hybridise them using the ovule germination method. The cross-fertilised ovule is excised 14–28 days after hybridisation and cultured in aseptic agar jelly solution. If the cross should still prove recalcitrant, the institution uses both the ovule germination method and colchicine treatment (i.e. polyploid inducement) to produce fertile 2n=44 plants in the next generation. The resulting crosses are:

C. persicum (2n = 48) × *C. hederifolium* var. *album* (2n = 34)
C. persicum (2n = 48) × *C. repandum* (2n = 20)
C. persicum (2n = 48) × *C. libanoticum* (2n = 30)
C. persicum (2n = 48) × *C. purpurascens* (2n = 34)
C. persicum (2n = 48) × *C. graecum* (2n = 84)
C. persicum (2n = 48) × *C. rohlfsianum* (2n = 96)

These hybrids are not yet influencing the market. The cross between *C. persicum* and *C. purpurascens* has, however, yielded some superbly fragrant plants, among them two cultivars, 'Uruwasi-no-kaori' and 'Kokō-no-kaori', which have been made public and are being produced commercially.

Species cyclamen in Japan

Although *C. coum* was first introduced to Japan in the early twentieth century, it is still an uncommon plant in gardens (MKNGP)

In the light of all the above, it need hardly be stated that the most popular cyclamen in Japan are *C. persicum* and its offspring. Mass production for winter display and the gift market made this species number one in the pot plant popularity league table, a position it held for decades until it was dislodged by the moth orchid *Phalaenopsis* during the fiscal flamboyance of the 'Bubble Boom'. Even so, *C. persicum* remains unchallenged as Japan's favourite plant for winter-time gifts.

The same cannot be said for the other cyclamen species. These beautiful plants are much-loved by Japan's alpine enthusiasts and our more expert and experimental plantsmen, but they are not very well known, nor are they considered particularly easy. A further inhibiting factor is their comparative slowness to achieve flowering size from seed, which is viewed by growers as an unacceptable commercial bottleneck. That said, seed of *C. coum* and *C. hederifolium* is becoming more widely available and these species will acclimatise themselves to gardens in many regions of Japan, especially when planted under conifers or the eaves of houses. Moreover, the exchange of ideas that has been taking place in recent years between plant lovers in Japan and Great Britain is doing much to foster a species cyclamen cult in Japan. These plants are, after all, suited in many ways to our traditional horticultural tastes: they are diminutive and charming, they have foliage as lovely as their flowers, and their needs, which might seem finicky, are easily met with some thought and experiment.

Cyclamen hederifolium with older shade-loving plants in a Japanese garden (NY)

The same bed in early spring, this time with *C. coum* (NY)

Cultivation

Facilities

We grow cyclamen in houses that are covered with polyvinyl, acrylic or glass. Fan heating is more suitable for acrylic and vinyl houses, whereas hot-water heating or a fan will do for glass. Houses of all kinds are double-glazed, usually with an internal layer of vinyl.

Various automatic watering systems have been used in the past. To begin with, pots were merely stood in trays that could be periodically filled with water; problems with stagnation and rots soon ensued, however, and especially in summer. These days, growers tend to prefer either capillary matting or a bath method in which the plants are stood in trays that can be filled with water and then drained via a plughole once the pots have taken up their fill.

Sowing

Depending on shipping time and cultivar, sowing takes place between the beginning of October and the end of February. For F_1 hybrids or early types that will be sold in 15-cm pots, seeds numbering around one and half times the number of plants wanted are sown. For the more reliable Japanese-selected cultivars or plants that will be sold in larger 18-cm pots, seeds numbering 1–2-times the number of plants ultimately desired are sown.

There are two methods of sowing: either use a tray or put one seed in a hole of a plugged tray. Cover the top with compost to a depth of 0.5–1 cm. As cyclamen seeds germinate in darkness, cover them with newspaper until they sprout. To attain the right germination temperature of 18–20°C, place the tray on a heated mat within a tunnel constructed of polyvinyl inside the house.

Potting

When the plants produce 2–3 leaves, re-pot them in a 7.5–10.5 cm diameter pot. Discard any weak or deformed plants at this stage. Plant the tuber with its top exposed.

Final potting

The timing of potting for sale depends on the climate and shipping time. For November shipping, the final potting would be between late June and the beginning of July. On the other hand, in the Tokyo area, the shipping peak is in December, so final potting is done between late August and the beginning of September. We change the potting season according to shipping time because it takes about three months from the initial appearance of the flower buds to blooming time. The appearance of those buds immediately follows the completion of the main foliage flush, a period of growth that is triggered by the final potting. So plants that are destined for shipping in November will need to be leafing out vigorously in July and August; whereas those that will go for sale in December and the New Year will need to be leafing out in their new pots in September.

Arranging leaves and dressing the plants

It is important to admit light to the top of the tuber to encourage foliage production, and consequently we like to arrange the cyclamen leaves. This 'dressing' also prevents the spread of *Botrytis cinerea* and improves the plant's general appearance. Any new leaves that are growing up in a tangle are separated and gently spread outwards; and any yellow leaves are promptly removed and treatment applied if *Botrytis* is found. These processes need to be carried out repeatedly until shipment.

Plant hormone treatments

We spray benzyladenine (BA) and gibberellin (GA) to increase the number of flowers. GA helps to develop flower buds and encourages all the flowers to open more or less simultaneously within a desired and predictable period. Nowadays, we apply both GA 0.75 at 1 ppm and BA 50 at 100 ppm between September and October. It is said that there is no need to use hormones when a plant is growing well; but we like to use them to give the plant a push when our Septembers are too hot.

Compost

There are many different types of cyclamen potting composts — almost as many as there are growers, some of whom guard the secret of their recipes jealously. Here are six examples:

1. 100% organic compost
2. 7 parts organic compost, 3 parts peat moss
3. 9 parts organic compost, 8 parts leafmould, 3 parts volcanic gravel
4. 5 parts metro mix (artificial organic compost), 2 parts leafmould, 3 parts red soil
5. 7 parts basic compost (i.e. 3 parts paddy field soil, 7 parts volcanic soil, 3 parts organic compost), 3 parts peat moss
6. 5 parts moss peat, 3 parts volcanic balls, 1 part palm fibre (coir), 1 part perlite

Our organic compost is made of volcanic soil, paddy field soil and 30–50% animal manure and left to ferment for 1–3 years. It becomes a very fertile medium that has a soft and porous granular texture. Red soil, which is used in the Kantō district, is a non-weathering (i.e. hard) clay soil found under volcanic ash. Coarse, solidified grains or pellets of it are selected and sold as the medium 'red balls'. Superficially, this medium resembles some forms of laterite and clay aggregate, although it is not inert and it may ultimately dissolve and degrade into sticky red clay. Paddyfield soil is gathered from spent paddyfields and so is silty to loamy and rich in both decomposed vegetable and animal material and the residues of rice crop feeds. Leafmould is composed of decomposed fallen leaves. Our moss peat tends to be pH neutral. Animal manure is gathered from pigs and cattle and usually combined with farmyard straw. Composted chaff is also used.

Up until the mid-1970s, florists' cyclamen cultivars were grown in clay pots using an organic compost. This medium has good porosity but retains moisture and releases nutrients slowly. Then, as plastic pots became popular, it became difficult to control water and feed on a mass production basis, so farmers started to use the mixed compost (example 5). However, a compost that includes moss peat tends to retain water, so the roots lacked oxygen. This was especially the case in summer for cyclamen growing in still and low-lying areas, whose roots would overheat and stagnate causing plant growth to decline or cease. To counteract these problems, growers have recently begun to use mixes that contain less peat and more bulky organic matter.

For the smaller cyclamen species and the cultivars and hybrids that resemble them, Japanese growers use a mixture of rough volcanic gravel or red balls with 10–20% each of grit and leafmould. This extremely open and fast-draining (effectively soilless) medium is common and now traditional in the world of Japanese alpine cultivation. It permits, indeed demands, frequent watering and feeding during the growing season, but it is also excellent at preventing excess damp and rots in both summer and winter.

Fertilisers

a) From germination to the final potting. Give 20-20-20 NPK diluted 1/4,000 once a week.
b) From potting to the end of July. After re-potting, the plant will grow rapidly, so growers tend to give both a slow-release base fertiliser and a quick-release non-organic liquid feed. When using a capillary system for watering, however, the base-fertiliser tends to be replaced by flushing the feed through the system in solution.
c) In cooler areas, the plants will keep growing throughout August. In this case, Japanese growers tend to change to a 10-13-10 NPK feed, which contains more phosphorus. On the other hand, if the plants are growing in high temperatures, only a weak fertiliser solution should be used and the plants' needs can be judged according to the colour of their leaves.
d) Up until the end of September, as flowering and shipping approach, the grower is aiming to encourage leaf, so fertiliser is freely given whenever temperatures permit. From October, however, growers reduce the amount of nitrogen given in order to encourage the production of flower buds.

Principal pests and diseases

The main pests of cultivated cyclamen in Japan are *Phytonemus pallidus*, *Frankliniella occidentalis*, *Mamestra brassicae* and aphids. These pests are particularly damaging from September onwards during Japan's long hot Indian summers. Other problems include *Rhizoctonia solani* in the sowing season and *Botrytis* in the growing season. The principal pathogens, however, are *Colletotrichum cyclamenae*, *Erwinia carotovota* subsp. *carotovota*, *Erwinia herbicola* and *Fusarium oxysporum* f. sp. *cyclaminis*. The damage done by *Erwinia* and *Fusarium* is fatal, and growers disinfect plants regularly to keep these diseases at bay. *Botrytis* is readily treated if diagnosed soon enough, but it can be a serious menace during Japan's long and humid autumns. At such times, and as the moment for shipping nears, it is necessary to enhance air circulation by opening windows or using a fan. Heating the glasshouse at the same time can also help to reduce humidity and with it the likelihood of *Botrytis* disease.

Informative exhibit at a meeting of the Japanese branch of the Cyclamen Society (NY)

Nationwide cyclamen shows

In 1976, a national cyclamen show was held under the aegis of the Tokyo Potplant Exhibition. The idea caught on and the show has since become an annual fixture organised, as it has been for many years now, by the Japan Flower Growers Association. Held each year in December, this is the only such show that the association supports and it is an important shop window for new cyclamen cultivars and cultivation techniques. Its format and aims are mirrored, however, in numerous regional cyclamen shows that are held each year in Japan. The Cyclamen Society now has a branch in Japan, which is currently the fastest-growing group in the Society, and shows for *Cyclamen* species are now being organised by this group.

The first meeting and show of the Cyclamen Society's Japanese group in Tokyo, 13 November 2010. There are now over 150 members (NY)

(d) New Zealand

Dorothy Hansen

[Adapted from an article first published in the Cyclamen Society's journal in 1994 — Ed.]

My love affair with dwarf cyclamen started many years ago. I had grown *C. hederifolium* for a long time and had pockets of them in various spots all over the garden, as well as a larger area at the base of a native rimu tree. A friend gave me some special tubers and I was quite delighted one day to find a beautiful white bloom with maroon blotches at the base. This plant was, of course, *C. coum* 'Album'. I thought it was just so pretty and interesting that from then on I was hooked. In the garden, we grow mostly *C. hederifolium*, *C. coum* of various colours and leaf patterns and, my favourite of all, *C. repandum*. They all grow well and easily in our conditions. The climate is very mild in our part of North Island, New Zealand, with temperatures not often much colder than 4°C and rarely exceeding 24°C; there is never frost.

We have cyclamen flowering outside for 12 months of the year. The hederifoliums start in December and usually, for some unknown reason, the whites are ahead of the pinks. They flower away merrily until joined by the coums in February or March, followed by the elegant repandums from May until their last blooms appear in November. Sometimes, an odd flower will still be showing in December. Then off we go again with the hederifoliums.

My biggest problem is finding enough shade for the plants in our hot sunny garden. I have a shade house where I grow most of the other types of cyclamen in pots. I intend experimenting, however, and will plant other species in different areas in the garden: I am sure that *C. cilicium*, *C. libanoticum*, *C. africanum*, *C. persicum* and *C. purpurascens* could do quite well outside in the right spot. It could be worth trying *C. rhodium* subsp. *rhodium* and subsp. *peloponnesiacum* too: if they take after *C. repandum*, they should stand the conditions; *C. graecum* will, I hope, enjoy being baked in our sunshine.

The only literature we have here on the growing of cyclamen, good and comprehensive as it is, is of course for English conditions, so it is a matter of experimenting and a case of trial and error. Conditions in parts of the South Island would be similar to those in England, but, although our North Island is roughly the same distance from the equator as is the Mediterranean, it is probably harder to replicate conditions that cyclamen enjoy here. The less-marked difference between summer and winter weather might not be a good thing for the plants' growth cycles. *Cyclamen africanum* is hardly dormant at all here, nor is *C. persicum*.

Not many nurseries or garden centres in New Zealand list cyclamen in their catalogues, and those that do list mainly *C. hederifolium* and *C. coum*, so special varieties are not easily available. Two specialist bulb or perennial nurseries in the north of the North Island have listed several cyclamen types that are sold as potted plants, not as dry tubers. One catalogue lists *C. africanum*, *C. cilicium*, *C. coum*, '*C. orbiculatum*', *C. graecum*, *C. hederifolium*, *C. libanoticum*, *C. persicum*, *C. repandum* and *C. rohlfsianum* — the last as 'rare and very few offered'. Some South Island nurseries have limited lists. To my knowledge, however, there is no specialist cyclamen nursery in New Zealand. Most people who have collections are, like me, purely hobbyists.

I have a friend who is a really expert grower of cyclamen and she has helped me enormously with plants, seeds and advice. She still has beautiful plants of *C. creticum*, *C. rhodium*, *C. intaminatum*, *C. cilicium*, *C. cyprium* and *C. rohlfsianum* that were imported from England many years ago. It is a joy to visit her shade house to see her beautiful healthy plants laden with flowers. Seed of most varieties germinates very readily here if sown fresh and proves difficult only very occasionally. Germination of the seed I have received from England is, however, quite spasmodic: sometimes good and sometimes not so good. I think perhaps it might help if it is put in the refrigerator for a few days before sowing. Most seedlings stay in growth for some time and seem loath to go into dormancy.

A lifetime of growing *Cyclamen* in New Zealand

Terry Hatch

I started growing *Cyclamen* over 50 years ago in the Dagenham parks department glasshouses Essex, England, where we grew about 200 plants of *C. persicum* every year for display. We used John Innes mix, clay pots and the old pot tapping reel on a stick to test for dryness before watering them: they had to be perfect for the 'Mayor's Parlour'. After finishing National Service in Cyprus, during which I grew some wild collected cyclamen plants, I set out for New Zealand in 1963 with my young wife and a copy of Lange and Blackwell's *New Zealand Native Plants*. We arrived on a typical Wellington day, very windy and no different from Essex on a dreadful day, but settled in South Auckland, which has a much milder climate. I worked in a local nursery that produced, among other things, 5,000–10,000 *C. persicum*. We used 12.5-cm plastic pots and a compost mix made up of peat, perlite and used coffee beans plus fertiliser. The plants were grown outside, protected by a huge lath house made of woven venetian blind material. Most of the plants were sold as house plants, but many ended up being planted out in shady gardens in the warmer parts of the country and possibly still have progeny popping up under trees today.

After three years, I started up my own nursery and grew mainly 'alpines', which flourished in the Mediterranean-type climate, i.e. dry, sunny, well-drained sites. I purchased *C. hederifolium* seeds from Thompson & Morgan and these flowered after ten months. The stocks soon expanded and became a good item of trade as this species grows well all over New Zealand and will self-seed into suitable sites. The tubers grow to a huge size and great age, producing amazing numbers of flower from November to May. I have selected very nice silver foliage forms over the years.

Cyclamen coum in its many variants grows well in cooler sites. I like to add a 10-cm layer of sawdust over them each year as they can dry out too much if near the surface, and this reduces the number of flowers. Further south at Taihape, there are keen growers who have carpets of *C. coum* growing under flowering cherry trees, a sight to behold. These plants do produce a lot more flower in a colder climate and do well in keen gardeners' rock gardens from Auckland to the far south of South Island.

I found *C. repandum* difficult in my old garden but with a recent move to a new site just 2 km away that has a cooler aspect, I now have these plants growing and spreading amazingly after 30 years of trying. They have huge leaves and masses of flowers and they are setting seed and producing self-sown seedlings. These plants are planted under *Podocarpus totora*, which has thick mats of roots and great wads of mycorrhiza that feed all and sundry, including vast clumps of *Clivia* in many colours and form, giant bulbs of *Crinum* species, tangles of *Polygonatum* species, Vireya rhododendrons and arisaemas: what an amazing range of plants originating from tropical to cold climates! *Cyclamen repandum* can be seen in a few gardens in most parts of New Zealand, but alas the keen collectors are fast declining.

Cyclamen graecum thrives in hot dry places and will do well when planted on the sunny side of large trees that keep the frost away. The corms grow huge and in very hot years have a great show of flowers; further south, the specialist collectors grow them in pots, as they do all the more tender species.

Cyclamen purpurascens will flower all year and is evergreen; it never has masses of flowers but what a scent!

For many years a nurseryman friend has been breeding *C. persicum*. Noel McMillan has grown this species since the 1950s and has developed many strains. He pioneered the amazing bearded cyclamen forms, which have taken up most of his life. Now in his 70s, he is still growing and selecting.

To sum up, cyclamen grow well in most areas of New Zealand that have a mild climate, as well as in coastal areas or gardens with mild microclimates, and do not need too much care. They will all grow to a ripe old age with some species requiring little more care than perhaps a top dressing of humus or grit.

4.4 EXHIBITING CYCLAMEN

Vic Aspland

The joys of exhibiting

The Show Secretary calls for attention, makes a few preliminary remarks, and then introduces the VIP of the day, who will present the cups and trophies. The successful exhibitors move to the front of the hall, to receive quiet words of congratulation, handshakes and trophies. You join the audience in a round of applause at each presentation. The ceremony concluded, the exhibitors break into groups and conversations scattered with references to subspecies and sinuses or petioles and peduncles ebb and flow. Scenes like this take place many times each year. You may have thought that the trophies really looked rather nice, and that you would quite like to win one sometime, but fear that your own plants might not be up to standard, and that, anyway, exhibiting is too difficult. After all, the exhibitors seem to be so knowledgeable. If you have thought like this, do not despair; they were all beginners at one time and that knowledge has been learned one step at a time. It really can be very easy to get your foot on the ladder to success.

A Cyclamen Society show, 20 October 2007. Specialist Jan Bravenboer (with camera) casts a critical eye (BM)

First steps

Probably the easiest species with which to begin your show career are those that flower in autumn, usually before the leaves emerge: *C. hederifolium* with its numerous forms, or *C. cilicium* and *C. graecum*. The cultivation of these plants has been covered elsewhere. As you will not have any leaves to bother about, you will have one less distraction while you are getting to grips with the basic principles of exhibition. With *C. hederifolium*, you may not necessarily even have to bother with year-round cultivation in a pot. Quite a few prize-winners have spent eleven months of the year growing in the open garden, before being lifted into a pot just for display. Why not walk round your garden in September or October and note any specimens that look show-worthy. Qualities to look for include: plenty of flowers that are open at the same time and nicely spaced across a circle; flowers that reach a uniform height (or possibly are higher towards the centre of the circle); flowers of good size and form (short dumpy petals are rarely aesthetically satisfying); flowers of good colour (plants with white flowers can often be much stronger growers than the pinks; deeper pinks are much in vogue at present, but deeper shades can fade badly as the flowers age). Mark your chosen plants carefully, so that they can be lifted into pots the following year, perhaps sometime in June or July. For exhibition, *C. cilicium* and *C. graecum* are better pot-grown throughout the year.

If your plants are already in pots, you will need to prepare them for exhibition, a job most easily done before the flower buds begin to develop. Tip away the old top-dressing and clean the pot. Stubborn salt deposits can be removed from clay pots by scrubbing with a plastic pan scrub dipped first in water, then in coarse sand. Even very stubborn deposits can be removed by careful work with a piece of old, soft, broken pot. Don't forget to clean inside the rim of the pot too! You may think that life will be much easier if you use plastic pots. Initially this is true and a wipe over with a damp cloth is all that is required, but as the pots age, the colour of plastic pots often fades and they become scratched, dull, and more difficult to clean. If you buy only one design of pot, however, you can drop your dull one into a new one just for show as modern designs also nest very well and the double rim will be hardly noticeable. Finally, having obtained a pristine pot, apply fresh clean top dressing and you are ready to go.

For showing, choose pots carefully: plastic (left) can fade, clay pots (right) often become encrusted with lime deposits (BM)

A show-quality plant of *C. hederifolium* forma *albiflorum* — a good species for beginners (MD)

Clay pots can be an attractive feature in themselves; in a competition they will probably carry no extra points but do enhance the overall appearance (BM)

When the flower buds begin to emerge through the top dressing, it helps to examine them regularly. If they emerge in clumps, it is possible at this stage to gently move them in different directions to obtain a uniform distribution. It is usually too late to do this when the buds begin to open. Regular turning of the pot will prevent the flowers from leaning in a particular direction. Some growers will do this once a week, whereas the most enthusiastic might give a quarter-turn daily! You are now almost ready. Most show schedules require a neat horizontal label for your plant. This is a chance to show your prowess with a pen, or to explore the various text styles available on your computer. On arrival at the show, give your plant a last look-over. If any faded flowers are present, carefully remove them; even a gap is better than a faded flower. With a little practice, you will learn to do this by grasping the stem firmly between thumb and finger and pulling away from the tuber; a slight tug may be necessary. If you have any doubt about the correct class, do not hesitate to ask the Show Secretary for advice, however busy he or she seems to be. All want to see your plant on the bench, and want it to receive the credit that it deserves.

You might already be growing a wide range of species in pots, and so might want to branch out and exhibit those that flower when in leaf. The general principles of preparation are merely an extension of those set out above. The appearance of your plant will be best if both leaves and flowers are symmetrically arranged and nicely spaced. The good news is that many plants arrange their leaves and flowers beautifully without any help from you, but any extra care that you take will not be wasted. The bad news is that some of your favourite plants might never be winners. You might have a deep regard for the first *C. mirabile* that you ever bought and it might flower beautifully, but if the leaves are rather dull and with a poor pattern, it will always come second to a specimen with equally good flowers and more interesting leaves. It is very difficult to define what exactly makes a 'good' leaf, but most exhibitors and judges know when they see one: a regular shape, a bold pattern with strongly contrasting colours, perhaps strong teeth around the edge or deep dissections; there are many factors involved. If you pick up an interesting-looking plant from a plant stall and the person next to you mutters a quiet "Oh!", you will know that you are on the right track. Again, inspect your plant carefully before placing it on the show bench and remove any damaged leaves, whether nibbled by pests, 'pinched' against the rim of the pot in transit or whatever; a gap is better than an unsightly leaf. It is essential that your plants are pest-free. Beware the greenfly that hide under the leaves in your cool greenhouse but emerge to bask in the warmth of the show hall, and beware the slug that roosts in your nice new top dressing and emerges to eyeball the judges!

Be brave. Take along some plants to your first show and you will be hooked. Seek out other exhibitors and ask them how they grow their plants; most are more than willing to pass on hints and tips, and you will have taken your first steps towards becoming an expert.

A show plant of *C. graecum* with a good balance of leaves and flowers (MD)

Moving on

Many show schedules incorporate graded competition: a proportion of the classes will be 'open', that is, open to all comers, whether they be nurserymen, the top growers in the country or you, but others will be restricted in some way to beginners or less experienced exhibitors. When you have gained some experience, won a few first prizes, and some of your plants are getting larger, you may want to move up a gear away from these restricted classes. The competition will be tougher, but the satisfaction you gain from that first-prize sticker will be greater. None of the basic principles change and the recipe for success is still the same: grow a good plant well and present it well.

There are numerous ways in which you can gain an edge over your rivals. In a close competition, good presentation can turn the decision in your favour. A plant in a small pot, with the leaves overhanging all around might be excellent in itself, but if the pot is placed in a larger one, then filled around with compost and neatly top dressed, it will make a much better exhibit with more balanced proportions. This procedure, called double potting, will also remove the need to clean the pot! It becomes more important to grow good forms of plants, but by now you will be getting your eye in and getting a feel for what makes a good plant.

When selecting plants though, remember all of the show classes. A *C. graecum* that forms a uniform cushion of stunningly beautiful leaves at the same time as a sparse display of small flowers could still be a winner as a foliage plant. For autumn foliage display, it may be better to remove the flowers, unless you want seed. If the latter is the case, keep your nerve. It is quite common for a car journey to a show to bring about pollination; if the flowers fall off, the seed pods will be less conspicuous and the overall effect better. In addition, the presence of the seed pods might be interpreted as a sign of superior cultivation. It may, however, also bring a flood of requests for seed. Similarly, some forms of *C. purpurascens* flower sparsely, so it may be worth removing the flowers for autumn shows if they have poor form or colour but the leaves are excellent. Being in flower does not, however, disqualify a plant from exhibition as a foliage plant, unless there are so many flowers that they detract from the foliage effect.

Returning to *C. graecum* for a moment; on young plants, the first leaf that develops each year may be much larger than those that follow. If this is left on, its disproportion can give a very unbalanced effect, so it is best removed when a few more leaves have developed.

Some species, *C. hederifolium*, *C. mirabile* and *C. alpinum* for example, can form petioles (leaf-stalks) that grow out from the tuber horizontally and then — often at the sides of the pot — turn up abruptly. This can produce a ring of leaves around the edge of the pot, with an isolated tuft of flowers in the centre and an unsightly bare area between the two. This tendency can be controlled by moving to a smaller size of pot, and then double potting just for a show. I have seen larger specimens of *C. hederifolium* showing this tendency and producing leaves with the flowers. In these cases, the problem had been turned to advantage, and the exhibitor had produced a floriferous white-flowered plant looking similar to a lemon meringue pie sitting in the centre of a green lace doily.

A similar problem is encountered with forms of *C. repandum* and *C. creticum*, which often form very long petioles. When they are grown in comparatively small pots, 5 or 10 cm of petiole can extend above the top of the pot and then flop untidily in all directions. Corrective measures include growing in a deeper pot,

The *C. repandum* group — this is *C. creticum* — readily become etiolated and flop irregularly, not ideal for the show bench (MD)

and keeping the plants in good light but as cool as possible to reduce this undesirable extension. Plunging is another possibility, but this can be a very difficult and frustrating task. Alternatively, shop around for more compact-growing specimens. Before I leave 'elongation' problems, I must mention etiolation. If plants are grown in a greenhouse, a problem can arise particularly during dull weather or if the house is in too shady a site. If temperatures are too high, the petioles and pedicels (flower stalks) will grow too vigorously. With insufficient light, the lower parts will become pale and weak and the stems might even flop over. A plant showing these symptoms will be heavily down-pointed. If the etiolation is not too extreme, it may be possible to 'save' such a plant by double potting. In more extreme cases, do not expect experienced judges to be fooled; they will notice the altered profile of the plant.

The big league

After a few more years of exhibiting, your entries will be confined to the open sections, where the competition can be severe and the judges unforgiving (but fair). Your rivals will include the top growers in the country, so you will need to bring all of your skill into play if you are to succeed. In addition to shows held by the Cyclamen Society, many shows held by the Alpine Garden Society and the Scottish Rock Garden Club also have classes for cyclamen, and here you will see plants grown and presented to the highest possible level. You may feel totally demoralised by the sight of plants in 36-cm pots, but don't worry, there are also classes with 19-cm size limits, so you won't need plants that are 30 or 40 years old to fill one of those.

Experienced travellers who have seen cyclamen species in the wild will tell you that very few plants in the wild bear as many flowers as those shown by the top growers. Indeed, some plants can produce too many flowers and/or leaves in cultivation to make the perfect exhibit. Too many leaves can become overcrowded and spoil the even distribution of the flowers. If there are too many flowers, they can become so overcrowded that they cannot open properly and their elegance is lost. In each case, the top exhibitor will remove the excess as growth progresses to produce an aesthetically satisfying result. Good spacing will also minimise the risk of fungal diseases, which may be a problem in inadequately ventilated glasshouses.

An amazing specimen of *C. alpinum* — but for the judges are there perhaps too many flowers, obscuring their individual form? (MD)

Finishing touches make all the difference. Here, some light-coloured chippings have been used to contrast with the leaves of a show plant of *C. graecum* (MD)

The top-dressing that you use can elevate your exhibits to the level of high art, and perhaps earn you the ultimate accolade of 'expert'. A number of uniformly graded dressings will look neat and tidy and are adequate for the exhibitor with too little time to spare. Sieved gravel of various kinds (2–3 mm is easy to use), sieved peat or fine composted bark will all give a workman-like result. Beware of flint and aquarium gravel! The rather garish yellow tone of the former does not give an attractive effect and the latter is unspeakable. Many cyclamen species grow in limestone areas, so it is worth looking out for limestone gravel. This seems to be available only in larger sizes; 8–10 mm seems to be fairly typical. Reference to a good book will give guidelines on the native habitats of the species and indicate the appropriate material. If you wish to attain the peak of art (and artifice), note that nature very rarely grades rock fragments uniformly, all sizes will be present; so for your top-dressing to be really natural-looking, you might need to obtain some larger pieces of rock and do some brisk work with a large hammer!

It will not necessarily be possible to reproduce the appearance of all cyclamen habitats. *Cyclamen creticum* is often found in limestone areas under pines, so a few pieces of limestone and a

A dark background, in this case bark chips, shows the leaves and flowers of this white *C. purpurascens* to advantage (MD)

light scattering of pine needles will produce perfection. The same treatment might be given to *C. repandum* and *C. balearicum*. However, *C. creticum* can also be found growing in chestnut groves. Unfortunately, chestnut leaves and seed cases are simply too large for a realistic effect in a pot. By contrast, almost anything goes for *C. hederifolium*. This species is so widespread in nature — from the south of France to southern Turkey — that it grows in an enormous variety of habitats. With the exception of sieved gravel, almost anything you dress it with will accurately represent one of its habitats.

Cyclamen purpurascens is a multi-purpose plant for exhibition and gives you a real chance to show your expertise. In Slovenia, it can often be found growing in beech woods. If used as a foliage plant, a simple dressing of dark peat can reproduce very well the appearance of much-decomposed leafmould, and can create an uncluttered and effective exhibit. If the plant is in flower, the same, with the addition of a few small beech leaves, can increase the authenticity. (I know of one exhibitor, now retired, who kept a small collection of leaves, preserved by soaking in glycerine, for such purposes.) Finally, in more mountainous areas, *C. purpurascens* grows in open limestone scree. If your plants bear the appellation 'Lake Garda', then again the addition of pine needles to the limestone would increase authenticity.

When exhibiting cyclamen, remember above all that you are doing it for fun. The angst-ridden exhibitor railing against an unfair judge's decision is a sorry figure who is best avoided. Searching for a good form can be exciting. Growing it well need not be too difficult. Yes, preparing it for exhibition can be a drudge, but that first-prize sticker makes it all worthwhile!

CHAPTER FIVE

Development of *Cyclamen persicum* cultivars

Diony de Bont & John Grimshaw

5.1 INTRODUCTION

Of the 23 species in the genus *Cyclamen*, only *C. persicum* has been subject to intensive breeding and selection, resulting in forms that are massively different from their wild progenitors. As a wild plant (see p. 111), distributed around the eastern Mediterranean (but not Persia!), it is naturally variable and this diversity has been exploited by florists in the development of horticultural forms. The flower colour of the species ranges from white to shades of pink and mauve, generally with a pink or crimson-magenta eye, but plants with pure white and deep purplish pink flowers occur occasionally. The flowers are scented, with long, narrow, twisted corolla-lobes. The foliage of wild *C. persicum* varies from plain green, to silver-veined and marked, to entirely silver in exceptional plants. *Cyclamen persicum* and *C. somalense* are the only *Cyclamen* species within the genus with non-coiling seed stems, a feature that has undoubtedly made its development much easier.

In this Chapter, we chart the development of the florists' cyclamen from its earliest introductions to the present day, as a subject for both selection and breeding and as an important horticultural crop. The cultivation of *C. persicum* cultivars is currently estimated to earn the horticultural industry approximately 100 million Euros annually in Europe. It is therefore not surprising that to the

Wild forms of *C. persicum*, from which all the 'house plant' cyclamen are derived. Seen here growing in Israel (MELJ)

horticultural industry and millions of consumers worldwide, the cultivars of *C. persicum* are known simply as cyclamen, eclipsing all the others.

Introduction of *C. persicum* to northern Europe

It seems certain that *C. persicum* has long been cultivated in the eastern Mediterranean, especially around monasteries and in graveyards, just as it is today. It was introduced to northern Europe in the early seventeenth century, as part of the wave of plant introductions from the Ottoman Empire around that time. The epithets *antiochicum*, *orientale*, *byzantium*, *indicum*, *aleppicum* and *e monte libani*, which are indicative of the origins of bulbous plants and their export from the Ottoman ports (even if the geography was sometimes a bit dubious), were all associated with the plant we now call *C. persicum* — a name that is itself an evocation of somewhere remote and exotic. The epithet *persicum* appears in the name *C. persicum vernale* in Jean and Vespasien Robin's *Manual* of 1624.

Cyclamen persicum soon became grown throughout northern Europe. According to the French botanist Joseph Pitton de Tournefort, the Morin family of Paris were growing two forms of the species by 1620. In 1629, John Parkinson published his *Paradisus Terrestris*, in which he described *Cyclamen Antiochenum Autumnale flore purpureo duplici*, 'the double-flowered sowebread of Antioch', apparently a *C. persicum* with flowers of 10 to 12 petals. Bachelier, an ambassador from the Pope, brought *C. persicum*, among several rare plants, to France in 1656, and 'the Cyclamen of Persia' was mentioned by Sir Thomas Hanmer in his *Garden Book* of 1659. In Germany, Johann Sigismund Elsholtz, physician, alchemist and botanist to the Elector of Brandenburg, mentioned what seem to be different forms of *C. persicum*, '*majus*' and '*vernum flore albicante, corolla purpurea odoratum*' in his *Garten Bau* of 1666. According to William Aiton's *Hortus Kewensis* (1789), *C. persicum* was brought to England from Cyprus in 1731 by Dr Sibthorp and had been grown since then in the Chelsea Physic Garden, providing the material on which Philip Miller based his *C. persicum*.

5.2 SELECTION AND BREEDING

The first cultivars

By the mid-eighteenth century, *C. persicum* was well established in cultivation, at least in specialist collections, and it had a small presence in the nursery trade. In 1739, Nicolaas Van Kampen from Haarlem, Holland, listed three varieties of the species, 'rose', 'white' and 'white with eye'. The nursery firm of Krelage in Haarlem continued to grow and develop the Van Kampen varieties until the end of the century, but in 1800, the entire collection was sold to a buyer in England. Although the colour range had been expanded, the flowers were still quite small but well scented. By 1837, several named selections, such as *albiflorum*, *odoratum*, *purpurascens* and *inodorum*, existed and two French breeders, Albert Truffaut (Versailles) and Claude Fournier, were working on cyclamen.

Serious work on developing the florists' strains of *C. persicum* seems to have started in about 1850, and this has led directly to the plants we know today. Breeders noted that by selfing plants carefully, forms would occur in which the purple ring extended further up the corolla-lobes, sometimes forming a coloured stripe along the centre. Through selection, this stripe was extended until the entire corolla was uniformly pink or purple. By 1853, French growers had been successful in selecting flowers in shades of pink, lilac and purple, although the flower size remained unaltered. What all the early breeders were trying to achieve was a greater range of flower colours and larger flowers with more variable flower forms. Scent and leaf variations were not important criteria in those days.

In 1868, A. Schmidt showed the new 'Wilhelm I' at an exhibition in Berlin. It was the first deep dark red cultivar. Cyclamen were still rare in these years, with only a few plants appearing at shows and exhibitions. Developments were also proceeding fast in England, and in 1869, the *Floral Magazine* illustrated the named selections 'Oriflamme' and 'Rubrum', which were both dark red, as well as the white form 'White Dedication', all developed by M. Wiggins of Isleworth. Many such new cultivar names appeared at about that time, but it was difficult to reproduce and keep the same characteristics in the following generations. Obviously the so-called varieties did not come true from seed and were not fixed cultivars. All were diploid.

The first large-flowered tetraploid cyclamen: *C.* 'Giganteum'

A big step forward was made in 1870 when Edmonds in Hayes, England, developed the cyclamen cultivar 'Giganteum', which had double the flower size of contemporary cultivars. In addition to its increased flower size, this whole plant was larger and stronger than any cyclamen seen before, indicating that it was the first tetraploid: doubling of chromosome number often results in the enlargement of flowers and leaves. The flower colour was white with a purple eye. This race also had great vigour and flowered in 14 months from seed, but with it came the first unscented flowers; B. Williams crossed 'Giganteum' with cultivars that had

One of the early cultivars of “C. giganteum”, ‘Rosy Morn’, drawn and engraved by John Nugent Fitch (1840–1927). From *The Floral Magazine* NS 1878, published by J. Reeve of Covent Garden (MD)

differently coloured flowers, and by 1884, had created a range of six different colours in the 'Giganteum' type. At about the same time, M. Wiggins bred the first edged cyclamen, 'Marginata', which was dark red with a white edge. It is evident that during the nineteenth century there was considerable exchange of plant material between breeders in different countries; for example, in this period when German horticulture was undergoing a revival, German florists spent time learning about cyclamen at English nurseries.

Salmon- and red-flowered cyclamen

The first cultivar that could be described as 'salmony' pink was 'Salmon Queen', bred by the seed grower Sutton & Sons (Reading, England) in 1894. It became an important parent. H. Kiausch (Berlin) made combinations between 'Salmoneum', 'Weiß mit Auge' (white with eye) and 'Rosa von Marienthal' (salmonish lilac), resulting in 1905 in the well known 'Rosa von Zehlendorf'. This was very popular and remained the most frequently grown cultivar for many years; it is still grown today. Kiausch improved 'Salmoneum' by crossing in several other cultivars and created 'Perle von Zehlendorf' in 1907.

'Rosa von Zehlendorf', a popular cultivar raised in Germany in the early twentieth century (SAH)

'Rosa von Marienthal' produced in Berlin by the firm of Kiausch (SAH)

'Perle von Zehlendorf', another German cultivar from Kiausch (SAH)

An early cultivar from The Netherlands, 'Vuurbaak' (SAH)

One of the most important targets in cyclamen breeding was to obtain the colour red. After obtaining a salmon-pink, all breeders were selecting for the darkest shades and making crosses to achieve this. Johannes Dlabka (Berlin-Zehlendorf) crossed 'Perle von Zehlendorf' with 'Dunkelrot' (deep red), which resulted in 1919 in the cultivar 'Leuchtfeuer' (='Bonfire'), which became known worldwide. In Aalsmeer (The Netherlands), the same colour was produced in the cultivar 'Vuurbaak' by crossing 'Carmin Salmoneum' with 'Perle von Zehlendorf'. 'Leuchtfeuer' and 'Vuurbaak' superseded 'Rosa von Zehlendorf' as the best-selling forms. The small-flowered red diploid 'Firefly' received an award in 1923. Bred by Sutton & Sons, this was the base for later diploid red varieties. These early red cultivars had the tendency to fade with the emergence of bluish tints, but this problem was solved by back-crossing with salmon-flowered cultivars.

'Papilio' and other special forms

In addition to the standard flower shapes, breeders have also been keen to develop flowers that have a 'new look'. Many novelties have appeared, only to disappear again as it proved impossible to maintain the type by selection or through inbreeding problems, or as they proved to be difficult to grow. Through the process of selection, however, some very distinct types occurred and became stabilised, leading not only to the more freakish-looking flowers of the 'Fimbriata', 'Rococo' and 'Cristata' groups, but also to the modern flower shape we know in today's florists' cyclamen.

'Rococo Yellow', a special Rococo with greenish-yellow edge (SAH)

One of the earlier mutations to appear resulted in the 'Fimbriata' flower type, a group of tetraploid varieties with a fringed flower edge. The first breeder to work seriously on Fimbriata cyclamen was Alwin Richter (Dresden-Striesen). He found a rose-coloured, fringed type in the late 1880s and subsequently he introduced three forms all with fimbriate petal margins: 'Leuchtendrot' (shining red) and 'Weiss mit Auge' (white with eye) in 1896 and 'Reinweiss' (pure white) in 1899.

A very important source for different flower forms was 'Papilio Brussels Best', bred by the Belgian Louis de Langhe (Saint-Gilles, near Brussels). In 1886, de Langhe's attention was drawn to one plant in a lot as it stood out because of its coarsely dentate, beautifully marbled leaves, rounded flower buds and flowers of unusual beauty. He separated this plant, pollinated it and improved the type with every succeeding year of selection so that many colours and flower forms appeared. The plants were remarkably

Early example of the 'Papilio' (Butterfly) varieties (SAH)

vigorous and floriferous. Finally finished, this form came on the market in 1896. 'Papilio', as De Langhe described it, had flowers with enlarged petals that were "held erect like the wings of a butterfly about to take to the air". The expanded surface of the petals of the flower made them very showy. Besides the flowers, the leaves were also very decorative with beautiful markings. The term papilio is used to describe a type of flower that has widened petals that are folded towards the edge and slightly hollow like a shell. The angle of the petals was about 45 degrees. The leaves were beautifully marked and dentate.

In 1900, from the 'Papilio' stock came 'Fimbriatum Carmineo Marginatum', later renamed 'Victoria', which was bred by Heineke (Bremen) and is still grown today. This spectacular new tetraploid cultivar had white flowers with a crimson eye and frilly crimson tips to the petals. The variety was not stable and even today there is always a percentage of plants without the frilly, coloured edge. In 1911, the French grower M. Eysser (St-Germain) introduced the purple 'Cattleya', a cultivar that is considered to be the most perfect 'Fimbriata'-type.

J. Schmidt (Erfurt) crossed 'Papilio' with 'Fimbriata' and this led to 'Rococo' (or 'Rokoko'), which was made available in several separate colours in 1904. 'Rococo' is a type of plant that has large flowers, which are extremely strongly ruffled and frilled, giving the effect of a double flower. The 'Rococo' types never became as popular as the fringed ones because their heavy, hanging flowers did not always open properly. The flowers are in fact too heavily fringed, which is why the petals hook into each other and do not open, just like a cabbage, and the plants are slow-growing. A very special type called 'Neue Gefranste' (new fringed), was bred by Franz Braukmann (Hildesheim, Germany), was easier in cultivation.

Cyclamen 'Neue Gefranste', bred by Braukmann (Germany) (SAH)

Another special form derived from 'Papilio' was 'Cristata', which has crested or bearded flowers. This crest results from outgrowths of tissue on the central portion of the outward surface of each petal. Its colour sometimes differs from that of the petal, giving a bicoloured effect.

Twisting of the petals was not wanted and breeders selected for flat, non-twisted petals.

It was around the 1920s that the typical elegant cyclamen flower form that we know today, with four (or more) petals pointing upwards and one petal bent downwards along the flower stem, appeared. Selection had resulted in the petals becoming increasingly broad, with the ultimate result that one petal could not turn upwards, being held back by the flower stem.

A 'Cristata'-type, 'Early Bush Pioneer' raised in 1897 (SAH)

Other developments in open-pollinated and non-F_1 hybrid cyclamen

Before the introduction of F_1 hybrids in the 1970s and their eventual dominance of the marketplace, breeders continued to work with open-pollinated cultivars and selections, resulting in numerous developments. For example, in 1926, Franz Braukmann produced the beautiful bicoloured 'Flamingo', whose flowers were deep salmon with light margins, as the result of a cross between 'Flamme' and 'Leuchtfeuer'. In 1948, the popular variety 'Apfelblüte' ('Apple Blossom') with delicate pink flowers arose from a cross between 'Lachshell' and 'Neu-Rosa' and was developed by Johannes Fisher (Dresden-Leuben).

In the 1930s, there was a tendency to select for smaller plants. An example was 'Ras Harting', produced in Aalsmeer, which was later followed by 'Ras Kuiper' and 'Ras Tas'. Working in The Netherlands, Professor Wellensiek began a research project in 1946 that tried to help breeders raise new cultivars. One of his main aims was to find out whether crosses between *C. persicum* and any of the other *Cyclamen* species or cultivated forms could be made successfully. This project, however, was not successful.

Leaf patterns

For many years, breeding for leaf patterns was of minor importance. It is possible to distinguish several main leaf patterns in cyclamen: completely green, marbled, silver-veined, silver-edged (with green centre), silver-centred (with green edge) and completely silver. There are many intermediate patterns. The first selection to be distinguished by its leaves was *C. persicum* f. *zonale*, which has clear

Example of a 'Cristata'-type, showing the raised crest in the centre of the petals (SAH)

silver zones and was first offered in 1896. A series of silver-leaved varieties was offered by J. Forsythe (Putteridge, UK) in 1929. In around 1937, a silver-leafed type arose from a cross between 'Salmoneum' and a small-flowered vermilion red 'Lubbenthal'-type at the Mekel nursery, in Brühl, Germany, where the firm of Struewe was also selecting for silver leaves.

Scented cultivars

Scent, being invisible, is a difficult characteristic to select for. This is the reason that fragrance has been lost in many cultivars. The wild *C. persicum* is well scented and the oldest cultivars were mostly scented, but scent was progressively lost in later selections. The fragrance is attractive and sweet and, on sunny days especially, it can be very pleasant. Scent can be an added value for sales.

In 1892, Ascherson discovered that there are two different kinds of perfumes in cyclamen. The corolla possesses the typical cyclamen scent, while another, less pleasant, scent is spread by the anthers. The typical cyclamen perfume is very sweet and can be lemony.

A long-stemmed variety suitable for cut flowers, 'Rheingaufeuer Schnitt' (SAH)

In 1911, Schmidt offered a scented race in three colours, and in 1937, Pregetter introduced a scented cyclamen that he crossed with his double forms. Chrestensen introduced the 'Bornthaler Duft' cyclamen, which were famous for their scent, in 1955.

Cut-flower cyclamen

Cyclamen selected for cut-flower production were sold from around 1896 onwards. Varieties for cut flowers use had long, strong stems and large flowers, preferably with unusual flower forms such as 'Rococo', 'Fimbriata' or bicolors. Cut flowers, if well treated, can last in a vase for two or three weeks.

Bosse (Friesdorf, Germany) started seriously breeding cut flower types, known as the 'Friesdorfer Schnitt-cyclamen' in 1957. Another German cultivar was 'Rheingaufeuer Schnitt', which had bright red flowers and strong stems of 40 cm, providing ideal cut flowers. In fact, since their first introduction, *Cyclamen* in Germany and in the East European and Scandinavian countries had always been quite important in cut-flower production. In the old days, growers in The Netherlands picked the first flowers of ordinary cyclamen when the plants were not ready for selling and sold them as cut flowers.

Double flowers

The oldest recorded double form is probably 'Aleppicum', a small-flowered white known in the early seventeenth century. By about 1850, double-flowered cyclamen were available, although they weren't uniformly double. Müller (Dresden-Striesen, Germany) had developed double-flowered cyclamen in almost every colour by 1880. Frederick de Bordet (Warsaw, Poand) discovered a semi-double plant in 1875 and after five years of breeding, he obtained a strain that was 50 to 75% double.

The double-flowered cultivar 'Monstruosum' arose at the nursery of M. Millet (France) around 1886. These plants were very robust and had ten petals instead of the normal five. This gives a double effect, although the flowers were not really double as usually understood as their stamens were still present and might also have doubled in number. Real double flowers have stamens that are converted to petals and are usually male sterile. It is also possible that extra petals can appear from the base of the stamens, leading to doubleness.

In 1915, John Vogt (Sturgis, Michigan) developed double cultivars that had 12, 16 and 20 petals, later even 64 petals! Some flowers attained diameters up to an enormous 12.5–17.5 cm. Pregetter (Austria) combined scent with double flowers. 'Kimono', a double-flowered series was launched by Sakata in the 1960s. Double-flowered *Cyclamen* were especially popular in France, where in the 1980s, Gaboriau had a good series that had large flowers with twisted petals and thick stems.

F_1 hybrids

In the 1970s, most seed companies started to develop F_1 hybrids from diploid material, but some companies, such as Morel, VEG, Beck and Clause made tetraploid F_1 Series.

Diploid varieties

The species *C. persicum* is of course diploid, but since the introduction of 'Giganteum', most new cultivars had been tetraploid. By back-crossing with the species, diploid types could be created again. Franz Braukmann, who had started breeding cyclamen as a hobby in 1900, was probably the first breeder who succeeded with this. In 1922, he bred the first pastel-coloured diploid cyclamen, whose parents are unknown. A striking variety was 'Flamme', which was deep salmon with light salmon edge but a darker centre. For many years, however, breeders ignored the 'Pastel' types.

Reichard (Berlin, Germany) bred a dwarf type of plant with many, relatively small flowers, in 1937; his work resulted in the so-called *parviflorum* 'Neu-Rosa' and 'Violetta', both of which turned out to be diploid. He called these varieties 'Fenster'-cyclamen (window-cyclamen).

Victor de Troyer (Ukkel, near Brussels, Belgium) is called the 'father of the diploid cyclamen'. He probably used diploid cyclamen like 'Flamme' from Braukmann among others, and began making crosses and selections in 1933. De Troyer was forced to stop his work in 1939 at the beginning of the Second World War and subsequently lost much of his material, but he continued after 1944. Around 1950, a new type of plant appeared that turned out to be diploid. The new diploid plants exceeded the old (tetraploid) varieties in flower size and thickness of the flower stems, but were also compact and flowered from their centres. In the beginning, these diploid plants were not completely uniform and were seen as populations rather than as cultivars. In the 1960s, de Troyer developed a range of varieties of different colours, including 'Zilverwit' (the whitest of all), 'Flamboyant', 'Mystique Melody' and many more. The Belgiun grower Marcel Stercks (Koningslo-Vilvoorde) also bred diploid cyclamen, and this was the reason why diploid cyclamen were called 'Belgian cyclamen' in those days.

'Mystique Melody', a unique bicolour variety from De Troyer (Belgium) (SAH)

In 1956, Lodder (Vleuten, The Netherlands) brought the 'De Troyer' types to The Netherlands and introduced them as 'Pastel cyclamen'. These pastel varieties were originally very variable and according to the rules could not be approved as cultivars, failing to meet the required 65% uniformity required by the NAK-S, the Dutch General Inspection Service of Ornamental Plants. Since 'Pastel' cyclamen were greatly in demand, a temporary exemption was made, allowing the seed to be sold freely. Growers such as Tas, Man, Lodder, Buis and Maarse re-selected the pastels and made them true to colour. 'Pastel' is a separate group of large-flowered diploid cyclamen and it contains a wide range of colours and shades, some of which do not exist in tetraploid cultivars. Besides plain-coloured varieties, there are many bicolors such as 'Appleblossom', whereas other varieties show flaming and edging. The 'Pastel' cyclamen have a great variation in flower form and beautifully marked leaves.

The first named 'Pastels' were probably introduced by Lodder in 1960 as a series in which the cultivars were named after composers. This series included cultivars with strong flower colours such as

'Pastel' Series, mixed colours named after famous composers, raised in the 1960s in The Netherlands (SAH)

scarlet and violet as well as pastel shades; examples are 'Bach' (deep rose), 'Beethoven' (purple violet) and 'Strauß' (salmon red). In total, there were 24 named 'Pastel' cultivars! 'Pastel' cyclamen tended to make long flower stems in the summer and therefore they were at their best in autumn and winter, when the stems were shorter. Tas (Aalsmeer, The Netherlands) created short-stemmed 'Pastel' cyclamen by crossing these varieties with tetraploids.

Victor de Troyer was able to distinguish tetraploid or diploid plants simply by looking at their leaves. Tetraploid cyclamen have thick leaves with dull markings, whereas the leaves of diploids are thinner, bright and clearly marked. De Troyer was also a perfectionist when it came to flower form. His ideal flower had about 7–9 flat petals overlapping each other in layers around the base and one petal that was held downwards along the stem. 'Pastel' cyclamen were distinguished by their strong growth, long shelf life and resistance to diseases.

In 1962, Prof. Susan Wellensiek (Wageningen, The Netherlands) published *The Descriptive List of Cyclamen Cultivars*. She collected plants from 22 breeders, which he compared and investigated. She concluded that most varieties were tetraploid with a chromosome number of 96 (some 92), a view that had also been reached by Prof. Dr Hans Kappert (Berlin) in 1941. Diploid cyclamen have 48 chromosomes. All of the German varieties turned out to be

One of the 'Pastel' varieties, 'Chopin' (SAH)

tetraploids, whereas the English and Dutch cultivars were either diploid or tetraploid. It turned out that there was a great deal of overlap in the material offered. Some cultivars were a mixture of tetraploid and diploid plants, with no morphological differences. Wellensiek's list provided an important basis for the identification and correct naming of the cyclamen cultivars known at that time. In the 1970s, open-pollinated diploid varieties were grown in countries with a moderate climate all over the world. These cultivars were often selected differently by breeders in different regions and under different climatic conditions. An example is Dauernheim in the United States, who bred very vigorous plants with big leaves and huge flowers with twisted petals. Dai Ichi in Japan produced the largest white form ever, with some plants carrying flowers with petals of 10 cm length (3 or 4 times the length of the petals of *C. persicum* seen in the wild!) and a total flower length sometimes measuring 20 cm! This 'Nomura White' was the largest of all cyclamen varieties ever and, unexpectedly, was diploid.

In the open-pollinated cultivars, every breeder distinguished his own selection with the name of the company, for example: 'Rasse Beck' (i.e. Beck's race), 'Rasse Fischer', 'Type Tas' or 'Ras Scholten'. Such selections are still being grown; for example, the open-pollinated cyclamen 'Ras Vollebregt' was selected by Vollebregt (Aalsmeer, The Netherlands), who should be mentioned as excellent cyclamen growers and selectors. The Dutch grower Scholten has been famous for his white cyclamen with big, round, flat flowers for many years, regularly getting the highest marks at the Dutch Hortifair for his 'Scholten Wit'.

Undulated cyclamen

De Troyer was also the breeder of a diploid waved cyclamen named 'Ondulatum', which he introduced in 1955. The flowers have a gently waved edge and the plants have strong, rather large leaves. In Aalsmeer, this type became known under names such as 'Barok' and 'Pastel Gefranjerd' (fringed pastel). After reselection, 'Ondulatum' was later the base for Sahin's 'New Wave', which became a series of twelve colours. 'Grandia' was a cultivar selected in the 1970s that had remarkably strongly ruffled flowers in salmon pink. It came about 50% true from seed.

Silver leaf

Occasional plants with completely silver foliage would appear in 'Mystique Melody' strains, but in the 1960s, De Troyer introduced the first diploid open-pollinated silver-leaved series, which was called 'Decora'. The name 'Rex' (derived from silver-leaved *Begonia rex* cultivars) is also used by some growers for silver-leaved varieties.

Miniature cyclamen

In the early days of cyclamen breeding, when the flowers were rather small, the aim was to breed for larger flowers, but the wheel eventually turned. In the 1960s, Wye College, University of London, initiated a cyclamen-breeding programme. They began to back-cross various large-flowered strains with the wild species. The aim of the program was to introduce a smaller plant with more refined flowers, attractively marked foliage and, most of all, scent. This led to the successful 'Wye College Hybrids'. This series was later re-selected and sold under the name 'Anglia' by Klaas Visser (Aalsmeer, The Netherlands).

About the same time, Prof. Wellensiek, in cooperation with Dr J. Doorenbos, began to make crosses between the modern cultivars (such as 'Sylphide') and species *C. persicum*. In 1972, these crosses resulted in a series of rather small-flowered cyclamen that became well known under the name 'Wellensiek cyclamen'; their names derived from the experimenters' children — 'Anneke', 'Willy'

Victor de Troyer with his cyclamen (SAH)

'Wellensiek' named after the professor who investigated the cytogenetics of *C. persicum* and its cultivars (SAH)

and 'Sonja'. The flowers and plants were rather large and today we would call them midis. Scholten (Aalsmeer, The Netherlands) improved the Wellensiek minis.

In 1973, the well-known 'Kleine Dresdnerin' saw the light; bred by V. E. G. (Weixdorf, Germany), it has since been sold worldwide by S&G.

A mixed batch of 'Miracle' miniature series, raised by Goldsmith (GSM)

A lot of breeding activity on small-flowered cyclamen occurred in Switzerland, with breeders such as Moll, Mauser, Nägeli, Wyss, Schlatter and Haller being involved. Early series of minis to be released onto the market were 'Romi' ('Roggli') and 'Schöne Helena' ('Belle Hélène') from Haller. The early miniature series were quite large-flowered and large-leaved varieties and were really midis rather than minis, but the Swiss breeders in particular selected for the smallest types. The smallest of all is 'Schlatter's Piccolino', which is typically cultured in 6–8-cm pots. The latest F_1 mini series are tending to be smaller plants with small leaves and flowers. These varieties are already smaller than the wild species.

Many of the open-pollinated miniature cyclamen are sweetly scented, as are some of the F_1 hybrids. This is especially true of the minis; for example, the F_1 Series 'Miracle' and 'Midori' both from Goldsmith, are considered to be well scented.

'Midori' another of the mini cyclamen types from Goldsmith (GSM)

Mini cyclamen were originally of minor commercial importance, but are becoming more and more valued. Young consumers tend to buy small-flowered cyclamen, as large- flowered ones are seen as old-fashioned plants. From the grower's point of view, minis are favourable, because the growing time is shorter and more plants can be grown in the space available. Miniatures flower in six months from seed, whereas large-flowered cyclamen need seven or eight months to flower. Minis take up less volume during transportation. Together, these factors make mini cyclamen commercially interesting for the grower.

F_1 hybrids

Starting in the 1970s, there was a need for more uniform-flowering cyclamen with a shorter growing time. Cyclamen had to be programmable for the modern growers and should flower all at once to make it possible to clear a crop within a short period of time. The auction system in The Netherlands makes this both possible and desirable. As in other crops, F_1 hybrids were the answer. Several breeders began to make uniform parents lines for the purpose of creating F_1 hybrids. In general, they used diploid 'Pastel'-types as these are easier to make uniform than tetraploids.

The first F_1 series on the market was the large-flowered 'Firmament' Series released by Royal Sluis (Enkhuizen, The Netherlands) in 1973. In 1974, Pannevis (Enkhuizen) started their (nameless) large-flowered F_1 Series with cultivars such as 'Rosamunde', 'Tzigane', 'Swan Lake' and 'Merry Widow'. This series was later sold under the name 'Pannevis'. This was the beginning of a turbulent development of new F_1 series in all flower sizes and in an increasing number of flower colours. All these series have been extended in subsequent years and 'Halios' (Morel, France), with 24 different colours, is the broadest series. Ten years later, in 1983, Royal Sluis again led the way with

'Rosamunde' one of the early F_1 cyclamen raised by Pannevis in 1974 (SAH)

The 'Halios' Series cyclamen at Morel Diffusion come in a wide colour range (MD)

Cyclamen trials at K. Sahin Zaden BV in the late 1980s (SAH)

The 'Laser' Series from Goldsmith have moderately sized flowers, classed as 'midis' (GSM)

the first F_1 miniature cyclamen, 'Sirius', whose flowers were a mixture of rose shades and white, followed by 'Royal Mini'. In the early 1990s, many mini F_1 Series came on the market. The 'Butterfly' series from Royal Sluis was the first large-flowered fringed F_1 hybrid introduced to the market in 1992.

A new group of hybrids, the midis, arose with a flower size in between that of the minis and the large-flowered varieties. Every seed grower introduced their own medium-sized F_1 hybrid, among which the best known are 'Laser' (Goldsmith) and 'Latinia' (Morel). These medium-sized F_1 cyclamen were not very popular at first, as their prices were the same as of the minis, although the plants were very early, very floriferous and many of them were sweetly scented. The scent has its origin in one parent that was a miniature. Retail dealers could not place midis, because they were neither large-flowered nor small. After several years, however, both the trade and customers became more accustomed to midis and demand increased.

The introduction of F_1 hybrids brought an end to the disease problems that had led to declining interest in cyclamen. In fact, these hybrids encouraged growers to grow ever larger quantities of cyclamen, often shifting from other crops, and over-production with falling prices resulted. The new large-scale growers did not have the experience and skill needed to grow good-quality plants; this resulted in lower-quality plants being marketed and damaged the reputation of cyclamen.

Midi cyclamen of the Morel 'Latinia' Series (MD)

Second-generation F_1 hybrids

The advantages of diploid hybrids over tetraploids eventually became evident. The first generation of F_1 hybrids sometimes left much to be desired with regard to characteristics such as stem length, uniformity and plant quality, but F_1 hybrids became more and more popular as improvements were made. During the 1990s and later, individual colour forms or indeed complete series were replaced with better selections. Varieties became more refined, more uniform and earlier flowering. Leaves became smaller, flower stems shorter and plants more compact, leading more and more to an ideal plant type. The rapid growth of the market share of F_1 hybrids has largely been at the expense of the open-pollinated cultivars. The F_1 hybrids are superior in growing time, uniformity, habit, floriferousness and vigour.

As well as standard varieties, the medium-flowered F_1 'Frou-Frou', which has waved flowers, arrived from Eyraudplants (St Laurent la Conche, France) early in the 2000s. The first silver-leaved series, 'Silverado' from Goldsmith (Andijk, The Netherlands), appeared in 2002; this is a mini at first only available in red and white. In 2006, this range was extended and a midi series called 'Sterling' (also Goldsmith) was introduced. A series with silver-centred and green-edged leaves, a leaf pattern that existed in some of De Troyer's lines, called 'Silver Heart' has recently been released by Goldsmith.

Morel trials of 'Latinia' Series (MD)

'Frou Frou', a second-generation F_1 hybrid with frilly flowers, raised by Eyraudplants (SAH)

A hanging basket containing the strikingly silvery-leaved 'Sterling', a midi cyclamen (Goldsmith) (GSM)

Attractive silver leaves are a feature of the mini cyclamen 'Silverado' (Goldsmith) (GSM)

'Silver Heart' has the silver zone in the centre (Goldsmith) (GSM)

The very large-flowered 'Magnum' series, introduced by Sahin in 1993 (SAH)

'Robusta', a large-flowered series from Goldsmith (2005) (GSM)

In 1993, Sahin introduced the 'Magnum' series, which has huge, often white, flowers. The 'Robusta' series, released by Goldsmith in 2005, are also large plants with extra-large flowers. As the 'Robusta' plants grow for a longer period vegetatively before forming a full-size plant, they take a little longer than usual to flower. 'Robusta' are grown in 15–20-cm pots.

Open-pollinated cultivars today

Since the F_1 hybrids entered the market, the open-pollinated, true from seed varieties have steadily decreased in quantity and popularity. Nevertheless, the open-pollinated cultivars have maintained a market share against the swelling stream of new F_1 hybrid series. Even so, there is a tendency for open-pollinated cyclamen breeders (such as Schoneveld (The Netherlands) and Eyraudplants (France)) to move into F_1 hybrids, and to reduce their activities with open-pollinated cyclamen.

Although young plants of open-pollinated cyclamen are cheaper than F_1 hybrids, the latter have many advantages, including uniformity in habit and flowering time, a higher percentage of saleable plants and a shorter cropping time. Despite this, there are still some cyclamen growers who grow open-pollinated cyclamen on a large scale, such as Vollebregt (The Netherlands), Scholten (The Netherlands), Chrestensen (Germany) and Eyraudplants (France). Eyraudplants offer fringed pastel and 'Sunkiss', a winter-flowering mini mixture containing 45% of flamed colours. They also list two F_1 series called 'Crystal', a compact large-flowered series, and 'Mini Star', a small-flowered series, as well as the open-pollinated series 'Micro Star', which has fringed, pastel-coloured flowers. Chrestensen still produces the old varieties 'Victoria', 'Rococo', 'Bornthaler Duft', 'Fringed', 'Luckenwalder Gefranst Schnitt', 'Barbarossa', 'Striata', 'Kleine Dresdnerin' and 'Pastel'. Scholten currently offers 'Pastel', 'Pastel flamed', 'Fringed', 'Mini mixed' and 'Standaard' (a large-flowered tetraploid variety). Peter Sprünken (Straelen, Germany) is still growing several fringed and cut-flower varieties.

In the 1970s, the famous Japanese singer Akira Fuse released the song 'Shikuramen no kaori', which translates as 'The scent of cyclamen'. This song became a big hit. Since then, *C. persicum* plants have become a best seller in Japan, despite this country's hot and humid climate which is not really suitable for growing cyclamen well. The Japanese have a special taste and have a good eye for fine forms. Around 30 cyclamen breeders are active in Japan, usually working from small nurseries. Of these, six are really important: Hayakawa, Takami, Takahashi, Kanazawa, Kage and Hayashi.

Working at the famous Yamatsuri nursery, Yoshihiro Kanazawa has bred many cultivars over the years, including 'Lipps Orange' (striking scarlet with light edge), 'Shining One Purple' (flamed purple), 'Chandelier' (ruffled pink), 'Papillon' (purple with white edge), the 'Chimo' Series (double mini) and 'Forest Fancy Candy'

Although an old variety, 'Striata' is maintained by the firm of Chrestensen, Germany (SAH)

(mini pink striated). The Yamatsuri nursery also offers a striped variety, 'Shooting Star Red'. A striking variety from Hara Nursery is 'Angel Wing', which has a very special rose-flamed flower without a darker eye, and Kage introduced F_1 'Piko' in midi and mini types with bicolour flowers (light edge). Several Japanese varieties have been brought to European cyclamen-breeding companies to improve their commercial value. Japanese breeders are not concerned about earliness, however, so some Japanese varieties are very late flowering.

Open-pollinated cyclamen provide opportunities for the development of special colours and forms that cannot be fixed in an F_1 hybrid. In about 1988, I (Diony de Bont) found a remarkable plant in an F_1 lot. Its corolla was lilac-coloured but it had a white, enlarged calyx looking like an extra whorl of small petals, giving the flowers a bicolor effect. This enlarged calyx prevents the petals from reflexing, so they spread horizontally, with the sepals overlapping them on the upper surface. As a result, the flower resembles a parachute or whirlygig. The characteristic was unstable within the plant, but interesting enough to continue breeding with. I crossed this plant with all kinds of colours and it was easy to carry over the character. Sahin introduced this type in 2001 under the name 'Whirlygig Series'. Although the characteristic remains unstable, when performing at its best, a Whirlygig cyclamen is smashing and very un-cyclamen-like. The bicolor effect produced by the white calyx is particularly effective and attractive in combinations with dark colours such as scarlet, purple, dark red and deep rose.

'Angel Wing', one of the many Japanese-raised varieties, this one from Hara Nursery (SAH)

The extraordinary 'Whirligig' is derived from a plant found by Diony de Bont among F_1 seedlings. The resulting Series was introduced in 2001 by Sahin (SAH)

5.3 CLASSIFICATION OF THE *C. PERSICUM* CULTIVARS

Cultivars of *C. persicum* can be classified in many different ways: by flower size, flower margins, leaf colour, ploidy level and so on.

Flower size

We can distinguish cyclamen plants of several flower sizes, usually defined as: extra large (magnum), large (maxi), midi, mini and micro.

The range of flower size in *C. persicum* cultivars, L to R: Micro, Mini, Midi, Large (Maxi), Extra Large (Magnum) (SAH)

'Sierra', a large-flowered type showing how one of the wide petals stays down, pressed on the flower stem (GSM)

Extra large plants, such as 'Magnum' or 'Robusta', have very large flowers (petals of 8–10 cm in length). These plants are best grown in large pots (15–20 cm).

Large plants include standard cyclamen types with petals of 6–7 cm in length. Their petals are twisted or flat and in some cases quite broad. In general, one petal points downwards along the stem. This type of cyclamen can be grown in 10–16-cm pots. Examples include most of the old tetraploid varieties: 'Pastel', 'Sierra', 'Concerto' and so on.

Midi cyclamen have medium-sized flowers with petals of 4–5 cm in length. Their petals are twisted and all point upwards. Examples are F_1 'Laser', F_1 'Latinia' and F_1 'Canto'. Midis can be grown in pots of 10–13 cm in diameter.

A 'Laser' Series variety from Goldsmith, a midi-sized cyclamen (GSM)

Mini cyclamen are the typical small-flowered types with twisted petals of 3–4 cm in length. Their flowers are usually scented as they are relatively close to the wild *C. persicum*. The earliest so-called mini cultivars, such as 'Wellensiek', were more of midi size than true minis. Real mini series include F_1 'Miracle', F_1 'Marvel' and F_1 'Metis'. Minis are best grown in pots of 6–12 cm in diameter.

A purple variety from the 'Marvel' Series, mini cyclamen more akin to the wild species and fragrant (GSM)

Micro cyclamen are a rather new group whose flowers are smaller than those of the minis: their petals are 2–3 cm long and their leaves are tiny and well proportioned. In general, micro cyclamen have scented flowers. Micro Series include 'Piccolino', F_1 'Micro' and F_1 'Midori'. Micros are best grown in 6–12-cm pots.

This white with a red eye is one of the 'Midori' Series, classified as a micro cyclamen and very like wild *C. persicum* (GSM)

Flower shape and margin

Smooth is the term used to describe the normal type of flower.

Fringed cultivars have a toothed petal margin, usually in the form of fine fimbriations or laciniations. These plants are often tetraploid cultivars such as 'Fimbriata' and 'Victoria'.

Waved cultivars, such as 'New Wave' and 'Nordseewellen', have a waved (i.e. undulate) petal margin.

Papilio cultivars have flowers whose petals are enlarged, like the wings of a butterfly.

A salmon pink 'Butterfly' ('Papilio') type (SAH)

Rococo cultivars have large flowers whose petals are extremely fringed and ruffled, giving a double effect; the ruffled edge is often greenish-yellow.

Anemone-flowered cultivars have petals that are not reflexed and stand horizontally, resulting in a large flower.

Crested cultivars, such as 'Cristata' and 'Barbarossa', have a bearded flower, resulting from an outgrowth of tissue at the base of the petals that is often fimbriate on its surface.

Whirlygig cultivars have flowers with an enlarged white calyx and a horizontal corona, giving a showy bicolour effect.

Bell-flowered cultivars, such as 'Bellisima', have a fringed flower bud that does not open.

Flower colour

Uni-coloured, also called self-coloured, cultivars (with or without a coloured eye) include most single-coloured forms.

Bi-coloured cultivars have a range of flower-colour patterns: -edged cultivars, such as 'Marginata', 'Mystic Melody', 'Verdi' and 'Fantasia', have petals with a light margin or a fine white line; -flamed cultivars, such as 'Apple Blossom', are white with a touch of colour; -striated cultivars, such as 'Striata' ('Harlequin'), are purple-red with pinkish-white stripes; and 'Victoria'-type cultivars, including 'Victoria' and 'Victoria Deco Mix', have a fringe (edge) that has a different colour from the rest of the petal (the same colour as the eye).

The very striking 'Victoria' in which the petals are outlined with the same colour as the basal 'eye' (SAH)

'Fantasia Violet Foncé' F_1 from Morel, France (SAH)

Leaf colour

Green-leaved varieties have lighter or dark green leaves without significant marbling. The old variety 'Sneeuwitje' even had dark brownish-green foliage. The underside of the leaves, especially in red and purple varieties, is purple coloured.

Marbled varieties include the majority of cyclamen, having leaves that are more or less marbled. The tetraploids are often dull-looking, where the diploids often have bright silver markings.

Silver-edged varieties are often just called silver leaf, as the pattern seen most often is one of mostly silver leaves: a broad margin is silver coloured, whereas the centre of the leaf is green. In former days, plants that had this leaf pattern were often called 'Rex', as in begonias. De Troyer used the name 'Decora'.

A scarlet-flowered version of 'Silver Heart' making a striking combination of red and silver (SAH)

Silver-centred cultivars are those whose leaves have a broad green margin and silver in the centre. They are not as common as silver-edged cultivars. This leaf pattern appears in the old 'Baardse's Wonder', in some of De Troyer's lines, in some minis and in some *Cyclamen* species. Goldsmith lately introduced the Mini Series 'Silver Heart', which has this leaf pattern.

Complete silverleaf cultivars are seldom seen. There was an old 'Mystic Melody' line that had a completely silver leaf.

Dentated leaf cultivars such as 'Rococo' can be recognised by their beautifully marbled and dentate leaves. 'Frou-Frou' has the same characteristic.

Ploidy level

Diploid cyclamen are like the wild species in that they have two sets of chromosomes: 2n = 48. Diploids have softer, thinner leaves than cyclamen of higher ploidy levels. In general, the small-flowered cyclamen are diploid, as are pastel cyclamen and most of the F_1 hybrids. Cyclamen whose flowers are white are generally diploid, although 'Kätchen Stoldt' is a white tetraploid exception.

Tetraploid cyclamen have four sets of chromosomes: 4n = 96. They are distinguishable by their thicker, stronger leaves. Old-fashioned varieties such as 'Vuurbaak' and 'Rosa von Zehlendorf' are tetraploids, as are some F_1 series such as 'Cheops' and 'Germanica'.

Aneuploid varieties have an irregular chromosome number: 4n = 90 or 92, 94, 95. Some of the old varieties are aneuploids.

Plant usage

Pot plants: the most usual application for cyclamen in cooler regions.

Bedding plants: in countries around the Mediterranean, cyclamen are very popular as bedding plants in the winter season. This use is becoming more frequent in northern Europe, especially in cities, but most do not survive hard frost.

Cut flowers: in the past, cyclamen were grown especially for cut flowers in Germany and Eastern European countries.

'Sierra' used as a bedding plant. Although not hardy outside in cold winter areas, these cyclamen are very ornamental for conservatories (GSM)

5.4 THE MARKET FOR CYCLAMEN

Until the 1970s, cyclamen were very popular as houseplants, but from then on their popularity has diminished for a number of reasons. First, cyclamen are known to be difficult to maintain as house plants. They regularly need water, not too much and not too little, and they can suffer from several diseases. As central heating became the norm, the situations under windows where cyclamen used to stand and would benefit from cool temperatures, became occupied by radiators. Further, there was a tendency for windowsills to disappear from rooms, so there was no place to put the plants any more. Then, there was a development of foliage plants, and flowering plants went out of fashion: cyclamen acquired an old-fashioned image as grandmothers' plants. In addition to these problems, the quality of the cyclamen plants on the market dropped because the advent of fast-growing F_1 hybrids enabled large growers without experience with cyclamen to grow them factory-wise on a large scale; this also led to overproduction, lower prices and reduced turnover for the growers. Now, growers are promoting cyclamen as a modern plant by presenting them in a fashionable way. In The Netherlands, 'De Cyclamen Commissie' promotes cyclamen as 'Boeiend Bloeiend' (fascinating flowering), whereas in Germany, growers cooperate to do their own promotion. Minis are more sought-after than large-flowered cultivars as they are seen as more trendy for young people.

Cyclamen are now very popular for outdoor decoration, particularly in towns and cities where they are better protected from inclement weather (GSM)

Cyclamen markets worldwide

In Europe, cyclamen must be cheap and they are grown rapidly with as little labour as possible. Big producers grow hectares of them for delivery to garden centres and supermarkets. Cyclamen are a mass product.

In Japan and China, the requirements are completely different. Cyclamen growers treat their product with the greatest care: every plant is manicured and leaves and flowers are hand-positioned in the right direction to produce a perfect plant for the customer. In Japan, cyclamen are seen as a special gift and are therefore very expensive. In China, cyclamen are grown to a very large size especially as gifts for the Chinese New Year, which is at the end of January. As the Chinese economy increases in size, so does the demand for cyclamen in that country.

Preferences for flowers of certain colours often follow the fashion, and there is also a difference in taste between countries. The southern European countries prefer strong colours, whereas Northern European countries go for lighter pastel shades. The Japanese prefer pink (cherry blossom) shades and in China, red is important. In Europe, the colours red and white are traditional favourites at Christmas. There is also a difference in the market for flower size. Southern European customers usually prefer mostly large flowers.

5.5 COMMERCIAL CULTIVATION OF CYCLAMEN

Until the middle of the nineteenth century, *C. persicum* was generally considered to be a half-hardy perennial, taking three to four years from sowing to flowering plant. Around 1826, John Willmot of Isleton, England, came up with the idea of eliminating the summer rest period by providing favourable conditions for continuous growth. This reduced the growing time before flowering to 15–18 months. The seeds were sown in August or September for flowering around Christmas of the next year. This change also led to an acceleration in breeding: it was now possible to grow a generation every two years. By about 1950, the sowing time had shifted to October or November for the following winter's crop. The first F_1 hybrids provided another breakthrough in the 1970s: their cropping was as short as roughly 6 months from sowing for minis, 6½ months for midis and 7 months for large-flowered varieties. The sowing time for Christmas flowering had now moved to May of the same year.

In Holland, up until the 1960s, cyclamen were grown only for flowering in November or December. Piet Tas (Aalsmeer, The Netherlands) created early-flowering forms of the pastel cyclamen by breeding and selection and succeeded in growing a flowering cyclamen in 6 months from potting. From then onwards, it was possible to grow two crops in a year and this turned the cyclamen world upside down. 'Type Tas' was a compact, short variety that could be grown year-round.

Before the 1960s, cyclamen were grown plunged in cold frames to achieve compact, robust plants. Cyclamen were also sometimes grown outside in summer and autumn under shading or fruit trees. In September or October the plants were then taken into the greenhouse to bring them into flower. Plants were typically grown in hand-made clay pots in a mixture of leafmould, loam, peat, sand and organic manure. The seedlings were first potted into a 6-cm pot and later on twice re-potted into larger pots.

A cyclamen nursery of the 1920s owned by C. Spaargaren (Kees Sahin's grandfather) of Aalsmeer, The Netherlands (SAH)

Modern growing systems: plants are automatically weighed and watered. Here seen at the Morel nursery in France (MD)

Since then, the cultivation of cyclamen has changed completely, often undergoing substantial mechanisation and automation. The plants are grown on concrete floors or rolling benches with ebb and flow systems. The potting compost is typically peat-based with added ingredients such as perlite and clay. A liquid fertiliser is applied via the irrigation system. Nurseries have increased in area as prices tended to decrease: it became necessary to grow large quantities in order to make a nursery profitable. Greenhouses have continuous roof and side windows that provide good cooling capability and air circulation.

In the past, cyclamen cultivation was very labour intensive because the plants had to be kept clean by taking away old leaves and flowers and any rotten parts. In The Netherlands, they used to say that cyclamen were "growing in your hands", indicating that good care resulted in good plants. Today, the need for this sort of labour has been reduced substantially by the use of fast-growing cultivars and by controlled climate conditions in which humidity is reduced and air circulation assisted by fans.

Seeds are germinated under controlled conditions in rooms where the temperature is kept constant at 18°C with high humidity and in absolute darkness. The seeds are sown in plugs, and after germination, the small plugs are transplanted into bigger plugs to avoid any growth disturbance. Later, these larger plugs are transferred into pots so that the plants can grow continuously. In the past, temperatures for seedling growth were kept moderate or cool, sometimes even close to 0°C in winter, but today, the plants are encouraged to grow fast for early delivery by providing higher temperatures. Research has shown that plants grow fastest with day and night temperatures of 18°C in the vegetative stage and 22°C day and night in the generative stage. Clay pots are seldom used and most growers have used plastic pots for many years.

Air circulation has always been of great importance to prevent diseases such as *Botrytis* (grey mould) causing heart rot and flower spots.

Seedlings at 15 weeks in the same trial bed (MD)

Seedling cyclamen 6 weeks from germination in the Morel nursery trial beds (MD)

5.6 BREEDING AND SEED PRODUCTION

Cyclamen are propagated from seed, which requires that the flowers are pollinated. Cyclamen breeders must carry out pollination by hand: pollen can be collected and transferred to the stigma of another flower on a breeder's thumbnail. The flower stem bends downwards after pollination, the corolla falls off and the capsule starts growing. The seed matures in about three months at which time the capsule becomes soft and finally opens to reveal brown seeds that are ready for harvesting. The seeds are then dried in a warm, dry room, are cleaned and then stored under stable moderate and dry conditions. The seeds can be stored for about five years without losing too much of their germination potential.

As outbreeding plants, cyclamen are normally pollinated in groups. F_1 hybrids can be self-pollinated only once or twice as too much self-pollination leads to inbreeding problems such as sterility, malformations and lost of vigour. The old way of breeding was to use the pollen from the best plant to pollinate the other plants of the group. Breeding is still a craft that requires a lot of work by hand. Selection must be done properly, pollination is done by hand, every seed capsule is picked separately and the plants must be kept clean.

To obtain uniform parent lines, the plants must be inbred and selected over a period of several years. Once the parents are uniform, it is time to make test crosses. If the test crosses prove to be good, the first small production can be made. It is not easy to maintain vitality in the parent lines, as in later generations the vigour and fertility may drop. Some seed companies maintain their parents in tissue culture and so avoid the inbreeding problem.

Seed production is very labour intensive. To produce an F_1 hybrid, flowers have to be emasculated in the bud stage by hand, then the stigma will become receptive and can be pollinated, again by hand, after three days. This is the reason why F_1 cyclamen seed is rather expensive and is sold by the unit. Some seed growers have moved their seed production abroad to reduce costs, but others prefer to keep full control and stay in their own country in Europe.

5.7 THE FUTURE OF CYCLAMEN

Yellow, orange and blue in cyclamen

In cyclamen, only flavonoids are responsible for the flower colour. The first creamy yellow colour found in cyclamen was present in the waved edge of 'Rococo' flowers. This type appeared as early as 1905. Breeders assumed this to be the beginning of a yellow-flowered cyclamen, but the waving and yellowish characteristics are linked, so both disappeared after crossing. The yellow colour in 'Rococo' is caused by chloroplasts (leaf colour pigments) and not by chromoplasts (flower colour pigments).

The mutant 'Yellow Kage', the first yellowish cyclamen ever, was bred in Japan around 1990. It was found as one single plant in a lot of 'Pure White' by T. Kage from Kyushu University. Improved stock selected from 'Yellow Kage' was called 'Neo Golden Boy' and got a mate in 'Neo Golden Girl', which was yellow with rose eye. Later, after re-selection, 'Neo Golden Girl' was introduced by Sahin as 'Wintersun', which was pale yellow, later fading to cream. Although striking, the plants were not vigorous and flowered late, so production ceased. No real yellow pigments exist anywhere in the genus *Cyclamen*, only the unstable chalcone, so a true yellow cyclamen is a long way off, if even possible.

Orange is also a most desirable colour that featured in some twentieth-century cultivar names, such as 'Orange' (Binnewies, released in 1934), 'Lachscharlachorange' (Fisher, released in 1934) and 'Orange As' (De Troyer). The colours of these cultivars are, however, far from real orange, more shades of salmon rose. The pigment pelargonidin that results in the colour orange is not present in cyclamen.

Blue is another colour that is not easy to achieve in cyclamen. Anthocyanins, the pigments causing blue flower colour, are present, but the pH within the cells of the petals is too low to make it visible. The Japanese breeder Hokko Sangyo bred 'Amethyst Blue', which comes the closest to having blue petals. It is propagated by tissue culture.

New trends

Real yellow, orange or blue cyclamen will remain a dream for both breeder and consumer. Perhaps one should be satisfied with so many colours and shades in cyclamen, but breeders will never be satisfied. There are always desires to be fulfilled, such as improved scent, new colours or flower forms, new leaf patterns, larger flowers and so on, in new combinations and forms. A shorter cropping time is always of interest and disease resistance is an important character for breeders to work on. Hardiness could be an extra value for cyclamen for garden use, but this can only be reached by crossing *C. persicum* with species such as *C. coum*, *C. purpurascens* and *C. hederifolium*. This has proved successful only when embryo-rescue rather than natural means are used.

A more heat-tolerant cyclamen that could survive the warm and dry climate of modern houses would be a great addition. Heat-tolerant cyclamen might also be grown in the summer and in southern countries. Perhaps crosses with *C. somalense* or *C. africanum* might result in more heat tolerance. The chromosome number of *C. africanum* is 34 or 68 and that of *C. somalense* is 96 (see Chapter 2, page 128), whereas that of diploid *C. persicum* is 48; these differences make crossings between these species difficult and no successful crosses have yet been made.

In 1946, Prof. Wellensiek started a breeding project to expand the breeding potential by making inter-specific crosses with *C. persicum*. This was unfortunately not successful, but in 1994, Hiroshi Ishizaka (Japan) made crosses between *C. persicum* 'Sonja' and *C. hederifolium*. This resulted in sterile plants, but after doubling the chromosome number, the amphidiploid plants were fertile.

In about 1995, Sahin made crosses between several cyclamen species and, using embryo rescue, obtained a hybrid between diploid *C. persicum* and *C. hederifolium*. This plant had relatively large lilac flowers and came true from seed. Unfortunately, this hybrid is not very hardy, but it flowers for a very long period and stops flowering only in the darkest winter months.

In 1995, Ishizaka obtained sterile hybrids from crosses between *C. persicum* and *C. purpurascens*. The resulting cultivar, 'Kaori No Mai', has lilac flowers and is propagated by tissue culture. The same interspecific cross has been made by Dr A. Ewald (D-Erfurt) who calls these plants *C.* × *perpuris* 'Odorella'. This half-hardy variety is propagated vegetatively and is promoted as being suitable for use as a garden plant, but although it persists in English gardens it is not a success. 'Odorella' is scented and available in four colours: white, pink, rose and light lilac.

Cyclamen 'Odorella', a hybrid between *C. persicum* and *C. purpurascens* (JB)

Cyclamen persicum as garden plant

Cyclamen persicum cultivars were always exclusively used as house plants. When the climatic conditions outside are moderate (especially in the autumn), however, modern varieties can do well and give colour to the garden during a dark period, when not many plants are available to cheer up the garden. They can be used outside in the spring, but at this time they must compete with numerous other plants such as bulbs, pansies and other bedding plants. Although *C. persicum* is not completely hardy, it can stand a few degrees of frost although the flowers may show some damage on their edges. Mr Tajima from Japan was probably the first person to introduce a (mini) cyclamen for the garden, in 1996, but planting cyclamen outside has now become a widespread practice.

The smaller varieties are ideal container plants to brighten the garden. These are midi-sized cyclamen of the 'Canto' Series (SG)

Around the Mediterranean, for example in Italy and the South of France, and in countries such as Japan, cyclamen are popular as bedding plants in the winter period. Series that are promoted as garden plants include 'Miracle' (Goldsmith), 'Mini Winter' (Schoneveld), 'Latinia', 'Halios', 'Metis' (all three from Morel), 'Canto' (S&G) and 'Darts' (Sakata). 'Mini Winter' has a more open habit that makes it less sensitive to *Botrytis*. In The Netherlands, Varinova recently bred a small-flowered F_1 series especially for outdoor use called 'Melody Outdoor'.

After 150 years of serious breeding, there is still a lot to wish for.

5.8 LIST OF CULTIVARS

Diony de Bont

The chronological list of cyclamen cultivars gives a good overview of the history and development of *Cyclamen*. The list is extensive, but it is not possible to compile a complete list. For some varieties, the year of introduction is unknown; these varieties are placed approximately in the chronological sequence. Series names are in bold type with their members indented below.

Year	Diploploid/ tetraploid	Variety or series	Colour	Flower size	Breeder	Country	Remarks
1629	Diplo	Double-flowered sowebread	White	Small	Unknown		
1739	Diplo	Rose, white, white with eye		Small	Nicolaas van Kampen	The Netherlands	
1836	Diplo	Unknown	Bright rose	Small	Claude Fournier	France	
1837	Diplo	'Albiflorum'	White	Small	Unknown		
1837	Diplo	'Odoratum'	Unknown	Small	Unknown		Scented
1837	Diplo	'Purpurascens'	Purple	Small	Unknown		
1837	Diplo	'Inodorum'	Unknown	Small	Unknown		
1868	Diplo	'Wilhelm I' (= 'Atrorubrum')	Deep dark red	Small	A. Schmidt	Germany	
1869	Diplo	'Oriflamme'	Dark red	Small	M. Wiggins	UK	
1869	Diplo	'Rubrum'	Dark red	Small	M. Wiggins	UK	
1869	Diplo	'White Dedication'	White	Small	M. Wiggins	UK	
1870	Tetra	'Giganteum'	Various	Large	Edmonds	UK	First large-flowered tetraploid
1871	Tetra	'Universum'	White with red eye	Large	M. Kopsel	Germany	
1873	Tetra	'Splendens'	Rose and red shades	Large	Richard Müller	Germany	*C. aleppicum* × *C. persicum*
1878	Tetra	'Brilliant'	Dark red	Large	Unknown	UK	
1878	Tetra	'Rosy Morn'	Light red	Large	Unknown	UK	
1878	Diplo	'Mont Blanc'	White	Large	H. Smith	UK	
1879	Tetra	'Crimson King'	Crimson	Large		UK	
1881	Tetra	'Crimhild'	Light red	Large	Riedel	Germany	'Giganteum' × 'Splendens'
1881	Tetra	'Rosa von Marienthal'	Salmony lilac	Midi	C. Stoldt	Germany	
1882	Tetra	'Leuchtend Dunkelrot'	Dark red		C. Stoldt	Germany	
1883	Tetra	'Rhum von Worms'	Deep dark red	Large	Riedel	Germany	
1883	Tetra	'Monstruosum'	Various	Large	M. Millet	France	Semi-double
1883	Tetra	'Marginata'	Dark red with white line	Large	M. Wiggins	UK	
1884	Tetra	'Dame Blanche'	White	Large	H. Smith	UK	
1885	Tetra	'Albert Victor'	Dark red	Large	M. Clarcke	UK	
1886	Tetra	'Sanguineum'	Deep red	Large	Ernst Benary	Germany	
1886	Tetra	'Leuchtend Hellrot'	Light red	Large	C. Stoldt	Germany	
1887	Tetra	'Princess of Wales'	White	Large	M. Clarcke	UK	First 'Cristata'
1889	Tetra	'Kätchen Stoldt'	White	Midi	C. Stoldt	Germany	Small-leaved, medium-flowered
1889	Tetra	'Fimbriata'	Various	Large	Alwin Richter	Germany	Fringed

Year	Diploploid/ tetraploid	Variety or series	Colour	Flower size	Breeder	Country	Remarks
1894	Tetra	'Salmon Queen'	Salmon	Large	Sutton & Sons	UK	First salmon
1896	Tetra	'Zonale'	Unknown	Large	Unknown	Unknown	Silver edge
1896	Tetra	'Papilio'	Various	Large	Louis De Langhe	Belgium	Special flower form
1897	Tetra	'Fimbriatum'	Various	Large	Karl Heineke	Germany	
1897	Tetra	'Bush Hill Pioneer'	White	Large	Hugh Low	UK	'Cristata'
1900	Tetra	'Fimbriatum Marginatum' (= 'Victoria')	White with red eye and edge	Large	Karl Heineke	Germany	
1900	Tetra	'Giganteum Salmoneum'	Salmon	Large	Thomas Fröbel	Switzerland	
1902	Tetra	'Rubin'	Dark red	Large	C. Stoldt	Germany	
1904	Tetra	'Silberblatt'	Various	Large	Meckel	Germany	
1904	Tetra	'Rococo'	Various	Large	J. Schmidt	Germany	
Unknown	Tetra	'Rococo-Erecta'	Various	Large	Ernst Binnewies	Germany	Improved 'Rococo'
Unknown	Tetra	'Neue Gefranste'	Various	Large	Franz Braukmann	Germany	'Rococo'-like
Unknown	Tetra	'Flore Pleno Gloria'	Various	Large	M. Heinemann	Germany	Double-flowered
1905	Tetra	'Rosa von Zehlendorf'	Palest pink with eye	Large	H. Kiausch	Germany	Important variety
1905	Tetra	'Brillantrosa'	Rose	Large	Tubbenthal	Germany	
1905	Tetra	'Triomphe de Lyon'	White, red eye and edge	Large	M. Rivoire	France	Same as 'Victoria'
1906	Tetra	'Rhum von Wandsbek' ('Glory of Wandsbek')	Deep salmon	Large	C. Stoldt	Germany	
1906	Tetra	'Rococo Schneekönigin'	White	Large	J. Schmidt	Germany	
1907	Tetra	'Perle von Zehlendorf'	Deep salmon	Large	H. Kiausch	Germany	Important variety
1908	Tetra	'Rococo Cattleya'	Purple	Large	J. Schmidt	Germany	
1908	Tetra	'Lachskarmin'	Salmon carmine	Large	C. Stoldt	Germany	
1909	Tetra	'Weisser Schwan'	White	Large	J. Schmidt	Germany	
1909	Tetra	'Deutsche Kaiserin'	Pink	Large	Tubbenthal	Germany	
1909	Tetra	'Striata'	White, purple stripes	Large	C. Stoldt	Germany	Also called 'Harlekijn' ('Harlequin')
1909	Tetra	'Rococo Gloria'	Pink shades	Large	J. Schmidt	Germany	
1910	Tetra	'Rosa von Wandsbek'	Salmon-rose	Large	C. Stoldt	Germany	
1911	Tetra	'Cattleya'	Purple	Large	M. Eysser	France	The most perfect 'Fimbriata'
1912	Tetra	'Schöne Dresdnerin'	Salmon	Large	E. Richter	Germany	
1913	Tetra	'M. M. Buckston'	Various	Large	Hurst & Son	UK	Irregular, multipetalled, 'Papilio'-type
1916	Tetra	'Lachs Purpur'	Deep salmon	Large	C. Stoldt	Germany	
1916	Tetra	'Carmin Salmoneum'	Bright salmon red	Large	A. Maarsse Kzn.	The Netherlands	
1919	Tetra	'Leuchtfeuer'	Red	Large	Johannes Dlabka	Germany	First red
1922	Tetra	'Lachscharlach'	Salmon scarlet	Large	Ernst Binnewies	Germany	
1922	Diplo	'Flamme'	Deep salmon with light salmon edge and middle part	Large	Franz Braukmann	Germany	First diploid

Year	Diploploid/ tetraploid	Variety or series	Colour	Flower size	Breeder	Country	Remarks
Unknown	Tetra	'Salmoneum Oculatum'	Salmon with red eye	Large	Unknown		
1923	Diplo	'Firefly'	Red	Small	Sutton & Sons	UK	First diploid red
1923	Tetra	'Harlekin' (or 'Harlequin')	White, purple stripes	Large	Johannes Dlabka	Germany	Same as 'Striata'
1923	Tetra	'Lehmania'	Various	Large	Walter	Switzerland	Star-shaped flower
Unknown	Tetra	'Anemoniflorum'	White, red eye and edge	Large	Haage & Schmidt	Germany	Plate-formed flower
1925	Tetra	'Rose van Aalsmeer'	Rose	Large	C. Spaargaren	The Netherlands	
Unknown	Tetra	'Lichtrood'	Vivid magenta red	Large	C. Spaargaren	The Netherlands	
1925	Diplo	'Sylphide'	Lilac	Large	Man	The Netherlands	Mutation from white
1926	Tetra	'Neulachsrosa'	Salmon rose	Large	C. Stoldt	Germany	
1926	Tetra	'Safranin mit Silbersaum'	Deep red with white line	Large	Ernst Binnewies	Germany	
1926	Tetra	'Flamingo'	Deep salmon with light margin	Large	Franz Braukmann	Germany	'Flamme' × 'Leuchtfeuer'
1927	Diplo	'Neurosa'	Rose	Large	Ernst Binnewies	Germany	
Unknown	Tetra	'Zartlachsrosa'	Soft salmon-rose	Large	Ernst Binnewies	Germany	
Unknown	Tetra	'Schneider's Rosa'	Rose	Large	Fritz Schneider	Germany	
1928	Diplo	'Goldlachs'	Gold salmon	Large	Franz Braukmann	Germany	
1928	Diplo	'Zartlachs'	Soft salmon	Large	Franz Braukmann	Germany	
1928	Diplo	'Silberlachs'	Silver salmon	Large	Franz Braukmann	Germany	
1928	Tetra	'Race Crot'	Various	Extra large	M. Fermont	France	Flowers 11 cm long
1928	Tetra	'Hermosa'	Palest pink with red eye	Large	Benary	Germany	
1928	Tetra	'Primadonna'	Salmon carmine	Large	Benary	Germany	
1928	Tetra	'Lachscharlachorange'	Salmon scarlet orange	Large	J. Fischer	Germany	
1929	Tetra	'Lachs mit Silberblatt'	Salmon	Large	Struewe	Germany	Silver leaf
1929	Tetra	'Rex'	Various	Large	Struewe	Germany	Silver leaf
1929	Tetra	'Hermione'	Amethyst with carmine eye	Large	Benary	Germany	
1929	Tetra	'Aurore Borealis'	Salmon with carmine eye	Large	Benary	Germany	
1929	Tetra	'Nordlicht'	Light carmine red	Large	Benary	Germany	
1930	Tetra	'Alpenglühen'	Salmon-orange with light edge	Large	Johannes Dlabka	Germany	
1930	Tetra	'Rex mit Silberblatt'	Deep salmon	Large	Struewe	Germany	Silver leaf
1930	Tetra	'Merveille Blanche' (White Wonder)	White	Extra large	M. Rivoire	France	Extreme large flowers
1930	Tetra	'Etincelant'	Vermillion	Large	M. Rivoire	France	
1931	Tetra	'Brilliantlachs'	Salmon	Large	Ernst Binnewies	Germany	
1932	Tetra	'Unique'	White with irregular red spots	Large	Suttons	UK	
1932	Tetra	'Mozart'	Cherry red	Large	Ernst Binnewies	Germany	

Year	Diploploid/ tetraploid	Variety or series	Colour	Flower size	Breeder	Country	Remarks
1932	Tetra	'Alpenglühen'	Salmon orange with light edge	Large	Johannes Dlabka	Germany	
1933	Tetra	'Gerhard Bubeck'	Orange red	Large	Gottlieb Bubeck	Germany	
1934	Tetra	'Orange'	Orange red	Large	Ernst Binnewies	Germany	
1936	Tetra	'Potsdam Strain'	Various	Large	Robert Blossfeld	Germany	
1936	Tetra	**'Triumph' Series:**	Various	Large	Sutton & Sons	UK	
		'Firebrand'	Orange-scarlet				
		'Hydrangea-pink'	Pink				
		'Mauve Queen'	Purple				
		'Salmon-Pink'	Salmon-pink				
		'Shell-pink'	Pale pink				
		'Scarlet'	Scarlet				
		'White'	White				
1937	Diplo	'Parviflorum'	Various	Mini	Reichardt	Germany	First selection on small-flowered
1937?	Diplo	'Violetta'	Violet	Large	Reichardt	Germany	
1937	Tetra	'Österreich'	Salmon	Large	Josef Pregetter	Austria	Fully double
1937	Tetra	'Salzburg'	Scarlet	Large	Josef Pregetter	Austria	Fully double
1937	Tetra	'Gruss aus Weiz'	Light salmon	Large	Josef Pregetter	Austria	Semi-double
1937	Tetra	'Kärnten'	Deep carmine	Large	Josef Pregetter	Austria	Semi-double, fringed
1937	Tetra	'Frau Pregetter'	Dark red	Large	Josef Pregetter	Austria	Semi-double
1937	Tetra	'Steiermark'	Rose with eye	Large	Josef Pregetter	Austria	Fringed
1937	Tetra	'Steirerkind'	Deep salmon	Large	Josef Pregetter	Austria	Semi-double
1937	Tetra	'Hofrat Rottenberger'		Large	Josef Pregetter	Austria	'Victoria'-type
1938	Tetra	'Prinses Juliana'	Rose	Large	D. Baardse	The Netherlands	Fringed 'Rose van Aalsmeer'
1938	Tetra	'Baardse's Wonder'	Deep salmon with light margin	Large	D. Baardse	The Netherlands	Same as 'Flamingo'
1939	Tetra	'Barbarossa'	Salmon, white beard	Large	Schmidt	Germany	'Cristata'-type
Unknown	Tetra	'Kirschlachs'		Large	Ernst Binnewies	Germany	
Unknown	Tetra	'Lichtlachs'	Light salmon	Large	Ernst Binnewies	Germany	
1941	Tetra	'Maarse's Purper'	Purple	Large	J. C. Maarse	The Netherlands	
1942	Tetra	'Erica'	Lilac	Large	Ernst Binnewies	Germany	
1942	Tetra	'Pink Pearl'	Salmon with light edge	Large	Ernst Binnewies	Germany	
1944	Tetra	'Helveticum'	Purplish red	Large	Gebr. Moll	Switzerland	
Unknown	Tetra	'Lavendel'	Lavender	Large	Ernst Binnewies	Germany	
1945?	Diplo	'Scholten Wit'	White	Large	H. Scholten	The Netherlands	Perfect white
1947	Tetra	'Andenken an Gottlieb Bubeck'	Bluish dark salmon	Large	Gottlieb Bubeck	Germany	
1948	Diplo	'Apfel Blüte' ('Apple Blossom')	Pink-flamed	Large	Johannes Fisher	Germany	
1948	Tetra	'Jubileum'	Purple	Large	G. Scholten	The Netherlands	Fringed
Unknown	Tetra	'Alexander Steffen'	Deep salmon	Large	Johannes Fisher	Germany	

Year	Diploploid/ tetraploid	Variety or series	Colour	Flower size	Breeder	Country	Remarks
1950?	Diplo	**'De Troyer' Series:**		Large	Victor de Troyer	Belgium	
		'Blanc Oculatum'	White with red eye				
		'Feu de Joie'	Orange red				
		'Flamboyant'	Rosy red				
		'Saumon Doré'	Gold salmon				
		'Mascotte'	Edged rose				
		'Meteor'	Red				
		'Mystique Melody'	Scarlet with light margin				
		'New Perle'	Salmon red				
		'Orange As'	Salmon orange				
		'Superba'	Rose				
		'Tip Top'	Scarlet				
		'Triomphe de Bruxelles'	Rose with light margin				
		'Zilverwit'	Silvery white				
1950 Unknown	Diplo	**'Decora' Series**	Various	Large	Victor de Troyer	Belgium	Silver leaf
1955	Tetra	**'Bornthaler Duft' Series**	Various	Large	N. Chrestensen	Germany	Scented
1955	Diplo	**'Ondulatum' Series**	Various	Large	Victor de Troyer	Belgium	Waved flower
1957	Tetra	'Friesdorfer Schnitt'	Various	Large	Bosse	Germany	Cut flower
Unknown	Tetra	'Heineke'		Large	Karl Heineke	Germany	Cut flower
Unknown	Tetra	'Weixdorf'		Large	V. E. G.	Germany	Cut flower
Unknown	Tetra	'Rheingaufeuer Schnitt'	Bright red	Large	Erfurter Samenzucht (Weigelt & Co)	Germany	Cut flower
1958?	Tetra	**'Parlament' Series**	Various	Large	Rudolf Roggli	Switzerland	Early
1960s	Diplo	**'Wye College Hybrids':**		Mini	Wye College	UK	Scented, named after butterflies, 'Wye Silver-leaved Scarlet' crossed with the species
		'Admiral'	Red				
		'Brimstone'					
		'Downland'					
		'Fritillary'					
		'Long Petal'					
		'Mauve'	Purple				
		'Orchid'					
		'Peacock'					
		'Ringlet'					
1964	Tetra	'Hallo'	Scarlet	Midi	Benary	Germany	Compact 'Leuchtfeuer'
1964	Tetra	**'Double' Series** ('Kimono')	Various	Large	Sakata	Japan	Double-flowered
1964	Tetra	'Cardinal'	Scarlet	Large	L. Daehnfeldt	Denmark	
1964	Tetra	'Albadonna'	White	Large	L. Daehnfeldt	Denmark	
1965	Tetra	'Rotkapchen'	Scarlet	Large	Hansen	Denmark	
1960–1970	Diplo	**'Pastel' Series:**	Various	Large	Lodder, Tas, Man, Maarsse, Buis	The Netherlands	Re-selected De Troyer types
		'Bach'	Deep rose				
		'Brahms'	Lilac				

Year	Diploploid/ tetraploid	Variety or series	Colour	Flower size	Breeder	Country	Remarks
		'Beethoven'	Purple				
		'Chopin'	Flamed salmon (appleblossom)				
		'Franck'	Flamed cherry red				
		'Händel'	Deep lilac rose				
		'Haydn'	Pink with purple eye				
		'Lehar'	Lilac rose				
		'Liszt'	Palest pink with eye				
		'Mozart'	Red with rose margin				
		'Ravel'	Deep salmon				
		'Rossini'	Flamed purple				
		'Schubert'	Soft lilac rose with purple eye				
		'Strauss'	Salmon red				
		'Strawinsky'	Wine red				
		'Tartini'	White				
		'Verdi'	Lilac-rose with white margin				
		'Vivaldi'	Salmon-pink with lighter margin				
1965?	Diplo	**'Romi' Series**	8 colours	Mini	Rudolf Roggli	Switzerland	
1967	Tetra	'Signal'	Scarlet	Midi	Schmidt	Germany	
1968?	Tetra	'Prinzessin'	White with fringed red edge	Large	Rudolf Roggli	Switzerland	Like 'Victoria'
1970	Diplo	**'Type Tas' Series**	Various	Large	Piet Tas	The Netherlands	Good, compact plant type
1970	Tetra	'Korfu'	Salmon	Large	Robert Mayer	Germany	Scented
1972?	Diplo	**'Silver Lustre' Series**	Various	Large	J. Fisher		Silver leaf, derived from De Troyer 'Decora' Series
1972	Diplo	**'Wellensiek' Series:**		Mini	Prof. Wellensiek	The Netherlands	
		'Anneke'	White				
		'Sonja'	White with eye				
		'Twinkie'	Dark salmon				
		'Willy'	Lilac				
1973	Diplo	**'Kleine Dresdnerin' Series:**		Mini	V. E. G.	Germany	
		'Anneli'	White with eye				
		'Betti'	Salmon rose				
		'Brigitte'	Pure white				
		'Gabi'	Salmon scarlet				
		'Heidi'	Deep lilac				
		'Kati'	Lilac with silver leaf				
		'Lilli'	Cherry red				
		'Rosemarie'	Soft lilac				
		'Steffi'	Carmine rose				
		'Susi'	Deep salmon rose				

Year	Diploploid/ tetraploid	Variety or series	Colour	Flower size	Breeder	Country	Remarks
1973	Diplo	**'Edelstein' Series**	7 colours	Mini	Gebr. Moll	Switzerland	
		'Amethyst'					
		'Rosasaphir'					
		'Rosaberil'					
		'Topas'					
Unknown	Diplo	**'Bambini' Series**	Various	Mini	Mauser	Switzerland	
1973 and on	Diplo	**F_1 'Firmament' Series:**		Large	Royal Sluis	The Netherlands	Named after constellations
		'Aquila'	Salmon scarlet				
		'Aries'	Unknown				
		'Columba'	Palest Pink with red eye				
		'Corona'	Deep rose				
		'Fornax'	Bright satiny red				
		'Dorado'	Salmon				
		'Lyra'	Unknown				
		'Orion'	Light rose with dark eye				
		'Pavo'	Salmon pink				
		'Pegasus'	Cherry red				
		'Phoenix'	Deep pink				
		'Scorpio'	Scarlet				
		'Taurus'	Deep salmon scarlet				
		'Tucana'	Purple violet				
		'Ursa'	Vivid rose				
		'Vela'	Light pink				
		'Virgo'	White				
1973 and on	Diplo	**F_1 'Concerto' Series:**	Various	Large	S&G	The Netherlands	Also called 'Pannevis' Series; 'Rosamunde' was introduced in 1973, the other colours later
		'Aïda'	Pale salmon-rose with dark eye				
		'Amalia'	Pink				
		'Apollo'	White				
		'Boheme'	Deep wine red				
		'Bolero'	Mid-lilac				
		'Bright Scarlet'	Bright scarlet				
		'Carmen'	Scarlet red				
		'Caruso'	Scarlet				
		'Copelia'	Salmon with eye				
		'Deep Scarlet'	Deep scarlet				
		'Esmeralda'	Dark salmon-red				
		'Estrella'	Salmon red				
		'Fidelio'	Pale violet-rose				
		'Figaro'	Rose				
		'Finlandia'	Pure white				
		'Flamenco'	Lilac flame				
		'Giselle'	Bright salmon-rose with dark eye				
		'Gypsy'	Salmon-red				

Year	Diploploid/ tetraploid	Variety or series	Colour	Flower size	Breeder	Country	Remarks
		'Iris'	White with eye				
		'Julia'	Deep rose				
		'Katya'	Magenta				
		'Lavender'	Lavender				
		'Leila'	Lilac				
		'Louisa'	Rose pink				
		'Lucia'	Salmon				
		'Maestro'	Coral flare				
		'Manon'	Deep salmon-rose with red eye				
		'Merry Widow' (= 'Lustige Witwe')	White with purple eye				
		'Nabucco'	Salmon-pink with red eye				
		'Norma'	White with rose-red eye				
		'Oberon'	Light lilac				
		'Ophelia'	Light salmon-rose with red eye				
		'Orpheus'	Scarlet-red				
		'Othello'	Bright scarlet-red				
		'Papageno'	Purple				
		'Pastel Flame Mix'					
		'Phantom'	Pastel lilac				
		'Pink with Eye'	Pink with eye				
		'Romeo'	Deep scarlet-red				
		'Rosamunde'	Rose-pink with dark eye				
		'Rose Pink'	Rose pink				
		'Scarlet'	Scarlet				
		'Swan Lake'	White				
		'Sylvia'	Violet				
		'Tosca'	Deep lilac				
		'Tzigane'	Salmon red				
1973	Tetra	**F_1 ' Germanica' Series:**		Large	V. E. G.	Germany	Named after towns in the former East Germany (DDR)
		'Arnstadt'	Pale salmon-pink				
		'Biel'	Dark lilac				
		'Eisenach'	White with pink eye				
		'Erfurt'	Bright red				
		'Gotha'	Pale rosy-pink with dark eye				
		'Jena'	Satiny violet				
		'Leuchtenburg'	Bright scarlet				
		'Weimar'	White				
Unknown	Tetra	**Large-flowered Series:**		Large	Dresdner	Germany Zierpflanzen	
		'Bob'	Salmon scarlet				
		'Dick'	Purple rose				

Year	Diploploid/ tetraploid	Variety or series	Colour	Flower size	Breeder	Country	Remarks
		'Fred'	Pure white				
		'Knut'	Light lilac				
		'Mark'	Dark lilac				
		'Roy'	White with eye				
		'Tom'	Salmon rose				
1973?	Diplo	**'Dauernheim' Series**	Various	Extra large	Dauernheim	USA	
Unknown	Diplo	**'Rochford' Series**	Various	Mini		UK	Derived from the 'Wye College' Series
1974?	Diplo	**'Piccolo' Series**	Various	Mini	W. Schlatter	Switzerland	
1974?	Diplo	**'Schöne Helena' Series** ('Belle Hélène')	5 colours	Mini	Haller	Switzerland	
1974?	Diplo	'Grandia'	Salmon pink	Large	Unknown		Strong ruffled flowers
1974?	Diplo	'Gerdes Opal'	Purple	Mini	Gerdes		Mini with silver leaves
1974?	Diplo	**'Nomura' Series**	Various	Extra large	Dai Ichi	Japan	
1974?	Diplo	**'Belles Provinces' Series:**	Various	Large	Gaboriau	France	
		'Alsace'	Orchid pink				
		'Anjou'	Salmon				
		'Artois'	Salmon-red				
		'Bearn'	Deep salmon with marbled leaves				
		'Bourgogne'					
		'Bretagne'	Deep salmon with marbled leaves				
		'Lorraine'	Deep salmon with silver leaves				
		'Normandie'	Bright-rose				
		'Picardie'	Brighty salmon with silver leaves				
		'Provence'	Violet				
		'Quebec'	Violet with silver leaves				
		'Tours'	Lilac				
		'Vendée'	Pale orchid-pink				
Unknown	Diplo	**'Mini' Series**	Various	Mini	Wilhelm Beck	Germany	
1976?	Diplo	**'Mini' Series:**		Mini	Fredi Nägeli	Switzerland	
		'Adagio'	Light salmon				
		'Allegreto'	Carmine shades				
		'Fortissimo'	Scarlet				
		'Largo'	Deep salmon				
		'Menuett'	Unknown				
		'Moderato'	Salmon shades				
		'Piano'	White with eye				
1977	Diplo	'Silver Cloud'	Mixture	Large	Hurst & Son	UK	Silver leaf, same as 'Decora' Series

Year	Diploploid/ tetraploid	Variety or series	Colour	Flower size	Breeder	Country	Remarks
Unknown	Diplo	**'Pastel Gefranjerd' Series**	Various	Large	Piet Vollebregt	The Netherlands	Waved flower ('Ondulatum' origin)
1979	Diplo	**'Baby' Series:**	Various	Mini	K. Sahin Zaden	The Netherlands	
		'Baby Charm'					
		'Baby Dream'					
		'Baby Face'					
Unknown	Diplo	**'Anglia' Series**	Various	Mini	Klaas Visser	The Netherlands	Re-selected Wye College Hybrids
Unknown	Diplo	'Coccinella'	10 colours	Mini	Mohr	Switzerland	
1981	Diplo	**'Mini' Series**	Various	Mini	Sakata	Japan	
1982	Tetra	**'New Double Mix' Series**	Various	Midi	Sakata	Japan	
1982	Tetra	F_1 'Vulkanfeuer'	Scarlet	Large	W. Beck	Germany	
1982	Tetra	**F_1 'Briljant' Series**	Various	Large	W. Beck	Germany	
1983	Diplo	F_1 'Sirius'	Mixture of rose shades and white	Mini	Royal Sluis	The Netherlands	First F_1 mini
1983	Tetra	**'Cheerful' Series**	Various	Midi	Sakata	Japan	Small-flowered 'Victoria' type
1983	Tetra	**'Kaori' Series**	Various	Mini	Sakata	Japan	
1983	Diplo	**'New Wave' Series**	Various	Large	K. Sahin Zaden	The Netherlands	Waved flower ('Ondulatum' origin)
1984		**'Misato' Series**	Various	Mini	Sakata	Japan	
1984	Diplo	**F_1 'Royal Mini' Series**	Various	Mini	Royal Sluis	The Netherlands	
1985	Diplo	**F_1 'Sierra' Series**	18 colours	Large	Goldsmith	The Netherlands	
1985	Tetra	**F_1 Series:**		Large	L. Clause	France	
		'Alouette'	Salmon red				
		'Bengali'	Scarlet				
		'Ibis'	Deep salmon				
		'Pastourelle'	White				
		'Polka'	Violet				
		'Ronde'	Pink with red eye				
		'Samba'	Rose				
		'Sylphide Cattleya'	Lilac				
1985	Diplo	**F_1 'Fairy Tale' Series:**		Large	De Ruiter Seeds	The Netherlands	
		'Aladdin'	Vivid red				
		'Bluebeard'	Purple				Later introduction
		'Cinderella'	Pale salmon-pink with dark eye				
		'Robin Hood'	Dark red				
		'Santa'	Pale satiny red				
		'Snow White'	White				Later introduction
		'Tinklebell'	Pale rose with dark eye				
1985	Diplo	**Friesen' Series:**	Various	Large	C. Stoldt	Germany	
		'Friesentraum'	White				
		'Friesenglut'	Scarlet red				
		'Friesenstar'	Salmon				

Year	Diploploid/ tetraploid	Variety or series	Colour	Flower size	Breeder	Country	Remarks
1986	Diplo	**F_1 'Butterfly' Series**	3 colours	Large	Royal Sluis	The Netherlands	First large-flowered fringed F_1
1987	Diplo	**F_1 'Zodiac' Series**	Various	Large	Royal Sluis	The Netherlands	
1987		**'Joyful' Series**	Various	Mini	Sakata	Japan	
1988	Diplo	**'Petite Wonder' Series**	3 colours	Mini	L. Daehnfeldt	Danmark	
1988	Tetra	**F_1 'Cheops' Series**	Various	Large	Morel	France	
1988	Diplo	**F_1 'Latinia' Series**	12 colours	Midi	Morel	France	
1988	Diplo	**F_1 'Halios' Series**	24 colours	Large	Morel	France	
1988	Diplo	**F_1 'Marvel' Series**	Various	Mini	Royal Sluis	The Netherlands	First F_1 mini series
1988	Diplo	**'Insel' Series:**		Mini	Wyss	Switzerland	Named after islands
		'Bali'	Salmon rose				
		'Bermuda'	Salmon				
		'Borneo'	White with rose eye				
		'Corsica'	Lilac with lighter margin				
		'Fiji'	Bright lilac-purple				
		'Hawaii'	Dark lilac				
		'Iceland'	White				
		'Java'	Dark red				
		'Mallorca'	Bright rose with dark eye				
		'Malta'	Cherry-salmon				
		'Mauritsius'	Dark salmon				
		'Samoa'	Red with silver margin				
		'Timor'	Pastel pink				
		'Tonga'	Luminous red				
1988?	Diplo	**F_1 'Serenade' Series**	Various	Mini	Takii	Japan	
1989	Diplo	**F_1 'Laser' Series**	18 colours	Midi	Goldsmith	The Netherlands	First F_1 midi; also sold under the name 'Turbo'
Unknown	Diplo	**'Sinfonia' Series**	8 colours	Mini	Fredi Nägeli	Switzerland	
Unknown	Diplo	**'Puppet' Series**	Various	Mini	Suttons	UK	
Unknown	Diplo	**'Weigelt's Ideal' Series**	11 colours	Mini	Erfurter Samenzucht (Weigelt & Co.)	Germany	
Unknown	Diplo	**'Zwerg Miniature' (= Dwarf Miniature) Series**	7 colours	Mini	Erfurter Samenzucht (Weigelt & Co.)		Germany
Unknown		'Flamenco Violett'	Purple	Mini	Gottlieb Bubeck	Germany	Silver leaf
Unknown	Diplo	**F_1 'Junior' Series**	Various	Midi	Clause	France	
1989	Diplo	**'Städte' Series (Stoldies):**	12 colours	Mini	Wyss	Switzerland	Named after Swiss towns
		'Basel'	Dark salmon pink				
		'Bern'	White with pink eye				

Year	Diploploid/ tetraploid	Variety or series	Colour	Flower size	Breeder	Country	Remarks
		'Brienz'	Pale rose-pink				
		'Davos'	Pure white				
		'Genf'	Cherry-red				
		'Luzern'	Lilac-rose				
		'Lugano'	Pale rose-pink with dark eye				
		'Solothurn'	Pale lilac				
		'St. Gallen'	Dark red				
		'Thun'	Rose-pink with dark eye				
		'Zug'	Pale salmon-pink				
		'Zürich'	Vivid red				
1989	Tetra	**'Luckenwalder' Series:**		Large	N. Chrestensen	Germany	Cut flower
		'Lola'	Deep salmon				
		'Lucie'	Silvery salmon				
		'Luise'	Dark salmon				
		'Lulu'	Salmon				
		'Luna'	Bright salmon				
1989	Diplo	'Pierce'	Delicate pink	Large	Hosogane	Japan	Darker under cool temperatures
1990?	Tetra	**'Napoli' Series**	Various	Large	Wyss	Switzerland	Cutflower
1990?	Diplo	**'Piccolino' Series**	4 colours	Micro	W. Schlatter	Switzerland	
1990?	Diplo	**'Nordseewellen' Series**	Various	Large	Stoldt		Waved flower
1990	Diplo	'Lips'	5 colours	large	Kanazawa	Japan	Flowers with pale edge
1990	Diplo	**'Papillon' Series**	4 colours	Large	Takahashi	Japan	White edge
1991	Diplo	**F_1 'Novella' Series**	Various	Midi	Pan American Seed	The Netherlands	
1991	Diplo	**F_1 'Minimate' Series**	6 colours	Mini	De Ruiter Seeds	The Netherlands	
Unknown		**F_1 'Senior' Series**	Various	Large	Clause	France	
1992?	Diplo	**'Piccolo 2000' Series**	14 colours	Mini	W. Schlatter	Switzerland	
1992?	Diplo	**'Piccolo 2000 Silberblatt' Series**	9 colours	Mini	W. Schlatter	Switzerland	Silver leaf
1992	Diplo	F_1 'Pastellette'	Pastel mixture	Large	De Ruiter Seeds	The Netherlands	
1992	Diplo	**F_1 'Dixie' Series**	Various	Mini	S&G	The Netherlands	
1992	Diplo	**'Sunkiss' Series**	14 colours	Mini	Eyraudplants	France	
1992	Diplo	**F_1 'Dressy' Series**	Various	Mini	Sakata	Japan	
1992	Diplo	**'Original' Series**	Various	Midi	Schoneveld	The Netherlands	Very florifourous
1993	Diplo	**F_1 'Maxi' Series**	Various	Large	K. Sahin Zaden	The Netherlands	
1993	Diplo	**F_1 'Magnum' Series**	Various	Extra large	K. Sahin Zaden	The Netherlands	
1993	Diplo	**F_1 'Midi' Series**	Various	Midi	K. Sahin Zaden	The Netherlands	
1993	Diplo	**F_1 'Mini' Series**	Various	Mini	K. Sahin Zaden	The Netherlands	
1993	Diplo	**F_1 'Intermezzo'**	9 colours	Midi	De Ruiter Seeds	The Netherlands	
1993	Diplo	**F_1 'Miracle' Series**	9 colours	Mini	Goldsmith	The Netherlands	
Unknown	Diplo	**'Damsbakken' Series**	Various	Mini	Damsbakken	Norway	
1993	Diplo	**F_1 'Gradation' Series**	Various	Midi	Sakata	Japan	Bicolours: flamed
1993	Diplo	**'Lingerie' Series**	5 colours	Large	Kanazawa	Japan	No eye

Year	Diploploid/ tetraploid	Variety or series	Colour	Flower size	Breeder	Country	Remarks
1994	Diplo	**F_1 'Darts' Series**	Various	Midi	Sakata	Japan	
1994	Diplo	'Wintersun'	Light yellow	Large	K. Sahin Zaden	The Netherlands	
1995	Diplo	**'Compact' Series**	Various	Mini	Schoneveld	The Netherlands	
1995	Diplo	**S_1 'Mini Star' Series**	11 colours	Mini	Eyraudplants	France	
1995?	Diplo	**'Isselzwerg' Series**	Various	Mini	Wilkes	Germany	Early
1995?	Diplo	**'Early Minex' Series:**	Various	Midi	Wyss	Switzerland	
		'Jack'	Deep salmon				
		'Jim'	Scarlet				
		'Jonny'	Light pink with eye				
1995?	Diplo	**F_1 'Romina' Series**	6 colours	Mini	Rudolf Roggli	Switzerland	
Unknown	Diplo	**'Apollo' Series**	12 colours	Large	C. Stoldt	Germany	
Unknown	Diplo	**'Brilliant' Series**	Various	Midi	Walz	Germany	
Unknown	Diplo	**'Fragrance' Series**	Various	Large	Johannes Fisher	Germany	
Unknown	Tetra	**'Valiant' Series:**	Various	Large	Johannes Fisher	Germany	
		'Charm'					
		'Huntsman'					
		'Pink Waves'	Pink				
		'Sunrise'	Orange-red				
Unknown		**'Isselquelle' Series**	Various	Large	Wilkes	Germany	Early
Unknown		**'Isselmündung' Series**	Various	Large	Wilkes	Germany	Late
Unknown	Diplo	**'Isselwelle' Series**	Various	Large	Wilkes	Germany	Waved flower
1996	Diplo	'Golden Boy'	Light yellow	Large	Kage	Japan	Improved 'Kage Yellow'
1996	Diplo	'Golden Girl'	Light yellow with rose eye	Large	Kage	Japan	
1996	Diplo	**F_1 'Colorado' Series**	Various	Large	De Ruiter Seeds	The Netherlands	
1996	Diplo	**F_1 'Rondo' Series**	13 colours	Large	S&G	The Netherlands	Compact
1996	Diplo	**F_1 'Royal' Series**	Various	Large	Royal Sluis	The Netherlands	
1996	Diplo	**F_1 'Melody' Series**	7 colours	Midi	De Ruiter Seeds	The Netherlands	
1997	Diplo	**F_1 'Quasar' Series**	Various	Large	W. Beck	Germany	
1997	Diplo	**S_1 'Crystal' Series**	15 colours	Large	Eyraudplants	France	
1997	Diplo	**F_1 'Melinda' Series**	8 colours	Midi	Ebbing-Lohaus	Germany	
1997	Diplo	**F_1 'Elmini' Series**	9 colours	Mini	Ebbing-Lohaus	Germany	
1997	Diplo	**'Prima Donna' Series**	5 colours	Large	Takahashi	Japan	Fringed 'Papillon'
1997	Tetra	'Victoria Crown'	Double 'Victoria'	Large	Takami	Japan	From tissue culture
1998	Diplo	**F_1 'Canto' Series**	Various	Midi	Novartis	The Netherlands	
1998	Diplo	**F_1 'Musical' Series**	11 colours	Mini	Varinova	The Netherlands	
1999	Diplo	**F_1 'Metis' Series**	Various	Mini	Morel	France	
1999	Diplo	**F_1 'Celia' Series**	10 colours	Large	Ebbing-Lohaus	Germany	
1999	Diplo	**F_1 'Majesty' Series**	Various	Large	Sakata	Japan	
1999	Diplo	**F_1 'Libretto' Series**	Various	Mini	S&G	The Netherlands	
1999	Diplo	**F_1 'Maxora' Series**	11 colours	Large	Varinova	The Netherlands	
1999		**'Solid' Series**	Various	Midi	Sakata	Japan	
2000	Diplo	**'Micro' Series**	Various	Micro	Schoneveld	The Netherlands	
2000	Diplo	**F_1 'Macro Pastel' Series**	14 colours	Midi	Schoneveld	The Netherlands	

Year	Diploploid/ tetraploid	Variety or series	Colour	Flower size	Breeder	Country	Remarks
2000	Diplo	**'Plumage' Series**	2 colours: red and purple	Large	Tanaka	Japan	Upper part of petal red, lower part white
2001	Diplo	**F_1 'Mini Winter' Series**	11 colours	Mini	Schoneveld	The Netherlands	
2001	Diplo	**F_1 'Original' Series**	12 colours	Midi	Schoneveld	The Netherlands	
2001	Diplo	**F_1 'Compact' Series**	18 colours	Mini	Schoneveld	The Netherlands	
2001	Diplo	**'Whirlygig' Series**	Mixture	Large	K. Sahin Zaden	The Netherlands	Special flower form with enlarged white calyx, giving bi-colour effect
2002	Diplo	**'Shooting Star' Series**	4 colours	Large	Kanazawa	Japan	Striped flowers
2002	Diplo	**F_1 'XL' Series**	15 colours	Large	Schoneveld	The Netherlands	
2002	Diplo	**F_1 'Midori' Series**	9 colours	Mini	Goldsmith	The Netherlands	
2002	Diplo	**F_1 'Silverado' Series**	5 colours	Mini	Goldsmith	The Netherlands	First silver-leaved mini F_1
2002	Diplo	**'Wink' Series**	3 colours: pink, purple and rose	Mini	Sakamoto	Japan	Double flower, faint scent; from tissue culture
2003	Diplo	**F_1 'Astrée' Series**	7 colours	Mini	Eyraudplants	France	
2003	Diplo	'Candy'	Striped	Mini	Hayashi	Japan	
2005	Diplo	**F_1 'Frou Frou' Series**	7 colours	Mini	Eyraudplants	France	First mini fringed F_1
2005		**'Can Can' Series**	Various	Midi	Sakata	Japan	Fringed
2005	Diplo	F_1 'Halios Victoria 50'	White with rose eye and edge	Large	Morel	France	50% fringed 'Victoria'-type
2005	Diplo	F_1 'Metis Victoria 'Deco Mix'	Mixture of rose shades and white	Midi	Morel	France	100% 'Victoria'-type
2005	Diplo	F_1 'Fantasia Purple'	Purple with white margin	Large	Morel	France	
2005	Diplo	**F_1 'Robusta' Series**	8 colours	Extra large	Goldsmith	The Netherlands	
2005	Diplo	**F_1 'Swan' Series**	18 colours	Large	Eyraudplants	France	
2005?	Diplo	'Angel Wing'	Flamed rose without eye	Large	Hara	Japan	Flamed without eye
2006	Diplo	**F_1 'Micro' Series**	13 colours	Micro	Schoneveld	The Netherlands	
2006	Diplo	**F_1 'Verano' Series**	11 colours	Mini	Schoneveld	The Netherlands	For the South of Europe
2006	Diplo	**F_1 'Sterling' Series**	10 colours	Midi	Goldsmith	The Netherlands	First silver-leaved midi F_1
2006	Diplo	F_1 'Fantasia Fuchsia'	Fuchsia with white margin	Midi	Morel	France	
2006	Diplo	**F_1 'Tianis' Series**	Various	Midi	Morel	France	
2006	Diplo	**F_1 'Miniwella' Series:**	Various	Midi	Florema	The Netherlands	Waved flower
2006	Diplo	'Miniwella Twinkle'	Various	Midi	Florema	The Netherlands	More compact
2006	Diplo	'Miniwella Lace'	Various	Midi	Florema	The Netherlands	More wavy flowers
2006?	Diplo	F_1 'Piko Midi'	2 colours	Midi	Kage	Japan	Light edge

Year	Diploploid/ tetraploid	Variety or series	Colour	Flower size	Breeder	Country	Remarks
2006?	Diplo	F_1 'Piko Mini'	2 colours	Mini	Kage	Japan	Light edge
Unknown	Unknown	'Amethyst Blue'	Bluish	Mini	Hokko Sangyo	Japan	From tissue culture
2007	Diplo	'Bellisima'	Mixture	Mini	Florema	The Netherlands	Fringed, bell-shaped flower
2007	Diplo	**F_1 'Picola'**	5 colours	Micro	Varinova	The Netherlands	
2007	Diplo	**F_1 'Melody Outdoor'**	4 colours	Mini	Varinova	The Netherlands	Especially bred for outdoors
2007	Diplo	**F_1 'Maxora Fringed'**	5 colours	Large	Varinova	The Netherlands	Waved flower edge
2008	Diplo	**F_1 'Silver Heart'**	5 colours; 2 mixes	Mini	Goldsmith	The Netherlands	Silver heart, green edge
2008	Diplo	**F_1 'Rainier'**	7 colours	Large	Goldsmith	The Netherlands	Green leaves
2008	Diplo	**F_1 'Winter Ice'**	3 colours: purple, salmon and white	Large	Goldsmith	The Netherlands	Green centre, broad silver edge
2008	Diplo	**F_1 'Da Vinci'**	6 colours	Mini	Schoneveld	The Netherlands	Compact, silver leaf
2008	Diplo	**F_1 'Picasso'**	7 colours	Mini	Schoneveld	The Netherlands	Silver leaf
2008	Diplo	**F_1 'Allure'**	9 colours	Midi	Schoneveld	The Netherlands	Compact; for warm seasons

Internet sites

www.cyclamen.com (Morel)
www.schoneveld.nl
www.goldsmith.com
www.florema.nl
www.kwekerijscholten.nl
www.eyraudplants.fr
www.ebbing-lohaus.de
www.varinova.com

CHAPTER SIX

Cyclamen in botanical illustration

6.1 CYCLAMEN IN THE OLD HERBALS

Martyn Denney & John Grimshaw

Knowledge of the medicinal virtues of plants has always been important in human societies and it is becoming clear that many wild animals also utilise plants for medicinal purposes. Much of this knowledge has been held by shamans, witchdoctors and physicians, and often passed on to the wider public rather sparsely, if at all. Word of mouth ensures its survival. In Europe, it was the Ancient Greeks who began recording this information and their influence on botany persists to the present day. The name *Cyclamen* itself has passed down to us from ancient times.

The manuscript herbals

Ancient documentation is fragmentary, and we have only echoes of the founders of botany, such as Aristotle, in the works of their followers. Of these, Theophrastus (372–287 BC) was pre-eminent, recording and classifying the plants known to him in his work *Enquiry into Plants*. He was followed, in around 65 AD, by the Greek physician Dioskurides (or Dioscorides), who, as a surgeon with the roman army of Emperor Nero, travelled widely in Italy, Gaul, Spain and North Africa. His knowledge was recorded in the herbal Περί ὑλῆς ιατρικῆς, usually known in the west by its Latin name of *De Materia Medica*, a work that was to form the basis of most European medicine until the Renaissance. A contemporary was Pliny the Elder (23–79 AD), who compiled a monumental *Natural History* recording all the facts he could gather on the natural world. Among his records was the statement that cyclamen flower twice a year, presumably conflating *C. hederifolium* with *C. repandum*, both widespread Italian species (Baumann, 1993).

Dioscorides

Dioscorides' herbal was a five-volume treatise that covered about 500 plants as well as therapeutically useful animal and mineral products. It documented the description and direct observations of plants, fruits, seeds and the effects that various drugs had on patients. *De Materia Medica* was the first extensive drug-affinity system. It included about 1,000 natural-product drugs (mostly plant-based), 4,740 medicinal usages for drugs, and 360 medical properties (antiseptic, anti-inflammatory, stimulants and so on). Book one describes the uses for aromatic oils, salves, trees and shrubs, and their products. Book two included uses for animals, parts of animals, animal products, cereals, pot herbs and sharp herbs. Book three detailed the properties of roots, juices, herbs and seeds. Book four continued to describe the uses for roots and herbs. Book five dealt with the medicinal uses for wine and minerals.

The earliest illustrated herbal manuscript is the *Juliana Anicia Codex* or *Codex Vindobonensis*. This wonderful illustrated work was made around 512 AD in Constantinople (present day Istanbul) and was given to Princess Juliana Anicia (daughter of Flavius Anicius Olybrius who had been Emperor of the West from 472 AD), in appreciation of her patronage in the construction of a district church in Constantinople. It contains the earliest existing illustration of cyclamen. Although often known as the Vienna Dioscorides or *De Materia Medica*, this Codex (Önb Cod. Med. Gr. 1), which is held in the Österreichische Nationalbibliothek in Vienna, doesn't contain a complete text of Dioscorides' five-volume work. Although Dioscorides' alphabetical herbal forms the greater part of the manuscript, this Codex is in fact a compilation of natural history writings, as was common throughout medieval times. We do not know which earlier manuscript was copied to form part of the *Juliana Anicia Codex*, or whether it was illustrated.

Pedanius Dioscorides was born in Anazarba near Tarsus in Cilicia (Turkey) and the cyclamen that he was principally familiar with was probably *C. graecum*, though his birthplace is just beyond the eastern extremity of its range. In 1655, the Codex was translated into English by John Goodyer, and this translation was eventually published in 1934 by Robert Gunther. We quote from it here:

194. KUKLAMINOS.
Greek Cyclamen
Cyclaminus [somme call it Cissanthemon, others call it Cissophyllon, somme Chelonion, others Ichthyotheron, somme call it Chuline, somme call it Zoroastris, or Trimphalites, Osthenes calls it Aspho, the Magi Miaspho, the Aegyptians Theske, the Romans Rapum terrae, (aut Umbilicus terrae, aut Orbicularis) and somme call it Arcar] hath leaves like unto Ivie, of a purple colour, divers, with whitish spots beneath and above, a stalke of four fingers high, bare & naked, on which are flowers like Roses, enclining to a purple colour, the roote black, much like the Rape roote, and sommewhat broader. Which being dranck with Hydromel, doth drive the flegme & water downeward, and doth move the Menstrua, being either pota or subdita. They say that if a woman great with childe doe goe over ye roote, that shee doth make abortion, and being tyed about her it doth hasten the birth. It is dranck with wine against deadly poysons, and especially against the Sea-Hare. And being applyed, it is a remedie against ye bitings of serpents and being mixed with wine it causeth drunckennesse. It doth also drive away the yellow Jaundice being dranck the quantitie of three drammes with Passum or diluted melicrate. But hee who drinckes it, must lye downe upon his bed in a warme howse, not open to the winde, and be covered with many cloaths that hee may sweate (the more easily). But the sweat that commes out, is found of a Gaully colour. The juice of it is put into the Nosthrills with Hony for to purge the head. And it is put up in wool into ye seate, for to bring downe the excrements of the belly. And the same being anointed on ye Navill, & ye lower part of the belly & ye hipp, doth mollify the belly, and doth work abortions. And the juyce anointed on with Hony is good for ye Oculi suffusi and hebetiores. It is mixed also with medicines that cause abortion. The juice being anointed on with Acetum doth restore the seate againe, when it is fallen downe. The roote being beaten is juiced, & squeesed out, the juice being sod to the consistence of Hony. The roote also doth cleanse ye skinne, & stayes the breaking out of pustles, & cures wounds with vinegar either of itself, or with hony. Being applyed it doth mollify the spleene (and abates it). It doth also take away the sunne-burning & ye Alopeciae. And the decoction thereof being fomented on is fitting for luxations, & the goute, and for little ulcers in the head, & for ye perniones. And it being sod in old oyle, it brings wounds to a cicatrix, the oyle being smeared on. The roote also being made hollow, is filled with oyle, & set over warme ashes, & sommetymes there is added unto it a little of Tyrrhenian wax, that it may be glewy, an excellent oyntment for such as are troubled with the Perniones. The roote of it also is layd up in store, beeing cut in pieces as that of ye Squill. It is sayd also that it is taken amongst love-procuring medicines being beaten, & soe made into Trochiscks. But it growes in shadie places, & especially under trees.

195. KUKLAMINOS ETERA.
Cyclamen
The other Cyclamen, which somme call Cissanthemon or Cissophyllon, hath leaves like unto Ivie but lesser, thick stalkes, knottie, rolling about ye trees that grow next to them, like the tendrells of vines. The flowers are white and odoriferous, but the fruit as the Acini uvae, like unto ivie, soft, single, sharpe in the tast, and viscous. The roote is unprofitable. It growes in rough places. The fruit of this being dranck as much as ye quantitie of one dragme with twoe Cyathi of white wine for fourtie daies doth melt ye spleene, & is a bringer downe of the belly by urine. It is also dranck for the Orthopnoea. And being dranck it doth purge away the rellicks after childe-bearing.

The illustration of *Cyclamen* in the *Juliana Anicia Codex* has upward-facing flowers and roots that look like those of *C. hederifolium* but has no leaves and is thought to be *C. graecum*. The illuminations in the Codex are thought likely to have been redrawn from early illustrations dating back as far as the second century BC, rather than from nature. A watercolour copy of the *Juliana Anicia Codex* illustrations on paper — including an almost exact copy of the *Cyclamen* — produced between 1458 and 1477 is held in the Botany Library at the Natural History Museum, London (Banksian mss. No. 63, fol. 175).

The first-century original manuscript of Dioscorides' herbal no longer exists, but it almost certainly wasn't illustrated. It is possible that when he described the second *Cyclamen*, which he called *KUKLAMINOS ETERA*, he was describing *C. hederifolium*, but that after his death, artists working on illustrated editions of his herbal, who had never seen the plant in nature, attempted to paint it literally from the description. In the *Juliana Anicia Codex*, this plant appears with the same tuber and roots as the first *Cyclamen* but with above-ground parts that are very close to ivy

Kyklaminos **from the** ***Juliana Anicia Codex*** **(*****Codex Vindobonensis*****) AD 512. The earliest known illustration of Cyclamen from an illuminated early sixth century version of Dioscorides herbal. It is believed to be** ***C. graecum*****. Reproduced with the permission of the Österreichische Nationalbibliothek, Vienna. ÖNB/Wien, Cod. med. gr. 1 fol. 164v**

Kyklaminos Etera from the *Juliana Anicia Codex* (*Codex Vindobonensis*) AD 512. Probably an artists impression of the plant based on Disocorides' description, possibly of *C. hederifolium*. Reproduced with the permission of the Österreichische Nationalbibliothek, Vienna. ÖNB/Wien, Cod. med. gr. 1 fol. 165v

Cyclaminos altera, Hederaceis foliis from *Plantarum seu Stirpium Icones* published by Christophe Plantin of Antwerp in 1581 to accompany Matthias de L'Obel's *Kruydtboeck*. This original woodblock was Christophe Plantin's attempt at illustrating the mythical plant that accompanied Dioscorides', description of *C. hederifolium* in medieval times. By permission of the Trustees of the Natural History Museum (London)

in appearance. In the printed editions of Dioscorides' herbal that appeared from the early sixteenth century to the mid-seventeenth century, authors such as Brunfels, Ruel, Matthioli, de L'Obel, d'Alechamps, Bauhin, Johnson (in the 1633 edition of Gerard's herbal), Panckow and Chabraeus wrote of and generally illustrated a 'Cyclamen altera' literally as a sprig of ivy, often alongside a true illustration of *C. hederifolium*. This very clearly demonstrates the way in which, for more than 1,150 years, plagiarised herbals were based upon unverified information. In this case, the erroneous illustrations derived from the simple mistake of a medieval artist, who drew from his own imagination having read Dioscorides' description of *C. hederifolium*.

Kyklameinos (left) and *Kyklameinos Etera* (right) from *Codex Neapolitanus*, sixth century AD. An early illustration of *C. graecum* with upward-facing flowers, and the erroneous literal depiction of Dioscorides' description of *C. hederifolium* with ivy leaves and flowers. Reproduced with the permission of the Biblioteca Nazionale di Napoli, Ms. Vind. Gr. 1 fol. 47r

The *Codex Neapolitanus* (Cod. Gr. 1) held in the Biblioteca Nazionale, Naples appears to have been based on the same earlier manuscripts as the *Juliana Anicia Codex* but was clearly not copied from the latter, with the illustrations being copied separately from the same earlier images. Neither the date nor the location of the *Codex Neapolitanus*' production are known with certainty, but the script suggests that it dates from the late sixth or early seventh century. Various authors have speculated as to where it was produced, with Constantinople, Rome, Ravenna and the Calabria-Sicily area being mentioned. It is possible that it was copied by a southern Italian who went to Constantinople to make the copy. From the arrangement of the illustrations and text, it is almost certain that the illustrations were drawn first — generally two at the top of each page — and the text added underneath afterwards. There are two illustrations of cyclamen in the *Codex Neapolitanus*, both superficially similar to those in the *Juliana Anicia Codex*.

The history of *Codex Neapolitanus* is somewhat sketchy. At one time, the codex belonged to the neapolitan Antonio Seripando, brother of Cardinal Girolamo, general of the Augustinians and one of the leaders of the Council of Trento. From about 1500, it was kept in the Augustinium convent of St. Giovanni a Carbonara. In 1718, during the reign of Charles VI of Hapsburg, the codex was taken to Vienna, and during that century, a series of engravings were made from its paintings by order of the Viennese Empress Maria Teresa. It is generally thought that only four copies were printed, though Collins (2000) limits this to just two. The codex was returned to Naples after the First World War and is now kept in the Biblioteca Nazionale di Napoli. In 1784, the botanist John Sibthorp, then Sherardian Professor of Botany at Oxford University, arrived in Vienna to study the *Juliana Anicia Codex* on his way to Greece to carry out field research for *Flora Graeca* (see p. 487). While in Vienna, he met Baron Nicholaus von Jacquin, the Director of the Vienna Botanic Garden and Professor of Botany and Chemistry at Vienna University. Jacquin introduced Sibthorp to the botanical artist Ferdinand Bauer and gave him one of the copies of the *Codex Neapolitanus* engravings. This now resides in the Sherardian Library of Plant Technology at the University of Oxford. Another copy of the engravings is held in the library of the Linnaean Society of London.

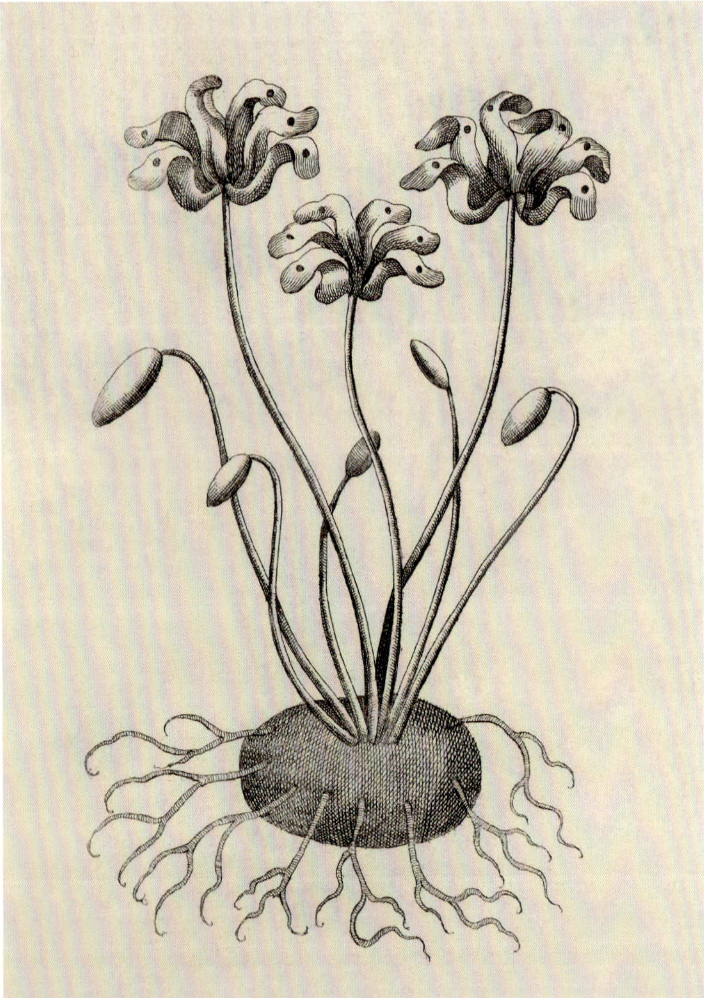

Kuklamenos from *Plantarum Disocoridis Icones*. An eighteenth century engraving of the *Codex Neapolitanus* painting of *C, graecum*. Only four copies were printed. Reproduced with the permission of the Sherardian Library of Plant Taxonomy, The Bodleian Library, University of Oxford. Sherard 443 fol. 190

In the Pierpont Morgan Library in New York, there is a mid-tenth century manuscript of Dioscorides' *De Materia Medica* in its alphabetical five-book form in Greek. It is likely to have been produced in Constantinople between 930 AD and 960 AD. Many of the illustrations were copied from the *Juliana Anicia Codex* but others were copied from other sources. In the case of cyclamen, *C. graecum* (Morgan MS. M.652, fol. 72v) and *C. Etera* (Morgan MS. M.652, fol 73v) are good copies from the *Juliana Anicia Codex*, though the flowers of the former are a little more open than those in the original.

The only illustrated copy of Dioscorides' five-volume Περί ὑλής ιατρικής translated into the old latin *De Materia Medica* is held in the Bavarian State Library in Munich (Clm 337). It could have been produced in the Benevento area of southern Lombardy, north-east of Naples in Italy, possibly in the third quarter of the tenth century. The illustrations were conceived as part of the overall work, but the space allocated to the illustrator by the scribe confined them to the top and bottoms of the two-column text. The illustrations are executed in a very limited palette and not always in a very naturalistic way, so that for example, plant leaves are sometimes blue. A cyclamen is shown with blue cordate leaves, basal roots and a curling seedpod (with attendant serpent!). The presence of the serpent indicates that the plant was believed to be a cure for snakebites. Clm337 also illustrates Cyclamen Etera with tiny ivy-like leaves and terminal umbels of non-descript flowers).

An early fifteenth century manuscript in the British Library (Ms. Add. 41623) has text that is principally from Dioscorides but that is remarkable for two things. The illustrations are clearly drawn from nature rather than being copied from earlier manuscripts, and the painting of a cyclamen reflects this and is possibly the earliest realistic image of the plant. The illustration shows a tuber with a floral trunk, six leaves and two flowers. The leaves are clearly shaped like those of *C. purpurascens*, which grows in the Belluno area, and have a pewter-coloured upper leaf surface and a red reverse. The flowers are truly reflexed with exerted pollen bearing anthers. Unusually, the source of the specimen is mentioned in some places in this manuscript. The town of Belluno in Northern Italy is mentioned in this respect four times, Mount Serva (nearby) is mentioned six times. Other locations include Monte Baldo, Callalta near Treviso and Monte Talvenal. It is thought that the manuscript was produced in the town of Belluno, and indeed an explanatory note says "...huius Ciuitatis Belluni" ("... of this city Belluno"). It is therefore often known as *Codex Bellunensis*.

Kiklaminu **(bottom right) from the tenth century Dioscorides *De Simplici Medicina Libri V*. The only existing illustrated copy of a sixth century latin translation of Dioscorides' herbal. Reproduced with the permission of the Bayerische Staatsbibliothek, München, Germany. BSB Clm 337 fol. 125**

Kiklamino – Dioscorides' *Kuklaminos Etera* — from the tenth century Dioscoridis *De Simplici Medicina Libri V*. The only existing illustrated copy of a sixth century latin translation of Dioscoridis' herbal. Reproduced with permission of the Bayerische Staatsbibliothek, München, Germany. BSB Clm 337 fol. 126

'Ciclamen' (bottom left) in the early fifteenth century *Codex Bellunensis*, probably the most realistic image of a cyclamen to be found in the manuscript herbals. © The British Library Board, BL Add. MS 41623 fol. 40v.

'Cyclaminos' from the sixth century, earliest known illustrated copy of the *Herbarius of Apuleius Platonicus*. Reproduced with the permission of the Universiteit Leiden Bibliotheken, The Netherlands. MS. VLQ 9 fol. 42r

Apuleius Platonicus and Liber Medicinae ex Herbis Femininis

The *Herbarius* of Apulieus Platonicus is thought to have been compiled in Latin in the fourth century, drawing upon both Greek and Latin sources and in particular Pliny and *Medicina Plinii*. The *Herbarius* that can be seen today consists of around 130 chapters, each devoted to a single plant with an illustration. The earliest surviving manuscript dates from the second half of the sixth century and was probably produced in southern Italy. It is held in the Isaac Vossius collection at the University of Leiden in The Netherlands (MS. VLQ 9). Each illustration is followed by a chapter of text that begins with synonyms in a number of languages, sometimes followed by details of the habitat and a numbered list of complaints that the plant can cure. Unlike the Dioscorides illustrations, the cyclamen image in the *Herbarius* shows a single stem arising from the tuber with leaves and flower stems branching from it. The flowers hang like lanterns similar to a *Galanthus* and, in the Dioscorides manuscripts, are not reflexed.

The *Herbarius* was often one of the principal works in compilation manuscripts that included the *Liber Medicinae ex Herbis Femininis*. The text of the latter was derived, at least in part, from Dioscorides Περί ὑλής ιατρικής and is likely to have been first produced in southern Europe before the sixth century. The illustrations often tend to be simpler and merely two-dimensional and were probably derived from a different unknown western European source. We have seen how it was normal for illustrations to be copied from earlier herbals rather than for new drawings to be made from nature. Through the next series of manuscripts, it is possible to trace characteristics that developed from the early ninth century over a period of some 400 years, for much of that period drifting further and further away from reality.

The early ninth century Plutei 73.41 is the earliest manuscript known to contain what was to become a very important compilation of texts that were repeated over the coming centuries. The compilation, held in the Biblioteca Medicea Laurenziana in Florence, Italy, contains 16 texts spanning 129 folios, but at their heart are 98 pages of the *Herbarius* of Apuleius Platonicus, 72 pages of the *ex herbis femininis* and 37 pages of the *Liber medicinae ex animalibus* of Sextus Placitus. Plutei 73.41 is thought to have been produced at Montecassino near Salerno in southern Italy. The illustrations are essentially unsophisticated pen-and-ink drawings with detail picked out in a single colour. In the case of cyclamen, the drawing is important as it contains elements that are to be reproduced in later manuscripts for the next 500 years. The branching flower and leaf stems, ball-like seed pods and three-petalled lantern-like flowers first seen in the sixth-century *Herbarius* VLQ 9 have been changed into a much more stylised two-dimensional design, with straightened radiating stems and, for the first time, curled basal stems (seed pods?) adjacent to the tuber.

'Cyclaminos' from the early ninth century manuscript compilation of the *Herbarius of Apuleius Platonicus* and the *ex herbis femininis*. Reproduced with the permission of the Biblioteca Medicea Laurenziana, Florence, Italy. MS. Plut.73.41 fol. 25v

The Bibliothèque Nationale de France in Paris has a manuscript compilation (Paris Lat. 6862) that dates from the middle of the ninth century. The last two thirds of the manuscript are a combination of the text and illustrations of the *Herbarius* and those of the *ex herbis femininis*, resulting in some plants having illustrations from both traditions. The manuscript originates in northern France, possibly around Tours or Metz, which were then under the influence of the court schools of Drogo of Metz and the emperor Charles the Bald. The illustration of a cyclamen is interesting for its hastate-patterned leaves with maroon

'Cyclaminos' from the middle-ninth-century compilation of the *Herbarius of Apuleius Platonicus* and the *Ex Herbis Femininis* which was produced in northern France. Reproduced with the permission of the Bibliothèque Nationale de France, Paris. MS. Lat. 6862 fol. 35r.

reverse, lantern-like flowers and wing-like roots. This is a far more accomplished illustration than that of Plut.73.41, though as it includes elements such as the ball-like seed pods and curled basal stems, it is probable that there is a common ancestry. The overall composition of the image is, however, unlike any other I have seen and it isn't possible to place it as a descendant of either the sixth-century VLQ 9 or Plut.73.41

The earliest known Latin version of the Apuleius Platonicus herbal produced in England is Oxford, Bod. MS. Ashmole 1431, held in the Bodelian Library, Oxford. This dates from the late eleventh century and was probably produced at St Augustine's Abbey in Canterbury. It has a relatively small page size, just 235 mm × 150 mm (9.25 inches × 5.9 inches), and was not an expensive production. The illustration of a cyclamen is clearly descended from the southern Italian Apuleius Platonicus Plut.73.41 manuscripts, and there are some elements that suggest a connection with Leiden VLQ.9. It is interesting that the following page of the manuscript (fol. 10v) contains an illustration not dissimilar to the *C. etera* found in the tenth century *De Materia Medica* Clm 337, but in Ashmole 1431 it is clearly named *Aristolochia rotunda*. Blunt and Raphael (1979) suggest that Ashmole 1431 is a descendant of the Cotton Vitellius C. III manuscript held in the British Library, but that manuscript has no illustration of a cyclamen.

There are two manuscripts of the Apuleius Platonicus herbal that were produced in Germany between the late-eleventh and mid-twelfth centuries. The British Library Harley 4986 manuscript, which dates from between the late-eleventh and early twelfth centuries, has been described as an illustrated pharmacopeial compilation. The text is in Latin and 44 of its 81 parchment folios were a "Pseudo-Apuleius Platonicus, De medicaminibus herbarum", with the rest being made up of a further 11 texts of a few pages each. The cyclamen illustration is crude and bears very little resemblance to nature with cordate unmarked lime-green leaves and pink Fleur-de-Lys-like flowers. The closely related Eton College ms. 204 is important for the bifolia inserted at the beginning. Folio 1v shows three rhizotomists collecting herbs, while folio 2r shows a pharmacist directing the weighing of drugs. The rest of the manuscript is not illuminated to the same fine standard but there are some fine paintings of plants. The *Herbe Ciclame* or *Herba Orbicularis* is similar to that in Harley 4986 and is most likely copied from it. It is, however, executed with more confidence and has been embellished by the artist. It has an elliptical tuber that looks like an 'all seeing eye' with a maroon outside and pink centre, the roots hanging down like a maroon fringe. The three petioles rise sinuously from the tuber and are topped by hirsute ovate leaves. Either side of the leaves are the pink Fleur-de-Lys-like flowers on pink pedicels.

'Ciclaminos' (top left) from the earlist known Latin version of the Apuleius Platonicus herbal produced in England. It dates from the late eleventh century and was probably produced at St Augustine's Abbey in Canterbury. Reproduced with the permission of The Bodleian Library, University of Oxford. Oxford, Bod. MS. Ashmole 1431 fol. 10r

According to Collins (2000), during the twelfth century there was a revival of interest in the *Herbarius* and its compilations. Two twelfth century manuscripts in the British Library in London are undoubtedly descendants of the Plut.73.41 ninth-century compilation. The Harley 5294 manuscript was probably produced in southern Italy in the second half of the twelfth century and has a pen and ink line drawing that is almost identical to, but more carefully executed than, those of its ancestor. The image of a cyclamen is completely uncoloured and is accompanied by a drawing of a man killing a serpent while meeting a priest holding a chalice at the entrance to a church.

Three rhizotomists collecting herbs. From a Latin Apulieus Platonicus herbal produced in Germany, possibly in the second half of the twelfth century. Reproduced by permission of the Provost and Fellows of Eton College. Eton MS. 204 fol. 1r

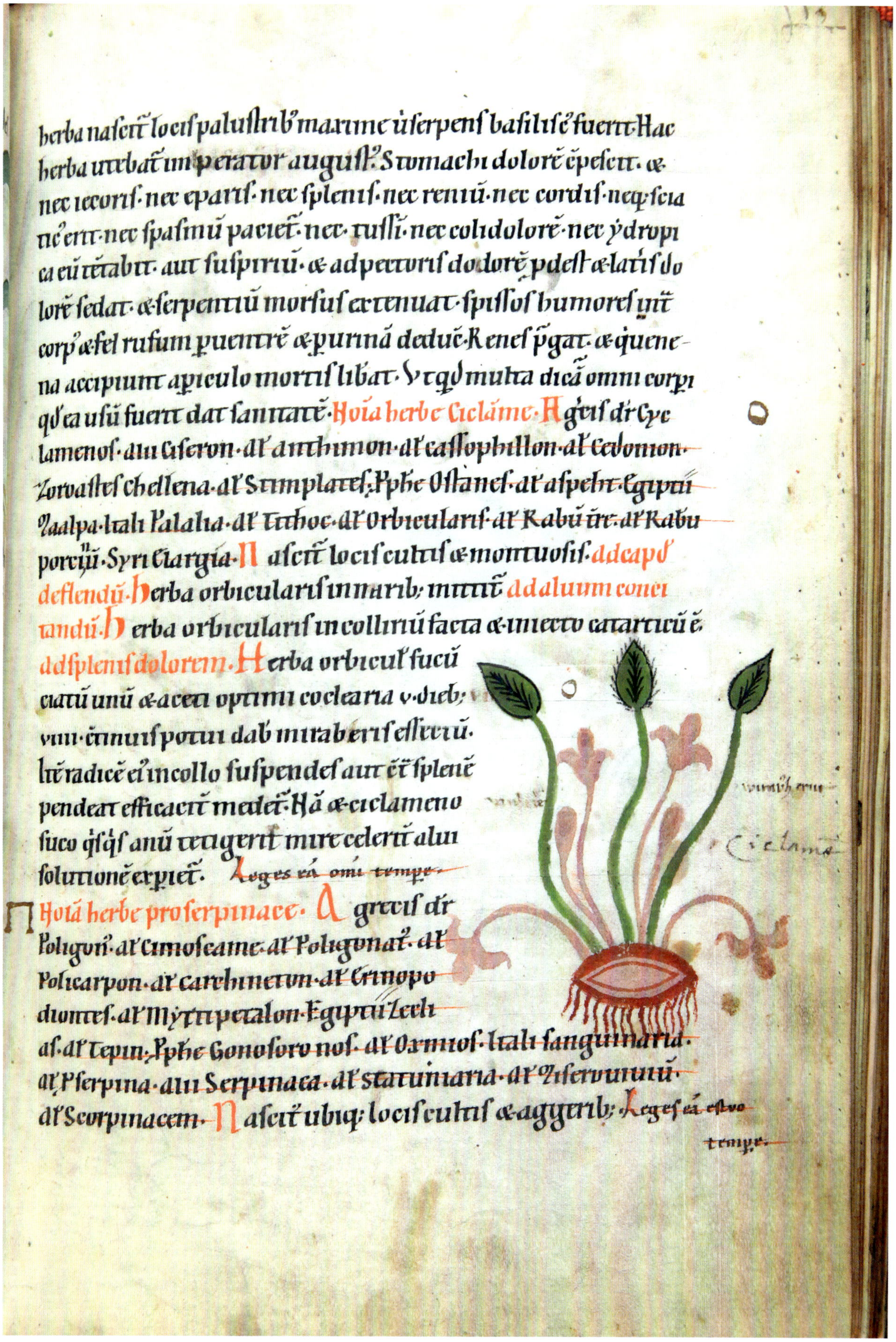

Ciclame from a Latin Apuleius Platonicus herbal produced in Germany, possibly in the second half of the twelfth century. Reproduced by permission of the Provost and Fellows of Eton College. Eton MS. 204 fol. 13r

Pen and Ink drawing of 'Cyclaminos' from the twelfth century Harley 5294 manuscript produced in southern Italy, showing a man killing a serpent while meeting a priest holding a chalice at the entrance to a church. © The British Library Board. MS. Harley 5294 fol. 15v

The second manuscript also held in the British Library, Harley 1585, is a more expensive production with entirely painted illustrations and text picked out in red. According to Collins (2000), it is thought to have been produced in the Stavelot area of eastern Belgium between 1145 and 1158. Collins also speculated on how a northern European manuscript could rely so heavily upon a predecessor from southern Italy: the Abbot of the imperial Benedictine Monastery of Stavelot at the time was Wibald, advisor to Lothair II, Conrad III and Frederick Barbarossa. Wibald was an erudite man of letters known to have been interested in the liberal arts, science and medicine; he was a teacher, patron of the arts and collector of manuscripts. Wibald travelled widely in his role as advisor to his imperial masters, and in 1137, he accompanied Lothair on a military expedition to southern Italy, where he was appointed Abbott of Montecassino. However, King Roger of Sicily threatened to destroy the monastery unless Wibald resigned the Abbacy so, after a mere 40 days, he returned to Stavelot. While in Montecassino, Wibald could have arranged to have Plut.73.41 copied or might have acquired an intermediate manuscript, like Harley 5294. After his return to Stavelot, he might then have commissioned a copy of the *Herbarius* and employed artists to illustrate it. The cyclamen image is far removed from nature, the artist having removed the dentations to the leaf margins and reduced the roots to triangular projections, while he has embellished the coiled basal stems with immature leaves. The two red leaves might indicate a connection with the Paris Latin manuscript 6862 or could simply be a product of the artist's imagination.

An almost identical image of a cyclamen appears in the Ashmole 1462 manuscript, held in the Bodelian Library, Oxford. Ashmole 1462 was produced in Northern England between 1190 and 1200 and is undoubtedly a descendant of Harley 1585. The colours used to paint the image are different, with the red replaced with a rust colour and the turquoise-green replaced with a slate-blue. The same image appears yet again in the British Library Sloane 1975 manuscript, but this illustration has been embellished to such an extent that the effect is completely different. Sloane 1975 was produced on vellum and is also thought to have been produced in Northern England between 1190 and 1200, though Collins (2000) suggests that both manuscripts could have originated in France. Sloane 1975 is a very sumptuously decorated herbal and the paintings of plants take stylisation to a new level, with parts of the plants being picked out with gold leaf and the whole set against a frame with a dark blue background.

In the first half of the thirteenth century, a small group of sumptuously produced manuscripts appeared. Two of these southern Italian Codices still exist, one in Vienna (ÖNB ms. 93) and the other at the Biblioteca Medicea Laurenziana in Florence (Plut.73.16). The latter contains the familiar *Herbarius* compilation with a large number of fully painted illustrations showing portraits of the authors of the component treatises (Plato, Hippocrates and Dioscorides), doctors in colourful flowing Greek robes treating patients, an extensive

A full-colour painting of 'Cyclaminos' from the twelfth century Harley 1585 manuscript produced in eastern Belgium. © The British Library Board. MS. Harley 1585 fol. 23v

Herbarius and an *Animalibus* showing amongst others, elephants, lions and wild boar. The nature of this manuscript had almost turned the original scientific treatise into a picture book. Collins (2000) comments that by this time, the pharmaceutical use of the *Herbarius* had been superseded by a number of treatises, many having been produced in Salerno during the twelfth century and others having been translated from Arabic texts, but none of these were illustrated. The cyclamen in Plut.73.16 is a restrained but striking image that owes allegiance to the sixth-century Leiden VLQ 9 *Herbarius*, the ninth-century Paris MS. Lat. 6862 and the twelfth-century Harley 1585, Ashmole 1462 and Sloane 1975, having the more natural stems of the first, the hastate-patterned leaves with a maroon reverse of the second, and the coiled basal stems of the last three.

An illumination of 'Cyclaminos' (right) picked out with gold leaf, from the twelfth century Sloane 1975 manuscript probably produced in northern England. © The British Library Board. MS. Sloane 1975 fol. 18v

The painting of *Cyclaminos* from the sumptuous Plut.73.16 manuscript produced in southern Italy possibly 1220–1250. Reproduced with the permission of the Biblioteca Medicea Laurenziana, Florence, Italy. MS. Plut. 73.16 fol. 49r

Tractatus de Herbis

In the British Library, there is a manuscript herbal *Tractatis de Herbis* (Egerton MS. 747), which is the earliest known example of this treatise. The author of the manuscript is unknown, but based his work on that of Dioscorides, the Apuleius Platonicus, Galen and *Macer Floridus*. To a significant extent, *Tractatis de Herbisi* is an enlarged, revised and illustrated version of the medical treatise of simples entitled *Liber de Simplici Medicini*, but known as the *Circa instans*, which was written in the first half of the twelfth century by Mattheus Platearius, part of a medical family from Salerno. Towards the end of the eleventh century the first medical school in Europe had been formed in Salerno near Naples. Blunt and Raphael (1979) describe Platearius's manuscript as "… the first dictionary of drug synonyms". According to Collins (2003), the *Tractatus de Herbis* contains excerpts from Isaac's treatise on diet, which was compulsory reading for students in Salerno from 1280, but its author does not seem to be aware of the ca. 1296 *Clavis Sanationis* or the 1317 *Opus Pandectarum Medicinae*. The proposition is, therefore, that the manuscript was written and illustrated sometime after 1280, possibly before 1296 but definitely before 1317. Egerton 747 illustrates a cyclamen whose tuber has roots arising from its sides and upper surface, and whose cordate leaves are held above pink reflexed flowers. It seems that at this time, herbal illustrators were correcting the mistakes that were introduced into herbals by copying from earlier works, and that wherever possible, they were consulting nature and making full use of the local flora. Certainly, *C. hederifolium* grows in the area around Salerno and Naples.

Slightly later, but clearly descended from Egerton 474, is the *Liber de Herbis et Plantis* Paris Lat. 6823, now held at the Bibliothèque Nationale de France. The *Tractatis de Herbis* text and illustrations were written and drawn on parchment by Manfredus de Monte Imperiale, possibly working in Naples around 1340. The illustrations are very similar to those of Egerton 747, but Manfredus used his own knowledge of plants to make the paintings more accurate. This is shown with the cyclamen drawing, which has leaves with a hastate pattern, the flowers are reflexed but they are shown in a 'stargazer' style and are upward-facing amongst the leaves. Similar too is the *Tractatis de Herbis* (ms. lat. 993), held at the Biblioteca Estense Universitaria in Modena. This too was derived from Egerton 747, and was produced in Bourg-en-Bresse in 1458. The illustration of a cyclamen in this manuscript is quite crude, with three plain green orbicular leaves on short petioles, each with a flower attached to one of its basal lobes.

Three manuscripts of the *Tractatis de Herbis* produced in Northern Italy in the hundred or so years between 1370 and the end of the fifteenth century introduced a new style of cyclamen image. The first of these was produced, Collins (2000) hypothesises in Padua, Northern Italy, between 1370 and 1380. The illustrations of the plants and especially the animals in the *Liber de animalibus* are exceptional. This manuscript is Ms. Masson 116 and is held at the Bibliothèque de l'Ecole des Beaux-Arts in Paris. The image of a cyclamen appears in the bottom right quadrant of an otherwise empty page, suggesting that text was to have been written but that the manuscript is incomplete. The second manuscript, British Library Sloane 4016, was produced 60–70 years later in 1440 and is believed to have been copied from Masson 116. In this manuscript, the text is secondary to the wonderful illuminations. The third manuscript, Bodleian Canon Ms. 408, dates from the late-fifteenth century and contains about 165 full-page illustrations of 348 mm × 245 mm), most with little or no text, though the cyclamen illustration has more than most. The cyclamen image in these manuscripts is an innovation. The tuber is almost orbicular with basal roots that appear to have almost been an afterthought. The petioles and pedicels are straight and radiate from the centre of the tuber. The three leaves have scalloped edges like some *C. hederifolium*, but the leaf laminas are covered in darker green spots. The four flowers are reflexed pink caps that hang stargazer-style around the pedicels, like an ink cap fungus. Although the flower appears to be entire, it does have five lobes that are joined for most of their length. The Masson 116 image has detail that wasn't transferred in the copy to Sloane 4016. Near the apex of the flower, there is a red circle that suggests that the artist had seen a cyclamen flower with a red blotch at the base of the petals. In the centre of this circle, the stamens and stigma are completely exerted. The flowers have been painted as if the petals reflex directly from the nectary so that there is no interior to the flower. There are red lines radiating from the apex of the flower that suggest the separation of the corolla lobes. In Sloane 4016, the apex of the flowers is rounded and there is no detail at the apex or lines to separate the corolla lobes. The scalloped nature of the leaf edges is much more pronounced. A paragraph of text has been added across the image between the tuber and leaves and the evidence is that the image was drawn first. In Bodleian Canon 408, the flowers are conical rather than convex and the apex is very clearly pointed. The leaves retain the more pronounced scalloped edges of Sloane 4016. The text that appears next to the image is an exact copy of that in Sloane 4016. It would be a fine painting, far larger than the others in the group, but unfortunately the central part of the plant has been obliterated by a repair.

The final group of *Tractatus de Herbis* manuscripts originate from Northern France in the fifteenth century. Paris MS. Fr. 9136 is a *Livre des Simples Médecines*, the French version of the Tractatus, and was a lavishly produced manuscript made for Louis de Bruges. The cyclamen has a slightly irregularly shaped tuber with roots from the sides and the base. The three leaves are cordate with dentate margins and are a dark blue green, with darker veins but no pattern. The four pink flowers arise directly from the tuber and are upward-facing slender trumpets that are almost reminiscent of some of the smaller *Colchicum* flowers. Very similar, is Paris MS. Fr. 623. This manuscript of the *Livre des Simples Médecines* was produced in the late fifteenth century and was probably copied from Paris MS. Fr. 9136. The image of a cyclamen is even more stylised. The five cordate leaves have similar dentate margins and a white contrasting mid-rib. The leaves are arranged in a line across the top of the tuber, interleaved with four flowers that look like pink golf tees.

The transition to printed herbals

Although printing is known to have taken place in China from the ninth century and moveable type was also known there, it wasn't successfully used for the printing of books until the fifteenth century in Europe. The limited character set of European languages was combined with the invention of the printing press in Mainz, Germany by Johannes Gensfleisch zur Laden zum Gutenberg around 1440. Over the following years, Gutenberg perfected the process and in the early 1450s started producing small texts, before printing approximately 180 copies of his Gutenberg (or 42-line) Bible between 1452 and 1454. In the next 20 years, printing spread rapidly across Europe, reaching Italy in 1465. The first printed book in English was produced in 1473 by William Caxton while he was in Bruges, Belgium; by 1476, Caxton had established a printing press in Westminster, producing the first printed books in England.

'Ciclamen' from *Tractatus de Herbis* MS Egerton 747. A late thirteenth century manuscript from Salerno, illustrating *C. hederifolium* with a new sense of realism. © The British Library Board. MS. Egerton 747 fol. 18v

'Ciclamen' from *Liber de Herbis et Plantis* c. 1340 Paris MS. Lat. 6823. This manuscript was copied from Egerton 747 but the illustrator Manfredus de Monte Imperiale used his own knowledge of plants to improve the illustrations. The leaves of the cyclamen clearly show a hastate pattern. Reproduced with the permission of the Bibliothèque Nationale de France, Paris. MS. Lat. 6823 fol. 33v

'Ciclame' (top left) from a *Tractatis de Herbis* c. 1440 MS. Sloane 4016 produced in northern Italy. This image appears in four manuscripts dating between 1370 and the late fifteenth century. In Sloane 4016 the text is secondary to the images of plants.

The very stylised image of 'Ciclamen' (right) from the *Livre des Simples Médecines* produced in northern France in the late fifteenth century. Reproduced with the permission of the Bibliothèque Nationale de France, Paris. MS. Fr. 623 fol. 52r

'Cyclaminos' from the *Editio Princeps Herbarium Apuleii Platonici* published in 1481 by Johannes Philippus de Lignamine, physician of Pope Sixtus IV. Though far removed from reality, this is probably the first printed image of a 'cyclamen'. The book was a printed version of the ninth century manuscript Montecassino *Codex Casinensis MS 97*. Reproduced with the permission of the Bayerische Staatsbibliothek, München, Germany. 4 Inc s.a 179 fol. 25v

'Cyclaminos' from the *Herbarium Apulei Platonici* late fifteenth century. This manuscript and its sister MS. Add. 17063) was almost certainly copied from the first printed *Editio Princeps Apulieus Platonicus* illustrated herbal. Reproduced with the permission of the British Library, London. MS. Add. 21115 fol. 24r

The first printed herbal was an unillustrated edition of the *De Viribus Herbarum* of *Macer Floridus* published in 1477 in Naples, Italy. A second edition, published in 1482, contained illustrations but these did not include a cyclamen.

The first printed illustration purporting to be of a cyclamen appeared in 1481 in the first illustrated printed herbal *Editio Princeps Herbarium Apuleii Platonici*, which was produced in Rome. Johannes Philippus de Lignamine, physician of Pope Sixtus IV had a printing press and, according to Payne (1904), having found the manuscript in the Monastery of Montecassino, he produced a printed edition, dedicating this to Cardinal Franscesco Gonzaga: "Ioannis Philippi de Lignamine Equitis Sicvli. S.N.D. Sixti. IIII. Familiaris ad Illvstrissimvm et Reverendissimvm dominvm do minvm. F.DF Conzaga Cardina. Mantvanvm". The illustrations are crude, and according to Blunt and Raphael (1979), there has been some debate as to whether they are woodblocks or from metal plates.

In the British Library, there are two virtually identical manuscripts of the *Herbarium Apulei Platonici* (Add. 17063 and Add. 21115). These are slim, quarto, economically produced vellum manuscripts, dating from the late-fifteenth century and probably originating from the area of Montecassino in Italy. Add. 21115 contains a dedicatory epistle from Johannes Philippus de Lignamine, physician to Pope Sixtus IV, to Cardinal Francesco Gonzaga, whose arms are on the first page. This dedication is very similar to that in the *Editio Princeps*.

The painting of a cyclamen in the manuscripts and the woodblock in the *Editio Princeps* — labelled *Herba orbicularis a Rapura* — are far removed from the plant we know. There are two leaf petioles,

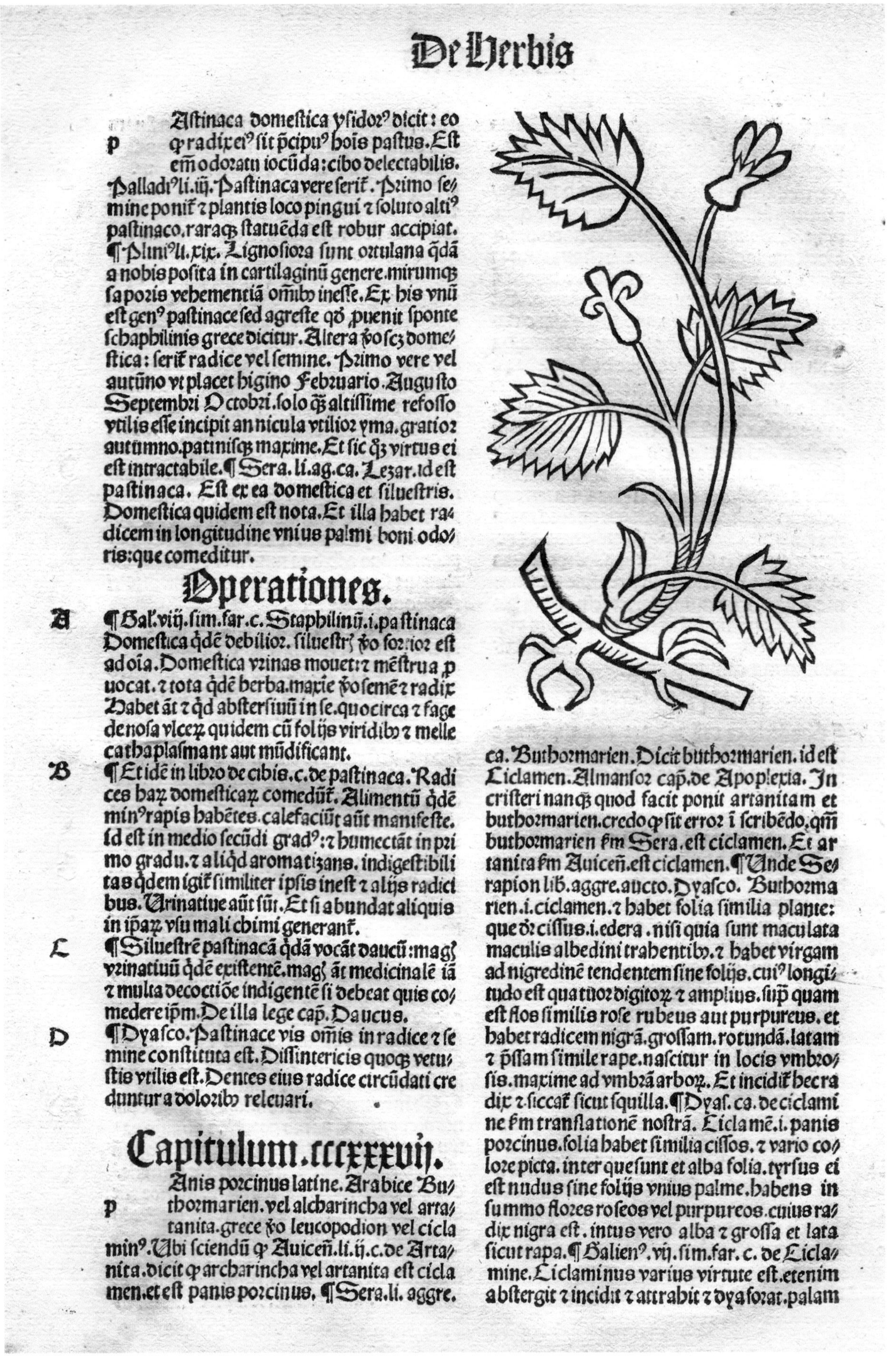

De Herbis

P Astinaca domestica ysidor⁹ dicit: eo q̄ radix ei⁹ sit pcipu⁹ hoīs pastus. Est enī odoratu iocūda: cibo delectabilis. Palladi⁹ li. iij. Pastinaca vere serit̄. Primo semine ponit̄ ⁊ plantis loco pingui ⁊ soluto alti⁹ pastinaco. raraq̄ꝫ statuēda est robur accipiat. ¶ Plini⁹ li. xix. Lignosiora sunt ortulana q̄dā a nobis posita in cartilaginū genere. mirumq̄ꝫ saporis vehementiā omībꝫ inesse. Ex his vnū est gen⁹ pastinace sed agreste qđ ꝓuenit sponte schaphilinis grece dicitur. Altera ꝟo scꝫ domestica: serit̄ radice vel semine. Primo vere vel autūno vt placet higino Februario. Augusto Septembri Octobri. solo q̄ꝫ altissime refosso vtilis esse incipit anniculo vtilior ymo. gratior autumno. patiniscꝫ maxime. Et sic q̄ꝫ virtus ei est intractabile. ¶ Sera. li. ag. ca. Lezar. id est pastinaca. Est ex ea domestica et siluestris. Domestica quidem est nota. Et illa habet radicem in longitudine vnius palmi boni odoris: que comeditur.

Operationes.

A ¶ Gal. viij. sim. far. c. Staphilinū. i. pastinaca Domestica q̄dē debilior. siluestris ꝟo fortior est ad oīa. Domestica vrinas mouet: ⁊ mēstrua ꝓuocat. ⁊ tota q̄dē herba. maxie ꝟo semē ⁊ radix Habet āt ⁊ q̄d abstersiuū in se. quocirca ⁊ fagedenosa vlcerꝫ quidem cū folijs viridibꝫ ⁊ melle cathaplasmant aut mūdificant.

B ¶ Et idē in libro de cibis. c. de pastinaca. Radices harꝫ domesticarꝫ comedūt̄. Alimentū q̄dē min⁹ rapis habētes. calefaciūt aūt manifeste. id est in medio secūdi grad⁹: ⁊ humectāt in primo gradu. ⁊ aliq̄d aromatizans. indigestibilitas q̄dem igit̄ similiter ipsis inest ⁊ alijs radicibus. Urinatiue aūt sūt. Et si abundat aliquis in ipsarꝫ vsu mali chimi generant̄.

C ¶ Siluestrē pastinacā q̄dā vocāt daucū: magis vrinatiuū q̄dē existentē. magis āt medicinalē iā ⁊ multa decoctiōe indigentē si debeat quis comedere ipm. De illa lege cap. Daucus.

D ¶ Dyasco. Pastinace vis oīs in radice ⁊ semine constituta est. Dissintericis quoq̄ꝫ vetustis vtilis est. Dentes eius radice circūdati creduntur a doloribꝫ releuari.

Capitulum. cccxxxvij.

P Anis porcinus latine. Arabice Buthormarien. vel alcharincha vel artanita. grece ꝟo leucopodion vel ciclamin⁹. Ubi sciendū q̄ Auiceñ. li. ij. c. de Artanita. dicit q̄ archarincha vel artanita est ciclamen. et est panis porcinus. ¶ Sera. li. aggre. ca. Buthormarien. Dicit buthormarien. id est Ciclamen. Almansor cap. de Apoplexia. In cristeri nanq̄ꝫ quod facit ponit artanitam et buthormarien. credo q̄ sit error ī scribēdo. q̄m buthormarien f̄m Sera. est ciclamen. Et artanita f̄m Auiceñ. est ciclamen. ¶ Unde Serapion lib. aggre. aucto. Dyasco. Buthormarien. i. ciclamen. ⁊ habet folia similia plante: que dr̄ cissus. i. edera. nisi quia sunt maculata maculis albedini trahentibꝫ. ⁊ habet virgam ad nigredinē tendentem sine folijs. cui⁹ longitudo est quatuor digitorꝫ ⁊ amplius. sup̄ quam est flos similis rose rubeus aut purpureus. et habet radicem nigrā. grossam. rotundā. latam ⁊ p̄ssam similē rape. nascitur in locis vmbrosis. maxime ad vmbrā arborꝫ. Et incidit̄ hec radix ⁊ siccat̄ sicut squilla. ¶ Dyas. ca. de ciclamine f̄m translatiōnē nostrā. Ciclamē. i. panis porcinus. folia habet similia cissos. ⁊ vario colore picta. inter que sunt et alba folia. tyrsus ei est nudus sine folijs vnius palme. habens in summo flores roseos vel purpureos. cuius radix nigra est. intus vero alba ⁊ grossa et lata sicut rapa. ¶ Galien⁹. viij. sim. far. c. de Ciclamine. Ciclaminus varius virtute est. etenim abstergit ⁊ incidit ⁊ attrahit ⁊ dyaforat. palam

Panis Porcinus or ciclamen from *Ortus Sanitatis* published in Strassbourg in 1497 by Johann Prüss. This clearly un-cyclamen-like image was first used to illustrate Vulgago or Ciclaminus in the 1487 editions of *Gart der Gesundheit* by Johannes von Cuba published in Strassbourg and Ulm (MD)

each with roughly four pairs of opposite leaflets and a terminal leaflet. The two flowers are on arching pedicels and resemble the centre of a composite flower nestling in the centre of its calyx. The plant illustrated in *Editio Princeps* is placed within a block frame and has an attendant serpent, indicating that it is proposed as a relief for snakebite. Medieval herbals were often subject to confusion over the identity of plants. Illustrations copied from one herbal to another suffered from artistic progression and it was common for plants to have numerous synonyms. The first synonym given for *Herba Orbicularis* is *Cyclaminos*, but Hunger (1935) maintains that *Herba Orbicularis* equates to *Aristolochia pallida*. I disagree as although it is not uncommon for there to be confusion between cyclamen and aristolochia, through many of the European manuscript herbals of *Apulieus Platonicus* from the ninth century onwards discussed earlier in this chapter, it is common to find *Herba Orbicularis* as a synonym of *Herba Cyclaminos* accompanied by an illustration that is clearly a *Cyclamen*. Unfortunately, the confusion over plant names and their accompanying illustrations managed to cross the divide from manuscript to printed herbals such as *Gart der Gesundheit* and *Hortus* Sanitatis. Well into the sixteenth century Johannes von Cuba in his *Kreutterbuch* (1533) and Theodor Dorsten in his *Botanicon* (1540) were offering woodcuts of *Ranunculus ficaria* with synonyms of Cyclaminus, Panis porcinus, Cyclamen, Arthanita, Erdtapffel and Schweinsbrodt.

The original manuscript that de Lignamine used as his exemplar for BL Add. 17063, Add. 21115 and the *Editio Princeps* was *Codex Casinensis MS 97*, produced in the early ninth century in Montecassino and still to be found in the Biblioteca del Monumento Nazionale di Montecassino at the Abbey of Montecassino. Between the fifteenth and twentieth centuries, the first 20 chapters of the manuscript were lost, so that the part containing a *Cyclamen* no longer exists.

It is tempting to believe that de Lignamine produced the British Library manuscript herbals Add. 17063 and Add. 21115 from *Codex Casinensis MS 97* then used these as the originals for the *Editio Princeps*, but Hunger offers fundamental reasons why this is not the case. He argues that both codices are in fact manuscript copies of the *Editio Princeps* and that Add. 21115, with its dedication, was possibly intended as a presentation copy for his patron Cardinal Franscesco Gonzaga, who in these early days of printing, may not have appreciated printed books. The most compelling reason for them being copies of the printed work is that they are faithful to the printer's errors that the *Editio Princeps* contains!

The *Gart de Gesundheit* was a herbal produced by Johan von Cuba in 1485, and subsequently published in various editions in the German cities of Augsburg, Strasbourg and Ulm. These manuscripts contained a chapter about a plant headlining as either Vulgago or Erdopfel, with the synonyms Panis Porcinis, Vulfago, Ciclaminus and Artanita. Vulgago was the Latin name for *Asarum*, and the editions produced in Augsburg in 1485, 1487 and 1488 illustrated the chapter with a clearly *Asarum*-like plant with an apparently creeping rootstock, cordate entire leaves and small *Asarum* flower. Unlike some other occurrences, the confusion of *Cyclamen* with *Asarum* is at least believable because of the superficial similarity in the leaves of *Asarum europaeum* and *Cyclamen*. The 1488 woodblock was a mirror image of that from 1485. The 1487 editions from Ulm and Strasbourg illustrated a plant with a creeping rootstock, sharply dentate leaves and diminutive flowers.

Le Grant Herbier, an illustrated herbal printed in Besançon, France by Pierre Metlinger between 1486 and 1488 contained a chapter headed 'De Ciclamen', with the synonym Pain a Porccina, but unfortunately this wasn't illustrated.

Hortus Sanitatis (or *Ortus Sanitatis*) was first published in Mainz by Jacob Meydenbach in 1491, with the section on plants titled *Tractatus de Herbis*, recalling the earlier manuscript herbals. In the editions of 1491, 1497, 1499 and 1500 the chapter on Panis Porcinis, with the synonyms Ciclamen and Artanita, was illustrated with a new woodblock of the same image that accompanied Vulgago in the Ulm and Strasbourg editions of *Gart de Gesundheit*.

It is interesting that in the editions of *Gart de Gesundheit*, *Le Grant Herbier*, and the 1491 *Hortus Sanitatis*, the frontispiece showed the same image of a group of apothecaries — albeit the woodblock was re-drawn in each case.

The printed herbals

Dioscorides' work remained important through the Dark and Middle Ages, being repeatedly copied. It was not questioned until the Renaissance, when commentaries on it appeared, and authors adapted its recommendations for their own conditions. The invention of moveable type enabled the printing of these, and earlier works, and the age of the printed herbal began. Herbals proliferated across Europe in the sixteenth century, improving not only in text but also in illustration, and it is to them that the tradition of accurate botanical illustration can be attributed.

The turning point from incomprehensible or crude illustration came in Brunfels' *Herbarum Vivae Eicones* of 1530, which included woodcuts of extreme accuracy and beauty, though none illustrated *Cyclamen*. The works of Fuchs, Matthiolus and de L'Obel stand out during this period.

Rather than discussing the works of these worthy (or less worthy) apothecaries individually, we will sum up the state of knowledge of each species during the sixteenth and early-seventeenth centuries. The herbals contain a great deal of material 'borrowed' from their predecessors; then as now, authors found it easier to rehash materials than do their own research. There is therefore a great deal of repetition with a little variation between authors.

The *Cyclamen* of the herbalists and apothecaries was *C. purpurascens*, known under various names and the only *Cyclamen* likely to be known to the herbalists of sixteenth-century Germany and Austria. It was used by western European herbalists to illustrate text, based on that of Dioscorides, which featured *C. graecum*. Leonhart Fuchs,

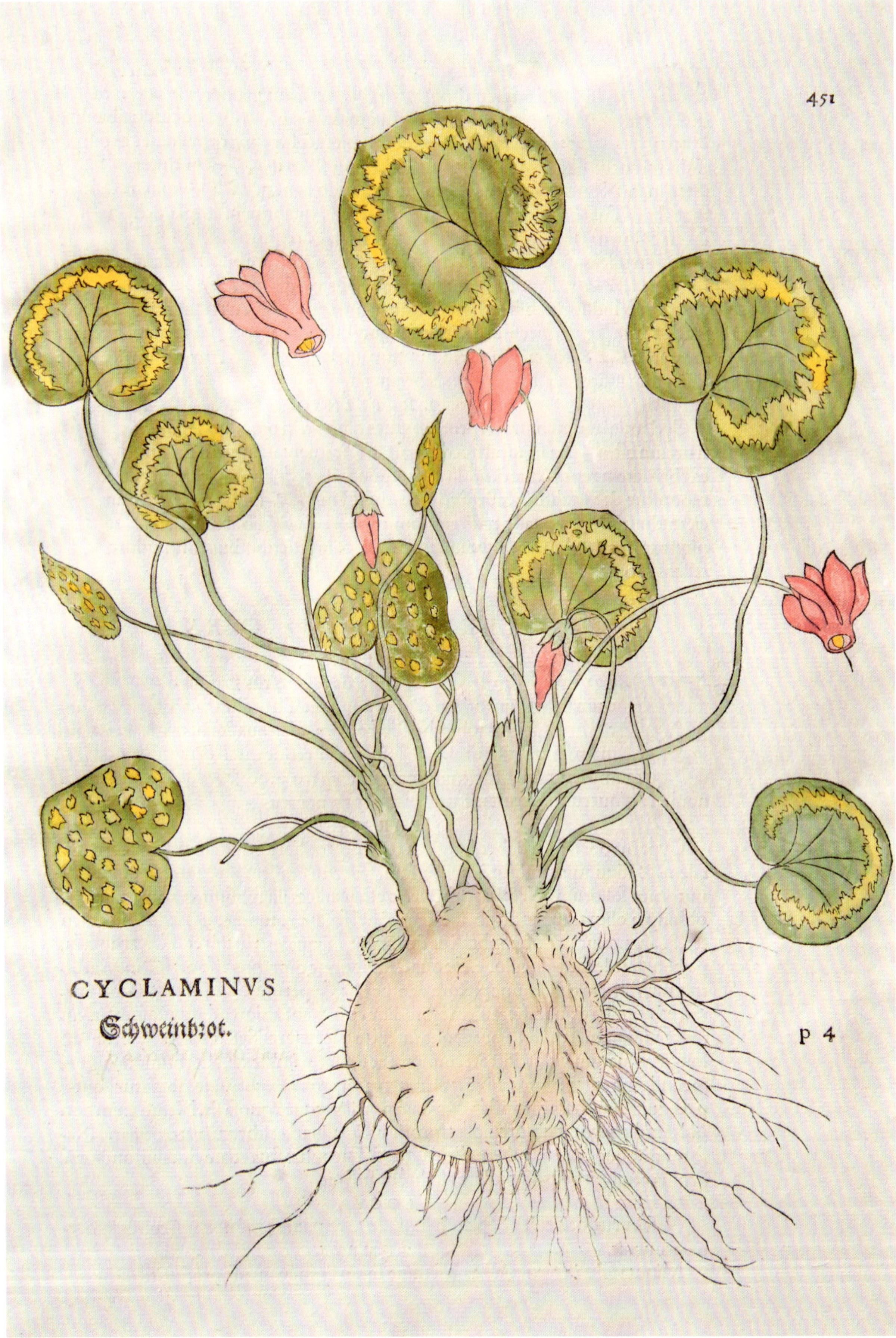

'Cyclamino' from Leonhart Fuchs' *De Historia Stirpium Commentarii Insignes* 1542. The iconic original woodblock image of *C. purpurascens* painted by Heinrich Füllmaurer, copied onto wood by Albrecht Meyer and engraved by Veit Rodolph Speckle (MD)

in his *De Historia Stirpium* of 1542, described it under the title *De Cyclamino*, as if it was the only one he knew. He noted that it flowered to the end of August and illustrated his text with a larger-than-life plate showing floral trunks and roots emerging from all over the surface of the tuber. Three leaves are spotted, recalling Dioscorides' perhaps fanciful description "… hath leaves like unto Ivie, of a purple colour, divers, with whitish spots beneath and above …", though Dioscorides is unlikely to have been describing *C. purpurascens*.

This truly iconic, earliest printed image of a true *Cyclamen*, was painted by Heinrich Füllmaurer, copied onto wood by Albrecht Meyer, and the woodblock engraved by Veit Rodolph Speckle. The original block with a massive image measuring 273 mm × 207 mm was probably used only three times: in the original 1542 Latin edition of Fuchs' herbal, in the first German edition of *New Kreuterbuch* in 1543, and then not until 1774–1778 when Salomon Schinz used it in his *Anleitung zu der Pflanzenkenntniss*. It was copied, with and without variance, and was re-drawn many times between 1542 and 1774, appearing in at least 50 volumes. One of the best re-drawings of Heinrich Füllmaurer's image was done by David Kandel for Hieronymus Bock's *Kreuter Buch*, first published in 1546 and then re-published in more than ten editions until 1630. At its smallest, the original was re-used by Fuchs in his duodecimal size *Plantarum Effegies* of 1549 and 1552 (minus one leaf) measuring just 62 mm × 36 mm. The most common use by Fuchs was a mirror image of 103 mm × 65 mm (with one missing leaf) that appeared in the less-expensive, reduced editions of his work from the 1543 Dutch *Den Nieuwen Herbarius* to the 1549 French *Commentaires Tres Excellens de l'Hystoire des Plantes*. This block was also used from 1553 to 1578 by the Dutch herbalist Rembert Dodoens in his *De Historia Stirpens* and by his translators Charles de l'Ecluse (French) and Henry Lyte (English).

With the publication of his *New Kreuterbuch* in 1543, Leonhart Fuchs realised that he had barely scratched the surface:

> "Before me there was no one of the more recent writers on botany who offered illustrations of more plants, and took care to display them more elegantly and splendidly, or related their histories with better method and order. However, although at the time I was satisfied with this edition, still not much later something began to displease me, mostly for two reasons. One, indeed, was that I had noticed that some of the plant histories were a little shorter than could satisfy the reader's expectations. The second was that, with passing time, I had acquired a wider and more complete knowledge of plants, and I felt that my statements about each had not been accurate enough. Therefore, so that those eager to learn more about plants might have richer and fuller histories, not only those in the first [book], but also of the others of which I have learned later, and to give myself a chance to correct those things which at the time I had unwisely left out. I began to ponder preparing another edition." (Meyer, Trueblood & Heller, 1999).

'Cyclaminus' from Hieronymus Bock's *De Stirpium*, 1552. David Kandel's reworking of the image of *C. purpurascens* painted by Heinrich Füllmaurer for Leonhart Fuchs (MD)

Fuchs' original 1542 *De Historia Stirpium* contained 511 images, whereas the three volumes of his Codex contain all these plus six images from the 1543 *New Kreuterbuch* and a further 1,012 new images making a total of 1,529 images according to Meyer, Trueblood and Heller (1999), but 1,525 according to Fuchs. Out of the 1,541 figures (sometimes more than one per block), 1,485 were of flowering plants. Whereas *De Historia Stirpium* and *New Kreuterbuch* make clear who the artists were, 893 of the plates in the Codex are unsigned. The remaining 636 are each signed by one of two artists: there are 413 by Jörg (Jerg Ziegler) but the artist who painted the other 223 is unknown. Likewise, we do not know who engraved the woodblocks for the Codex or just how many of these had been completed by the time of Fuchs' death in 1566. Fuchs struggled to have his manuscript published, and in April 1563, he complained to his friend Camerarius the Younger about his continuing problems: "First of all, I have indeed completed the most elaborate and sumptuous work of all on the History of Plants, divided into three volumes of great bulk, containing 1,504 illustrations. Two years ago an agreement was reached between me and Michael Isingrin's widow, whose written promise about the matter I hold. But such is the perfidy of people that she has now shamelessly withdrawn from her contract. She wants to sell the plates, for a good price, to another printer, to whose fancy the publication of this work would appeal. But so far no one has been found for me." (Meyer, Trueblood & Heller, 1999). After Fuchs' death, his son Friderich attempted to have the manuscript published but he could find no-one willing to invest the 3,000 florins necessary.

The Codex had a chequered history and disappeared for long periods, eventually being acquired by the Austrian National Library in 1954. It now comprises nine volumes (31.5 cm × 20.8 cm), known as the *Vienna Codex* 11117–11125. *Vienna Codex* 11121 contains three illustrations of 'Cyclamen'. Folio 155 is a magnificent painting of the image used in *De Historia Stirpium*, labelled *Cyclaminus prima maior*. As one would expect, it is a mirror image of the woodblock print of 1542 and it is tempting to speculate whether it is indeed the original. Two leaves have been modified by Fuchs, with annotations, making them cordate rather than orbicular. Folio 157, *Cyclaminus prima minor*, is clearly a more delicate plant with a small round tuber and short floral trunk. The below-ground parts of the petioles and pedicels are uncoloured, and this simple but clever device brings life to the image. It is possible that it is intended to depict *C. repandum*, but the leaves could equally be *C. hederifolium*. The tuber further confuses, showing the delicacy of *C. repandum* but with roots appearing from all over the surface as in *C. hederifolium* but the flowers lack auricles. Leonhart Fuchs' *De Stirpium Historia* is remarkable in several ways, not least because, along with the works of Otho Brunfels, it represents a new realism with plants being drawn from nature rather than the artist's imagination or copied from earlier volumes. It is therefore both interesting and disappointing that for folio 159 of his Codex, Fuchs depicts *Cyclaminus altera* Dioscoridis, but where earlier authors generally showed a plant with a cyclamen tuber, the below-ground parts of Fuchs plant are completely devoid of this. The illustration is probably of *Clematis vitalba*.

According to Meyer, Trueblood and Heller (1999), after Friderich Fuchs gave up trying to publish his father's manuscript, it passed into the hands of a printer from Ulm named Wagner and then into obscurity, until it surfaced in 1732 when a Viennese book dealer, Matthias Bayer, offered it for sale. During this period of obscurity, however, Dominique Chabrée (Dominicus Chabraeus), a doctor, produced a three-volume illustrated herbal *Historia*

330 The thirde Booke of

other groweth high, and wrappeth it selfe about shrubbes and plantes, and it hath no notable roote, and it is called Cyclaminus altera.

The Description.

1 Cyclaminon (which we may cal round Sowbread) hath broade leaues spread vpon the grounde with peaked corners lyke to Ivy leaues, and slightly dented round about the edges: and of a swart or darke greene colour aboue, yet powdered or garnished with white speckes or spots, and the middle part of the sayde leafe is somewhat white: but that syde of the leafe whiche is next the grounde, is purple colour, but sometimes deeper and sometimes lighter. The flowers hang vppon tender stalkes, nodding or beckning downewardes, and their leaues turning vpwardes or backwardes, in colour lyke to the purple violet, but not so faire: and of but a little or no sauour. There folowe small knoppes with seede, growing vpon small stalkes that are winded or turned two or three tymes about. The roote is turned rounde lyke to a Turnep, or Bulbus roote, and somewhat flat or pressed downe, with diuers hearie stringes by it, and it is blacke without, and white within, & in withering it gathereth wrinckles.

Cyclaminus orbicularis.
Sowbread.

2 The second Cyclaminon, or Sowbread, his leaues be also broade and nothing peaked or angled, but in a manner rounde, and nothing speckled vppon, or at least wayes very harde to be perceiued: they be also of a sadde or blackish greene colour, but vnderneath of a red purple colour. The flowers are lyke to the first, but of a better sauour. The roote is somewhat smaller.

3 The third kinde also hath leaues without corners, but they be somwhat dented or snip rounde about the edges: these leaues also are speckled, and blackish in the middle. The flower is of a deeper purple, and of a most pleasant sauor. But the roote is smaller than any of the rest.

The Place.

Sowbread groweth in moyst and stony shadowy places, vnderneath trees, hedges, and bushes, and in certayne wooddes, but not euerywhere. It groweth about Artoys and Vermandoys in Fraunce, & in the forest of Arden, and in Brabant. It is also common in Germanie and other Countries. But the thirde kinde is the dayntiest, and yet not strange in Italie.

The Tyme.

The kindes of Sowbread do flower in Autumne about September, afterwardes springeth vp the leaues, which are greene all the winter. The seede wareth ripe about sommer next folowing.

The Names.

1 The first is called in Greeke κυκλάμινος ἢ ἰχθυόθηρον: in Latine Cyclaminus, Rapum terræ, Tuber terræ, and Vmbilicus terræ: of Apuleius Orbicularis, Palalia, Malum terrę, Rapum porcinum, and Panis porcinus: in shoppes Cyclamen, and Arthanita: in English Sowbread: in Frenche *Pain de pourceau*: in Italian Pan

Cyclaminus orbicularis from Henry Lyte's *A Nievve Herball or Historie of Plantes wherein is contained the whole discourse and perfect description of all sortes of Herbes and Plantes: their divers et sundry kindes*, 1578. The first English translation of Rembert Dodoens' *De Stirpium Historia*. This woodblock of *C. purpurascens* was first used by Leonhart Fuchs in the 1543 first Dutch edition of his herbal *Den Nieuwen Herbarius* (MD)

Cyclaminus prima maior (*C. purpurascens*) from Leonhart Fuchs manuscript herbal (1544–1563). As a painting, this is a mirror image of the woodblock that appeared in Fuchs' 1542 *De Historia Stirpium*, but note that Fuchs has marked the image with intended modifications. Reproduced with the permission of the Österreichische Nationalbibliothek, Vienna. ÖNB/Wien, Cod. 11121 fol. 155

Cyclaminus prima minor from Leonhart Fuchs', manuscript herbal (1544–1563). An image of a cyclamen with confusing morphological details. It was subsequently published as an engraving of *Cyclamen parva radice* in 1651 and 1666 by Dominique Chabrée. Reproduced with the permission of the Österreichische Nationalbibliothek, Vienna. ÖNB/Wien, Cod. 11121 fol. 157

Cyclaminus altera Dioscoridis from Leonhart Fuchs manuscript herbal (1544–1563). Since medieval times herbalists illustrated this mythical plant described by Pedanius Dioscorides, often with highly imaginative plants. The illustration provided by Fuchs is probably *Clematis vitalba.* Reproduced with the permission of the Österreichische Nationalbibliothek, Vienna. ÖNB/Wien, Cod. 11121 fol. 159

Plantarum Universalis published in Yverdon, Switzerland in 1650 and 1651. This work was an expansion of *Historiae Plantarum Generalis Novae et Absolutiss* (1619) by Jean Bauhin (Johann Bauhin, 1541–1613) and Johann Heinrich Cherler (1570–1610). Amongst its many small woodblocks, it used 21 engraved from drawings in Fuchs' manuscript and these included the image of *Cyclaminus prima minor*, which was included under the name *Cyclaminus parva radice* (small tuber). Chabrée also illustrated five other *Cyclamen* using woodblocks that were variations on images originally used by Fuchs (1542), Dodoens (1574), de L'Obel (1581) and Camerarius/Matthioli (1586). He used them again, but with larger woodblocks, in his *Stirpium Icones et Sciagraphia*, published in Geneva in 1666. Chabrée said of *Cyclaminus pava radice* that it originated from Monte Salève near Geneva and in the Rhaetian Alps, and flowered in June and July; this suggests that it is synonymous with *C. purpurascens*, though the illustration is certainly not that species.

Another prolific writer was Pietro Andrea Matthioli, born into a medical family in Siena, Italy in 1501. His first treatment of Dioscorides was published in Italian in 1544 and then in Latin in 1554 as '*Medici Senensis Commentarii, in Libros Sex Pedacii Dioscoridis*'. It was illustrated with images that were laid out in such a way as to entirely fill the wooden block and thus they are very rectangular. *Cyclamen purpurascens* appears as 'Cyclaminus' described as *Cyclaminus folia habet hederæ, purpurea* … . This woodblock is quite unremarkable and lacks the delicacy of Fuchs' illustration, but it appeared in many editions, twice (in 1562 and 1572) as a mirror

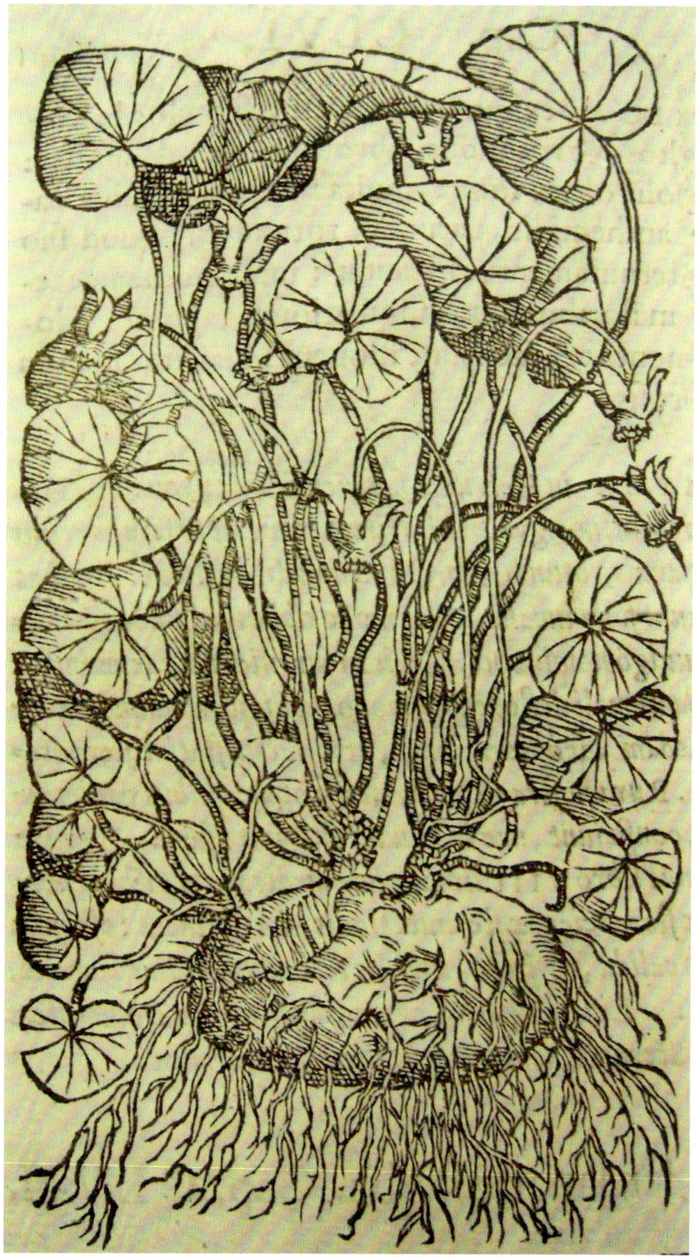

Cyclaminus folia habet hederæ, purpurea (*C. purpurascens*) from Pietro Andrea Matthioli's *Commentarii Secundo Aucti* (1559). By permission of the Trustees of the Natural History Museum (London)

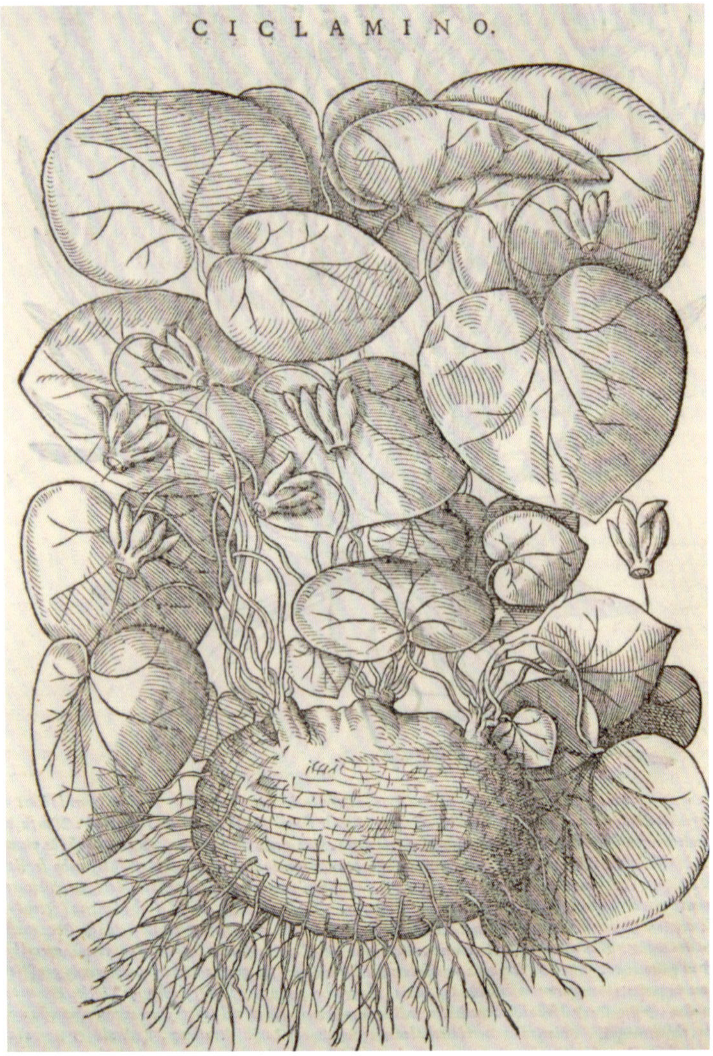

'Ciclamino' from Pietro Andrea Matthioli's *I Discorsi* (1568). This depiction of *C. purpurascens* by Giorgio Liberale of Udine, engraved by Wolfgang Meyerpeck from Germany first appeared in 1563 in Matthioli's *Neu Kreutterbuch*. By permission of the Trustees of the Natural History Museum (London)

image. In 1561, 1567 and 1584, a small mirror image (a mere 57 mm × 34 mm) was used by Antoine du Pinet to illustrate his duodecimal-sized *Historia Plantarum*. The image was also copied by Jacques d'Alechamps in his *L'Histoire Generale des Plantes* of 1586, 1615 and 1653, and finally as late as 1710, in William Salmon's *Botanologia – The English Herbal or History of Plants*. A new set of much larger figures (210 mm × 148 mm) appeared in later editions of Matthioli's treatment that were published in 1562, 1563, 1565, 1568 and posthumously in 1583, 1585 and 1604.

These figures were the work of Giorgio Liberale of Udine and Wolfgang Meyerpeck from Germany and took the Matthioli illustrations into a different league. They were intricately shaded in such a way that colouring was unnecessary, though coloured copies do exist, one such being the splendid 1563 *New Kreüterbuch*, held in the Botany Library of the Natural History Museum in London. Nine years after Matthioli's death (in 1577) editions of his works were re-issued under the editorship of Leonhart Fuchs' friend, the German botanist Joachim Camerarius the Younger, who re-illustrated them with drawings by Conrad Gessner and himself. These introduced us to a peculiar plant with an extremely extended floral trunk that appeared variously under the names *Cyclaminus macrorrhizos*, *Cyclaminus longius radicatus* and *Cyclamen radice oblonga*. Through nine editions, the woodblock was reversed and inverted, and in the 1598 and 1674 editions of Matthioli's *Opera*, a part of it was accompanied by the woodblock of *C. purpurascens* first used in 1590 by Jacob Theodor (Tabernæmontanus) in his *Kreuterbuch* and *Eicones Plantarum* and by John Gerard in his *Herball* of 1597. *Cyclaminus macrorrhizos*

Cyclamen (*C. purpurascens*) from Jacob Theodor's *Kreuterbuch* (1590). This image was later used in editions of Pietro Andrea Matthioli's herbal and by John Gerard in his 1597 *Herball or Generall Historie of Plantes*. By permission of the Trustees of the Natural History Museum (London)

was also copied in Dominique Chabrée's *Stirpium Icones* of 1666 (where it appeared with four variants of illustrations from Dodoens, de L'Obel and Fuchs' and Theodor Zwinger's *Theatrum Botanicum Krauter Buch*, published in 1696.

In the 1530s, Brunfels had mentioned cyclamen as if it was not known to him, save through reference to Dioscorides, though the cyclamen of Dioscorides was possibly *C. graecum*. However, variations on Dioscorides' text were used with illustrations of *C. purpurascens* by many authors, including Ruel, Matthioli, Catellana, Amato and Cordus. By 1540, Egenolph (a publisher in Frankfurt, and a ruthless pirater of others' illustrations) was able to write of *cyclaminos folia hederae, purpurea et varia*, making it clear that there was no longer a single cyclamen. Names proliferated, such that de l'Obel (in *Stirpium Icones*, published by Christophe Plantin in1581) could call it *Cyclamen vulgare Officinarum orbiculato folio* (the common cyclamen of the apothecaries with round leaves), giving the alternative names *Panis porcinum*, *Rapum terrae*, and *Arthanita*.

The translation of these names might best be left to William Turner, who produced his *Herball* between 1551 and 1568. His section on cyclamen is headed 'Of Sowesbread' and continues:

"Sowesbread is called in Greek *Ciclaminos*, is also in Latin *ciclaminus*, *rapum terrae*, *umbilicus terrae*, et *panis porcinus*, of some, *tuber terrae*, in Duche *Sewbrodt*, in French *pain de porceau*. I have not sene it in England, wherefore I knowe no usual name for it, but least it should be brought in to England or be found in anye place in England I name it Sowesbread or rape violet because the flowers is lyke a violet and the roote is lyke a rape."

Regarded as the father of English botany, Turner became Dean of Wells and was deeply involved with the Reformation, counting Bishops Latimer and Ridley as friends. Unlike them, however, he escaped the attentions of Mary I by going into exile abroad, explaining why his *Herball* (in English) was published in Cologne. The period of exile also gave him the opportunity to roam the continent, picking up botanical instruction and information, as well as enabling him to make field observations:

"I have seen *Cyclaminus* both in Italy and also in Germany: but there was great difference between them for the Italian was thryse as big as the Duche was, and much longer and lyker unto ivy. The Duche *cyclamineus* … had leaves as round as *Asarum* hath …"

As for habitat notes:

"*Cyclaminum* groweth much in shadowy places and moste under trees."

As an instructor in herbalism, Turner provides several 'vertues' for cyclamen, amongst which may be noted the following: "It is perilous for women with chylde to go over this roote. It is also a remedy against serpents if it be layd against the place. It is dronken against deadlye venom and specially against the fishe, which is called in Latin *Lepus marinus*, that is to say, the Sea Hare. It is put in wol to the fundament to drive forth the excrements of the belly."

The fullest exposition of the 'vertues' of cyclamen comes from John Gerard, the barber-surgeon whose *Herball or Generall Historie of Plantes* of 1597 is one of the great books of Tudor England. Gerard stands condemned both as a plagiarist (the work is based on a previous translation of a herbal by Dodoens by a Dr Priest, though Gerard alleged that this manuscript had 'perished') and as a fantasist (the descriptions of the Barnacle tree bearing live geese, and the vegetable lamb). Regardless, his book is an invaluable source of information about the plants grown in England — Gerard gardened in Holborn — in the late-sixteenth century. The notes on wild plants show that Gerard was genuinely interested in the plants he wrote about. Although lengthy, it is worth quoting the passage on the medicinal properties of cyclamen for both the charm of the language, and the amusement to be derived from it:

"*The Nature*

Sowbread is hot and drie in the third degree.

The Vertues

A. The root of Sowbread dried into powder and taken inwardly in the quantitie of a dram and a halfe, with mead or honied water, purgeth downeward tough and grosse flegme, and other sharpe humours.

B. The same taken in wine as aforesaid, is very profitable against all poison, and the bitings of venomous beasts, and to be outwardly applied to the hurt place.

C. The pouder taken as aforesaid, cureth the iaundise and stoppings of the liver, taketh away the yellow colour of the bodie, if the patient after the taking hereof be caused to sweat.

D. The leaves stamped with honie, and the juice put in the eies, cleereth the sight, taketh away al spots and webs, pearl or haw, and all the impediments of the sight, and is put into that excellent ointment called *Unguentum Arthanitae*.

E. The roote hanged about women in their extreme travel with childe, causeth them to be delivered incontinent, and taketh away much of their paine.

F. The leaves put into the place hath the like effect, as my wife hath proved sundrie times upon divers women, by my advise and commandment, with good successe.

G. The iuice of Sow-bread doth open the Hemorrhoids, and causeth them to flow being applied with wooll or flocks.

H. It is mixed with medicines that consume or waste away knots, the king's evill, and other hard swellings: moreover it clenseth the head by the nostrils, it purgeth the belly being anointed therwith, and killeth the child. It is a strong medicine to destroy the birth, being put up as a pessarie.

I. It scoureth the skin and taketh away sun-burning, and all blemishes of the face, pilling of the haire, and marks also that remaine after the small pocks and mesels: and given in wine to drinke, it maketh a man drunke.

K. [*sic*] The decoction thereof serveth as a good and effectuall bath for members out of ioint, the gout, and kibed heeles.

L. The root being made hollow and filled with oile, closed with a little wax, and rosted in the hot embers, maketh an excellent ointment for the griefs last rehearsed.

M. Being beaten and made up into trochisches, or little flat cakes, it is reported to be a good amorous medicine to make one in love if it be inwardly taken.

The Danger

It is not good for women with childe to touch or take this herbe, or to come neere unto it, or stride over the same where it groweth for the naturall attractive vertue therein contained is such, that without controversie they that attempt it in maner aforesaid, shall be delivered before their time: which danger and inconvenience to avoid, I have (about the place where it groweth in my garden) fastened sticks in the ground, and some other stickes I have fastened also crosse-wise over them, lest any woman should by lamentable experiment finde my words to be true, by their stepping over the same."

So, to Gerard, cyclamen was a veritable panacea, although one feels the modern warning "do not try this at home" is rather appropriate. His notes serve to remind us of the horrors of pre-scientific medicine, and the forgotten ailments of the past: a kibe is "an ulcerated chilblain, especially on the heel" (Concise Oxford Dictionary). We should not be too quick to mock, however, since Dr J. Rupreht has reported on the toxicity of *C. purpurascens* in *The Cyclamen Society's Journal* (1994), noting particularly its danger as an abortifacient (see also Chapter 1, page 17).

Gerard's *Herball* was re-issued in 1636 with extensive additions and commentary by Thomas Johnson, one of the most able botanists of the day and not one to put up with Gerard's nonsense lightly. This was his comment on *The Danger*, given above:

"I iudge our Author something too womanish in this, that is, led more by vain opinion than by any reason or experience, to confirme this his assertion, which frequent experience shows to be vaine and frivolous, especially for the touching, striding over, or coming neere to this herbe."

Gerard's *Herball* of 1597 contains descriptions and illustrations of two distinct *Cyclamen*, *C. purpurascens* and *C. hederifolium*; *C. repandum* and *C. balearicum* were added by Johnson in 1636. The descriptions are useful, and there is a particularly charming passage on the seed ripening process (under *C. purpurascens*):

"...these knoppes [the seed heads] are wraped after a few days in the small stalkes, as thred around a bottome, where it remaineth so defended from the iniurie of Winter close upon the ground, covered also with the green leaves aforesaid, by which means it is kept from the frost, even from the time of his seeding, which is in September until Iune: at which time the leaves do fade away, the stalkes and seed remaining bare and naked, whereby it injoyeth the Sun (whereof it was long deprived) the sooner to bring them to maturitie."

In addition, Johnson included the Christophe Plantin illustration of ivy used for the strange *C. altera* first encountered in the *Juliana Anicia Codex* (see illustration on p. 418).

No such anomalies are to be found in the treatment of *Cyclamen* in the next great English plant book, the *Paradisi in Sole Paradisus Terrestris. A Garden of all sorts of pleasant flowers which our English ayre will permit to be noursed up* by John Parkinson (1629). The title is a pun on his name, which translates as 'the park in the sun; the earthly paradise'. Although originally an apothecary, Parkinson became botanist to Charles I and Queen Henrietta Maria (to whom his book is dedicated). In the *Paradisus*, the commentary on medicinal virtues is relegated to a minor paragraph. It is first and foremost a gardening book, describing the plants known to Parkinson and giving suggestions on how to grow them: a forerunner of innumerable similar tomes. As a historical record of British horticulture in the Jacobean period, it is invaluable. Parkinson's later book, *Theatrum Botanicum* (1640) is more herbalist in scope, and lacks the spark and interest of the *Paradisus*. It adds nothing to our knowledge of cyclamen.

For the cyclamen enthusiast, the *Paradisus* starts well. The title page (shown on p. 537) is occupied by an engraving of the Garden of Eden, the original paradise: a rather hirsute Eve is picking strawberries, Adam is doing something obscure with a tree bearing strategically placed foliage. They are surrounded by choice and rare plants: there is a banana in fruit, dates and grapes as well as apples, even a prickly pear, and a pineapple. The ground flora is most interesting, with a fine tessellated *Colchicum*, various tulips, a double *Hepatica*, lilies, carnations, *Tagetes*, and saffron crocus being conspicuous. In the centre, below the tree of life itself, is a *C. hederifolium* bearing three flowers (complete with auricles) and two curling pedicels.

Parkinson placed his account of *Cyclamen* adjacent to that of *Erythronium dens-canis*, saying:

"The likenesse of the flowers, and the spotting of the leaves of the *Dens Caninus*, with these of *Cyclamen* ... maketh me joyne it next thereunto: as also that after the bulbous roted plants I might begin with the tuberous that remaine... ." thereby clearly showing that he was aware of the distinction between monocots and dicots.

He described and illustrated ten taxa, equivalent to *C. purpurascens*, *C. hederifolium*, *C. repandum*, *C. creticum* and probably *C. balearicum* and *C. persicum*. There are descriptions of *C. purpurascens* under the names *Cyclamen Aestivum* and *Cyclamen Romanum rotundifolium*, whereas both white (*Cyclamen autumnale hederaefolio flore albo*) and narrow-leaved (*Cyclamen autumnale hederaefolio angustifolium*) forms of *C. hederifolium* are described, the illustration of the latter showing quite a desirable hastate leaf.

Parkinson described plants in the order of their flowering, so he begins with *C. repandum*, known to him as *Cyclamen vernum flore purpureo* the "Purple flowred Sowebread of the Spring". His description is perfectly accurate: "[the leaves] at their first coming up are folded together; among which come the flowers, of a reddish-purple colour and very sweete... ".

Cyclamen from John Parkinson's *Paradisi in Sole Paradisus Terrestris* (1629) (MD)

He continues to describe the process of the coiling down of the seed pod, this being the first opportunity to do so in his book: "winding his footestalke, and coyling itselfe like a cable". A particularly memorable description of the appearance of the coiled pedicel and ripening pod is given for *C. purpurascens*: "standing almost like the head of a Snake that is twined or folded within the body thereof." Under *C. repandum*, Parkinson discusses how a young plant develops from seed "which being presently sowne, will growe first into round rootes, and afterwards from the shoot small leaves." This is very similar indeed to the first description of the germination process of a cyclamen seed given by Carolus Clusius in 1583, and one suspects that Parkinson's account of *Cyclamen* is based quite heavily on the great botanist's works. Charles de l'Ecluse, generally known as Clusius, was perhaps the most important of the sixteenth-century botanists and gardeners: it is regrettable that there seems to be no English translation of his work. As an example, he noted that a cyclamen flower was composed of a single 'leaf', i.e. a corolla, but that it was deeply divided into five lobes.

Together with the "Purple flowred Sowebread of the Spring", Parkinson knew, or knew of, a white *Cyclamen*, *Cyclamen Vernum flore albo*, which is probably *C. balearicum* from the Montpellier region of France. Clusius had certainly grown it, noting that it flowered in April and May of 1606, but that it perished in the frost of the *rigidissima hiems* (the 'most severe' winter) of 1608. Clusius' illustration of the plant is consistent with *C. balearicum*.

Parkinson says that the spring-flowering cyclamen come from: "the Pyrenaean Mountains in Italy [perhaps the Appenines were meant!], and in Candy, and about Mompelier in France." The reference to Candy is particularly interesting, this being the former name for Crete (now remembered by cyclamen enthusiasts in the shape of the Cretan endemic *C. graecum* subsp. *candicum* (Ietswaart, 1990)). He was referring to the plant he called *Cyclamen Vernum Creticum flore albo*, "White Candy Sowebread of the Spring", which he said had larger leaves (than *Cyclamen vernum albo flore*), "with more corners at the edges" (i.e more strongly toothed) and "more eminent [conspicuous] spots on them", together with larger and longer flowers. It is very tempting to think that this is the modern *C. creticum*.

Cyclamen hederifolium was known to northern Europeans from the mid-sixteenth century, usually as the ivy-leaved cyclamen, and is fortunately easy to trace through the herbals. It is interesting that it was noted as being variable in both leaf pattern and flower colour (e.g. by Clusius in 1583), and Parkinson describes how its characters, including the strength of its scent, vary when plants are grown from seed. He noted that it came from "divers places in France and Italie", while the white form was "reported to grow in the Kingdom of Naples".

The origin of plants was a subject of interest to the herbalists and early botanists, particularly because they (Thomas Johnson in particular) were trying to piece together local floras but also to gather information on wider distributions. The question of whether cyclamen were native to England was debated following Gerard's (1597) assertion that "It is reported to mee by men of good credit, that *Cyclamen* … groweth upon the mountains of Wales, on the hills of Lincolnshire, and in Somersetshire…". Johnson (1636) notes rather tartly "I cannot learne that this growes wilde in England." Parkinson was more forthcoming: "I have very curiously enquired of many, if ever they found them in any parts of England, neare or farther off from the places where they dwell: but they have all affirmed, that they never found, or ever heard of any that have found any of them. This only they have assured, that there groweth none in the places, where some have reported them to grow."

Cyclamen vulgare Officinarum orbiculato folio from *Plantarum seu Stirpium Icones* published by Christophe Plantin of Antwerp in 1581 to accompany Matthias de L'Obel's *Kruydtboeck*. This woodblock of *C. purpurascens* was subsequently used by several authors including Rembert Dodoens, Charles de L'Ecluse (Clusius), Johann Theodor de Bry, Thomas Johnson (in the 1636 edition of Gerard's *Herball*) and Michael Bernhard Valentini. By permission of the Trustees of the Natural History Museum (London)

Both Clusius (1601) and Parkinson (1629) discuss other cyclamen that are harder to identify. It is clear that there was an extensive trade in bulbs from Ottoman Turkey and that all sorts of plants were being received in the north, some becoming established, others perishing but not before being recorded in print or illustration. In many cases, it is difficult to be sure whether or not the author had ever actually set eyes on the plant described. One such is the plant discussed by Parkinson as *Cyclamen Antiochenum Autumnale flore purpureo duplici*, the "Double-flowred Sowebread of Antioch". It is very tempting to think that this is *C. persicum*, perhaps in the form of semi-selected plants. It "hath his leaves somewhat round [like the leaves of *C. purpurascens*] but with lesse notches or corners, and full of white spots on them"; the illustrated leaf certainly could be from *C. persicum*. Some stalks bore two or three large flowers, with ten or twelve lobes, but it is also clear that others had normal flowers. The flower is described as a "faire Peach colour, (like the flowers of *C. repandum*) deeper at the bottome." Others were similar but white, flowering in spring; the single ones could flower in either autumn or spring. According to Parkinson, these came from "Antioch in Syria", which might further support a view that this was *C. persicum*.

One of the problems involved in identifying such plants is that the woodcuts in the *Paradisus* are usually appallingly crude copies of the delicate work of earlier authors. Single block cuts are roughly copied into a multi-species plate: much detail has been lost in the process. It is the major flaw in this seminal book.

Gerard had re-used woodcuts from the works of others. In his 1597 edition, *C. purpurascens* was taken from the 1590 editions of the *Kreuterbuch* and *Eicones Plantarum* of Jacob Theodor (Tabernæmontanus), whereas that of *C. hederifolium* (also used in Johnson's 1636 edition of Gerard) was from Rembert Dodoens *Purgantium Aliarumque* of 1574. In the 1636 edition, it is interesting to observe the variation in quality of illustration of the four species depicted. The *C. hederifolium*, the *C. purpurascens* and C. vernum (=*C. repandum*), all taken from the 1581 *Plantarum seu Stirpium Icones* of Matthias de L'Obel, show the quality of German engraving at the time, whereas the C. vernum album (putative *C. balearicum*) from Charles de L'Ecluse (Carolius Clusius) *Curae Posteriores* of 1611 is beautiful and accurate. On the continent, Christophe Plantin of Antwerp was publisher for the work of Dodoens, Clusius and de l'Obel, and the blocks were used rather freely in their works and those of others. The books often appeared in various editions and in different languages, sometimes only the pictures were gathered together. The same illustration would be used in many variations. As an example, the *C. hederifolium* of Rembert Dodoens *Purgantium Aliarumque* in its original form was used in various editions of at least 14 volumes between 1574 and 1719 by Rembert Dodoens, Mathias de l'Obel, Christophe Plantin, John Gerard, Charles de L'Ecluse, John Parkinson, William Salmon and Michael Bernhard Valentini; a mirror image was used in the 1586, 1615 and 1653 editions of *Historia Generalis Plantarum* of Jacques d'Alechamps; a

Cyclamen vernum from *Plantarum seu Stirpium Icones* published by Christophe Plantin of Antwerp in 1581 to accompany Matthias de L'Obel's *Kruydtboeck*. This woodblock of *C. repandum* was subsequently used by several authors including Charles de L'Ecluse (Clusius), Thomas Johnson (in the 1636 edition of Gerard's *Herball*) and slightly modified by Dominique Chabrée and Michael Bernhard Valentini. By permission of the Trustees of the Natural History Museum (London)

Cyclamen vernum album (*C. balearicum*) from Thomas Johnson's 1636 edition of John Gerard's *Herball or Generall Historie of Plantes*. This image was first used by Charles de L'Ecluse (Clusius) in his *Curae Posteriores* (1611) (MD)

Cyclamen folio Hederæ from *Plantarum seu Stirpium Icones* published by Christophe Plantin of Antwerp in 1581 to accompany Matthias de L'Obel's *Kruydtboeck*. This woodblock of *C. hederifolium* was first published in Rembert Dodoens 1574 *Purgantium Aliarumque* and subsequently used extensively by other authors including John Gerard in his *Herball or Generall Historie of Plantes* of 1597. By permission of the Trustees of the Natural History Museum (London)

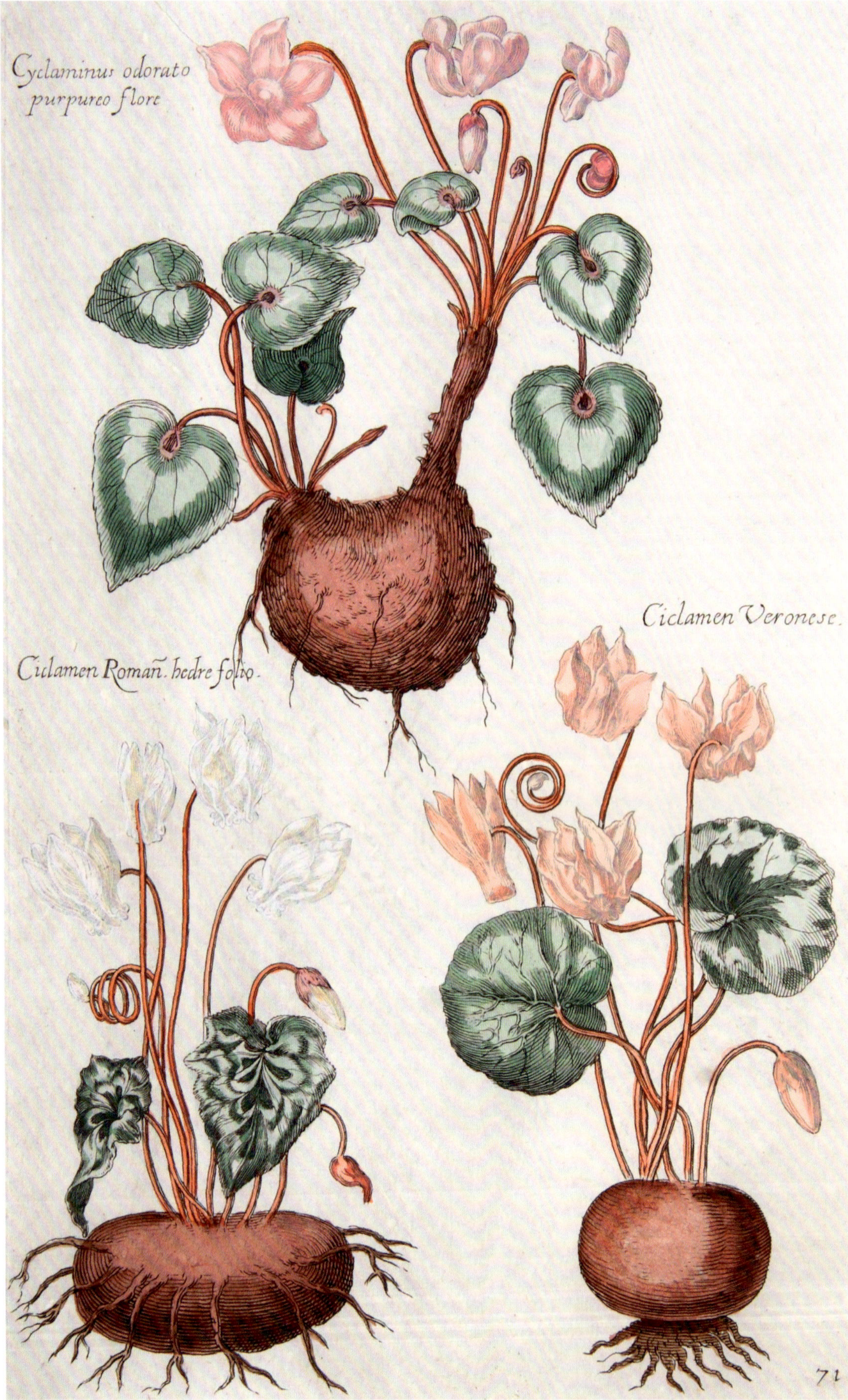

Cyclamen from Johann Theodor de Bry's *Florilegium Novum* (1612). The image of *Cyclaminus odorato purpureo flore* (*C. purpurascens)* was first published in 1581 by Christophe Plantin to accompany Matthias de L'Obel's *Kruydtboeck* (MD)

reduced-size mirror image appeared in the 1666 *Stirpium Icones* of Dominique Chabrée (Dominicus Chabraeus); a mirror image of the left half of the block only, appeared in the 1644 *De Historia Plantarum* of Theophrastus; and a re-worked mirror image was used by Thomas Panckow in his *Herbarium portatile* of 1654 and 1679.

The shift away from the strictly herbalist approach to books on plants began in the latter part of the sixteenth century. By this time, relatively peaceful and prosperous conditions pertained across much of Europe, and it was possible for botanists to travel, gardeners to correspond and merchants to trade. The result was a massive influx of new plants, the first to revolutionise gardening. It is difficult now to think of European gardens before about 1550, when the basic materials were the wild flowers of the meadows and woods, with a handful of more exotic species: carnations, roses, *Iris germanica* and hollyhocks perhaps. Plants such as the common double red peony (*Paeonia officinalis* 'Rubra Plena'), which appeared in about 1550, and crown imperial (*Fritillaria imperialis*) (1576) must have astonished the gardeners of the day. The result was also that the herbalists were confronted with a body of plants about which they had no traditional medicinal knowledge, and they were at a loss when attempting to write about these plants in their books. As Gerard (1597) wrote about the crown imperial: "The vertue of this admirable plant is not yet knowne…". As a gardener however, he was well able to describe the plant.

Thus, the history of botanical writing and publishing diverges from the strictly medicinal herbals and towards books of horticultural or botanical record. Herbals continued to be published, and still are — Nicholas Culpeper's book of 1652 has (inexplicably) remained in print to the present day — but they became of much less importance. From about 1600 onwards, botanico-horticultural books diversified, some into the coffee-table equivalents of their day, such as Johann Theodor de Bry's *Florilegium Novum* (1611), Emanual Sweert's *Florilegium* (1612), Crispin van de Passe's *Hortus Floridus* (1614), Daniel Rabel's *Theatrum Florae* (1622) and Pierre Vallet's *Le Jardin du Roy Tres Chrestien, Loys XIII, Roy de France et de Navare…* (1623), all with beautiful illustrations but little or no text, or the extreme of vanity publishing of Besler's *Hortus Eystettensis*, recording the plants in a single garden. These may be termed florilegia, picture books of flowers; such remain in the mainstream of horticultural publishing. Others recorded botanical and other information. Some combined botany with herbalism, such as Pierre Bulliard's *Herbier de la France* (1780–1793), probably the first botanical work to be printed in colour without retouching by hand, and François Pierre Chaumeton's *Flore Médicale* (1816). Charles de L'Ecluse (Carolius Clusius) combined botany with his accounts of travels in Spain or elsewhere, whereas others were primarily concerned with passing on horticultural information, as was John Parkinson. It is these works that are of most interest to modern gardeners.

By the end of the sixteenth century, the art of wood engraving was past its peak and becoming supplanted by engraving on copper. To complete this survey of significant early flower books, four works that were illustrated by copperplate engravings of remarkable quality deserve particular mention.

In 1613, Basilius Besler published *Hortus Eystettensis*, an enormous royal folio volume (57 × 46 cm) that, it has been remarked, requires a wheelbarrow to carry around. It is a record of the plants growing in the garden and diocese of Johann Konrad von Gemmingen (1593/5–1612), Prince Bishop of Eichstatt in northern Bavaria. A truly extraordinary book, it was the creation of a team of unattributed draughtsmen and engravers under the direction of Besler, an apothecary in Nuremberg. The work was mostly carried out in Nuremberg, the Prince Bishop sending boxes of flowers in season. Looking through the book, one is struck by the high proportion of extremely familiar plants and one can see that the garden at Eichstatt must have been a plantsman's paradise. Produced in various editions, some had the plates hand coloured, others did not. We prefer the uncoloured versions, which show the clear lines of the draughtsman and engraver, but it is the coloured copies that have been reprinted in recent years. Cyclamen are represented by *C. repandum*, *C. purpurascens* (three illustrations), and a magnificent portrait of a tuber of *C. hederifolium* bearing 43 flowers, plus buds and seed pods. As in all the plates in the book, the cyclamen plants are illustrated in carefully designed layouts that are beautifully rendered, but one cannot help feeling that they are laid out for pressing, spreadeagled upon the page. Despite this, the result is beautiful; it is no wonder that Besler's work is reproduced almost as frequently as the illustrations of Redouté.

Quite the opposite to the formality of *Hortus Eystettensis* is a book published just one year later, the *Hortus Floridus* of Crispin van de Passe. This is a true florilegium, a book of flower pictures printed in a landscape format. The plants are arranged by the season in which they flower. They are viewed from their own level, and are shown emerging from the ground, sometimes with a flat Dutch landscape in the background. The copperplates are exquisitely rendered. It was intended to be a sophisticated colouring book, and notes on colouring the plates were given together with the descriptions of the plants. Published originally in Arnhem in 1614, an English edition was produced at Utrecht in 1615, the publisher no doubt perceiving a good market among the flower-loving English! Only two cyclamen are illustrated, *C. purpurascens* and *C. hederifolium*. Both have a plate to themselves, while a further plate is used to illustrate their tubers. Both are fine illustrations, but the *C. hederifolium* suggests that it has been lifted and replanted, with the geniculate petioles above ground and the whole plant a bit loose and sprawling. On the other hand, the *C. purpurascens* depicted is a beautiful plant, with abundant glossy leaves, three flowers and two buds; it is attended by a hawk moth and a beetle. The tubers are also accurately depicted: flattened with roots from the sides for *C. hederifolium*, rounded and rooting all over, with floral trunks, for *C. purpurascens*. The colouring instructions say: "The roote of this Italian sow breade is of a chestnut coloure, the uppermost spriggs above are somewhat lighter … the shaddow is a black red."

Cyclamen from Daniel Rabel's *Theatrum Florae* (1622). A short, but fabulous book containing just 57 plates of mainly bulbous plants plus two frontispieces and no text published at the height of tulip mania. Rabel was portraitist to Marie of Medici before becoming botanical illustrator to Louis XIII (MD)

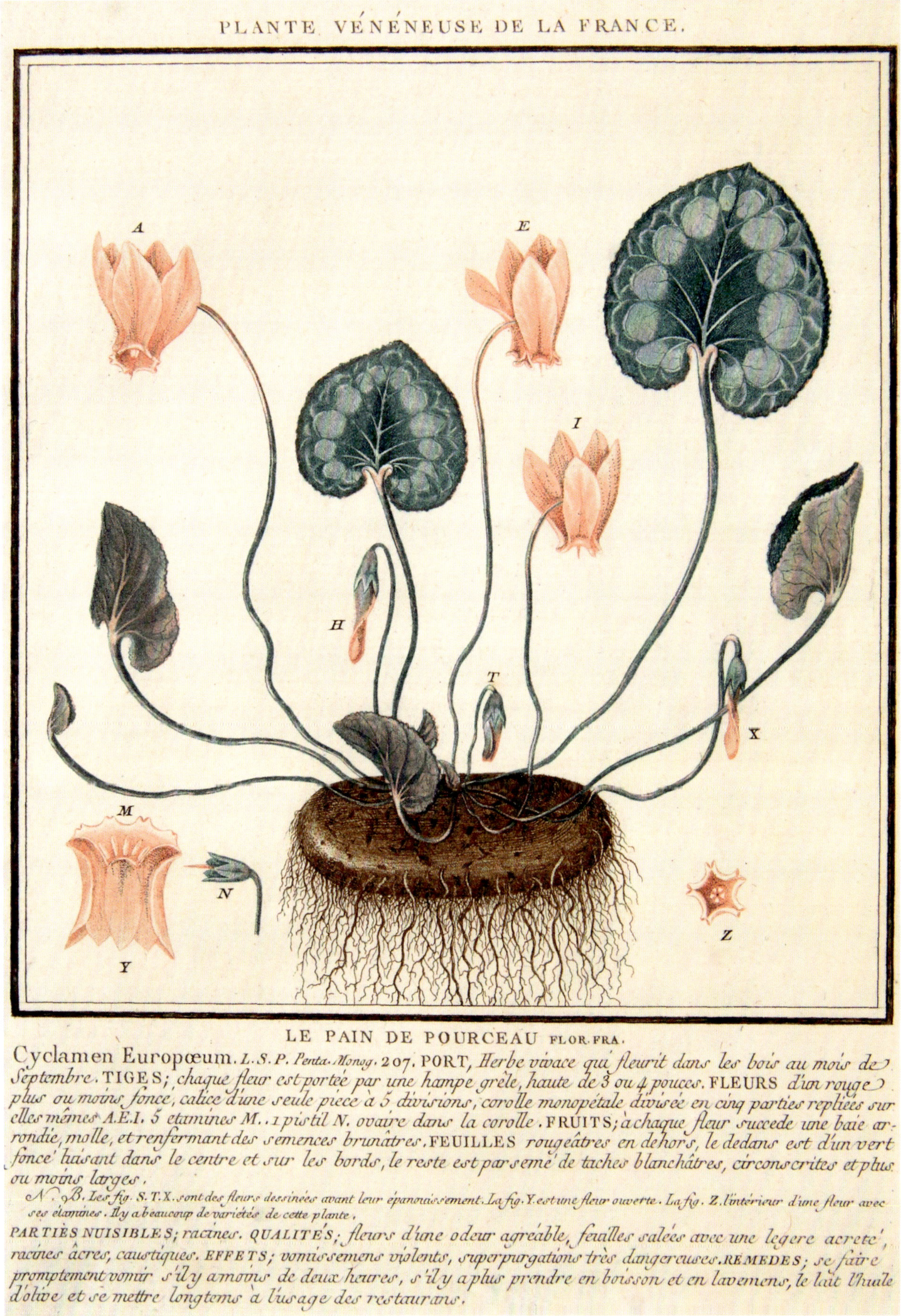

Cyclamen Europæum (*C. purpurascens*) from Jean Baptiste François Pierre Bulliard's *Herbier de la France* 1780–1793. Engraved and colour printed without retouching by Bulliard. The image was formed with a copperplate line engraving in black, followed by three tint plates to add colour to the flowers and leaves (MD)

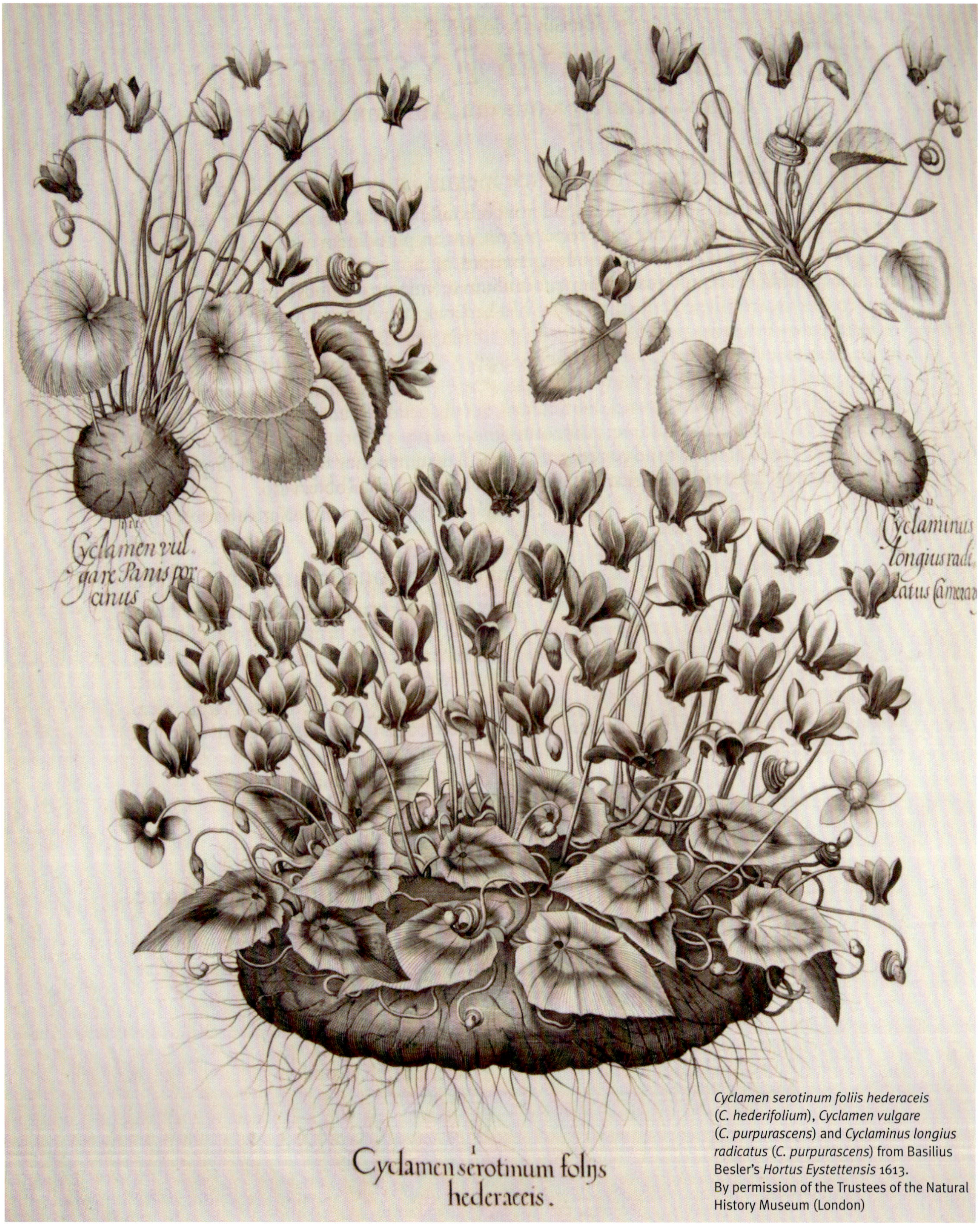

Cyclamen serotinum foliis hederaceis (*C. hederifolium*), *Cyclamen vulgare* (*C. purpurascens*) and *Cyclaminus longius radicatus* (*C. purpurascens*) from Basilius Besler's *Hortus Eystettensis* 1613.
By permission of the Trustees of the Natural History Museum (London)

The exceptionally fine engraving of *Cyclaminos Italica rotundifolia* (*C. purpurascens*) from Crispin van de Passe's *Hortus Floridus* (1614) (MD)

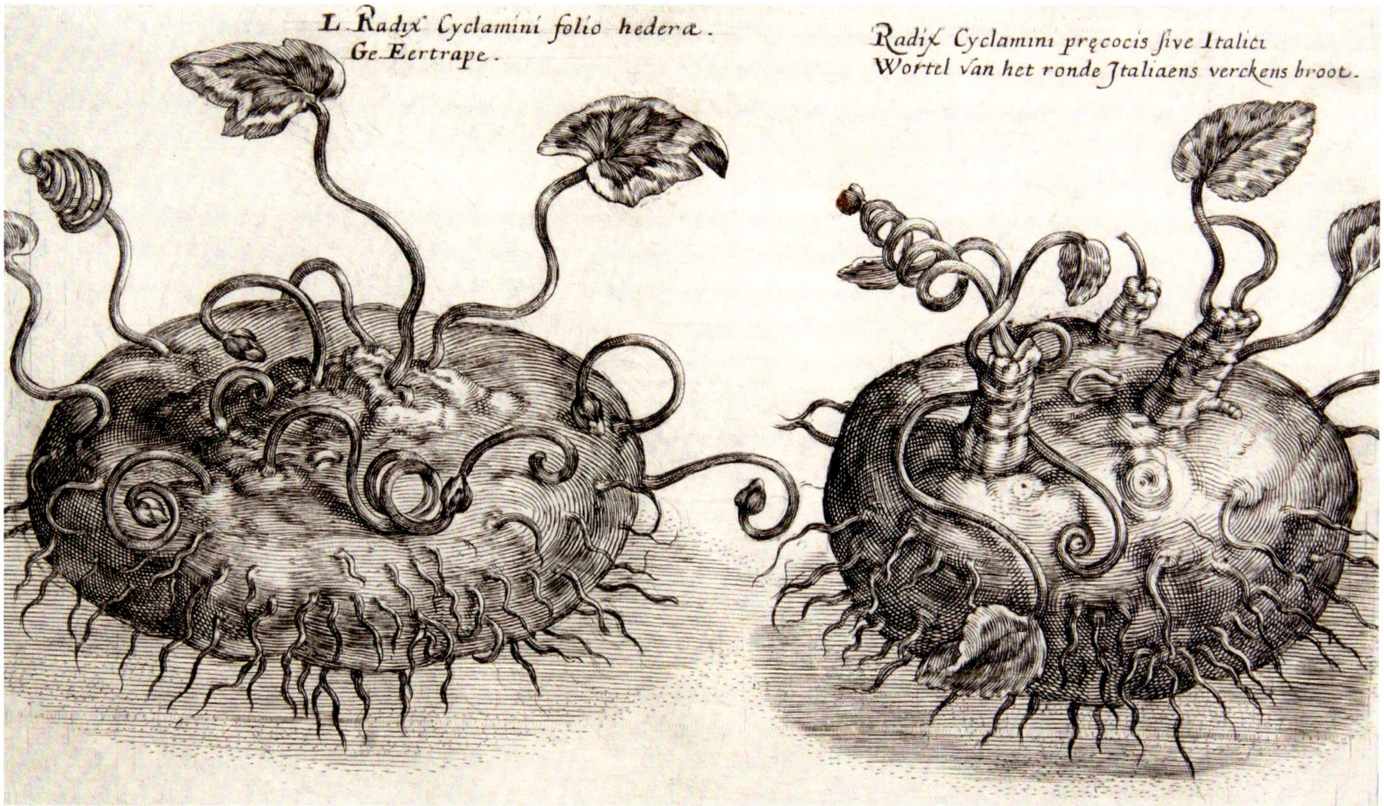

Tubers of *Cyclamini folio hedera* (*C. hederifolium*) and *Cyclamini precocis sive Italici* (*C. purpurascens*) from Crispin van de Passe's *Hortus Floridus* (1614) (MD)

Forms of *C. hederifolium* and *C. purpurascens* from Johann Wilhelm Weinmann's *Phytanthoza Iconographia* (1737). A mezzotint engraving, printed in colour and then hand finished. Many of the illustrations in *Phytanthoza Iconographia* were by Georg Dionysius Ehret (MD)

The pharmacist and botanist Johann Wilhelm Weinmann produced *Phytanthoza Iconographia* 123 years later. Although he ran a pharmacy business, his interest was in botany. He established a small botanic garden in Regensburg, Germany, and he started publishing botanical notes in the early 1720s. In 1728, he was introduced to Georg Dionysius Ehret and hired him to produce 1,000 engravings within a year. Ehret achieved just half that number and was dismissed, but he was to become one of the foremost floral illustrators of the eighteenth century. Weinmann hired other illustrators to complete the work, and a Regensburg physician, Dr Johann Georg Nicolaus Dieterichs, to write the text. This was no coffee-table book: the text (in both Latin and German) included sections on the character of each plant, its culture, and its medical uses; four pages were devoted to cyclamen. As was common at the time, the book was published in parts between 1737 and 1745. There were criticisms that some of the images were faked, but it was an impressive publication nonetheless. It was produced as a folio volume 38 cm × 25.4 cm (15 inches × 10 inches) but some of the plates were double-size. These larger plates included that of the six cyclamen that were illustrated: *Cyclamen hederaceum angulosum flore albo*, *Cyclamen rotundi folium flore albo*, *Cyclamen bipennium folio longiori anguloso*, *Cyclamen folio cordato*, *Cyclamen bipennium flore carneo* and *Cyclamen folio rotundiori vulgatius*. In reality, these are four forms of *C. hederifolium* and two of *C. purpurascens*. The plates are mezzotint engravings: a form of intaglio in which the copper plate is first roughened to hold ink in the pits and then the image is created by burnishing smooth parts of the surface so that they print lighter. In this way, it is possible to print tones and mid-tones. The illustrations were printed in colour and then hand-finished.

Another 60 years later, Robert John Thornton, who had trained in medicine at Guy's Hospital in London, conceived the idea of producing a monumental work the *New Illustration of the Sexual System of Linnaeus*. It was to be issued in parts between 1797 and 1807. The third section of the work was to be the *Temple of Flora*, planned as a series of 90 images in elephant folio 55.3 cm × 42.9 cm (21¾ inches × 16⅞ inches). Thornton commissioned a number of artists and painters to prepare the plates, including Sydenham Edwards, Abraham Pether, Philip Reinagle and James Sowerby. Unfortunately, by 1807 when just 32 plates (or 28 according to some, as the bibliography is somewhat confusing) had been completed, Thornton ran out of money and the work was cut short. In 1811, he obtained an Act of Parliament that allowed him to run the 'Royal Botanic Lottery' to raise funds with a first prize of the original paintings, but this was unsuccessful. Thus, in spite of the publication of a quarto edition of the book in 1812, Thornton was almost penniless when he died in 1837. His magnificent illustration of *C. persicum* was one of just two plates painted by Abraham Pether and is a hand-coloured and colour printed aquatint with stipple and line engraving by William Elmes. According to Buchan (1951), there were three states of this cyclamen plate:

> "In the first state the top of the castle is indistinct and has no pinnacles on the towers, and this is the first feature to inspect. The hillside is pure aquatint; the shading behind the cyclamen flowers is lightly cross-hatched, while the tree trunk to the right has only a few lines on it. In the second state the castle is more prominent and five distinct sharp pinnacles have been added, while many extra etched lines are to be seen — notably behind the cyclamen flowers; on the tree trunk; and under the cyclamen leaves on the left, which themselves stand out more sharply. The principal change in the third state is the addition of the aquatint to the sky on the left, so that only a streak of light remains above the mountains, while in the earlier states the light reached the top corner. The leaves of the cyclamen [now have] ... light and dark patches, the coarse-grained aquatint has been added to the middle distance. Much additional aquatinting has been applied to other parts of the plate. The most easily noticed difference, however, are the changes in the castle between states one and two, and in the sky between states two and three."

We hope that in this section we have provided some indication of the information about cyclamen present in Medieval, sixteenth- and seventeenth-century herbals and florilegia. An increasing number of both manuscript and printed herbals have been digitised and are available on the internet. This is of great benefit to researchers and is excellent for conservation, as it reduces the wear of precious manuscripts and books.

'The Persian cyclamen' (*C. persicum*) from Robert John Thornton's *Temple of Flora* (1804). By permission of the Trustees of the Natural History Museum (London)

Librarians rightfully guard their treasures, but without exception, we have found that they welcome keen researchers and freely offer their help, advice and time. We are particularly grateful to the librarians of Magdalen College and the Sherardian Library of Plant Taxonomy, University of Oxford; The Royal Horticultural Society's Lindley Library, London; and Science Library, Wisley; The Royal Botanic Gardens, Kew; the Linnean Society of London; and the Botany Library of the Natural History Museum, London for having allowed us to peruse most of these works in the original version. We are grateful to librarians at the British Library, Bodleian Library, Eton College Library and in a number international institutions for their willingness to provide images from their most precious and rarest works.

Minta Collins' *Medieval Herbals* (2000) provides an incredibly thorough treatment of manuscript herbals that dwarfs our specialist interest in cyclamen and has been an inspiration, leading us to cyclamen illustrations around Europe. A complete overview of the old herbals is given by Agnes Arber in her magisterial *Herbals* (3rd Edition, 1986), to which we are indebted for much of the background information cited here. H. Walter Lack's *Garden of Eden* (2001) is a copiously illustrated review of botanical illustration from Dioscorides to the present day, and contains an invaluable bibliographic background.

6.2 THE SEVENTEENTH CENTURY TO THE PRESENT

Vic Aspland

Woodcut produced for John Parkinson's *Paradisi in Sole Paradisis Terrestris* (1629). By permission of the Trustees of the Natural History Museum (London)

Parkinson's *Paradisi* marked a crucial turning point in the history of the illustration of cyclamen. The woodcut of cyclamen contained in that work demonstrated the fact that probably seven new species had been introduced in the mere 32 years since Gerard's *Herball*. Although the highest quality woodcuts could reach impressive standards of artistry, in the hands of all but the most accomplished artist they were limited by the technology involved. To put it simply, the design was drawn on the face of a plank of wood, and all other areas *except* the design were cut away. The raised areas were inked and the block brought into contact with a sheet of paper to print the design. Woodcuts were developed in Europe in about 1400, and were ideal for use with the printing press, which had been perfected in the 1450s, as they could be set with movable type to put words and pictures on the same page.

To follow the illustration of cyclamen beyond Parkinson takes one on an interesting journey through the advance of technology, the wider dissemination of cyclamen in cultivation, and increasing knowledge of these plants. The next technological step to follow woodcuts was the development of etchings in about 1500 and line engravings in about 1600. The two processes are inextricably entwined after 1600, as it was not uncommon for the two techniques to be applied to one printing plate. In particular, if an etching became worn after many printings (or was too 'thin' initially), it could be refreshed or enhanced by judicious engraving.

Line engraving raised the technology and artistry of illustration to a new level. The lines forming the image were cut into a copper or zinc plate using a sharp tool called a burin or graver. The finished plate was inked and wiped clean, so that the only ink left was in the cut lines. The plate and paper were then brought into contact under pressure. Because the ink was held in grooves, the process needed much greater pressure than printing using woodcuts to transfer the ink to the paper. In addition, since the engraving was in effect the opposite of the woodcut, it could not be used in conjunction with movable type: illustrations were usually printed on separate pages, or illustrations and text had to be printed on the same page by a two-step process. The process allowed the creation of much finer lines than were possible with woodcuts, and consequently the inclusion of much finer detail. The engraver was a specialist who worked from a watercolour painting by a botanical artist.

Etchings were produced by covering the surface of the metal plate with a layer of wax, and drawing the illustration using a specialised needle. Finally, the plate was immersed in an acid bath, in which the acid dissolved the metal where the wax had been removed. After cleaning, the plate was used for printing in the same way as an engraving. The smaller force applied to the needle allowed the formation of fine, flowing curves, but the correct acid 'cut' was a matter of skilled judgment.

The two new processes were rapidly adopted for the illustration of botanical and other subjects because of the finer detail that they made possible, and because the metal plates were harder-wearing than the wood blocks. In Britain, however, it was a very long time before either process was used to illustrate a cyclamen; I am aware of nothing until the late-eighteenth century. British horticulturalists were not, however, idle in the meantime.

Sir Thomas Hanmer, a Royalist during the English Civil war, had estates in Wales, and was a keen horticulturalist. He completed his *Garden Book* in 1659, with the intention of supplying knowledge that he thought was not widely available. By a quirk of fate, however, the book was not published until 1933, after a remarkably long time lag. His notes for January record that "These only Beare flowers in this cold Moneth: viz. WYNTER ACONITE, BLACK HELLEBORE, WYNTER CYCLAMENS". He grew plants in pots too, noting: "Sett potted Anemones and Beares eares and Cyclamens and other weake rootes out of violent raines and Snowes … that they may not receave too much wett at a tyme, nor the earth bee washt away." Sage advice that is equally applicable to the present day. Amongst the plants listed as flowering in April were "Spring cyclamens", a tantalising phrase that might represent late-flowering *C. coum* or perhaps *C. repandum*. His final reference for August records under "Remembrances for this moneth, Autumnall Cyclamens", indicate that he was growing *C. hederifolium* too.

In 1769, Philip Miller, Gardener to the Worshipful Company of Apothecaries at their Botanic Garden in Chelsea, authored *The Gardener's Kalendar*, a comprehensive account of the gardening knowledge of the time. He was clearly thoroughly familiar with a number of *Cyclamen* species. In the sections on plants in flower in the open air, he lists the "round-leaved Spring Cyclamen" (clearly *C. coum*) for January, February and March, and for December, he lists the "red-flowered Spring Cyclamen", so it would seem that selection of colour forms had already begun! Among the plants listed as flowering in the greenhouse, he mentions "Persian Cyclamen" in January and "Aleppo Cyclamens" in March and April. It would appear that forms of *C. persicum* were already becoming established as a favourite greenhouse plant!

In recent years, cyclamen enthusiasts became quite excited when news emerged of autumn-flowering forms of *C. persicum*, so it may come as a surprise that Miller mentions the flowering of Persian cyclamen in October and November, and of Aleppo Cyclamen in December, without considering this to be in any way special. Elements of the book are very modern. For example, he was unsure of the plural, scattering 'cyclamen' and 'cyclamens' at random throughout the work. All gardeners are busy people, and he was no less so, including the reminder for August: "You may now sow the seeds of … cyclamen", followed in September by the somewhat plaintive, "You may yet sow the seeds of… cyclamen." Other sage advice is as true today as it was when he wrote it: for May — "Transplant bulbs as blow in the autumn as autumnal cyclamen or sow-bread provided their leaves are

Cyclamen hyeme & verre florens folio anguloso ample flore albo basi purpurea, Persicum dictum Ten. (*C. persicum*) from Philip Miller (1760) *Figures of the most Beautiful, Useful, and Uncommon Plants described in The Gardeners Dictionary*. In the 8th edition of the latter (1768), Miller replaced polynomial names with the binomial *C. persicum*. By permission of the Trustees of the Natural History Museum (London)

decayed." (Blow was the term of the time for flowering.); June — "This is the proper season for taking up and transplanting the roots of cyclamens ... and such other bulbous-rooted flowers, whose leaves have decayed, some of which do not succeed if kept any long time out of the ground." The truth of the latter observation will be recognised by anyone who has bought dry cyclamen tubers and failed to establish them.

Autumnal cyclamens, clearly a reference to *C. hederifolium*, are listed as flowering in the open from August to October. Miller's book does not contain extensive descriptions of the plants themselves, so in some cases, we have to infer what Miller was writing about. It was clearly written by someone who knew his plants, so considered that little description was necessary for like-minded people who grew them. In this respect, the book was moving away from the library of the gentleman and into the hands of the gardener. It also makes it clear that at least three species: *C. coum*, *C. hederifolium* and *C. persicum* were becoming valuable components of the garden flora.

As in present times, when one political diary prompts the publication of another, Miller's horticultural work prompted others. 1782 saw the publication of a book by Thomas Mawe, somewhat cheekily titled *Every Man His Own Gardener. Being a New, and Much More Complete GARDENER's KALENDER*. As far as the plants of interest to us were concerned, this was a pretty thin claim, the only reference to cyclamen reading, under June: "transplant cyclamens; the leaves are now decayed; that is, take up the roots and part them: then new prepare the mould, and plant them again." The text that follows, does not make it clear whether he is referring to *C. coum* or *C. persicum*, as he states that "This plant generally begins to flower in February or March, according to its situation." Which just goes to show that you can't judge a book by the introductory blurb.

Mawe was not to know that a revolution in the documentation of knowledge relating to plants was on the horizon: the publication of *Curtis's Botanical Magazine*, which commenced in 1787. Its intention was clearly stated on the title page: "Intended for the use of Such LADIES, GENTLEMEN, and GARDENERS, as wish to become scientifically acquainted with the Plants they cultivate." Clearly, Curtis considered gardeners to be a race apart, and neither ladies nor gentlemen! The normal practice in the production of plant illustrations was for an artist to prepare a watercolour painting of the featured plant; then another craftsman would produce the printing plate, at this time an engraving or etching. (Note that in many recent publications, the term 'engraving' is used indiscriminately to cover both engravings and etchings, and that the painter of the original illustration usually gets far more credit than the craftsman whose plate put the ink on the paper!) In the *Botanical Magazine*, the plates were then hand-coloured, a laborious process considering that up to 6,000 plates were produced each month in the early days.

Cyclamen coum features in the second issue of the magazine, published in March 1787, but that plate is omitted here as it has been reproduced so often elsewhere (see page 514). Instead, I will mention the plate of *C. persicum* published in 1788. The original watercolour is the work of Sydenham Edwards, who worked for Curtis for many years, and the copper printing plate was made by Francis Sansom. The plate embodies the best artistic principles; the flowers and leaves nicely fill the page, and the smooth, sinuous curves of the stems enhance the composition. Note that modelling and texture is added to the basic outlines by the additional engraving of more-or-less parallel lines; those on the petals are particularly effective.

The accompanying text states that, "It is generally cultivated in pots, in light undunged earth, or in a mixture of loam and lime rubbish, and kept in frames, or on the front shelf of the green-house, where it may have plenty of air in the summer, but guarded against too much moisture in winter.", a recipe that will be recognisable to modern cultivators. Possibly of even greater interest, however, is the following statement: "Flowers early in the spring, and is admirably well adapted to decorate the parlour or study." This must surely be one of the first recommendations of this species as a pot plant for the house, rather than the greenhouse. I am sure that Curtis would have been astounded by the prophetic nature of his words, given the enormous popularity of *C. persicum* in recent times. Rather further from the mark is the attribution of this species to "the East-Indies", a somewhat diffuse geographical area stretching roughly from Sri Lanka to the Philippines via the Malay Archipelago!

A few years ago, I was excited to come upon a second illustration of this plant, printed from a plate also produced by Sansom and published in *The New Botanic Garden* (t. 19) in 1806. Two versions of this are shown here. There are indeed differences of detail, but the overall 'feel' of the plates is the same; could the craftsman have produced the second version from memory, or perhaps worked from the same original but wanted to insert differences for his own satisfaction?

A little later, the slightly different technique of stipple engraving was well-demonstrated in F. P. Chaumeton's *Flore Medicale*, a multi-volume herbal, the first edition of which was published in Paris in 1814. The cyclamen illustration in this book was painted by Pierre Jean Françoise Turpin, a highly praised artist considered to be as accomplished as Redouté. As usual, the engraver gets much less credit, being simply recorded as 'Lambert'. Rather than creating lines for modelling, in this case the engraving tool was repeatedly 'pecked' into the copper plate to produce many small hollows. When the print was made, these ink-filled hollows produced the many spots that so effectively convey the patterning on the leaves. As in the *Botanical Magazine*, the prints were then hand-coloured.

Cyclamen persicum from *Curtis's Botanical Magazine* 2: t. 77 (1788). Line engraving, hand-coloured. Painted by Sydenham Edwards and engraved by Francis Sansom (VAA)

Cyclamen persicum and *Crocus vernus*. From *The New Botanic Garden* Plate 19 (1812) (VAA)

As left, variation; the colour of '*C. vernus*' changed to yellow, so becoming *C. luteus*! (MD)

Cyclamen persicum from Chaumeton's *Flore Medicale*, illustration by P. Turpin. Stipple engraving, hand coloured (VAA)

Detail from *Flore Medicale* (VAA)

The early part of the nineteenth century was characterised by industrial, technical and social changes, which took place with increasing rapidity. The Industrial Revolution brought rapid technical advances (the cylinder press, for example, was invented in 1846) and a number of existing printing techniques were popularised over a short timescale, so it is not possible to place their wider use in a distinct chronological order. The period brought increasing wealth, and an increasingly large middle class, who had money to spend on more than just subsistence. Naturally, entrepreneurs appeared, happy to produce things on which money could be spent. If people would buy newspapers, why not gardening magazines? But before these could be produced cheaply in quantity, a number of technical problems had to be solved.

This period saw an advance on the woodcut: the wood engraving. Rather than cutting the design in the face of a plank, the engraving was cut into the end grain of a block of fine-grained wood (commonly box). This gave a much harder-wearing block and made much finer detail possible, but still left a technical problem: the wood engraving could not be set with moveable type so as to allow the printing of text and illustrations in one operation. In this period of breakneck progress, it was soon discovered that the woodblock could be used as a mould to produce a metal plate, on which the design was raised rather than recessed, and so could be set with type. This was the great breakthrough that made the gardening magazine possible. Strangely, the metal plates (and the prints too) continued to be called wood engravings.

For a brief time, magazines came and went thanks to takeovers, mergers and so on, just as in the present time. *The Journal of Horticulture, Cottage Gardener and Country Gentleman* has a title redolent of the time; it was first published weekly in 1861 and sold for three pence. One of my favourite illustrations appeared in Volume V of 1863. Only 8 cm × 10 cm, it bears the caption "Cyclamen Atkinsii" and is accompanied by an explanation of the plant's origin. In the words of the raiser: "After many ineffectual attempts to produce a good cross between *C. coum* or *C. vernum*, and *C. persicum*, combining the neat habit of the two former with the colour and larger petals of the latter, having at the same time the foliage dark, yet relieved with a lighter band, or marbled, I at length succeeded in raising the hybrid now figured, from seeds produced by a variety of *C. coum* impregnated with *C. persicum*." He also reported that: "Amongst the seedlings, it was found that every plant deviating from the marking of the foliage of the seed-bearing parent, produced white or blush flowers, whilst those retaining its plain dark leaf have invariably bloomed in different shades of the colour of that species." The further description states that the plant represented by the figure was exhibited with

Cyclamen atkinsii from *The Journal of Horticulture, Cottage Gardener and Country Gentleman* (1863). Wood engraving. Compare with the *Flore des Serres* lithograph of 1880 (see p. 473) (VAA)

about 70 fully expanded flowers and that the flowers were: "of a French white, marked with a deep crimson blotch at the base of each segment". The dimensions reported indicate that the plant was much larger than normal in all its parts. Given our present knowledge that polyploid forms are often larger and more vigorous than diploid ones, it is tempting to wonder whether pollination with *C. persicum* had stimulated the production of polyploid seedlings of *C. coum*. Note the attribution of plain leaves to *C. coum*. Later authors have recorded this plant as derived from a cross between *C. coum* and *C. ibericum* (merely another form of *C. coum*, introduced from a different area). Perhaps a bee actually made this cross behind Atkins's back; after the passage of so much time, we can only speculate upon whose version of events is correct. I remember that when I first began to buy cyclamen, perhaps 40 years ago, the offerings in the trade were described thus: '*C. coum*: plain leaves, flowers white, pink or crimson' and '*C.* × Atkinsii: marbled leaves, flowers white, pink or crimson'. Although long considered to be invalid, the name *C.* × Atkinsii can still be found occasionally in bulb catalogues; the bulb trade is far more reluctant to change names than are botanists and keen growers!

While the wood engraving catered for the 'cheap and cheerful' end of the market, there was an explosion in the popularisation of techniques that could cater for upmarket needs. Lithography is said to have been invented by Senefelder in Munich in 1797. In this process, limestone slabs were ground to a flat surface, and the artist drew the design directly onto the stone using a grease pencil. The porous stone was then moistened with water, and an oil-based ink applied with a roller. The ink only adhered to the grease pencil drawing. The print was made by pressing a sheet of paper against the stone surface. A disadvantage of the process was that the stone only produced a print in a single colour; the artist needed to draw on a separate stone for each colour! The printer had a similar problem: the paper had to be very accurately aligned with each stone. That both of these problems were surmounted is a tribute to the high skill of both craftsmen. A very fine example of the technique is a plate of *C. macrophyllum* (=*C. africanum*) published in 1853, in the journal *Flore des Serres et Jardins de l'Europe*. It includes seven colours, and the thickness of the inks in the leaf give it a wonderful three-dimensional quality. In this particular plate, the plant parts have a hard outline, paralleling the style of the hand-coloured engraving, but this is omitted from many lithographs, giving a much softer outline, particularly to petals. At the top of the leaves, it can be seen that the register of the colour and line impressions does not quite correspond. This is attributable to the fact that the paper had to be very slightly dampened before each impression was made, which could lead to the paper stretching slightly.

The production of the journal is a story in itself. Louis Benoit van Houtte was a Belgian horticulturalist who founded a nursery that grew to be probably the largest in Europe at the time. He decided to publish a journal featuring coloured plates of the best

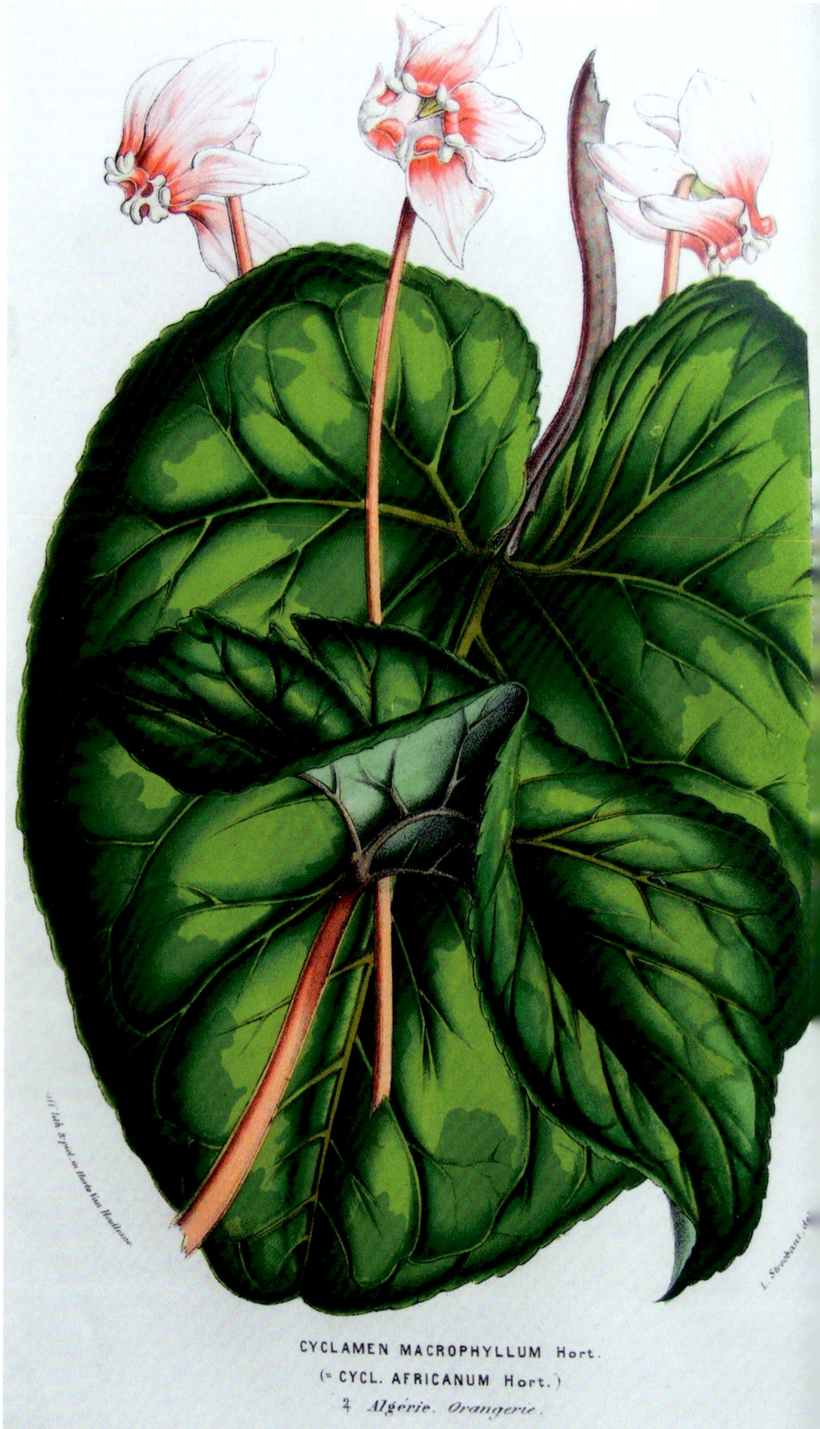

Cyclamen macrophyllum from *Flore des Serres* (1853). Lithograph and original painting by L. Stoobant. This has outstandingly rich application of colour, with high glaze (VAA)

Detail from *Flore des Serres* (VAA)

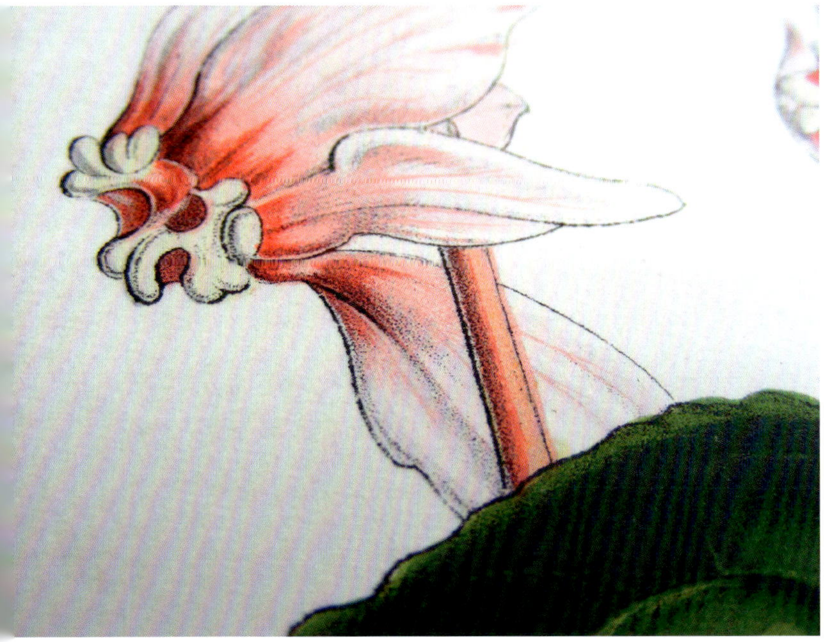

Detail from *Flore des Serres* (VAA)

plants in cultivation at the time (which, unsurprisingly perhaps, were available from his nursery!). He was dissatisfied with the quality of the printing available for this momentous work, so, being a decisive person, founded his own printing shop on the nursery site, so that he could personally control the quality of the product. This was clearly a successful ploy, for the journal ran to 23 volumes, with over 2,000 coloured plates produced.

Job obsolescence is a familiar theme today, but was equally common in the past. The introduction of lithography allowed the artist to create his design directly onto the stone, whereas previously the engraver or etcher created a facsimile of the artist's watercolour painting on a zinc or copper printing plate. This development, coupled with advances in press technology brought cost reductions and eliminated the need for the engraver. In 1845, *Curtis's Botanical Magazine* passed out of the hands of the Curtis family, and the new owner very quickly adopted lithography for the illustrations. Interestingly, he did not go the whole hog and use chromolithography, but continued to have the outline printed plates coloured by hand.

Stepping slightly back in time, an alternative technique was jostling for popularity, the chromoxylograph. This had much in common with the lithograph: a separate plate was needed for each colour, but the working medium was wood rather than stone. My chosen example is interesting for a number of reasons. This illustration of *Cyclamen coum vernum* was published in 1872 in David Wooster's *Alpine Plants* and represents a melding of two techniques. The outline, here printed in grey, was produced by a line engraving on a metal plate. The engraved lines are particularly strong in the petioles and there is stippled and line texturing in the leaves and flowers. The six colours were then added using individual wood blocks, one for each colour. This is a superb example of the craft, executed by the company of Benjamin Fawcett in Driffield, Yorkshire. The accurate register of the different blocks is a tribute to the skill of Frank Lydon, the craftsman who made them all, and perhaps also to the carefully chosen machine-made wove paper.

While the printing itself is outstanding, the technology of the ink used for this particular plate is less so. In different impressions, although the colour of the leaves and petioles is uniform, the flower colour varies widely, from a deep, glowing magenta, through paler tones and also to a remarkable orange. If the orange-flowered version represented a living plant, it would cause quite a stir in the cyclamen world! While the colour consistency is, shall we say, a little suspect, the accompanying text is more so. It states that: "Although the tubers of *C. coum* are remarkably acrid, they form the principle food of the wild swine of Sicily" and later "*C. coum* grows wild in many parts of Italy and Germany." It would seem that his information was supplied by less botanically inclined travellers. On the other hand, Wooster seems likely to have grown the plants, for he advises that, "The seeds should be sown as soon as they are ripe", a tip always passed on to beginners with cyclamen.

Cyclamen coum from *Alpine Plants* by David Wooster (1872). Chromoxylograph. Different flower colour versions exist (VAA)

Different colour version of this plate (VAA)

Detail and different colour version of the same plate (VAA)

Different colour version of the same plate (VAA)

Another of my favourite plates from the *Flore des Serres* stable, number 851, published in 1880, is captioned CYCLAMEN ATKINSI, Hort.

To my eye, the comparison with the wood engraving mentioned on p. 469 is striking. Both of the depicted plants are in shallow pans, and the overall arrangement of leaves and flowers has the same 'feel'. I often wonder if the one illustration inspired the other. It is also notable that the flowers in this later plate conform to the description by Atkins quoted above. Closer inspection shows that the lightness of the flower petals is enhanced by the hair-thinness of the outline lines, or on many flowers, by their total absence. I count a total of eight colours on this plate, for which the accurate alignment of the individual colour blocks is a remarkable technical achievement. This plate also tells us that in the 19 years between the publication of the two plates, *C.* × *atkinsii* had found its way from England to Belgium, where it had been grown to a standard that was well up to modern show standards.

C. × *atkinsii* from *Flore des Serres*, Plate 851 (1880). Chromolithograph (VAA)

Detail from Plate 851 (VAA)

Detail from Plate 851 (VAA)

Cyclamen persicum 'Roi des Noirs', from *Revue Horticole* (1892). Chromolithograph. No outline stone was used! Note the rich colours (VAA)

Five details from the plate of 'Roi des Noirs' from *Revue Horticole* (1892) (VAA)

Van Houtte's product might have been superb, but he had healthy competition from a French journal, *Revue Horticole*, which was founded in 1829 and published bimonthly. This also included chromolithographs of superb quality. The illustration of *Cyclamen persicum* 'Roi des Noirs' published in 1892 was produced without an outline plate, divorcing it completely from the engraving style. This, coupled with the wonderful richness of the inks, gives a wonderful impression of a vibrant, living plant. A close examination of the vignette to the lower left clearly shows the influence of the porous texture of the limestone printing plate. Ironically, the plates were printed almost in van Houtte's back yard, by the company of G. Severeyns in Brussels. I was interested to discover that the editor of *Revue Horticole* at the time, Edouard André, won a competition to design Sefton Park near Liverpool in 1867, and thereafter became a famous garden designer, working all over Europe. The journal continued its successful run until 1935.

Cyclamen cyprium from William Robinson's supplement to *The Garden* (1876). Chromolithograph? Poor quality, in effect a cheap and cheerful early 'free gift' (VAA)

Detail from above (VAA)

Detail from above (VAA)

My next choice, an illustration of *C. cyprium* was produced in 1876. Compared with some of the masterpieces already mentioned, this chromolithograph is definitely in the second rank. The texture of the colour is very 'flat' and the paper is of poorer quality. So why did I choose it? Because it was produced as a supplement to *The Garden*, founded by William Robinson. This highly successful magazine ran from 1871 to 1927. Plant portraits were a regular feature, from 1880 painted by Henry Moon, who became well-known for his naturalistic plant portraits. The *C. cyprium* plate featured here is unattributed, but is probably not one of Moon's productions as the style is unlike that of his other work. William Robinson himself was highly influential in gardening circles, and was one of the leading advocates of the naturalistic gardening style, as opposed to the highly formal and regimented style so popular in the Victorian period.

Robinson must have been a man of enormous energy, as in April 1903 he launched a further gardening journal, *Flora and Sylva*, which was published monthly. This was much influenced by the Arts and Crafts movement; it was printed on heavy-grade paper and featured superb illustrations. Robinson's aspirations were set out in his 'Foreword' written in December 1903, which appears in the bound copy of volume 1: "Witness of the hot chase after process illustrations, small type, tin-shine paper, smudge lithographs, tomb-stone weights, and the less delightful features of modern books, the spirit of old things began to move in me and led at last to my going to the printer, who did not at first see my meaning." He further explained that he "...went with the flower drawings to the best colour printer in Europe; to the paper mills that still make real paper, and found surviving a wood-engraver who understood my good artist's drawings, and so began." He was clearly not a fan of the increasing flood of books that were becoming available, made possible by the introduction of the powered printing press during the 1880s, and yearned for a mythical halcyon past in which books were lovingly hand-crafted by artisans. (But when books were so produced, one might add, their price was unaffordable by the masses!) His publication was illustrated by quality chromolithographs, supported by a greater number of wood engravings.

The August 1903 issue is of interest to cyclamen enthusiasts as it contains a review article covering the cyclamen in cultivation at that time, a total of ten species. *Cyclamen libanoticum* is depicted in a chromolithograph, and it is recorded as occurring at high elevation on Mount Lebanon. Of interest to modern cultivators is the comment: "It is a hardy kind, of easy culture, effective in autumn for its leaves, and in spring for its early flowers." At the present time, this species is grown almost exclusively as a pot plant in the cold greenhouse, but I have found it to be hardy in the garden to at least −8°C in a north-facing situation. This plate is unambiguously attributed to Henry Moon.

Cyclamen libanoticum from *Flora and Sylva* Vol 1 (1903). Chromolithograph with stipple embellishment. After printing, the plate was run through textured rollers to give it a heavy surface texture (VAA)

Although the powered press made bulk printing much more practicable, it brought new challenges for the illustrator. The higher pressure exerted by the rotary cylinders in the press meant that conventionally finished stones did not produce a clean impression on the paper. It became necessary to polish the stones to a much smoother, shiny surface, and also to use a smoother, slightly glossy paper. The use of smoother stones caused a loss of the subtle texture associated with the 'graininess' of the earlier types, but the artist rose to the challenge. Texture was re-introduced into the print by the use of stippling: fine dots and dashes applied to the stone with a very fine pen. As a further measure, to conceal the glossy surface of the paper, the print was calendared. This process involved passage under a textured roller to impress small irregularities in the surface, so as to simulate the appearance of hand-made paper. Although very effective in some of the plates, I do not consider this to be an improvement to this particular image! The technique smacks more of 'arty-crafty' than Arts and Crafts.

There is much in the accompanying *Flora and Sylva* article that is of interest to the modern cultivator. In the list of "the sorts in cultivation", *C. alpinum* is briefly described as "A dwarf plant recently found on the Taurus Mountains of Asia Minor, at a high elevation." This gives the journal a really up-to-date edge: Dammann & Co. in Naples had only recently received tubers from Turkey in about 1892. We have to presume that Robinson (if he actually grew *C. alpinum*) also obtained his plants from Italy. The story of the loss of this species from cultivation, its rediscovery in 1956 by Davis & Polunin, the assignment of the name *C. trochopteranthum*, and the return in 2002 to the former name has been recorded elsewhere in this book.

It is easy to forget that travel in Turkey over 100 years ago was difficult. Because of this, knowledge of the variation within an individual species was poor. If different forms of a species were introduced from different areas, they could all too readily acquire different names and considerable confusion could arise. And yet, an observant cultivator might grope towards the truth more readily than a botanist. The sections of the *Flora and Sylva* review on *C. Atkinsii*, *C. coum* and *C. ibericum* demonstrate the point. Extracts read as follows: "*C. Atkinsii*, A hybrid between *C. coum* and *C. ibericum*, raised by the late Mr. Atkins of Painswick. It is only a fine form of *ibericum*, from which it is barely distinguished by its larger foliage, sometimes shaded with whitish-green above, and by its larger flowers, white, rose, red, lilac or purple." This description is clearly at odds with Atkins' original description. I have not been able to trace the originator of the alternative parentage, but it looks as if garden cultivation and open pollination rapidly degraded the original production. This possibility is supported by the comments on *C. ibericum*: "This is only a finer form of *C. coum* from the Iberian Caucasus, and a plant so like *C. Atkinsii* that the two strains frequently do duty for one another." To add a final touch, it is recorded of *C. coum* that: "There are several varieties in cultivation." The difficulties of obtaining accurate distribution records are demonstrated by that reported for *C. coum*: South Europe, Asia Minor, Caucasus, Greece and Syria. In those days, there were no Cyclamen Society Expeditions to put dots on maps! The distribution for *C. neapolitanum* (now, of course, *C. hederifolium*) is better: Italy, Corsica, Greece and the south of France, and a very modern comment is added: "There are various forms and too many synonyms."

Unfortunately, Robinson's high aspirations did not receive sufficient support, and a host of new printing techniques were bringing down costs (and quality too, he might have said) at a rapid rate, so the last edition of *Flora and Sylva* rolled off the press in December 1905. The plate of *C. libanoticum* was the first, and only, cyclamen to be illustrated. One of the technical factors leading to the demise of Robinson's publication was the increasing popularity of the halftone process, first used in a newspaper in the USA in 1873. This technique had rapidly become popular in the UK, and by the 1890s, was being used increasingly for the printing of pictures based upon photographs. The process depended upon two principles. The first was that a photographic image could be broken up into a series of dots of variable size by the use of a screen. At first, the screen was made by scribing parallel lines on two glass plates and gluing them together at right angles. The image, now consisting of dots, was transferred to the surface of a sensitised metal plate (the chemistry need not concern us) and the plate etched. A black spot on the original image generated a large, deep pit in the printing plate, and a white spot no pit. Thus when inked, the single plate printed a continuous range of tones, from white, through greys to black, using just black ink. The advantages were un-opposable: the process could be highly automated, it was ideal for use with high-speed, high pressure presses, and was thus cheap. I can understand Robinson's dismay on seeing the loss of the craft skills and artistry of the engraver, etcher and lithographer. It is thus ironic that a new gardening journal was launched in 1929 under the editorship of E. H. M. Cox, illustrated entirely by the halftone process. Its title: *New Flora and Sylva*!

Although the halftone process gave a product that was 'cheap and cheerful' and ideal for newspapers, magazines and the like, the quest for a similarly economical colour process was soon underway. This process was ultimately derived from the original technique by adding to the use of the halftone screen three colour filters — red, blue and green — which were used in turn when the original artwork was photographed. This allowed the production of three printing plates in addition to the black one, corresponding to the three primary colours. The alignment of the screen was rotated slightly each time, so that when printed in cyan, magenta and yellow (and, of course, black), the dots did not overlap. If the printed dots are sufficiently close, and sufficiently small, the eye adds these ink colours together to give the other colours. Yellow + magenta, for example gives red. In spite of many technical improvements in the process, this colour-printing method is used almost universally to this day. If you examine almost any printed material with a magnifying glass (including this book), the coloured spots and 'rosettes' created by the screen rotation are easily seen. Naturally, the quality of the product depends on many factors, perhaps the most important being the size of the dots; 60–120 dots per inch is sufficient for newspapers, but a density of 130–300 dots per inch is needed for better-quality products. An increased dot count alone will not ensure a better product though; the finer the dot count, the smoother the paper surface needs to be, whether this is achieved by surface coating or mineral-filling.

It must be said that the quality of the colour reproduction could be 'variable'. An example is the illustration of *C. europaeum* (now *C. purpurascens*) by Henry Correvon published in *Alpenblumen* in 1917. The plates are based on watercolours by Philippe Robert and are very distinctive. Fortunately, for my purposes, the cyclamen plate is one of the better ones. The flowers and leaves are artistically arranged, yet retain the character of the plant, and the colours of the flower, leaf and tuber are accurately reproduced. What lets this plate down in my eyes is the peculiarly dull toning of the background, which is rendered more obvious by contrast with the high gloss of the paper. The production of muddy, dull colours was a hazard of the process, and occurred when the dots of translucent ink were printed on top of each other, rather than side by side. Perhaps the artist was unfortunate in choosing for his background wash precisely the colour that would give the poorest result in the three-colour process? Common names are listed below the illustration: 'Cyclamen d'Europe', 'Violett des Alpes', 'Erdscheibe', 'Alpenveilchen' and 'Common Cyclamen'. This multiplicity of names is surely an excellent argument for the use of Latin binomials?

Cyclamen purpurascens (*europaeum*) from *Alpenblumen* (1917). Three-colour halftone. Demonstrates good and bad aspects (VAA)

I return to *New Flora and Sylva*, in which an article on cyclamen appeared in the July 1933 edition, penned by F. C. Stern. He opens with the observation: "Cyclamen species are a fascinating race of plants." While we would all agree with his sentiments, his choice of words is a little odd, perhaps. The quality of the halftone plates is best described as 'variable'. The illustrations of *C. coum album* and *C. hederifolium* are dark but adequate (although the latter is captioned *C. repandum*!), but the plate of *C. persicum* in Cyprus is impressively poor.

Stern's article contains a perplexing comment on *C. coum* and *C. ibericum*:

> "These two species are easily mixed up, though they are quite easily distinguished. *C. coum* has small kidney-shaped leaves, rounded at the top, of a dark green colour with no markings. The flower-stem when the seed-head is formed rolls up like a cork-screw. *Cyclamen ibericum* has the same shaped leaves but rather more pointed at the top; the leaves though dark green have always silver markings on the stem. Another difference is that the flower stem of *C. ibericum* when in seed does not roll up."

Detail from above (VAA)

Cyclamen coum from *New Flora and Sylva* (1933). Halftone. Poor contrast; a merely adequate product (VAA)

Detail from above (VAA)

Was Stern actually growing a plant that we no longer have today, or did he simply complete his article by slotting in (incorrect) information that was not based on his own experience? Perhaps he should have thought more deeply about the implications of his comment that, "It is difficult to keep these two from hybridising if they are grown near each other in the garden." These two names seem to have given gardeners a problem over more than 70 years, and their purported hybrid even more, for Stern records that: "There is a hybrid between them called *C.* × *Atkinsonii*" (sic). This fascinating article is full of certainties and observations that give pause for thought when compared with our modern knowledge. Those interested in discussions on climate change might like to ponder the significance of the observation on *C. hederifolium*: "The pink flowers come out about September 1 and last till November."

Whereas Robinson's *Flora and Sylva* was brought to an end by technical progress and economic pressures, the fate of *New Flora and Sylva* was sealed by the Second World War, the last issue appearing in August 1940.

By the 1930s, the general economic climate and the advance of technology were also making the position of *Curtis' Botanical Magazine* precarious. It became increasingly difficult to find diligent hand-colourists prepared to work on runs of several hundred plates (and difficult to pay them too!). The war brought an end to the long deliberations, and 1948 saw the first volume of the *Botanical Magazine* to be illustrated using four-colour halftone.

By the middle of the twentieth century, the photograph coupled with the halftone process and its various colour derivations reigned almost supreme, but some nice examples of original artworks were reproduced in *Collins Guide to Bulbs* published in 1961. The two plates of cyclamen in that book are reproductions of fine watercolours painted by Rear-Admiral Paul Furse, a famous plant explorer with a major interest in bulbs. They demonstrate the fact that a good artist can cram more information into a given space than a photographer. The essence of the plants is captured, and the artist has filled all areas of the plate. It is unfortunate that in some editions, the quality of the printing is not as good as might be wished.

Detail from image right (VAA)

Several *Cyclamen* species are shown in this Paul Furse plate, published in *Collins' Guide to Bulbs* (1961) (VAA)

Cyclamen libanoticum by Ann V. Webster, from *Curtis's Botanical Magazine* (1957). A good-quality 'modern' screen print. Compare with the plate from *Flora and Sylva* on page 477 (VAA)

Happily, high-quality reproduction of original artworks still continues in the *Botanical Magazine*, which holds fast to traditional values while using modern printing techniques. A good example is the plate of *C. libanoticum* published in 1957 (and the comparison with the *Flora and Sylva* illustration is interesting). The flower colours, for example, are remarkably fresh, and capture the naturally vibrant quality of the living plant. Recognising the rather untidy appearance of the plant, in cultivation at least, the artist has filled the available space in a nicely balanced fashion, and the detail is good thanks to the high dot frequency. In more recent times, the quality of reproduction has been maintained, but having loved the prints made in the past on slightly creamy-coloured matt papers, I am still slightly shocked by the very glossy surface and brilliant whiteness of the paper used in more recent times. It is sad, in a way, that Curtis's magazine finally had to succumb to the advance of technology, but although the crafts of the etcher, engraver and lithographer are no longer employed, that of the watercolour artist, encapsulating the essence of the living plants, remains. Long may it survive!

The 25 superb watercolours by Pandora Sellars and Christabel King reproduced in this book are a twenty-first century landmark in the history of cyclamen illustration.

Detail from above (VAA)

CHAPTER SEVEN

A Celebration of cyclamen in art

Martyn Denney & Brian Mathew

In the early stages of the preparation of this book, it was decided that artwork should play a significant role, hence the commissioning of superb new watercolours by Pandora Sellars and Christabel King.

As we have seen in Chapter 6, which covers botanical illustration, cyclamen have been highly popular with artists over the centuries. That chapter explored, more or less chronologically from the early Herbals onwards, the way in which cyclamen have been depicted and the techniques used to prepare and print the images. The examples presented there were selected specifically to illustrate the various points under discussion.

Recognising that there are many other superb examples of cyclamen paintings, we have brought together some of these, albeit a rather random selection, in this chapter as a celebration of the rich and varying styles of talented artists who have depicted cyclamen over the years.

Cyclamen in Ottoman Turkey

Certain plants feature very prominently in illustrations from Ottoman Turkey, most notably tulips, narcissus, hyacinths, roses and carnations. These appear frequently on the superb Iznik tiles of the thirteenth and later centuries, sometimes in forms that suggest there had already been much selection and perhaps hybridisation of cyclamen. Cyclamen appear infrequently, in fact illustrations of them from Ottoman times are hard to find. Nevertheless, the research by Yıldız Demiriz for her book *Osmanlı Kitap Sanatında Doğal Çiçekler* (*Naturalistic Flowers in Ottoman Manuscripts*), published in 2005, revealed some interesting early images. One work she consulted was an *Astrolojik Almanak* by Mehmet Üsküdari, dated 1620, a copy of which is in the Chester Beatty Library, Dublin. Each of the pages of astrological charts is adorned in the margin with flowers — mostly those listed above — and on one page by a cyclamen. Although somewhat stylised in that the leaves are shown to be attached to the flower stalk, the flower itself is unmistakable and most likely represents a wild form of *C. persicum*, which has pink flowers with long, narrow twisted petals. There are other *Cyclamen* illustrations in Yıldız Demiriz's book, sourced from various eighteenth-century Ottoman works, and most are similar to the one reproduced here; one shows a pure white form, probably also of *C. persicum*.

Cyclamen in flower paintings

Cyclamen feature frequently in sumptuous still-life paintings produced over the centuries by great flower artists. Although the bouquets are in a sense artificial in that they consist of a great many different species and varieties that flower at different times of the year, they are nevertheless painted with great accuracy and are immediately recognisable. As a representative of this genre, we could do no better than to reproduce Jan Brueghel the Elder's 'Flowers in a Wooden Vase'. This painting depicts about 130 different flowers, including cyclamen, in the bouquet; a complete cyclamen plant lies on the table to the left of the vase, with tuber, leaves and flowers clearly displayed.

Cyclamen illustrated in an Ottoman Turkish calendar, the *Astrolojik Almanak* by M. Üsküdari (1620). ©The Trustees of the Chester Beatty Library, Dublin

Jan Brueghel the Elder (1568–1625). Oil on panel. About 130 different flowers are depicted, with a whole cyclamen plant lying on the table in the bottom left corner. Reproduced by courtesy of the Kunsthistorisches Museum, Vienna

Detail from above, showing *C. hederifolium* with tuber

"Ciclamen, panis porcinis, arthanita, erdapfl, sowbrot or schweinbrot together with *Campanula celesta* or *Gentiana minor*" from *Arzneipflanzenbuch*, a manuscript on paper (31 cm × 21 cm). Reproduced with the permission of the Bavarian State Library, Munich (BSB Cod. Icon 26 fol. 7r)

Cyclamen from a sixteenth century plant book

On paper 31cm × 21 cm. The *Arzneipflanzenbuch* manuscript was probably produced in Augsburg between 1520 and 1530, though Blunt and Raphael suggest that it could have been produced in northern Italy by a German artist around 1500. The text is in Latin, German and Italian and the book features an illustration that includes a painting of a cyclamen entitled "Ciclamen, panis porcinis, arthanita, erdapfl, sowbrot or schweinbrot together with *Campanula celesta* or *Gentiana minor*".

John Sibthorp, Ferdinand Bauer, James Edward Smith and James Sowerby and the creation of *Flora Graeca*

Martyn Denney

John Sibthorp was born on 28 October 1758 in Oxford, where his father, Dr Humphrey Sibthorp, younger son of John Sibthorp of Canwick Hall, Lincolnshire, was the Sherardian Professor of Botany at the University of Oxford. Little is known of Sibthorp's early life, but he was educated at Magdalene College School, Oxford and then at Lincoln Grammar School. He went on to study medicine at Oxford, graduating with a B.A. in 1777 and an M.A. in 1780, the year his mother died. Sibthorp's mother had been wealthy in her own right following the death of her own father, and left him a handsome inheritance including an estate in Devon. Having been elected Radcliffe travelling fellow of University College, Sibthorp went to the University of Edinburgh to study medicine. After graduating as M.B. at Oxford in 1783, he continued his studies at Montpellier, where he was elected a member of the Academy of Sciences. Subsequently, Sibthorp studied botany at the Jardin du Roi on the left bank of the River Seine in Paris. He spent the next three years continuing his education in medicine and botany in Edinburgh, Montpellier and Paris, returning to Oxford in October 1783. In 1784, Sibthorp's uncle died and his father, on succeeding to the Canwick property, resigned the Sherardian professorship. John Sibthorp succeeded his father in this position.

Looking for a way to enhance his international reputation and having considerable personal financial resources, Sibthorp conceived the idea of producing a Flora of the western part of the area occupied by the Ottomans at that time. He would be documenting an area known to the first-century Greek herbalist Pedanius Dioscorides from Cilicia (now in Turkey, but at that time part of the Roman Empire) upon whose five-volume work most herbals from the sixteenth, seventeenth and eighteenth centuries were still based, but he would include many new plants. Unusually for the period, Sibthorp decided to base his Flora on original fieldwork and to collect his own specimens, rather than to simply regurgitate the material of others. In preparation, while on his way to 'Greece' in 1784, Sibthorp visited the Imperial Library in Vienna to study the *Juliana Anicia Codex*, a Dioscorides manuscript produced in Istanbul around 512 AD. Whilst in Vienna, Sibthorp met with Baron Nicholaus von Jacquin, the Professor of Botany and Chemistry at Vienna University, who introduced him to the botanical artist Ferdinand Bauer.

Ferdinand Bauer was born in Feldsberg, Austria (now Valtice in the Czech Republic) on 20 January 1760, the youngest son of Lukas Bauer, court painter to the Prince of Lichtenstein. Although Lukas died when Ferdinand was just one year old, his talents lived on in his three sons: Francis becoming botanical painter to King George III at Kew, and Joseph becoming curator of the Prince of Lichtenstein's art collection. After their father's death, the three orphaned Bauer brothers were taken in by Father Norbert Boccius, Prior of the monastery and hospital of *Fratrum Misericordiae* (Merciful Brothers) in Feldsberg, who taught them to draw. A keen botanist, Boccius engaged the brothers to work on illustrations for his *Liber Regni Vegetabilis Retinens Plantas ad Vivum Pictas*, a 14-volume florilegium with 2,750 illustrations. In 1780, the three brothers moved to Vienna where Franz and Ferdinand were employed as botanical artists by Jacquin. At that time, Jacquin had just published his *Florae Austriacae* and would have been working on his *Icones Plantarum Rariorum*, which was to be published in three volumes between 1781 and 1793.

Sibthorp was so impressed by Ferdinand Bauer's work that he travelled to Feldsberg to see the paintings that Bauer had done for Boccius, and employed the young artist to accompany him on his expedition on a salary of £80 per annum (later increased to £100). While Sibthorp collected plants and made notes, Bauer pressed the herbarium specimens and made sketches of them. Many collections were made each day, so for speed and to limit the pigments and paper that he had to carry, Bauer had developed a system of recording colours on pencil sketches using a colour chart. Initially this consisted of just 140 colours but later versions extended to about 1,000 colour shades. Paper was handmade and expensive in the eighteenth century and it had to be kept dry in inclement weather. Consequently, even the paper required for the pencil sketches was purchased during the trip.

John Sibthorp and Ferdinand Bauer left Vienna in March 1786 for Trieste, from where they sailed to Venice. They travelled south through Italy via Bologna to Florence, and then by a circuitous route via Livorno (Leghorn) to Pisa. The following extract is from Sibthorp's diary for 30 March 1786 (MS Sherard 214 fol 6.):

> "The 30th. Left Leghorn having engaged a Muleteer with two Mules & a Horse to attend me. The Horse proved so bad that he fell down with me three times. This retarded our Journey & we arrived late at Pisa. On the way observed Cyclamen N.S. 73. Cheiranthus Cheiri. 74 (on the Walls of Pisa) Euphorbia Helioscopia 75."

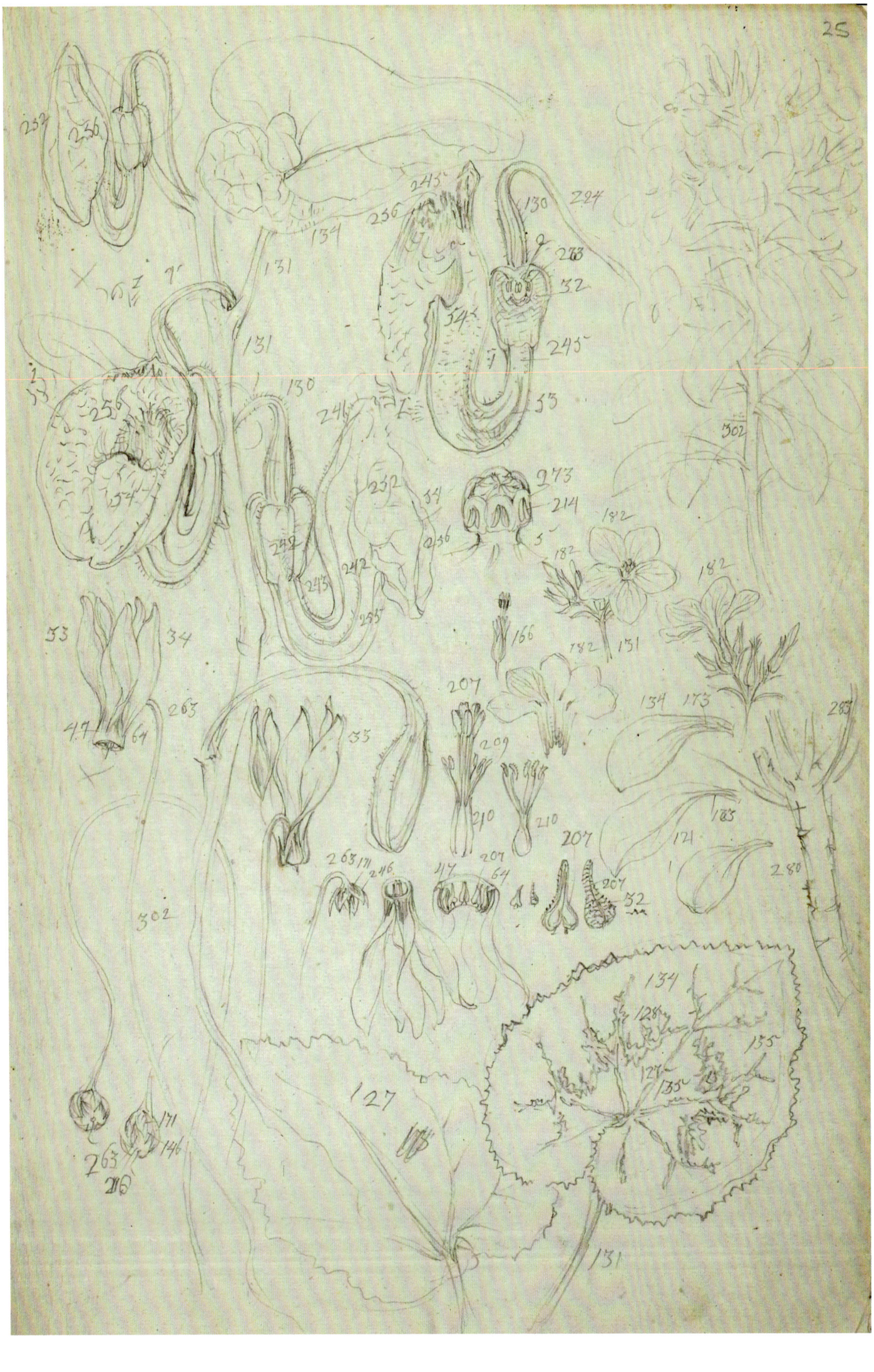

Ferdinand Bauer's field sketch of *C. persicum* drawn on the island of Cyprus in May 1787. It was subsequently used to create Bauer's painting of *C. latifolium*. Reproduced with the permission of the Sherardian Library of Plant Taxonomy, The Bodleian Library, University of Oxford. (MS Sherard 247 fol 25r)

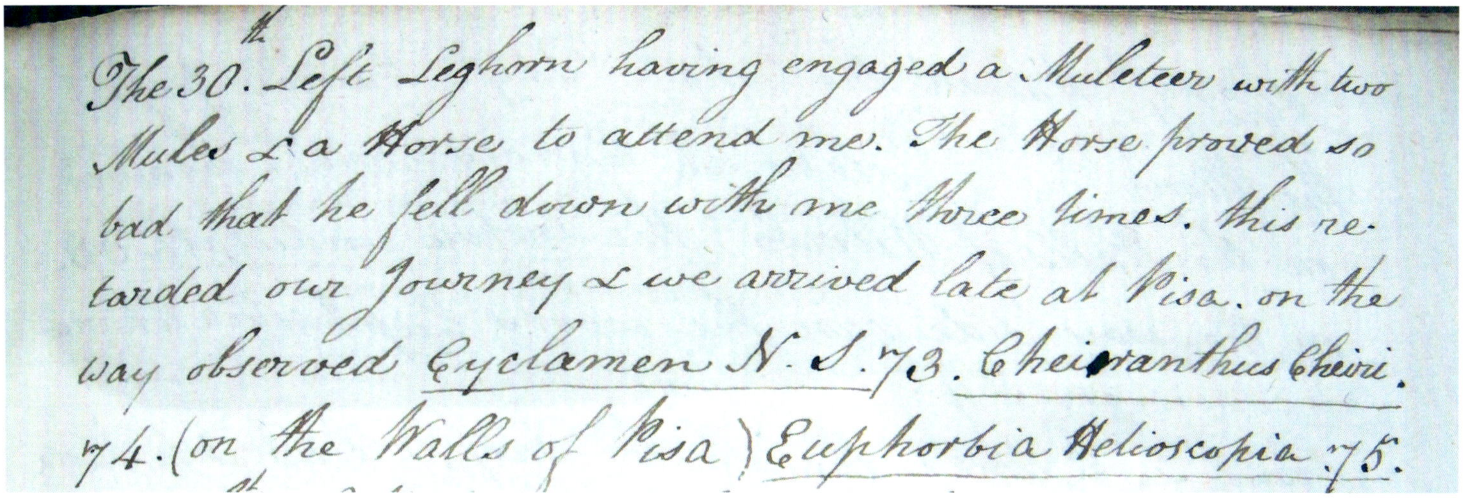
The 30th. Left Leghorn having engaged a Muleteer with two
Mules & a Horse to attend me. The Horse proved so
bad that he fell down with me three times. this re-
tarded our Journey & we arrived late at Pisa. on the
way observed Cyclamen N. S. 73. Cheiranthus Cheiri.
74. (on the Walls of Pisa) Euphorbia Helioscopia 75.

An extract from John Sibthorp's diary for 30 March 1786, noting the discovery of *C. repandum*. 'Leghorn' is Livorno, Italy. Reproduced with the permission of the Sherardian Library of Plant Taxonomy, The Bodleian Library, University of Oxford (MS Sherard 214 fol 6)

The new species of *Cyclamen* referred to was *C. repandum*. The numbers are those of herbarium specimens that were collected by Sibthorp but unfortunately they have not survived.

From Pisa, Sibthorp and Bauer continued south through Sienna, San Lorenzo and Rome to Naples, from where they sailed for Crete via the Straits of Messina and Milos. They spent much of the summer of 1786 on the island, before travelling north to Athens and then on to Constantinople (Istanbul), where they spent the winter. Had their visit to Crete been earlier in the season, perhaps Sibthorp would have encountered and recorded *C. creticum*. In the spring of 1787, they sailed east via Lesbos, Chios, Kos and Rhodes to Cyprus, where they spent five weeks studying the fauna and flora.

While in the Paphos area on the 10 and 11 May 1787, Sibthorp noted in his diary (MS Sherard 215):

> "I observed … the prickly Chicory, the Silene fructosa, the Cyclamen cypricum & the Ruta graveolans grew on the Rocks in the Road from Pafos to Iktinia."

Sibthorp's reference to *C. cypricum* clearly refers to *C. persicum* as it is the only spring-flowering *Cyclamen* species on Cyprus. This record was described by Lack as "… another major discovery by Sibthorp …" because at the time, *C. persicum* was known only in cultivation and was thought to have originated from Persia.

Sibthorp collected herbarium material and living plants, some of which were grown in the Royal Gardens at Kew, and these were attributed to Sibthorp by William Aiton in his *Hortus Kewensis* of 1789:

> "*persicum* 3. C. foliis oblongo-ovatis cordatis crenatis
> Cyclamen persicum *Mill. dict. Curtis Magaz.* 44
> Persian Cyclamen.
> *Nat.* of Island of Cyprus. *John Sibthorp*, M.D.
> *Cult.* 1731. *Mill. dict. Edit.* 1. *n.* 5.
> *Fl. February–April.*"

Leaving Cyprus, Sibthorp travelled to Athens, Thessaloniki and Mount Athos, before heading for Patras in the Peloponnese to board a ship, which reached England in December 1787. He returned to Greece in March 1794, visiting Olympos, Mount Athos and some of the islands before returning to Athens. In 1795, he toured the Peloponnese, Zante, Kephalonia and Prevesa, where he caught a 'cold' that turned out to be tuberculosis. Sibthorp travelled by ship to Ancona in Italy and then returned to England overland. He went to his estate in Devon to recuperate but feeling no better, travelled to Bath, where he died on 8 February 1796. He was buried in Bath Abbey.

In his will, Sibthorp left his books, along with an estate at South Leigh, Oxfordshire, to Oxford University. The proceeds were used first to publish his *Flora Graeca* and then to endow a chair in Rural Economy. Sibthorp's executors appointed Dr James Edward Smith to complete *Flora Graeca*. Smith first published the *Prodromus* — a two-volume octavo unillustrated Flora that secured the publication of Sibthorp's new species — and thus it is Smith's name that appears after *C. repandum*. The first volume of the *Prodromus* appeared in 1806 and the second in 1813. Smith completed the first volume of *Flora Graeca* in 1806, and the second (which contains *Cyclamen*) in 1813. Further volumes followed, six being completed before Smith died in 1828 and a seventh shortly afterwards in 1830. Volumes 8, 9 and 10 were edited by John Lindley and published between 1833 and 1840. Only 30 copies of the complete work were printed and the total cost was over £30,000.

Ferdinand Bauer produced 966 paintings from his field sketches, and James Sowerby was engaged to produce engravings from them. Sowerby was well known as a botanical engraver. He produced the illustrations for L'Héritier's monograph on geraniums *Geranologia* and, from 1787, he had produced engravings for *Curtis's Botanical Magazine* (including the well-known engravings of *C. coum* (see page 514) and *C. persicum*) (see page 465), but he is perhaps best

Ferdinand Bauer's drawing of *C. latifolium* (*C. persicum*) for *Flora Graeca*, after his own field sketches. Reproduced with the permission of the Sherardian Library of Plant Taxonomy, The Bodleian Library, University of Oxford (MS Sherard 244 vol. 4 fol. 176)

known for his illustrations in *The English Botany; or Coloured Figures of British Plants*, which were accompanied by descriptions by James Edward Smith and published in 36 volumes from 1790 until Sowerby's death in 1822. It is therefore unsurprising that Smith chose Sowerby for the *Flora Graeca* work. Sowerby's engravings on copper plates of the two *Cyclamen* were precise, faithful copies of Bauer's paintings and therefore the plates in *Flora Graeca* are mirror images of the paintings.

The beautiful engravings of *Cyclamen* appeared in volume 2 of *Flora Graeca* published in 1813, but there are descriptive problems with both of them. *Cyclamen repandum* was collected in Italy, and doesn't grow in Greece; it shouldn't therefore appear in either the *Prodromus* or in *Flora Graeca*. This species was, however, new to science and we have to be grateful for Bauer's wonderful painting and for Sowerby's engraving, which is the 'type' for the species. *Cyclamen latifolium* is clearly *C. persicum*, but in the *Prodromus*, Smith erroneously refers to the engraving that appears in *Flora Graeca* within the description of *C. hederifolium* and the accompanying text in the *Flora* is consistent with this, clearly stating: "C. hederifolium. *Prodr.* v.1.128; excluso synonymo Aitoni."

James Sowerby's 1813 engraving of *C. repandum* from *Flora Graeca*, after a painting by Ferdinand Bauer. This engraving was designated the lectotype of *C. repandum* by Compton & Culham, *Journal of the Cyclamen Society* 27(2): 70 (2003). Reproduced with the permission of the Sherardian Library of Plant Taxonomy, The Bodleian Library, University of Oxford (MS Sherard 761 v2, tab 186.)

'Cyclamen Europæum' (*C. purpurascens*) from Baron Nicholas Joseph Jacquin's *Floræ Austriacæ* Vol. 5, tab 401 (1778). The same illustration was used by Joseph Jacob Plenck in his 1788 *Icones Plantarum Medicinalium*. Both works were published in Vienna. By permission of the Trustees of the Natural History Museum (London)

'*Cyclamen*' from *Lust- und Artzney-Garten des Königlichen Propheten Davids. Das ist: Der Gantze Psalter* by Johann Gerharden (1675). Copper-plate engraving by Georg Christoph Eimmart the younger. The plates were printed on both sides; on the front with an emblem that has a legend in Latin and German verse; on the reverse, with a plant, that has a legend in German alone. In general, the emblems were used to clarify some of the terms found in the Psalter. The plants were for decoration. The psalms were translated by Wolfgang Helmhard Hohberg, presumably from Latin. Hohberg was an Austrian nobleman who was born in Ober-Thumritz, Austria, in 1612. The greater part of the music is by Hieronymus Gradenthaler and the paraphrase of the Psalms is by J. L. Prasch. The manuscript was printed in Regensburg by Fischer and Freysinger (MD)

'Cyclamen Bipenninum Flore Carneo' from *Nauwkeurige Beschryving der Aard-gewassen* by Abraham Munting, Utrecht and Leiden (1696). Abraham Munting was head of the Botanical Garden at Groningen and Professor of Botany at Groningen University from 1658. The book was translated into Latin and published by Franz Kiggelaer in Amsterdam and Leiden as *Phytographia Curiosa, Exhibens Arborum, Fruticum, Herbarum & Florum Icones* in a smaller folio format in 1702. All of the engravings show plants inserted into a landscape, with the plant names shown on trailing ribbons or engraved into the side of stone blocks or plant pots. There are eight engravings of *Cyclamen*, five showing a single plant with the tuber suspended over a landscape, and a further three that show two plants, one above the other (MD)

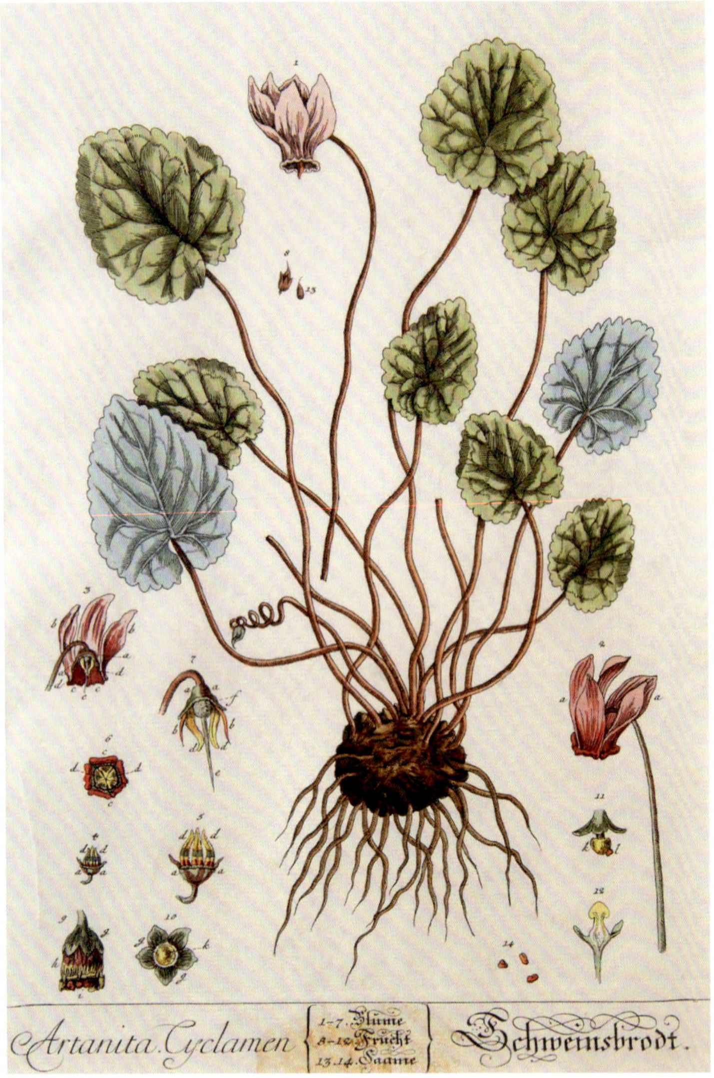

'Artanita Cyclamen' (*C. purpurascens*) from *Herbarium Blackwellianum* published in 1757 by Christopher Jacob Trew (MD)

'Cyclamen Europeum' (*C. hederifolium*) from '*La Botanique mise à la portée de tout le monde ou collection des plantes d'usage dans la médecine, dans les alimens et dans les arts*' by Nicolas François Regnault (1774). Drawing and hand-coloured copper-plate engraving by Genevieve de Nangis Regnault. This print has been designated as the lectotype for *C. hederifolium* (see *The Cyclamen Society Journal*, June 2008) (MD)

Elizabeth Blackwell's *C. purpurascens*

Elizabeth Blackwell was the daughter of a successful merchant from Aberdeen. At the age of 28, she secretly married her cousin, Alexander, and settled in Aberdeen, where Alexander maintained a medical practice. Although his education was sound, a question arose concerning Alexander's credentials, which resulted in the young couple's fast move to London to escape charges that Alexander was practicing under a bogus license. Alexander decided to establish his own printing house, even though he did not belong to the guild and had not served the required apprenticeship as a printer. Consequently, he was subject to heavy fines that forced him to close shop. Elizabeth's dowry was exhausted, and Alexander's free-spending habits and fines meant that a substantial debt was incurred, and it could not be paid. Alexander found himself in a debtor's prison.

Elizabeth must have been desperate at this point: her husband was in jail and she had no source of income, but she had a household to support and now a child to care for. Fortunately, she learned that a new herbal was needed to depict and describe exotic plants from the New World. She decided that she could illustrate it, and Alexander, given his medical background, could write the descriptions of the plants. The project received the support of the Society of Apothecaries and several leading doctors. As she completed drawings, Elizabeth would take them to her husband's cell, where he supplied the correct names in Latin, Greek, Italian, Spanish, Dutch and German. Unlike her husband, Elizabeth was unskilled in botany. To compensate for this, she was aided by Isaac Rand, then curator of the Chelsea Physick Garden, where many of these new plants were under cultivation. The *Herbal* was issued in weekly parts between 1737 and 1739, each part containing four illustrated plates and

'Cyclamen linearifolium' from Augustin-Pyramo de Candolle's *Icones Plantarum Galliae Rariorum* (1808). One example of some of the fanciful plants that have been illustrated, though they never existed in nature. Candolle described it as the 'Grasblätteriges Cyclamen' (grass-leafed cyclamen). *Cyclamen linearifolium* also featured in Leopold Trattinick's *Archiv der Gewachskunde* (1818) alongside *C. europaeum*, *C. europaeum* var. *descendens* and *C. coum*, the last of which was illustrated using the same image as the 1787 Plate 4 from *Curtis's Botanical Magazine*. By permission of the Trustees of the Natural History Museum (London)

a page of text. Not only did Elizabeth do the drawings, she also engraved the copper printing plates for the 500 images and text, and hand-coloured the printed illustrations. The complete work was published in two volumes and entitled "*A Curious Herbal containing five hundred cuts of the most useful plants, which are now used in the practice of physick, to which is added a short description of ye plants and their common uses in physick*".

Income from the book bought Alexander's release from prison: he did not, however, handle his freedom well. Debts accumulated again, forcing the couple to sell some of the publication rights to the book. It was re-issued in 1757 in Nuremberg in a revised and enlarged format as *Herbarium Blackwellianum* with German text by Christopher Jacob Trew, a botanist and physician. The engravings were done by N. F. Eisenberger.

Cyclamen persicum from *The Botanical Cabinet*. Conrad Loddiges & Sons had a nursery in Hackney, East London. They started publishing an illustrated catalogue of the nursery's plants from around 1780, and this grew into *The Botanical Cabinet*. The monthly magazine featured ten hand-coloured illustrations and ran from 1817 to 1833, totalling 2,000 plates. The plates were drawn by the author George Loddiges himself, his daughter Jane, his brother William, local engraver George Cooke and two members of his family (Edward Cooke, who became a leading Victorian artist, and W. I. Cooke), apprentice engravers T. Boys and William Miller (who later became principal engraver to the artist J. M. W. Turner), and a Miss Rebello. All of the plates were engraved on copper by George Cooke. The magazine featured four *Cyclamen*: *C. hederifolium*, drawn by T. Boys, *C. persicum* and *C. coum*, drawn by George Loddiges and *C. repandum* drawn by George Cooke (MD)

'Cyclamen neapolitanum' (*C. hederifolium*) from *The Botanical Register*. A hand-coloured copper-plate etching by G. Barclay, after a painting by Miss S. A. Drake. *The Botanical Register* was started by Sydenham Edwards in 1815, four volumes being issued before he died on 8 February 1819. A further nine volumes were issued under the editorship of the publisher, James Ridgeway, and a further 19 (up until 1847) after Dr John Lindley was appointed editor. This print appeared in 1838 during John Lindley's term of office (MD)

'Cyclamen d'Europe' (*C. persicum*) from *Phytographie Médicale, Histoire des Substances Héroiques et des Poisons* by Joseph R. Roques. Drawing and hand-finished colour stipple engraving by Edouard Hocquart. The work was published in 36 parts between 1821 and 1825. Hocquart, a Professor of Mathematics and Botany at the College d'Ath, drew and engraved the fine plates, which were exquisitely colour-printed with some hand finishing. All of the very rich colours were applied directly to the printing plate for the printing of each sheet, this laborious technique yielding remarkably beautiful results (MD)

'Cyclamen' from *Recueil de Planches de Botanique de L'Encyclopédie* Vol. 1, plate 100 (1823). (Plates to accompany *Encyclopédie Méthodique*: *Botanique 1783–1808* by Jean Baptiste Pierre Antione de Monet de Lamarck). The individual flower parts in this engraving were copied from Joseph Pitton de Tournefort's *Institutiones Rei Herbariæ*, which was published in Paris in 1700. They were also used by Egbert Buys in his *Nieuw en Volkomen Woordenboek van Konsten en Weetenschappen*, published in Amsterdam in 1770 (MD)

'Cyclamen europeaum' (*C. purpurascens*), drawing and hand-coloured copper engraving by Friedrich Guimpel from *Getreue Darstellung und Beschreibung der in der Arzneykunde gebräuchlichen Gewächse wie auch solcher, welche mit ihnen verwechselt werden können*, Vol. 13, tab. 8 (1837).
This journal was originally published by Friedrich Gottlob Hayne. After Hayne's death in April 1832, it was carried on by Drs J. F. Brandt and J. T. C. Ratzeburg. There were 13 parts published between 1805 and 1837. Friedrich Guimpel was born in Berlin on 1 August 1774 and was a botanical artist and engraver who became Professor at the Berlin Academy of Arts. It was his work for Karl Ludwig Willdenow on *Hortus Berolinensis* that made his reputation, and it is likely that it was through Willdenow that Guimpel met Hayne (MD)

Cyclamen hederifolium from *British Phænogamous Botany* Vol. 6, plate 505 (1843). This journal was published by Whittaker, Treacher and Co., London, for William Baxter, Curator of the Oxford Botanic Garden, in order to "... supply the lovers of Botany with a set of Figures which may be relied on for accuracy". Baxter chose one species from each of the 500 genera of flowering plants then known, so that they would fit into the six volumes he produced between 1834 and 1843. Baxter said of this print: "The accompanying plate is from a very beautiful drawing by Mr Isaac Russell, botanical draughtsman and glass painter, of Oxford, from a specimen kindly communicated to me by Mr Walter W. Reeves, of Farnham, Surrey, from its station near Sandhurst, Kent." The copper plate was engraved by Charles Mathews (MD)

'Cyclamen vernum' (*C. repandum*) chromolithograph by Guillaume Severeyns from *Neerlands's Plantentuin* Vol. 1, plate 3 (1865) by Dr Cornelius Anthonie Abraham Oudemans. Printed in Groningen by J. B. Wolters. Oudemans was Professor of Medicine and Botany at the Atheneum in Amsterdam; later, in 1877, he was appointed Professor of Botany and Pharmacology at the University and director of the Botanical Garden. Oudemans' book had a total of just 52 plates and was published in three volumes between 1865 and 1867. Volume 3 (1867) contained *C. coum*, *C.* × *atkinsii* and *C. ibericum* (Plate 44) and *C. repandum* (Plate 45), both of the chromolithograph plates were by Emrik & Binger (MD)

The first of two plates showing cyclamen from Joseph Kratz's *Primulaceen* (1861). The book covers *Primula*, *Auricula* and *Cyclamen*. Kratz was at heart a 'splitter' who described many 'species' named by others plus a few of his own, the majority of which have never been accepted. This plate shows: *C. autumnale*, *C. (macrophyllum) africanum*, *C. subhastatum*, *C. silicicum*, *C. europaeum*, *C. neapolitanum*, *C. europaeum album*, *C. hederifolium latifolium*, *C. coum pulcherrimum*, *C. coum carneum* and *C. repandum*. By permission of the Trustees of the Natural History Museum (London)

The second of two plates showing cyclamen from Joseph Kratz's *Primulaceen* (1861). This plate shows: *C. aleppicum*, *C. persicum album*, *C. persicum*, *C. persico-europaeum rubrum flore-pleno*, *C. persico-europaeum erectum flore-pleno*, *C. aleppicum maculatum elegans*, *C. ibericum*, *C. Atkinsii*, *C. Roebbelenianum*, *C. hybridum guttatum* and *C. hederifolium ficariaefolium*. By permission of the Trustees of the Natural History Museum (London)

Primulaceen by Joseph Kratz

Joseph Kratz was almost certainly a nurseryman from Tübingen in Germany. In 1861 he published '*Primulaceen. Beschreibung der in unsern Gärten eingebürgerten Gattungen Primula, Auricula und Cyclamen. Ein handbuch für freunde dieser lieblichen Zierpflanzen*' (Primulaceae. A description of genera naturalised in our gardens *Primula*, *Auricula* and *Cyclamen*. A guide for friends of these lovely ornamental plants). Kratz described 52 different species, subspecies and forms or varieties and whilst he wasn't responsible for the majority of these cyclamen names himself, he was at heart a 'splitter'. In the form he gave them, just five of the names are accepted today (*C. coum* Mill., *C. persicum* Mill., *C. hederifolium* Ait., *C. africanum* Boiss. & Reut., and *C. repandum* Sm.) though in addition he listed a small number of taxa that were nearly correct: *C. silicicum* Boissier (a mis-spelling of *C. cilicium* Boiss. & Heldr.), *C. europaeum purpurascens* Mill. (*C. purpurascens* Mill.) and *C. hederifolium graecum* Lk. (*C. graecum* Link.).

Kratz described two new taxa of his own:

Cyclamen Röbbelenianum, supposedly the impossible (outside a laboratory) cross between *C. persicum* and *C. macrophyllum*. The latter is often considered a synonym of *C. africanum* and although Kratz described it as coming from south Europe and north Africa, it was named by F. W. Sieber in 1823 in a review of Cretan specimens held in the Tournefort herbarium in Paris and the name could therefore refer to what is known today as *C. confusum* (Grey-Wilson) Culham, Jope & Moore.

Cyclamen hederifolium grandiflorum which he described as: 'Flower. Very large, the petals wide 1½ inches long, white with a hint of pink, the throat dark pink; Leaves: 2-3½ inches across with markings as *C. hederifolium* but with brighter middle zones' – presumably a horticultural selection or cultivar.

Whilst his work contained only two engravings of cyclamen, a brief look in *Index Londinensis* shows that Joseph Kratz was second only to Alexis Jordan (see next page) in the number of cyclamen taxa he illustrated.

'Cyclamen europaeum L.' (*C. purpurascens*) from *Flora von Deutschland, Österreich und der Schweiz in Wort und Bild für Schule und Haus* by Otto Wilhelm Thomé Vol. 4, tab. 473 (1885) (MD)

Otto Wilhelm Thomé's *Flora von Deutschland, Österreich und der Schweiz in Wort und Bild für Schule und Haus*

Professor Doctor Otto Wilhelm Thomé (1840–1925) was a German botanist and botanical artist from Cologne. His Flora was published in four volumes, with a total of 572 botanical illustrations, in 1885 in Gera, Germany. Professor Thomé provided the text, whereas the plates were the work of Walter Müller (sometimes spelt Mueller).

In the same year that Otto Wilhelm Thomé was born, Diederich Franz Leonhard von Schlechtendal, Professor of Botany and Director of the Botanical Gardens at the Martin Luther University of Halle-Wittenberg, began the publication of a 24-volume work also called *Flora von Deutschland*, which contained 2,400 illustrations. Schlechtental's *Flora von Deutschland* was reprinted several times, and a fifth edition that was produced between 1880 and 1887 by Ernst Hans Hallier was expanded to 30 volumes. Hallier re-used Walter Müller's illustration of *C. purpurascens* in a simpler form (as Plate 1983) along with two other much cruder plates of *C. hederifolium* and *C. repandum* (numbered plates 1984 and 1985), which do not carry Walter Müller's initials.

Art Nouveau fan design by A. V. Meckel from *Le Journal de la Décoration*, a French art quarterly periodical published by Armand Guérinet (c. 1890–1900). It also appeared in *Dekorative Vorbilder eine Sammlung von Figurlichen Darstellungen*, published by Verlag Hoffman Stuttgart (1893) (MD)

Alexis Jordan: the ultimate splitter

Alexis Jordan was the ultimate 'splitter'. In his *Icones ad Floram Europæ Novo Fundamento Instaurandam Spectantes*, published in 1903, he illustrated 27 *Cyclamen* taxa with paintings by C. Delorme. Of these, 26 were new species described by Jordan, the other plant being *C. subhastatum*, which had previously been described by Reichenbach in *Flora Germanica*. None of the 27 names have been accepted. Jordan divided his species into a number of Grex and these correspond to species that we recognise, as follows:

Grex *C. Coi* Mill. (*C. coum* Mill.)
- *C. zonale* Jord. (Plate 504)
- *C. brevifrons* Jord. (Plate 504 bis.)
- *C. apiculatum* Jord. (Plate 505)

Grex *C. verni* Ait. (*C. repandum* Sm.)
- *C. stenopetalum* Jord. (Plate 506)
- *C. spectabile* Jord. (Plate 507)
- *C. eucardium* Jord. (Plate 508)
- *C. lobospilum* Jord. (Plate 509)
- *C. ilicetorum* Jord. (Plate 510)
- *C. rarinævum* Jord. (Plate 511)

Grex *C. europæi* L. (*C. purpurascens* Mill.)
- *C. breviflorum* Jord. (Plate 512)
- *C. cyclophyllum* Jord. (Plate 513)
- *C. umbratile* Jord. (Plate 514)
- *C. lilacinum* Jord. (Plate 515)
- *C. holochlorum* Jord. (Plate 516)

Grex *C. hederaefolii* Ait. (*C. hederifolium* Ait.)
- *C. angulare* Jord. (Plate 517)
- *C. subhastatum* Reichenb. (Plate 518)
- *C. sabaudum* Jord. (Plate 519)
- *C. insulare* Jord. (Plate 520)
- *C. albiflorum* Jord. (Plate 521)
- *C. ædirhizum* Jord. (Plate 522)

Grex *C. africani* et *macrophylli* Hort. (*C. africanum* Boiss. & Reut.)
- *C. venustum* Jord. (Plate 523)
- *C. subrotundum* Jord. (Plate 524)
- *C. pachybolbum* Jord. (Plate 525)
- *C. algeriense* Jord. (Plate 526)

Grex *C. graeci* Link (*C. graecum* Link)
- *C. velutinum* Jord. (Plate 527)

Grex *C. persici* Mill. (*C. persicum* Mill.)
- *C. tunetanum* Jord. (Plate 528)
- *C. albidum* Jord. (Plate 529)

'Cyclamen apiculatum' (*C. coum*) Plate 505 from Alexis Jordan's *Icones ad Floram Europæ Novo Fundamento Instaurandam Spectantes* Vol. 3 (1903). By permission of the Trustees of the Natural History Museum (London)

'Cyclamen lobospilum' (*C. repandum*). Plate 509 from Alexis Jordan's *Icones ad Floram Europæ Novo Fundamento Instaurandam Spectantes* Vol. 3 (1903). Jordan gave the type locality as Bastia, Corsica. By permission of the Trustees of the Natural History Museum (London)

'Cyclamen lilacinum' (*C. purpurascens*), Plate 515. Jordan gave the type locality as Biella in the Piedemont, Italy. 'Cyclamen holochlorum' (*C. purpurascens*), Plate 516. Jordan gave the type locality as Evian-les-Bains, France. From Alexis Jordan's *Icones ad Floram Europæ Novo Fundamento Instaurandam Spectantes* Vol. 3 (1903). By permission of the Trustees of the Natural History Museum (London)

'Cyclamen subrotundum' (*C. africanum*). Plate 524 from Alexis Jordan's *Icones ad Floram Europæ Novo Fundamento Instaurandam Spectantes* Vol. 3 (1903). Jordan gave the type locality as Africa borealis: Philippeville (Skikda, north-eastern Algeria). By permission of the Trustees of the Natural History Museum (London)

CHAPTER EIGHT

Cyclamen in popular culture

8.1 CYCLAMEN ON POSTAGE STAMPS

Eugene Kozloff

The use of illustrations of animals, flowers and other biological subjects on postage stamps began in earnest before the middle of the twentieth century, when it became practical to print small multicoloured images. Some countries and dependencies had, however, produced engraved stamps featuring notable examples of fauna and flora much earlier. Canada was in fact the first (in 1851), to depict an animal, its famous beaver, and the same image was used on some twentieth-century issues. Between 1854 and 1912, when it became part of modern Australia, Western Australia repeatedly featured the black swan on its stamps. Lebanon's cedar is the only plant species to have enjoyed such a long run, but this did not begin until 1925.

Before about 1950, most stamp collectors, other than scholarly specialists, were happy to assimilate just about any postal issue that came their way, and this attitude has persisted to at least some extent. Today, however, such an enormous quantity of new issues is being produced each year that many collectors are compelled to restrict their new acquisitions to particular geographic or political entities, such as Mexico or the Middle East, or to topics such as automobiles, pharmacy, composers of music or art. Butterflies, birds, reptiles, mushrooms, garden flowers, and plants of curative value are just a few of the many subjects pursued by collectors whose interests are related to horticulture, natural history, special areas of biology, and medicine. This article, dealing only with stamps that feature images of cyclamen, should be of interest to those who enjoy connecting certain species of the genus, countries or regions of the world.

The history of cyclamen on stamps did not begin until 1938, when Germany issued a set of nine charity stamps, each bearing the image of an old Austrian castle, in the foreground of which is a flower. These stamps, one of which illustrates *C. purpurascens*, could be used for postage, but they were also expected to raise money to aid families disadvantaged by a harsh winter.

The next cyclamen stamp to appear, in Switzerland in 1947, also showed *C. purpurascens* and also was linked to charitable concerns. More than 30 postage stamps featuring various cyclamen species have appeared since then. Florists' cultivars of *C. persicum* are naturally prominent among them, but for fanciers of cyclamen, the more appealing images will probably be those featuring species native to the countries that have issued the stamps. It should be borne in mind that some of the images reproduced here do not accurately portray particular species. If a stamp is not labelled, and if the species does not grow wild in the country that has issued the stamp, it is assumed that the illustration shows *C. persicum*.

In the interests of brevity and to make it easy to locate them, the stamps on which cyclamen are illustrated are listed in a table according to the year in which they were issued, and with the numbers assigned to them in the Stanley Gibbons or Scott catalogues, which are revised and enlarged annually. The Stanley Gibbons catalogue (abbreviated SG) is routinely referred to in

Issues of postage stamps that are illustrated by cyclamen

Date	Issuing country or territory	Species depicted	Stanley Gibbons (SG) or Scott (Sct) catalogue numbers
1938	Germany	*C. purpurascens*	SG 667; Sct B127
1947	Switzerland	*C. purpurascens*	SG J123; Sct B169
1948	Austria	*C. purpurascens*	SG 1106; Sct B243
1952	San Marino	*C. persicum* (florists' cultivar)	SG ★446, ★449, ★450, 451; Sct ★C82, ★C85, ★C86, C87
1953	San Marino	*C. persicum* (florists' cultivar)	SG 470; Sct 342
1958	Greece	*C. graecum* (as *C. greacum*)	SG 798; Sct 631
1959	Israel	*C. persicum* (as *C. latifolium*)	SG 162; Sct 205
1960	Czechoslovakia	*C. purpurascens* (as *C. europaeum*)	SG 1192; Sct 1014
1961	Albania	*C. persicum* (florists' cultivar)	SG 679; Sct 595
1964	Lebanon	*C. persicum* (florists' cultivars)	SG 814, ★815; Sct 421, ★422
	Poland	*C. persicum* (florists' cultivar)	SG 1528; Sct 1279
1965	Albania	*C. persicum* (florists' cultivar)	SG 920; Sct 834
1966	Hungary	*C. purpurascens* (as *C. europaeum*)	SG 2168; Sct 1741
	German Democratic Republic	*C. persicum* (florists' cultivar)	SG E910; Sct 844
1971	Liechtenstein	*C. purpurascens*	SG 532; Sct 481
1973	Hungary	*C. purpurascens* (as *C. europaeum*)	SG 2821; Sct 2241
1975	Berlin[1]	*C. purpurascens*	SG B496; Sct 9NB121
1980	Tunisia	*C. africanum* (as *C. europaeum*)	SG 961; Sct 773
1982	Italy	*C. persicum* (florists' cultivar)	SG 1753; Sct 1510
	Turkish Republic of Northern Cyprus	*C. persicum*	SG 110; Sct 102

C. purpurascens, Germany, 1938

C. purpurascens, Switzerland, 1947

C. purpurascens, Austria, 1948

C. persicum, San Marino, 1952

C. persicum, San Marino, 1953

C. graecum, Greece, 1958

C. purpurascens, Czechoslovakia, 1960

C. persicum, Israel, 1959

C. persicum, Albania, 1961

Date	Issuing country or territory	Species depicted	Stanley Gibbons (SG) or Scott (Sct) catalogue numbers
1983	Union of Socialist Soviet Republics	*C. coum* ("elegant cyclamen", possibly referring to *C. elegans*)	SG 5334; Sct 5151
1984	Turkey[2]	*C. pseudibericum* (as *C. pseudoibericum*)	SG 2670; Sct 2284
1985	Cambodia (Kampuchea)	*C. persicum* (florists' cultivar)	SG 634; Sct 599
	North Korea	*C. persicum* (florists' cultivar)	SG N2530; Sct 2510
1990	Republic of Cyprus	*C. cyprium*	SG 788; Sct 769
	Syria	*C. persicum* (florists' cultivar)	SG 1758; Sct 1198
	Turkey[3]	*C. pseudibericum* (as *C. pseudoibericum*)	SG ⋆3075; Sct ⋆2467
1991	Algeria	*C. africanum*	SG 1060; Sct 938
	Monaco	*C. persicum* (florists' cultivar)	SG 2036; Sct 1749
1993	Russia	*C. persicum* (florists' cultivar)	SG 6400; Sct 6135
1998	North Korea	*C. persicum* (florists' cultivar)	SG N3800; Sct (not yet assigned)
	Tunisia	*C. persicum*	SG 1402; Sct 1178
	Uganda	*C. graecum*	SG 1991; Sct 1566e
2005	Japan	*C. persicum* (florists' cultivar)	SG 3358; Sct (not yet assigned)

[1]Issued for use in the American, British and French occupation sectors of Berlin.

[2]Issued also as a postcard (see page 524); see also year 1990, Turkey.

[3]Re-issue of SG 2670, Sct 2284 in 1984, with surcharge for rate increase.

The only cyclamen stamps not reproduced in this chapter are a few in which the pertinent image appears more than once in the same series, or that differ from other issues only in colour or because of a surcharge. These stamps are, however, listed in this table and are indicated by an asterisk (⋆).

Of the commonly recognised species of the genus *Cyclamen*, the following have not yet been shown on postage stamps: *C. balearicum*; *C. alpinum* (=*C. trochopteranthum*), *C. cilicium*, *C. intaminatum*, *C. mirabile* and *C. parviflorum*, all endemic to Turkey; *C. creticum*; *C. hederifolium*; *C. libanoticum*, endemic to Lebanon; *C. repandum*; *C. rohlfsianum*, endemic to Libya; and *C. somalense*, endemic to Somalia.

C. persicum, Lebanon, 1964

C. persicum, Albania, 1965

C. purpurascens, Hungary, 1966

C. purpurascens, Liechtenstein, 1971

C. persicum, Poland, 1964

C. persicum, German Democratic Republic, 1966

C. purpurascens, Hungary, 1973

C. *purpurascens*, Berlin, 1975

C. africanum, Tunisia, 1980

C. persicum, Italy, 1982

C. persicum, Turkish Republic of Northern Cyprus, 1982

C. coum, Union of Socialist Soviet Republics, 1983

C. pseudibericum, Turkey, 1984

C. persicum, Cambodia (Kampuchea), 1985

C. persicum, North Korea, 1985

C. cyprium, Republic of Cyprus, 1990

C. persicum, Syria, 1990

C. africanum, Algeria, 1991

C. persicum, Monaco, 1991

C. persicum, Russia, 1993

C. persicum, North Korea, 1998

C. persicum, Tunisia, 1998

C. graecum, Uganda, 1998

C. persicum, Japan, 2005

Three 'personalised' cyclamen stamps designed by Jan Bravenboer for postage from The Netherlands the UK, Australia, New Zealand and many other countries or dependencies of the British Commonwealth; the Scott catalogue (abbreviated Sct) is used in the USA and most of Latin America.

Obviously, stamp dealers whose commerce regularly extends beyond the borders of a single country are likely to be conversant with at least SG and Sct, and perhaps with other cataloguing systems, such as Yvert (France) and Michel (Germany). It is important to understand that the numbers used in one catalogue are routinely different from those in another, but knowing the year of issue should enable enthusiasts to track down any stamp in which they are interested.

Most of the stamps that feature cyclamen belong to a set of two or more flower issues published at the same time. Complete sets are usually easier to find than individual stamps, especially if unused stamps are desired; this is because not many dealers are enthusiastic about separating stamps that belong together and that are generally offered for sale that way. The SG or Sct numbers given in the table, as well as those in other catalogues, will, of course, lead a collector to a complete set of stamps as well as to a single cyclamen stamp.

Acknowledgements

Although I was able to locate most of the cyclamen stamps in a private collection, I would have missed some issues if I had not been helped by the late Graham Simpson and, more recently, by Robert Coveny, a botanist at Royal Botanic Garden, Sydney, Australia. It was a pleasure to correspond with both of these fanciers concerning our mutual interest in cyclamen and in the postage stamps that illustrate them. Graham, who was originally expected to write this chapter, had already pulled together a substantial collection, and he very kindly shared duplicates and colour reproductions. Among the discoveries made by Robert was the stamp issued by Germany in 1938, believed to be the first to feature a cyclamen. He also provided Stanley Gibbons' numbers for many issues. The help of my friend and professional colleague Craig Staude in preparing accurate copies of stamp images is warmly appreciated.

8.2 CYCLAMEN AND CERAMICS, POTTERY AND GLASSWARE

It is hardly surprising that such a familiar flower as cyclamen has been used frequently as a source of imagery on ceramics, pottery and glassware — it is a much-loved plant that is instantly recognisable, even if reproduced in an impressionistic style. Coupled with this popularity is the simplicity of form of cyclamen flowers and leaves, which is undoubtedly attractive to those designers who seek less complex floral subjects.

Spode Ware, *Curtis's Botanical Magazine* and Cyclamen[1]

Pam Wooliscroft & Brian Mathew

Several *Cyclamen* species have been illustrated in *Curtis's Botanical Magazine*, the first nearly 225 years ago. One of them provides an interesting link between this venerable journal and the equally historic ceramics firm of Spode. Botanical subjects were used on Spode ceramic wares from the early 1800s onwards, and it is clear that many of the designs were influenced by watercolours taken from the *Botanical Magazine*, which first appeared in 1787.

One of the earliest illustrations in the *Magazine*, Plate number 4 in the very first issue, was of *C. coum*, drawn by James Sowerby,

Spode plate, part of a dessert service first produced c. 1812 (BM)

[1]Editor's footnote. A more extensive and well-illustrated article on this subject by Pam Woolliscroft, Curator of the Spode Museum, appears in *Curtis's Botanical Magazine* Vol. 19, part 3 (2002).

Cyclamen coum from *Curtis's Botanical Magazine* (Vol. 1: t. 4, 1787) used by the Spode ceramics factory to adorn Spode pottery (Kew)

and this is one of the images selected by Spode to illustrate their ceramics. In fact, it was this illustration that instigated research (by PW) into the influence of the *Magazine* on Spode designs, which came about through a sequence of unrelated events. In 2000, one of us (BM), then President of the Cyclamen Society and Editor of *Curtis's Botanical Magazine*, was visiting Valerie Finnis, the founder and secretary of the Merlin Trust. She was displaying some dessert plates on a table and one of these bore an image of *C. coum*. Although reproduced wholly in green, including the flowers, it was immediately recognisable as illustration no. 4 from the magazine. The plate was one piece of a dessert service given to Valerie Finnis by her late mother-in-law, Lady Charles Montagu Douglas-Scott. Following this encounter, the research began, and it soon became clear that other images from the *Botanical Magazine* were also used in this particular service.

As a background to this story, it is interesting to provide an outline of the techniques involved in utilising published artwork in the decoration of pottery. The images from the *Botanical Magazine* were transferred on to the pottery by a process known as 'transfer printing' at the biscuit ware stage (i.e. after the first firing). Before the printing could begin, the pattern was first engraved onto a sheet of copper. The ceramic colour was then applied to the warmed copper plate and the surplus removed, leaving the colour only within the engraved design. Specially prepared tissue paper was then laid carefully in place and pressed onto the copper plate in a hand-powered press. The paper was then removed, taking the colour with it. It was then cut to the shape of the pottery, applied colour-side down to the biscuit ware and rubbed hard by hand with a stiff brush. Once the colour had been successfully transferred from the paper, the pot was washed, removing the thin paper and leaving the colour behind. The next process was a low-temperature firing to 'harden on' the colour.

The Spode factory records show that the *C. coum* image was used under the pattern no. 1678, the first recorded use of which was in about 1812. It is interesting to note that the engraver did not follow the original exactly, presumably attempting to improve it slightly, for whatever reason. The upper-most flower in the original illustration is a side view, whereas on the plate it has been turned up to show the central 'eye' of the flower. Other than this, the ceramic version is nearly identical. The cyclamen image also appears also on an oval plate ('Devonia shape') in the same set. Several other *Botanical Magazine* images depicting various plants were used in this service. The original hand-engraved copper plates for these 'Botanical Studies' still exist at the Spode factory.

A selection of pieces showing *C. repandum* from the Portmeirion® 'Botanic Garden' range of ceramics. Photo: Wayne Mountford, Portmeirion

Portmeirion pottery

The *C. repandum* design in Susan Williams-Ellis's Portmeirion® Pottery 'Botanic Garden' range of ceramics (see page 515) was introduced in 1972. The illustration is taken from *La Flore des Dames* by Albert Jacquemart, a tiny book of only 14 × 7 cm that was published in Paris in 1840 and described as a 'Botanique a l'usage des Dames et des Jeunes Personnes'. The artwork for this painting of *C. repandum* is attributed to A. Duménil, although it also bears a remarkable similarity to that of Sydenham Edwards, which had been published earlier in *Curtis's Botanical Magazine*, tab. 1001 (1808) (unfortunately mis-identified there as *C. hederifolium*). Good paintings and engravings were often used and re-used in various publications, much as photographs are today. Portmeirion also produce a range of glassware — vases, goblets and tumblers — that feature a cyclamen image.

Cyclamen repandum from *Curtis's Botanical Magazine* t. 1001 (1808) has some similarity with the image used on Portmeirion pottery (Kew)

Other ceramics and pottery

China dish from Germany, 1950. Collection of Jan Bravenboer (JB)

Plate and set of hand-made bowls from Cyprus. Collection of Brian and Jo Walker (BM)

Porcelain bread boards from Germany, c. 1930. Collection of Jan Bravenboer (JB)

Bohemian tea set, 1890–1919. Collection of Jan Bravenboer (JB)

Schnapps flask and measure from the Austrian Tirol. Collection of Edwin Rose (EW)

Item from large dinner set of china. Collection of Peter Jones (PJ)

Bavarian porcelain cobalt tea set. Collection of Jan Bravenboer (JB)

Small container with lid decorated with cyclamen. Collection of Melvyn Jope (BM)

Tiles

The cyclamen image has also been utilised for the adornment of ceramic tiles. A few of these are illustrated here to show widely differing interpretations.

Tiles from Germany in the 'Jugendstil', Art Nouveau period, 1890–1910. Collection of Jan Bravenboer (JB)

Drawing by Phil Cornish of a tile decorating a Victorian fireplace. The design is clearly intended to trace the outline of a swan; the flowers and buds are coloured deep blue (BM)

Cyclamen on Belgian tiles, c. 1920s. The art-deco design probably represents *C. persicum*. Collection of John Grimshaw (JG)

Glassware

The cyclamen image has also been used to adorn glassware and there is perhaps no more elegant example of this than a vase made by the Daum factory, c. 1902.

Daum made great use of botanical subjects, both wild subjects from the flora of Lorraine and cultivated plants from the gardens of Nancy, in particular the world-famous nursery of Victor Lemoine who was noted for raising many new hybrids and notably lilacs. Henri Bergé (1868–1936), one of the artists for Daum, recorded the names of the plants depicted, so it is likely that this 'cyclamen d'Europe' represents *C. purpurascens*. Thanks go to the library of the Fine Arts Museum of Nancy for information about the Daum glass factory and in particular the cyclamen vase.

Vase aux cyclamens. Daum vase, c. 1902. Acquired by the Musée des Beaux Arts de Nancy from Daum Cristalleries in 1983 (INV. 83.1.6). The gold-rimmed vase is made of pale rose-tinted glass on which cyclamen leaves and flowers have been painted in purple enamel with a gold background decoration. Height 29.8 cm, top diameter 17 cm. Photo by Gilbert Mangin (GM)

Glass plate, bought in an antique market in Cirencester. Collection of Brian and Margaret Mathew (BM)

Glass bowl engraved with cyclamen flowers. Collection of Melvyn Jope (BM)

The Pen-y- Llwyn Trophy: an Alpine Garden Society award, based on a nineteenth-century claret jug. The trophy was made of Caithness crystal glass and silver by J. A. Campbell, silversmith. Cyclamen design by Margaret Todd, engraved by Robert Ellison (BM)

8.3 CYCLAMEN ON JEWELLERY AND METALWARE

Fabergé tiara

A fabulous piece of jewellery made in the early twentieth century by the legendary firm of Carl Fabergé was exhibited by Wartski's of London in c. 1997 in an exhibition entitled 'One hundred tiaras: an evolution of style 1880–1990'. It is made of diamonds and gold in the form of a band of cyclamen leaves and flowers (allegedly of *C. coum*) and can be removed from its gold frame to allow conversion into a necklace. Geoffrey C. Munn in *Tiaras. A History of Splendour* (Antique Collectors' Club, 2001) describes this as "possibly the most bewitching of all the tiaras made by Fabergé." He attributes it to the workshops of the chief jewellery workmaster to Fabergé, Albert Holmström, c. 1903.

Belt buckle, found in an antiques shop in Surbiton, Surrey. With thanks to Jill Marsden (BM)

A tiara in the form of a band of *Cyclamen* tied with ribbon, purchased from Fabergé for the Hon. Mrs Wilson Fox. Private Collection/Bridgeman Art Library (Bridgeman)

Small tin decorated with cyclamen. Collection of Jan Bravenboer (JB)

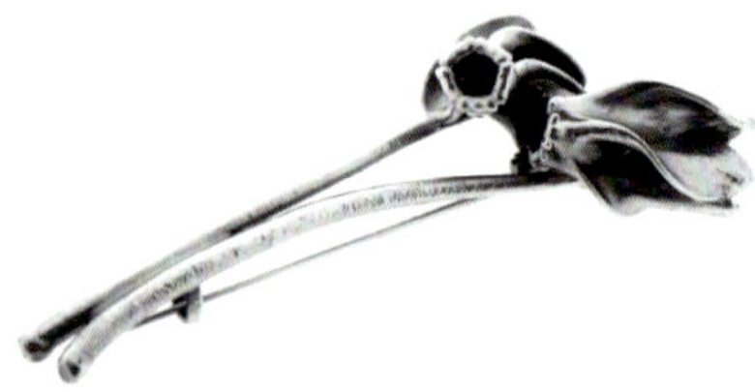

Brooch by Asimi, made from a cyclamen flower coated in copper, then plated in silver. Each brooch is made from an individual flower, so all of the brooches in this floral range are unique. Reproduced by courtesy of Asimi Limited, Edinburgh (Asimi)

Another diadem, illustrated as Plate 270 in Munn's book, is in the form of cyclamen flowers and leaves set with diamonds and mounted in platinum. This Art Nouveau piece was made in the workshops of Audianne in April 1899.

Other items

Art Nouveau silver cigarette case, c. 1900 (E-Bay)

Small jewellery box, c. 1890. Seen in an antiques shop, Esher, Surrey (JB)

8.4 OTHER USES OF THE CYCLAMEN IMAGE AND NAME

Brian Mathew

Cards

The cyclamen image has appeared on countless cards of all types and we have selected just a few to illustrate the range.

Fifteenth-century playing card. Reproduced from *The Art of Botanical Illustration* by Wilfred Blunt and William T. Stearn (Antique Collectors Club, 1994) (ACC)

Perhaps one of the oldest examples of the cyclamen image being used on cards is a fifteenth-century playing card. According to Wilfred Blunt and William T. Stearn, writing in *The Art of Botanical Illustration*, "The earliest metal engravings of flowers appear to be those found upon playing cards made about 1440–1450 by the Meister der Spielkarten and other German artists. These curious prints ... are of particular interest in that they precede even the first known botanical woodcuts."

Jill White has a collection of the delightful woven silk cigarette cards that were issued by J. Wix and Sons Ltd with Kensitas cigarettes in the 1930s, and one of these shows a cyclamen. Like others in the set, it carries a short verse, is fully washable, and shows the cyclamen as an emblem for diffidence — its hanging head symbolic of modesty or shyness. The cyclamen card was number 10 in a series of 40 flower cards, the second such series (the first had 60 cards). It can be found in two sizes, small and medium; some cards in the series were also issued in postcard size. Each card came with a folder and cultural tips and, for 1 shilling, a decorative album to take the full series could be supplied. Although cards from the first series can be quite easy to find and collect, and are relatively inexpensive, examples of the second series can cost up to £50 each.

Woven silk cigarette card issued with Kensitas cigarettes in the 1930s. Collection of Jill White (BM)

Prayer card, France, 1902. Collection of Jan Bravenboer (JB)

German coffee card, Toledo Spice Co. Collection of Jan Bravenboer (JB)

Postcard, Germany, dated 1902. Collection of Jan Bravenboer (JB)

Christmas card showing *C. persicum*. Collection of Keith Fry (KF)

Postcard, Germany, 1910. Collection of Jan Bravenboer (JB)

Postcard, Italy, dated 1913. Collection of Jan Bravenboer (JB)

Postcard, Germany, dated 1904, depicting *C. persicum*. Collection of Jan Bravenboer (JB)

Christmas card depicting *C. purpurascens* (as *C. europaeum*). Collection of Keith Fry (KF)

Postcard, France, c. 1930. Collection of Jan Bravenboer (JB)

Postcard depicting *C. purpurascens*, Germany, 1951. Collection of Jan Bravenboer (JB)

Turkish postcard and stamp showing *C. pseudibericum* (JB)

Telephone cards, Cyprus. Collection of Jan Bravenboer (JB)

Embroidery

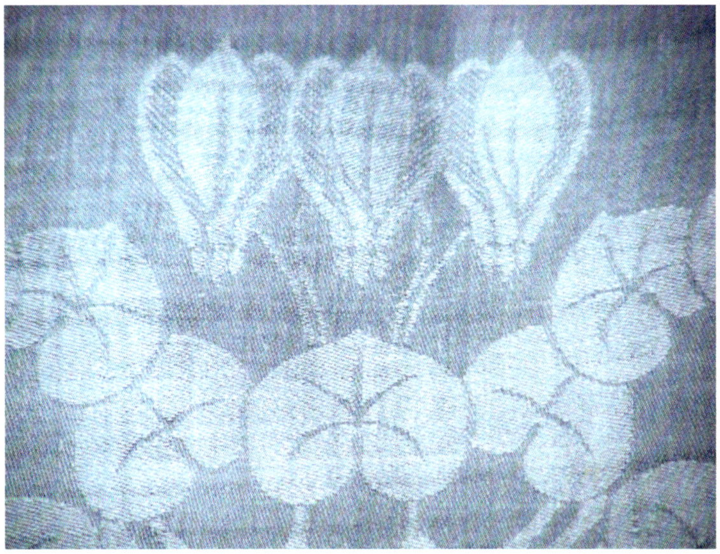

Cyclamen table napkin. Collection of Vic and Janet Aspland (VAA)

Embroidered icon seen in a church in Crete (JF)

Cyclamen and ships

It was rather by chance that we first heard of a ship called HMS Cyclamen, but this led to similarly named vessels being encountered. There appear to have been a succession of three British ships and one French warship bearing the name. The first of the British vessels was a drifter, built in 1911 and then hired for use by the Royal Navy between 1915 and 1919.

At about this time, the naval sloops known as the "Arabis" Class were built and named after plants, and these included an HMS Cyclamen. She was built in 1916 and served in the Royal Navy until 1932 when she was sold to Metal Industries, a ship-breakers firm in Charlestown, Fife. HMS Cyclamen was one of the vessels involved in the bombardment of Khasab and Qidah on 21 April 1930.

The sloop HMS Cyclamen (middle left) en route for the Persian Gulf.
Taken from the magazine *The Graphic* (1921)

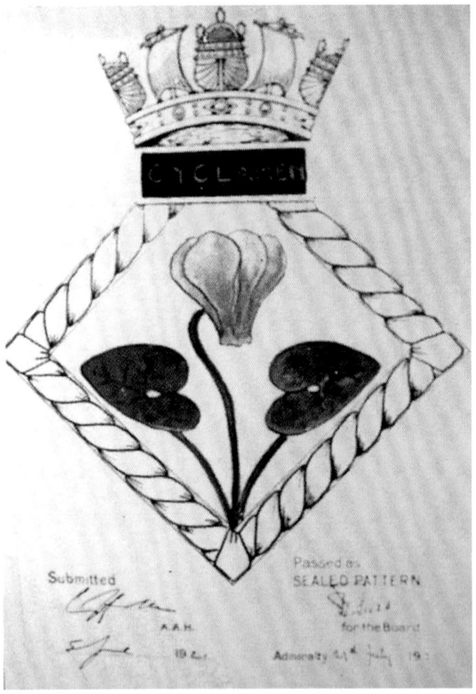

A badge for HMS Cyclamen submitted in 1921, presumably for the "Arabis" Class sloop of that name. Reproduced courtesy of the Imperial War Museum (IWM)

Postcard of the crew of HMS Cyclamen, date and place unknown

The third British ship bearing the name HMS Cyclamen was a successful World War II corvette, built for convoy protection and anti-submarine work. Along with numerous other ships named after plants (several bulbous ones including *Crocus*, *Fritillary* and *Narcissus*), and collectively called the "Flower" Class, K83 was built just at the start of the war, quickly and cheaply for this

Photo of crew of HMS Cyclamen, Muscat 1930, and inscription on reverse

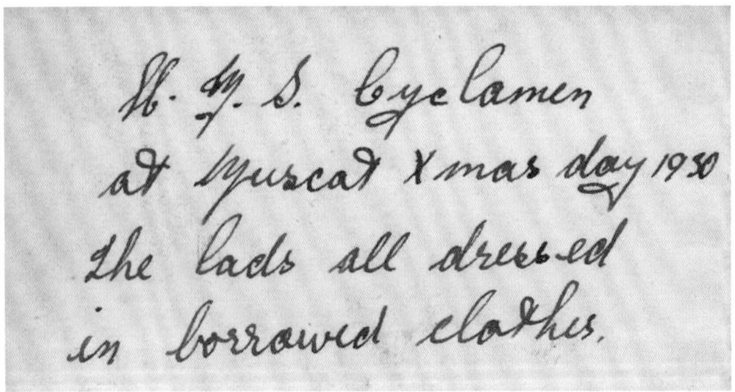

The corvette K83, HMS Cyclamen, photographed in October 1942. Reproduced courtesy of the Imperial War Museum (IWM)

specific task. The dates indicated just how quickly this ship was built: the keel of HMS Cyclamen was laid on 30 November 1939 in Aberdeen (J. Lewis & Sons), she was launched on 20 June 1940 and commissioned on 30 September 1940! *Cyclamen* was sold in 1947 to the firm of Chr. Salvesen of Leith and re-named the *Southern Briar*. Under this name, she was modified and used as a towing vessel and whaler between 1949 and 1963. In 1956–1957, she was involved with the National Institute of Oceanography in tagging whales in the Bellinghausen Sea. Eventually, in 1996, she was sold to a breaker's yard in Belgium, although it is reported that she was wrecked en route, off the shore of Denmark, when the towing cable broke.

The French naval vessel bearing the name MSC Cyclamen (M674) was a minesweeper, in service from 1953 to 1983, during which time she took part in the Suez Canal War of 1956.

MSC Cyclamen M674. Postcard, date unknown

Currency

Cyclamen have also been used to enhance both bank notes and coinage, some examples of which are shown below.

Cyprus £10 note showing, among various items of natural history, two cyclamen, perhaps *C. graecum* and *C. persicum*. Collection of Jan Bravenboer (JB)

Half-cent coin from Cyprus. Collection of Jan Bravenboer (JB)

500,000 Turkish Lira proof of silver coin, obverse (KF)

Bank note from Israel showing *C. persicum*. Collection of Jan Bravenboer (JB)

Council emblem

The municipal insignia of the town of Kiryat Tiv'on between Haifa and Nazareth illustrates a cyclamen, probably *C. persicum*. The local council website provides the following explanation: "The symbol of Kiryat Tiv'on is the cyclamen, which grows between rocks and reflects the closeness of the town to nature, its effort to maintain a unique community atmosphere, and balance between landscape preservation and adherence to rules and regulations concerning development and the environment."

Town insignia of Kiryat Tiv'on, Israel

Even on furniture ...

A piece of furniture seen on display in the Great Hall at Knightshayes Court, Devonshire was decorated along its front with flower paintings, including a cyclamen. The Burges bookcase was on loan from the Ashmolean Museum, Oxford.

and finally ... anyone for the cyclamen rally?

The Cyclamen Rally in Cyprus was first organised in 2004 by the Mercedes Benz Club as a way of promoting good relations between Cypriots on the two politically separate parts of the island. The rally was such a success that it has become an annual event, now organised by the Cyprus Federation of Classic Vehicles. The name of the rally is appropriate as the classic cars parade their elegance in the autumn, at a time when the native cyclamen commence flowering. There are three native species in Cyprus: *C. cyprium*, *C. persicum* and *C. graecum*.

Poster for the *Cyclamen Rally* in Cyprus. Thanks to the Cyprus Federation of Classic Vehicles

CHAPTER NINE

Cyclamen literature – a select bibliography

Richard Bailey

To create any select bibliography is no mean task: no two people will ever agree on which works should, or should not, be included. Then there is the problem of how to present the information: in alphabetical order by author, in chronological order of publication, or perhaps works should be classified according to their content — taxonomy, ecology, history, use or cultivation and so on. Should availability or language be criteria for inclusion? How far back should one go?

When so much information is available online, what can a hard-copy bibliography offer that is not available electronically? Reference to sources so obscure or so ancient that they have not yet been scanned by Google or the British Library? The opportunity to study in some remote locality without electricity or internet access? An impartial assessment of the works listed? Or, perhaps just the pleasure of handling a real book?

But my decisions in compiling this bibliography were ultimately made on the basis of a simple question: what would be most interesting and useful to this book's diverse body of readers?

The sequence that I chose reflects first the fact that this book has been compiled by the Cyclamen Society, and for those wishing to source an impressive array of factual material on cyclamen the publications of the Society are a great place to begin. Following these publications are other periodical publications that are important in the context of cyclamen studies. Floras are another useful source of information.

These resources are followed by other publications grouped into four historical sections. Within these four time periods, references are listed in alphabetical order of author. In each period, there are two parts: books and periodicals. Within each section, I have tried to select items that are indicative of the main thrust of interest in cyclamen at the time, and which themselves have good reference lists. Finally, there is a 'Miscellanea' for any works that do not readily fit elsewhere.

Type descriptions (i.e. the original descriptions of a species) are not listed, but reference is made (p. 534) to *Index Kewensis* and *The International Plant Names Index* (IPNI).

Finally, remember that the taxonomy and nomenclature of the genus is constantly being revised. Nomenclature used in works that are more than a few years old is unlikely to be correct throughout.

In compiling this account of cyclamen literature, I would like to acknowledge both the staff of the R.H.S. Lindley Library, for help well beyond the call of duty, and my wife, Susan, for help and support.

To summarise, the arrangement of the list of publications is as follows:

9.1 Cyclamen Society publications

9.2 Periodicals that have particular relevance to cyclamen

9.3 Floras
- a. Floras covering a region of several countries
- b. Floras covering a single country

9.4 Books and periodicals in historical sequence, arranged in the following time periods:
- a. Pre-1800. Most of the works listed are Herbals, or comparable works. No periodical articles from this period are listed.
- b. 1800 to 1900. The 'pre Hildebrandian' era. Hildebrand's seminal work, *Die Gattung Cyclamen* L., published in 1898, marked a turning point in our knowledge of the genus. Works that are listed from this period include books and some articles from periodicals.
- c. 1901 to 1975. Roughly, the period between the publication of Hildebrand's book and the foundation of the Cyclamen Society in 1977. Again, works that are listed from this period include books and some journal articles.
- d. Post 1975. This is the era of the Cyclamen Society, which has generated enormous interest in the genus amongst amateurs and professionals, academics and gardeners. Both books and journal articles from this period are listed.

9.5 'Miscellanea Cyclamenologica': sundry works.

9.1 CYCLAMEN SOCIETY PUBLICATIONS

The 36 volumes of the Cyclamen Society's journals contain a wealth of personal observations made by enthusiasts (BM)

Cyclamen. The Journal of the Cyclamen Society
(1977–present) ISSN 01433571.

Editors: 1977–1990 Colonel J. A. Mars; 1991–2000 I. A. Aird; 2001–2010 Dr J. C. B. Clennett; 2011 onwards M. Denney. Assistant Editor: 1992–2008 Dr Helena M. Wiesner.

The cyclamen society journal has appeared under several titles in its 30-odd years of publication. The Society was founded on 27 January 1977 and its first publication, *The Cyclamen Society Newsletter* No. 1 appeared in June that year; it was edited by James Mars and composed of nine pages that were duplicated in A4 format. Later, this first newsletter was reissued in A5 format to match Newsletter No. 2, which was published in December 1977. These two newsletters were reprinted, with the addition of a cover and lists of the Society's officers and committee, as the first two parts of what became, in 1978, *The Cyclamen Society's Journal*, now known as *Cyclamen The Journal of the Cyclamen Society* and commonly abbreviated to *Cyclamen*.

The journal contains an immense amount of information. It includes reports on the Society's field studies, taxonomic information and articles on particular species and geographical regions. Perhaps most valuable for many readers is the wide range of anecdotal information, observations and speculation — information of a kind all too seldom recorded in print. An index to volumes 1–24 (1977–2000) and the Conference Reports (see below) were published in 2001.

The following alphabetical list notes some of the interesting items that have been published in the *Journal of the Cyclamen Society*, which I have selected to cover a range of topics:

Affre, L., Thompson, J. D. & Debussche, M. (1996). *Cyclamen balearicum*: the ecology and reproductive ecology of natural populations. *Journal of the Cyclamen Society* 20: 47–52.

Bailey, R. H. (1986). José Quer y Martinez and the occurrence of cyclamen in mainland Spain. *Journal of the Cyclamen Society* 10: 50–55.

Brickell, C. D. (1979). Notes on the taxonomy of *Cyclamen cilicium*, *C. coum* and their close relatives. *Journal of the Cyclamen Society* 2: 49–54.

Ciotir, C., Yesson, C., Culham, A. & Cristea, N. (2011). DNA barcoding of *Cyclamen* species, another approach to species identification. *Journal of the Cyclamen Society* 35: 26–29.

Clennett, C. (1999). Pollen studies in *Cyclamen*. *Journal of the Cyclamen Society* 23: 20–24.

Clery, R., Lelievre, D. & Wiltshire, T. (2004). On the scent of cyclamen. *Journal of the Cyclamen Society* 28: 24–31.

Compton, J., Culham, A. & Spencer, M. (2008). Typification of some cyclamen names. *Journal of the Cyclamen Society* 32: 22–30.

Compton, J., Culham, A. & Thornton-Wood, S. (2002). Cyclamen diversity project: Phase 1 report. *Journal of the Cyclamen Society* 26: 41–42.

Culham, A. & Bota, P. (2009). The taxonomic status of *Cyclamen* subgenus *Psilanthum* Schwarz. *Journal of the Cyclamen Society* 33: 22–25.

Debussche, M. (2003). Frost and cyclamen under Mediterranean climatic conditions. *Journal of the Cyclamen Society* 27: 14–18.

Debussche, M. & Thompson, J. D. (1999). Distribution, ecology and floral variation of *Cyclamen* species in France. *Journal of the Cyclamen Society* 23: 52–55.

Goodwin, N. (1984). A year of cyclamen. *Journal of the Cyclamen Society* 8: 50–53.

Greilhuber, J. (1989). Taxonomic relevance of karyotype structure in *Cyclamen* L. subgen. *Psilanthum* Schwarz. *Journal of the Cyclamen Society* 13: 57–62.

Kakuzen, H. (1983). How I cultivate my cyclamen in Japan. *Journal of the Cyclamen Society* 7: 47–48.

Killingback, A. & Pritchard, M. (2009). Lebanese cooking with cyclamen. *Journal of the Cyclamen Society* 33: 29–30.

Mathew, B. (1995). Cyclamen serendipity. *Journal of the Cyclamen Society* 19: 14–16

Mathew, B. (2007). Illustrations of cyclamen from the past. *Journal of the Cyclamen Society* 31: 50–53.

Meikle, R. D. (1989). Variation in natural populations of *Cyclamen coum* sensu lato. *Journal of the Cyclamen Society* 13: 63–64.

Sparkes, B. B. (1977). Wye College small flowered scented cyclamen. *Journal of the Cyclamen Society* 1: 23–27.

Wallis, B. & Wallis, R. (2011). *Cyclamen elegans* in the Caspian forests. *Journal of the Cyclamen Society* 35: 16–19.

Wiltshire, T. (2011; 2012). Cyclamen cultivars. Part 1: *Journal of the Cyclamen Society* 35: 23–25; Part 2: 35: 66–67; Part 3: 36: 28–29.

Yesson, C. & Culham, A. (2008). Cyclamen and climate change. *Journal of the Cyclamen Society* 32: 10–12.

Cyclamen Society conference reports

Reports of the Society's first three conferences, held in 1978, 1980 and 1982, were published as separate booklets (edited by R. H. Bailey). Subsequent reports appeared in *Cyclamen*. Notable articles in each issue are listed:

Cyclamen. Proceedings of the conference of the Cyclamen Society held at Wye College, Kent, 30 September–1 October 1978 (1979). [23 pp.].
Meikle, R. D. (1979). Some notes on *Cyclamen*. p. 3–12.

Cyclamen 1980. Proceedings of the second conference of the Cyclamen Society, October 1980 (1981). 27 pp.
Chapman, G. P. The relevance of chromosome studies to improving Cyclamen, p. 12–16.
Ward, S. The cytology of the genus *Cyclamen*. p. 17–25.
Mathew, B. (1980). Encounters with wild *Cyclamen*. p. 3–9.

Cyclamen 1982. Proceedings of the third conference of the Cyclamen Society, April 1982 (1983). 58 pp.
Maatsch, R. Habitat and variation in wild populations of *Cyclamen persicum* and their relation to the production of cultivars and florists' hybrids. p. 37–44.

Cyclamen Society expedition collection numbers and field notes

Three booklets, compiled by Martyn Denney, have been published covering field studies carried out between 1987 and 2006.

1987–1996. Published 1998. Expeditions to Greece, Israel and Turkey.

1997–1999. Published 2000. Expeditions to Greece and Turkey.

2000–2006. Published 2007. Expeditions to Greece, Corsica, Sicily and Turkey.

These booklets comprise notes, made in the field and lightly edited, describing each site visited and identifying the plants collected by number. A few private visits by members to important sites have been included if work of interest to the Society was undertaken and the information was subsequently passed to the Society. All Cyclamen Society fieldwork has been carried out with the permits necessary for the collection, export and import of plants.

Cyclamen of Greece is the latest booklet, an account of all the species occurring in Greece (BM)

A bibliography of the genus *Cyclamen* L.

A brief bibliography comprising 12 unnumbered pages was compiled by Martyn Denney in 1999. Approximately 230 references were drawn from many sources but with emphasis on work published in the Society's journal. I have selected the articles listed below as being of special interest:

Other publications

Frank, E. (1999). *The genus Cyclamen* L. Translation from the German of Hildebrand (1898) (see p. 537).

Mathew, B. & Özhatay, N. (2001). *The Cyclamen of Turkey*. 33 pp. ISBN 0-9537526-1-5. A guide to the species of cyclamen growing in Turkey.

Mathew, B. & Özhatay, N. (2001). *Turkiye'nin Siklamenleri*. ISBN 0-9537526-2-3. Turkish language version of the above, translated by Sema Atay.

Moore, P. & Jope M. (2011). *The Cyclamen of Greece*. 40 pp. ISBN 978-0-9537526-3-8. A guide to the species of *Cyclamen* growing in Greece.

9.2 PERIODICALS THAT HAVE PARTICULAR RELEVANCE TO CYCLAMEN

Several periodicals publish articles or notes on *Cyclamen* reasonably frequently. A few articles of especial interest are detailed below, but the majority are not listed in this bibliography and readers are referred to the indexes of these periodicals.

Alpine Gardener. The Bulletin of the Alpine Gardening Society

(1930–present, quarterly) ISSN 0002-6476.

Formerly the *Quarterly Bulletin of the Alpine Garden Society*, this journal took its present name in 2001. An index to volumes 1–61 (1930–1993) was published in 1994 (ISBN 0-90004861-6); a further index covering 1930 to 2005 was produced both on paper (ISBN 978 0 90 004876) and on disk. Until 2006, the indexes to each volume were included in each year's part 4; from 2007, a separate index was produced for each volume. Most references to cyclamen are to notes in the 'Alpine Anthology' (or its predecessor 'Alpine Commentary'), descriptions of area studies, general garden information, or reports on shows and awards. There are only a few volumes in which there is no reference to the genus. A supplement, published in March 2012, details awards to plants made by the R.H.S. Joint Rock Garden Plant Committee. The first list (for 2009–2010) records the award of a Cultural Commendation to a plant of *C. mirabile*. Further supplements are expected.

Articles of particular interest include:

Anon. (1944). *Cyclamen europaeum* var. *album*. *Quarterly Bulletin of the Alpine Garden Society* 12: 182–183.

➢ An early description of this plant, which was introduced to gardens only in the 1980s.

Bird, R. (1988). Cyclamen in the blood. *Quarterly Bulletin of the Alpine Garden Society* 56: 312–317.

Buglass, F. W. (1978). Cyclamen — leaf form and tubers. *Quarterly Bulletin of the Alpine Garden Society* 46: 240–241.

➢ Part of a larger article on the paintings of Gerard Parker.

Mountfort, C. C. (1946). Cyclamen. Peter Moore and the Tilebarn Nursery. *Quarterly Bulletin of the Alpine Garden Society* 14: 8–25.

➢ A useful summary of knowledge at the time.

Various authors (1950). *Pedicularis*. *Quarterly Bulletin of the Alpine Garden Society* 18: 105–106; 196–197; 307–309.

➢ Three notes on propagation, including vegetative propagation.

Schwarz, O. (1951). Taxonomy and nomenclature of Cyclamen. *Quarterly Bulletin of the Alpine Garden Society* 19: 396–404.

Curtis's Botanical Magazine
(1787–present, quarterly) ISSN (print) 1355–4905; ISSN (online) 1467-8748.

The complete set of *Curtis's Botanical Magazine*, from 1787 to the present (BM)

Founded by William Curtis as *The Botanical Magazine*, this periodical was re-titled *Curtis's Botanical Magazine* in 1801; from 1984, it was called *The Kew Magazine* until taking up its present title again in 1995. A note describing the changes in name, pagination and citation appeared in the 1995 issue. Until 1984, articles in this journal comprised text and a painting featuring just one plant, subsequently longer more general articles appeared. A consolidated index to volumes 1–164 was published in 1956 and a handwritten index to volumes 165–184 is held in the R.H.S. Lindley Library, Vincent Square, London.

William Curtis, founder of *The Botanical Magazine* (BM)

In volumes 1–164, cyclamen are featured in plates numbered 4 (reporting *C. coum* from Germany and Italy), 44 (in which *C. persicum* is said to be "a native of the East Indies"), 1001 (which notes that *C. repandum* can be propagated by root cuttings), 5758 and 8989. Re-numbered plates 29, 192, 296, 307, 417, 437 (the last recording "satin white"-flowered *C. alpinum*), 450, 579, 603 and 783 in volumes 165–184 also feature cyclamen. Other interesting articles include:

Grey-Wilson, C. (1986). *Cyclamen purpurascens*. *Curtis's Botanical Magazine* 3: 104–109.

Grey-Wilson, C. & Wilford, R. (1998). *Cyclamen colchicum*. *Curtis's Botanical Magazine* 15: 180–185.

Garden. The Journal of the Royal Horticultural Society

(1846–present, monthly) ISSN 0308-5457 (a new series, re-numbered as volume 1, was started in 1866)

A consolidated index to this journal has been published for 1846–1985, including indexes to the *Proceedings* from 1939. Recent volumes are indexed in each year's December issue. Many cyclamen articles have been published, mostly short notes or details of plants given awards. Reports on trials at Wisley are published in *Extracts from the Proceedings of the R.H.S.* (see below). A few articles of particular interest include:

Aspland, V. (2005). The cyclamen in the churchyard. *Garden* 130: 734–735.

Bourne, V. (2004). Cyclamen at work. *Garden* 129: 32–35.
➢ An account of work at Tile Barn Nursery.

Bowles, E. A. (1949). Cyclamens in the garden. *Garden* 84: 325–332.

Frank, E. (1991). Expeditions of the Cyclamen Society. *Garden* 116: 552–558.

Grey-Wilson, C. (1998). Cyclamen revolution. *Garden* 123: 639–641.

Kelly, J. (1993). Cyclamen in all their variety. *Garden* 118: 115–121.

Meikle, R. D. (1965). Some notes on *Cyclamen coum*. *Garden* 90: 296–298.

Oldfield, S. & Halliwell, B. (1984). Cyclamen conservation and cultivation. *Garden* 109: 516–518.

Palmer, L. (1960). Cyclamens. *Garden* 85: 210–217.

Palmer, L. & Synge, P. M. (1965). The winter flowering cyclamen. *Garden* 90: 293–296.

Streeter, F. (1950). *Cyclamen persicum*. *Garden* 75: 185.

Toogood, A. & Gilkerson, R. (1997). Cyclamen at Wisley. *Garden* 122: 16–18.

IndexKewensis and *The International Plant Names Index* (*IPNI*)

This substantial work should, perhaps, not be included under journals, but it scarcely fits elsewhere.

Index Kewensis lists the full published references for descriptions of new flowering plant taxa (originally genus and species, but all taxa, from family down, since 1971). It is the prime printed source for tracing the type descriptions of new and old plant names. The first volumes were published in 1895 and regular supplements have appeared since, the most recent in 1996 and all in hard copy format. Since 2000, the work of cataloguing plant names has been taken on by the *International Plant Names Index* and references to all type descriptions are published online at www.ipni.org

Rock Garden. The Journal of the Scottish Rock Garden Club

(1937–present, quarterly) ISSN 0265-5500

Most volumes are indexed, and early issues have a contents page. Recent references are mostly to plants from shows or area studies. Issue Number 1 (before the journal had volume numbers) had an article by G. P. Porter entitled simply: Cyclamen.

Two other articles of interest appeared together:

Dryden, K. (1984). The cultivation of hardy cyclamen. *Rock Garden* 18: 332–337.

Johnstone, R. (1984). Cyclamen cultivation. *Rock Garden* 18: 323–333.

Rock Garden Quarterly. Bulletin of the North American Rock Garden Society

(c.1942– present) ISSN 1081-0765

Most volumes are indexed. By comparison with British publications, this journal contains few references to cyclamen, but the articles listed give an idea of the problems of and possibilities for growing cyclamen in North America. This journal is not widely available in the UK.

Chase, H. (2005). Growing small bulbs in Newfoundland. *Rock Garden Quarterly* 63: 15–36

Lonsdale, J. (2001). The magic of cyclamen. *Rock Garden Quarterly* 59: 243–253.
➢ An Englishman's account of growing cyclamen in Pennsylvania. See also the same author's article in *Journal of the Cyclamen Society* 23: 55–57 (1999).

Firak, G. R. (2005). *Cyclamen purpurascens*. *Rock Garden Quarterly* 63: 56–58.
➢ Also mentions other species. See also the same author's article in *Journal of the Cyclamen Society* 24: 18–19 (2000).

Grothaus, M. (1997). Life with bulbs in an Oregon garden. *Rock Garden Quarterly* 55: 83–85.

Schellinger, K. (1996). Cyclamen for Minnesota gardens. *Rock Garden Quarterly* 54: 51–54.

Royal Horticultural Society, Extracts from the Proceedings

First published in 1838, the *Extracts* (which were preceded by the *Transactions*) record R.H.S. awards to plants and the results of R.H.S. trials at Wisley and elsewhere. Reports on trials of *C. persicum* cultivars have appeared in 1973 (vol. 98: p. 103–105), 1981 (vol. 106: p. 81–84), 1989 (vol. 114: p. 130–132) and 1997 (vol. 121: p. 84–85). Until 1990, articles published in the *Extracts* are included in indexes to *The Garden* (entries for the *Extracts* are in parentheses or italicised in these indexes), subsequently in the *Extracts* themselves. From 2004, the *Extracts* have been published only on line.

9.3 FLORAS

Works covering the flora of several countries

Blamey, M. & Grey-Wilson, C. (1988). *Wildflowers of the Mediterranean*. Harper Collins, London. Republished in 2004 by A. & C. Black, London. 560 pp. ISBN 0-7136-7015-0.

➢ A good general Flora giving a broad picture of where to find some ten species of cyclamen.

Holubec, V. & Krivka (2006). *The Caucasus and its Flora*. Loxia, Prague. 388 pp. ISBN 80-9025413.

➢ Six good photographs but limited data on distribution.

Meikle, R. D. & Sinnott, N. H. (1978). Cyclamen. In: *Flora Europaea* vol. 3. ed. T. G. Tutin *et al.*, pp. 25–26. Cambridge University Press, Cambridge.

North, C. (1997). *A Botanical Tour Round the Mediterranean*. New Millennium, London. xiv + 502 pp. ISBN 1-8584 5075-6

Polunin, O. (1980). *Flowers of Greece and the Balkans*. Oxford University Press. xv + 592 pp. ISBN 0-1921-7626-9.

➢ Includes references for Albania, Bulgaria and most of the former Yugoslavia as well as Greece.

Post, G. E., revised by J. E. Dinsmore (1933). *Flora of Syria, Palestine and Sinai*. Volume 2. American Press, Beirut. xviii + 928 pp.

Sheasby, P. (2007) *Bulbous Plants of Turkey and Iran*. Alpine Garden Society, Pershore, U.K. 280 pp. ISBN 978-090-004877-7.

➢ Well illustrated but scant data on distribution.

Works covering the flora of a single country

Adler, W., Oswald, K. & Fischer, R. (1994). *Exkursionflora von Österreich*. Eugen Ulmer, Stuttgart and Wien. 1181 pp. ISBN 3-8001-3461-6.

➢ Records the presence of *C. purpurascens* in Austria.

Brun, B., Brun, L., Conrad, M. & Gamisson, J. (1975). *La Nature en France: Corse*.

Horizon de France. 220 pp. ISBN 2-85027-070-9.

➢ Report on the Cyclamen Society's field study in Corsica in *Cyclamen* 27: 50–57, provides more detail.

Clement, E. J. & Foster, M. C. (1994). *Alien Plants of the British Isles*. B.S.B.I., London. xviii + 590 pp. ISBN 0-9011-5823-2.

➢ Presents a fascinating summary of cyclamen species that are naturalised in Britain.

Coste, H. (1903). *Flore Descriptive et Illustrée de la France*: vol. 2. Paul Klinksieck, Paris. 627 pp.

Feinbrun-Dothan, N. (1978). *Flora Palaestina* vol. 3. Israel Academy of Sciences and Humanities, Jerusalem. 481 pp. ISBN 965-208-000-4.

➢ Limited information only; more detail can be found in the report of the Cyclamen Society's field study in Israel in *Cyclamen* 14: 15–16.

Fielding, J. & Turland, N. (2005). *Flowers of Crete*. Royal Botanic Gardens, Kew; xix + 650 pp. ISBN 1-842-079-X.

Fournier, P. (1961, 2nd ed. 1977). *Les Quatres Flores de la France*. (2nd ed. in two volumes, vol. 1 text; vol. 2 plates). Lechevalier, Paris. xlviii + 1105 pp. ISSN (2nd ed.) 2-7205-0493-9.

Hegi, G. (1927). *Illustrierte Flora von Mittel Europa*. Vol. 5, part 3. J. F. Lehmanns Verlag, München. pp.1567–2250.

Hess, H. E., Landolt, E. & Hirzel, R. (1976). *Bestimmungschlüssel zur Flora der Schweiz*. Birkhauser, Basel and Stuttgart. 657 pp. ISBN 3-7643-0656-4.

➢ Records *C. hederifolium* and *C. purpurascens* from Switzerland.

Kit Tan (2001). *Endemic Plants of Greece: The Peloponnese*. Gad Publishers, Copenhagen K, Denmark. ISBN 87-12-03857-1.

➢ A sumptuous work with 111 colour plates.

Mathew, B. & Özhatay, N. (2001). *The Cyclamen of Turkey*. A guide to the species of *Cyclamen* growing in Turkey.

➢ Many reports of the Cyclamen Society's work in Turkey, giving more detail, can be traced in the index to *Journal of the Cyclamen Society*, see p. 530.

Meikle, R. D. (1978). *Cyclamen*. In: Davis, P. H. (Ed.) *Flora of Turkey and the East Aegean Islands*. Vol. 6. Edinburgh University Press, Edinburgh. xviii + 825 pp. ISBN 0-8522-4336-7.

Meikle, R. D. (1985). *Flora of Cyprus*. Vol. 2. Bentham Moxon Trust, Kew. xiii + 1134 pp. ISBN 0-9504-8764-3.

Moore, P. J. M. & Jope, M. (2011). *The Cyclamen of Greece*.

➢ Many detailed reports of the Cyclamen Society's work in Greece can be traced in the index to *Journal of the Cyclamen Society*.

Pignatti, S. (1982). *Flore d'Italia*. Vol. 2. Edagricola, Bologna. 732 pp. ISBN 88-206-2311-0.

Pobedimova, E. G. (1967). *Cyclamen*. In: B. K. Shishkin & E. G. Bobrov (eds.). *Flora of the USSR*. Vol. XVIII. Israel Program for Scientific Translation, Jerusalem; viii + 600 pp.

➢ Includes details of the many species in the 'Coum Group' described by Pobedimova.

Romo, A. M. (1994). *Flores Silvestres de Baleares*. Editoriales Rueda, Madrid; 412 pp. ISBN 84-7207-073-5.

Turland, N. J., Chilton, L. & Press, J. R. (1993). *Flora of the Cretan Area*. HMSO, London. xii + 439 pp. ISBN 0-11-310043-4.

➢ Basically a check list with distribution maps using a 10 km square base.

9.4 THE HISTORICAL SEQUENCE

Pre-1800

Books

Most herbals (and the majority of works published before 1800 are herbals) include illustrations of and reference to cyclamen, often taking information and illustrations from earlier titles. The number of references to cyclamen indicates significant interest in the genus from physicians and apothecaries throughout Europe. This is a time when the study of plants was almost exclusively the sphere of physicians and apothecaries and usually, therefore, tied to the officinal uses of plants and, necessarily, their identification. The nomenclature used in these works is basically pre-Linnaean and often difficult to reconcile with present names, so many illustrations are not easy to identify.

The titles listed here are a selection from various countries covering nearly three centuries. When searching indexes of this period, it is important to look not just for cyclamen but also for variations on the name (e.g. ciclamen), various forms of sowbread (e.g. pain porceau, pan porcino or sewbrot) and earth apple (e.g. erdapfel, rapum terrae or terrae malum).

Bauhin, C. (1621). *Pinax Theatri Botanici*. J. Regis, Basel. Lists many names, including *C. vernum luteum*.
➢ Many later herbals take illustrations and information from this work.

Besler, C. (1613). *Hortus Eystettensis*.
➢ Excellent illustrations of four plants, including *C. longius radicatis* (also called *C. odoratus*) with an unusually long floral trunk, probably *C. purpurascens*.

Boerhaave, H. (1720). *Index Alter Plantarum Quae in Horto Academica Lugdano-Batavo*. Vol. 2. Lugdano-Batavo [Leiden]. An updated edition of *Horti Academici Lugdano-Batavi Catalogus* (1687).
➢ Includes reference to two double-flowered plants *C. vernum album flore pleno* and *C. orbiculato folio inferne purpurascente flore pleno* (both probably *C. coum*).

Bock, H. (1539). *Krätterbuch*. Strassburg. (The 1630 edition was examined for this note.)
➢ Reference to some twelve species under antique names from *C. major* to *C. fibre odorato*. One illustration of a plant that has non-auriculate flowers and marbled leaves includes two spotted leaves reminiscent of *C. rhodium* subsp. *peloponnesiacum*. The tuber and most of the leaves, however, resemble *C. graecum*. Could this have been drawn from mixed material?

Brunfels, O. (1530–1536). *Herbarium Vivae Eicones*. J. Schott, Strassburg.
➢ Vols. 1 and 2 contain discussion of the cyclamen known at the time; probably the earliest reference to cyclamen since classical times.

Clusius, C. (1601). *Rariorum Plantarum*. Officina Plantiniana, Antwerp.
➢ Illustrates and describes three species: probably *C. hederifolium*, *C. purpurascens* and *C. repandum*.

Cornut, J. (1635). *Canadensium Plantarum Historia*. S. Lemoyne, Paris.
➢ Illustrates a double cyclamen (probably *C. graecum*) with the name *C. orientale*.

Deville, N. (probably 1670). *Histoire des Plantes de Europe et des Plus Visites*. Duplain, Lyon.
➢ One illustration, probably of *C. coum*, with a description and uses. This book has a complex history of editions; I examined the 1737 edition.

Dodoens, R. (1554). *Crüydeboeck*. van der Loë, Antwerp.
➢ The second edition of 1644 includes more information than the original.

Gerard, J. (1597). *Herball*. J. Norton, London. Revised by Thomas Johnston (1633).
➢ A facsimile reprint of the 1633 edition was published in 1975 by Dover Publications, New York. ISBN 0-486-23147-X. Four plants are illustrated and described.

L'Obel, M. (1676). *Plantarum Seu Stirpium Historia*. C. Plantini, Antwerp.
➢ Illustrates and describes *C. hederifolium* and mentions "*C. vernum flore rubro*". A later work by L'Obel, *Plantarum Seu Stirpium Icones* (1681), includes several illustrations taken from earlier publications.

Lonitzer, A. (1565). *Botanicum. Plantae Historiae*. C. Enolphus, Francoforti.
➢ Descriptions and illustrations owe much to Bock's *Krätterbuch*.

Mattioli, P. A. G. (1586). *Neu Volkommenes Krauter-buch*. J. J. Decter, Basel.
➢ Illustrations and discussion of two plants. One, *C. macrorrhizos*, with a very long, bifurcated floral trunk with roots and swellings along the whole length.

Morison, R. (1715). *Plantarum Historia Universalis Oxoniensis*. Vol. 3. Oxford & London.
➢ Both the illustrations and the text owe a great deal to earlier work but provide a useful overview of knowledge of and interest in cyclamen at the time.

Munting, A. (1696). *Naauwkeurige Beschriving der Aardgewassen*. 2 volumes. Leyden & Utrecht.
➢ Includes reference to a double, yellow-flowered cyclamen: *C. vernum montanum creticum flore luteo pleno* (and to other fabulous plants, including a double red narcissus!).

Parkinson, J. (1629). *Paradisi in Sole. Paradisus Terrestris*. Facsimile reprint published in 1904 by Methuen, London.

Frontispiece from John Parkinson's *Paradisi in Sole. Paradisus Terrestris* (1629). By permission of the trustees of the Natural History Museum (London)

➢ There is even a cyclamen illustrated on the title page! Some ten cyclamen are described including the peach-coloured, double-flowered *C. antiochenum* (now accepted as almost certainly *C. persicum*); probably the first reference to a double cyclamen in English (but what is meant by peach coloured?).

Porta, R. (1650). *Phytognomica*. I. Berthelin, Rothmagi [Rouen].

➢ One illustration of a round tuber, which looks corky, with roots all over and non-auriculate flowers. Might this have been drawn from memory?

Salmon, W. (1710). *Botanologia. The English Herbal or History of Plants*. Rhodes and Taylor, London.

➢ One of the longest and most interesting entries in all of the herbals comprising five folio pages with two illustrations. Includes notes on cultivation and suggests that "they are sooner propagated by roots slipped of" [than by seed] and that "one stock is enough to produce a great many because it is an officinal herb".

Turner, W. (1551). *A New Herball*. S. Mierdman, London. Reprinted in 1989 by Cambridge University Press. ISBN 052144548-5.

➢ Concentrates on uses, no illustrations.

1800 to 1900

At the beginning of the nineteenth century, any interest in cyclamen was still predominantly that of physicians and apothecaries. Interest in the genus as conservatory plants burgeoned by the mid nineteenth century, but the cultivation and breeding of *C. persicum* dominated the scene with hardy species playing only a minor role. In the later years of this century, good, solid morphological and taxonomic work — crowned and summarised by Hildebrand's monograph in 1898 — began as universities, especially in continental Europe, increasingly turned to research.

Books

De Candolle, A. P. (1808). *Icones Plantarum Galliae Rariorum* Vol. 1 (the only volume published). Paris.

➢ Describes and illustrates *C. linearifolium*: an auriculate species with narrow linear leaves. It has been suggested that this 'species' was the result of mixing leaves and flowers on a herbarium sheet (see Aspland (1985) in *Journal of the Cyclamen Society* 9: 39–44).

Edwards, F. C. (1894). *Cyclamen and How to Grow Them*. Manchester.

➢ Deals with *C. persicum*. A semi popular work giving some idea of the importance of cyclamen.

Hildebrand, F. (1898). *Die Gattung Cyclamen L. eine Systematische und Biologische Monographie*. Gustav Fischer, Jena.

Die

Gattung Cyclamen L.

eine systematische und biologische Monographie

von

Dr. Friedrich Hildebrand,

Professor der Botanik an der Universität zu Freiburg i. B.

Mit 6 lithogr. Tafeln.

JENA.

Verlag von Gustav Fischer

1898.

Hildebrand's superlative monograph of 1898 (BM)

Friedrich Hildebrand (1835–1915) (JB)

➢ A starting point for the taxonomy and morphology of cyclamen. The thirteen species that were recognised at the time (listed as *C. persicum*, *C. europaeum*, *C. repandum*, *C. balearicum*, *C. cilicicum*, *C. ibericum*, *C. coum*, *C. alpinum*, *C. neapolitanum*, *C. afrikanum*, *C. cyprium*, *C. graecum* and *C. rohlfsianum*) are described, and the book provides general information on the biology and distribution of the genus. The descriptions and drawings of each species provide detailed morphological information unequalled at the time (or even since). An English translation, *The Genus Cyclamen L.*, translated by Erna Frank and including a brief biographical note on Hildebrand, was published by the Cyclamen Society in 1999 (ISBN 0-9537526-0-7).

Jonghe, J. (1844). *Monographie du Genre Cyclamen*. Brussels.

Articles in periodicals

Many articles and short notes on cyclamen have been published in horticultural journals, often in answer to readers' queries. Many journals also included references to cyclamen, usually *C. persicum* cultivars, in lists of "new plants" or in notes from shows. An extensive list of illustrations of *C. persicum* cultivars, from 1827 until 1935, is included in Doorenbos, 1950 (see p. 540).

'A foreman of a London nursery' (1843). On the cultivation of cyclamens. *Floricultural Cabinet* 11: 235.

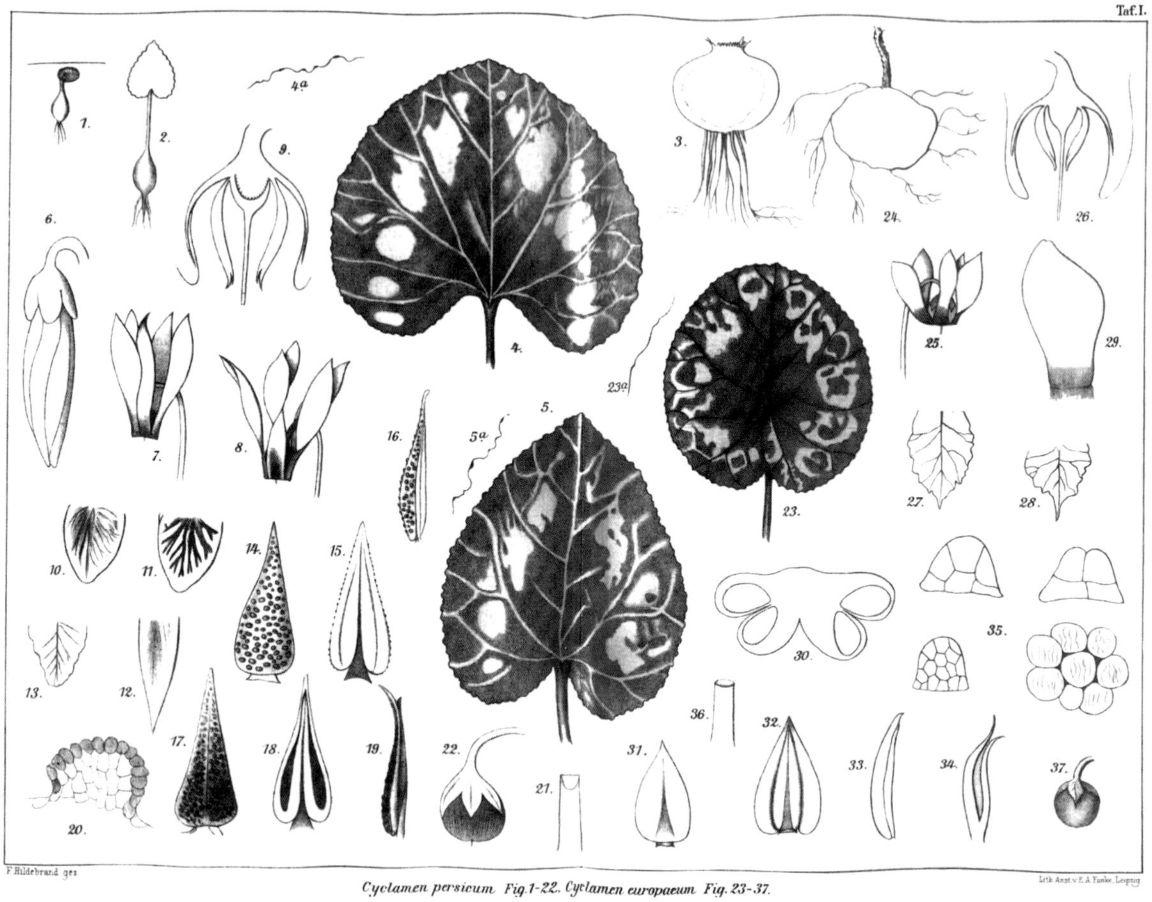

One of the plates from Hildebrand's monograph, giving an indication of the extraordinary detail of his observations (BM)

André, E. (1886). Origine des Cyclamens à fleurs doubles. *Rev. Hort.* 58: 515.

Baker, J. G. (1883). A synopsis of the species of Cyclamen. *Gard. Chron.* n. ser. 19: 307–308; 339–340; 468.

C. B. B. (1837). On the culture of cyclamen. *Floricultural Cabinet* 5: 58.
➣ A short note typical of many published in florists' magazines of the time.

C. W. S. (1879). Flower forcing in market gardens. *Garden* 15: 43–44.
➣ About one-third of this article is devoted to cyclamen, describing the techniques used by a nurseryman who was cultivating 12,000–15,000 plants annually. Cyclamen were grown "in pans and boxes placed on a bed of cocoa nut fibre".

Dauthenay, H. (1900). Nouvelles variétés de cyclamens à grandes fleurs. *Rev. Hort.* 72: 390–391.
➣ Includes an excellent coloured lithograph.

Dyer, W. T. (1887). The cultural evolution of *Cyclamen latifolium*. *Proc. Roy. Soc. B* 61: 135.

Dyer, W. T. (1897). Progress of Cyclamen. *Gard Chron.* ser. 3, 21: 316–318; 330–332; 350.
➣ Illustrates some early cristate forms of *C. persicum*.

Gordon, G. (1843). On the varieties of Cyclamen. *Gard. Chron.* 3: 660.

G. W. (1880). Cyclamen at Ealing Dean. *Garden* 18: 665–666.
➣ An account of *C. persicum* at a London nursery.

Hildebrand, F. (1896). Über die eigentümliche haarbildungen auf den knollen einiger arten von *Cyclamen*. *Bot. Zeit.* 54: 133–138.

T. P. T. (1875). A good method of growing the Persian cyclamens. *Garden* 7: 98.

Wilmot, J. (1826). On the culture of the *Cyclamen persicum*. *Gardener's Magazine* 1: 386–388; followed by a letter answering a query in 2: 377–378.

1901 to 1975

Basic scientific and cultural work continued in the first half of the twentieth century, albeit interrupted by two world wars. Early in this period, Mendel's work on genetics was rediscovered and the importance of chromosomes in inheritance recognised, giving a fresh approach and impetus to breeding and research. The discovery of the structure of DNA in 1955 and the recognition of its importance to taxonomy (as opposed to genetics) came too late to be of importance to studies of cyclamen in this period, work in that area came later. Many works still concentrated on *C. persicum*.

Books

Blasdale, W. C. (1952). *Cyclamen persicum. Its Natural and Cultivated Forms*. Stanford University Press, Stanford, California.
➣ A semi-popular work by a professor of chemistry whose scope is defined by its title. An important book in its time.

Maatsch, R. (1971). *Cyclamen*. Paul Parey, Berlin.
➣ An update of Schneider and Maatsch (1950).

Meyer, F. (1952). *Alpenveilchen. Ein Einblick in die Gattung Cyclamen, im Besonderen in die Gärtnerische Entwicklung des Cyclamen persicum und seine Kultur*. Eugen Ulmer, Stuttgart.
➣ Covers distribution of those species recognised at the time of publication, more than half of the book covers the development of techniques for cultivating *C. persicum*.

Puttock, A. G. (1958). *The Cyclamen*. John Gifford Ltd., London.
➣ A modest work with some nice illustrations, mostly black and white. Includes a note on early Dutch work on the *in vitro* cultivation of cyclamen embryos.

Saunders, D. E., revised by Meikle, R. D. & Grey-Wilson, C. (1973). *Cyclamen. A Gardener's Guide to the Genus*. Alpine Garden Society, London.
➣ The original version of this book was first published in *Quart. Bull. Alpine Gard. Soc.* 27: 18–25. Since this first revision, the work has been further revised by Christopher Grey-Wilson, most recently in 2007. ISBN 0-9-00048-1.

Seghers, N. (1916, 2nd ed. 1925). *Monographie du Genre Cyclamen. Histoire-evolution, Culture et Multiplication*. Brussels, (2nd ed. Alph. Leys, Ostende).
➣ A scarce volume, largely concerned with the large-flowered hybrids of *C. persicum*, allowing an interesting insight into early twentieth century methods of cultivation. It includes an early account of the development of these hybrids.

Schneider, F. & Maatsch, R. (1950). *Cyclamen*. Paul Parey, Berlin and Hamburg.
➣ Largely concerned with the cultivation of *C. persicum*; there are some interesting illustrations both from earlier works and of contemporary plants.

Wellensiek, S. J., Doorenbos, J., van Bragt, J. & Legre, R. A. H. (1961). *Cyclamen. A Descriptive List of Cultivars*. Publication 200, Laboratorium voor Tuinbouwplantenteelt, Landbouwhogeschool Wageningen.
➣ The first list published by the International Registration Authority for *C. persicum* hybrids. Includes information on the chemistry of flower colour in cyclamen, the history of *C. persicum* and the genetics and cytology of the species. A list of 19 references is included. Well illustrated.

Articles in periodicals

De Haan, I. & Doorenbos, J. (1951). The cytology of the genus cyclamen. *Meded. Landbouwhogeschool Wageningen* 51: 151–170.

Doorenbos, J. (1950). Taxonomy and nomenclature of cyclamen. *Meded. Landbouwhogeschool Wageningen* 50: 19–34.

Doorenbos, J. (1951). History of the 'Persian' cyclamen. *Meded. Landbouwhogeschool Wageningen* 51: 33–66.

Doorenbos, J., Legro, R. A. H. & van Bragt, J. (1962). Description of *Cyclamen* cultivars and its importance for breeding. *Adv. Hort. Sci. and Appl.* 2: 326–331.

Glasau, F. (1939). Monographic der Gattung Cyclamen auf morphologischcytologischer grundlage. *Planta* 30: 507–550.

Hill, A. W. (1920). Studies in seed germination. Experiments with Cyclamen. *Ann. Bot.* 34: 417–429.

➢ A frequently cited paper describing propagation from cotyledon and leaf cuttings — not, it appears, entirely successful. Perhaps more work is needed to investigate this?

Neuray, G. (1973). Bud formation in *Cyclamen persicum*. *Acta Horticulturae* 31: 77–79.

Legro, R. A. H. (1959). The cytological background to *Cyclamen* breeding. *Meded. Landbouwhogeschool Wageningen* 59: 1–51.

Lepper, L. (1975). Nachtrag zur cytologie von *Cyclamen* L. *Z. Friedrich Schiller Univ. Math.-Naturwiss. Reihe* 24: 426–429.

Le Texnier (1909). Le cyclamen. *Rev. Hort.* nuov. ser. 9: 408–410.

➢ A brief summary of the introduction and development in Europe. Re-printed in a series of pamphlets: *Essais sur l'Histoire de quelques Fleurs d'Ornement* using the author's name, F. Le Tesnier, by Paul Pigelet et fils, Orleans.

Schwarz, O. (1938). Cyclamen studien. *Gartenflora* n.s. 1938: 11–38.

Schwarz, O. (1955). Systematische monographic der gattung *Cyclamen* L. Teil 1. *Feddes Repert. Sp. Nov. Regni Veg.* 58: 234–283.

Schwarz, O. (1964). Systematische monographie der gattung *Cyclamen* L. Teil 2. *Feddes Repert. Sp. Nov. Regni Veg.* 69: 73–103.

Stearn, W. T. (1970). The enigmatic *Cyclamen indicum* of Linnaeus. *Israel Journ. Bot.* 19: 266–270.

➢ Reprinted in *Journal of the Cyclamen Society* 3: 22–28 (1979).

Valenta, V. (1947). A monograph of abnormities [abnormalities] in *Cyclamen persicum* Mill. *Studia Botanica Čechoslovaca* 8: 125–202.

➢ Contains a useful bibliography listing many eastern European works seldom referred to elsewhere. For other information on this topic see *Journal of the Cyclamen Society* 17: 36 (1993) and 29: 45–46 (2005).

Wendelbo, P. (1961). Studies in Primulaceae III. On the genera related to *Primula* with special reference to their pollen morphology. *Årbok Univ. Bergen, Mat-Naturv. Ser.* 19: 1–31.

Widmer, R. E., Stephens, L. C. & Angell, M. V. (1974). Giberellin accelerates flowering of *Cyclamen persicum* Mill. *Hort. Science* 9: 476–477.

Post-1975

By now new ideas and techniques, in particular micropropagation, the use of DNA in taxonomic studies and the development of F_1 hybrids in plant breeding, had begun to revolutionise both botany and horticulture. For cyclamen, however, the most significant single event in this period was undoubtedly the foundation of the Cyclamen Society. The Society has sponsored the largest ever programme of field research into cyclamen, a programme that has extended our knowledge of the ecology and distribution of the genus, collected material for laboratory research here in Britain and introduced new plants to gardens.

Work on the cytology of cyclamen has been continued in Europe and elsewhere and, related to this, new techniques aimed at producing interspecies hybrids have been studied in Japan. *In vitro* reproduction of *C. persicum* has also been studied both in Europe and Japan. At the University of Reading, material collected by the Cyclamen Society has been used in taxonomic studies, including DNA sequencing, and in studies of climate change sponsored by the Society.

Breeding programmes have produced a vast assortment of new types, including F_1 hybrids, in *C. persicum*, which is one of the most important pot-plant crops in commerce — on a visit to Hong Kong, I saw plants grown in The Netherlands on sale in Kowloon flower market. Despite the claims of nurserymen, however, I am sceptical that a hardy *C. persicum* has yet been produced. Little breeding has been attempted with species other than *C. persicum*, new cultivars coming more often from selections made in gardens or from wild populations.

Books

Bavcon, J. (2009). *Common Cyclamen (*C. purpurascens *Mill.) and its Diversity in Slovenia.* Department of Biology, University Botanic Garden, Ljubljana. ISBN 978-961-90262-8-1.

➢ A local study of *C. purpurascens*, its ecology and variability; includes a classification of the wild 'forms'.

Bongartz, W. (1999). *Cyclamen.* Thalacker Median, Braunschweig. ISBN 3-87815-116-0.

➢ In German, principally considers *C. persicum* cultivars, although there is a wide range of information. Well illustrated, with an excellent list of references, mostly in German.

Eames, A. & Potter, R. (1998). *Cyclamen Essentials. Grower Digest 2.* Grower Books, Swanley, Kent. ISBN 1-899372-08-3.

➢ A concise, clearly written guide for commercial growers of *C. persicum* cultivars. Nonetheless, it is useful to growers of any species for its comprehensive list of pests and diseases, although many of the chemical controls listed are no longer available, at least to amateur growers.

Grey-Wilson, C. (1988). *The Genus Cyclamen.* A Kew Magazine Monograph. Royal Botanic Gardens Kew and Christopher Helm. ISBN 0-88192-111-4.

A recent work devoted to one species of cyclamen and confined to one study area, Slovenia (BM)

The most recent comprehensive monograph of *Cyclamen*, by Christopher (Kit) Grey-Wilson (BM)

➢ A useful work presenting the state of knowledge at the time, beautifully illustrated with watercolours by Mary Grierson.

Grey-Wilson, C. (1997, revised edition 2002). *Cyclamen, a Guide for Gardeners, Horticulturists and Botanists*. B. T. Batsford, London. ISBN 0-7134-7864-0.

➢ A comprehensive monographic treatment, presenting a new classification and describing several new taxa. Many colour photographs. The 2002 revised edition (ISBN 0-7134-8760-7) provided enhanced coverage with many, larger photographs. A wealth of information brought together.

Hendrikx, P. (2002). Cyclamen: *Handbook voor Liefhebbers*. Procyklòs, Hooten, The Netherlands, ISBN 90-9014176-6.

➢ A comprehensive guide to cyclamen species, their identification and cultivation. Nice illustrations.

Nightingale, G. (1982, revised edition 1984). *Growing Cyclamen*. Croom Helm, London; Timber Press, Beaverton, U.S.A., ISBN 0-70991805-4; paperback edition 1987, ISBN 0-7470-2402-2.

➢ Covers cultivation, in all situations, of the range of *Cyclamen* species; includes some interesting information on the history of the genus and other incidental topics.

Rampinini, G. (1991). *La Coltivazione del Ciclamino. Non tutto, ma quasi. Guida Pratica alla Coltivazione del Ciclamino*. Pentagono Editrice, Milan. (no ISBN assigned).

➢ A guide to the commercial production of *C. persicum* with descriptions of cultivars, pests and diseases. Well illustrated with good coloured photographs.

Schwenkel, H.-G. (ed.) (1999). *Reproduction of* Cyclamen persicum *Mill. Through Somatic Embryogenesis using Suspension Culture Systems*. European Commission Directorate General for Research, Luxembourg. ISBN 912494- 02-4-91-0.

➢ A multi-authored volume comprising papers reporting basic research on *Cyclamen* cultivation *in vitro*, excellent lists of references, but not for the faint hearted. A brief article on this theme appeared in the *Journal of the Cyclamen Society* 24: 50–55 (2000).

Sønderhousen, O. (1933). *Cyclamen Vildarter*. Jyllandsgruppen, Alpine Garden Society. (no ISBN assigned).

➢ An introduction to the genus, some of the information on distribution is questionable.

Szczepaniak, S. (2006). *Cyklameny*. PWRiL, Poznan. ISBN 83-09-01829-0.

➢ Basically covers *C. persicum*. Useful bibliography of German and eastern European publications not often seen in Britain.

van der Plas-Haarsma, M. (1987). *Cyclamen in Trade*. Traffic Rapport Nr. 5, 57 pp. Traffic (Nederland), Zeist.

➢ A thorough compilation of data on trade in *Cyclamen* that provides a useful baseline for further work. For further information on trade in Turkey see: *Journal of the Cyclamen Society* 25: 12–14 (2001).

Widmer, R. E. & Lyons, R. E. (1985). *Cyclamen persicum. Handbook of Flowering Plants*. Vo1. 2. CRC Press. A book for the market grower.

Articles in periodicals

Affre, L., Thompson, J. D. & Debussche, M. (1995). The reproductive biology of the Mediterranean endemic *Cyclamen balearicum* Willk. (Primulaceae). *Bot. Jour. Linn. Soc.* 119: 309–330.

Affre, L., Thompson, J. D. & Debussche, M. (1997). Genetic structure of continental and island populations of the Mediterranean endemic *Cyclamen balearicum* (Primulaceae). *Amer. Jour. Bot.* 84: 437–451.

Anderberg, A. A. (1994). Phylogeny and subgeneric classification of *Cyclamen* L. (Primulaceae). *Kew Bulletin* 49: 455–466.

Anon. (1983). Hardy cyclamen. *Gardening from Which?* 1983: 283–287.

Bailey, R. H. (1991). Has there ever been a yellow Cyclamen? *Arch. Nat. Hist.* 18: 231–235.

Bennett, S. T. & Grimshaw, J. M. (1991). Cytological studies in *Cyclamen* subg. *Cyclamen* (Primulaceae). *Plant Syst. Evol.* 176: 135–143.

Clennett, J. C. (2002). An analysis and revision of *Cyclamen* L. with emphasis on subgenus *Gyrophoebe* O. Schwarz. *Bot. Jour. Linn. Soc.* 138: 473–481.

Compton, J., Clennett, J. C. & Culham, A. (2004). Nomenclature in the dock. Overclassification leads to instability: a case study in the horticulturally important genus *Cyclamen* (Myrsinaceae). *Bot. Jour. Linn. Soc.* 146: 339–349.

Greilhuber, J. (1989). Karyotype structure and evolution in *Cyclamen* L. subgen. *Psilanthum* Schwz. (Primulaceae). *Flora* 183: 103–113.

Grey-Wilson, C. (1991). Cyclamen — a reappraisal. *New Plantsman* 13: 1–20.

Grey-Wilson, C. (2002). Sprenger's alpine cyclamen. *Plantsman* (N.S.) 1: 173–177.

Ishizaka, H. (2003). Cytogenetic studies in *Cyclamen persicum*, *C. graecum* (Primulaceae) and their hybrids. *Plant Syst. Evol.* 239: 1–14.

Ishizaka, H. & Uematsu, J. (1995). Interspecific hybrids of *Cyclamen persicum* Mill. and *C. purpurascens* Mill. produced by ovule culture. *Euphytica* 82: 31–37.

Ishizaka, H. & Uematsu, J. (1995). Amphidiploids between *Cyclamen persicum* Mill. and *C. purpurascens* Mill. induced by treating ovules with colchicines *in vitro* and sesquidiploids between the amphidiploid and the parental species induced by conventional crosses. *Euphytica* 86: 211–218.

Sealey, J. L. (1981). Cyclamen for western gardens. *Pacific Horticulture* 142: 21–26.

Thulin, M. & Warfa, A. M. (1989). *Cyclamen* (Primulaceae) in tropical Africa. *Pl. Syst. Evol.* 166: 249–252.
➢ Includes the type description of *C. somalense*.

Yesson, C. & Culham, A. (2006). A phyloclimatic study of *Cyclamen*. *BMC Evolutionary Biology* 6: 72.

Yesson, C. Toomey, N. H. & Culham, A. (2009). Cyclamen: time, sea and speciation biogeography using a temporally calibrated phylogeny. *J. Biogeogr.* 36: 1234–1252.

9.5 MISCELLANEA CYCLAMENOLOGICA

Sundry ephemera, of the type known to librarians as 'grey literature', also feature *Cyclamen*. The following is a small selection. Dates are usually unknown; the item might be subject to frequent, even continual, revision and the place of publication, author or publisher might change. n.d. = no date provided.

Anon (2009). An untitled, coloured cartoon showing cyclamen being bulldozed away illustrating an article about settlements in the West Bank lands east of Jerusalem. The only cartoon I know illustrating cyclamen. *The Guardian* 26th July 2009: 26.

Ashwood Nurseries (n.d.) *Cyclamen. A Cultural Guide*. Kingswinford, West Midlands.

Bailey, R. H. (1988). A visit to some Turkish cyclamen. *Extra. Newsl. of the Dept. of Extra-Mural Studies, University of London.*

Colegrave's Seeds (1981). Cyclamen culture. *Grower Information Bulletin* C4.

Culham, A. (2005). *Back to the Future*. Walker Institute, University of Reading. Considers the effects of climate change on cyclamen.

Cyclamen Society (n.d.). *Cyclamen from Seed. A Growers Guide.*

Cyclamen Society (n.d.). *Cyclamen Species. Descriptions and Cultural Advice, for Growing Outdoors or Under Glass.*

Department of the Environment (n.d.). *Endangered Plants*. London.
➢ A leaflet advising travellers of plants requiring permission for importation into Britain under the Convention on International trade in Endangered Species (CITES) regulations.

Grant, M. (2012). Is it a bulb, corm or tuber? *Journal of the Royal Horticultural Society* 137: 17.
➢ Discusses the differences between the structures — and ends by saying, that for cyclamen "there is no consistent view from plant morphologists"!

Royal Horticultural Society (2004 onwards). *Cyclamen as Pot Plants* (Horticultural Advisory Services leaflet).

Scott, M. A. (1983). *Miniature Cyclamen.* Leaflet No. 5. Efford Experimental Horticulture Station, Ministry of Agriculture, Fisheries and Food.

References by chapter

CHAPTER 1

Habermehl, G. (1985), *Mitteleuropäische Giftpflanzen und ihre Wirkstoffe*, Springer-Verlag, Dordrecht.

Ministry of Agriculture Fisheries and Food, UK (MAFF) (1979). Bulletin 161, British Poisonous Plants; HMSO, London.

Schwartz-Tzachor, R., Eisikowitch, D. & Dafni, A. (2005). Flower characteristics and breeding system of two phenological ecotypes of *Cyclamen persicum* Mill. (Myrsinaceae) in Israel. *Oesterreichisches Botanisches Wochenblatt* 274 (1): 127–134.

Steinegger, E. & Hansel, R. (1988). *Lehrbuch der Pharmakognosie und Phytopharmazie*. Springer-Verlag, Dordrecht.

CHAPTER 2

2.4 Conservation

Affre, L., Thompson, J. D. & Debussche, M. (1995). The reproductive biology of the Mediterranean endemic *Cyclamen balearicum* Willk. (Primulaceae). *Botanical Journal of the Linnean Society* 118: 309–330.

Andrieu, E., Debussche, M., Munoz, F. & Thompson, J. D. (2011). How does herbivory affect individuals and populations of the perennial herb *Paeonia officinalis*? *Flora* 206 (6): 544–549.

Andrieu, E., Thompson, J. D. & Debussche, M. (2007). The impact of forest spread on a marginal population of a protected peony (*Paeonia officinalis* L.): the importance of conserving the habitat mosaic. *Biodiversity & Conservation* 16: 643–658.

Grey-Wilson, C. (2002). *Cyclamen. A guide for Gardeners, Horticulturists and Botanists*; 2nd Edition. B. T. Batsford Ltd, London.

Hildebrand, F. (1898). *Die Gattung Cyclamen L., eine Systematische und Biologische Monographie*. Gustav Fischer, Jena.

IPCC (2007). Fourth assessment report on climate change. Available online: http//www.ipcc.ch/publications_and_data/publications_ipcc_fourth_assessment_report_synthesis_report.htm

Jope, M. (2006). Caterpillar damage. *Journal of the Cyclamen Society* 31 (1): 28.

Rupreht, J. (1995). Names for cyclamen. *Journal of the Cyclamen Society* 19: 52–53.

Saez-Royuela, C. & Telleria, J. L. (1986). The increased population of the Wild boar (*Sus scrofa* L.) in Europe. *Mammal Rev*. 16: 97–101.

Sala, O. E., Chapin, F. S., Armesto, J. J. *et al.* (2000). Biodiversity — global biodiversity scenarios for the year 2100. *Science* 287: 1770–1774.

Yesson, C. & Culham, A. (2006). A phyloclimatic study of *Cyclamen*. *BMC Evolutionary Biology* 6: 72.

CHAPTER 3

3.3 Taxonomic comments on subgenus Psilanthum

Affre, L. & Thompson, J. D. (1997). Population genetic structure and levels of inbreeding depression in the Mediterranean island endemic *Cyclamen creticum* (Primulaceae). *Biological Journal of the Linnean Society* 60: 527–549.

Bota, P. M. P. G. (2006). *A Taxonomic and Molecular Investigation in Cyclamen subgenus Psilanthum Schwarz*. M.Sc. dissertation, University of Reading.

Gorer, R. (1963).Wild flowers on Rhodes. *Gardeners' Chronicle* 154: 388–389.

Schwarz, O. & Lepper, L. (1975). Zwei neue *Cyclamen* aus dem östlichen Mittelmeergebiet. *Feddes Repertorium* 86 (9–10): 491–497

Yesson, C., Toomey, N. H. & Culham A. (2008). *Cyclamen*: time, sea and speciation biogeography using a temporally calibrated phylogeny. *Journal of Biogeography* 36 (7): 1223–1252.

Other relevant references can be found in section 3.4 (listed on p. 544).

Further resources:

For those that are interested, a number of websites now archive old literature, so it is possible to see, free of individual charge, the descriptions for some of these *Cyclamen*:

1. *Cyclamen repandum* was first described by Smith in *Floræ Græcæ Prodromus* (Sibthorp & Smith, 1806; viewable free of charge at http://biodiversitylibrary.org/page/6114456) although a much fuller description and illustration appeared later in *Flora Graeca* (Sibthorp & Smith, 1836; http://www.bodley.ox.ac.uk/users/millsr/isbes/FG/FGE1/target354.html and http://www.bodley.ox.ac.uk/users/millsr/isbes/FG/FGE1/target355.html).
2. *Cyclamen balearicum* Willkomm (1875; http://biodiversitylibrary.org/page/9603730) is an old name of long standing for plants in Majorca.

3.4 Ecology and Evolution of subgenus Psilanthum

Affre, L. & Thompson, J. D. (1998). Floral trait variation in four *Cyclamen* (Primulaceae) species. *Plant Systematics & Evolution* 212: 279–293.

Affre, L. & Thompson, J. D. (1999). Variation in self-fertility, inbreeding depression and levels of inbreeding in four *Cyclamen* species. *Journal of Evolutionary Biology* 12: 113–122.

Affre, L., Thompson, J. D. & Debussche, M. (1995). The reproductive biology of the Mediterranean endemic *Cyclamen balearicum* Willk. (Primulaceae). *Botanical Journal of the Linnean Society* 118: 309–330.

Affre, L., Thompson, J. D. & Debussche, M. (1997). Genetic structure of continental and island populations of the Mediterranean endemic *Cyclamen balearicum* (Primulaceae). *American Journal of Botany* 84: 437–451.

Anderberg, A. A. (1994). Phylogeny and subgeneric classification of *Cyclamen* L. (Primulaceae). *Kew Bulletin* 49: 455–467.

Anderberg, A. A., Trift, I. & Källersjö, M. (2000). Phylogeny of *Cyclamen* L. (Primulaceae): evidence from morphology and sequence data from the internal transcribe spacers of nuclear ribosomal DNA. *Plant Systematics & Evolution* 220: 147–160.

Aspland, V. (1994). Expedition to Crete 1994. *Journal of the Cyclamen Society* 18: 42–44.

Aspland, V. (2005). The Cyclamen Church. *Journal of the Cyclamen Society* 29: 15–17.

Bailey, R. H. (1991). The Society's expedition to Rhodes and Kárpathos, 3–17 April 1991. *Journal of the Cyclamen Society* 15: 46–50.

Bailey, R. (1992). *Cyclamen creticum* on Kárpathos. *Journal of the Cyclamen Society* 16: 41–43.

Briquet, J. & Litardière, R. de (1938). *Prodrome de la Flore Corse*, tome III, Lechevallier, Paris: p. 202–203.

Clennett, J. C. B. (2002). An analysis and revision of *Cyclamen* L. with emphasis on subgenus Gyrophoebe O. Schwarz. *Botanical Journal of the Linnean Society* 138: 473–481.

Clennett, J. C. B. (2004). Classification changes in cyclamen — a guide for the grower. *Journal of the Cyclamen Society* 28: 10–11.

Compton, J. A. & Culham, A. (2003). New combinations in *Cyclamen* subgenus *Psilanthum*. *Journal of the Cyclamen Society* 27: 70–72.

Compton, J. A., Clennett, J. C. B. & Culham, A. (2004). Nomenclature in the dock. Overclassification leads to instability: a case study in the horticulturally important genus *Cyclamen* (Myrsinaceae). *Botanical Journal of the Linnean Society* 146: 339–349.

Debussche, M. (2003). Frost and cyclamen under Mediterranean climatic conditions. *Journal of the Cyclamen Society* 27: 14–18.

Debussche, M., Debussche, G. & Affre, L. (1995). La distribution fragmentée de *Cyclamen balearicum* Willk. en France: analyse historique et conséquence des activités humaines. *Acta botanica Gallica* 142: 439–450.

Debussche, M., Debussche, G. & Grandjanny, M. (2000). Distribution de *Cyclamen repandum* Sibth. & Sm. subsp. *repandum* et écologie en Corse et France continentale. *Acta Botanica Gallica* 147: 123–142.

Debussche, M., Garnier, E. & Thompson, J. D. (2004). Exploring the causes of variation in phenology and morphology in Mediterranean geophytes: a genus-wide study of *Cyclamen*. *Botanical Journal of the Linnean Society* 145: 469–484.

Debussche, M., Grandjanny, M., Debussche, G. & Affre, L. (1996). Ecologie d'une espèce endémique et rare à distribution fragmentée: *Cyclamen balearicum* Willk. en France. *Acta Botanica Gallica* 143: 65–84.

Debussche, M., Grandjanny, M., Debussche, G., Mus, M., Torres, N. & Fraga, P. (1997). Ecologie d'une espèce endémique en milieu insulaire: *Cyclamen balearicum* Willk. aux îles Baléares. *Anales Jardín Botánico de Madrid* 55: 31–48.

Debussche, M., Moore, P. & Moore, N. (2006). Confusing cyclamen on Andros. *Journal of the Cyclamen Society* 30: 29–31.

Debussche, M. & Quézel, P. (1997). *Cyclamen repandum* Sibth. & Sm. en Petite Kabylie (Algérie): un témoin biogéographique méconnu au statut taxinomique incertain. *Acta Botanica Gallica* 144: 23–33.

Debussche, M. & Thompson, J. D. (2000). Les populations à fleurs blanches de *Cyclamen repandum* Sibth. & Sm. en Corse. *Bull. Soc. Bot. Centre-Ouest* 31: 81–88.

Debussche, M. & Thompson J. D. (2002). Morphological differentiation among closely related species in disjunct distributions: a case study of Mediterranean *Cyclamen* L. subgen. *Psilanthum* Schwarz (Primulaceae). *Botanical Journal of the Linnean Society* 139: 133–144.

Debussche, M. & Thompson, J. D. (2003). Habitat differentiation between two closely related Mediterranean plant species, the endemic *Cyclamen balearicum* and the widespread *C. repandum*. *Acta Oecologica* 24: 35–45.

Debussche, M. & Thompson, J. D. (2004). The classification of subgenus *Psilanthum* … revisited (yet again!). *Journal of the Cyclamen Society* 28: 12–16.

Denney, M. (1994) Cyclamen on Rhodes. *Journal of the Cyclamen Society* 18: 48–49.

Denney, M. (1996). The Cyclamen Society expeditions to Crete. *Journal of the Cyclamen Society* 20: 36–43.

Denney, M. (1999). Cyclamen Society expedition to the Dodecanese, 1998. *Journal of the Cyclamen Society* 23: 15–19.

Denney, M. (2003). Corsica field study. *Journal of the Cyclamen Society* 27: 50–57.

Denney, M. (2006). CITES update. *Journal of the Cyclamen Society* 30: 56–57.

Fry, K. (1992). *Cyclamen creticum* in Crete. *Journal of the Cyclamen Society* 16: 44.

Gamisans, J. & Marzocchi, J. F. (1996). *La Flore Endémique de la Corse*. Edisud, Aix-en-Provence.

Gielly, L., Debussche, M. & Thompson, J. D. (2001). Geographic isolation and evolution of Mediterranean endemic *Cyclamen*: insights from chloroplast *trn*L (UAA) intron sequence variation. *Plant Systematics and Evolution* 230: 75–88.

Greilhuber, J. (1989). Karyotype structure and evolution in *Cyclamen* L. subgen. *Psilanthum* Schwz. (*Primulaceae*). *Flora* 183: 103–113.

Grey-Wilson, C. (1988). *The Genus Cyclamen. A Kew Magazine Monograph*. Christopher Helm/Timber Press, London, Portland.

Grey-Wilson, C. (1997). *Cyclamen. A Guide for Gardeners, Horticulturists and Botanists*. B. T. Batsford Ltd, London.

Hildebrand, F. (1898). *Die Gattung Cyclamen L., eine Systematische und Biologische Monographie*. Gustav Fischer, Jena.

Johnstone, R. (1992). The Society's Peloponnese expedition, April 1992. *Journal of the Cyclamen Society* 16: 45–47.

Lavergne, S., Thompson, J. D., Garnier, E. & Debussche, M. (2004). The biology and ecology of narrow endemic and widespread plants: a comparative study of trait variation in 20 congeneric pairs. *Oikos* 107: 505–518.

Mathew, B. (1999). White-flowered *Cyclamen repandum* from the Peloponnese. *Journal of the Cyclamen Society* 23: 50–51.

Moore, P. (1993). Peloponnese expedition, April 1993. *Journal of the Cyclamen Society* 17: 50–53.

Moore, P. (1996). *Cyclamen repandum* subsp. *peloponnesiacum* forma *vividum*. *Journal of the Cyclamen Society* 20: 13–14.

Quézel, P. (1956). Contribution à l'étude des forêts de chênes à feuilles caduques d'Algérie. *Mémoires de la Société d'Histoire Naturelle d'Afrique du Nord* 1: 1–57.

Rosselló, J. A., Alomar, G. & Mus, M. (1996). *Flora endemica de Balears*. FODESMA, Consell Insular de Mallorca, Palma.

Schwarz, O. (1964). Systematische Monographie der Gattung *Cyclamen* L., II. *Feddes Repertiorum Specierum Novarum Regni Vegetabilis* 69: 73–103.

Selvi, F. & Stefanini, P. (2005). *Biotopi Naturali e Aree Protette Nella Provincia di Grosseto*. Provincia di Grosseto ed., Grosseto.

Strobelberger, J. S. (1620). *On Rare Plants Which Spontaneously Grow in France, and Especially in Narbonne Region, Languedoc and Provence, with Indications on Their Sites and Other Interesting Details*. Beithmann, Jena. [In Latin]

Ten Hoope, M. (2004). Another 13 days in the Peloponnese. *Journal of the Cyclamen Society* 28: 62–64.

Thompson, D. (2006). Cyclamen in Western Crete. *Journal of the Cyclamen Society* 30: 58–62.

Thompson, J. D., Gaudeul, M. & Debussche, M. (2009). Conservation value of sites of hybridization in peripheral populations of rare plant species. *Conservation Biology* 24 : 236–245.

Thompson, J. D. (2005). *Plant Evolution in the Mediterranean*. Oxford University Press, Oxford, 293 pp.

Wiltshire, T. & Dale, H. (1995). Cyclamen Society expedition to Crete 1995. *Journal of the Cyclamen Society* 19: 54–55.

3.5 Field studies

Affre, L. & Thompson, J. D. (1997). Variation in the population genetic structure of two *Cyclamen* species on the island of Corsica. *Heredity* 78: 205–214.

Debussche, M. & Thompson, J. D. (2002). Morphological differentiation among closely related species with disjunct distributions: a case study of Mediterranean *Cyclamen* L. subgen. *Psilanthum* Schwarz (Primulaceae). *Botanical Journal of the Linnean Society* 139: 133–144.

Grey-Wilson, C. (1988). *The Genus Cyclamen. A Kew Magazine Monograph*. Christopher Helm/Timber Press, London, Portland.

Grey-Wilson, C. (2002). *Cyclamen. A guide for Gardeners, Horticulturists and Botanists*; 2nd Edition. B. T. Batsford Ltd, London.

Meikle, R. D. (1978). *Cyclamen* in Davis, P. H. (ed.), *Flora of Turkey* 6: 128–134. University Press, Edinburgh.

Royal Horticultural Society's Colour Chart (1995). RHS, London.

Ternek, Z. (1957). The Lower Miocene (Burdigalian) Formations of the Adana Basin, their relations with other formations, and oil possibilities. *Bulletin of the Mineral Research and Exploration Institute of Turkey* 49: 60–80.

CHAPTER 4

4.2 Pests and diseases

Bravenboer, J. (2006). Downy or false mildew on cyclamen. *Cyclamen* 30 (1): 7.

Clennett, J. C. B. (1990). The control of pests and diseases in cyclamen. *Journal of the Cyclamen Society* 14 (2): 18–27.

Clennett, J. C. B. (1996). Vine weevil in cyclamen. *Cyclamen* 20 (1): 16–20.

Cowles, R. S. (1995). Black vine weevil biology and management. *Journal of the American Rhododendron Society* 49 (2): 83.

Cowles, R. S. (2003). Practical black vine weevil management. *Journal of the American Rhododendron Society* 57 (4): 219.

Helm, H. (2001). Root weevils: troublesome *Rhododendron* pests. *Journal of the American Rhododendron Society* 55 (4): 195.

Scopes, N. & Ledieu, M. (Eds) (1983). *Pest and Disase Control Handbook*. BCPC Publications, Croydon.

Van Tol, R. (2007). Novel strategies for vine weevil control. *Cyclamen* 31 (1): 22–27.

Whitehead, R. (Ed.) (2007). *The UK Pesticide Guide*. BCPC Publications, Alton.

4.3a North America

Bailey, L. H. (1904). *Cyclopedia of American Horticulture*. Macmillan, New York, p. 424–426.

Bailey, L. H. (1925). *The Standard Cyclopedia of Horticulture*. Macmillan, New York, p. 933–935.

Blanchan, N. (1909). *The American Flower Garden*. Doubleday Page & Company, New York.

Goodwin, N. (2005). *Montrose: Life in a Garden*. Duke University Press, Durham.

Goodwin, N. & Allen, L. (2001) *A Year in our Gardens; Letters by Nancy Goodwin & Allen Lacy*. The University of North Carolina Press, Chapel Hill.

Griffiths, D. (1934). *Bulbs from Seed*. Circular No. 311, United States Department of Agriculture, Washington, D.C., p. 9–10.

Hamblin, S. (1927). *Lists of Plant Types For Landscape Planting*. Harvard University Press, Cambridge.

Lawrence, E. (1942) *A Southern Garden*. The University of North Carolina Press, Chapel Hill.

Lawrence, E. (1957). *The Little Bulbs*. Criterion Books, New York.

Lawrence, E. (1961). *Gardens in Winter*. Harper and Brothers, New York.

Meehan, T. (1864). *The Gardeners Monthly and Horticultural Advertiser*. W. G. P. Brinckloe, Philadelphia.

Sedgwick, M. C., assisted by Cameron R. (1907). *The Garden Month by Month*. Garden City Publishing Company, Inc., Garden City, New York.

Wilder, L. B. (1927). *Adventures in My Garden and Rock Garden*. Doubleday, Page & Company, Garden City, New York.

4.3b Australia

Aitkin, R. & Looker, M. (2002). *The Oxford Companion to Australian Gardens*. Australian Garden History Society, Oxford University Press, Oxford.

Brookes, M. & Barley, R. (1992). *Plants Listed in Nursery Catalogues in Victoria 1855–1889*. Ornamental Plant Collection Association, Royal Melbourne Botanic Gardens, Victoria.

Grey-Wilson, C. (2002). *Cyclamen. A Guide for Gardeners, Horticulturists and Botanists*. Second Edition. B. T. Batsford Ltd, London.

Hitchmough, J. (1989). *Garden Bulbs for Australia and New Zealand*. Viking O'Neil, Australia.

Law-Smith, J. (1991). *The Garden Within*. National Trust of Australia (Victoria), Melbourne.

Price, S. (1986). *The Urban Woodland*. Lothian Publishing Co., Australia.

Polya, R. (1981). *Nineteenth Century Plant Nursery Catalogues of South-eastern Australia — A Bibliography*. La Trobe University Library, Melbourne.

Various authors. Plant catalogues and letters. Caroline Simpson Library, Sydney.

CHAPTER 5

Anon (1955). L'Avenir Horticole no 1.

Anon (1985). Thema: moderne *Cyclamen*-kultur. Deutscher Gartenbau 39 (3): 98–116.

Anon (1998). The Commercial Greenhouse Grower. 16 pp.

Bailey, R. H. (Ed) (1982). *Cyclamen 1982: Proceedings of the Conference of the Cyclamen Society*. The Cyclamen Society.

Blasdale, W. C. (1952). *Cyclamen persicum — its Natural and Cultivated Forms*. Stanford University Press, Stanford.

Böhmig, F. (1953). *Die Gattung Cyclamen*. Neumann Verlag, Berlin.

Bongartz, W. (1999). *Cyclamen*. Thalacker Medien, Braunschweig.

Boonstra, H. (1993). *Teelt van Cyclamen*. DLV Bloemisterij, Aalsmeer. 38 pp.

Doorenbos, J. & Boer, M. (1951). *De Herkomst van het Nederlandse Handelssortiment van Cyclamen persicum*. Wageningen.

Edwards, F. (1894). *Cyclamen* and how to grow them. Manchester Guardian.

Forsythe, J. W. (1929). Seed catalogue. Putteridge, Luton.

Grey-Wilson, C. (2002). *Cyclamen. A guide for Gardeners, Horticulturists and Botanists*. 2nd Edition. B. T. Batsford Ltd, London.

Grey-Wilson, C. (1988). *The Genus Cyclamen*. Christopher Helm Ltd, Bromley.

Kappert, H. (1941) Die Bedeutung der Polyploidie in der CyclamenzUchtung. *Züchter* 13: 106–144.

Maatsch, R. (1971). *Cyclamen*. 2nd Edition. Verlag Paul Parey, Berlin.

Nightingale, G. (1982). *Growing Cyclamen*. Croom Helm Ltd, London.

Oskam, W. (1964). Diploide *Cyclamen* aus Belgien. *Deutsche Gartenbörse* nr 3 (18-1-64).

Puttock, A. G. (1958). *The Cyclamen*. John Gifford Ltd, London.

Raalte, D. van, (1950). *Handboek der Bloemisterij* II. C. Misset, Doetinchem.

Rampinini, G. (1991). *La coltivazione del ciclamino* (non tutto, ma quasi). Pentagono Editrice, Italy.

Schneider, F. & Maatsch, R. (1937). *Cyclamen*. Verlags Paul Parey, Berlin.

Seghers, N. (1930). *Monographie du genre Cyclamen*. Alph. Leys, Ostende.

Takamura, T., Tomihama, T. & Miyajima, I. (1995). Inheritance of yellow-flowered characteristic and yellow pigments in diploid *Cyclamen* cultivars. *Scientia Horticulturae* 64 (1): 55–63.

Wellensiek, S. J., Doorenbos, J., Van Bragt, J. & Legro, R. A. H. (1961). *Cyclamen, a Descriptive List of Cultivars*. Veenman en Zonen BV, Wageningen.

CHAPTER 6

6.1. Old herbals

Manuscripts

Bayerische StaatsBibliothek, Munich. Clm 337 fol. 125–126. *De Material Medica*. 10th century.

Biblioteca Estense Universitaria, Modena. Ms. Lat. 993 (alfa.l.9.28). *Tractatus de Herbis*. 1458.

Biblioteca Medicea Laurenziana, Florence. Plut.73.16 fol. 49r. 13th century.

Biblioteca Medicea Laurenziana, Florence. Plut.73.41 fol. 25v. 9th century.

Biblioteca del Monumento Nazionale di Montecassino, Abbey of Montecassino, Italy. Codex Casinensis 97. *Herbarius of Apuleius Platonicus*. Early 9th century.

Biblioteca Nazionale, Naples. Naples Cod. Gr 1 fol. XLVII *Codex Neapolitanus* 6th–7th century.

Bibliotheck der Rijksuniversiteit, Leiden. Isaac Vossius Coll. MS. VLQ 9 fol. 42r. *Herbarius Apuleius Platonicus*. 6th century.

Bibliothèque de l'Ecole des Beaux-Arts, Paris. Ms. Masson 116. *Tractatus de Herbis*. 1370–1380.

Bibliothèque Nationale de France, Paris. Paris fr, 623 fol. 52r. *Livre des Simples Médecines*. 15th century.

Bibliothèque Nationale de France, Paris. Paris lat. 6823 fol. 33v. c. 1340.

Bibliothèque Nationale de France, Paris. Paris lat. 6862 fol. 35r. 9th century.

Bibliothèque Nationale de France, Paris. Paris lat. 9136 Platearius, Livre des Symples Medichinos. 15th century.

Bodleian Library, Oxford. Ashmole 1431 fol. 10r. Herbarius of Pseudo-Apuleius. 1070–1100.

Bodelian Library, Oxford. Ashmole 1462 fol. 19r. Pseudo-Apuleius, Herbarium. 1190–1200.

Bodelian Library, Oxford. Canon Mss. 408 fol. 55v. 15th century.

Bodleian Library (Sherardian Library of Plant Taxonomy), Oxford. Sherard 443 fol 190. Plantarum Disocoridis Icones. 18th century.

British Library, London. Add. 17063 fol. 14r. Apuleius Herbarium. c. 1480.

British Library, London. Add. 21115 fol. 24r. Herbarium Apulei Platonici. c. 1480.

British Library, London. Add. 41623 fol. 40v. Codex Bellunensis. 15th century.

British Library, London. Egerton. 747 fol. 18v. Tractatus de herbis. 1280–1317.

British Library, London. Harley 1585 fol. 23v. Pseudo-Apuleius Platonicus, Herbarium. 1145–1158.

British Library, London. Harley 4986 fol. 9r. Pseudo-Apuleius Platonicus. 11th–12th century.

British Library, London. Harley 5294 fol. 15v. Pseudo-Apuleius Platonicus. 12th century.

British Library, London. Sloane 1975 fol. 18v. Liber Platonis Apoliensis. 1190–1200.

British Library, London. Sloane 4016 fol. 31r. Icones Plantarum. 1440.

Eton College Library, Windsor. Eton ms. 204 13r Apulei platonici de Medicaminibus Herbarum. 12th century.

Natural History Museum (Botany Library), London. Banksian ms. No. 63 fol. 175. 1458–1477.

Österreichische Nationalbibliothek, Vienna. Cod. med. gr. 1 fol. 164v–165v Περί ὑλής ιατρικής Juliana Anicia Codex. 512.

Österreichische Nationalbibliothek, Vienna. Cod.11121 fol. 155, 157, 159 Codex Vindobonensis Palatinus/Codex Fuchs. Leonhart Fuchs. 1544–1563.

Pierpont Morgan Library, New York. Morgan MS. M.652 fol. 72v and 73v. 930–960.

Printed Works

Alechamps, J. d' (1586). *Historiae Generalis Plantarum*. Gulielmum Rovillium, Venice.

Bauhin, J. & Cherler, J. (enlarged and illustrated by Dominicus Chabreaus) (1650–1). *Historia Plantarum Universalis*. Ebroduni, Yverdon (Switzerland).

Besler, B. (1613). *Hortus Eystettensis*.

Bock, H. (1546). *Kreuter Buch*.

Brunfels, O. (1530). *Herbarum Vivae Eicones*. Hans Schotten, Strasszburg.

Bry, J. T., de (1611). *Florilegium Novum*. Oppenheim.

Bulliard, J. B. F. P. (1780–1793). *Herbier de la France*. Paris.

Chabraeus, D. (1666). *Stirpium Icones et Sciagraphia*. Philip Gamoneti & Iac de la Pierre, Geneva.

Chaumeton, F. P. (1816). *Flore Médicale*. Panckoucke, Paris.

Cuba, J. von (1485). *Gart der Gesundheit*. Johann Schönsperger, Augsburg.

Cuba, J. von (1487). *Gart der Gesundheit*. Johann Schönsperger, Augsburg.

Cuba, J. von (1487). *Gart der Gesundheit*. Johann Grüninger, Strassburg.

Cuba, J. von (1487). *Gart der Gesundheit*. Konrad Dinckmut, Ulm.

Cuba, J. von (1488). *Gart der Gesundheit*. Johann Schönsperger, Augsburg.

Cuba, J. von (1533). *Kreutterbuch von allem Erdtgewachs*. Christian Egenolph, Frankfurt.

Culpeper, N. (1652). *The English Physician; or, An Astrologo-physical Discourse of the Vulgar Herbs of this Nation*. William Bentley, London.

Dodoens, R. (1553). *De Stirpium Historia*. John Loe, Antwerp.

Dodoens, R. (translated by L'Ecluse, C. de) (1557). *Histoire des Plantes*. Jean Loe, Antwerp.

Dodoens, R. (1574). *Purgantium Aliarumque*. Christophe Plantin, Antwerp.

Dodoens, R. (translated by Lyte, H.) (1578). *A Nievve Herball, or Historie of Plantes*. Gerard Dewes, Pawles Churchyard at the Signe of the Swanne, London.

L'Ecluse, C. de (Carolus Clusius) (1583). *Rariorum Stirpium per Pannonias Observatarum Historiae*.

L'Ecluse, C. de (1601). *Rariorum Plantarum Historia*. Christophe Plantin, Antwerp.

L'Ecluse, C. de (1611). *Curae Posteriores*.

Fuchs, L. (1542). *De Historia Stirpium Commentarii Insignes*. Michael Isingrin, Basel.

Fuchs, L. (1543). *New Kreüterbůch*. Basel.

Fuchs, L. (1543). *Den Nieuwen Herbarius*. Michiel Hingrin, Basel.

Fuchs, L. (1549). *Commentaires Tres Excellens de l'Hystoire des Plantes*. Jacques Gazeau, Paris.

Fuchs, L. (1549). *Plantarum Effigies*. Balthazar Arnoullet, Lyon.

Gerard, J. (1597). *The Herball or General Historie of Plantes*. John Norton, London.

Gerard, J. (enlarged by Johnson, T.) (1636). *The Herball or General Historie of Plantes*. Adam Islip, Joice Norton & Richard Whitakers, London.

Gonzaga, F. (1481). *Herbarium Apuleii Platonici mit Widmungsbrief des Autors an Kardinal Francesco Gonzaga bzw. an Kardinal Giuliano della Rovere*. Johannes Philippus de Lignamine, Rome.

L'Obel, M. de (1581). *Kruydtboeck oft Beschrÿuinghe*. Christophe Plantin, Antwerp.

L'Obel, M. de (1581). *Plantarum seu Stirpium Icones*. Christophe Plantin, Antwerp.

Matthioli, P. A. (1554). *Medici Senensis Commentarii, in Libros Sex Pedacii Dioscoridis*. Vincent Valgris, Venice.

Matthioli, P. A. (1562). *Senensis Medici, Commentarii Denuo Aucti in Libros Sex Pedacii Dioscoridis*. Gabriel Coter, Lugduni.

Matthioli, P. A. (1563). *New Kreüterbuch*.

Matthioli, P. A. (edited by Camerarius, J.) (1586). *De Plantis Epitome*.

Matthioli, P. A. (1598). *Opera*.

Metlinger, P. (1486–1488) *Le Grant Herbier*. Besançon, France.

Meydenbach, J. (1491) *Ortus Sanitatis*. Mainz

Panckow, T. (1654). *Herbarium Portatile*. Berlin.

Parkinson, J. (1629). *Paradisi in Sole Paradisus Terrestris*. London.

Parkinson, J. (1640) *Theatrum Botanicum*. Thomas Cotes, London.

Passe, C. van de (1614). *Hortus Floridus*. Ioannem Iassonium, Arnhem.

Pinet, A. du (1561). *Historia Plantarum*. Gabriel Coter, Lugduni.

Prüss, J. (1497) *Ortus Sanitatis*. Strassburg.

Prüss, J. (1499) *Ortus Sanitatis*. Strassburg.

Prüss, J. (1500) *Ortus Sanitatis*. Strassburg.

Rabel, D. (1622). *Theatrum Florae*. Petrum Firens, Paris.

Ruel, J. (1516). *Pedacij Dioscoridis Anazarbei de Medicinali Material Libri Quinqi*. Henry Stephan.

Ruel, J. (1536). *De Natura Stirpium Libri Tres*. Simon Colin, Paris.

Salmon, W. (1710). *Botanologia — The English Herbal or History of Plants*.

Schinz, S. (1774) *Anleitung zu der Pflanzenkenntniss*. Bansenhauses.

Sweert, E. (1612). *Florilegium*. Anthony Kempner, Frankfurt am Main.

Theodor, J. (1590). *Eicones Plantarum*. Fankfurt am Main.

Theodor, J. (1591). *Kreuterbuch*. Fankfurt am Main.

Theophrastus Eresii (1644). *Historia Plantarum*. Henry Laurent, Amsterdam.

Turner, W. (1551). *A New Herball*. Steven Mierdman, London.

Vallet, P. (1623). *Le Jardin du Roy tres Chrestien, Loys XIII, Roy de France et de Navare…* . Paris.

Weinmann, J. W. (1737). *Phytanthoza Iconographia, sive, Conspectus Aliquot Millium…* . Hieronymum Lentz, Regensburg.

Zwinger, T. (1696). *Theatrum Botanicum Krauter Buch*. Jacob Philipp Richtern, Frankfurt.

Reference works

Arber, A. (1912). *Herbals, Their Origin and Evolution: A Chapter in the History of Botany, 1470–1670*. Cambridge University Press, Cambridge.

Baumann, H. (translated by Stearn, W. T. & Stearn, R. S.) (1993). *Greek Wild Fowers and Plant Lore in Ancient Greece*. The Herbert Press, London

Blunt, W. & Raphael, S. (1979) *The Illustrated Herbal*. Frances Lincoln, London.

Blunt, W. & Stearn, W. T. (1994). *The Art of Botanical Illustration*. Antique Collectors Club, Woodbridge.

Collins, M. (2000). *Medieval Herbals: The Illustrative Tradition*. The British Library, London.

Collins, M. (2003). Introduction in *A Medieval Herbal — A facsimile of British Library Egerton MS 747*. British Library, London.

Grigson, G. (with bibliographical notes by Buchanan, H.) (1951). *Thornton's Temple of Flora*. William Collins, London.

Gunther, R. T. (1934). *The Greek Herbal of Dioscoride Illustrated by a Byzantine A.D. 512, Englished by John Goodyer A.D. 1655, Edited and first printed A.D. 1933*. Hafner Publishing Company, London.

Hunger, F. W. T. (1935). *The Herbal of Pseudo-Apuleius*. E. J. Brill, Leyden, The Netherlands.

Ietswaart, J. H. (1990) Infraspecific delimitation of the *Cyclamen graecum* complex. *Journal of the Cyclamen Society* 14 (2): 49–53.

Lack, H. W. (2001). *Garden of Eden*. Taschen, Cologne.

Meyer, F., Trueblood, E & Heller, J. (1999) *The Great Herbal of Leonhart Fuchs: De Historia Stirpium Commentarii Insignes*. Stanford University Press, Stanford.

Payne, J. F. (1904). *The Fitz-Patrick Lectures for 1903: English Medicine in the Anglo-Saxon Times*. Clarendon Press, Oxford.

Rupreht, J. (1994). Toxicity of *Cyclamen purpurascens*. *Journal of the Cyclamen Society* 18 (1): 15–17.

Singer, C. J. (1921). Greek biology and its relation to the rise of modern biology. In: *Studies in the History and Method of Science* Vol. II. by Singer, C. J. Oxford University Press, Oxford.

CHAPTER 7

John Sibthorp, Ferdinand Bauer, James Edward Smith & James Sowerby and the creation of Flora Graeca

Aiton, W. (1789). *Hortus Kewensis; or, A Catalogue of the Plants Cultivated in the Royal Botanic Garden at Kew*. George Nicol, Pall Mall.

Bauer, F. (1786–1787). Sketches of plants for Flora Graeca. Bodleian Library, University of Oxford: Sherardian Library of Plant Taxonomy, MS Sherard 247 fol. 25r.

Bauer, F. (1788–1792) Flora Graeca drawings v.4: 176–177. Bodleian Library, University of Oxford: Sherardian Library of Plant Taxonomy, MS Sherard 244.

Harris, S. (2007). *The Magnificent Flora Graeca*. Bodleian Library, University of Oxford.

Lack, H. W. & Ibáñez, V. (1997). Recording colour in the late eighteenth century botanical drawings: Sydney Parkinson, Ferdinand Bauer and Thaddäus Haenke. *Curtis's Botanical Magazine* 14 (2): 87–100.

Lack, H. W. & Mabberley, D. J. (1999). *The Flora Graeca Story*. Oxford University Press, Oxford.

Mabberley, D. J. (1999). *Ferdinand Bauer – The Nature of Discovery*. Merrell Holberton & The Natural History Museum, London.

Sibthorp, J. (1786–1787). Sibthorp Diaries. Bodleian Library, University of Oxford: Sherardian Library of Plant Taxonomy, MS Sherard 214–215.

Smith, J. E. (1806). *Florae Græcae Prodromus sive Plantarum Omnium Enumeratio, qu as in Provinciis Aut Insulis Græciae invenit Johannes Sibthorp, M.D.* v1:127–128. Richard Taylor, London.

Smith, J. E. (1813). *Flora Græca sive Plantarum Rariorum Historia, quas in Provinciis Aut Insulis Græcia legit, Investigativit et Depingi Curavit Johannes Sibthorp, M.D.* v2: 71–72; tab 185–186. Richard Taylor, London. Bodleian Library, University of Oxford: Sherardian Library of Plant Taxonomy, MS Sherard 761.

Names of Cyclamen Index

Accepted names for the 23 species of *Cyclamen* and their subspecies, varieties and botanical forms (f.) are shown in bold. Photographs and illustrations are indicated by italic page numbers. Bold page numbers indicate the major references for a name.

Cyclamen persicum cultivars and series

S

T

U

V

General index

Birth and death dates are given for historical persons. Where known, the country in which cyclamen breeders and seed merchants are based is given in parentheses. Images are indicated by an italic page number.

D

E

H

N

O

P

T